Werner Heisenberg

GESAMMELTE WERKE / COLLECTED WORKS

Werner Heisenberg

GESAMMELTE WERKE / COLLECTED WORKS

Abteilung / Series A

Original Scientific Papers
Wissenschaftliche Originalarbeiten

Abteilung / Series B

Scientific Review Papers, Talks, and Books
Wissenschaftliche Übersichtsartikel, Vorträge und Bücher

Abteilung / Series C

Philosophical and Popular Writings
Allgemeinverständliche Schriften

The *Gesammelte Werke / Collected Works* of Werner Heisenberg
are published jointly by R. Piper GmbH & Co KG Verlag, München, Zürich
and Springer-Verlag, Berlin, Heidelberg, New York, Tokyo.
The volumes of Series A and B are issued by Springer-Verlag,
those of Series C by R. Piper Verlag.

Werner Heisenberg

GESAMMELTE WERKE
COLLECTED WORKS

Edited by

W. Blum, H.-P. Dürr, and H. Rechenberg

Series A / Part I

Original Scientific Papers

Wissenschaftliche Originalarbeiten

Springer-Verlag

Berlin Heidelberg New York Tokyo 1985

Dr. Walter Blum
Professor Dr. Hans-Peter Dürr
Dr. Helmut Rechenberg

Max-Planck-Institut für Physik und Astrophysik, Werner-Heisenberg-Institut für Physik
Föhringer Ring 6, D-8000 München 40, Fed. Rep. of Germany

Annotators:

Professor Dr. Subrahmanyan Chandrasekhar

Laboratory for Astrophysics and Space Research, The Enrico Fermi Institute
The University of Chicago, 933 East 56th Street, Chicago, IL 60637, USA

Professor Dr. Charles P. Enz

Département de Physique Théorique, Section de Physique, Université de Genève
32, Boulevard d'Yvoy, CH-1211 Genève 4, Switzerland

Professor Dr. Friedrich Hund

Charlottenburger Straße 19, D-3400 Göttingen, Fed. Rep. of Germany

Professor Dr. Bartel Leendert van der Waerden

Wiesliacher 5, CH-8053 Zürich, Switzerland

ISBN 3-540-13400-X Springer-Verlag Berlin Heidelberg New York Tokyo
ISBN 0-387-13400-X Springer-Verlag New York Heidelberg Berlin Tokyo

Library of Congress Cataloging in Publication Data. (Revised for series A/part I) Heisenberg, Werner, 1901-1976. Gesammelte Werke = Collected works. Includes bibliographies. Contents: ser. A.Original scientific papers - ser. B. Scientific review papers, talks, and books. 1. Physics-Collected works. I. Blum, W. (Walter), 1937-. II. Dürr, H.-P. (Hans-Peter), 1929-. III. Rechenberg, H. (Helmut), 1937-. IV. Collected works. V. Title. QC3.H33 1984 530 84-10633

Reproduction and offset printing: Beltz Offsetdruck, 6944 Hemsbach
Bookbinding: J. Schäffer OHG, 6718 Grünstadt
2153/3150-543210

Preface

Werner Heisenberg is one of the greatest scientists of our century. His work extends over a period of fifty years, ranging from turbulence theory, the establishment of quantum mechanics, its fundamental applications in atomic and solid state physics, to the theory of cosmic ray phenomena and of elementary particles. He recognized early the epistemological significance of the new discoveries. He was able to place the radical changes in the foundation of physics of this century in the historical context of natural philosophy. His thoughts on language as *the* medium to grasp scientific truth, artistic truth, religious truth, truth in general, reached many auditors and readers, scientists as well as non-scientists.

In the 75 years of his life the political and social structure of his home country, of Europe and the world over underwent drastic changes. He grew up in Imperial Germany, made his great contributions to quantum mechanics during the period of the Weimar Republic and was engaged in nuclear physics when the potentates of the Third Reich tried to discredit relativity and quantum theory as "degenerate" science; in World War II he participated in the German effort to develop a nuclear reactor. After the war, he devoted himself mainly to the physics of elementary particles. In addition, he acted in many official capacities: for the promotion of research, the reconstruction of science in the Federal Republic of Germany, and the advancement of international collaboration. His public speeches demonstrate the responsibility of the scientist in a world transformed by science.

The different parts of his far-ranging work are intimately connected. For this reason the *Collected Works* contain all the published papers and essays, without exception. We also print all the completed but unpublished manuscripts of the intellectual estate. This includes some articles that had been prepared for publication (e.g., the report to the Eighth Solvay Conference which had been scheduled for October 1939 but cancelled because of the beginning of war). It also includes a manuscript of a systematic natural philosophy, which is printed here for the first time under the title *"Ordnung der Wirklichkeit"*, and two unpublished essays on similar themes. In addition, we reproduce the more than twenty originally classified technical reports on the reactor project, and about a dozen other completed unpublished manuscripts and obituaries. More information will be given with the individual papers.

We have incorporated in the *Collected Works* neither the correspondence and memoranda of Heisenberg, nor his incomplete sketches and manuscripts. Lecture notes, unless published, have been omitted as well. This material is also contained in the *Werner-Heisenberg-Archiv*, at present located at the *Werner-Heisenberg-Institut für Physik* in Munich, and later on will be administrated by the *Bibliothek und Archiv der Max-Planck-Gesellschaft* in Berlin.

We have arranged the writings of Heisenberg in three series. Series A contains the original scientific papers, Series B the scientific review papers, reports and books, as well as the texts and abstracts of his talks at scientific conferences. Series C presents the lectures, essays and writings directed to a wider public. A complete bibliography of Heisenberg, and a biographical sketch by David C. Cassidy and Helmut Rechenberg are included.

The complete edition of this wide-ranging work could only be achieved by the close collaboration of a scientific book publisher and a general publisher. Series A and B are issued by Springer-Verlag, and Series C by R. Piper Verlag.

The scientific papers of Series A have been organized by the editors under 14 topics; for each topic we have found an internationally renowned colleague to write an annotation to Heisenberg's papers on that topic. The purpose of these introductory notes is to provide the reader a review of the origin, content and impact of these contributions. The editorial language of Series A and B is English, so that we have rendered in English the titles of the papers which Heisenberg published in German or other languages. Furthermore, we have in Series B occasionally favored the English version of a given text if it is available. Finally, all the annotations have either been written in, or translated into, English.

The writings in Series C have also been organized into topics, not necessarily confined to a single volume. Within each topic grouping in Series A and C, and within Series B, the ordering of the papers is strictly chronological. The time sequence is determined by the date of completion or submission of the papers or manuscripts.

The texts are as a rule given in photographic reproduction. Several publications were rearranged on the page in order to be more readable in the *Collected Works;* however, we have always noted the original page numbers at the margin. Some essays, especially those taken from newspapers and manuscripts, as well as the classified war reports on the reactor project, have been newly set in print.

Editing such an extensive work could not have been accomplished without assistance from many sides. We thank the members of the Heisenberg family, especially Mrs. Elisabeth Heisenberg, for essential help in preparing the *Collected Works.* The documents presented to the Werner-Heisenberg-Archiv have proven to be of inestimable value for the practical work of the editors.

Both publishers have, in spite of their different traditions, collaborated well with each other and the editors in their aim to present the complete *Collected Works* in an adequate and unified form. We would like to thank Wolf Beiglböck, Ernst Reinhard Piper, Klaus Stadler and their helpers for their fine cooperation, and especially Klaus Piper for taking the early initiative.

David C. Cassidy, who has been working for many years on the bibliography and biography of Werner Heisenberg, has assisted us since the very beginning of the project. We would also like to thank him for his cooperation with the bibliography and the *Biographical Data.*

We are indebted to the *Alexander-von-Humboldt-Stiftung* and the *Max-Planck-Gesellschaft zur Förderung der Wissenschaften e.V.* for a substantial subsidy of the publication.

München, July 1985 *W. Blum, H.-P. Dürr, and H. Rechenberg*

Editors' Note

The present first volume of Series A contains Heisenberg's original scientific contributions to four topics: hydrodynamic stability and turbulence theory (1922–1948), atomic and molecular structure (1922–1925), quantum mechanics (1925–1927), and applications of quantum mechanics (1926–1933). They were written (apart from some papers on the statistical theory of turbulence) by the young Werner Heisenberg and indicate the broad scope of his scientific education and creativity.

It is our great pleasure to thank Professors Subrahmanyan Chandrasekhar, Friedrich Hund, Bartel Leendert van der Waerden and Charles P. Enz, who have undertaken the task of writing valuable introductions to the papers of Heisenberg. We hope that these annotations will assist the reader to know the origin and historical background and guide his understanding of the contents as well as of the later influence of Heisenberg's contributions to the topics in question.

For the few editorial notes (other than the annotations) one of us (H.R.) is responsible.

München, July 1985 *W. Blum, H.-P. Dürr, and H. Rechenberg*

Contents

Group 3: **Quantum Mechanics (1925–1927)**

Group 4: **Applications of Quantum Mechanics (1926–1933)**

Biographical Data

Werner Heisenberg (1901–1976)

1. Youth (1901–1920)

Werner Karl Heisenberg was born in Würzburg, Germany, on 5 December 1901. His father, August Heisenberg (1869–1930), stemmed from a family of master craftsmen in Osnabrück, and had studied classical philology in Munich; at the time of Werner's birth he held the dual positions of teacher at the *Altes Gymnasium* in Würzburg and *Privatdozent* for Greek philology at the University of Würzburg. His mother was Anna Wecklein, the daughter of Nikolaus Wecklein (1843–1926), a classical philologist and rector of the *Maximilians-Gymnasium* in Munich. They married in 1899 and their first son, Erwin, was born in Munich in 1900; he later became a chemist and worked in industry (he died in 1965).

The family returned to Munich in April 1910 when August Heisenberg was called (in January) to succeed his teacher Karl Krumbacher in the chair for medieval and modern Greek philology at the university. On 18 September 1911 Werner entered the *Maximilians-Gymnasium,* still under his grandfather's rectorship and at that time the best humanistic gymnasium in the Bavarian capital. Werner studied diligently and rapidly and was regarded as an outstanding, ambitious, self-confident pupil. He excelled particularly in mathematics; far exceeding the teaching program, he taught himself differential and integral calculus, worked with elliptic functions, and studied abstract number theory. Besides his academic achievements Heisenberg learned to play the piano and by the age of 13 he was playing master compositions. He remained an excellent player throughout his life.

The First World War and its consequences intruded into Heisenberg's boyhood life. His father, a reserve infantry officer, was immediately called to duty and remained away from the family for nearly the entire war. An increasing shortage of food and fuel prevailed throughout Germany and forced the occasional closing of schools. Because of his family's lack of food, Werner signed up for the war assistance service in the summer of 1918 and helped to bring in the harvest with schoolmates on a farm near Miesbach in Upper Bavaria.

The loss of the war and the abdication of the monarchy generated revolutionary unrest throughout Germany. In Bavaria a socialist republic came to power on 7 November 1918; it developed (in Munich) into a Soviet Republic *(Räterepublik)* on 7 April 1919 that was crushed in early May 1919 by troops dispatched by the Reich Government. During the fighting in Munich and the ensuing restoration of moderate socialist rule, Heisenberg, along with many of his schoolmates, voluntarily served in support of one of the republican units, Cavalry Rifle Command No. 11 *(Kavallerie-Schützen-Kommando No. 11).* He carried messages, guarded the street from the roof of the Catholic seminary across from the university, and read Plato's *Timaeus* in Greek. Shortly afterwards Heisenberg became associated, as the elected leader of a small group of younger boys, with the New Boy Scouts *(Bund deutscher Neupfadfinder),* a representative of the postwar German youth movement that strove for renewal of personal and social life in Germany. Until the prohibition of independent youth groups in 1933 Heisenberg spent nearly all of his leisure time with his group, hiking and camping within Germany and on trips to neighboring countries (e.g. to Finland in summer 1923). He loved hiking, skiing, mountain climbing, and enjoyed the beauty of nature all his life.

2. Student and Postdoctoral Years (1920–1925)

After a brilliant performance on his final gymnasium examinations *(Abitur),* for which he was accepted as a scholar of the prestigious Maximilianeum Foundation, Heisenberg entered the University of Munich during the winter semester of 1920/21. At first he planned to study pure mathematics, but a discomforting meeting with the famous professor of that subject, Ferdinand von Lindemann, resulted in his turning to theoretical physics. Arnold Sommerfeld, professor of that field, immediately recognized his talents and admitted him to his *Seminar,* composed of advanced students and postdoctoral researchers. Among the *Seminar* members at that time were Gregor Wentzel and Wolfgang Pauli. Otto Laporte and Karl Bechert arrived a little later.

During his three years of university study Heisenberg attended Sommerfeld's lectures covering all fields of theoretical physics, including the qantum and relativity theories, as well as the special lectures offered by Karl Herzfeld. He showed less interest in the lectures and laboratory exercises in experimental physics provided by Wilhelm Wien. For one of his two minors he attended mathematics courses offered by Lindemann, Alfred Pringsheim, Artur Rosenthal and Aurel Voss. For his second minor he studied astronomy with Hugo von Seeliger.

In Sommerfeld's *Seminar* Heisenberg studied the latest literature in physics and carried on intensive discussions of the problems arising from it with his teacher and a group of excellent fellow students. Under Sommerfeld's direction he delved into the problems of atomic theory, a difficult and abstract discipline that tended to attract the best and most dedicated young theorists of the day. As early as his first semester Heisenberg presented a quantum-theoretical analysis of the anomalous Zeeman effect, the results of which he utilized a year later in his first publication (submitted in November 1921). At the same time he examined problems in hydrodynamics, a subject on which he published his second paper (on Kármán vortices, 1922) held his first public lecture in September 1922 (at a topical conference in Innsbruck), and wrote his doctoral dissertation (finished in July 1923).

In June 1922 Sommerfeld brought his star pupil to Göttingen where Niels Bohr presented a series of lectures on quantum theory and atomic structure – the "Bohr Festival" (12–22 June 1922). There Heisenberg met the leading representatives of atomic physics in Germany and Europe, among them Max Born, Paul Ehrenfest, James Franck, Hendrik Anthony Kramers and Alfred Landé. Bohr and Born began to take notice of the young physicist. When Sommerfeld left for a guest professorship at the University of Wisconsin, Madison, during the winter of 1922/23, Born invited Heisenberg to Göttingen as his personal assistant (November 1922–March 1923). In Göttingen Heisenberg learned the rigorous mathematical methods of the Hilbert school and worked with Born on the many-body problem in atomic theory, especially on the energy states of the helium atom.

During the summer semester of 1923 Heisenberg completed his dissertation in Munich. In it he successfully treated the problem of the onset of turbulence, which had resisted all previous efforts made by mathematicians and physicists. Sommerfeld was very pleased with this result and the mathematical methods Heisenberg employed. However, Wilhelm Wien, the examiner in experimental physics, came to quite a different conclusion in his field during the final orals (on 23 July 1923) and wanted to fail the candidate. Heisenberg thus received his doctorate "cum laude" rather than "magna" or "summa cum laude", the two highest marks.

In October 1923 Heisenberg became Born's assistant at the University of Göttingen. He continued to work with Born on atomic and molecular models. By means of a modification of the Bohr-Sommerfeld quantum rules he obtained certain advances in the explanation of the anomalous Zeeman effect. He got his *Habilitation* at the University of Göttingen on 28 July 1924. Although only 22 years old, he thus became a *Privatdozent* and fully qualified as a university lecturer.

Heisenberg made his first visit to Bohr's institute in Copenhagen during the spring of 1924 (15 March to early April). With the help of a Rockefeller fellowship (International Education Board) he resided in Copenhagen during the seven months from middle of September 1924 to early April 1925, and during the fall of 1925 (September-October), and otherwise returned to Born's institute in Göttingen. At Bohr's institute Heisenberg met a number of talented young physicists from a variety of nations, among them Christian Møller (Denmark), Svein Rosseland (Norway), G.H. Dieke (Holland), Ralph Fowler (England), Ralph de Laer Kronig and David Dennison (both from U.S.A.). During his visits he learned to speak Danish and English and worked intensively with Bohr and his closest collaborator Kramers on the most difficult problems of atomic theory. This work led to Heisenberg's papers on resonance fluorescence (submitted in November 1924), the dispersion of light by atoms (with Kramers, December 1924) and the structure of complex spectra and their Zeeman effects (April 1925).

During his years of study and postdoctoral research Heisenberg had received in Munich and Göttingen with Sommerfeld and Born a thorough education in all aspects of theoretical physics, while simultaneously familiarizing himself with the main problems of atomic and quantum theory. With Bohr in Copenhagen he deepened his understanding of the foundations of quantum physics. As his friend Pauli wrote (to Kramers on 27 July 1925), Heisenberg "learned a little of philosophical thinking" and moved "noticeably away from the purely formal". Heisenberg himself later summarized the influence of his teachers thus: "From Sommerfeld I learned optimism, from the Göttingen people mathematics, and from Bohr physics".

3. The Development of Quantum Mechanics (1925–1927)

The early 1920s witnessed fundamental difficulties in atomic physics. The quantum theory of atomic structure, founded by Bohr and largely developed by Bohr and Sommerfeld, did not describe the properties of complicated atoms and molecules. Moreover, the discovery of the Compton effect at the end of 1922 focussed attention on the problem of the nature of radiation. Its interpretation in the light-quantum hypothesis contradicted classical radiation theory, and the radical attempt by Bohr, Kramers and Slater in early 1924 to resolve the difficulty by assuming only statistical conservation of energy and momentum was refuted by the experiment of Walther Bothe and Hans Geiger in April 1925.

Heisenberg was growing ever more concerned with these and other difficulties in atomic theory. His works on the anomalous Zeeman effect, only successful in part, and his unsuccessful calculation of the helium states with Born had sensitized him by early 1925 to the "crisis" of current theory. Nevertheless, his latest calculations in Copenhagen on dispersion theory and on complex spectra, especially the principle of "sharpened" correspondence applied in these works, seemed to point toward a future satisfactory theory.

With characteristic optimism the Göttingen *Privatdozent* took on a new and difficult problem at the beginning of May 1925, the calculation of the line intensities in the hydrogen spectrum. Heisenberg began with a Fourier analysis of the classical hydrogen orbits, intending to translate them into a quantum theoretical scheme – just as he had done with Kramers for the dispersion of light by atoms. But the hydrogen problem proved much too difficult, and he replaced it with the simpler one of an anharmonic oscillator. With the help of a new multiplication rule for a quantum-theoretical Fourier series he succeeded in writing down a solution for the equations of motion for this system. On 7 June 1925 he went to the island of Helgoland to recover from a severe attack of hay fever. There he completed the calculation of the anharmonic oscillator, determining all the constants of the motion. He made use, in particular, of a modified quantum condition that was later called by Born, Pascual Jordan and himself a "commutation relation", and he proved that the new theory yielded stationary states (conservation of energy). Returning to Göttingen on 19 June 1925 Heisenberg composed his fundamental paper "Über die quantentheoretische Umdeutung kinematischer und mechanischer Beziehungen" ("On a Quantum Theoretical Reinterpretation of Kinematic and Mechanical Relations"), which was completed on 9 July 1925. In this paper, the starting point for a new quantum mechanics, Heisenberg announced as the leading philosophical principle of quantum mechanics that only observable quantities are allowed in the theoretical description of atoms. Heisenberg reported his new results during visits shortly thereafter with Paul Ehrenfest in Leiden and with Ralph Fowler in Cambridge.

After Born and Jordan managed in August and September 1925 to develop the mathematical content of Heisenberg's work into a consistent theory with the help of infinite Hermitian matrices (Z. Phys. *34,* 858, 1925), Heisenberg participated, starting in September 1925, in the completion and application of the new "matrix mechanics", culminating in the long "three-man-paper", by Born, Heisenberg and Jordan, submitted on 16 November 1925. Further developments followed rapidly: Pauli calculated the stationary states of the hydrogen atom in October 1925; Cornelius Lanczos in Frankfurt and Born and Norbert Wiener in the USA extended the method of operator mechanics to describe continuous motions (December 1925); and Paul Adrien Maurice Dirac in Cambridge developed independently of the Göttingen school a different scheme based upon Heisenberg's July paper, the method of *q*-numbers (November 1925), in which many-electron atoms and the relativistic Compton effect could be handled successfully (spring 1926). In addition Heisenberg and Jordan utilized electron spin and matrix mechanics to solve the old problems of hydrogen fine structure and the anomalous Zeeman effect (April 1926); and finally Heisenberg discovered the phenomenon of quantum-mechanical resonance (June 1926), which played a decisive role in his subsequent calculation of the term system of the helium atom (July 1926).

In May 1926 Niels Bohr offered Heisenberg a position at his institute in Copenhagen as Lector and successor to his assistant Kramers. There Heisenberg delivered lectures at the university (in Danish) on contemporary physical theories, directed beginning students, helped guest researchers with their problems, and discussed with Bohr the most important results of quantum mechanics. In the summer and fall of 1926 the main topic of discussion was wave mechanics, the quantum atomic theory that Erwin Schrödinger began introducing in January 1926. The complete mathematical equivalence between Göttingen's matrix mechanics or Dirac's *q*-number scheme and Schrödinger's wave mechanics was proved by Jordan and

Dirac in December 1926, after preparatory work by Schrödinger (March 1926), Pauli (April 1926), and Carl Eckart (June 1926). However, Schrödinger's physical interpretation of the square of the wave amplitude as the continuously distributed charge density of the electron was rejected by Born, Bohr and Heisenberg and replaced on Born's proposal by the interpretation that it is the probability for finding the electron at each location (June 1926). In close contact with Pauli, and in intense discussion with Bohr, Heisenberg analyzed what he termed the "perceptual content of the quantum-theoretical kinematics and mechanics". As a result of the analysis he presented in March 1927 the so-called "indeterminacy" or "uncertainty relations", which limit the simultaneous measurement of canonically conjugate variables, such as the position and momentum of a particle. Bohr, on the other hand, pondered the simultaneous use of the physical pictures of particles and waves, which resulted in his general principle of "complementarity" announced in fall of 1927.

Born's statistical interpretation of Schrödinger's wave function, Heisenberg's uncertainty relations, and Bohr's complementary principle formed the basis of the physical interpretation of the new quantum mechanics, as explicated by Bohr in his lectures at the Volta Conference in Como (September 1927) and at the Solvay Congress in Brussels (24–29 October 1927). This "Copenhagen Interpretation" of quantum mechanics, as it was later called, found acceptance by most physicists, but not by all: Albert Einstein in particular raised serious objections to it at the 1927 and 1930 Solvay conferences and later, for example, in his paper with Boris Podolsky and Nathan Rosen (Phys. Rev. *47,* 777, 1935).

4. Professor in Leipzig (1927–1942)

Through his fundamental contributions to quantum mechanics Heisenberg soon became known far beyond the circle of the Munich, Göttingen and Copenhagen physicists as a leading representative of the new, successful atomic theory. At the Meeting of the German Scientists and Physicians in Düsseldorf (19–25 September 1926) he delivered the main address on quantum mechanics. In October 1927 Heisenberg, then only 25 years old, was called from Copenhagen to the chair for theoretical physics at the University of Leipzig, where Peter Debye had just accepted the chair for experimental physics. In spring 1929 Friedrich Hund joined Heisenberg's theoretical physics institute, and in 1931 the mathematician Bartel Leendert van der Waerden, who developed group theoretical methods in quantum mechanics, came to Leipzig as professor of mathematics.

Together Heisenberg and his colleagues created in Leipzig a leading center for research on atomic and quantum physics that attracted numerous talented students and collaborators. Felix Bloch, Rudolf Peierls and Edward Teller were among Heisenberg's first students. Hans Euler, Léon Rosenfeld, Victor Weisskopf, Carl Friedrich von Weizsäcker and Gian Carlo Wick also studied under Heisenberg. As part of their research program Debye and Heisenberg set up an annual symposium (until 1933) on current research in physics, the famous Leipzig University Week *(Universitätswoche).* The first of these took place on 18–23 June 1928 with, among others, Paul Dirac speaking on his relativistic electron equation and Fritz London on the theory of the chemical bond.

Heisenberg's research in Leipzig concentrated first on applications and extensions of quantum mechanics. In May 1928 he provided an explanation of ferromagnetism from first principles in which he showed that the quantum mechanical exchange integral that had played a decisive role in his solution of the helium problem could account for the strong molecular magnetic field in the interior of ferromagnetic materials. A little later Felix Bloch supplemented Heisenberg's theory of ferromagnetism with the invention of spin waves. During the period 1928–1931 Heisenberg and his pupils presented a series of basic contributions to the theories of molecules and metals. At the same time Heisenberg collaborated with Wolfgang Pauli, who held the chair in theoretical physics at the E.T.H. (Swiss Polytechnic) in Zurich since spring 1928, on the formulation of relativistic quantum field theory. In March and September 1929 they completed two basic papers that allowed a systematic treatment of interacting quantum fields, but that also indicated the existence of severe difficulties and divergences in the theory, such as infinite self energies of electrons. The discovery of the neutron in early 1932 occasioned Heisenberg's development of the neutron-proton model of the nucleus (independently of Dimitri Iwanenko), in which he introduced the concept of nuclear exchange force and the formalism of isotopic spin (June and July 1932). In recognition of his work on quantum mechanics and its applications Heisenberg received the Nobel Prize for physics in 1933 (for the year 1932).

Besides his intensive research during the early years in Leipzig, Heisenberg undertook a number of trips abroad. In spring of 1929 he delivered a series of lectures at the University of Chicago on "The Physical Principles of the Quantum Theory" which were published as his first book. He made the return trip with Dirac, who had also come to the United States, through the American West (August 1929) to Japan and India (September–November 1929). In 1932 he lectured at the Ann Arbor summer school. Heisenberg participated as well in numerous conferences, such as the Solvay congresses of 1927 (24–29 October), 1930 (20–25 October) and 1933 (22–29 October) in Brussels, the centenary meeting of the British Association in London (29 September 1931) and the nuclear physics conference in Rome (October 1931).

Throughout the 1930s Heisenberg very frequently visited Niels Bohr in Copenhagen and occasionally Wolfgang Pauli in Zurich; his close relations with both men and their colleagues were for him irreplaceable in the difficult period.

Quantum mechanics led to entirely new conceptions in the description of the microscopic world and to a revision of some basic physical and philosophical notions, such as the principle of causality. After the completion of the theory by Born, Bohr and Heisenberg, the latter two, as did others, devoted considerable attention to obtaining many of the intellectual consequences of the new theory. These consequences often went far beyond the realm of physics, extending into chemistry, biology, and even into social and ethical phenomena. To many laymen, however, the new ideas seemed as strange and disturbing as the revisions brought about earlier by the relativity theory – if not more so. In order to explain the new conceptions to nonspecialists and to relieve the public of any bewilderment, which – as with relativity theory – grew dangerously infected with ideology, Heisenberg wrote for and spoke often before wide audiences of laymen interested in science during the 1930s. His widely read first collection of essays, *Wandlungen in den Grundlagen der Naturwissenschaft* (1935), contained two such addresses.

Despite the political situation in Germany, which also affected his collaborators, Heisenberg's physics continued to flourish. The main focus of his research concentrated during the 1930s on the analysis of problems in relativistic quantum field theory. A major component of this work involved the analysis of elementary processes in high energy cosmic-ray interactions, where the new elementary particles positron and meson were discovered in 1932 and 1937, respectively. For the solution of problems in field theory and the explanation of cosmic-ray phenomena, fundamentally new ideas appeared to Heisenberg unavoidable. In cosmic-ray physics he argued in favor of the existence of particular multiple processes, which he called "explosion showers", that could be comprehended only by means on a nonlinear quantum field theory (June 1936, May 1939). He further claimed that the divergence difficulties in quantum field theory had to be resolved by assuming a fundamental length that limited the validity of field theory at very small distances (June 1938).

In the years following 1933 Heisenberg became the main spokesman in Germany for modern theoretical physics. Although the number of his students decreased, Heisenberg presented systematic lectures on all fields of theoretical physics, including the politically disfavored relativity – both the special and the general theories. He defended relativity theory even in an article published in the Nazi Party newspaper *Völkischer Beobachter* on 26 February 1936. Together with two colleagues in experimental physics, Hans Geiger in Berlin and Max Wien in Jena, Heisenberg authored a memorandum to the Reich Minister for Education and Science *(Reichsminister für Erziehung und Wissenschaft)* opposing the official discrediting of theoretical physics and emphasizing its great importance for the training of needed physicists. His opponents, who gathered around Philipp Lenard and Johannes Stark, sought to dismiss relativity and quantum theory as "Jewish" and "non-German".

After Arnold Sommerfeld reached retirement age in 1934, attacks by representatives of the so-called "German physics" greatly increased when Heisenberg was named by the faculty of the University of Munich as the first candidate to succeed him. On 25 July 1937 an article entitled "Weiße Juden in der Wissenschaft" ("White Jews in Science") and signed by Stark appeared in *Das schwarze Korps,* the journal of the SS. It contained wild accusations against modern theoretical physics and theoreticians, such as Heisenberg, and closed with the demand that Heisenberg and his colleagues be made to "disappear" like the Jews. Heisenberg responded with a letter to the head of the SS, Heinrich Himmler, urgently requesting a termination of the campaign against him. After a long series of investigations and interrogations by the SS, Heisenberg attained his goal in a letter from Himmler on 21 July 1938. The campaign ceased, but the Munich chair went in December 1939 to the aerodynamicist Wilhelm Müller. Heisenberg managed, however, to have an article on the significance of modern theoretical physics published in the organ of the Reich student's organization. The article, entitled "Die Bedeutung der 'modernen' theoretischen Physik" ("The meaning of 'modern' theoretical physics"), finally appeared in 1942 after considerable delay, and was eagerly read by many young people.

In spite of the official discrediting of theoretical physics in the Third Reich, Heisenberg and Hund managed to keep an unusually stimulating atmosphere in their institute, made even more lively by many foreign guests. Besides the German members of the institute (E. Bagge, S. Flügge, B. Kockel and H. Voltz), there were a number of visitors from America (R. S. Mulliken, J. C. Slater, J. H. Van Vleck, C. Zener), Italy (U. Fano, G. Gentile), Japan (K. Ariyama, Y. Fujioka, S. Kikuchi, S. Tomonaga, K. Umeda, S. Watanabe), Scandinavia (B. O. Grönblom, H. Wergeland), Hungary (L. Tisza) and Yugoslavia (I. Supek), to name only a few. Heisenberg also continued his relations with foreign physicists through lectures and travels to scientific conferences. In 1937 he attended the Galvani Conference in Bologna (18–21 October), in April 1934 and March 1938 he traveled to England (to Cambridge and Manchester, respectively), and in summer of 1939 he went to America where he presented his latest results on the theory of cosmic-ray showers at the Symposium on Cosmic Rays at the University of Chicago (27–30 June). Although he received several tempting offers of positions in America, he decided to return to Germany.

In April 1937 Heisenberg married Elisabeth Schumacher, daughter of Hermann Schumacher, a professor of economics at the University of Berlin. In early 1938 the twins Maria and Wolfgang were born; the family was later increased by Jochen (1939), Martin (1940), Barbara (1942), Christine (1944) and Verena (1950). For a secure place of refuge during the imminent war the Heisenbergs bought in 1939 a summer house in Urfeld on Lake Walchen, south of Munich, that had previously belonged to the painter Lovis Corinth. When Heisenberg transferred to Berlin in 1942 to assume directorship of the Kaiser Wilhelm Institut für Physik, his family did not accompany him but moved to Urfeld rather than to Berlin, for safety against bombing raids.

5. The War Years (1939–1945)

The outbreak of war on 1 September 1939 directly affected Heisenberg's scientific career. He was called to military service, and to his surprise was ordered to report to the Army Weapons Bureau (*Heereswaffenamt)* in Berlin. There he and other leading German atomic scientists, the so-called Uranium Club *(Uranverein),* were asked to investigate whether the fission of uranium, discovered in December 1938 by Otto Hahn and Fritz Strassmann in Berlin, could be used for gaining energy on a large scale. Within two months Heisenberg completed a comprehensive report in which he developed the theory of a chain reaction with uranium fission by neutrons (6 December 1939); he followed it soon by a second report refining the theory and taking into account recently improved data (29 February 1940).

The Army Weapons Bureau designated the *Kaiser-Wilhelm-Institut (KWI) für Physik* in Berlin-Dahlem as the center of its uranium research. Peter Debye, director of the institute and a Dutch citizen, was prevented as a foreigner from working on the secret research. He placed himself on leave and accepted a guest professorship at Cornell University in January 1940. Kurt Diebner assumed Debye's office as head of the Berlin uranium project. At the suggestion of C. F. von Weizsäcker and especially Karl Wirtz, two of Debye's former collaborators now working on the project, Heisenberg acted as scientific advisor. He thus traveled weekly to Berlin from Leipzig and wrote several reviews on the progress of research at the Berlin institute (December 1940–February 1942). Other, non-secret research begun under Debye (for example on X-ray analysis and on nuclear and low temperature physics) continued at the *KWI für Physik* under the formal supervision of Max von Laue, vice-director. During 1941 and 1942 Heisenberg also directed a seminar at the institute in which he and other members and guests of the institute discussed the recent progress and problems of cosmic-ray physics. These talks were published in 1943 in the collection *Kosmische Strahlung* (Cosmic Radiation) dedicated to Sommerfeld on his 75th birthday.

In spring of 1940 Heisenberg began experiments with Robert and Klara Döpel at the University of Leipzig in which they studied possible arrangements of uranium and neutron moderator substances, since for uranium fission slow neutrons turned out to be most effective. These experiments substantiated the usefulness of heavy water as moderator (28 October 1941) and indicated in early 1942 that a spherical arrangement of natural uranium and heavy water led to a small multiplication of the irradiating neutrons (confirmed in June 1942). Thus the prerequisites for the construction of a functioning reactor were fulfilled. On 4 June 1942 a meeting was held between Albert Speer, Minister for Armament and War Production *(Reichsminister für Rüstung und Kriegsproduktion),* and other officials and military leaders and the scientists, including O. Hahn, W. Heisenberg, K. Diebner, P. Harteck and K. Wirtz. There it was decided that the uranium project should continue with the goal of constructing a nuclear reactor, but not an atomic bomb. As a result the Army Weapons Bureau returned the *KWI für Physik* to the Kaiser Wilhelm Society *(Kaiser-Wilhelm-Gesellschaft),* but the reactor project remained secret. Heisenberg, the chief scientific advisor to the institute since 1940 and theoretical director of the successful Leipzig experiments, was named head of the leading reactor research group in Germany and appointed on July 1942 director *at* the *KWI für Physik* (to substitute for the absent director Debye). He was assisted by many of Debye's assistants, including Horst Korsching and Karl Wirtz, as well as by Erich Bagge and Fritz Bopp. Kurt Diebner, previous head of the Berlin project, went with several collaborators to the army laboratory at Gottow where he set up his own uranium experiments. C. F. von Weizsäcker accepted a professorship for theoretical physics at the University of Strasbourg.

The work on the uranium project did not exhaust Heisenberg's activities during the war years. He occupied himself particularly with problems in cosmic-ray physics and developed in several papers a new approach to the theory of elementary particles based on the concept of the scattering matrix. Three papers, received in September and October 1942 and May 1944, respectively, were published in the *Zeitschrift für Physik.* He regularly gave lecture courses at the universities of Leipzig and Berlin (after summer 1942) on topics of theoretical physics. And he traveled, as far as the war restrictions allowed, to foreign countries, e. g. to Budapest (April 1941), to Zurich and Bern (November 1942) and to Leiden and Utrecht (October 1943), to meet with colleagues and friends and to deliver talks on his (non-secret) scientific research or on more general questions of physics and science. A particular visit was paid in October 1941 to Niels Bohr in Copenhagen. Without violating the secrecy of his work on the uranium project, Heisenberg tried to convey the message that in Germany the construction of the atomic bomb was not being considered. The failure of this visit cast a shadow on the future relations between Heisenberg and his teacher and friend.

Despite the early successes, which at first gave the German reactor project a slight edge over Fermi's American project, progress was very slow. Although the basic scientific and technological problems had been overcome and only the task remained of assembling enough uranium and heavy water to create an energy producing reactor, the project was increasingly hindered by the war conditions. Germany could barely provide enough uranium and not nearly enough heavy water, deficiencies made more acute by the fact that at least three groups performed competing experiments: besides the largest group in Berlin, a group in Hamburg under Paul Harteck and the group in Gottow under Diebner. The increasing air raids on Berlin forced the transfer of all apparatus to a bunker in the Dahlem institute, and finally, in late 1944, the entire project was shipped out of Berlin and housed in a rock bunker under the castle chapel in the southern German town of Haigerloch. The *KWI für Physik* had already been transferred earlier (since 1943) to the nearby town of Hechingen. In the first months of 1945 the model reactor B_8 nearly went critical (the point at which the reaction is self-sustaining). But the lack of a few hundred kilograms of uranium (out of 1.5 tons) and of several hundred liters of heavy water prevented success before war's end.

At the end of April 1945 members of an American science intelligence unit, the Alsos Mission, reached Tailfingen (where Otto Hahn's *KWI für Chemie* had moved), Hechingen and Haigerloch just ahead of the advancing French troops. The German atomic scientists gathered there – Bagge, Hahn, Korsching, von Laue, von Weizsäcker and Wirtz – were taken prisoner. Heisenberg, however, had gone a few days earlier by bicycle to Urfeld to join his family. He was taken prisoner there on 3 May 1945 by a small unit of the Alsos Mission led by Colonel

Boris T. Pash. Heisenberg and Walther Gerlach, Reich Commissioner *(Reichsbeauftragter)* for the reactor project, along with Diebner and Harteck, soon joined the Tailfingen-Hechingen group in France. All ten scientists were transferred to Castle Facquerel in Belgium and finally to England (on 23 July 1945), while the Americans dismantled the Haigerloch reactor and transported all of the apparatus, uranium, heavy water, and numerous documents to the United States.

6. The Period of Reconstruction and Renewal (1946–1958)

In April 1945 Soviet troops reached Berlin, the Reich collapsed and the Russians began sending to the Soviet Union the remaining apparatus in the Dahlem institute, the institute library, and the several German scientists they found there (among them Ludwig Bewilogua who had built the low temperature laboratory and had run the high voltage generator during the war). The Berlin *KWI für Physik* ceased to exist, while the Hechingen division barely survived with a handful of scientists – including Fritz Bopp, the crystallographer Georg Menzer and the spectroscopist Hermann Schüler – and a handful of experiments unrelated to nuclear physics.

Heisenberg was held along with the other prominent members of the Uranium Club at the country estate Farm Hall near Cambridge, England, well treated and equally well watched. During their stay at Farm Hall the atomic bomb was dropped on the Japanese city of Hiroshima on 6 August 1945. From news reports the German scientists learned that the Allies had, in their efforts to gain energy from atomic fission, obviously progressed much further than the Germans and had achieved military uses, an application that the Germans had believed to lie only far in the future. Their many discussions about the consequences of military applications of nuclear energy contributed to their later stance against nuclear weapons for the German Army.

At Farm Hall Heisenberg also set the stage for his future research. After years of difficulty, political pressure and reduced exchange of ideas he took the opportunity to devote himself in leisure to a wide range of physical problems. Included were the still unresolved problems of explaining superconductivity, which he discussed with von Laue, and the application of statistical methods to the description of turbulence, discussed with von Weizsäcker. He also decided to remain at the *KWI für Physik* if possible, despite lucrative offers to come to the United States.

On 3 January 1946, the German scientists were released from Farm Hall and returned to Western Germany. Heisenberg went with Hahn, von Laue, von Weizsäcker and Wirtz to Göttingen in the British Zone, where Max Planck and Ernst Telschow were seeking to continue the Kaiser Wilhelm Society and its institutes. Heisenberg devoted himself upon his return to two large tasks: the reconstruction of the *KWI für Physik* as a center for experimental and theoretical research in physics and the renewal of research in Germany. In particular he pondered a suitable vehicle for mediating between scientific research and the future German government in order to insure technical prosperity and the protection of science as the source of technical progress.

During the immediate post-war years Germans had to rebuild practically from the ground up. With the collapse of economic and political organization and the division of Germany into four occupation zones, the reconstruction of research in Germany was extremely difficult.

The conditions were perhaps most favorable in the British Zone, where the authorities, especially the liaison officer Colonel Bertie Blount, appeared quite open to the needs of the German scientists. As early as 1 January 1946 scientists and officials formed the German Scientific Council *(Deutscher Wissenschaftlicher Rat)* under the chairmanship of the Nobel Prize chemist Adolf Windaus. It mediated between the British military government and German scientific institutes and enabled, among other activities, the rebuilding of the Physico-Technical Institute *(Physikalisch-Technische Reichsanstalt)* in Braunschweig and the refounding of the German Physical Society *(Deutsche Physikalische Gesellschaft)* within the British Zone. The old British Hanoverian university city of Göttingen, largely untouched by the war, became once again a scientific center. During a meeting with Göttingen scientists in October 1946 the British authorities permitted the installation in Göttingen of several Kaiser Wilhelm institutes on the grounds of the former Aerodynamical Research Institute *(Aerodynamische Versuchsanstalt)* which had been completely dismantled as war-related. The institutes for physics, physical chemistry and medical research, previously all in Berlin, were thus revived. Finally the Kaiser Wilhelm Society was renamed and reinstituted in the three western zones as the Max Planck Society with seat in Göttingen; after 16 February 1948 all its institutes in western Germany were renamed Max Planck institutes (MPI).

In July 1946 Heisenberg began reequipping the institute for physics, whose direction he assumed after the official director Debye declined an offer to return. Two departments were set up initially: one for theoretical physics under von Weizsäcker, the other for experimental physics under Wirtz. Research was limited by the directives of the Allied Control Commission in that Germans were forbidden to work in a variety of fields considered war related, especially areas of nuclear physics involving slow neutrons and nuclear fission. Despite the limitations Heisenberg and his colleagues, many of whom had worked with him earlier in Berlin, did not lack research topics. Foremost remained the study of cosmic-ray physics, begun in Leipzig, continued in Berlin, and now picked up once again in Göttingen. Most of the institute resources were directed at first into theoretical research, since financing did not suffice for experimental studies. Only gradually did the institute catch up with the rapid developments occurring elsewhere in cosmic-ray and elementary particle physics at that time. The second edition of *Kosmische Strahlung,* published in 1953, indicated a near approximation to international standards in this subject. During the early post-war years Heisenberg also presented a theory of superconductivity (1946–1948), a statistical theory of turbulence (1946–1948), and examined the theoretical properties of elementary particles (after 1946).

Besides the direction of his institute and his own research Heisenberg devoted himself with great energy to the revival of scientific research in Western Germany and especially to his conception of governmental science policy. Learning from his experiences in the Third Reich, his observations of the close cooperation between scientists and government in Great Britain, and believing that a modern industrial state required a central science policy, Heisenberg sought a direct role for the government of the new Federal Republic of Germany in forming a national policy of support for science and technology and a role for science advisors to the chancellor. Such conceptions found expression in the German Research Council *(Deutschen Forschungsrat),* founded on 9 March 1949 by the Max Planck Society and the surviving academies of science in West Germany. It was composed of 15 leading scientists with Heisenberg as president. The new council, supported by Chancellor Konrad Adenauer, represented German science in international affairs and directly in the chancellor's office. Among its successes were the acquisition of Marshall Plan money for the support of German research, the admission of the Federal Republic to the International Union of Scientific Councils and to

UNESCO, the provision of federal responsibility for science in the West German constitution, and the assurance of mutual support between German industry and science. Yet the institution of the German Research Council contradicted the long tradition in Germany that science fell under the authority of the cultural ministers of each state *(Land)*. Hence the council was challenged increasingly by the Emergency Association of German Science *(Notgemeinschaft der Deutschen Wissenschaft)*, revived in January 1949 by the ministers of culture and the university rectors. To avoid rivalry and the damaging overlap of interests, Heisenberg, with much regret, gave up his idea of a federal science body. The two bodies, i. e. the Research Council and the Emergency Association, were joined into the present-day German Research Association *(Deutsche Forschungsgemeinschaft)* in August 1951. The task of representing German science was assumed by the senate of the new body, composed mainly of the old Research Council. Heisenberg was elected to the presidential committee of the new Research Association; he also directed after February 1952 its Commission for Atomic Physics *(Kommission für Atomphysik)* which coordinated nuclear research (with the exception of research in nuclear fission) nationwide.

Heisenberg's constant involvement in German science policy found its counterpart in his involvement in the organization of research in his own institute. On 1 July 1947 a new astrophysical department was added under the direction of Ludwig Biermann, who had earlier worked at the Babelsberg observatory and the University of Hamburg. Biermann and his collaborators concentrated in Göttingen on cosmic plasma physics and on quantum theoretical problems in astrophysics. They soon attained significant results, in particular the prediction of the so-called "solar wind" from an analysis of the behavior of comet tails (Biermann and Rhea Lüst) and a new formation of the basic equations of plasma physics (Arnulf Schlüter). Their work required, as did other topics treated at Heisenberg's institute, the increased application of electronic data handling. Since such tools were not available at that time in Germany, the Max Planck Society began its own development program. Heinrich Billing, since 1 June 1950 head of the computer group at the Institute for Instrument Studies *(Institut für Instrumentenkunde)* in Göttingen and after 1957 a member of the *MPI für Physik*, developed and constructed the programmed calculating machines G1 and G2; the former was used after October 1952 at the MPI in Göttingen. Billing's breakthroughs in computer technology were unfortunately not realized by German industry.

The work produced by the Göttingen MPI during the years of reconstruction won a growing international reputation, and the institute's director Heisenberg attained increasing recognition as the leading representative of German science in the international arena. The reestablishment of international relations, also of great importance to Heisenberg, began officially with his invitation to lecture at the British universities in Cambridge, Edinburgh and Bristol in December 1947. During the following years he repeatedly visited Niels Bohr in Copenhagen and in summer 1950 he participated in the International Congress of Mathematicians in Cambridge, Massachusetts (30 August–6 September). In 1954 he served as the West German delegate to the conference on Atoms for Peace in Geneva. When in 1952 the European Council for Nuclear Research came into being, Heisenberg headed the German delegation and participated in the decision to locate the large European research center for high energy physics, CERN, in Geneva, Switzerland. The Scientific Policy Committee of CERN, responsible for planning the research program, elected Heisenberg as its chairman.

Immediately after his return to post-war Germany Heisenberg emphasized in a programmatic speech before Göttingen students the role of "science as a tool of mutual understanding between peoples" (13 July 1946). He plunged into the task of getting German scientists reaccepted as members of the international family of scientists and of renewing personal relations after the isolation and alienation in the previous period of the Third Reich and World War II. In this endeavor he assigned a particularly important role to the reestablished Alexander von Humboldt Foundation, whose first president he was appointed on 10 December 1953 by Chancellor Adenauer. The purpose of the foundation is to enable young scholars and scientists from around the world to collaborate with German colleagues while guests at German research institutes. Numerous Humboldt fellows were invited to Heisenberg's institute in Göttingen and later in Munich. He held the president's office until a severe illness forced him to resign in October 1975.

On 5 May 1955 the Paris Accords came into effect, whereby the Western Allies granted the Federal Republic full sovereignty and full membership in the NATO alliance. All restrictions on West German research were thus removed and, after ten years of restraint, the development of German nuclear energy resumed in full. In October 1955 Adenauer created a Federal Ministry for Atomic Questions *(Bundesministerium für Atomfragen),* forerunner of the wider ranged, present-day Ministry for Research and Technology *(Bundesministerium für Forschung und Technologie).* Heisenberg served as a leading member of the German Atomic Commission *(Deutsche Atomkommission)* set up by the new ministry and composed of scientists, industrialists, and politicians with the purpose of advising the ministry on nuclear energy policy. Heisenberg also directed the committee on nuclear physics within the Atomic Commission, served as a member of the Bavarian Atomic Commission, and acted as the main impetus toward the construction of Germany's first nuclear reactor, a research model set up at Garching near Munich in October 1957.

While Heisenberg supported the development of nuclear energy for peaceful uses, he and other scientists equally energetically opposed the plans of Chancellor Adenauer to equip the West German army with tactical nuclear weapons. Adenauer met with stiff resistance from Hahn, Heisenberg, von Weizsäcker and other atomic scientists, eighteen of whom issued a public declaration (basically formulated by von Weizsäcker and Heisenberg) from Göttingen on 12 April 1957, opposing research on or possession of nuclear weapons by West Germany. The West German army has since remained non-nuclear.

Heisenberg's main scientific interest in the early fifties increasingly focused on the search for a consistent quantum field theory of elementary particles. After an unsuccessful attempt at obtaining a nonlocal theory, he became concerned after 1952 with the investigation of nonlinear field equations, in which the mathematical space of states was extended beyond that used since the early days of quantum mechanics. In this theory, which Heisenberg and his collaborators developed in a series of papers between November 1953 and December 1956, the conditions of relativistic invariance could be immediately introduced and finite results could always be obtained without the use of supplementary subtraction or normalization procedures. After Heisenberg had demonstrated the consistency of his ideas and methods in the case of a model field theory, the so-called Lee model (October 1957), he entered into a close collaboration with Pauli that yielded in early 1958 the proposal of a nonlinear spinor equation designed to describe the properties and the behavior of all known elementary particles (dubbed the "world formula" by eager journalists). While Heisenberg expounded the equation in April 1958 and on several later occasions, Pauli withdrew his support.

Heisenberg's efforts to obtain a consistent quantum field theory for all elementary particles harmonized with the philosophical views that he presented at that time in many public lectures and lecture series (e.g., in the Gifford Lectures delivered in the winter term 1955/56 at St. Andrew's University in Scotland).

In September 1958 Heisenberg's *MPI für Physik* moved from Göttingen into a large new building in Munich near the English Garden, which solved the problem of overcrowding at the old institute. Differing from original plans, Karl Wirtz and his reactor group did not move to Munich, but left the MPI in March 1957 and settled at the nuclear research center of the *Kernreaktor Bau- und Betriebsgesellchaft mbH.* close to Karlsruhe. C. F. von Weizsäcker did not go to Munich either, as he accepted in June 1957 a call to a philosophy chair at the University of Hamburg. Yet he remained a regular guest in the Munich MPI during his semester vacations.

7. The Munich Years (1958–1976)

When Heisenberg moved to Munich in 1958, he characteristically broadened the program of his institute, which ultimately led to the establishment of a series of new institutes. First, the former astrophysics department under Ludwig Biermann was raised to an *Institut für Astrophysik* which, together with the *Institut für Physik,* formed Heisenberg's *Max-Planck-Institut für Physik und Astrophysik.* The program of the Göttingen institute was divided between the two institutes. The groups for experimental and theoretical elementary particle physics (and related fields) constituted, together with a group on experimental plasma physics (under Gerhard von Gierke), the physics institute. The astrophysics institute housed a department for electronic computers (Billing), one for theoretical astrophysics (Reimar Lüst) and one for theoretical plasma physics (Schlüter). With the addition of experimental plasma physics the institute compensated for the loss of Wirtz' reactor group and kept an open option on contributing to fundamental research in nuclear fusion. Yet it soon overflowed the available laboratory space in Munich and in June 1960 began moving to Garching, to form the present *MPI für Plasmaphysik* (move completed in 1968). Heisenberg remained, however, in the scientific administration of the new institute. The *MPI für extraterrestrische Physik* under Reimar Lüst also emerged (in 1964) from Heisenberg's institute and settled on the grounds of the Garching unit, but it remained an official sub-division of the mother institute.

While Heisenberg played an active role in these extensions of his institute, he remained convinced that the large experimental and theoretical tasks in elementary particle physics could only be solved by international efforts. He thus continued to aid and advise the European laboratory CERN near Geneva, which began operations in January 1959. Many members of his institute performed experimental work at CERN. During the later expansion of CERN Heisenberg actively supported the construction of the storage rings, which he dedicated on 16 October 1971. He also provided essential support during the early planning stages for the electron synchrotron (DESY) near Hamburg, whose first high-energy accelerator went into operation with a 6 GeV electron beam in February 1962.

Among Heisenberg's theoretical researches in Munich the development and evaluation of the nonlinear spinor theory occupied center place. In March 1959 he completed with his collaborators Hans-Peter Dürr, Heinrich Mitter, Siegfried Schlieder and Kazuo Yamazaki a long paper enunciating the principles of the theory together with several consequences; they obtained, in particular, its resonance states, one of which, the η-meson, was found more than a year later. Further investigations yielded an organizing of nucleons and hyperons with the help of the spurion concept (December 1964) and a calculation of the electromagnetic fine structure coupling constant (January 1965). Heisenberg wrote an introductory textbook on nonlinear spinor theory, which was published in 1966, and he retained his belief in it until the end of his life. He argued that all experimental researches on the structure of matter during the past decades had yielded either the already known elementary particles, or new ones having the same qualities. Moreover, the assumption of successive layers of further, ever more elementary objects, such as quarks, would not allow the achievement of the main goal of a fundamental theory, i. e., to explain the dynamical behavior of matter; it would rather relegate the problem to the next deeper level. It appeared to Heisenberg that his nonlinear spinor theory agreed in spirit with Plato's representation of the structure of matter by simple geometrical forms. He proposed to replace Plato's form by a highly symmetrical field equation. All properties of matter should follow from this field equation and the imposed conditions, although the detailed computations involved nonlinear and often very complicated approximation procedures.

In Heisenberg's later work the foundations of physics, which guarantee the unity of the description of nature, tended to merge with the conceptions of Plato's world view. He tirelessly presented his ideas about the intimate connection between physics and philosophy to wider audiences. He saw this connection verified in the historical development of quantum physics, describing parts of his own role in that development in his recollections *Der Teil und das Ganze* (1969, in English: *Physics and Beyond,* 1971). But the connection was also reflected in Heisenberg's standpoint on questions of art, and even on questions of religion and society.

On 31 December 1970 Heisenberg resigned the directorship of the *MPI für Physik und Astrophysik,* the institute that he had directed for almost 30 years. Yet he still came regularly to his office and continued to work on scientific papers and on more general articles, lectures and the second edition of his book on the nonlinear spinor theory. He participated at selected conferences, such as the symposium in honor of Dirac's 70th birthday in Miramare near Trieste (18–25 September 1972) and the colloquium celebrating the 200th anniversary of the Brussels Academy (16–17 May 1973). He traveled to the United States for the last time in April 1973. In the middle of 1973 he fell seriously ill. He slowly improved and a year later he appeared to have fully recovered. But in July 1975 he suffered a severe relapse. He died at his home in Munich on 1 February 1976.

Heisenberg received numerous national and international awards in recognition of his work and his influence on science and society. He received, besides the Nobel Prize for physics, the Barnard Medal of Columbia University (New York, 1929), the Matteucci medal of the *Accademia Nationale dei Lincei* (Rome, 1929) the Planck medal of the German Physical Society (1933), the Copernicus prize of the University of Königsberg (1943), the Hugo Grotius medal (1956), the order *Pour le mérite für Wissenschaften und Künste* (1957), the *Kulturpreis* of the City of Munich (1958), the Niels Bohr medal (Copenhagen, 1970) and the Guardini prize (Munich, 1973). In addition he received high national and international honors. He was a member of more than 30 scientific societies, including the Saxonian Academy (Leipzig, 1930), the *Kaiserlich Leopoldinisch-Carolinische Akademie der Naturforscher* (Halle, 1933), Norwegian Academy of Sciences (Oslo, 1936), Göttingen Academy of Sciences (1938), the Royal Swedish Academy (Uppsala, 1938), the *Société Philomatique* (Paris, 1938), the Royal Dutch Academy of Sciences (Amsterdam, 1939), the Prussian Academy of Science (Berlin, 1943), the *Accademia Nazionale dei Lincei* (Rome, 1947), the Bavarian Academy of Sciences (Munich, 1949), the Royal Danish Academy of Sciences (Copenhagen, 1951), the Pontificial Academy of Sciences (1955), the Royal Society of London (1955), the American Academy of Arts and Sciences (Boston, 1958) and the US National Academy of Sciences (Washington, 1961). At the very beginning of Heisenberg's studies Sommerfeld reminded him of the words of Schiller: „Wenn Könige bauen, haben Kärrner zu tun." ("When kings build, wagoners have to work.") By this he meant that his pupil had first to work along side the wagoners. Little did he suspect how quickly Heisenberg would begin to build and to contribute, as did few others in our century, to the construction of modern theoretical physics.

David C. Cassidy and Helmut Rechenberg

Group 1

Hydrodynamic Stability and Turbulence (1922–1948)

Hydrodynamic Stability and Turbulence (1922–1948)

An Annotation by S. Chandrasekhar, Chicago

Part I

Heisenberg's long paper (No. 2, p. 31 *), devoted in part to a study of the stability of plane-parallel laminar flows, is, by any standards, an important and fundamental contribution to the subject. In this paper, besides describing the essential steps that must be taken to solve the underlying mathematical problem, Heisenberg devises for the first time the use of "inner" and "outer" approximations with suitable matching conditions for the solution of ordinary differential equations (of order two or higher) with a turning point – a method later to be described by the initials "W.K.B.". And beneath the mathematical developments, one can discern the operation of a powerful physical insight. Heisenberg himself described later the physical and the mathematical ideas in this early work of his in an address to the International Congress of Mathematicians in 1950 [1]. Perhaps a somewhat more technical account of these ideas may be useful as an introduction to his paper.

The problem, Heisenberg considers, is the classical one of the stability of plane laminar flow (in the x-direction, say) confined between two parallel planes (which we may assume to be at $y = \pm 1$, by measuring distances in units of the separation, d, between the two planes) with a velocity profile $w(y)$ (measured in units of the velocity U at, say, $y = 0$). Restricting oneself to two-dimensional disturbances (which by Squire's theorem entails no loss of generality) and expressing the perturbations, u and v, in the velocities in the directions x and y in terms of a stream function, $\psi(x, y)$, in the manner

$$u = \frac{\partial \psi}{\partial y} \quad \text{and} \quad v = -\frac{\partial \psi}{\partial x}, \tag{1}$$

and putting

$$\psi(x, y) = \varphi(y) \exp[\mathrm{i}\,\alpha(x - ct)], \tag{2}$$

one obtains the Orr-Sommerfeld equation,

$$(w - c)(\varphi'' - \alpha^2 \varphi) - w'' \varphi = -\frac{\mathrm{i}}{\alpha R}(\varphi'''' - 2\alpha^2 \varphi'' + \alpha^4 \varphi), \tag{3}$$

where primes denote differentiations with respect to y and

* All page numbers refer to pages in this volume. (Editors)

$$R = Ud/\nu \tag{4}$$

is the Reynolds number of the flow (ν, denoting the kinematic viscosity).

In the inviscid limit of zero viscosity and infinite Reynolds number, equation (3) reduces to Rayleigh's equation,

$$(w-c)(\varphi'' - \alpha^2\varphi) - w''\varphi = 0\,. \tag{5}$$

In the context of this equation, Rayleigh had shown that a necessary and sufficient condition for instability is that $w(y)$ has a point of inflexion in the interval, $-1 < y < +1$. For Poiseuille flow,

$$w(y) = 1 - y^2\,, \tag{6}$$

there is no point of inflexion, and the inviscid flow is stable. But the flow *is* unstable when viscosity is present and the Reynolds number exceeds a certain critical value, R_c. The problem is to ascertain this critical Reynolds number for the onset of instability, and Heisenberg proceeds as follows.

First, Heisenberg seeks a solution of the Orr-Sommerfeld equation by making the substitutions,

$$\varphi = \exp[\textstyle\int g(y)\,dy] \tag{7}$$

and

$$g(y) = (\alpha R)^{1/2} g_0(y) + g_1(y) + \mathrm{O}[(\alpha R)^{-1/2}]\,. \tag{8}$$

These substitutions provide approximations to two solutions, φ_1 and φ_2, which are rapidly varying when αR is large. (The method of approximation sought here is the same as the Wentzel-Kramers-Brillouin approximation familiar in the quantum theory.) To obtain approximations to two additional solutions, φ_3 and φ_4, Heisenberg considers an expansion in powers of $(\alpha R)^{-1}$ so that in the lowest order, these approximations are the solutions of Rayleigh's equation (5). These "inviscid" approximations are further expanded in powers of α^2. This is the weakest point in Heisenberg's analysis. In the more recent investigations, one uses instead the exact solutions obtained by numerical methods.

Heisenberg next considers the manner of connecting the approximate solutions obtained on either side of the turning point y_0 where $w(y_0) - c = 0$. He accomplishes this matching by introducing a "stretched variable",

$$\eta = (y - y_0)(\alpha R)^{1/3}\,, \tag{9}$$

and expanding φ_1 and φ_2 in powers of $(\alpha R)^{-1/3}$. This procedure leads to what one now calls the "inner" expressions of the solutions. In dealing with φ_1 and φ_2 only the first approximation is required; and Heisenberg obtains this approximation from Hopf's solution [2] for linear profiles. When $\alpha \ll 1$, $\varphi_3 = w - c$ and no connecting relation is needed. But in the case of φ_4, two terms of the inner expansion are required; and they must be matched on either side of y_0 to

$$\varphi_4 \simeq (w-c)\int_{-1}^{y} dy\,(w-c)^{-2}\,. \tag{10}$$

The outer approximations to the solutions were then used to obtain the corresponding approximations to derive the dependence of the characteristic value R_c on α. By this procedure Heisenberg established (a) that velocity profiles which are unstable in the absence of viscosity (i.e., velocity profiles which satisfy Rayleigh's criterion of having an inflexion point of the interval, $-1 < y < +1$) remain unstable for all sufficiently large values of R; and (b) that velocity profiles which are monotonic in the interval, $-1 < y < +1$, without inflexion points (and hence stable in the absence of viscosity) remain stable in the presence of viscosity. And finally, Heisenberg attempted for the first time to compute the entire curve of marginal stability for Poiseuille flow. Because of the approximate nature of his solutions, he was able to obtain only rough approximations to the upper branch of the curve of marginal stability (see Fig. 2 of his 1924 paper). Nevertheless, using rough estimates, he was able to sketch the correct qualitative behaviour of the entire curve; and he further estimated that

$$\alpha \simeq 0.7 \quad \text{and} \quad R_c = O(10^3)\,. \tag{11}$$

For many years following Heisenberg's investigation, his methods and his results were matters of controversy; but they were eventually justified by C. C. Lin [3] in 1945. Finally, for comparison, we may note that the presently accepted values for marginal stability are

$$\alpha = 1.097 \quad \text{and} \quad R_c = 5.772 \times 10^3\,. \tag{12}$$

In spite of the limitations and approximations which Heisenberg used (the most serious being his restriction to small wave-numbers in his outer expansions), his treatment of the Orr-Sommerfeld equation has had a major impact on all subsequent developments. For a review of these later developments, see Drazin and Reid [4].

Part II

In the second part of his 1924 paper, Heisenberg considers fully turbulent flow and attempts to derive the profile of the mean velocity of the flow which then prevails. For this purpose, he considers a mean-field approximation in which the motion is governed by the same Orr-Sommerfeld equation in which the mean flow replaces $w(y)$ and allowance is made for the action of Reynolds' stresses. The basic idea is to expand the stream function in a Fourier series, which for flows symmetrical about the y axis is of the form,

$$\psi = \sum_{n=-\infty}^{+\infty} \varphi_n(y) \exp[\mathrm{i}n(\beta t - \alpha x)] \tag{13}$$

where $\varphi_{-n} = \varphi_n^*$ and $\varphi_0' = \bar{w}$ is the mean flow which results from the finite amplitude of the motions prevailing in the nonlinear domain. By going to a second approximation, Heisenberg seeks to relate the mean velocity distribution with that which follows from Blasius' law, $w \sim \eta^{1/3}$ (where η denotes the distance from the wall) for the boundary layer. The significance of this part of his investigation is difficult to assess; but the approach itself has influenced and inspired later work (cf., Meksyn and Stuart [5] and Herbert [6]).

Part III

Already at the time Heisenberg was writing his 1924 paper, his primary interests were centred on the basic contradictions that beset the "old" quantum theory of Bohr and Sommerfeld. It is in this area that Heisenberg was soon to make his far-reaching contributions. But after the end of World War II, he briefly returned to his earlier interest in hydrodynamic turbulence, stimulated by his then close association with von Weizsäcker. However, this time, Heisenberg turned to the statistical theory of turbulence and the scaling laws of Kolmogoroff and Onsager based on the notion of the cascade of energy from large eddies (small wave numbers) via the inertial range to small eddies (large wave numbers) in which energy is dissipated by viscosity. Heisenberg incorporated these notions in an "elementary theory" (Nos. 3, 5, and 6) which for a time played a significant role in the developments. The basic ideas of Heisenberg's theory can be formulated as follows.

Considering the rate of dissipation of energy by eddies with wave numbers of less than a particular k, Heisenberg distinguishes between the energy directly dissipated in the form of molecular motion and thermal energy and the energy communicated in the form of kinetic energy to all eddies with wave numbers exceeding the specified k. According to the equations of hydrodynamics, we can write for the dissipation into thermal energy, per unit volume, the expression,

$$\varepsilon_k(\text{thermal}) = \varrho\nu\overline{|\operatorname{curl} v|^2} = 2\varrho\nu\int_0^\infty F(k)\,k^2 dk\,, \tag{14}$$

where $F(k)$ denotes the spectrum of turbulence (derived from a Fourier analysis of the prevailing velocity field) and ν is the coefficient of kinematic viscosity. In analogy with this equation Heisenberg writes for the energy transferred to all eddies with wave numbers exceeding a given k by eddies with wave numbers less than this k, the expression,

$$\varepsilon_k(\text{mechanical}) = 2\varrho\nu_k\int_0^k F(k)\,k^2 dk\,. \tag{15}$$

In writing this expression, the assumption is made that the process of energy transfer between the sets of eddies considered can be visualized in terms of a suitably defined coefficient of "*eddy viscosity*", ν_k. Heisenberg further assumes that ν_k can be expressed as an integral over the spectrum in the form,

$$\nu_k = \kappa\int_k^\infty \frac{dk}{k^{3/2}}\sqrt{F(k)}\,, \tag{16}$$

where κ is a numerical constant of order unity. The justification for this assumption, apart from the fact that it is of the right dimensions, is that for a power law, $F(k) \sim k^{-n}$, Eq. (16) is equivalent to the expression,

$$\nu_k \sim l_k v_k \sim v_k/k\,, \tag{17}$$

derived on the picture of the eddies with wave numbers exceeding k, describing a mean free path l_k of order $1/k$ with a certain mean speed, $v_k \sim k^{-(n-1)/2}$.

Combining equations (14), (15), and (16), we have

$$\varepsilon_k = 2\varrho\left\{\nu + \kappa \int_k^\infty \frac{dk}{k^{3/2}}\sqrt{F(k)}\right\}\int_0^k F(k)\,k^2 dk\,. \tag{18}$$

Using this equation, Heisenberg discussed the spectrum of the turbulence both when the conditions are stationary and an equilibrium spectrum obtains and when the conditions are nonstationary and the turbulence is decaying.

When the equilibrium spectrum obtains,

$$\varepsilon_k = \text{constant}\,, \tag{19}$$

and the solution of the resulting equation is given by (cf. Chandrasekhar [7])

$$F(k) = F_0\left(\frac{k_0}{k}\right)^{5/3}\frac{1}{[1+(k/k_s)^4]^{4/3}} \tag{20}$$

where k_0 is a measure of the largest eddies present and

$$k_s = k_0\left(\frac{3\kappa}{4\nu}\sqrt{\frac{F(k_0)}{k_0}}\right)^{3/4}. \tag{21}$$

Heisenberg missed obtaining this exact solution; he gave instead an interpolation formula (see Eq. (28) in his paper (No. 3, p. 88) together with a demonstration that Kolmogoroff's $k^{-5/3}$-law emerged from his considerations.

For decaying turbulence,

$$\varepsilon_k = -\frac{\partial}{\partial t}\int_0^k F(k,t)\,dk\,, \tag{22}$$

and the resulting equation for the spectrum allows a solution of the functional form

$$F(k,t) = \text{constant}\left(\frac{t_0}{t}\right)^{1/2} f\left(\frac{k\sqrt{t}}{k_0\sqrt{t_0}}\right), \tag{23}$$

where f can be deduced from solving an ordinary differential equation (cf. Chandrasekhar, *loc. cit.*, Eq. (4)). The solution so obtained has been discussed and used in the literature (cf. Batchelor [8]).

To summarize, then, while hydrodynamic stability and hydrodynamic turbulence were at best marginal to Heisenberg's principal interests, his contributions to this area have nevertheless had a major impact on subsequent developments.

References

1 W. Heisenberg: "On the Stability of Laminar Flow", in *Proc. Int. Congress of Mathematicians, Cambridge, Massachusetts, U.S.A., August 30 – September 6, 1950, Volume II* (American Mathematical Society, Providence, R.I., 1952) pp. 292 – 296 (reprinted as No. 36 in *Volume B* of *Collected Works*)

2 L. Hopf: Der Verlauf kleiner Schwingungen auf einer Strömung reibender Flüssigkeit. Ann. Phys. (4) **44**, 1 – 60 (1914)

3 C. C. Lin: On the stability of two-dimensional parallel flows. Parts I, II, III, Q. Appl. Math. **3**, 117 – 142, 218 – 234, 277 – 301 (1945/46)

4 P. G. Drazin, W. H. Reid: *Hydrodynamic Stability* (Cambridge University Press, Cambridge 1981) pp. 124 – 251 (Chapter IV)

5 D. Meksyn, J. T. Stuart: Stability of viscous motion between parallel planes for finite disturbances. Proc. Roy. Soc. (London) **A 208**, 517 – 526 (1951)

6 T. Herbert: Zur nichtlinearen Stabilität ebener Potentialströmungen. Z. Angew. Math. Mech. **57**, T187 – T189 (1977)

7 S. Chandrasekhar: On Heisenberg's elementary theory of turbulence. Proc. Roy. Soc. (London) **A 200**, 20 – 33 (1949)

8 G. K. Batchelor: *The Theory of Homogeneous Turbulence* (Cambridge University Press, Cambridge 1953) pp. 161 – 168 (§ 7.5)

Editorial Note

In the following pages the six original papers of Heisenberg on the theory of turbulence are reproduced. Paper No. 1 (p. 27) constitutes the apprentice work of the second semester student, who treated a problem arising from A. Sommerfeld's course on hydrodynamics (in the summer semester 1921) and managed to calculate for the first time all parameters determining the dimensions of vortex streets in turbulent fluids between two parallel plates. Sommerfeld was very pleased with the performance of his student, which might have contributed to his suggestion (in the summer of 1922) that Heisenberg should deal in his doctoral thesis with a more difficult fundamental problem, namely, the origin of turbulence. Heisenberg soon found a method of deriving the critical Reynolds number, at which a laminar flow between two parallel plates becomes unstable, and presented it as a contribution to the *Hydro-aerodynamische Konferenz*, held from 10 to 13 September 1922 in Innsbruck (see paper No. 1 of Volume B). He then developed the mathematical techniques and worked out a description of turbulent flows, which he incorporated in his dissertation "Über Stabilität und Turbulenz von Flüssigkeitsströmen" (University of Munich, July 1923). A slightly improved version (including information obtained from discussions with L. Prandtl in Göttingen) he submitted in February 1924 to *Annalen der Physik*; this is paper No. 2 reprinted below (p. 31). [A detailed account of Heisenberg's early work on turbulence theory and the subsequent controversial debate on the results of his thesis may be found in J. Mehra and H. Rechenberg: *The Historical Development of Quantum Theory, Volume 2*, (Springer, New York, Heidelberg, Berlin 1982) Section 1.7.]

The postwar work on the theory of statistical turbulence (papers Nos. 3, p. 82, 5, p. 115, and 6, p. 120) originated from discussions with Carl Friedrich von Weizsäcker at Farm Hall in England, the place of their common internment (from July 1945 to the end of that year). Von Weizsäcker had been interested for several years in the formation of the solar system from a rotating gas cloud in turbulent motion [see his paper "Über die Entstehung des Planetensystems", Z. Astrophys. **22**, 319 (1943)]. The little joint note of Heisenberg and von Weizsäcker, containing a qualitative discussion of the shape of spiral nebulae and reprinted below as paper No. 4 (p. 112), emerged from this astronomical problem. At the same time, von Weizsäcker also composed a detailed paper on turbulent rotating gas masses, based on mathematical methods suggested by Heisenberg [see his paper: "Die Rotation kosmischer Gasmassen", Z. Naturforschung **3a**, 524 (1948), especially footnote 6 on p. 527]. What did interest Heisenberg primarily in the renewed work on the turbulence problem was the fact that turbulence represented an essentially nonlinear phenomenon, and its statistical treatment seemed to provide methods which could be applied as well to another, in his opinion highly nonlinear, problem: multiparticle production in high energy collisions between elementary particles (see his lectures of 27 June 1966 on "Nonlinear Problems in Physics" at the International School of Nonlinear Mathematics and Physics, Munich, reprinted as No. 61 in Volume B). The latter problem would soon afterwards, i.e., by 1949, move into the center of Heisenberg's scientific work (see the papers of Group 11).

Physikalische Zeitschrift *23*, 363–366 (1922)

$$a_1{}^2 = \frac{100\,K}{r\,(100 + 100\,r\,K)} \sim \frac{6{,}52}{r\,(100 + 6{,}52\,r)}.$$

$r =$	$10\,r$	r	$0{,}1\,r$	$0{,}01\,r$	$0{,}001\,r$
Durchmesser	127000 km	12750 km	1275 km	128 km	13 km
$a_1 =$	0,063	0,25	0,81	2,55	8,1

a_1 ergibt sich für Merkur zu 0,41; Venus 0,246; Erde 0,248; Mars 0,344.

Wir sehen, daß Merkur mit $0{,}38 = a$ nahe an dem Grenzpunkte steht ($a_1 = 0{,}41$)[1]. Dagegen stehen die anderen Planeten weit von der kritischen Stelle ab.

Wenn das Gesetz für die Umdrehung nicht nur zwischen Sonne und Planeten, sondern auch zwischen den Planeten und ihren Monden gilt, so können wir für diese die typischen Abstände ausrechnen. Für unsern Erdmond ergibt sich dabei $a_1 = 0{,}408$; die wirkliche Entfernung ist 0,00258. Er steht also vollständig in dem Bereiche, wo jede Umdrehung unmöglich ist. Für die anderen Monde ergeben sich durchaus die gleichen Verhältnisse.

1) Umdrehung unmöglich.

(Eingegangen 27. Juli 1922.)

Die absoluten Dimensionen der Kármánschen Wirbelbewegung.

Von Werner Heisenberg.

In einer bekannten Arbeit[1]) gelang es Kármán, den Widerstand W einer ebenen Platte, die senkrecht zu ihrer Ebene im Wasser mit der Geschwindigkeit U bewegt wird, zu berechnen auf Grund von Daten, die dem Strömungsbild selbst entnommen waren.

Die Kármánsche Untersuchung beschäftigte sich mit dem Strömungsbild in einigem Abstand hinter der Platte. Das Experiment zeigte, daß sich dort eine regelmäßige Anordnung von Wirbelfäden, eine „Wirbelstraße" ausbildet (Fig. 2), die, hinter der Platte zurückbleibend, mit langsamerer Geschwindigkeit als jene im Wasser vorschreitet. Die relativen Dimensionen dieser Anordnung, d. h. das Verhältnis des Wirbelabstandes (l) zur Breite (h) der Wirbelstraße (Fig. 1) wurden auf Grund einer Stabilitätsuntersuchung ermittelt; die Geschwindigkeit u und eine lineare Abmessung des Wirbelsystems

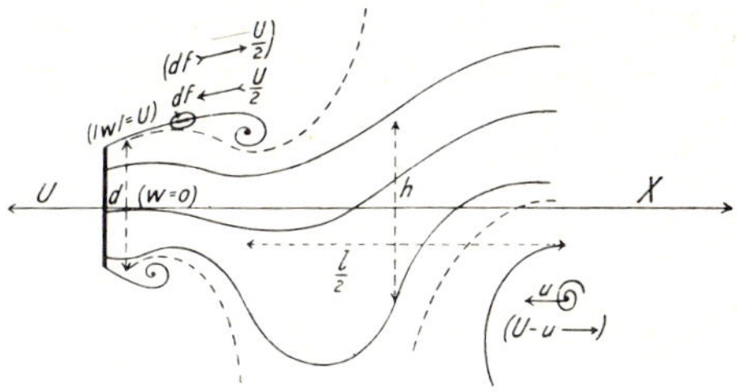

Fig. 1.

(etwa der Abstand l zweier aufeinanderfolgender Wirbel gleichen Drehsinns) mußten jedoch experimentell gesucht werden, damit der Widerstand W wirklich angegeben werden konnte. Die Frage nach dem Zustandekommen des Wirbelsystems blieb unbeantwortet.

Nach dem Helmholtzschen Satze, daß in einer reibungslosen Flüssigkeit keine Wirbel entstehen können, lag es zunächst nahe, die Viskosität für die Bildung des Wirbelsystems verantwortlich zu machen; auch Kármán hielt (l. c.) ein Eingehen auf die Vorgänge in der „Grenzschicht" der Platte für nötig, um die Unbekannten u und l des Wirbelschweifes zu berechnen. Nach den Berechnungen von Oseen[1]) haben wir jedoch anzunehmen, daß der Einfluß der Reibung gerade in unmittelbarer Nähe der Platte geringer ist, als in einiger Entfernung hinter derselben. Ferner wurde von Jaffé[2]) gezeigt, daß auch in einer reibenden Flüssigkeit im allgemeinen keine Wirbel entstehen können, daß also der Grund der überall in der Hydrodynamik auftretenden Wirbelbildung nicht in der Reibung gesucht werden kann. Es liegt nach Jaffé vielmehr dann ein Grund zur Wirbelbildung vor, wenn in den äußeren Kräften oder den Geschwindigkeiten der Flüssigkeit Unstetigkeiten auftreten. In unserem Falle sind an der Platte sicherlich solche Unstetigkeiten vorhanden. Es soll also im Folgenden der Versuch gemacht werden, aus einer Betrachtung dieser Unstetigkeiten die Größen u und l bzw. $\frac{u}{U}$ und $\frac{l}{d}$ ($d =$ Plattendicke) zu ermitteln.

Die einzige[3]), bisher bekannte Strömungsform des Wassers an ebenen Platten, bei der die Strömungsgeschwindigkeit am Rand der Platte nicht unendlich wird, ist die Helmholtz-

1) Kármán, Göttinger Nachr. 1911, S. 509 u. 1912, S. 547 und Kármán u. Rubach, diese Zeitschr. **13**, 49, 1912.

1) Oseen, Ann. d. Phys. **46**, 231 u. 646, 1915.

2) Jaffé, diese Zeitschr. **21**, 541, 1920.

3) Dies trifft, worauf mich Herr Prof. Jaffé freundlicherweise aufmerksam machte, nicht ganz zu. A. R. Richardson (Phil. Mag. **38**, 433, 1919) gibt eine mathematisch sehr interessante andere Lösung an, die allerdings kaum physikalische Bedeutung haben dürfte.

Physikalische Zeitschrift 23, 363–366 (1922)

Kirchhoffsche unstetige Potentialbewegung[1]) (Fig. 1).

Die mehrfach festgestellte praktische Unmöglichkeit dieser Strömung, die wegen ihrer Unstetigkeitsflächen als Wirbelströmung aufzufassen ist, hat ihren Grund jedenfalls in der Labilität der auftretenden Wirbelschichten. Immerhin liegt die Vermutung nahe, daß der unstetigen Potentialbewegung wenigstens in unmittelbarer Nähe der Platte Geltung zukommen müsse.

In der Tat zeigen photographische Abbildungen übereinstimmend, daß das wirkliche Strömungsbild in unmittelbarer Nähe der Platte der Kirchhoffschen Potentialbewegung außerordentlich ähnlich ist. Wir wollen also über die Entstehung des Kármánschen Wirbelsystems folgende Annahme versuchen (vgl. Fig. 1)[2]):

In unmittelbarer Nähe der Platte bildet sich zunächst die unstetige Potentialbewegung aus. In einigem Abstand von der Platte rollen sich die Unstetigkeitsflächen bzw. Wirbelschichten dieser Potentialbewegung, wegen ihrer Labilität, auf und liefern in weiterem Abstande von der Platte die Kármánsche Wirbelanordnung.

Diese Annahme kann insofern nur eine Annäherung an die Wirklichkeit darstellen, als die in einigem Abstand auftretenden Wirbel auch die Strömung in nächster Nähe der Platte nicht unbeeinflußt lassen werden. Ferner können wir sie nur für die Vorderseite und die Kanten der Platte gelten lassen, da wir im „toten" Wasser jedenfalls eine ziemlich undefinierte Wirbelbewegung (nicht aber Wirbelbildung) haben.

So einfach und plausibel nun der so formulierte Ansatz zunächst scheint, so lassen sich doch, wie mir Herr Prof. Prandtl freundlichst mitgeteilt hat, gewichtige Einwände gegen ihn erheben, vgl. dessen Zusatz am Schluß des Artikels. Herr Prof. v. Kármán hat für diese Einwände die folgende Form gefunden: „Wendet man den Ansatz wörtlich auf den Bobileffschen Fall des symmetrischen Keils an, so liefert er für alle Keilwinkel denselben Widerstandskoeffizienten (nämlich stets denselben, wie für die ebene Platte)". Es bleibt also die Frage offen, weshalb gerade in unserem Falle der Ansatz der Wirklichkeit weitgehend entspricht. Jedenfalls wird die weitere Rechnung zeigen, daß bei der Platte die oben gemachte Annahme zu bemerkenswert guter Übereinstimmung mit der Erfahrung führt.

1. Bedingung für die Geschwindigkeit des Wirbelsystems.

Zur Berechnung der Unbekannten u und l bieten sich ohne Schwierigkeit zwei Gleichungen dar. Zunächst folgt aus dem Satz von der Erhaltung des Wirbelmomentes, den wir auf jede der beiden sich einrollenden Wirbelschichten anwenden, daß das pro Zeiteinheit an jedem der beiden Plattenränder entstehende Wirbelmoment gleich dem pro Zeiteinheit durch das Wirbelsystem abtransportierten Wirbelmoment der einen oder anderen Drehrichtung sein muß. Dieses Wirbelmoment hat, wie eine einfache Überlegung ergibt, die Größe:

$$\frac{\zeta}{l}(U-u), \qquad (1)$$

wenn ζ das Moment des einzelnen Wirbelfadens darstellt. Um andererseits das am Plattenrand entstehende Wirbelmoment zu berechnen, bemerken wir: Bei der Helmholtzschen unstetigen Potentialbewegung ist im „toten Wasser" hinter der Platte (diese als feststehend gedacht) die Geschwindigkeit 0, an der von der Platte ausgehenden „freien" Oberfläche ist sie U, also ist der Geschwindigkeitssprung an der Unstetigkeitsfläche U (vgl. Fig. 1). Zerlegen wir die Unstetigkeitsfläche in ihre Elemente df, die wir als Wirbelfäden auffassen, und bilden das Geschwindigkeitsintegral auf geschlossenem Wege um solch einen Wirbelfaden df herum (vgl. Fig. 1), so finden wir, daß sein Moment $U \cdot df$ beträgt. Ferner ist dieses Element df keineswegs in Ruhe relativ zur Platte, denn dann hätten wir im toten Wasser die Geschwindigkeit $-\frac{U}{2}$, außen $+\frac{U}{2}$. Das Element entfernt sich vielmehr auf der Unstetigkeitsfläche mit der Geschwindigkeit $\frac{U}{2}$ von der Platte. Wir schließen daraus: das pro Zeiteinheit am Plattenrande entstehende Wirbelmoment ist

$$U \cdot \frac{U}{2}. \qquad (2)$$

Wir erhalten aus (1) und (2) die Gleichung:

$$U \cdot \frac{U}{2} = \frac{\zeta}{l}(U-u). \qquad (3)$$

2. Bedingung für den Wirbelabstand.

Eine zweite Beziehung folgt daraus, daß es eine gewisse Wassermenge gibt (vgl. Fig. 2), die stets innerhalb der beiden Wirbelreihen vorwärts

1) Vgl. etwa Lamb, Lehrbuch der Hydrodynamik, IV, § 76, S. 113.

2) In Fig. 1 sind die Geschwindigkeiten ($w = u - iv$) zum Teil in doppelter Weise eingetragen: Die eingeklammerten Angaben beziehen sich auf Geschwindigkeiten relativ zur Platte, die nicht eingeklammerten auf Geschwindigkeiten relativ zur Flüssigkeit im Unendlichen.

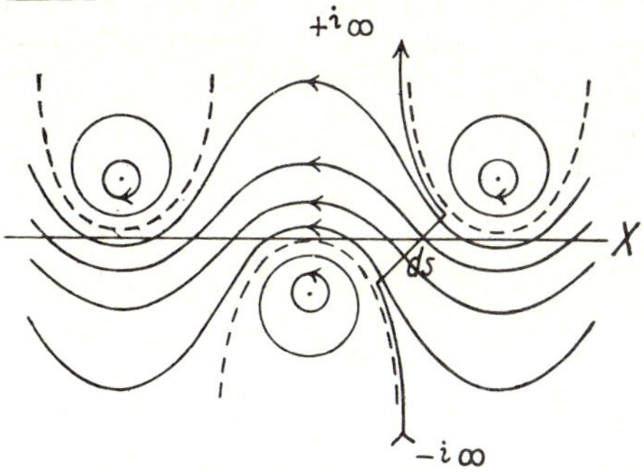

Fig. 2.

geschoben wird. Während die Flüssigkeit außerhalb der Wirbelstraße im Mittel in Ruhe ist (d. h. ein Flüssigkeitsteilchen entfernt sich hier im Laufe der Zeit nicht beliebig weit von seinem Anfangsort), fließt innerhalb der Straße ein kontinuierlicher Flüssigkeitsstrom vorwärts. Die in diesem Strom beförderte Wassermenge muß gleich derjenigen sein, die durch die bewegte Platte ständig nach vorne verdrängt wird. Letztere beträgt $U \cdot d$.

Um erstere zu berechnen, schreiben wir nach Kármán das komplexe Potential der Wirbelströmung an:

$$\chi = \frac{i\zeta}{2\pi} \lg \frac{\sin (z_0 - z)\frac{\pi}{l}}{\sin (z_0 + z)\frac{\pi}{l}}, \qquad (4)$$

$$z_0 = \frac{l}{4} + \frac{hi}{2}$$ (h Abstand der beiden Wirbelreihen).

$z = x + iy$ (die X-Achse läuft, wie in Fig. 1 gezeichnet, parallel den Wirbelreihen und halbiert den Raum zwischen denselben). Um die gesuchte Wassermenge zu erhalten, integrieren wir auf einem an sich willkürlichen Wege vom Linienelement ds (u Geschwindigkeitsvektor, $w = u - iv$, J bedeutet „Imaginärer Teil von"):

$$\int_a^b u_n\, ds = \int_a^b J\,(w\,dz) = J \int_a^b \frac{d\chi}{dz}\, dz = J\,(\chi)\Big|_a^b.$$

Die Grenzen a, b des Integrals sind die beiden die gesuchte Wassermenge begrenzenden Stromlinien (in Fig. 2 punktiert), wobei die besondere Lage der Punkte a, b auf den Stromlinien belanglos ist. Nun sind in unserem Falle die begrenzenden Stromlinien die einzigen, die (in der Y-Richtung) ins Unendliche reichen. Wir erstrecken also das Integral einfach von $z = +i\infty$ nach $z = -i\infty$. In Fig. 2 ist der Integrationsweg von $-i\infty$ entlang einer punktierten Stromlinie in die Nähe der Wirbel geführt, dann quer über das Strombett und von da wieder entlang der Grenzstromlinie nach $+i\infty$. Die pro Zeiteinheit durch den Strom gehende Flüssigkeitsmenge wird demnach:

$$J\,(\chi_{+i\infty} - \chi_{-i\infty}).$$

Nun ist nach (4):

$$\chi_{\alpha+\beta i} = \frac{i\zeta}{2\pi} \lg \frac{e^{i\frac{\pi}{l}\left(\frac{l}{4}+\frac{hi}{2}-\alpha-\beta i\right)} - e^{-i\frac{\pi}{l}\left(\frac{l}{4}+\frac{hi}{2}-\alpha-\beta i\right)}}{e^{i\frac{\pi}{l}\left(\frac{l}{4}+\frac{hi}{2}+\alpha+\beta i\right)} - e^{-i\frac{\pi}{l}\left(\frac{l}{4}+\frac{hi}{2}+\alpha+\beta i\right)}}.$$

Daraus für:

$\beta \to \infty$, $\alpha = 0$ (α geht nicht in den imaginären Teil von χ ein)

$$\chi_{+i\infty} = \frac{i\zeta}{2\pi} \lg \left(-e^{+\frac{i\pi}{2} - \frac{h}{l}\cdot\pi}\right) = -\frac{3\zeta}{4} - \frac{i\zeta h}{2l},$$

$$\chi_{-i\infty} = \frac{i\zeta}{2\pi} \lg \left(-e^{-\frac{i\pi}{2} + \frac{h}{l}\pi}\right) = -\frac{\zeta}{4} + \frac{i\zeta h}{2l}.$$

Also wird die Flüssigkeitsmenge (das Vorzeichen lassen wir außer acht, da es uns nur auf den absoluten Betrag ankommt):

$$|J\,(\chi_{+i\infty} - \chi_{-i\infty})| = \frac{\zeta \cdot h}{l}. \qquad (5)$$

Wir erhalten daraus als zweite Beziehung:

$$U \cdot d = \frac{\zeta \cdot h}{l}. \qquad (6)$$

3. Schlußfolgerung.

Zur numerischen Auswertung von (3) und (6) benutzen wir noch die von Kármán l. c. angegebenen Beziehungen:

$$\frac{\zeta}{l\sqrt{8}} = u \qquad (7)$$

$$\text{und } \frac{h}{l} = 0{,}283\cdots \qquad (8)$$

Aus Gleichung (3) folgt dann:

$$\tfrac{1}{2} U^2 = u \cdot \sqrt{8}\,(U - u),$$

$$1 = 2 \cdot \sqrt{8} \cdot \frac{u}{U}\left(1 - \frac{u}{U}\right),$$

$$\frac{u}{U} = 0{,}229. \qquad (9)$$

Die andere Wurzel $\frac{u}{U} = 0{,}771$ ist aus einem sogleich anzugebenden Grunde auszuschließen.

Gleichung (6) geht wegen (7) und (8) über in:

$$U \cdot d = u \cdot \sqrt{8} \cdot l \cdot 0{,}283.$$

Daraus folgt nach (9):

$$\frac{l}{d} = \frac{1}{\sqrt{8} \cdot 0{,}283 \cdot 0{,}229} = 5{,}45; \quad (10)$$

$$\frac{h}{d} = 1{,}54. \quad (11)$$

Die andere Wurzel für $\frac{u}{U}$ würde einen viel kleineren Wert für $\frac{l}{d}$ und $\frac{h}{d}$ ergeben. Da aber die Diskontinuitätsflächen der Potentialströmung nach außen gerichtet sind, so ist es sehr verständlich, daß diese zweite Möglichkeit nicht realisierbar ist.

Kármán gibt (l. c.) als empirische Werte an:

$$\frac{u}{U} = 0{,}20; \quad \frac{l}{d} = 5{,}5.$$

Als Wert für den spezifischen Widerstandskoeffizienten ψ_w, definiert durch:

$$W = \psi_w \cdot L \cdot d \cdot U^2 \cdot \varrho$$

(L = Länge der Platte senkrecht zur Zeichenebene, ϱ = Dichte der Flüssigkeit),

ergibt sich aus den theoretischen Zahlen für $\frac{u}{U}$ und $\frac{l}{d}$:

$$\psi_w = 0{,}90.$$

Auch dieser Wert scheint mit neueren Messungen in bester Übereinstimmung zu stehen.

München, Institut für theoretische Physik.

(Eingegangen 18. Juli 1922.)

Bemerkung zur vorstehenden Arbeit, von L. Prandtl.

Die von Herrn Heisenberg bereits erwähnten Einwände, die ich gegen seine Ansätze zu machen habe, seien im folgenden, da sie für die Leser der vorstehenden Arbeit von Interesse sein dürften, kurz formuliert:

1. Der Widerstand der Platte ist rund doppelt so groß wie der, den die Kirchhoffsche Formel liefert. Hinter der Platte herrschen kleine Geschwindigkeiten, also volum-konstanter Druck bis an die Kanten heran. Dem etwa doppelt so großen Druckabfall von der Mitte der Vorderseite bis zur Kante entspricht auch ein etwa verdoppeltes $\frac{U^2}{2}$; also ist die sekundliche Wirbelproduktion $= \alpha \frac{U^2}{2}$ zu setzen, wo α bei der ebenen Platte ungefähr = 2 ist (genau genommen ist auch die Bewegung auf der Rückseite der Platte von Einfluß auf die Wirbelproduktion, doch ist dieser Einfluß nur geringfügig).

2. Es ist nicht ausgeschlossen, daß ein merklicher Teil der positiven und negativen Wirbelmomente durch Vermischung in der turbulenten Zone hinter der Platte sich gegenseitig vernichtet und daher in dem abziehenden Wirbelsystem nicht mehr vorhanden ist. Ist der vernichtete Bruchteil $= \beta$, so wäre an Stelle von Gl. (3) zu setzen:

$$(1 - \beta)\, \alpha \frac{U^2}{2} = \frac{\zeta}{l}(U - u).$$

3. Der Schluß, daß die zwischen den Wirbeln nach vorn geförderte Flüssigkeitsmenge gerade $= U \cdot d$ sein soll, ist nicht zwingend. Mit der Platte rückt in der als ruhend gedachten Flüssigkeit auch die Grenze des Wirbelgebiets vor, und auch dieser Platz muß durch die Strömung ausgefüllt werden. Durch eine Zusammenziehung des Wirbelgebiets bei der Formulierung der Wirbelreihe kann eine entgegengesetzte Wirkung eintreten, so daß also auch Gl. (6) mit einem unbestimmten Faktor, der größer oder kleiner als 1 sein kann, behaftet ist.

Meines Erachtens ist die an sich recht instruktive Heisenbergsche Rechnung nur geeignet, im Zusammenhalt mit den Versuchsergebnissen über die hier erwähnten Berichtigungszahlen Aufschlüsse zu liefern.

Göttingen, 29. Juli 1922.

Die Verbiegungen in einzelnen Punkten unterstützter kreisförmiger Platten.

Von A. Nádai.

1. Im Folgenden werden die Formänderungen und die Spannungen kreisförmiger Platten angegeben, die durch Kräfte quer zu ihrer Ebene belastet und auf ihrem Umfang in einzelnen Punkten unterstützt sind. Bei einer dünnen ebenen Platte ist wegen ihrer geringen seitlichen Steifigkeit die Möglichkeit eines ungleichmäßigen Aufliegens auf den Rand und einer dadurch hervorgerufenen Vergrößerung des Durchhanges unter der Wirkung von seitwärts gerichteten Kräften leicht gegeben. Es dürfte von Interesse sein, einmal überhaupt die

1924. № 15.

ANNALEN DER PHYSIK.

VIERTE FOLGE. BAND 74.

1. *Über Stabilität und Turbulenz von Flüssigkeitsströmen;*

von Werner Heisenberg.

Einleitung.

Das Turbulenzproblem, das ganz allgemein den Gegenstand der folgenden Untersuchungen bilden soll, ist im Laufe der Zeit in so vielen Arbeiten von so vielen verschiedenen Gesichtspunkten aus behandelt worden, daß es nicht unsere Absicht sein kann, einleitend über die bisherigen Resultate eine Übersicht zu geben. Wir verweisen zu diesem Zweck auf eine Arbeit von Noether[1]) über den heutigen Stand des Turbulenzproblems, in welcher auch die meisten Literaturangaben zu finden sind.

Für unseren Zweck genügt es, den gegenwärtigen Stand des Turbulenzproblems in ganz groben Umrissen zu skizzieren: die bisherigen Untersuchungen zerfallen in zwei Teile: die einen von ihnen befassen sich mit der Stabilitätsuntersuchung irgendwelcher laminaren Bewegung, die anderen mit der turbulenten Bewegung selbst.

Die ersteren führten anfangs zu dem negativen Resultat, daß alle untersuchten Laminarbewegungen stabil seien. v. Mises[2]) und Hopf[3]) bewiesen auf Grund eines Ansatzes von Sommerfeld[4]) die Stabilität des der Couetteschen Anordnung entsprechenden linearen Geschwindigkeitsprofils, Blumenthal[5]) gelangte bei einem von Noether zur Diskussion gestellten Profile 3. Grades zu demselben Ergebnis. Dagegen gelang es später Noether[6]), ein labiles Profil anzugeben — allerdings

1) F. Noether, Zeitschr. f. angew. Math. u. Mech. 1. S. 125, 1921.

2) R. v. Mises, Beitrag z. Oszillationsprobl.: Heinr. Weber-Festschrift. 1912. S. 252.

3) L. Hopf, Ann. d. Phys. 44. S. 1. 1914.

4) A. Sommerfeld, Atti d. IV. congr. int. dei Mathem. Rom 1909.

5) O. Blumenthal, Sitzungsber. d. bayr. Akad. d. Wiss. S. 563. 1913.

6) F. Noether, Nachr. d. Ges. d. Wiss. Göttingen 1917.

ein Profil, das schon im Falle der reibungslosen Flüssigkeit labil ist und auch in Wirklichkeit nie als stationärer Bewegungszustand realisierbar ist. In neuerer Zeit aber hat Prandtl[1]) gezeigt, daß es in der Tat Profile gibt, die erst bei Berücksichtigung der Reibung labilen Charakter besitzen.

Die andere Gruppe von Arbeiten, die in jüngster Zeit durch die Rechnungen[2]) von v. Karman, Latzko u. a. große Erfolge erzielt hat, untersucht die turbulente Bewegung selbst und geht dabei auf halbempirischem Wege mit Benutzung der Ähnlichkeitsgesetze vor. Theoretisch fußt sie fast durchweg auf dem Boden der Prandtlschen Grenzschichttheorie. Das für uns wichtigste Resultat dieser Arbeiten ist das aus dem empirischen Blasiusschen Widerstandsgesetz (die Zitate bei Schiller[5])) folgende sogenannte $y^{1/7}$-Gesetz der turbulenten Geschwindigkeitsverteilung.

Ein Hauptziel der erstgenannten Arbeiten, der Stabilitätsuntersuchungen, bildete stets die Ermittelung der kritischen Reynoldsschen Zahl. Eine befriedigende Berechnung dieser Zahl ist bisher nicht gelungen und es muß auch als fraglich betrachtet werden, ob eine solche durch Stabilitätsuntersuchungen geleistet werden kann. Die Versuche von Ekman[3]), Ruckes[4]) und Schiller[5]) zusammen mit dem negativen Ergebnis Hopfs beim linearen Geschwindigkeitsprofil legen vielmehr den Gedanken nahe, daß die kritische Reynoldssche Zahl nicht den Punkt angibt, an dem die Laminarbewegung anfängt labil zu werden, sondern den, an dem zum erstenmal die turbulente Bewegung als stationärer Zustand möglich ist. Vom Standpunkt der Theorie aus müssen wir also darauf gefaßt sein, unter Umständen zwei kritische Reynoldssche Zahlen zu finden, die eine entsprechend

---

1) L. Prandtl, Physik. Zeitschr. **23.** S. 19. 1922 und O. Tietjens, Dissert. Göttingen 1922.

2) Vgl. bes. Zeitschr. f. angew. Math. u. Mech. **1.** S. 233f. 1921.

3) V. Ekman, Turbulent motions of Liquids. Arch. f. Mat. och fysik **6.** S. 12. 1919.

4) W. Ruckes, Dissert Würzburg 1907. Vgl. auch den Vortrag von W. Wien, Über turbul. Bewegungen. Phys. Zeitschr. **8.** 1904 und Verh. d. deutsch. phys. Gesellsch. **9.** 1907.

5) L. Schiller, Rauhigkeit und kritische Zahl. Physik. Zeitschr. **3.** S. 412. 1920.

dem Beginn der Turbulenz, die andere entsprechend dem Niederbrechen der Laminarbewegung.

Auch die vorliegende Untersuchung soll in die zwei verschiedenen Teile, die Behandlung des Stabilitätsproblems einerseits, die der turbulenten Bewegung andererseits zerfallen.

Das Ziel des ersten Teiles soll sein, alle bisherigen Untersuchungen unter einem einheitlichen Gesichtspunkt zusammenzufassen, d. h. möglichst allgemein die Bedingungen dafür aufzustellen, wann ein Profil labilen oder stabilen Charakter trägt und die Methoden anzugeben zur Lösung der Stabilitätsgleichung für jedes beliebige Geschwindigkeitsprofil und zur Berechnung der für labile Profile sich ergebenden kritischen Reynoldsschen Zahl. Dieses Ziel kann natürlich nur durch Anwendung von Näherungsverfahren unvollkommen erreicht werden. Dennoch hoffen wir, die qualitativ wesentlichen Gesichtspunkte durch solche Rechnungen klarstellen zu können. Die Untersuchung irgend eines beliebigen Profils scheint zwar zunächst physikalisch sinnlos zu sein, da nur bestimmte Profile in Wirklichkeit vorkommen. Indem wir aber irgendein Profil als endliche Störung eines anderen auffassen können, wie dies z. B. Noether a. a. O. getan hat, wir andererseits später die Untersuchungen auf das zunächst unbekannte Grundprofil der turbulenten Bewegung ausdehnen müssen, so scheint die Untersuchung eines beliebigen Profils doch von großer Wichtigkeit.

Als Anwendung der Methoden soll das Parabelprofil durchgerechnet werden.

Im zweiten Teile wollen wir versuchen, unter gewissen stark idealisierenden Voraussetzungen Differentialgleichungen für die turbulenten Bewegungen abzuleiten und aus diesen qualitative Aufschlüsse zu erhalten über einige Eigenschaften der turbulenten Geschwindigkeitsverteilung.

I. Teil: **Die Stabilitätsgleichung.**

§ 1. **Festlegung des mathematischen Problems.**

Die wesentlichste Beschränkung, die wir unseren Rechnungen auferlegen, besteht in der ausschließlichen Betrachtung von ebenen Laminarbewegungen und nur ebenen Störungen derselben. Unter Zugrundelegung eines rechtwinkligen Koordinatensystems X, Y, Z nehmen wir also an, daß die Ge-

Annalen der Physik (4) 74, 577–627 (1924)

schwindigkeit in der Z-Richtung Null und alle übrigen Größen von z unabhängig seien. Weiter aber werden wir auch nur die Stabilität solcher Laminarbewegungen prüfen, die zwischen zwei geraden, parallelen Wänden verlaufen. Dabei nehmen wir an, daß die Wände der X-Achse parallel laufen, daß also auch die zu untersuchende Laminarbewegung eine Geschwindigkeitskomponente nur in der X-Richtung besitzt. Diese Geschwindigkeit w in der X-Richtung wird in irgendeiner Weise von y abhängen; über die Funktion $w = w(y)$ behalten wir uns vor, später einige Annahmen über Stetigkeit, Symmetrie usw. zu machen; sonst soll diese Funktion jedoch zunächst ganz willkürlich sein.

Würden wir $w = a\,y$ setzen, so würden unsere Ansätze gerade mit den von Hopf beim Couetteschen Fall untersuchten identisch. Auf die Frage, ob sich die untersuchten Profile $w = w(y)$ als stationäre Bewegungen realisieren lassen, gehen wir zunächst nicht ein (vgl. u. S. 581).

Bevor wir die (bereits von Sommerfeld a. a. O. aufgestellte) Stabilitätsgleichung noch einmal in Kürze aus den Stokesschen Differentialgleichungen herleiten, führen wir in bekannter Weise dimensionslose Variable ein. Sei h eine charakteristische Länge (etwa der Abstand beider Wände), U eine charakteristische Geschwindigkeit des Profils, μ die Zähigkeit, ϱ die Dichte, $\frac{U \cdot h \cdot \varrho}{\mu} = R$ die Reynoldssche Zahl, so führen wir statt x, y, u, v, t, p (unter u, v werde die Geschwindigkeit in der x- bzw. y-Richtung verstanden, t sei die Zeit, p der Druck) neue Varibale x_0, y_0, u_0, v_0, t_0, p_0 ein vermöge der Relationen

$$(1)\quad \begin{cases} x_0 = \dfrac{x}{h}; \quad y_0 = \dfrac{y}{h}; \quad u_0 = \dfrac{u}{U}; \quad v_0 = \dfrac{v}{U}; \quad t_0 = t \cdot \dfrac{U}{h}; \\ \qquad\qquad p_0 = p \cdot \dfrac{h}{\mu\,U}. \end{cases}$$

Lassen wir nachträglich den Index 0 weg, so lauten die Stokesschen Gleichungen:

$$(2)\quad \begin{cases} \dfrac{\partial u}{\partial t} + u\dfrac{\partial u}{\partial x} + v\dfrac{\partial u}{\partial y} = \dfrac{1}{R}\left(-\dfrac{\partial p}{\partial x} + \dfrac{\partial^2 u}{\partial x^2} + \dfrac{\partial^2 u}{\partial y^2}\right), \\ \dfrac{\partial v}{\partial t} + u\dfrac{\partial v}{\partial x} + v\dfrac{\partial v}{\partial y} = \dfrac{1}{R}\left(-\dfrac{\partial p}{\partial y} + \dfrac{\partial^2 v}{\partial x^2} + \dfrac{\partial^2 v}{\partial y^2}\right). \end{cases}$$

Weil wir Inkompressibilität voraussetzen, schreiben wir

$$(3)\quad u = \frac{\partial \psi}{\partial y}, \quad v = -\frac{\partial \psi}{\partial x}.$$

Durch Elimination von p erhalten wir bekanntlich:

$$\frac{\partial}{\partial t} \Delta \psi + \frac{\partial \psi}{\partial y} \frac{\partial}{\partial x} \Delta \psi - \frac{\partial \psi}{\partial x} \frac{\partial}{\partial y} \Delta \psi = \frac{1}{R} \Delta \Delta \psi . \tag{4}$$

Unter Δ ist hier das Differentiationssymbol

$$\frac{\partial^2}{\partial x^2} + \frac{\partial^2}{\partial y^2}$$

verstanden.

In Gleichung (4) ist noch nichts über unser spezielles Problem, die Stabilitätsuntersuchung einer gewissen Laminarströmung, enthalten. Dementsprechend wird Gleichung (4) auch die Grundlage für die Rechnungen des Teils II bilden. Um speziell zur Stabilitätsuntersuchung überzugehen, teilen wir die Bewegung und damit auch das Vektorpotential ψ in eine Grundströmung und darüber überlagerte kleine Schwingungen. Wir machen also den Ansatz:

$$\psi = \Phi(y) + \varphi(y) e^{i(\beta t - \alpha x)}, \tag{5}$$

$$\frac{\partial \Phi}{\partial y} = w(y) = w . \tag{6}$$

Gehen wir mit diesem Ansatz in (4) ein, lassen dabei alle Glieder weg, die φ nicht enthalten (da wir die Gleichung (4) für $\varphi = 0$ als erfüllt ansehen), lassen ferner alle in φ quadratischen Glieder weg (da wir φ als klein voraussetzen), so lautet die entsprechende Differentialgleichung für φ:

$$(\varphi'' - \alpha^2 \varphi)\left(w - \frac{\beta}{\alpha}\right) - \varphi w'' = \frac{i}{\alpha R} (\varphi'''' - 2\alpha^2 \varphi'' + \alpha^4 \varphi) . \tag{7}$$

Daß wir Gleichung (4) für $\varphi = 0$ als erfüllt ansehen, bedeutet physikalisch, daß wir nur solche Grundströmungen w betrachten, die entweder vermöge äußerer Kräfte wirklich stationär sind, oder deren zeitliche Veränderung langsam gegen die der kleinen Schwingungen erfolgt.

Gleichung (7) ist in dieser Allgemeinheit schon von Noether a. a. O. abgeleitet. Sie ist eine gewöhnliche Differentialgleichung für φ von der vierten Ordnung. Dem entspricht es, daß die Funktion φ vier Grenzbedingungen erfüllen muß; es müssen u und v, daher auch φ und φ' an den beiden Wänden verschwinden. Setzen wir $\beta/\alpha = c$, so daß c im wesentlichen die Wellengeschwindigkeit bedeutet, so läßt sich das mathematische Problem folgendermaßen formulieren:

Es sind die Lösungen der Gleichung

$$(7\,\mathrm{a})\quad (\varphi'' - \alpha^2 \varphi)(w - c) - \varphi w'' = \frac{i}{\alpha R}(\varphi'''' - 2\alpha^2 \varphi'' + \alpha^4 \varphi)$$

mit der Nebenbedingung, daß an den begrenzenden Wänden (etwa $y = 1$ und $y = -1$) $\varphi = 0$ und $\varphi' = 0$ sein soll, zu untersuchen. Zu jedem Wert von α und R soll der zugehörige Wert von c und β berechnet werden; α sei der Einfachheit halber stets positiv. Je nachdem der imaginäre Teil von β positiv, Null oder negativ ist, haben wir es mit einer stabilen, ungedämpften oder labilen Schwingung zu tun. Es sind die Bedingungen für das Profil w aufzusuchen, unter denen Gleichung (7a) nur stabile, bzw. unter denen sie auch labile Schwingungen zuläßt.

Bevor wir zu den Lösungsmethoden übergehen, sei noch auf eine besondere Eigenschaft der Gleichung (7a) hingewiesen. Im Limes der reibungslosen Flüssigkeit $R = \infty$ verwandelt sich Gleichung (7a) in eine Differentialgleichung 2. Grades für φ.

$$(8)\qquad (\varphi'' - \alpha^2 \varphi)(w - c) - \varphi w'' = 0\,.$$

Dem entspricht es, daß für $R = \infty$ nur noch zwei Grenzbedingungen zu befriedigen sind, welche besagen, daß die Normalkomponente der Geschwindigkeit, also v oder φ an beiden Wänden verschwinden soll, nicht mehr aber φ'.

Die Bedingungen für die Lösbarkeit von Gleichung (8) sind schon von Rayleigh[1]) ausführlich untersucht worden. Man kann, um eine einfache Bezeichnung einzuführen, „schwingungsfähige" und „nichtschwingungsfähige" Grundströmungen unterscheiden, je nachdem Gleichung (8) eine Lösung mit reellem c besitzt, die den Grenzbedingungen genügt, oder nicht.[2]) Gibt es Lösungen mit komplexem c, so wird, wie sich später zeigen wird, auch bei Berücksichtigung der Reibung die Stabilitätsfrage für diese Schwingungen schon durch (8) entschieden, die Schwingungen sind dann immer *labil.*

Man wird aber darüber hinausgehend zu der Vermutung geführt, daß das Profil w dann und nur dann unter Einfluß

1) Lord Rayleigh, Papers I. S. 361; III. S. 575, 594; IV. S. 203.

2) Hierbei genügt es aber *keineswegs*, das Profil durch Tangentenpolygone anzunähern, das Resultat hinsichtlich der möglichen Schwingungen würde dadurch völlig gefälscht.

der Reibung labile oder ungedämpfte Schwingungen zulassen wird, wenn es zu den schwingungsfähigen Grundströmungen gehört.

Diese Vermutung liegt um so näher, als sie sich an allen bisher untersuchten Profilen bestätigt hat.[1]) Trotzdem wird sie keineswegs dadurch begründet, daß Gleichung (8) im Limes $R = \infty$ aus Gleichung (7a) hervorgeht, denn es ist z. B. in den Arbeiten von Oseen[2]) bewiesen, daß der Grenzübergang $R = \infty$ in den Differentialgleichungen schon mehrfach zu falschen Resultaten, besonders hinsichtlich der Grenzbedingungen der reibungslosen Flüssigkeit, geführt hat und daß man daher den Grenzübergang nur an den Integralen der Gleichung (7 a) vornehmen darf. Außerdem ist es von vorneherein keineswegs zu entscheiden, ob die Reibung die ungedämpften Schwingungen von (8) im Sinne einer Dämpfung oder einer Anfachung abändert.

Wir werden im folgenden versuchen, unsere oben ausgesprochene Vermutung zu beweisen, indem wir zeigen, daß die *schwingungsfähigen* Systeme oberhalb eines bestimmten Wertes der Reynoldsschen Zahl im allgemeinen *labilen* Charakter, alle *nichtschwingungsfähigen* dagegen *stabilen* Charakter besitzen.

Durch diesen Satz wird das Problem der Stabilität eines Profils ganz wesentlich vereinfacht; denn für sehr kleine Werte von α lassen sich bekanntlich die Lösungen von (8) direkt hinschreiben.

§ 2. Die Lösungsmethoden und das allgemeine Verhalten der Integrale von (7a).

Die wichtigste Eigenschaft der Gleichung (7a), die eine näherungsweise Darstellung ihrer Lösungen erlaubt, besteht darin, daß R als sehr groß betrachtet werden kann. Es wird sich nämlich zeigen, daß, falls eine Labilitätsgrenze existiert, diese im allgemeinen bei sehr hohen Werten von R liegt. Da es andererseits physikalisch ganz unwahrscheinlich ist, daß bei kleinen Werten von R noch einmal Labilität des betreffenden Profils eintreten kann, so reicht es für unsere Zwecke zunächst aus, R als sehr groß anzusehen.

1) Vgl. auch L. Prandtl, Phys. Zeitschr. a. a. O.

2) Vgl. z. B. C. W. Oseen, Beiträge z. Hydr. Ann. d. Phys. **46.** S. 231 u. 623. 1915.

Diese Annahme macht es möglich, die Lösungen von (7a) anzunähern durch Entwicklung nach negativen Potenzen von R bzw., wie sich zeigen wird, von $\sqrt{\alpha R}$. Ferner werden wir α als klein annehmen und die Lösungen gegebenen Falles nach positiven Potenzen von α^2 entwickeln.

Die beiden Entwicklungsmethoden nach $(\alpha R)^{-1/2}$ einerseits, nach α^2 andererseits, scheinen sich insofern zu widersprechen, als im ersten Falle αR als groß, im zweiten α^2 als klein angenommen wird. Der Widerspruch wird jedoch dadurch behoben, daß R im allgemeinen als außerordentlich groß betrachtet werden kann, so daß z. B. für $R = 2000$, $\alpha = 1/_{10}$; $\alpha R = 200$, $\alpha^2 = 1/_{100}$ wird, was für eine gute Konvergenz beider Entwicklungen völlig hinreicht. Die Konvergenzeigenschaften dieser Näherungsmethoden sind aber noch genauer zu betrachten. Die Untersuchung lehrt, daß die Reihen nach $(\alpha R)^{-1/2}$ im allgemeinen divergieren, daß sie aber den bekannten Charakter der semikonvergenten Reihen tragen, d. h. daß die Glieder zuerst abnehmen, dann wieder wachsen und daß man die bestmögliche Annäherung erhält, wenn man die Reihe beim kleinsten Glied abbricht. Unser Näherungsverfahren hat also ähnliche Konvergenzeigenschaften, wie z. B. die in der Astronomie verwendeten Reihen der Störungstheorie, deren Verhalten Poincaré, Meth. nouv. d. l. mec. cel. II ausführlich schildert.

Eine erhebliche Schwierigkeit für die Verwendung der semikonvergenten Entwicklungen liegt aber darin, daß sie in der Umgebung eines bestimmten Punktes ihre Gültigkeit verlieren, so daß es nicht ohne weiteres entschieden werden kann, in welcher Weise die Näherungslösungen auf beiden Seiten des Punktes aneinandergefügt werden müssen, um ein *bestimmtes* Integral der Gleichung (7a) *beiderseits* anzunähern. Wir werden hierauf in § 3 ausführlich eingehen.

Die Entwicklung nach positiven Potenzen von α^2 scheint im allgemeinen wirklich zu konvergieren. Für spezielle Profile läßt sich diese streng beweisen (z. B. für das lineare Profil); eine Untersuchung über die Frage, unter welchen Bedingungen für das Profil w diese Konvergenz wirklich stattfindet, haben wir jedoch nicht ausgeführt.

Wir beginnen mit der Ableitung der Näherungslösungen der Gleichung (7a):

(7a) $(\varphi'' - \alpha^2 \varphi)(w - c) - w'' \varphi = \frac{i}{\alpha R}(\varphi'''' - 2\alpha^2 \varphi'' + \alpha^4 \varphi).$

Zu diesem Zwecke setzen wir zunächst

$$\varphi = e^{\int g\,dy}, \quad g = \sqrt{\alpha R}\, g_0 + g_1 + \frac{1}{\sqrt{\alpha R}} g_2 + \cdots \tag{9}$$

Die Entwicklung wollen wir auf die beiden höchsten Glieder in $\sqrt{\alpha R}$ beschränken. Es folgt:

$$\alpha R g_0^2 (w - c) + \sqrt{\alpha R}(g_0' + 2 g_0 g_1) \cdot (w - c) = i \alpha R g_0^4 + i \sqrt{\alpha R}(4 g_0^3 g_1 + 6 g_0^2 g_0').$$

Hierbei ist vorausgesetzt, daß α^2 und w'' von der Größenordnung 1 oder jedenfalls $\ll \sqrt{\alpha R}$ seien. Es ergibt sich jetzt durch einfache Rechnung:

$$g_0 = \sqrt{-i(w - c)}, \quad g_1 = -\frac{5}{2}\frac{g_0'}{g_0}, \quad \int g_1\,dy = -\frac{5}{2}\lg g_0. \tag{10}$$

Wir erhalten also zwei partikuläre Integrale der Gleichung (7a):

$$\varphi_{1,2} = (w - c)^{-5/4} e^{\pm \int_{y_0}^{y} \sqrt{-i\alpha R (w - c)}}. \tag{11}$$

Der Punkt y_0 soll dadurch festgelegt werden, daß für $y = y_0$ $w = c$ sei. y_0 kann also unter Umständen auch komplex sein. Das Vorzeichen der Wurzel soll so gewählt werden, daß für $w - c = -a e^{i\omega}$, $\quad -i\alpha R(w - c) = \alpha R \cdot a e^{i\left(\omega + \frac{\pi}{2}\right)}$ die Wurzel wird:

$$(\alpha R a)^{1/2} \cdot e^{i\left(\frac{\omega}{2} + \frac{\pi}{4}\right)}.$$

Bemerkenswert an diesen beiden Integralen ist, daß α^2 in φ in dieser Näherung nicht vorkommt (d. h. nur in der Kombination αR, die man, wie aus Gleichung (7a) hervorgeht, in gewissem Sinne als die eigentliche Reynoldssche Zahl bezeichnen könnte).

Wie wir später sehen werden, bestimmen die beiden Integrale (11) das Verhalten von φ in der Grenzschicht und das Nichtvorkommen von α^2 in (11) bedeutet physikalisch, daß wir nur solche Schwingungen betrachten, deren Wellenlänge groß ist gegen die Grenzschichtdicke — was bei den empirisch beobachteten labilen Schwingungen sicher der Fall ist.

Annalen der Physik (4) 74, 577–627 (1924)

Zu einem vollständigen Lösungssystem von (7a) brauchen wir jedoch noch zwei weitere Integrale; naturgemäß werden wir dazu diejenigen nehmen, die durch Entwicklung nach Potenzen von $(\alpha R)^{-1}$ aus den Integralen der Gleichung (8) hervorgehen.

Zu diesem Ende lösen wir zunächst Gleichung (8) durch Entwicklung nach Potenzen von α^2. Wir setzen also:

$$(\varphi'' - \alpha^2 \varphi)(w - c) - \varphi w'' = 0, \tag{8}$$

$$\varphi = \varphi^{(0)} + \alpha^2 \varphi^{(1)} + \alpha^4 \varphi^{(2)} + \ldots \tag{12}$$

Daraus folgt:

$$\begin{aligned} \varphi^{(0)''}(w - c) - \varphi^{(0)} w'' &= 0. \\ \varphi^{(1)''}(w - c) - \varphi^{(1)} w'' &= \varphi^{(0)} \cdot (w - c), \\ \varphi^{(2)''}(w - c) - \varphi^{(2)} w'' &= \varphi^{(1)} \cdot (w - c), \\ \ldots \ldots \ldots \ldots \ldots \end{aligned}$$

Es ergeben sich durch die Methode der Variation der Konstanten die beiden Integrale:

$$(13) \left\{ \begin{aligned} \varphi_{3\,(R=\infty)} &= (w - c)\left(1 + \alpha^2 \int \frac{dy}{(w-c)^2} \int dy\,(w-c)^2 + \ldots\right), \\ \varphi_{4\,(R=\infty)} &= (w - c) \int \frac{dy}{(w-c)^2}\left(1 + \alpha^2 \int dy (w-c)^2 \int \frac{dy}{(w-c)^2} \right. \\ &\qquad\qquad \left. + \ldots\right). \end{aligned} \right.$$

Diese Integrale sind nun noch um Größen der Ordnung $(\alpha R)^{-1}, \ldots$ usw. zu korrigieren, wenn sie der Gleichung (7a) genügen sollen.

Ohne die entsprechende Reihenentwicklung im einzelnen anzuschreiben, geben wir als Resultat φ mit den Größen der Ordnung $(\alpha R)^{-1}$ an:

$$(14) \left\{ \begin{aligned} \varphi_3 &= (w - c)\left\{1 + \alpha^2 \int \frac{dy}{(w-c)^2} \int dy\,(w-c)^2 + \alpha^4 \ldots \right. \\ &\qquad \left. + \frac{i}{\alpha R} \int \frac{dy}{(w-c)^2} \frac{d^3}{dy^3}(w - c) + \ldots \right\}, \\ \varphi_4 &= (w - c) \int \frac{dy}{(w-c)^2} \left\{1 + \alpha^2 \int dy\,(w-c)^2 \int \frac{dy}{(w-c)^2} + \alpha^4 \ldots \right. \\ &\qquad \left. + \frac{i}{\alpha R} \frac{d^3}{dy^3}(w - c) \int \frac{dy}{(w-c)^2} + \ldots \right\}. \end{aligned} \right.$$

Mit (11) und (14) ist ein vollständiges Lösungssystem der Gleichung (7a) näherungsweise gefunden.

Ehe wir dieses Lösungssystem zur Erfüllung der Grenzbedingungen anwenden, wird es zweckmäßig sein, wenn wir uns den physikalischen Sinn der vier Integrale φ_1, φ_2, φ_3, φ_4 klar machen und einige Resultate vorwegnehmen, die wir erst später begründen können.

Die Integrale φ_1, φ_2 haben die Eigenschaft, bei den uns interessierenden großen Werten von R sehr rasch veränderlich zu sein, wie aus dem Exponenten der Ordnung $\sqrt{\alpha R}$ hervorgeht. Wenn also etwa φ_1 an der einen Wand von der Größenordnung 1 ist, so wird es in einigem Abstand von der Wand exponentiell verschwinden (an sich könnte es auch noch außerordentlich groß werden, doch dies wird naturgemäß durch die Grenzbedingungen verhindert). Das hat zur Folge, daß sich, von der unmittelbaren Umgebung der Wände abgesehen, φ aus φ_3 und φ_4 allein zusammensetzt, d. h. dem Verhalten von φ in der reibungslosen Flüssigkeit sehr ähnlich ist.

Daß α^2 in φ_1, φ_2 nicht explizit vorkommt (vgl. Gleichung (11)), wohl aber in φ_3, φ_4 (vgl. Gleichung (14)), muß physikalisch offenbar so gedeutet werden, daß bei der Annahme, α^2 sei nicht sehr groß ($\alpha^2 < \sqrt{\alpha R}$, vgl. Gleichung (7a) und (9)) die Wellenlänge zwar als unendlich groß gegen die Grenzschichtdicke, nicht aber gegen die Kanalbreite betrachtet werden kann. Der charakteristische Unterschied zwischen den „Grenzschichtintegralen" φ_1, φ_2 einerseits, den der reibungslosen Flüssigkeit entsprechenden Integralen $\varphi_{3,4}$ andererseits, spricht sich also in dem Auftreten bzw. Nichtauftreten von α^2 bezeichnend aus.

Was die Konvergenz der Entwicklung nach Potenzen von α^2 betrifft, so können wir hoffen, daß sie für Werte von α^2 von der Größenordnung 1 noch sehr wohl hinreicht, um eine gute Näherung zu ermöglichen, denn für ein lineares Profil gehen die Reihen für φ_3, φ_4 über in solche vom Typus der Potenzentwicklung von $\operatorname{Cos} \alpha$, die in der Umgebung $\alpha = 1$ noch sehr rasch konvergiert.

Das Strömungsbild, das wir nach allen diesen Schlüssen erwarten müssen, entspricht den Ansätzen, die man in der Prandtlschen Grenzschichttheorie macht. Mit Ausnahme der unmittelbaren Umgebung der Wände gehorcht die Bewegung sehr nahe den Differentialgleichungen der reibungslosen Flüssigkeiten. An den Wänden selbst aber haftet eine Grenzschicht, deren Dicke von der Größenordnung $(\alpha R)^{-1/2}$ ist. In dieser nimmt die Geschwindigkeit u gegen die Wand zu rasch gegen Null ab, während v auch außerhalb der Grenzschicht schon nahezu Null ist.

§ 3. Die Übergangssubstitutionen.

Wenn wir den Verlauf der Integrale von (7a) von der einen begrenzenden Wand bis zur anderen studieren wollen, so müssen wir damit rechnen, daß an einer Stelle $y = y_0$ im Kanal $w - c = 0$ ist (oder wenigstens der reelle Teil von $w - c$ Null ist), daß also die Wellengeschwindigkeit dort mit der Geschwindigkeit der Grundströmung übereinstimmt. In diesem Punkte verlieren die Näherungsformeln (11) und (14) für die Integrale von (7a) ihre Gültigkeit.

Es ist daher nötig, die Übergangssubstitutionen für $\varphi_1, \varphi_2, \varphi_3, \varphi_4$ zu kennen, die beim Übergang von $\Re(w - c) > 0$ zu $\Re(w - c) < 0$ anzuwenden sind. Zu diesem Ende entwickeln wir w und φ in der „Umgebung" des kritischen Punktes y_0 nach Potenzen von $(\alpha R)^{-1/3}$ und setzen deshalb $y - y_0 = \eta \cdot (\alpha R)^{-1/3}$. Ferner nehmen wir an, daß der imaginäre Teil von y_0 von *kleinerer* Größenordnung als $(\alpha R)^{-1/3}$ sei. Ist er nämlich von *höherer* Größenordnung, so verstehen sich die Übergangssubstitutionen von selbst, weil dann im ganzen Bereich reeller y nirgends ein „kritischer Punkt" auftritt. Ist er von *derselben* Größenordnung, so läßt sich das Verhalten von φ und w leicht aus den beiden eben angeführten Grenzfällen interpolieren. Wir können zunächst sogar $\Im(y_0) = 0$ setzen, weil sich φ in unserem Falle $(\Im(y_0) \ll (\alpha R)^{-1/3})$ nach Potenzen von $\Im(y_0)$ entwickeln läßt und fürs erste nur das Verhalten von φ für $\Im(y_0) = 0$ gebraucht wird.

Wir setzen also jetzt $\varphi = \varphi_0 + \varepsilon \varphi_1 + \varepsilon^2 \varphi_2 + \dots$, $\varepsilon = (\alpha R)^{-1/3}$; $w - c = \varepsilon \cdot a \eta + \varepsilon^2 b \eta^2 + \dots$

Dann entsteht aus (7a):

$$\varphi_0'''' + \varepsilon \varphi_1'''' + \dots$$
$$= -i[(\varphi_0'' + \varepsilon \varphi_1'' + \dots) a \eta + \varepsilon \varphi_0'' b \eta^2 - 2 \varepsilon \varphi_0 b] + \dots$$

Also in erster Näherung:

$$\varphi_0'''' = -i \varphi_0'' \eta a; \tag{15}$$

in zweiter Näherung:

$$\varphi_1'''' = -i[\varphi_1'' \eta a + \varphi_0'' b \eta^2 - 2 \varphi_0 b]. \tag{15a}$$

Für die Integrale φ_1, φ_2 Gl. (11) entnehmen wir aus

(15), daß sie sich im „kritischen Gebiete" (η Größenordnung 1) wie die von Hopf beim *linearen* Profile gefundenen Integrale, d. h. wie gewisse Zylinderfunktionen verhalten. Wir können also jetzt schon schließen, daß die Übergangssubstitutionen für φ_1, φ_2 aus (11) bis auf Größen der Ordnung $(\alpha R)^{-1/3}$ dieselben, wie beim linearen Profil sein müssen:

$$(16) \qquad \begin{cases} \varphi_1 \to \varphi_1 + i\,\varphi_2 \\ \varphi_1 - i\,\varphi_2 \to \varphi_1 \end{cases}$$

entsprechen einem Übergang von

$$\mathfrak{Re}\,(w - c) < 0 \to \mathfrak{Re}\,(w - c) > 0.$$

Für das Studium der Integrale φ_3, φ_4 in der Umgebung von $y - y_0 = 0$ reichen aber die bisherigen einfachen Rechnungen nicht aus, denn für diese würde die Näherungslösung (15) lauten $\varphi'' = 0$; aus (14) wissen wir aber, daß im Limes $R \to \infty$ φ_4' an der Stelle $y - y_0 = 0$ im allgemeinen logarithmisch unendlich wird. Die Gleichungen (15) und (15a) sind also in dieser Form ungeeignet zur Erfassung dieser Singularität.

Wir setzen jetzt vielmehr $\alpha = 0$ und $w''' = 0$ (d. h. wir brechen die Entwicklung von w mit dem 2. Gliede ab), integrieren aber sonst (7a) zunächst *exakt*. Dabei bemerken wir, daß $\varphi = w - c$ ein partikuläres Integral dieser vereinfachten Gleichung (7a) sein muß, und für φ machen wir daher den aus der Theorie der linearen Differentialgleichungen geläufigen Ansatz: $\varphi = (w - c) \int \psi\, d\, y$. Dann folgt aus

$$\varphi'''' = -\, i\, a\, R\, (\varphi''\, (w - c) - w''\, \varphi)$$

für $\varphi = (w - c) \int \psi\, d\, y$:

$$\psi'''\,(w - c) + 4\,\psi''\, w' + 6\,\psi'\, w''$$
$$= -\, i\, \alpha\, R\, (2 w'\,(w - c)\,\psi + (w - c)^2\, \psi'),$$

was nach nochmaliger Integration übergeht in:

$$(17) \quad \psi''\,(w - c) + 3\,\psi'\, w' + 3\,\psi\, w'' = -\, i\, \alpha\, R\, ((w - c)^2\, \psi - C).$$

C ist eine Integrationskonstante. Führt man jetzt wieder ein:

$$\eta = (y - y_0)\,(\alpha R)^{1/3}, \quad \varepsilon = (\alpha R)^{-1/3}, \quad w - c = \varepsilon\, a\, \eta + \varepsilon^2\, b\, \eta^2,$$
$$\psi = \psi_0 + \varepsilon\, \psi_1 + \dots,$$

so ergibt sich:

$$(17\text{a}) \quad \begin{cases} \psi_0''\, a \cdot \eta + 3\,\psi_0'\, a = -\, i\, (a^2\, \eta^2\, \psi_0 - C) \\ \psi_1''\, a\, \eta + 3\,\psi_1'\, a + \psi_0''\, b\, \eta^2 + 6\,\psi_0'\, b\, \eta + 6\,\psi_0\, b \\ \qquad = -\, i\, (2 a\, b\, \psi_0\, \eta + a^2\, \eta^2\, \psi_1)\,. \end{cases}$$

Diese Differentialgleichungen enthalten natürlich immer noch sämtliche Lösungen von (7a). Da wir speziell φ_4 studieren wollen (φ_3 verhält sich für $\alpha = 0$ im Punkte $y = y_0$ regulär), so suchen wir *diejenige* Lösung von (17a) aus, die sich in einigem Abstand von y_0, also für große $(w - c)\,\alpha R$, wie $\frac{1}{(w-c)^2}$ verhält; denn wir wissen ja aus (14), daß φ_4 in einigem Abstand von y_0 durch

$$(w - c)\int \frac{d\,y}{(w-c)^2}$$

gegeben ist. Es wird also nach (17a)

$$\psi_0 = \frac{C}{a^2\,\eta^2}$$

und

$$\psi_1''\,\eta + 3\,\psi_1' = -\,i\left(\frac{2\,b}{a^3}\,C\eta + a\,\eta^2\,\psi_1\right). \tag{18}$$

ψ_1 ist wieder dadurch völlig bestimmt, daß es sich „im Unendlichen" wie $-\frac{2\,b\,C}{a^3\,\eta}$ verhalten soll.

Wenn wir nun nach den Übergangssubstitutionen für φ_4 (und φ_3) fragen, so ist damit folgendes gemeint: In den asymptotisch gültigen Darstellungen (14) für φ_3, φ_4 kommt immer das Integral $\int \frac{d\,y}{(w-c)^2}$ vor, das seinen Sinn verliert, wenn es über die Stelle $y = y_0$ $(w - c = 0)$ hinweg zu erstrecken ist. In Wirklichkeit tritt aber eben an Stelle von $\frac{1}{(w-c)^2}$ in der Nähe des kritischen Punktes die Funktion ψ. Es kommt also alles auf das Verhalten von ψ (insbesondere ψ_1) in der kritischen Umgebung an. Ist dieses Verhalten, also auch die Größe des Integrals $\int \psi\,d\,y$ (erstreckt über die kritische Stelle hinweg) bekannt, so ist diese Kenntnis gleichbedeutend mit der Kenntnis der Übergangssubstitutionen für φ_3, φ_4.

Die durch (18) und die Grenzbedingung im Unendlichen charakterisierte Lösung ψ_1 lautet:

$$\left\{\begin{aligned} \psi_1(\eta) = -\,\frac{\pi\,b\,C}{3\,a^2\,\eta}\Big\{&H^{(1)}_{2/3}\left(\tfrac{2}{3}(-\,i\alpha_0\,\eta)^{3/2}\right)\int_{+\infty}^{\eta} H^{(2)}_{2/3}\,\eta^2\,d\eta \\ &-H^{(2)}_{2/3}\int_{-\infty}^{\eta} H^{(1)}_{2/3}\,\eta^2\,d\eta\Big\}. \end{aligned}\right. \tag{19}$$

Hierin treten die Hankelschen Zylinderfunktionen vom Index $\frac{2}{3}$ und dem Argument $\frac{2}{3}(-i\alpha_0\eta)^{3/2}$ auf $(\alpha_0 = a^{1/3})$. Dabei ist das Vorzeichen von $(-i\alpha_0\eta)^{3/2}$ so zu nehmen, daß $(-i\alpha_0\eta)^{3/2}$ für $\eta = \frac{r e^{\frac{\pi i}{2}}}{\alpha_0}$ positiv wird. Eine nähere Untersuchung von (19) lehrt, daß sich ψ_1 in der ganzen oberen Halbebene und sogar noch in der unteren Halbebene teilweise, nämlich für $\eta = r e^{i\xi}$ in den Grenzen

$$-\frac{i\pi}{6} < \xi < \frac{7i\pi}{6}$$

im Unendlichen wie $-\frac{2bC}{a^3\eta}$ verhält, wenn α_0 oder a positiv ist. Ist a negativ, so vertauschen sich obere und untere Halbebene und es gilt

$$\lim_{r\to\infty}\psi_1(re^{i\xi}) = -\frac{2bC}{a^3\eta}, \quad \text{wenn} \begin{cases} \frac{5i\pi}{6} < \xi < \frac{13\pi i}{6} \\ a < 0, \end{cases} \quad \text{oder} \begin{cases} -\frac{i\pi}{6} < \xi < \frac{7\pi i}{6} \\ a > 0. \end{cases} \tag{20}$$

Daraus entnehmen wir das wichtige Resultat:

$$\int_{-\infty}^{+\infty}\psi_1\,d\eta = \begin{cases} \frac{2bC}{a^3}i\pi & a > 0, \\ -\frac{2bC}{a^3}i\pi & a < 0. \end{cases} \tag{21}$$

Es sind also jetzt die Übergangssubstitutionen für φ_3, φ_4 bis auf Größen der Ordnung $(\alpha R)^{-1/3}$ genau auch für endliche Werte von α gefunden; denn wir wissen jetzt — und das ist hinreichend — was wir nach (21) unter dem Integral $\int\frac{dy}{(w-c)^2}$, erstreckt von $w - c < 0$ nach $w - c > 0$, zu verstehen haben.

Auch die Formeln (16) *lassen sich aus* (17) *noch einmal ableiten.* Denn den asymptotischen Lösungen (11) der Gleichung (7a) entsprechen die Integrale

$$\frac{1}{\eta}H^{(1),(2)}_{2/3}\left(\tfrac{2}{3}(-i\alpha_0\eta)^{3/2}\right) \tag{19a}$$

der homogenen Gleichung (18) $(C = 0)$. Die Aufgabe, die Übergangssubstitutionen der „asymptotischen" Integrale (11) und (14) zu finden, ist hiermit in der für uns erforderlichen Genauigkeit (bis auf Größen der Ordnung $(\alpha R)^{-1/3}$) gelöst.

§ 4. Erfüllung der Grenzbedingungen und die Stabilität der dem Lösungssystem I entsprechenden Schwingungen.

Nachdem unsere bisherigen Betrachtungen von der Art des Profils ganz unabhängig waren — bis auf einige, die singulären Punkte betreffenden Einschränkungen, die dem Profil auferlegt werden mußten — wollen wir, um uns nicht in einer übergroßen Anzahl verschiedener Möglichkeiten zu verlieren, die Natur der Grundströmung weiter spezialisieren. Die Betrachtungen gelten jedoch viel allgemeiner. Wir nehmen also an, daß die begrenzenden Wände durch die Formeln $y = +1$; $y = -1$ dargestellt sind, daß ferner die Wand $y = 1$ gegen die andere eine Relativgeschwindigkeit in der positiven X-Richtung besitze (von der Größe $w(+1) - w(-1)$) und daß die Laminarströmung an den Wänden hafte (was der Versuchsanordnung von Couette entspricht); daß schließlich im Gebiete $-1 < y < +1$, also in der Flüssigkeit, $\Re(w - c)$ einmal und nur einmal Null sei. Außerdem werden wir im ganzen Gebiete Stetigkeit für w und die Differentialquotienten von w voraussetzen und darüber hinausgehend noch annehmen, daß die Funktionen w, w', w'' usw. stets normale Größe besitzen, d. h. z. B. nicht an einzelnen Stellen groß werden von der Ordnung $(\alpha R)^{1/2}$.

Ferner betrachten wir zunächst für die folgenden Rechnungen α als so klein und αR als so groß, daß wir mit genügender Genauigkeit setzen können

$$(14\text{a}) \qquad \begin{cases} \varphi_3 = w - c, \\ \varphi_4 = (w - c) \displaystyle\int_{-1}^{y} \frac{dy}{(w - c)^2} . \end{cases}$$

Die Festlegung der unteren Grenze des Integrals in φ_4 bedeutet offenbar keine Beschränkung der Allgemeinheit unserer Lösungen. Wir setzen dadurch vielmehr φ_4 als diejenige lineare Kombination von φ_3 und φ_4 fest, die an der Stelle $y = -1$ verschwindet. Sollte dort auch $w - c$ verschwinden, so tritt offenbar an Stelle von φ_4 die Funktion $\varphi_3 = w - c$, die jetzt für $y = -1$ Null wird.

Um die Grenzbedingungen zuerst an der Wand $y = -1$

zu befriedigen, bilden wir zwei Aggregate f_1, f_2 aus $\varphi_1, \varphi_2, \varphi_3, \varphi_4,$ für die wirklich $\varphi = \varphi' = 0$ ist für $y = -1$.

$$(22)\quad \begin{cases} f_1 = \varphi_4 + \dfrac{1}{g_0(-1)[w(-1)-c] - \frac{9}{4} w'(-1)} \left[\dfrac{\varphi_3}{w(-1)-c} - \dfrac{\varphi_1}{\varphi_1(-1)}\right], \\ f_2 = \varphi_4 - \dfrac{1}{g_0(-1)[w(-1)-c] + \frac{9}{4} w'(-1)} \left[\dfrac{\varphi_3}{w(-1)-c} - \dfrac{\varphi_1}{\varphi_1(-1)}\right]. \end{cases}$$

Hierbei verstehen wir unter g_0 von jetzt ab die Wurzel $\sqrt{-i\alpha R(w-c)}$, nicht, wie in (10), $\sqrt{-i(w-c)}$, um uns das Anschreiben des Faktors $\sqrt{\alpha R}$ zu ersparen.

Um die Grenzbedingungen auch an der anderen Wand zu erfüllen, muß man zwei Konstanten A und B so zu bestimmen suchen, daß

$$A f_1(+1) + B f_2(+1) = 0,$$
$$A f_1'(+1) + B f_2'(+1) = 0.$$

Die Bedingung für die Möglichkeit einer solchen Bestimmung lautet:

$$(23)\qquad \begin{vmatrix} f_1(+1) & f_2(+1) \\ f_1'(+1) & f_2'(+1) \end{vmatrix} = 0.$$

Durch diese Bedingung wird c bzw. β festgelegt, wenn R und α gegeben sind. Es handelt sich also jetzt darum, die Gleichung (23) nach c aufzulösen und das Vorzeichen des imaginären Teils von β zu bestimmen. Gleichung (23) bildet das vollständige Analogon zur Sommerfeldschen Turbulenzgleichung für das lineare Profil.

Aus (16) entnehmen wir:

$$(24)\quad \begin{cases} f_1(+1) = \varphi_4(+1) + \dfrac{1}{g_0(-1)[w(-1)-c] - \frac{9}{4} w'(-1)} \left[\dfrac{\varphi_3(+1)}{w(-1)-c} - \dfrac{\varphi_1(+1) + i\varphi_2(+1)}{\varphi_1(-1)}\right], \\ f_1'(+1) = \varphi_4'(+1) + \dfrac{1}{g_0(-1)[w(-1)-c] - \frac{9}{4} w'(-1)} \left[\dfrac{\varphi_3'(+1)}{w(-1)-c} - \dfrac{\varphi_1'(+1) + i\varphi_2'(+1)}{\varphi_1(-1)}\right], \end{cases}$$

Annalen der Physik (4) *74*, 577–627 (1924)

$$
(24)\quad \begin{cases}
f_2(+1) = \varphi_4(+1) - \dfrac{1}{g_0(-1)\,[w(-1)-c] + \frac{9}{4}\,w'(-1)} \left[\dfrac{\varphi_3(+1)}{w(-1)-c} - \dfrac{\varphi_2(+1)}{\varphi_2(-1)}\right], \\[2ex]
f_2'(+1) = \varphi_4'(+1) - \dfrac{1}{g_0(-1)\,[w(-1)-c] + \frac{9}{4}\,w'(-1)} \left[\dfrac{\varphi_3'(+1)}{w(-1)-c} - \dfrac{\varphi_2'(+1)}{\varphi_2(-1)}\right].
\end{cases}
$$

Diese Werte von f_1, f_2, f_1', f_2' setzen wir in (23) ein, nehmen aber zuerst noch eine Abschätzung der Größe der einzelnen Glieder vor, um die Rechnung nicht durch Anschreiben unwesentlicher Glieder unnötig zu komplizieren. Zu diesem Zwecke bemerken wir, daß im allgemeinen entweder $\varphi_2(+1) \gg \varphi_2(-1)$ oder $\varphi_2(+1) \ll \varphi_2(-1)$ sein wird. Dies bewirkt der Faktor $\sqrt{\alpha R}$ im Exponenten von φ_1, φ_2 in (11), wenn nicht gerade

$$
\Re e \int_{+y_0}^{-1} \sqrt{-i(w-c)}\, dy = \Re e \int_{y_0}^{+1} \sqrt{-i(w-c)}\, dy
$$

sein sollte, was wir ausschließen.

Welcher der beiden Fälle eintritt, läßt sich nicht vorher entscheiden; im allgemeinen sind beide möglich und liefern beide Lösungen. Im Falle eines schiefsymmetrischen Profils gibt der eine Fall die zu dem des anderen symmetrischen Lösungen. Jedenfalls verhalten sich die beiden Möglichkeiten in prinzipieller Hinsicht ganz analog und es genügt daher, eine von beiden zu untersuchen. Wir nehmen also an:

$$
\varphi_2(+1) \ll \varphi_2(-1),
$$

d. h. (vgl. S. 585) der Punkt $w = c$ soll näher an $w = w(+1)$ als an $w = w(-1)$ liegen.

Daraus folgt, daß $\varphi_1(-1) \sim \frac{1}{\varphi_2(-1)}$ außerordentlich klein, also $\frac{1}{\varphi_1(-1)}$ sehr groß ist. Es bleiben daher in f_1 und f_1' nur die Glieder stehen, die $\varphi_1(-1)$ im Nenner haben, in f_2 und f_2' fallen die Glieder, die φ_2 enthalten, fort.

Aus (23) erhalten wir so:

$$
(25)\quad \begin{cases}
[\varphi_1(+1) + i\varphi_2(+1)] \\
\left[\varphi_4'(+1) - \dfrac{\varphi_3'(+1)}{\left[g_0(-1)[w(-1)-c] + \frac{9}{4}w'(-1)\right][w(-1)-c]}\right] \\
= [\varphi_1'(+1) + i\varphi_2'(+1)] \\
\left[\varphi_4(+1) - \dfrac{\varphi_3(+1)}{\left[g_0(-1)[w(-1)-c] + \frac{9}{4}w'(-1)\right][w(-1)-c]}\right].
\end{cases}
$$

Auch in dieser Form ist die Gleichung für c noch recht kompliziert. Wir vereinfachen daher (25) zunächst weiter, indem wir jetzt nicht nur Größen der Ordnung $e^{-\sqrt{\alpha R}}$, sondern auch solche der Ordnung $(\alpha R)^{-1/2}$ streichen.

Zu diesem Zwecke stellen wir fest, daß $g_0(+1)$ von der Ordnung $(\alpha R)^{1/3}$ ist — wir schließen also zunächst aus, daß $w(+1) - c$ sehr klein ist —, daß ferner:

$$
\varphi_1'(+1) + i\varphi_2'(+1) = -\frac{5}{4}\,\frac{w'(+1)}{w(+1)-c}\{\varphi_1(+1) + i\varphi_2(+1)\} \\
+ g_0(+1)\{\varphi_1(+1) - i\varphi_2(+1)\}.
$$

Von den Gliedern der Gleichung (25) behalten wir daher nur noch diejenigen bei, die mit dem Faktor $g_0(+1)$ multipliziert sind.

So entsteht das einfache Resultat:

$$
[\varphi_1(+1) - i\varphi_2(+1)]\,\varphi_4(+1) = 0
$$

oder

$$
(26)\qquad \left[e^{2\int_{y_0}^{y=+1}\sqrt{-i\alpha R(w-c)}\,dy} - i \right] \int_{-1}^{+1} \frac{dy}{(w-c)^2} = 0.
$$

Diese Gleichung besitzt zwei völlig verschiedene Lösungssysteme:

$$
\text{I.}\quad e^{2\int_{y_0}^{+1}\sqrt{-i\alpha R(w-c)}\,dy} = i, \qquad \text{II.}\quad \int_{-1}^{+1}\frac{dy}{(w-c)^2} = 0.
$$

Das System I stellt das vollkommene Analogon dar zu den Lösungen, die Hopf für das lineare Profil erhalten und a. a. O. § 4 ausführlich diskutiert hat. In der Tat zeigt sich, daß die dem System I entsprechenden Schwingungen stets stabilen Charakter tragen.

Denn aus

$$e^{2\int_{y_0}^{+1}\sqrt{-i\alpha R(w-c)}\,dy} = i \text{ folgt:}$$

(27) $$2\int_{y_0}^{+1}\sqrt{-i\alpha R(w-c)}\,dy = \pi i(\tfrac{1}{2} + 2n),$$

wo n eine positive (vgl. S. 585) ganze Zahl ist. Es ist leicht zu sehen, daß diese Gleichung nur erfüllt sein kann, wenn $\alpha c = \beta$ einen positiv imaginären Teil besitzt. Also sind die durch (27) charakterisierten Schwingungen tatsächlich gedämpft, der Betrag der Dämpfung ist von der Größenordnung $w(+1) - c$ und braucht daher keineswegs klein zu sein.

§ 5. Das Lösungssystem II und die Bedingungen für die Labilität eines Profils.

Die Lösungen im System II sind identisch mit den Lösungen der Rayleighschen Gleichung (8) und genügen der Bedingung:

(28) $$\int_{-1}^{+1}\frac{dy}{(w-c)^2} = 0$$

oder (vgl. die Bemerkung zu (14 a) S. 592) ganz allgemein

$$\varphi_4 = (w-c)\int^{y}\frac{dy}{(w-c)^2} = 0 \text{ für } y = +1 \text{ und } y = -1.$$

Die letztere Form unterscheidet sich von der ersteren in gewissen Ausnahmefällen, die nachher besprochen werden sollen; außerdem stellt (28) natürlich nur eine erste Näherung ($\alpha = 0$) dar. Für die Lösungen von (28) müssen jetzt vier verschiedene Möglichkeiten unterschieden werden: Entweder 1. besitzt (28) Lösungen mit komplexem c. Dann ist das Profil immer *labil*, da stets auch der konjugiert komplexe Wert zu c eine Lösung darstellt; oder 2. es gibt Lösungen von (28) mit *reellem* c. Dann bezeichnen wir mit Prandtl a. a. O. das Profil als „*schwingungsfähig*". Dies kann nach (21) nur vorkommen, wenn an der Stelle ($w = c$) $w'' = 0$ ist, wenn also das Profil entweder einen Wendepunkt besitzt oder aus linearen

Stücken zusammengesetzt ist. Oder 3. es gibt *reelle* Werte von c, die wenigstens den *reellen* Teil von

$$\int_{-1}^{+1} \frac{dy}{(w-c)^2}$$

zu Null machen. Oder endlich 4. Es tritt keiner dieser drei Fälle ein, (28) besitzt keine Lösung. Im Fall (3) und (4) bezeichnen wir das Profil als „*nicht schwingungsfähig*". Wir behaupten nun, daß Fall 1. stets *Labilität*, Fall 3. und 4. stets *Stabilität*, Fall 2. im allgemeinen *Labilität* des zugrunde gelegten Profils ergibt. Für 1. und 4. haben wir dies im vorhergehenden schon bewiesen. Im Fall 3. setzen wir $c = c_r + i\,c_i$, wo c_r denjenigen reellen Wert von c bedeutet, für den der reelle Teil von

$$\int_{-1}^{+1} \frac{dy}{(w-c)^2}$$

verschwindet. Dann wissen wir aus § 3, daß für $c_i \leqq 0$ der imaginäre Teil des Integrals $\frac{2b}{|a^3|}\pi i$ wird, für $c_i \gg |(\alpha R)^{-1/3}|$ aber gleich $-\frac{2b}{|a^3|}\pi i$. Also muß es aus Stetigkeitsgründen (vgl. § 3) eine Stelle $c_i > 0$ geben, an welcher auch der *imaginäre* Teil des Integrals (28) verschwindet. Die vier so charakterisierten Lösungen von (28) liefern also eine Größe c mit positiv imaginärem Bestandteil, also *stabile* Schwingungen.

Der Fall 2. schließlich erfordert etwas ausführlichere Rechnungen. Bevor wir sie ausführen, bemerken wir, daß zu 2. auch zwei Lösungstypen von (28) gehören, die sich nicht in der Form

$$\int_{-1}^{+1} \frac{dy}{(w-c)^2} = 0$$

darstellen lassen. Wenn nämlich $w(+1) = w(-1)$ ist, so lautet eine Lösung von (28) $\varphi = w - w(+1)$; in der Tat ist hier $\varphi = 0$ für $y = +1$ und $y = -1$. Ferner kann es vorkommen, daß z. B. w' für $w = w(+1)$ unendlich wird. Dann ist auch

$$\varphi = \big(w - w(+1)\big) \int_{-1}^{y} \frac{dy}{[w - w(+1)]^2}$$

eine Lösung von (8), die den Grenzbedingungen genügt. Wir wollen hierauf nicht weiter eingehen, da dieser Fall im Teil II

Annalen der Physik (4) *74*, 577–627 (1924)

ausführlicher diskutiert werden wird. Wesentlich ist aber noch zu bemerken, daß der Unterschied zwischen Fall 2. und 3. sehr groß ist und daß es z. B. in *keiner Weise genügt, nach Rayleigh ein krummes Profil durch ein Polygon anzunähern.*

Zur Veranschaulichung dieses Unterschieds ist

$$\Re e\,(J(c)) = \Re e \int_{-1}^{+1} \frac{d\,y}{(w-c)^2}$$

als Funktion von c qualitativ in Fig. 1 dargestellt, wo die ausgezogene Kurve dem krummen, die gestrichelten Kurven dem aus linearen Stücken bestehenden Profile entsprechen. Man sieht, daß jeder Knick eine neue Wurzel $\Re e\,(J) = 0$ verursacht, weil J im Punkte $c = w_{\text{Knick}}$ beim geknickten Profil sich wie $\dfrac{1}{w - c_{\text{Knick}}}$ verhält. Dies entspricht dem bekannten Rayleighschen Satz, daß es soviel Schwingungswurzeln wie Knicke gibt. Trotzdem besitzt das krumme Profil *keine* Schwingungswurzel. — Nach dieser Bemerkung kehren wir zu unserer Behauptung zurück, daß die schwingungsfähigen Profile im allgemeinen bei Berücksichtigung der Reibung labil werden.

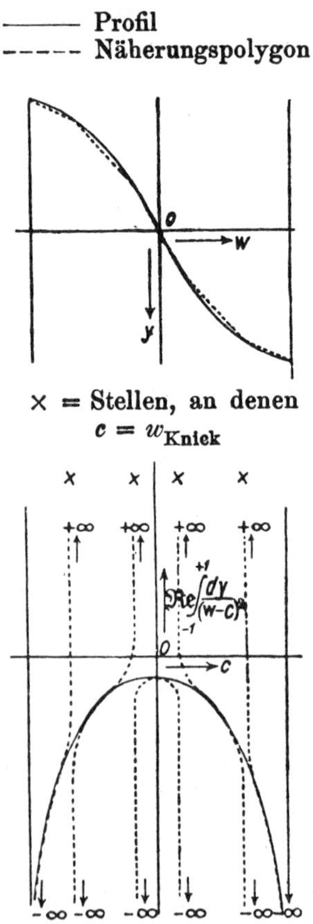

Fig. 1.

Zum Beweise dieser Labilität kehren wir zur Gleichung (25) und zu den genaueren Lösungen im System II zurück. Da wir wissen, daß c bis auf Größen der Ordnung $(\alpha\,R)^{-1/2}$ reell ist, können wir

$$\varphi_2\,(+1) \gg \varphi_1\,(+1)$$

annehmen. Vernachlässigen wir ferner die Glieder der Ordnung $(\alpha\,R)^{-1/2}$ in (25), so erhalten wir nach leichten Umformungen:

$$\int_{-1}^{+1} \frac{d\,y}{(w-c)^2} = \frac{1}{g_0\,(-1)\,[w\,(-1) - c]^2} - \frac{1}{g_0\,(+1)\,[w\,(+1) - c]^2}\,.$$

Setzen wir weiter $c = c_0 + \delta$, wo c_0 die Nullstelle von J, δ eine kleine Größe der Ordnung $(\alpha R)^{-1/2}$ bedeutet. α nehmen wir der Einfachheit halber positiv an, dann können wir auf der rechten Seite c durch c_0 ersetzen und die linke in eine Taylorsche Reihe nach δ entwickeln. So ergibt sich, wenn wir die Taylorsche Reihe mit dem 2. Gliede abbrechen, was wir als hinreichende Näherung voraussetzen:

$$J(c) = J(c_0) + \delta \frac{dJ}{dc}_{(c = c_0)}$$

und aus (25) wegen

$$J(c_0) = 0\,; \quad g_0 = \sqrt{-i\alpha R(w - c)}$$

(über das Vorzeichen vgl. S. 585).

$$\delta \frac{dJ}{dc}_{(c=c_0)} \cdot \sqrt{2\alpha R} = \frac{1 - i}{[c_0 - w(-1)]^{5/2}} + \frac{1 + i}{[w(+1) - c_0]^{5/2}}\,.$$

Hieraus folgt wegen $c_0 - w(-1) > w(+1) - c_0$ (c_0 soll näher an $w(+1)$ liegen!), daß der imaginäre Teil von δ, also auch der von c und von β dasselbe Vorzeichen hat, wie $\frac{dJ}{dc}_{(c=c_0)}$ und daß Schwingungen, welche einem negativen Werte von $\frac{dJ}{dc}$ entsprechen, *labilen* Charakter tragen. Wenn also unser teilweise lineares Profil noch die Eigenschaft hat, daß $\frac{dJ}{dc} < 0$ ist an der Stelle $w = c$, so ist es *labil*. Diese Bedingung $\frac{dJ}{dc} < 0$ ist aber sehr häufig, z. B. immer dann erfüllt, wenn der Punkt $w = c$ in der Nähe der einen Wand (z. B. $y = +1$) liegt und das Profil von der Stelle $w = c$ bis zum Rand linear verläuft.

Zusammenfassend folgern wir:

Die Labilität bzw. Stabilität eines Profils kann bei allen von uns bisher betrachteten Profilen schon durch das Verhalten derselben bei reibungsloser Flüssigkeit entschieden werden. Profile, die im letzteren Falle ungedämpfter Schwingungen fähig sind, werden bei Berücksichtigung der Reibung unter bestimmten Voraussetzungen labil. Die letzteren Profile müssen, wie oben gezeigt, ganz spezielle Eigenschaften haben, z. B. entweder teilweise aus *linearen* Stücken zusammengesetzt sein, oder sie müssen einen Wendepunkt $w'' = 0$ besitzen (vgl. oben).

Diese Profile der Gattung 2. sind aber zugleich die einzigen, die noch ein physikalisches Interesse beanspruchen können. Denn nur sie besitzen ein Verhalten hinsichtlich ihrer Stabilität, das etwa den Vermutungen von Reynolds entspricht. Wir werden im folgenden zeigen, daß diese Profile wirklich im allgemeinen eine „kritische" Reynoldssche Zahl besitzen (mit Ausnahme der geknickten Profile).

§ 6. Die Reynoldssche Zahl der Stabilitätsgrenze; numerische Rechnung am Parabelprofil.

Wenn also ein Profil gegeben ist, das bei reibungsloser Flüssigkeit ungedämpfte Schwingungen zuläßt und mit Reibung labil ist, so taucht die Frage auf: Welches ist der kleinste Wert der Reynoldsschen Zahl, für den Labilität eintritt? Um diese Frage zu beantworten, reichen die vereinfachten Gleichungen (25), (26) usw. nicht aus. Wir müssen zur Gleichung (23) und den Formen (11) und (14) für die Integrale φ_1, φ_2, φ_3, φ_4 zurückkehren. Es ist aber natürlich ganz unmöglich, allgemein für ein beliebiges Profil w die kritische Reynoldssche Zahl als Funktion von w und Integralen über w darzustellen; unsere Aufgabe soll es nur sein, den Weg anzugeben, auf dem man zur kritischen Geschwindigkeit gelangt und dann an einem speziellen Beispiele die Rechnung durchzuführen.

Da in unseren letzten Rechnungen α und R nur in der Kombination αR vorgekommen sind (weil wir α^2 als klein annahmen), können diese Rechnungen auch bestenfalls nur einen kritischen Wert für αR, nicht aber für R allein liefern. Wir müssen daher zunächst das Verhalten der Wurzeln von (23) bei wachsendem α^2 untersuchen. Von den Wurzeln von (23) interessieren uns hierbei nur die im Lösungssystem II, welche der Gleichung $\varphi_4(+1) = 0$ genügen.

An Stelle von (28) müssen wir also die Gleichung

$$(28\text{a})\left\{\begin{aligned} \varphi &= (w-c)\int\frac{dy}{(w-c)^2}\left(1+\alpha^2\int dy\,(w-c)^2\int\frac{dy}{(w-c)^2}+\ldots\right)\\ \varphi &= 0 \text{ für } y=-1,\ y=+1 \end{aligned}\right.$$

diskutieren. Besteht das Profil, wie bei Rayleigh, aus linearen Stücken, so gibt es (vgl. S. 598) stets eine Wurzel von (28a) für jeden Knick und diese Wurzeln bleiben für *jeden* Wert von α^2 bestehen. Das geknickte Profil liefert also *keinen*

Maximalwert von α^2 und kann daher auch niemals auf eine kritische Reynoldssche Zahl führen.[1])

Anders ist dies, wenn (vgl. S. 597) eine Lösung von (28) bzw. (28a) mit reellem c deswegen möglich ist, weil entweder irgendwo im Profil $w'' = 0$ ist oder weil $w(+1) = w(-1)$, $\varphi = w - w(+1)$ eine Lösung von (28) darstellt. Diese letzteren Lösungstypen liefern nämlich immer nur für einen *ganz bestimmten* Wert von α^2 eine Lösung von (28a). Für $w'' = 0$ ist c schon festgelegt dadurch, daß eben $w = c$ sein soll für $w'' = 0$; also definiert die Gleichung (28a) einen *ganz bestimmten* Wert α^2. Für den Fall $w(+1) = w(-1)$ aber gibt es offenbar nur für $\alpha^2 = 0$ eine Lösung von (28a).

Für diese Art von Lösungen der Gleichung (28) bzw. (28a) die im Limes $R = \infty$ durch *einen ganz bestimmten Wert* von α^2 charakterisiert sind. werden wir erwarten, daß sich bei Berücksichtigung der Reibung α auch nur um geringe Beträge von seinem „bestimmten" Wert entfernen kann, das Auftreten eines *Maximalwertes* (und im Falle $w'' = 0$ auch eines *Minimalwertes*) für α ist bei diesen Profilen sehr verständlich. *Alle Schwingungen also, deren Wellenlänge kleiner ist, als eine bestimmte kritische Wellenlänge, sind in solchen Fällen für alle Werte von αR gedämpft.*

Nachdem eine obere Grenze für α^2 gefunden ist, wird man die ungefähre Größe für die untere Grenze von αR zu bestimmen suchen. Eine einfache Untersuchung der Gleichung (25) lehrt, daß wesentliche Veränderungen im imaginären Teil von c erst eintreten, wenn der Exponent von e in der Näherungsdarstellung (11) in $\varphi_1(+1)$, $\varphi_2(+1)$ auf Werte der Größenordnung 1 gesunken ist. Ist dies aber der Fall, so haben wir den kritischen Wert, bei dem der imaginäre Teil von c von negativen zu positiven Werten übergeht, sehr bald erreicht, wie auch das Zahlenbeispiel zeigen wird. Nehmen wir an, daß w zwischen $w = c_0$ (c_0 = reeller Teil von c) und $w(+1)$ wesentlich linear verläuft, so lautet daher die Bedingung für die ungefähre Größe von αR:

$$(29) \qquad \frac{(\alpha R)^{1/2}\,[w(+1) - c_0]^{3/2}}{w'(+1)} \sim 1$$

1) Hierbei ist immer noch vorausgesetzt, daß R und αR groß und $\alpha \ll R$ sei. Möglicherweise treten also kritische Reynoldssche Zahlen dann auf, wenn diese Voraussetzungen nicht mehr gelten. Die betreffenden Reynoldsschen Zahlen R dürften aber dann bei so kleinen Werten liegen, daß sie physikalisch sicher keine Bedeutung haben.

oder

(29a) $$(\alpha R)^{1/3} \sim \frac{w'(+1)^{2/3}}{w(+1) - c_0}.$$

Da $w(+1) - c_0$ in den uns interessierenden Fällen klein sein wird, so kann man aus (29) vermutungsweise auf große kritische Reynoldssche Zahlen schließen. Zugleich bemerken wir hier, daß es für einen bestimmten Wert von R *stets* nicht nur einen Maximalwert, sondern auch einen *Minimal*wert für α der labilen Schwingungen geben wird. Dies folgt daraus, daß wir ja einen Minimalwert von αR (nicht R) gefunden haben.

Als numerisches Beispiel für unsere bisherigen allgemeinen Rechnungen wählen wir das Parabelprofil, weil es physikalisch am interessantesten ist. Es gehört zur Klasse der „schwingungsfähigen Profile" vom Typus $w(+1) = w(-1)$.

Wir wollen auch hier nur die ebene Bewegung betrachten, d. h. nicht die Poiseuillesche Strömung in Rohren, sondern die Strömung, die zwischen zwei parallelen ruhenden Wänden ($y = +1$, $y = -1$) unter Einfluß eines konstanten Druckgradienten herrscht. Wir setzen also

(30) $$w = 1 - y^2.$$

Aus der Symmetrie von w und $w - c$ läßt sich ableiten, daß φ eine gerade Funktion von y sein muß.[1]) Von den Lösungen der Gleichung (7a) greifen wir also zwei symmetrische partikuläre Integrale heraus und suchen die Grenzbedingungen an der einen Wand, etwa $y = -1$, zu befriedigen. Die an der anderen Wand sind dann von selbst erfüllt. Als das eine dieser symmetrischen Integrale kann offenbar einfach φ_3 genommen werden. Für das andere wählen wir:

$$\frac{\varphi_1(y)}{\varphi_1(0)} + \frac{\varphi_2(y)}{\varphi_2(0)}.$$

1) Zerlegt man φ in einen in y geraden und einen ungeraden Teil, so muß, wegen der Symmetrie von $w - c$ und w, jeder Teil von φ für sich der Differentialgleichung (7a) und den Grenzbedingungen genügen. Für die allgemeine Stabilitätsuntersuchung des Profils $1 - y^2$ ist es daher ausreichend, die beiden Fälle „φ gerade" und „ungerade" getrennt und nur diese beiden Fälle zu behandeln. Es ist aber leicht zu sehen, daß die Annahme symmetrischer Schwingungen, d. h. „φ ungerade" zu keiner Lösung von (28) und daher nicht zu labilen Schwingungen führt. Die Annahme „φ gerade" genügt also zur Stabilitätsuntersuchung. Dies ist insofern bemerkenswert, als hiernach alle symmetrischen Schwingungen stabil und nur unsymmetrische Störungen labil sind.

Wie aus (29a) hervorgeht, wird c bei unserem Profil in Nähe der kritischen Geschwindigkeit klein sein von der Ordnung $(\alpha R)^{-1/3}$, wir werden also in den folgenden Rechnungen Glieder von höherer als erster Ordnung in c streichen. Ferner stellen wir fest, daß $\varphi_2(0) \gg \varphi_1(0)$ sein wird, so daß sich in der Umgebung von $w = 0$ und $w = c$ die zweite symmetrische Funktion φ einfach auf $\varphi_1(y)$ reduziert. Aus (16) geht dann hervor, daß wir zur Befriedigung der Grenzbedingungen für $y = -1$ die beiden Integrale φ_3 und $\varphi_1 - i\varphi_2$ zur Verfügung haben. Gleichung (23) geht damit über in

$$\begin{vmatrix} \varphi_3(-1) & \varphi_1(-1) - i\varphi_2(-1) \\ \varphi_3'(-1) & \varphi_1'(-1) - i\varphi_2'(-1) \end{vmatrix} = 0 . \tag{31}$$

In φ_3 ist hierbei als untere Grenze für die auftretenden Integrale immer $y = 0$ zu nehmen, um die Symmetrie von φ_3 zu gewährleisten. In φ_3 werden wir ferner nur bis zu Größen der Ordnung α^4 entwickeln und bei der Entwicklung nach $(\alpha R)^{-1}$ mit den Gliedern der Ordnung $(\alpha R)^{-1}$ abbrechen. Wir schreiben nun (31) in der Form

$$\frac{\varphi_1'(-1) - i\varphi_2'(-1)}{\varphi_1(-1) - i\varphi_2(-1)} = \frac{\varphi_3'(-1)}{\varphi_3(-1)} . \tag{32}$$

Setzt man (11) und (14) ein, so ergibt sich

$$\left\{ \begin{aligned} &\frac{e^{2\int_{y_0}^{-1}\sqrt{-i\alpha R(w-c)}\,dy} + i}{e^{2\int_{y_0}^{-1}\sqrt{-i\alpha R(w-c)}\,dy} - i}\sqrt{i\alpha R c} = -\frac{9}{2c} \\ &\quad + \frac{\frac{\alpha^2}{c^2}\left[\int_0^{-1} dy\,(w-c)^2 + \alpha^2 \cdots + \frac{i}{\alpha R}\cdots\right]}{1 + \alpha^2\int_0^{-1}\frac{dy}{(w-c)^2}\int_0^y dy\,(w-c)^2 + \alpha^4 \cdots + \frac{i}{\alpha R}\cdots} . \end{aligned} \right. \tag{33}$$

Da c sehr klein wird, nehmen wir in erster Näherung w von o bis c als linear an: $w \sim 2(y+1)$. Dann wird

$$\int_{y_0}^{-1} \sqrt{c - w}\, dy = -\frac{1}{3} c^{3/2}. \tag{34}$$

Annalen der Physik (4) *74*, 577–627 (1924)

Setzen wir $z = \frac{1}{3} c^{3/2} \cdot (2 \alpha R)^{1/2}$, so entsteht aus (33)

$$\frac{e^{-(1+i)z} + i}{e^{-(1+i)z} - i} \cdot \frac{3z(1+i)}{2} = -\frac{9}{2} + \frac{\alpha^2}{c} \frac{\int\limits_0^{-1} dy (w-c)^2 + \dots}{1 + \alpha^2 \int \dots}. \tag{35}$$

Diese Gleichung ist zu (26) völlig analog. Uns interessiert vor allem der Grenzwert R bzw. z, bei dem die labilen Schwingungen sich in stabile verwandeln, also der imaginäre Teil von c gerade *Null* ist. Dieser Grenzwert z wird natürlich noch von α abhängen. Wir nehmen also jetzt c als *reell* an und erhalten dadurch den Grenzwert von z bzw. R als Funktion von α. Den Minimalwert von R auf dieser $R(\alpha)$-Kurve wird man als charakteristische Reynoldssche Zahl für das Parabelprofil ansprechen. Die Durchrechnung zeigt nun, daß man durch die Form (35) der Stabilitätsgleichung nur den oberen Teil der Kurve $R = R(\alpha)$ (in der Fig. 2 ausgezogen) erhält — was ja nach den Betrachtungen von § 6 zu erwarten war, daß man aber den unteren Teil der Kurve $R = R(\alpha)$ nur berechnen kann, wenn man für φ_1, φ_2 andere Näherungen [vgl. (19a)] als die asymptotischen Formeln (11) verwendet. Die kritische Reynoldssche Zahl bezeichnet eben gerade das Gebiet, in welchem die asymptotischen Formeln ihre Gültigkeit verlieren. Da dieser Umstand zu sehr komplizierten numerischen Rechnungen führen würde und wir (vgl. § 7) allgemein der hier charakterisierten Art von Labilität keine wesentliche physikalische Bedeutung beimessen können, haben wir zur Berechnung des unteren Teiles (in Fig. 1 gestrichelt) der Kurve $R = R(\alpha)$ rohe Schätzungen benutzt, die keinen Anspruch darauf machen können, quantitative Ergebnisse zu liefern. Das qualitative Verhalten der Kurve wird aber wohl richtig wiedergegeben. Wir entnehmen also aus der Fig. 2:

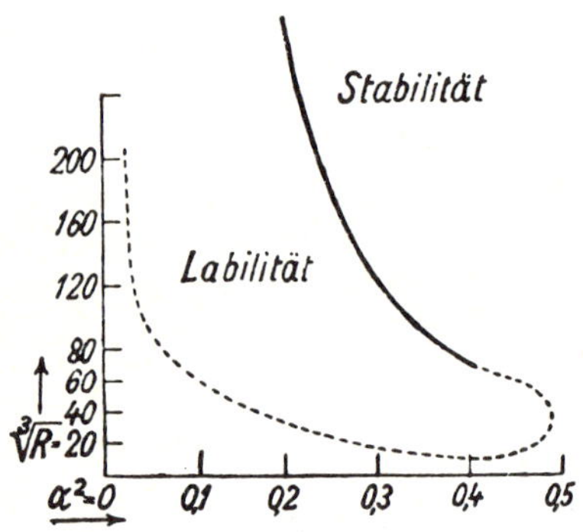

Fig. 2.

1. Es gibt sowohl einen Maximalwert von α, wie einen Minimalwert von R, nach deren Über- bzw. Unterschreitung die Labilität aufhört.

2. Für einen *bestimmten* Wert von R existiert sowohl ein *Maximal*- wie auch ein *Minimal*wert von α; innerhalb dieser Werte herrscht *Labilität*, außerhalb *Stabilität*.

3. Der Maximalwert von α liegt etwa bei $\alpha = 0{,}7$ ($\alpha^2 = \frac{1}{2}$). Der Minimalwert von R bei Größen der Ordnung 10^3. Eine einigermaßen genaue Berechnung dieses Minimalwertes aus der Figur ist nicht möglich.

§ 7. Physikalische Diskussion der Resultate des Teiles I.

Fassen wir noch einmal ausführlich alle Ergebnisse zusammen, die wir für das Stabilitätsproblem gefunden haben. In erster Linie hat sich im Verlaufe der Rechnung gezeigt, daß man allgemein die Frage nach der Stabilität eines Profils bei reibender Flüssigkeit entscheiden kann, wenn man, wie Lord Rayleigh, den Grenzfall reibungsloser Flüssigkeit (Gleichung 8) behandelt. Profile, die dann, d. h. für $R = \infty$ labil sind, bleiben dies auch für hinreichend große endliche Werte von R (§ 4) — wie dies wohl von vornherein zu erwarten war. Ebenso erweisen sich Profile, die im reibungslosen Falle *keiner* Schwingungen fähig sind, als *stabil* (§ 4) und Profile, die bei der Untersuchung durch Gleichung (8) *ungedämpfte* Schwingungen zulassen, im allgemeinen als *labil* (§ 5). Dieser letztere Fall ist offenbar der einzige, der physikalisch gegenüber der reibungslosen Hydrodynamik etwas Neues bedeutet. Wir müssen aber hervorheben, daß er, entgegen dem, was man aus Rayleighs Arbeiten zunächst schließen möchte, einen *Ausnahmefall* darstellt. Wenn man von den Möglichkeiten $w(+1) = w(-1)$, $w'_{\text{Rand}} = \infty$ (§ 5) absieht, so ist es für das Eintreten dieses Ausnahmefalls *notwendige* Bedingung, daß irgendwo in der Flüssigkeit $w'' = 0$ ist. Die von Rayleigh eingeführten geknickten, aus linearen Stücken bestehenden Profile gehören zu diesen Ausnahmeprofilen; hieraus ist aber nur zu schließen, daß eben gekrümmte Profile zum Zwecke der Stabilitätsuntersuchung *nicht* nach Rayleigh durch Polygone angenähert werden dürfen (S. 598). Auch für überall gekrümmte Profile ($w'' \neq 0$) findet man allerdings (§ 5) bei Benützung der Differentialgleichungen *mit* Reibungsgliedern eventuell Schwingungen, die für jeden Wert von αR gedämpft sind, bei denen aber im Limes $R = \infty$ der Betrag der Dämpfung wie $(\alpha R)^{-1/3}$ gegen *Null* geht; man hat also auch hier für $R = \infty$ ungedämpfte Schwingungen. Diese Schwingungswurzeln gehen aber *verloren*, wenn man die vereinfachte Differentialgleichung (8) ohne Reibungsglieder zugrunde legt. Insofern liegt also auch hier *keine* Ausnahme der Regel vor, nach welcher dann, wenn die reibungslose Gleichung (8) *ungedämpfte* Schwingungen zuläßt, die Berücksichtigung der Reibung eine *Anfachung* ergibt. Die Möglichkeit ungedämpfter Schwingungen für die Gleichung (8) muß aber, wie oben gesagt, als Ausnahmefall betrachtet werden. Das Parabelprofil gehört zu diesen Ausnahmeprofilen (§ 6). Fragen wir bei den „labilen Profilen" weiter nach dem Wertebereich für R, innerhalb dessen Labilität stattfindet, so zeigt sich, daß *auf „kritische Reynoldssche Zahlen" auch im allgemeinen nur Profile der letzten Klasse führen können, d. h. nur solche, die ohne Reibung ungedämpfte Schwingungen zulassen. Von diesen lieferten aber die Rayleighschen „geknickten Profile" dennoch keine* (vgl. auch S. 601, Anm. 1) *kritische Reynoldssche Zahl.* Bei diesen gibt es nämlich wohl einen Minimalwert von αR, nicht aber einen Maximalwert von α, also auch keinen Minimalwert von R für die Labilität (§ 6). *Nur diejenigen Profile der letzen Klasse, bei denen im reibungslosen Falle nur ein bestimmter Wert von* α *auf ungedämpfte Schwingungen führt* (z. B. die Typen $w'' = 0$ für $w = c$), $w(+1) = w(-1)$, *ergeben einen Maximalwert von* α *und einen Minimalwert von* αR, *also auch einen Minimalwert für* R. Für einen bestimmten Wert von R gibt es daher bei diesen Profilen sowohl einen *Maximal-* wie einen *Minimalwert* von α für die labilen Schwingungen. Alle diese Resultate sind in Übereinstimmung mit den bisherigen Stabilitätsuntersuchungen hydrodynamischer Profile.[1])

Es fragt sich jetzt, wie sich experimentell diese mathematischen Resultate äußern werden. Es erscheint verwunderlich, daß sich empirisch die stabilen (z. B. das Couettesche[2])

1) Vgl. die in der Einleitung zitierten Arbeiten.

2) Vgl. jedoch hierzu die interessanten Stabilitätsuntersuchungen der Couetteschen Bewegung gegen dreidimensionale Störungen. G. J. Taylor, Stability of a viscous liquid contained between two rotating cylinders. Phil. Transact. of the Royal Society London **223.** S. 289–343. 1922.

und die labilen (z. B. das Poiseuillesche) Profile ganz gleich verhalten. Oberhalb einer bestimmten Reynoldsschen Zahl tritt bei hinreichenden Störungen Turbulenz ein; macht man die Störungen möglichst klein, so läßt sich das Laminarprofil bei beliebig hohen Reynoldsschen Zahlen erhalten. Besonders die letztere Tatsache, die von Ekman (a. a. O.) am Parabelprofil geprüft worden ist, scheint bei den labilen Profilen im Widerspruch zur Theorie zu stehen. Es ist aber leicht zu erkennen, daß dieser Widerspruch nur scheinbar ist.[1] Je kleiner nämlich die äußeren Störungen sind, um so länger dauert es (besonders bei hohen Reynoldsschen Zahlen, da die Anfachung von der Ordnung $(\alpha R)^{-1/2}$ ist (vgl. § 5), bis sie die Bewegung merklich beeinflussen; bei der Strömung in Rohren wird man also diesen Zeitpunkt immer so weit hinausschieben können, daß die betreffende Flüssigkeitsmenge, um deren Stabilität es sich handelt, das Rohr schon wieder verlassen hat, wenn ihre Labilität in Erscheinung tritt. Die von uns berechnete Reynoldssche Zahl könnte also nur an in sich geschlossenen Rohren geprüft werden, bei denen immer dieselbe Flüssigkeitsmenge strömt.[2] Die Versuche von Schiller (a. a. O.) andererseits, durch die gezeigt wird, daß es unterhalb einer bestimmten Reynoldsschen Zahl *nur* Laminarbewegung gibt, können durch Stabilitätsuntersuchungen überhaupt nicht erfaßt werden, da man hier gar nicht als ursprüngliche Bewegung die laminare hat, sondern da es sich nur um die Existenz oder Nichtexistenz einer turbulenten Bewegungsform dabei handelt. Jedenfalls ist aus allen diesen Betrachtungen nur zu schließen, daß sich durch Stabilitätsbetrachtungen allein das Turbulenzproblem eben durchaus *nicht* lösen läßt.

Immerhin können uns die vorhergehenden Untersuchungen auch für das eigentliche Ziel, die Berechnung der turbulenten

---

1) Auf diese Möglichkeit, die Ekmanschen Versuche als eine Art Anlaufeffekt aufzufassen, hat mich Hr. Professor Prandtl hingewiesen, dem ich hierfür und für viele andere wertvolle Hinweise an dieser Stelle den herzlichsten Dank aussprechen möchte.

2) Vielleicht sind aber die Störungen der Stabilität, die Ruckes a. a. O. bei ziemlich kleinen Reynoldsschen Zahlen beobachtet, durch die Labilität nach § 7 hervorgerufen. Dies wäre dann sehr wohl denkbar, wenn die kritische Reynoldssche Zahl nach §7 unter derjenigen liegt, bei welcher zum ersten Male Turbulenz (vgl. Teil II) möglich ist.

Bewegung, wesentliche qualitative Ergebnisse liefern. Wenn wir nämlich die turbulente Bewegung auffassen als eine bestimmte Grundströmung mit überlagerten *ungedämpften* Schwingungen, so können wir aus unseren Rechnungen schließen, daß der Minimalwert R, für den dieser Bewegungstypus möglich ist, wahrscheinlich ebenfalls bei Werten der Größenordnung 10^3 liegt; daß die Wellenlänge der ungedämpften Schwingungen bei $2\pi\, h/2$, nämlich α bei Werten der Ordnung 1 liegt und daß α für ein gegebenes R zwischen bestimmte, eventuell sehr enge Grenzen eingeschlossen ist; daß ferner diese Schwingungen, wie aus der Kleinheit von $w\,(+1) - c$ zu schließen ist, den Charakter einer Wandstörung haben. Diese qualitativen Ergebnisse sind von der speziellen Form der Grundströmung ganz unabhängig. Über solche qualitativen Anhaltspunkte hinaus aber liefern die bisherigen Rechnungen nichts, was uns der eigentlichen Lösung des Turbulenzproblems näherbrächte.

II. Teil: **Die turbulente Bewegung.**

§ 1. **Formulierung des mathematischen Problems.**

Diejenige Reynoldssche Zahl, die man für gewöhnlich als die „kritische" bezeichnet, die z. B. in den Versuchen von Schiller gemessen wird und die den Eintritt der Turbulenz bei hinreichend großen Störungen angibt, hat mit Stabilitätsfragen und mit der Laminarströmung nichts zu tun; sie ist durchaus eine charakteristische Konstante der *turbulenten* Bewegung. Ebenso lassen das Blasiussche Widerstandsgesetz und die aus ihm abgeleitete bekannte Folgerung, daß die turbulente Geschwindigkeit in der Nähe einer Wand mit der $^1/_7$ ten Potenz der Wandentfernung anwächst, deutlich erkennen, daß die sogenannte „turbulente" Bewegung ihre eigenen wohl definierten Gesetzmäßigkeiten hat und daß sie eine zweite mögliche Bewegungsform der reibenden Flüssigkeiten darstellt. Zur Lösung des Turbulenzproblems kann man also nur versuchen die turbulente Bewegung ihrer Undefiniertheit zu entkleiden und sie soweit zu idealisieren, bis sie der mathematischen Analyse durch die Stokesschen Gleichungen zugänglich wird.

Das „Turbulenzproblem" der Hydrodynamik ist ein Problem der energetischen, nicht der dynamischen Stabilität. Es gibt zwei verschiedene Bewegungsformen der reibenden

Flüssigkeit, von denen jede einen bestimmten Wertebereich der Reynoldsschen Zahl besitzt, innerhalb dessen sie möglich ist. Die Laminarströmung ist möglich von $R = 0$ bis $R = \infty$, wird aber unter Umständen oberhalb eines bestimmten Wertes von R dynamisch labil. Die turbulente Bewegung dagegen existiert erst oberhalb eines bestimmten kritischen Wertes von R, ist dann aber immer[1]) energetisch stabiler als die Laminarbewegung. In demjenigen Bereich von R, in welchem beide Bewegungsformen möglich sind, kann man daher stets die Flüssigkeit aus dem laminaren Zustand durch hinreichend große Störungen, in den turbulenten Zustand „herunterfallen" lassen.

Um die turbulente Bewegung näherungsweise mathematisch behandeln zu können, fassen wir wieder die Strömung zwischen zwei parallelen Wänden ins Auge und machen zunächst folgende Annahmen[2]): Die Strömung soll a) um die X-Achse symmetrisch, die begrenzenden Wände in Ruhe sein; sie soll b) periodisch in der X-Richtung mit der Periode $2\pi/\alpha$[3]) und c) periodisch in der Zeit mit der Periode $2\pi/\beta$ sein. d) Alle Störungen sollen sich mit der Geschwindigkeit β/α relativ zur X-Achse fortpflanzen, d. h. wenn wir die Bewegung in eine Fourierreihe entwickeln, so sollen in den Exponenten von e nur Vielfache von $i\,(\beta\, t - \alpha\, x)$ vorkommen.[4])

Die Fourierentwicklung der Stromfunktion soll also lauten:

$$(36)\quad \begin{cases} \psi = \varphi_0(y) + \varphi_1(y)\, e^{i(\beta t - \alpha x)} + \overline{\varphi_1(y)}\, e^{-i(\beta t - \alpha x)} \\ \qquad\qquad + \varphi_2(y)\, e^{2i(\beta t - \alpha x)} + \dots \end{cases}$$

Die mathematische Aufgabe besteht dann in der Ermittelung der (nach a) ungeraden Funktionen $\varphi_0, \varphi_1, \varphi_2 \dots$

1) Vgl. F. Noether, a. a. O.

2) Auch dieser Ansatz, der eine einfache Verallgemeinerung des Sommerfeldschen Stabilitätsansatzes darstellt, wurde zur Untersuchung der turbulenten Bewegung selbst von F. Noether: Zur Theorie der Turbulenz, Jahresber. d. deutsch. Math. Vereins **23**. S. 138. 1914 ohne weitergehende Folgerungen angegeben.

3) Zur Annahme eines *bestimmten* α berechtigt uns das Ergebnis von Teil I, daß α zwischen bestimmte Grenzen eingeschlossen ist, die insbesondere in der Nähe des Minimalwertes von R sehr eng sind.

4) Wir brauchen nicht hervorzuheben, daß die wirklichen Bewegungsvorgänge zweifellos viel komplizierter sind; trotzdem kann man wohl erwarten, daß diese Ansätze qualitative Aussagen über die Turbulenz gestatten.

($\overline{\varphi}_1$, $\overline{\varphi}_2 \ldots$ konjugiert zu φ_1, φ_2). Der Grad der Konvergenz der Reihe (36) ist zunächst völlig unbekannt, die Konvergenzfrage kann erst entschieden werden nach der Berechnung von φ_0, $\varphi_1 \ldots$ Wollen wir die Genauigkeit bis zur n^{ten} Näherung treiben, d. h. wollen wir $\varphi_0 \ldots \varphi_n$ berechnen, so bekommen wir offenbar $(n+1)$ simultane Differentialgleichungen für die $(n+1)$ unbekannten Funktionen $\varphi_0 \ldots \varphi_n$. Im folgenden werden wir teils der ersten, teils der zweiten Näherung bedürfen. Wir gehen also mit dem Ansatz (36) in Gleichung (4) ein, vergleichen beiderseits die Koeffizienten der periodischen Funktionen und brechen mit dem Gliede $e^{2i(\beta t - \alpha x)}$ und daher mit φ_2, $\overline{\varphi}_2$ ab. Für φ_0' schreiben wir (wie in (6) für Φ') w. So entstehen drei simultane Differentialgleichungen (die gleichzeitig entstehenden konjugierten Gleichungen brauchen wir nicht hinzuschreiben):

$$(37)\quad \left\{\begin{aligned}
&\frac{d^2}{dy^2}\Big[(\varphi_1'\overline{\varphi}_1 - \overline{\varphi}_1'\varphi_1) + 2(\varphi_2'\overline{\varphi}_2 - \overline{\varphi}_2'\varphi_2)\Big] = \frac{i}{\alpha R}w'',\\
&(\varphi_1'' - \alpha^2\varphi_1)\left(w - \frac{\beta}{\alpha}\right) - \varphi_1 w'' - \varphi_2'(\overline{\varphi}_1'' - \alpha^2\overline{\varphi}_1)\\
&\qquad - 2\varphi_2(\overline{\varphi}_1''' - \alpha^2\overline{\varphi}_1') + 2\overline{\varphi}_1'(\varphi_2'' - 4\alpha^2\varphi_2)\\
&\qquad\qquad + \overline{\varphi}_1(\varphi_2''' - 4\alpha^2\varphi_2')\\
&\qquad\qquad = \frac{i}{\alpha R}(\varphi_1'''' - 2\alpha^2\varphi_1'' + \alpha^4\varphi_1),\\
&(\varphi_2'' - 4\alpha^2\varphi_2)\left(w - \frac{\beta}{\alpha}\right) - w''\varphi_2 + \frac{1}{2}(\varphi_1'\varphi_1'' - \varphi_1\varphi_1''')\\
&\qquad\qquad = \frac{i}{2\alpha R}(\varphi_2'''' - 8\alpha^2\varphi_2'' + 16\alpha^4\varphi_2).
\end{aligned}\right.$$

Von diesen Gleichungen läßt sich die erste zweimal integrieren und liefert:

$$(37\text{a})\quad \varphi_1'\overline{\varphi}_1 - \overline{\varphi}_1'\varphi_1 + 2(\overline{\varphi}_2\varphi_2' - \overline{\varphi}_2'\varphi_2) = \frac{i}{\alpha R}(w' - C - C_1 y).$$

C und C_1 bedeuten willkürliche Integrationskonstanten.

Da die linke Seite von (37a) und w nach Forderung (a) ungerade in y sind, so muß in unserem Falle C verschwinden.

Gehen wir von der zweiten auf die erste Näherung zurück, so reduziert sich unser Gleichungssystem auf zwei simultane Differentialgleichungen für w und φ_1:

$$(38)\begin{cases} \varphi_1'\bar{\varphi}_1 - \bar{\varphi}_1'\varphi_1 = \dfrac{i}{\alpha R}(w' - C_1 y), \\ (\varphi_1'' - \alpha^2\varphi_1)\left(w - \dfrac{\beta}{\alpha}\right) - w''\varphi_1 = \dfrac{i}{\alpha R}(\varphi_1'''' - 2\alpha^2\varphi_1'' \\ \qquad + \alpha^4\varphi_1). \end{cases}$$

Einschaltungsweise wollen wir uns jetzt überlegen, was an Stelle von (38) tritt, wenn wir nicht eine um die X-Achse symmetrische Strömung (Forderung a), d. h. die Strömung einer Flüssigkeit unter Druckgefälle zwischen zwei ruhenden Wänden, sondern eine um die X-Achse antisymmetrische Strömung (d. h. Strömung zwischen zwei relativ zueinander bewegten Wänden ohne Druckgefälle wie beim Couetteschen Fall) betrachten. Forderung (*b*) und (*c*) sollen bestehen bleiben. Der Ansatz (36) wird dann nicht mehr genügen, da φ_1, φ_2 usw. bei beliebigem β/α keine geraden Funktionen mehr sind; wir müssen, um das ganze Strömungsbild ungerade zu erhalten, auch noch die symmetrischen Schwingungen der Form $e^{i(-\beta t - \alpha x)}$ in den Ansatz für ψ aufnehmen d. h. ψ muß beginnen mit den Gliedern

$$\varphi_0 + \varphi_1(y)e^{i(\beta t - \alpha x)} + \varphi_1(-y, e^{-i(\beta t + \alpha x)} + \overline{\varphi_1(y)}\, e^{-i(\beta t - \alpha x)} + \ldots$$

Das hat zur Folge, daß in ψ schließlich alle Glieder der Form $e^{i(m\beta t - n\alpha x)}$ auftreten (Wegfallen der Forderung (*d*)).

An Stelle von (38) ergibt sich:

$$(39)\begin{cases} \overline{\varphi_1(y)}\,\varphi_1(y)' - \overline{\varphi_1(y)'}\,\varphi_1(y) + \overline{\varphi_1(-y)}\,\varphi_1(-y)' \\ \qquad - \overline{\varphi_1(-y)'}\,\varphi_1(-y) = \dfrac{i}{\alpha R}(w' - C), \\ (\varphi_1'' - \alpha^2\varphi_1)\left(w - \dfrac{\beta}{\alpha}\right) - w''\varphi_1 = \dfrac{i}{\alpha R}(\varphi_1'''' - 2\alpha^2\varphi_1'' \\ \qquad + \alpha^4\varphi_1). \end{cases}$$

Die beiden Gleichungen des Systems (38) bzw. (39) haben eine einfache anschauliche Bedeutung.

Die zweite Gleichung ist nichts anderes, als unsere frühere „Stabilitätsgleichung“ (7), welche die Amplitude der einer Grundströmung w überlagerten Schwingung bestimmt und die Grundlage für unsere Untersnchungen im Teil I bildete. Die erste

Annalen der Physik (4) *74*, 577–627 (1924)

Gleichung aber stellt den Impulssatz dar. Die linke Seite dieser Gleichung gibt nämlich im wesentlichen den durch die turbulente Wirbelung im Mittel übertragenen Impuls an[1]), das Glied mit w' rechts stellt die laminare Impulsübertragung dar, die Konstante C bzw. $C_1 y$ ist die „Konstante“ des Impulssatzes.

An den Wänden ist wegen der Grenzbedingungen $\varphi_1 = \varphi_1' = 0$. Also ist dort $w' = C$ bzw. $= C_1 y_{\text{Wand}}$; an den Wänden überwiegt also der laminare Impulstransport den turbulenten, w'_{Wand} wird im allgemeinen sehr groß sein (vgl. den nächsten Paragraphen). In der Kanalmitte aber, d. h. im ganzen Kanal außerhalb der unmittelbaren Wandnähe, ist w' von der Größenordnung 1, also sehr klein gegen C bzw. $C_1 y$. Hier wird daher der turbulente Impulstransport den laminaren völlig überwiegen.

Der Bauart der Systeme (38) und (39) entspricht es, daß wir eine triviale Lösung derselben sofort angeben können, nämlich

$$\varphi_1 = 0, \qquad w' = C \quad \text{bzw.} \quad w' = C_1 y,$$

d. h. wir kommen so zur Laminarbewegung zurück.

Unser Ziel muß es jetzt sein, über die nichttrivialen Lösungen von (38) und (39) bestimmte Resultate zu erhalten.

§ 2. Die turbulente Bewegung in Wandnähe und das Widerstandsgesetz.

Das wichtigste Ergebnis über das Verhalten von w in unmittelbarer Nähe der Wände ist das von v. Kármán (a. a. O.) durch Dimensionsbetrachtungen aus dem Blasiusschen Widerstandsgesetz abgeleitete Gesetz, daß w in der Nähe einer Wand wie $\eta^{1/7}$ anwächst (wenn η die Entfernung von der Wand darstellt). Wir wiederholen kurz den Kármánschen Gedankengang, da wir uns dadurch allgemein darüber orien-

1) Wir meinen hier den mittleren Impuls in der X-Richtung, der bei unserem Problem in der Y-Richtung übertragen wird. Der Impuls in der X-Richtung ist im wesentlichen gleich u, die Geschwindigkeit des den Impuls tragenden Teilchens in der Y-Richtung ist v, also der in der Zeiteinheit übertragene Impuls $u \cdot v$, im Mittel $\overline{u \cdot v}$, was im Falle (36) ergibt:

$$\overline{u \cdot v} = -i\alpha(\Phi_1 \varphi_1' - \Phi_1' \varphi_1).$$

tieren können, was wir — auch ohne das Blasiussche Gesetz zu kennen — für das Verhalten von w in Wandnähe zu erwarten haben.

Die an einer Wand angreifende Schubspannung τ (d. h. also der Widerstand) muß sich, wie aus Dimensionsbetrachtungen leicht zu ersehen ist, darstellen lassen in der Form:

$$\tau = \varkappa \cdot \mu \cdot \frac{U}{h} \cdot f(R),$$

wo $\varkappa$ eine bestimmte dimensionslose Konstante bedeutet.

Wenn wir speziell ein Potenzgesetz annehmen, so ist

$$\tau = \varkappa \cdot \mu \cdot \frac{U}{h} \cdot R^{\xi} = \varkappa \cdot \mu \cdot \frac{U}{h} \cdot \left(\frac{U \cdot h \cdot \varrho}{\mu}\right)^{\xi}. \tag{40}$$

Aus (40) folgt umgekehrt:

$$U = \left[\frac{\tau}{\varkappa} \cdot \left(\frac{h}{\mu}\right)^{1-\xi} \cdot \varrho^{-\xi}\right]^{\frac{1}{1+\xi}}.$$

Die Geschwindigkeitsverteilung in Wandnähe muß sich dann durch eine Gleichung der Form darstellen:

$$w = \frac{1}{U}\left[\tau \cdot \varrho^{-\xi}\left(\frac{h}{\mu}\right)^{1-\xi}\right]^{\frac{1}{1+\xi}} f\left(\frac{\eta}{h}\right)$$

(w ist hier wieder [vgl. (1) und (6)] dimensionslos gewählt und enthält daher U im Nenner, η bedeutet den Abstand von der Wand).

Nehmen wir wieder in 1. Näherung ein Potenzgesetz an (σ sei eine dimensionslose Konstante):

$$w = \frac{\sigma}{U}\left[\tau\, \varrho^{-\xi} \cdot \left(\frac{h}{\mu}\right)^{1-\xi}\right]^{\frac{1}{1+\xi}} \left(\frac{\eta}{h}\right)^{\varepsilon}.$$

Fordert man jetzt, daß die Geschwindigkeitsverteilung in unmittelbarer Nähe der Wand nur von τ, μ, ϱ, nicht aber von h abhängen soll, was physikalisch sehr plausibel ist, so folgt:

$$\frac{1-\xi}{1+\xi} = \varepsilon. \tag{41}$$

Für $\xi = \frac{3}{4}$, wie es dem Blasiusschen Gesetze entspricht, ergibt sich also $\varepsilon = \frac{1}{7}$.

Um uns über die physikalische Bedeutung dieses Resultats klar zu werden, bemerken wir: $w \sim \eta^{1/7}$ bedeutet, daß $w' = dw/d\eta$ am Rande unendlich ist, daß also w sich un-

Annalen der Physik (4) *74*, 577–627 (1924)

endlich an die Wand anschmiegt. Nun ist es aber klar, daß in Wirklichkeit w' an der Wand nicht unendlich sein kann, denn w' bedeutet ja gerade im wesentlichen die Schubspannung an der Wand und ist daher nach (40) bis auf einen von R unabhängigen Zahlenfaktor gleich R^{ξ} zu setzen.

(42) $$w'_{\text{Rand}} \sim R^{\xi}.$$

w' an der Wand ist also bei den uns interessierenden großen Werten von R sehr groß. Die Geschwindigkeitsverteilung $w \sim \eta^{1/7}$ wird ihrer Ableitung entsprechend nur im Grenzfall unendlich großen Wandabstandes oder reibungsloser Flüssigkeit ($R = \infty$) in Strenge gültig sein. Noch einfacher läßt sich dieser Sachverhalt überblicken, wenn wir das gefundene Gesetz $w \sim \eta^{1/7}$ in der Form schreiben $\eta \sim w^7$. Aus der Tatsache endlicher Schubspannung wissen wir, daß das erste Glied der Potenzentwicklung $\eta(w)$ von der Form $\gamma_1 w$ sein muß. Dieses Glied ist aber sehr klein, nämlich im wesentlichen gleich dem reziproken Werte von w', also von der Ordnung $R^{-\xi}$ (vgl. 42).

Die Meinung des abgeleiteten Gesetzes $w \sim \eta^{1/7}$ ist also offenbar die, daß die Reihenentwicklung von $\eta(w)$ anfangen soll mit den Gliedern

(43) $$\eta = \gamma_1 w + \gamma_7 w^7 + \ldots$$

wo γ_1 außerordentlich klein ist, und daß daher das erste Glied $\gamma_1 w$ für einigermaßen große Werte von w gegen das zweite $\gamma_7 w^7$ gestrichen werden kann.

Nach den vorhergehenden Darlegungen erwarten wir unabhängig von der Gültigkeit des $\frac{1}{7}$-Gesetzes für das Aussehen der Grundströmung der turbulenten Bewegung in der Mitte kleine Krümmungen, in Wandnähe Anschmiegen der Grundströmung an die Wände.

Für ein solches Profil gelten die Untersuchungen des Teiles I nicht unmittelbar, denn wir haben dort w', w'' usw. als endlich vorausgesetzt. Doch lassen sich diese Untersuchungen leicht auf solche Profile wie das hier vorkommende verallgemeinern (vgl. § 5, S. 597). Insbesondere wird die Lösung der reduzierten Gleichung (8) (also $\lim R = \infty$) mit Befriedigung der Grenzbedingungen hier besonders einfach; das soeben charakterisierte Profil gehört nach § 5, S. 597 zu den „schwingungsfähigen", es existiert eine Lösung von (28) mit reellem c; dies ist außerordentlich wichtig, zeigt es doch, daß die „turbulenten Profile" immer „labil" sind nach § 5, oder anders ausgedrückt, daß gerade die Abweichung (42) vom laminaren Widerstandsgesetz die Möglichkeit eines „labilen" Profils, also der Turbulenz gibt.

Die Lösung von (28) heißt für $\alpha = 0$

$$\varphi = [w - w(+1)] \int_{-1}^{y} \frac{dy}{[w - w(+1)]^2} \,. \tag{44}$$

Daß φ für $y = -1$ Null wird, haben wir durch Wahl der unteren Grenze des Integrals erreicht; daß es aber auch für $y = +1$ Null wird, wegen der Wahl $c = w(+1)$ ist leicht zu sehen aus folgender Umformung:

$$\varphi = [w - w(+1)] \int_{w(-1)}^{w} \frac{dw}{[w - w(+1)]^2 \cdot w'} \,.$$

Das Integral der rechten Seite wird nämlich im Punkte $w = w(+1)$ von *niedrigerer* Ordnung unendlich als $\frac{1}{w - w(+1)}$, da w' dort (im Limes $R = \infty$) unendlich wird. Also ist $\varphi_1 = 0$ für $y = +1$. Durch (44) haben wir im Limes reibungsloser Flüssigkeit die Amplitude der turbulenten Schwingungen dargestellt und aus den Grenzbedingungen den Wert für β/α, nämlich $\beta/\alpha = w(+1)$ abgeleitet. Es ist aber selbstverständlich, daß auch die zu (44) symmetrische Lösung

$$\varphi = [w - w(-1)] \int_{+1}^{y} \frac{dy}{[w - w(-1)]^2} \tag{44a}$$

den Grenzbedingungen genügt; also gilt hier $c = w(-1)$. Bei der Couetteschen Anordnung schließen wir daher aus (44) und (44a), daß zwei zueinander symmetrische Schwingungssyteme existieren, deren Wellengeschwindigkeit bzw. mit den Geschwindigkeiten der beiden Wände ($w(+1)$ und $w(-1)$) übereinstimmen.

Für die symmetrische Strömung zwischen zwei *ruhenden* Wänden dagegen ist $w(+1) = w(-1) = 0$. Aus (44) und (44a) schließen wir dann, daß jedes Integral der Form

$$\varphi = w \int \frac{dy}{w^2}$$

den Grenzbedingungen genügt. Aus der Forderung (a), daß

Annalen der Physik (4) 74, 577–627 (1924)

φ ungerade sein soll, ergibt sich aber, daß wir als untere Grenze des Integrals $\int \frac{dy}{w^2}$ $y = 0$ wählen müssen; also

$$\varphi = w \int_0^y \frac{dy}{w^2} \,. \tag{44b}$$

Im Falle symmetrischer Strömung gilt also insbesondere für β / α:

$$w(+1) = w(-1) = \frac{\beta}{\alpha} = 0 \,. \tag{44c}$$

Von der turbulenten Grundströmung interessiert uns zunächst am meisten die Art der Singularität derselben an den Wänden, also wenn wir annehmen $w \sim \eta^\varepsilon$, der Exponent ε. Wir wollen zu zeigen versuchen, daß aus den Differentialgleichungen (38) bzw. (39) im Limes $R = \infty$ wenigstens in unmittelbarer Wandnähe tatsächlich ein solches Potenzgesetz mit dem Exponenten $\varepsilon = \frac{1}{7}$ folgt. Dabei ist allerdings der Konvergenzbereich der benutzten Potenzreihen nicht sicher gestellt, so daß die Schlüsse, soweit sie auf die Gestalt des Profils in einigem Abstand von der Wand angewendet werden, unsicher sind. Wir entwickeln φ_1 und w in der Umgebung von $\eta = 0$ nach ganzen und positiven Potenzen von η — dies ist möglich für jedes endliche — und dann umgekehrt η nach ganzen Potenzen von w. So werden wir direkt zu der Formel (43) für $\eta(w)$ geführt.

Wir behaupten — und dies ist das wichtigste Resultat, das wir später brauchen — φ_1 läßt sich in 1. Näherung darstellen durch eine Reihe der Form

$$\varphi_1 = \alpha_2 \eta^2 + \alpha_5 \eta^5 + \alpha_8 \eta^8 + \ldots,$$

wo $\alpha_2, \alpha_8 \ldots$ reelle, $\alpha_5, \alpha_{11} \ldots$ rein imaginäre Konstanten sind; ferner ist w von der Form:

$$w = \beta_1 \eta + \beta_7 \eta^7 + \ldots \tag{45}$$

Diese Behauptung läßt sich für die Differentialgleichungen (38) direkt durch Ansatz von φ und w mit unbestimmten Koeffizienten beweisen, wenn die Glieder $\alpha_0, \alpha_1, \alpha_2, \alpha_3$ und β_0, β_1 gegeben sind. Suchen wir also vorerst diese Glieder zu bestimmen. Zunächst muß φ_1 und φ_1' für $\eta = 0$ wegen der Grenzbedingungen Null sein, also fängt die Reihe für φ_1 mit $\alpha_2 \eta^2$ an ($\alpha_0 = \alpha_1 = 0$). Daß ferner das folgende Glied $\alpha_3 \eta^3$

wegfällt, d. h. sehr klein wird gegen die übrigen Glieder, können wir nachträglich verifizieren. Einschaltungsweise wollen wir hier zu diesem Zwecke durch einmalige näherungsweise Integration der zweiten Gleichung (38) beweisen, daß α_3 von der Größenordnung αR wird. Für $\alpha^2 = 0$ lautet Gleichung (38)

$$\varphi_1''\left(w - \frac{\beta}{\alpha}\right) - \varphi_1 w'' = \frac{i}{\alpha R}\varphi_1'''',$$

woraus folgt:

$$(46)\qquad \varphi_1'\left(w - \frac{\beta}{\alpha}\right) - \varphi_1 w' = \frac{i}{\alpha R}\varphi_1''' + A.$$

Die Konstante A ist hierin von derselben Größenordnung, wie die linke Seite von (46) in der Mitte des Kanals, also höchstens von der Größenordnung 1 (vgl. I, § 2). φ_1''' am Rande ist daher wegen der Grenzbedingungen von der Ordnung αR. Dasselbe gilt für α_3.

Nehmen wir also einstweilen α_3 als klein an und suchen dies später zu rechtfertigen. Von den Konstanten β_0, β_1 ist die erste β_0 wegen Forderung (a) § 1 gleich Null.

Die Konstanten α_2 und β_1 sind zunächst willkürlich[1]) und wir haben auch keine Möglichkeit, sie aus der Lösung der Differentialgleichungen (38) und (39) in Wandnähe abzuleiten. Diese Möglichkeit bestände erst, wenn es uns gelänge, die Lösung (45) analytisch bis zur anderen Wand fortzusetzen. Diese Aufgabe jedoch ist mathematisch außerordentlich kompliziert, schon deswegen, weil, wie sich zeigen wird, die vereinfachten Gleichungen (38), (39) nicht ausreichen, um φ_1 und w in der Mitte des Kanals zu bestimmen. Obwohl wir demnach auf die Lösung dieser Aufgabe verzichten müssen, können wir doch erwarten, daß wir durch die bloße Entwicklung von φ_1 und w in der Nähe *einer* Wand mit unbestimmten Koeffizienten α_2, β_1 doch diejenigen qualitativen Eigenschaften von w und φ_1 in Wandnähe erhalten können, die erfahrungsgemäß ganz unabhängig sind von dem Verhalten der Flüssigkeit in der Kanalmitte wie z. B. das Gesetz $w \sim \eta^{1/7}$.

1) Wir wollen α_2 als reell annehmen; dies bedeutet keine Beschränkung der Allgemeinheit, denn φ_1 ist nur bis auf einen Faktor der Form $e^{i\chi}$ bestimmt, da man in (36) den Anfangspunkt der Zeit beliebig wählen kann.

Gehen wir mit dem Ansatz:

$$\varphi_1 = \alpha_2 \eta^2 + \alpha_4 \eta^4 + \alpha_5 \eta^5 + \dots$$
$$w = \beta_1 \eta + \beta_2 \eta^2 + \beta_3 \eta^3 + \dots$$

in Gleichung (38) ein, wobei wir die zweite Gleichung durch (46) ersetzen. Wir nehmen also wieder α als sehr klein an, was hier nur bedeutet (vgl. I, § 2), daß die Wellenlänge der Schwingungen groß sein soll gegenüber der Grenzschichtdicke. Außerdem setzen wir nach (44c)

$$\frac{\beta}{\alpha} = 0 .$$

Für die erste Gleichung (38) schreiben wir ferner

$$(47) \quad \begin{cases} - i\alpha R(\varphi_1' \bar{\varphi}_1 - \bar{\varphi}_1' \varphi_1) = 2\alpha R(\varphi_{1i}' \varphi_{1r} - \varphi_{1i} \varphi_{1r}') \\ \qquad = w' - C_1 y . \end{cases}$$

Hierin bedeutet φ_{1r} den reellen, φ_{1i} den imaginären Teil von φ_1.

Aus (46) und (47) folgen jetzt die Rekursionsformeln

$$(48) \qquad n(n-1)(n-2)\alpha_n = -i\alpha R \sum_{s=2}^{n-2} (n-2s)\alpha_{n-1-s}\beta_{s-1} ,$$

$$(49) \qquad n\beta_n = 2\alpha R \sum_{s=2}^{n-2} (n-2s)\alpha^i_{n-s}\beta^r_s ;$$

$$\text{dazu} \qquad \beta_1 = C_1 y_{\text{Rand}} ,$$
$$2\beta_2 = C_1 .$$

Hierin bedeutet α_s^r, α_s^i den reellen bzw. imaginären Teil von α_s.

Aus (48) folgt zunächst

$$\alpha_4 = 0 .$$

Aus (49) ergibt sich dann:

$$\beta_3 = \beta_4 = \beta_5 = \beta_6 = 0 .$$

Auch β_2 können wir näherungsweise Null setzen.

Aus (49) folgt nämlich

$$\beta_2 = \frac{\beta_1}{2 y_{\text{Rand}}} ;$$

das Glied $\beta_2 \eta^2$ ist also für sehr kleine η gegen das erste Glied $\beta_1 \eta$ zu vernachlässigen; für größere η aber überwiegen die höheren Glieder $\beta_7 \eta^7$ usw. völlig.

Nehmen wir also auch $\beta_2 = 0$ an und rechnen so die höheren Glieder der Reihen für β_1 und w aus.

Es folgt

$$\alpha_5 = -i\alpha R \frac{\alpha_2 \beta_1}{3\cdot 4\cdot 5}; \quad \alpha_6 = \alpha_7 = 0; \quad \alpha_8 = -(\alpha R)^2 \frac{\alpha_2 \beta_1^2}{3\cdot 5\cdot 6\cdot 7\cdot 8};$$

$$\alpha_9 = \alpha_{10} = 0;$$

$$\alpha_{11} = -(\alpha R)^3 \left(\frac{\alpha_3^3 \beta_2}{2\cdot 7\cdot 9\cdot 10\cdot 11} - \frac{\alpha_2 \beta_1^3}{3\cdot 5\cdot 6\cdot 8\cdot 9\cdot 10\cdot 11}\right);$$

$$\alpha_{12} = \alpha_{13} = 0.$$

$$\beta_7 = -(\alpha R)^2 \frac{\alpha_2^2 \beta_1}{70}; \quad \beta_8 = \beta_9 = \beta_{10} = \beta_{11} = \beta_{12} = 0;$$

$$\beta_{13} = (\alpha R)^4 \left(\frac{\alpha_2^2 \beta_1^3}{5\cdot 5\cdot 6\cdot 8\cdot 11\cdot 13\cdot 14} - \frac{\alpha_2^4 \beta_1}{7\cdot 10\cdot 11\cdot 13}\right).$$

Die von uns behauptete Darstellung (45) für w ist also bewiesen und es läßt sich leicht zeigen, daß auch von den weiteren Gliedern β nur jedes sechste Glied einen endlichen Wert hat.

Für die Darstellung von η als Potenzreihe nach w folgt hieraus:

$$\eta = \gamma_1 w + \gamma_7 w^7 + \gamma_{13} w^{13} + \cdots$$

$$\gamma_1 = \frac{1}{\beta_1}, \qquad \gamma_7 = (\alpha R)^2 \frac{\alpha_2^2}{70\,\beta_1^7}, \tag{50}$$

$$\gamma_{13} = (\alpha R)^4 \left(\frac{\alpha_2^4 \cdot 3\cdot 3\cdot 17}{7\cdot 10\cdot 10\cdot 11\cdot 13\cdot \beta_1^{13}} - \frac{\alpha_2^2}{5\cdot 5\cdot 6\cdot 8\cdot 11\cdot 13\cdot 14\,\beta_1^{11}}\right).$$

Die Glieder γ_2 bis γ_6, γ_8 bis γ_{12}, γ_{14} usw. sind alle gleich Null.[1])

Die Entwicklung (50) stimmt nun in der Tat völlig mit (43) überein und auch ohne Kenntnis der Konstanten α_2 und β_1 scheinen wir so zu dem von Kármán halbempirisch abgeleiteten Gesetz $\eta \sim w^7$ zu gelangen. Die Koeffizienten γ_1 und γ_7 lassen sich jedoch nicht berechnen. Umgekehrt können wir vielleicht aus dem empirischen Befund für die Koeffizienten β_1 und α_2 schließen, daß γ_7 von der Größenordnung 1, γ_1 von der Ordnung $(\alpha R)^{-3/4}$, also $\alpha_2 \sim (\alpha R)^{13/8}$ ist. Nachträglich bestätigen wir so auch unsere frühere Behauptung $\alpha_3 \eta \ll \alpha_2$. Fragt man sich nach der Größenordnung der Werte von w, für die das

1) Diese Potenzreihe $\eta(w)$ läßt sich natürlich auch direkt aus (46) und (47) ableiten, ohne den Umweg über die Reihe von $w(\eta)$, wenn w als unabhängige Variable eingeführt wird. Doch sind die dazu nötigen Rechnungen etwas komplizierter.

3. Glied in (50) klein ist gegen das zweite, für die also das w^7-Profil wirklich gilt, so ergibt sich $w \sim \beta_1^{-1/8}$, also $\sim R^{-1/8}$. Aus den Differentialgleichungen (38) folgt nach alledem das Profil $w \sim \eta^{1/7}$ zunächst nur qualitativ; darüber, daß das $\frac{1}{7}$-Profil fast bis zur Kanalmitte hin beobachtet ist, geben unsere Rechnungen keinen Aufschluß. Das war jedoch auch nicht zu erwarten, denn die übrigen Konstanten, die in das Gesetz eingehen, hängen auch von dem Verhalten der Flüssigkeit an der gegenüberliegenden Wand ab.

Fassen wir einschaltungsweise noch einmal kurz zusammen, welche Vernachlässigungen wir bei der Ableitung von (50) aus (48) und (49) begangen haben und suchen dadurch zu bestimmen, mit welcher Genauigkeit die aus (50) gezogenen Schlüsse richtig sind. Zunächst haben wir das System (38) statt (37) benutzt, also Größen der Ordnung φ_2/w gestrichen. Ferner haben wir $\alpha_3 = 0$, $\beta/\alpha = 0$, $\beta_2 = 0$ gesetzt und damit Größen der Ordnung

$$\frac{\alpha_3\,\eta}{\alpha_2}, \qquad \frac{\beta}{\alpha\,w}, \qquad \frac{\beta_2\,\eta}{\beta_1} \qquad \text{bzw.} \qquad \frac{\beta_1}{\beta_7\,\eta^6}$$

vernachlässigt. Die Genauigkeit unserer Rechnungen wird durch das größte der hier vernachlässigten Glieder bestimmt sein. Einfache Größenordnungsbetrachtungen, die wir hier nicht durchführen, machen es wahrscheinlich, daß von diesen Gliedern φ_2/w das größte ist, daß aber auch dieses Glied mit $R \to \infty$ gegen Null geht. Durch Wahl eines hinreichend großen Wertes für R wird man also die Genauigkeit der aus (48),(49) und (50) abgeleiteten Resultate beliebig weit treiben können.

Was das Blasiussche Widerstandsgesetz betrifft, so läßt sich dieses natürlich umgekehrt nach dem oben geschilderten Verfahren mittels Dimensionsbetrachtungen aus dem Gesetz $\eta \sim w^7$ herleiten, wenn man, wie geschehen, annimmt, daß das Verhalten von w in Wandnähe von der Kanalbreite unabhängig ist. Die Frage aber, ob auch diese letztere physikalisch sehr plausible Annahme aus den Differentialgleichungen (38) bzw. (39) folgt, müssen wir aus den oben angegebenen Gründen (Unmöglichkeit der analytischen Fortsetzung) unbeantwortet lassen.

Eine bemerkenswerte direkte Folgerung über das Widerstandsgesetz können wir jedoch aus (38) und (39) durch Größenordnungsbetrachtungen ziehen. Im Kanal, die unmittelbare

Wandnähe und den Punkt $y = 0$ [vgl. u. (66)] ausgeschlossen, kann man nämlich statt der ersten Gleichung (38) wegen der Größe von C_1 schreiben (vgl. S. 612)

(51) $$i\alpha R(\varphi_1' \bar{\varphi}_1 - \bar{\varphi}_1' \varphi_1) = C_1 y.$$

Da nun die Amplitude φ_1 nicht mit $R \to \infty$ gegen unendlich gehen kann — dies würde allen unseren Rechnungen den physikalischen Sinn nehmen —, so folgt, daß C_1 höchstens groß von der Ordnung αR ist, daß also der Exponent ξ von Gleichung (40) < 1 sein muß (was ja auch in gewisser Weise aus (41) hervorgeht). Daraus folgt [vgl. Gleichung (42) und (40)], daß das in der Hydraulik gewöhnlich angenommene Widerstandsgesetz $\tau = \text{const} \cdot U^2$ eine obere Grenze für alle denkbaren Widerstandsgesetze der Turbulenz darstellt, die *unabhängig* ist *von der Wandbeschaffenheit.* Man kann vermutungsweise schließen, daß das Gesetz $\tau \sim U^{7/4}$ nur gültig ist für glatte Wände — nur für diese haben wir ja auch $\eta \sim w^7$ erhalten —, daß sich aber das Widerstandsgesetz für rauhe Wände immer mehr dem quadratischen annähert.[1]) Denn für rauhe Wände wird die Amplitude φ_1 von R unabhängig und von der Größe der Wandstörungen sein — außerdem werden für rauhe Wände die Grenzbedingungen nicht mehr verursachen, daß φ_1 in erster Näherung reell ist, wie es der Gleichung (44) entspricht.

An den Schlüssen dieses Paragraphen ändert sich nichts, wenn man an Stelle der Differentialgleichungen (38) die Gleichungen (39) zugrunde legt.

§ 3. Die turbulente Bewegung außerhalb der unmittelbaren Wandnähe.

Für die Bewegung in der Kanalmitte ist es wesentlich, daß sich φ_1 hier aus denjenigen beiden Integralen von (7a) zusammensetzt, die schon bei der reibungslosen Flüssigkeit, also für Gleichung (8) auftreten (vgl. I, § 2). Die wichtigste hieraus folgende Eigenschaft von φ_1 ist die, daß es, bis auf Größen der Ordnung φ_2 und $(\alpha R)^{-1}$ der Bedingung

(52) $$\varphi_1' \bar{\varphi}_1 - \bar{\varphi}_1' \varphi_1 = \text{const.}$$

genügt.

---

1) Vgl. die genaueren Untersuchungen von v. Kármán a. a. O. und die experimentellen Untersuchungen von Schiller, dieselbe Zeitschrift **3.** S. **2.** **1923.**

Dies ergibt sich vermöge des Abelschen Satzes aus der Tatsache, daß, bis auf Größen der Ordnung φ_2 und $(\alpha R)^{-1}$ φ_{1r} und φ_{1i} (der reelle bzw. imaginäre Teil von φ_1) Lösungen der Differentialgleichung (8) sind.

Daraus ist zu schließen, daß die Gleichungen (38) bzw. (39) nicht ausreichen, um die Bewegung in der ganzen Kanalbreite festzulegen, daß wir vielmehr zu Gleichung (37) und zu dem ihr im Couetteschen Fall entsprechenden Gleichungssystem zurückgreifen müssen.

Dies bringt im allgemeinen eine Komplizierung der mathematischen Aufgabe mit sich. Nur im Couetteschen Fall läßt sich die Aufgabe verhältnismäßig leicht lösen, weil die erste Gleichung (39) bis auf Größen der Ordnung w'/C also $(\alpha R)^{-2/3}$ und φ_2^2, vgl. (37), *mit Gleichung* (52) *übereinstimmt.* Während also Gleichung (52) ihrer Ableitung aus dem Abelschen Satz gemäß nur bis auf Größen der Ordnung φ_2 richtig ist, soll hier im Couetteschen Fall noch (39) bis auf Größen der Ordnung φ_2^2 gelten. Dieser Forderung wird Genüge geleistet, wenn wir setzen:

(53) $$\varphi_2 = 0.$$

Diese Gleichung ist also als Lösung unserer zugrunde gelegten Differentialgleichung für den Couetteschen Fall anzusehen.

Aus (53) würde für φ_1 aus (37) folgen

(54) $$\varphi_1''' \varphi_1 - \varphi_1'' \varphi_1 = 0.$$

Nun gilt für den Couetteschen Fall nicht das System (37), sondern ein komplizierteres, das wir hier nicht anschreiben wollen. Wir geben aber von ihm an, daß es ebenso wie (37) für $\varphi_2 = 0$ zur Lösung (54) führt und daher zum Ergebnis:

(55) $$\varphi_1 = a\, e^{\gamma y} + b\, e^{-\gamma y}$$

Hier sind a, b, γ irgendwelche komplexe Konstanten. Für w folgt dann aus der 2. Gleichung (39) bzw. aus ihrer reduzierten Form (8):

(56) $$w - c = a_1\, e^{\gamma_1 y} + b_1\, e^{\gamma_1 y}.$$

Da nun $w - c = 0$ sein soll an der einen Wand, da andererseits w ungerade sein soll um $y = 0$, so folgt, daß w einfach durch Nullwerden von γ_1 und geeignetes Wachsen von a_1 und b_1 zum linearen Profil ausarten muß.

Wir erhalten also das wichtige Resultat, daß beim Couetteschen Fall das Grundprofil w der turbulenten Bewegung in der ganzen Kanalbreite wesentlich linear verläuft (— jedoch stark abweichend vom laminaren Profil; es wird viel flacher sein, als das laminare —) [*daß es sich aber* (vgl. II. § 2) *am Rand wieder wie* $\eta^{1/7}$ *an die Wände anschmiegt*].

Gehen wir nun zum komplizierteren Fall einer Strömung zwischen zwei ruhenden Wänden, also genau zum System (37) über. Zur Lösung müssen wir uns naturgemäß mit rohen Näherungen begnügen. Zunächst können wir in (37) die rechten Seiten aller drei Gleichungen, nämlich die Reibungsglieder, streichen — dies wird durch die Überlegungen von I, § 2 vollauf gerechtfertigt. Dann setzen wir [vgl. Gleichung (44c)] $\beta/\alpha = 0$.

Für φ_1 erhalten wir so an Stelle der zweiten Gleichung (37)

$$(57)\quad \begin{cases} \varphi_1''w - w''\varphi_1 - \alpha^2 w\varphi_1 - \varphi_2'(\bar\varphi_1'' - \alpha^2\bar\varphi_1) - 2\varphi_2(\bar\varphi_1''' \\ \quad - \alpha^2\bar\varphi_1') + 2\bar\varphi_1'(\varphi_2'' - 4\alpha^2\varphi_2) + \bar\varphi_1(\varphi_2''' - 4\alpha^2\varphi_2') = 0. \end{cases}$$

Entwickeln wir φ_1 als Lösung der Gleichung (57) noch Potenzen von α^2 einerseits, von φ_2 andererseits, bemerken ferner, daß φ_1 ungerade sein soll (vgl. 44b) und schreiben $\varphi_1 = \varphi_{10} + \varphi_{11}$, so ergibt sich bei Berücksichtigung nur der linearen Glieder:

$$(58)\qquad \varphi_{10} = a\,w\int_0^y \frac{dy}{w^2}.$$

$$(59)\quad \begin{cases} \varphi_{11} = a\,w\displaystyle\int_0^y \frac{dy}{w^2}\int_0^y dy(\alpha^2 w\varphi_{10} + \varphi_2'\varphi_{10}'' + 2\varphi_2\varphi_{10}''' \\ \qquad\qquad - 2\varphi_{10}'\varphi_2'' - \varphi_{10}\varphi_2'''). \end{cases}$$

Hierein ist φ_1 natürlich nur bis auf einen konstanten Faktor a bestimmt, den wir hier, was keine Beschränkung bedeutet, reell angenommen haben.

Gehen wir mit diesem Werte von φ_1 in die vereinfachte 1. Gleichung (37), nämlich:

$$(60)\qquad \varphi_1\bar\varphi_1'' - \bar\varphi_1\varphi_1'' = \text{const.}$$

ein, so ergibt sich, wenn φ_{2i} den imaginären Teil von φ_2 bezeichnet:

$$(61)\quad (\varphi_{2i}'\varphi_{10}'' + 2\varphi_{2i}\varphi_{10}''' - 2\varphi_{10}'\varphi_{2i}'' - \varphi_{10}\varphi_{2i}''')\int_0^y \frac{dy}{w^2} = \text{const.}$$

Jetzt genügt aber φ_{2i}, wie aus der dritten Gleichung (37) und daraus, daß φ_1 in erster Näherung reell ist, folgt, der Gleichung

$$\varphi_{2i}'' w - \varphi_{2i} w'' = 0 . \tag{62}$$

also

$$\varphi_{2i} = b w \int_0^y \frac{dy}{w^2} = c \varphi_{10} . \tag{63}$$

Gehen wir mit diesem Werte von φ_{2i} in (63) ein, bedenken wir ferner, daß für $y = 0$ die linke Seite von (61) und damit die Konstante der rechten Seite Null ist (dies bedeutet für die Konstante der rechten Seite von (60) nur, daß sie in erster Näherung Null, d. h. klein von der Ordnung $\varphi_1 \varphi_2^2$ bzw. $\alpha^2 \varphi_1 \varphi_2$ oder $\alpha^4 \varphi_1$ ist), so erhalten wir

$$\varphi_{10} \varphi_{10}''' - \varphi_{10}'' \varphi_{10}' = 0, \tag{64}$$

was vollständig mit (54) übereinstimmt.

Diese Gleichung wird nun allerdings, ebenso wie (54), in der Umgebung des Punktes $y = 0$ trivial, sie ist, weil φ eine ungerade Punktion von y ist, dort identisch erfüllt. Sie kann also dort auch keine Bestimmung von w liefern. Dies führt beim symmetrischen Profil (64) zu einer merkwürdigen Unstetigkeit im Punkte $y = 0$. (Beim ungeraden Profil ist eine solche Unstetigkeit aus den Differentialgleichungen nicht zu erkennen). Intergriert man nämlich (37a), so erhält man, wie oben gezeigt, nach einmaliger Integration die Gleichung:

$$2\alpha R [\bar{\varphi}''_1 \varphi_1 - \varphi_1'' \bar{\varphi}_1 + 2 (\bar{\varphi}_2'' \varphi_2 - \varphi_2'' \bar{\varphi}_2) + \ldots] = w'' - C \tag{65}$$

[wo C (vgl. S. 612) bei Gültigkeit des Blasiusschen Gesetzes von der Größenordnung $(\alpha R)^{3/4}$, also jedenfalls sehr groß ist]. Die linke Seite von (65) verschwindet aber mit φ_1 und φ_2 (die ja ungerade Funktionen von y sind) an der Stelle $y = 0$. Dort muß also gelten

$$w_{y=0}'' = C . \tag{66}$$

Dies bedeutet, daß $w_{y=0}''$ sehr groß ist ($\sim (\alpha R)^{3/4}$), daß also w an der Stelle $y = 0$ einen scharfen Knick [1]) (Krümmungsradius

1) Auf diesen Knick hat mich freudlicherweise Hr. Prof. Prandtl auf Grund des empirischen Materials hingewiesen. Der Knick scheint dabei empirisch nicht so scharf zu sein, wie ihn die Rechnung ergibt, was sich wohl zwangslos dadurch erklärt, daß eben die Annahme (a) S. 609 über die Symmetrie auch der *Wirbel* und *Störungen* nicht genau der Wirklichkeit entspricht.

$\sim (\alpha\, R)^{-3/4}$) aufweist. In geringer Entfernung von diesem Punkt muß w nach (64) wieder wesentlich linear verlaufen.

Wir erhalten das Resultat:

Auch bei der Strömung zwischen zwei ruhenden Wänden — was sich wohl auch auf die Strömung im Rohr übertragen läßt — verläuft das Profil näherungsweise über die ganze Kanalbreite linear, in der Mitte jedoch besitzt es einen scharfen Knick (an die Wände schmiegt es sich mit dem $y^{1/7}$-Gesetz an). (Vgl. Fig. 3.)

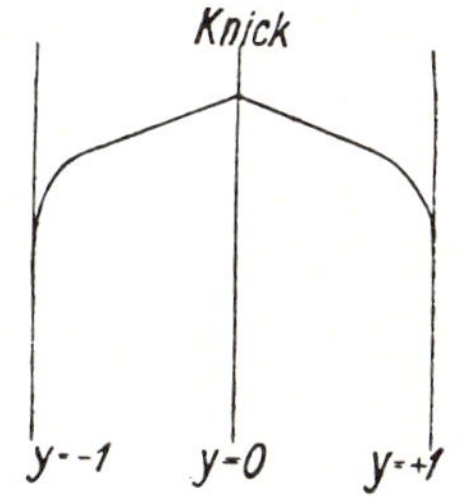

Fig. 3.

Die physikalische Ursache des scharfen Knicks ist die, daß der Gradient der *turbulenten* Impulsübertragung für $y = 0$ aus Symmetriegründen verschwindet, daß also, weil der Gradient der gesamten Impulsübertragung über die Kanalbreite *konstant* ist, der Gradient der *laminaren* Impulsübertragung, d. h. w'' dort sehr groß sein muß.

§ 4. Schlußbemerkungen und Zusammenfassung der physikalischen Ergebnisse.

Unsere Untersuchungen haben noch zwei wesentliche Lücken: Erstens liefern sie nicht den Übergang von dem $\eta^{1/7}$-Profil in das im mittleren Teil gültige lineare Profil und zweitens beschränken sie sich auf große Werte von R, liefern also auch nicht den Minimalwert von R, falls ein solcher existiert, für den die turbulente Bewegung noch möglich ist. Die erste dieser beiden Lücken ist am schwersten auszufüllen (vgl. S. 617) und wir können auch kein Verfahren angeben, das in befriedigender Weise die hier gestellte Aufgabe lösen würde. Man kann versuchen, die beiden von der Wand und der Kanalmitte herkommenden Näherungen aneinander zu stückeln, was durch die Bedingung geschehen müßte, daß an der betreffenden Ansatzstelle φ_1, φ_1', φ_1'', φ_1''', w und w' stetig sein sollen; die Konvergenz der Entwicklungen (45) und (50) reicht aber kaum aus, um so eine einigermaßen definierte Näherung zu garantieren. Jedenfalls hängt das Endresultat, das Profil w, noch stark von der Art der Zusammensetzung der beiden Näherungen ab. Es muß schließlich auch als fraglich betrachtet werden, ob eine solche exakte Durch-

Annalen der Physik (4) 74, 577–627 (1924)

führung der Ansätze (S. 609) wesentlich neue physikalische Resultate in Übereinstimmung mit der Erfahrung liefern würde, da diese Ansätze zweifellos eine sehr starke Idealisierung der Wirklichkeit bedeuten.

Die Ausfüllung der zweiten Lücke dagegen bietet prinzipiell gar keine Schwierigkeiten, alle Hilfsmittel dazu sind schon im Teil I enthalten, und wenn einmal das Profil w vollständig bekannt ist, so reichen die im Teil I geschilderten Methoden prinzipiell aus, um nach Art von I, § 6 den Minimalwert von R zu berechnen, für den die turbulente Bewegung möglich ist. Man könnte z. B. die kritische Reynoldssche Zahl für ein Profil berechnen, das nach dem oben erwähnten Verfahren durch Anstückelung der beiden Näherungen gewonnen ist oder man könnte für diese Untersuchung das empirisch beobachtete Profil zugrunde legen und so die Reynoldssche Zahl auf halbempirischem Wege ausrechnen. In jedem Falle wird man wohl — das haben die Untersuchungen im Teil I wahrscheinlich gemacht und direkte, hier nicht angegebene Rechnungen bestätigt — zur selben Größenordnung der kritischen Reynoldsschen Zahl, nämlich $R \sim 10^3$ kommen; der genaue Wert von R wird allerdings noch zu sehr von der Art abhängen, wie das Profil gewonnen wurde, als daß wir Vergleiche mit der Erfahrung ziehen könnten. Deshalb haben wir hier noch keine derartige Berechnung von R durchgeführt.

Fassen wir zum Schluß zusammen, was als physikalisches Ergebnis aus unseren Untersuchungen über das Turbulenzproblem zu schließen ist: Im Teil I haben wir erkannt, daß die laminare Bewegung und ihre Stabilitätsverhältnisse keine wesentliche Bedeutung für das Turbulenzproblem und die kritische Reynoldssche Zahl besitzen. Im Teil II aber haben wir die turbulente Bewegung selbst untersucht und können daraus einige Angaben über den turbulenten Bewegungszustand machen: Allgemein besitzt die Geschwindigkeitsverteilung über den ganzen Kanal einfachsten Charakter, sie ist — je nach den Versuchsbedingungen — linear oder konstant (§ 3), in der Mitte besitzt sie bei der (symmetrischen) Strömung zwischen zwei ruhenden Wänden einen scharfen Knick; an den Wänden schmiegt sie sich im $\eta^{1/7}$-Profil den Wänden an (§ 2). Darüber, daß das $^1/_7$-Profil bis weit in das Kanalinnere hinein Gültigkeit besitzt, geben die Rechnungen keinen Aufschluß. Die turbulenten Schwingungen verlaufen beim Couetteschen Fall im Innern des Kanals nahezu harmonisch (§ 3, Gl. (53)), in der Nähe der Wände treten alle Oberschwingungen auf. Die Geschwindigkeit der Wellen stimmt mit der Wandgeschwindigkeit überein (§ 2 Gl. 44—44c), beim Couetteschen Fall gibt es zwei Gruppen von turbulenten Schwingungen, deren eine hinsichtlich ihrer Fortpflanzungsgeschwindigkeit mit der einen Wand übereinstimmt, während die andere die Geschwindigkeit der anderen Wand besitzt. Die turbulenten Störungen tragen also äußerlich den Charakter einer Wandstörung. Dabei ist aber hervorzuheben, daß diese Störungen unabhängig von Wandrauhigkeit und ähnlichen Einflüssen als freie Schwingungen existenzfähig sind. Die Amplitude der turbulenten Wellen nimmt gegen die Wände (folgt aus (44) § 2) stark zu, um erst in unmittelbarster Wandnähe zu Null zu gehen.

Die Wellenlänge der auftretenden Schwingungen (I, § 8) ist der Größenordnung nach gleich, eher etwas größer als die Kanalbreite. Der Minimalwert der Reynoldsschen Zahl (I, § 8), für welchen noch die Turbulenz möglich ist, liegt größenordnungsmäßig bei 10^3. Als Widerstandsgesetz (§ 2, Gl. (50)) scheint sich für glatte Wände aus dem Profil $\eta^{1/7}$ unter gewissen Voraussetzungen das Blasiussche $\tau \sim u^{7/4}$ zu ergeben. Für rauhe Wände nähert es sich wahrscheinlich (§ 2) dem hydraulischen Gesetz $\tau \sim u^2$ an. — Das Ziel der vorliegenden Arbeit bildete weniger die Aufstellung dieser zum großen Teil schon bekannten Gesetzmäßigkeiten, als vielmehr der Nachweis, daß sich alle bisherigen, teilweise scheinbar sich widersprechenden Ergebnisse durch einfache Grundannahmen einheitlich mathematisch beschreiben lassen.

Meinem hochverehrten Lehrer, Herrn Professor Sommerfeld, möchte ich für die Anregung zu dieser Arbeit und für vielfache Förderung den herzlichsten Dank aussprechen.

München, Institut für theoret. Physik.

(Eingegangen 20. Februar 1924.)

Zeitschrift für Physik *124*, 628–657 (1948)

Zur statistischen Theorie der Turbulenz*.

Von

W. HEISENBERG.

Max Planck-Institut für Physik, Göttingen.

Mit 4 Figuren.

(Eingegangen am 16. Dezember 1946.)

Die in der vorhergehenden Arbeit von v. WEIZSÄCKER dargestellte Auffassung von der Turbulenz wird mathematisch mit Hilfe der üblichen Methode der Fourierzerlegung behandelt. Dabei wird das Spektrum der turbulenten Bewegung bis zu den kleinsten Wellenlängen, d. h. bis in den laminaren Bereich hinein abgeleitet, die mittleren Druckschwankungen und die Korrelationsfunktionen berechnet. Schließlich wird versucht, die für die Energiedissipation in der statistischen turbulenten Bewegung charakteristische Konstante aus den hydrodynamischen Gleichungen herzuleiten.

In der von G. J. TAYLOR[1] und v. KÁRMÁN[2] entwickelten statistischen Theorie der Turbulenz wird die unregelmäßige turbulente Bewegung einer Flüssigkeit durch einige charakteristische Funktionen beschrieben, zwischen denen einfache mathematische Relationen bestehen: die „spektrale" Verteilung der Energie auf Wellen verschiedener Wellenlänge, die Korrelationen zwischen den Geschwindigkeiten an Punkten eines vorgegebenen räumlichen oder zeitlichen Abstandes und dgl. Die Arbeiten von G. J. TAYLOR enthalten ausführliche empirische und theoretische Angaben über diese Funktionen.

v. WEIZSÄCKER[3] hat in der vorausgehenden Arbeit die wichtigste dieser Funktionen, nämlich die spektrale Verteilung der Energie, für den Grenz-

* Richard BECKER zum 60. Geburtstag gewidmet.

[1] TAYLOR, G. J.: Proc. Roy. Soc. A **151**, 421 (1935); **156**, 307 (1936); **164**, 15 (1938); **164**, 476 (1938).

[2] KÁRMÁN, TH. v.: Journ. Aero. Sci. **4**, 131 (1937).

[3] WEIZSÄCKER, C. F. v.: ZS. f. Phys. im Erscheinen. Die v. WEIZSÄCKERsche und die vorliegende Abhandlung sind in enger Zusammenarbeit in der Zeit des gemeinsamen Aufenthalts in England 1945 entstanden. Erst nach Abschluß der Arbeiten hat uns Herr G. J. TAYLOR freundlicherweise mitgeteilt (Frühjahr 1946), daß wesentliche Gedanken dieser Arbeiten schon von KOLMOGOROFF: Compt. Rend. Acad. Sc. UdSSR. **30**, 301 (1941); **32**, 16 (1941) und ONSAGER (Phys. Rev., **68**, 286, 1945) gefunden und veröffentlicht worden sind. Vgl. einen Bericht von G. K. BATCHELOR am VI. Int. Kongr. f. angew. Mechanik, Paris **1946**. Etwa um die gleiche Zeit erhielten wir ferner Kenntnis von einer Arbeit von PRANDTL und WIEGHARDT, die ähnliche Gedanken enthält und die inzwischen in den Göttinger Akademie-Berichten erschienen ist (Nachrichten d. Akad. d. Wissenschaften in Göttingen, Math.-Physikal. Klasse, aus dem Jahre **1945**, S. 6). Die vorliegende Arbeit kann also nur als Ergänzung und Vervollständigung dieser früheren Untersuchungen gelten.

fall großer Reynoldsscher Zahlen auf Grund von Ähnlichkeitsbetrachtungen abgeleitet. Die folgenden Abschnitte sollen zunächst die v. WEIZSÄCKERschen Überlegungen in die gewohnte Sprache der Fourierzerlegung übertragen und mit Hilfe dieser Übertragung das Abbrechen des Spektrums bei großen Frequenzen infolge der molekularen Zähigkeit studieren. Dann sollen die Folgerungen für die Korrelationsfunktionen und die Druckschwankungen gezogen werden und schließlich soll eine Ableitung der Grundkonstante der Energiedissipation versucht werden.

1. *Darstellung der v. Weizsäckerschen Überlegungen in der Sprache der Fourierkomponenten.*

Bei hinreichend großen Reynoldsschen Zahlen findet die Energiedissipation bei der turbulenten Bewegung in der Weise statt, daß die großen Turbulenzelemente dadurch Energie verlieren, daß für sie die Energie- und Impulsübertragung durch kleine Turbulenzelemente wie eine zusätzliche Zähigkeit (vgl. z. B. PRANDTL[1]) wirkt. Im stationären Betrieb wird also dauernd Energie von größeren in kleinere Turbulenzelemente übergeführt, wobei der Spektralbereich einer bestimmten Wellenlänge stets von größeren Wellen her soviel Energie erhält, wie er nach kleineren Wellen abgibt. Zur Aufrechterhaltung dieses Gleichgewichtes ist eine bestimmte Energieverteilung notwendig, die dann, wenn die molekulare Reibung vernachlässigt wird, nach v. WEIZSÄCKER durch das Gesetz

$$\rho \frac{\overline{v^2}}{2} = \rho \frac{v_0^2}{2} = \rho \int F(k)\, dk \qquad (1)$$

$$F(k) \sim k^{-5/3}$$

dargestellt wird. $\left(k = \frac{2\pi}{\text{Wellenlänge}}\right.$ bedeutet die Wellenzahl, $v_0 = \sqrt{\overline{v^2}}$ ein Maß für die mittlere Geschwindigkeit[2]$\Big)$. Dieses Spektrum $F(k) \sim k^{-5/3}$ ist nach zwei Seiten begrenzt. Bei kleinen Wellenzahlen, d. h. großen Wellenlängen, wird irgendwie die Strömung aufhören, als isotrop turbulent gelten zu können. Denn die größten Turbulenzelemente sind durch die Geometrie der Apparaturen gegeben, die die Turbulenz erzeugen. Dieses Ende des Spektrums bei kleinem k kann also überhaupt nicht Gegenstand einer rein statistischen Theorie sein. Bei großen k dagegen wird das Spektrum durch die molekulare Zähigkeit begrenzt. Bei großen k wird schließlich die molekulare Zähigkeit größer werden als die scheinbare turbulente Zähigkeit, das Spektrum wird dort dann sehr rasch abfallen.

[1] PRANDTL, L.: Strömungslehre, 3. Aufl., S. 105 u. f. Braunschweig: Vieweg 1942.

[2] Unser v_0 unterscheidet sich von dem v_0 in der v. WEIZSÄCKERschen Arbeit um einen Zahlenfaktor der Größenordnung 1.

Zeitschrift für Physik *124*, 628–657 (1948)

Für die Rechnungen wollen wir die folgenden Bezeichnungen verwenden: Die Geschwindigkeit $\mathfrak{v}$ soll in einem Normierungsvolumen V in eine Fourierreihe entwickelt werden

$$\mathfrak{v} = \sum_{\mathfrak{k}} \mathfrak{v}_{\mathfrak{k}} e^{i\mathfrak{k}\mathfrak{r}} \left(\mathfrak{k}_x = \frac{2\pi}{Lx} n_x, \ldots; n_x, n_y, n_z \text{ ganze Zahlen}\right). \qquad (2)$$

Dabei ist $\mathfrak{v}_{\mathfrak{k}} = \mathfrak{v}^*_{-\mathfrak{k}}$, und die Anzahl der „Eigenschwingungen" zwischen k und $k + \Delta k$ ist durch $\frac{4\pi k^2 \Delta k V}{(2\pi)^3}$ gegeben[1]. Dann wird

$$\frac{\overline{v^2}}{2} = \frac{1}{2} \sum_{\mathfrak{k}} \left|\mathfrak{v}_{\mathfrak{k}}^2\right| = \frac{1}{2} \int 4\pi k^2\, dk \frac{V}{(2\pi)^3} \left|\mathfrak{v}_{\mathfrak{k}}^2\right| = \int F(k)\, dk, \qquad (3)$$

also

$$F(k) = (2\pi)^{-2} k^2 V \left|\mathfrak{v}_k^2\right|. \qquad (4)$$

Aus div $\mathfrak{v} = 0$ folgt

$$(\mathfrak{v}_{\mathfrak{k}}\, \mathfrak{k}) = 0. \qquad (5)$$

Der Koeffizient der molekularen Zähigkeit werde μ genannt, der mittlere Energieverlust durch Reibung ist dann, wegen div $\mathfrak{v} = 0$ unter der Voraussetzung ruhender Begrenzungsflächen:

$$S = \mu\, \overline{(\mathrm{rot}\, \mathfrak{v})^2} \qquad (6)$$

$$= \mu \sum_{\mathfrak{k}} \left|[\mathfrak{v}_{\mathfrak{k}}\, \mathfrak{k}]^2\right| = \mu \int F(k)\, 2k^2\, dk. \qquad (7)$$

Wenn das Spektrum in einem großen Bereich dem Gesetz $F(k) \sim k^{-5/3}$ folgt, so wird die Gesamtenergie durch die größten Turbulenzelemente bestimmt. Wir können etwa annehmen, das Gesetz $k^{-5/3}$ gelte bis herunter zu einer kleinsten Wellenzahl k_0, für kleinere k sei $F(k) = 0$. Dann ist

$$v_0^2 = 2 \int_0^\infty F(k)\, dk = 2 \int_{k_0}^\infty \frac{C}{k^{5/3}}\, dk = 3\, C\, k_0^{-\frac{2}{3}}, \qquad (8)$$

[1] Dieses Verfahren ist etwas weniger anschaulich, aber mathematisch bequemer, als die übliche Entwicklung nach sin und cos. Es läuft formal auf die Grenzbedingung hinaus, daß $\mathfrak{v}, \frac{\partial \mathfrak{v}}{\partial x}, \ldots$ an einer Begrenzungsfläche des Volumens V die gleichen Werte haben soll wie an der gegenüberliegenden Begrenzungsfläche.

also $$C = \frac{v_0^2}{3} k_0^{2/3}$$ und

$$F(k) = \frac{v_0^2}{3} \frac{k_0^{2/3}}{k^{5/3}} \quad \text{für} \quad k \geq k_0, \tag{9}$$

$$\overline{|v_k^2|} = \frac{v_0^2 k_0^{2/3} \cdot (2\pi)^3}{6\pi V k^{11/3}}. \tag{10}$$

V. WEIZSÄCKER betrachtet den Energieverlust S_k desjenigen Teiles des gesamten Spektrums, dessen Wellenzahlen unter k liegen. Für diese Turbulenzelemente wirken die Turbulenzelemente kleinerer Wellenlänge $\left(< \frac{2\pi}{k}\right)$ wie eine zusätzliche Zähigkeit. Man kann also allgemein schreiben

$$S_k = (\mu + \eta_k) \int_0^k F(k') \cdot 2k'^2 \, dk', \tag{11}$$

wobei η_k die zusätzliche turbulente Zähigkeit bezeichnen soll und durch das Zusammenwirken aller Turbulenzelemente mit Wellenlänge $< \frac{2\pi}{k}$ zustande kommt. Dimensionsmäßig ist η_k nach PRANDTL das Produkt aus Dichte, Mischungsweg und Geschwindigkeit, wobei der Mischungsweg vergleichbar sein wird mit dem Durchmesser der betreffenden Turbulenzelemente, während die Geschwindigkeit der Turbulenzelemente etwa durch $v_0 \left(\frac{k_0}{k}\right)^{\frac{1}{3}}$ gegeben ist. In Anlehnung an die v. WEIZSÄCKERsche Arbeit wird man daher setzen:

$$\eta_k = \varkappa \rho \int_k^\infty dk' \sqrt{\frac{F(k')}{k'^3}}. \tag{12}$$

($\varkappa$ ist ein Zahlfaktor).

Der Ausdruck unter dem Integralzeichen ist im wesentlichen durch Dimensionsbetrachtungen festgelegt; man könnte sich aber natürlich denken, daß etwa die Wellen k' in der Nähe von k mit etwas anderen Gewichten in das Integral eingehen, als die Wellen mit großen k-Werten, d. h. der Integrand könnte noch von der dimensionslosen Zahl k'/k abhängen. Wegen der einheitlichen Form des Spektrums $F(k) \sim k^{-5/3}$ kann man aber alle diese Unbestimmtheiten in den Zahlenfaktor $\varkappa$ aufnehmen und dem Integral willkürlich die exakte Form (12) geben. Dieses Verfahren ist einwandfrei im Gebiet des $k^{-5/3}$-Gesetzes, wird jedoch ungenau an den Enden des Gebietes, an denen die Geometrie oder die molekulare Reibung das Spektrum verändert. Aber auch an der letzteren Grenze wird (12) noch eine gute Approximation sein, die zum mindesten qualitativ das Eingreifen der Reibung richtig darstellt.

Die Konstante $\varkappa$ in Gl. (12) muß durch die hydrodynamischen Gleichungen exakt festgelegt sein, sie hat in allen Fällen, in denen von statistischer isotroper Turbulenz gesprochen werden kann, den gleichen Zahlwert und hängt in keiner Weise von der Geometrie der Strömung ab. Die theoretische Bestimmung dieser wichtigen Zahl wird im Abschnitt 5 versucht werden. Dort wird sich auch zeigen, daß die turbulente Energiedissipation tatsächlich als ein Doppelintegral vom Typus $\int_0^k dk' \ldots \int_k^\infty dk'' \ldots$ geschrieben werden kann, wobei der Integrand die von k' nach k'' pro Zeiteinheit übergeführte Energie bedeutet [Gl. (89)]. Dieses Integral ist komplizierter als der vereinfachende Ausdruck (13), der aus (11) und (12) hervorgeht, aber für die folgenden Betrachtungen können (11) und (12) als hinreichende Näherungen gelten. Für S_k erhält man so:

$$S_k = \left(\mu + \rho\varkappa \int_k^\infty dk'' \sqrt{\frac{F(k'')}{k''^3}}\right) \int_0^k F(k') \cdot 2k'^2\, dk'. \tag{13}$$

Der entscheidende Schritt der v. WEIZSÄCKERschen Überlegung ist die Feststellung, daß dieser Ausdruck für $k \ll k_0$ von k unabhängig sein muß:

$$S_k = S = \text{const (für } k \gg k_0), \tag{14}$$

weil ja die Gesamtenergie zum allergrößten Teil im langwelligen Gebiet des Spektrums liegt, also der Energie„durchsatz" von k unabhängig werden muß.

Die Gl. (13) kann als Bestimmungsgleichung des turbulenten Spektrums $F(k)$ aufgefaßt werden, die für das Gebiet großer Reynoldsscher Zahlen das $k^{-5/3}$-Gesetz und für noch größere k-Werte das Abklingen des Spektrums infolge der molekularen Zähigkeit liefern muß.

2. *Die Gestalt des Spektrums in der Gegend der kleinsten Turbulenzelemente.*

Wir setzen zunächst $\frac{\mu}{\rho} = \nu$ und differenzieren (13) nach k. Dann ergibt sich

$$\left(\frac{\nu}{\varkappa} + \int_k^\infty dk' \sqrt{\frac{F(k')}{k'^3}}\right) F(k) \cdot k^2 = \sqrt{\frac{F(k)}{k^3}} \int_0^k F(k')\, k'^2\, dk'. \tag{15}$$

Dann definieren wir neue Variable x und w durch die Gleichungen:

$$x = \lg \frac{k}{k_0}, \quad F(k) = F(k_0)\, e^{-w}; \quad w = w(x). \tag{16}$$

Damit geht (15) über in:

$$e^{\frac{7}{2}x-\frac{w}{2}}\left(\frac{\nu}{\varkappa}\sqrt{\frac{k_0}{F_0}}+\int_x^\infty e^{-\frac{w}{2}-\frac{x}{2}}\,dx\right)=\int_0^x e^{3x-w}\,dx. \quad (17)$$

Die Konstante $\frac{\nu}{\varkappa}\sqrt{\frac{k_0}{F_0}}=\frac{\nu k_0\sqrt{3}}{\varkappa v_0}$ [vgl. Gl. (9)] ist im wesentlichen die reziproke Reynoldssche Zahl der Gesamtströmung und daher stets sehr klein; wenn die Reynoldssche Zahl selbst klein wäre, so könnte die Strömung überhaupt nicht turbulent sein. Durch nochmaliges Differenzieren entsteht aus (17)

$$\left(\frac{7}{2}-\frac{1}{2}\frac{dw}{dx}\right)\left(\frac{\nu}{\varkappa}\sqrt{\frac{k_0}{F_0}}+\int_x^\infty e^{-\frac{w}{2}-\frac{x}{2}}\,dx\right)=2e^{-\frac{w}{2}-\frac{x}{2}}. \quad (18)$$

In dieser Gleichung kann man das Integral $\int_x^\infty dx\, e^{-\frac{w+x}{2}}$ näherungsweise auswerten, indem man $w(x)$ in der Nähe von x entwickelt: $w(x_1)=$ $=w(x)+(x_1-x)\frac{dw}{dx}+\cdots$ und mit dem zweiten Gliede abbricht. Da die Exponentialfunktion rasch abfällt, erhält man so eine gute Näherung:

$$\int_x^\infty dx\, e^{-\frac{w+x}{2}}\approx 2\frac{e^{-\frac{w+x}{2}}}{1+\frac{dw}{dx}}. \quad (19)$$

Durch Einsetzen in (18) ergibt sich schließlich:

$$\left(7-\frac{dw}{dx}\right)\left(\frac{\nu}{2\varkappa}\sqrt{\frac{k_0}{F_0}}\,e^{\frac{w+x}{2}}+\frac{1}{1+\frac{dw}{dx}}\right)=2. \quad (20)$$

Man erkennt aus (20) sofort den allgemeinen Verlauf des Spektrums. Für nicht zu große x und w kann das erste Glied in der Summe vernachlässigt werden und man erhält

$$\left(7-\frac{dw}{dx}\right)=2\left(1+\frac{dw}{dx}\right),\ \text{d.h.}\ \frac{dw}{dx}=\frac{5}{3};\ F(k)=F_0\left(\frac{k}{k_0}\right)^{-5/3}, \quad (21)$$

wie es nach der v. Weizsäckerschen Theorie sein muß. Für große x und w dagegen überwiegt das erste Glied, daher muß dann

$$\frac{dw}{dx}=7,\ F\sim\text{const}\cdot k^{-7} \quad (22)$$

Zeitschrift für Physik *124*, 628–657 (1948)

sein. In der Gegend der kleinsten Turbulenzelemente fällt das Spektrum daher sehr rasch, nämlich mit der 7. Potenz der Wellenzahl, ab.

Nur im Übergangsgebiet von (21) nach (22) sind numerische Rechnungen nötig, um die Lösung von (20) zu ermitteln. Da für kleinere x, d. h. für das Gebiet

$$\begin{aligned} 1 \ll x \ll \frac{3}{4} \lg \frac{2\varkappa}{\nu} \sqrt{\frac{F_0}{k_0}} \\ w = \frac{5}{3} x \end{aligned} \tag{23}$$

gesetzt werden kann [damit ist dann nicht nur (18) sondern auch (17) in ausreichender Genauigkeit erfüllt], kann man von Punkt zu Punkt fortschreitend $\frac{dw}{dx}$ nach (20) aus w berechnen und damit w für höhere x ableiten. Dabei genügt es, die numerische Rechnung für einen bestimmten großen Wert der Konstanten, sagen wir $\frac{2\varkappa}{\nu}\sqrt{\frac{F_0}{k_0}} = a$ durchzuführen. Für einen anderen Wert b kann man dann w durch eine einfache Ähnlichkeitstransformation erhalten:

$$w_b(x) = w_a\left(x + \frac{3}{4} \lg \frac{a}{b}\right) - \frac{5}{4} \lg \frac{a}{b}, \tag{24}$$

wie man durch Einsetzen in (20) und (23) erkennt.

Die Abb. 1 gibt das Ergebnis der numerischen Rechnung für $\frac{2\varkappa}{\nu}\sqrt{\frac{F_0}{k_0}} = 1000$ wieder. Die numerische Integration zeigt, daß für diesen Wert im Gebiet

$$\begin{aligned} x > 5 \\ w(x) \approx 7x - 21{,}85 \end{aligned} \tag{25}$$

ist. Allgemeiner wird also in diesem Gebiet des k^{-7}-Gesetzes:

$$w(x) = 7x + 3{,}0 - 4 \lg\left(\frac{\varkappa}{\nu}\sqrt{\frac{F_0}{k_0}}\right), \tag{26}$$

d. h.

$$F(k) = 0{,}0496 \cdot \frac{F_0^3}{k_0^2} \left(\frac{\varkappa}{\nu}\right)^4 \left(\frac{k_0}{k}\right)^7. \tag{27}$$

Eine brauchbare Interpolationsformel, die in den beiden Grenzfällen richtig ist und auch im Übergangsgebiet keine großen Fehler gibt, lautet:

$$F(k) = F_0 \left(\frac{k_0}{k}\right)^{5/3} \left[1 + \left(\frac{k}{k_s}\right)^{\frac{8}{3}}\right]^{-2}. \tag{28}$$

Definiert man $L_0 = \frac{\pi}{k_0}$ als den „Durchmesser der größten Turbulenzelemente" und führt als Reynoldssche Zahl der Gesamtströmung

$$R_0 = \frac{\rho v_0 L_0}{\mu} \tag{29}$$

ein, so wird nach (9), (27) und (29)

$$k_s = 0{,}16\, k_0\, (R_0\, \varkappa)^{3/4}\,. \tag{30}$$

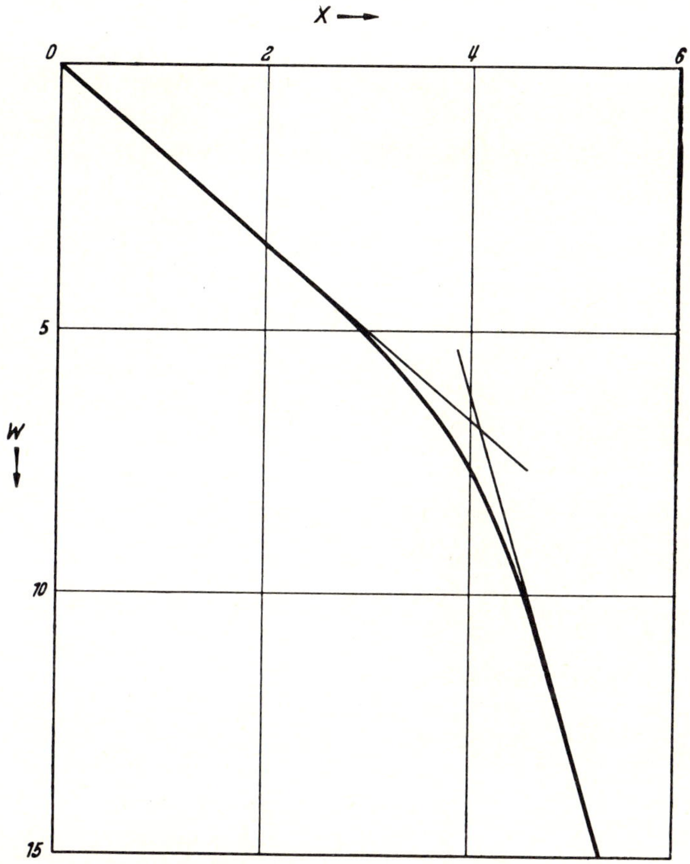

Abb. 1. Darstellung der Funktion $w(x)$.

Man kann $L_s = \frac{\pi}{k_s}$ als den „Durchmesser der kleinsten Turbulenzelemente" bezeichnen und erhält

$$L_s = 6{,}25\, L_0\, (R_0\, \varkappa)^{-3/4}\,. \tag{31}$$

Durch (9), (28) und (30) ist die Gestalt des Spektrums im ganzen k-Bereich festgelegt. Bei den wirklichen Strömungen wird allerdings der Ver-

Zeitschrift für Physik *124*, 628–657 (1948)

lauf des Spektrums bei kleinen k-Werten ($k \sim k_0$) ein anderer sein, da dort die Geometrie der Versuche, z. B. die Gestalt der Gitter, mit denen die Turbulenz erzeugt wird, eine Rolle spielt. Man wird dann, um den Vergleich mit der Erfahrung sinngemäß durchführen zu können, eine Größe k_0 einführen derart, daß etwa im Bereich des $k^{-5/3}$-Gesetzes (also für $k_s \gg k \gg k_0$) die Formel

$$F(k) = \frac{v_0^2}{3\,k_0}\left(\frac{k_0}{k}\right)^{5/3} \tag{32}$$

richtig wird. Die so definierte Größe k_0 sagt dann nicht direkt etwas über den Verlauf des Spektrums bei den kleinsten k-Werten aus. Wohl aber wird das Spektrum im allgemeinen in der Gegend $k \sim k_0$ stark vom $k^{-5/3}$-Gesetz abweichen.

Auch für $k \gg k_s$ wird das Spektrum nicht unbegrenzt die Form k^{-7} beibehalten. Denn die bekannten Untersuchungen von BURGERS[1] machen es sehr wahrscheinlich, daß es bei hinreichend kleinen Reynoldsschen Zahlen schließlich gar keine turbulenten Bewegungen mehr gibt. Andererseits fällt das k^{-7}-Gesetz so rasch ab, daß das Gebiet $k \gg k_s$ praktisch keine Rolle spielt. Ein etwas größerer Fehler wird, insbesondere im Übergangsgebiet, durch die Ungenauigkeit der Gl. (13) selbst entstehen; aber es lohnt wohl nicht, schon jetzt die sehr viel komplizierteren Gleichungen des 5. Abschnittes auf das hier gestellte Problem anzuwenden. Die richtigen Gleichungen würden jedenfalls zu etwas anderen Zahlfaktoren in (27), (30) und (31) führen.

Für den Vergleich mit der Erfahrung braucht man die Energieverteilung über die Wellenzahlen einer bestimmten Richtung, etwa über k_x, da die Spektren experimentell von SIMMONS[2] und DRYDEN[3] durch die zeitlichen Schwankungen der Geschwindigkeit in einem Luftstrom gemessen worden sind, der mit einer konstanten relativ zu v_0 großen Geschwindigkeit U am Meßort vorbeigeführt wird. Dieses Spektrum hat auch eine verschiedene Form, je nachdem es sich um die Fourierentwicklung von $\mathfrak{v}_x$ oder $\mathfrak{v}_y$ handelt. Experimentell wird zunächst das Spektrum für $\mathfrak{v}_x$ gebraucht, wir wollen aber auch das für $\mathfrak{v}_y$ ableiten, da es später bei der Berechnung der Korrelationsfunktionen notwendig ist. Da nach Gl. (5) $(\mathfrak{v}_{\mathfrak{k}}\,\mathfrak{k}) = 0$ ist, wird

$$\overline{\mathfrak{v}_{\mathfrak{k}x}^2} = \frac{v_k^2}{2}\left(1 - \frac{k_x^2}{k^2}\right). \tag{33}$$

[1] BURGERS, J. M.: Verh. d. Kgl. Nied. Akad. d. Wiss. **17**, Nr. 2, 1 (1939); **18**, Nr. 1, 1 (1940).

[2] SIMMONS u. SALTER: Proc. Roy. Soc. A **165**, 73 (1938).

[3] DRYDEN, SCHUBAUER, MOCK u. SKRAMSTAD: Nation. Adv. Comm. Aeron., Nr. 581 (1938); DRYDEN, H. L.: Proc. V. Intern. Congr. f. Applied. Med. Cambridge (Mass.), S. 362 (1938).

Das Spektrum von $\mathfrak{v}_x$ in k_x, das wir mit $F_x(k_x)$ bezeichnen wollen, wird daher

$$F_x(k_x) = \iint\limits_{-\infty}^{+\infty} \frac{d k_y \, d k_z}{4 \pi k^2} \, \frac{1}{2} \left(1 - \frac{k_x^2}{k^2}\right) F(k)$$

$$= \frac{1}{4} \int\limits_{k_x}^{\infty} \frac{d k}{k^3} (k^2 - k_x^2) \, F(k). \tag{34}$$

In ähnlicher Weise wird das Spektrum für $\mathfrak{v}_y$:

$$F_y(k_x) = \iint\limits_{-\infty}^{+\infty} \frac{d k_y \, d k_z}{4 \pi k^2} \, \frac{1}{2} \left(1 - \frac{k_y^2}{k^2}\right) F(k)$$

$$= \frac{1}{8} \int\limits_{k_x}^{\infty} \frac{d k}{k^3} (k^2 + k_x^2) \, F(k). \tag{35}$$

Aus (34) und (35) folgt für

a) $k_0 \ll k \ll k_s$:

$$F(k) = F_0 \left(\frac{k_0}{k}\right)^{5/3}, \quad \text{also}$$

$$F_x(k_x) = \frac{9 F_0}{110} \left(\frac{k_0}{k_x}\right)^{5/3} \quad \text{und} \tag{36}$$

$$F_y(k_x) = \frac{6 F_0}{55} \left(\frac{k_0}{k_x}\right)^{5/3}.$$

b) $k \gg k_s$:

$$F(k) = F_0 \frac{k_0^{5/3} k_s^{\frac{16}{3}}}{k^7},$$

$$F_x(k_x) = \frac{F_0}{126} k_0^{5/3} k_s^{\frac{16}{3}} k_x^{-7},$$

$$F_y(k_x) = \frac{2 F_0}{63} k_0^{5/3} k_s^{\frac{16}{3}} k_x^{-7}. \tag{37}$$

Als brauchbare Interpolationsformel [die aber im Übergangsgebiet etwas weniger genau als (28) ist] kann man wieder setzen:

$$F_x(k_x) = \frac{9 F_0}{110} \left(\frac{k_0}{k_x}\right)^{\frac{5}{3}} \left[1 + \left(\frac{k_x}{k_s^{xx}}\right)^{\frac{8}{3}}\right]^{-2} \tag{38}$$

mit

$$k_s^{xx} = 0{,}645 \; k_s \tag{39}$$

Zeitschrift für Physik *124*, 628–657 (1948)

und
$$F_y(k_x) = \frac{6 F_0}{55}\left(\frac{k_0}{k_x}\right)^{5/3}\left[1+\left(\frac{k_x}{k_s^{y\,x}}\right)^{\frac{8}{3}}\right]^{-2} \tag{40}$$

mit
$$k_s^{y\,x} = 0{,}793\; k_s\,. \tag{41}$$

Bevor der Vergleich mit der Erfahrung im einzelnen durchgeführt wird, soll die Frage gestellt werden, bei welchen kritischen Reynoldsschen Zahlen der Übergang vom $k^{-5/3}$-Gesetz in das k^{-7}-Gesetz – also, wenn man so will, der Übergang von der eigentlich turbulenten zur laminaren Bewegung – erfolgt. Als die kritische Reynoldssche Zahl hierfür kann man etwa den Ausdruck

$$R_s = \frac{\rho\, v_s\, L_s}{\mu} \tag{42}$$

ansehen, wobei nach v. WEIZSÄCKER $v_s = v_0\left(\frac{k_0}{k_s}\right)^{\frac{1}{3}}$ ist. Aus (29), (30) und (31) erhält man dann

$$R_s = \frac{10{,}2}{\varkappa}\,. \tag{43}$$

Der Zahlwert von $\varkappa$ wird später noch diskutiert werden. Jedenfalls findet also der Übergang bei einem bestimmten numerischen Wert der Reynoldsschen Zahl statt, wie aus allgemeinen Ähnlichkeitsüberlegungen zu erwarten ist.

In Abb. 2 sind die Messungen des Spektrums $F_x(k_x)$ von SIMMONS (l. c.) mit der Theorie verglichen. Es handelt sich um Intensitätsmessungen an einen Luftstrom, der mit den Geschwindigkeiten $U = 456$ cm/sec (o), 608 cm/sec (×), 1060 cm/sec (□) am Meßort vorbeiströmt und durch ein Gitter von 7,6 cm Maschenweite turbulent gemacht worden ist, die Messung erfolgte 2,1 m hinter dem Gitter. Die Meßpunkte SIMMONS' sind nur im rechten Teil der Abb. einzeln eingetragen, in der linken Hälfte ist der ungefähre Streubereich der Meßpunkte durch einen senkrechten Strich angegeben. Abszisse ist k in cm^{-1} (in logarithmischem Maßstab), Ordinate $F_x(k_x)$ ebenfalls logarithmisch, in willkürlichen Einheiten. Wenn man annimmt, daß U/v_0 in allen drei Meßreihen den gleichen Wert hat, was durch andere Messungen von TAYLOR bestätigt wird, so erhält man bei einer geeigneten Wahl dieses Verhältnisses die drei in die Abb. eingezeichneten Kurven. Qualitativ werden die SIMMONSschen Meßpunkte durch die Kurven gut dargestellt, insbesondere auch das Auseinanderlaufen der drei Meßreihen im kurzwelligen Teil des Spektrums. Im einzelnen aber gibt es auch erhebliche Abweichungen; man erkennt aus der Figur, daß der Gültigkeitsbereich des $k^{-5/3}$-Gesetzes hier so klein ist, daß eine zuverlässige Nachprüfung nicht möglich ist. Dies liegt an der Kleinheit der Reynoldsschen Zahl R_0. Bei $k = 1\ \mathrm{cm}^{-1}$ ist der Durchmesser der Turbulenzelemente

3 cm, also etwa halb so groß wie die Maschenweite des Gitters; in diesem Gebiet ist die Turbulenz noch nicht völlig isotrop, das $k^{-5/3}$-Gesetz kann also noch nicht gelten. Schon bei $k = 4\ \text{cm}^{-1}$ aber macht sich der Einfluß der molekularen Zähigkeit bemerkbar, die Intensität sinkt stärker ab.

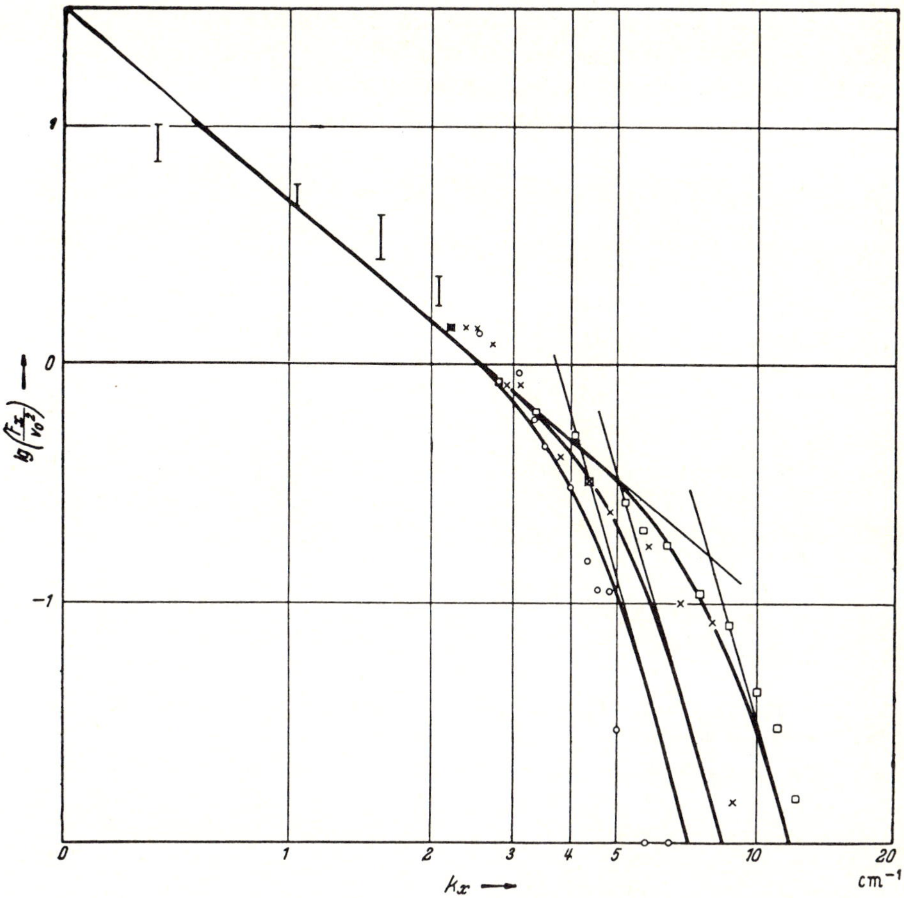

Abb. 2. Die turbulente Energieverteilung als Funktion der Wellenzahl.

Auch die verwandten Messungen von DRYDEN (l. c.), die sich über ein großes Spektralgebiet erstrecken, sind bei so kleinen Reynoldsschen Zahlen durchgeführt, daß die Gültigkeit des $k^{-5/3}$-Gesetzes kaum nachgeprüft werden kann. Es wäre also wünschenswert, ähnliche Messungen mit sehr viel größeren Reynoldsschen Zahlen durchzuführen. Für das Verhältnis U/v_0 erhält man aus der Angleichung der theoretischen Kurven an die

Zeitschrift für Physik *124*, 628–657 (1948)

Meßpunkte $U/v_0 = 53\,\varkappa$, wenn man L_0 mit der Maschenweite des Gitters identifiziert. Dieser Wert paßt gut zu Messungen[1] dieses Verhältnisses bei ähnlichen Versuchen, wenn man annimmt, daß $\varkappa$ etwa 0,5 ist.

Eine andere und wohl genauere Bestimmung von $\varkappa$ erhält man aus dem zeitlichen Abklingen der Turbulenz, das schon von TAYLOR (l. c.) theoretisch vollständig behandelt worden ist. Für den gesamten Energieverlust pro cm³ und sec: S ergibt sich aus (9), (13) und (14)

$$S = \rho \varkappa \frac{\sqrt{3}}{8} v_0^3 k_0. \tag{44}$$

Also muß für die zeitliche Abklingung von v_0 gelten:

$$\frac{d}{dt}\left(\frac{v_0^2}{2}\right) = -\varkappa \frac{\sqrt{3}}{8} v_0^3 k_0, \tag{45}$$

mit der Lösung[2]

$$v_0(t) = \frac{v_0(0)}{1 + \frac{\sqrt{3}}{8} \varkappa k_0 v_0(0)\, t}. \tag{46}$$

TAYLOR, der im wesentlichen diese Gl. abgeleitet hat, berichtet über Messungen von SIMMONS, bei denen $\frac{U}{v_0(t)}$ als Funktion von $t = \frac{x}{U}$ (x = Abstand des Meßpunktes vom Gitter) ermittelt wurde. Aus (46) folgt:

$$\frac{U}{v_0(t)} = \frac{U}{v_0(0)} + \frac{\sqrt{3}}{8} \varkappa \pi \frac{x}{L_0} = \frac{U}{v_0(0)} + 0{,}68 \varkappa \frac{x}{L_0}. \tag{47}$$

Wenn man $v_0 = u' \sqrt{3}$ setzt ($u' = \sqrt{\overline{v_x^2}}$ nach TAYLOR) und L_0 mit der Maschenweite identifiziert, so folgt aus den TAYLORschen Messungen $\varkappa = 0{,}85$, aus den entsprechenden Messungen von DRYDEN ein etwas kleinerer Wert. Wegen der Unsicherheit in dem für L_0 einzusetzenden Wert ist diese Bestimmung aber wohl noch um etwa 50% unsicher.

3. *Die Korrelationsfunktionen.*

TAYLOR und v. KÁRMÁN (l. c.) haben die Korrelationen studiert, die zwischen den Geschwindigkeiten an zwei Punkten in gegebenem Abstand bestehen. Die beiden Korrelationsfunktionen $R_1(x)$ und $R_2(x)$, die hierbei die Hauptrolle spielen, sind definiert als

---

[1] Vgl. TAYLOR, G. J.: Proc. Roy. Soc. A **164**, 486 (1938).

[2] Anm. bei der Korrektur: Bei dieser Lösung ist k_0 = const vorausgesetzt, was sicher für größere Zeiten nicht zutrifft. Das Problem der Abklingung wird genauer untersucht in einer im Erscheinen begriffenen Arbeit des Verfassers (Proc. Roy. Soc. A).

$$R_1(x) = \frac{\overline{v_x(P_1)\, v_x(P_2)}}{\overline{v_x^2}}, \qquad R_2(x) = \frac{\overline{v_y(P_1)\, v_y(P_2)}}{\overline{v_y^2}}, \tag{48}$$

wobei der Punkt P_2 um die Strecke x in der x-Richtung gegen den Punkt P_1 verschoben ist.

Diese Funktionen stehen nach TAYLOR in einfachem Zusammenhang mit den Spektren:

$$\left.\begin{aligned} R_1(x) &= \frac{\int_0^\infty d k_x F_x(k_x) \cos k_x x}{\int_0^\infty d k_x F_x(k_x)}, \\ R_2(x) &= \frac{\int_0^\infty d k_x F_y(k_x) \cos k_x x}{\int_0^\infty d k_x F_y(k_x)} \end{aligned}\right\}. \tag{49}$$

Mit Hilfe der Gl. (34) und (35) geht (49) über in

$$R_1(x) = \frac{3 \int_0^\infty d k\, F(k)\, (\sin k x - k x \cos k x)\, k^{-3} x^{-3}}{\int_0^\infty d k\, F(k)},$$

$$R_2(x) = \frac{\frac{3}{2} \int_0^\infty d k\, F(k)\, (k^2 x^2 \sin k x + k x \cos k x - \sin k x)\, k^{-3} x^{-3}}{\int_0^\infty d k\, F(k)}. \tag{50}$$

An diesen Ausdrücken erkennt man unmittelbar die Richtigkeit der v. KÁRMÁNschen Beziehung:

$$R_2 = R_1 + \frac{x}{2} \frac{d R_1}{d x}. \tag{51}$$

Die Formeln (50) können näherungsweise ausgewertet werden in den beiden Grenzfällen: $x \ll \frac{1}{k_s}$ und $\frac{1}{k_s} \ll x \ll \frac{1}{k_0}$. Wenn $x \ll \frac{1}{k_s}$ ist, entwickelt man die Integranden zweckmäßig nach Potenzen von x. Die ersten Entwicklungsglieder führen dann auf die Größe

$$\overline{k^2} = \frac{\int_0^\infty d k\, F(k)\, k^2}{\int_0^\infty d k\, F(k)}, \tag{52}$$

Zeitschrift für Physik *124*, 628–657 (1948)

die aus (13), (44) und (29) leicht berechnet werden kann:

$$\overline{k^2} = \frac{\sqrt{3}}{8\pi} \varkappa R_0 k_0^2. \tag{53}$$

So erhält man:

$$\text{Für } x \ll \frac{1}{k_s}: \quad \begin{aligned} R_1(x) &= 1 - \frac{x^2 \overline{k^2}}{10} + \cdots \\ R_2(x) &= 1 - \frac{x^2 \overline{k^2}}{5} + \cdots. \end{aligned} \tag{54}$$

TAYLOR hat eine Länge λ durch die Gleichung

$$\lambda^2 = \frac{5}{\overline{k^2}} \tag{55}$$

definiert und als ein Maß für die Größe der kleinsten Turbulenzelemente bezeichnet. Nach (53) wird

$$\lambda = 2{,}71 \frac{L_0}{\sqrt{R_0 \varkappa}} = 0{,}434 L_s (R_0 \varkappa)^{\frac{1}{4}}. \tag{56}$$

Es muß betont werden, daß λ nicht identisch ist mit der Größe L_s [Gl. (31)], die wir als ,,Durchmesser der kleinsten Turbulenzelemente" bezeichnet haben, und daß λ auch anders als L_s von L_0 und v_0 abhängt. Ein Vergleich von (31) und (56) zeigt, daß für hinreichend große Reynoldssche Zahlen die Länge $L_s \ll \lambda$ wird.

Im entgegengesetzten Grenzfall $\frac{1}{k_s} \ll x \ll \frac{1}{k_0}$ erhält man aus (50):

$$\begin{aligned} R_1(x) &= 1 - 0{,}643\, (k_0 x)^{2/3} + \cdots \\ R_2(x) &= 1 - 0{,}858\, (k_0 x)^{2/3} + \cdots. \end{aligned} \tag{57}$$

Dabei hängen diese ersten Entwicklungsglieder noch nicht von der speziellen Form des Spektrums in der Nähe von k_0 ab; erst für $x \sim \frac{1}{k_0}$ wird die Form des Spektrums in der Nähe von k_0 wichtig, dort kann man aber das Problem nicht mehr mit rein statistischen Methoden behandeln. Die Formeln (54) und (57) geben also wohl eine vollständige Beschreibung der Korrelationen, soweit sie als Folge statistischer isotroper Turbulenz betrachtet werden können.

Die Formeln (54) und (57) zeigen auch deutlich, daß die Korrelationsfunktion nicht in allen Strömungen die gleiche Form hat, daß vielmehr bei Veränderungen der Parameter die inneren und die äußeren Teile der

Funktion verschiedenartige Ähnlichkeitstransformationen durchmachen. Dieser Punkt ist besonders von TAYLOR[1] gegenüber einer andersartigen Vermutung von v. KÁRMÁN (l. c.) betont worden.

Zum Vergleich mit der Erfahrung sind in Abb. 3 die Messungen von $R_1(x)$ und $R_2(x)$ durch SIMMONS eingetragen (Kreise bzw. Punkte), ferner die nach der genauen Formel (50) berechneten theoretischen Kurven.

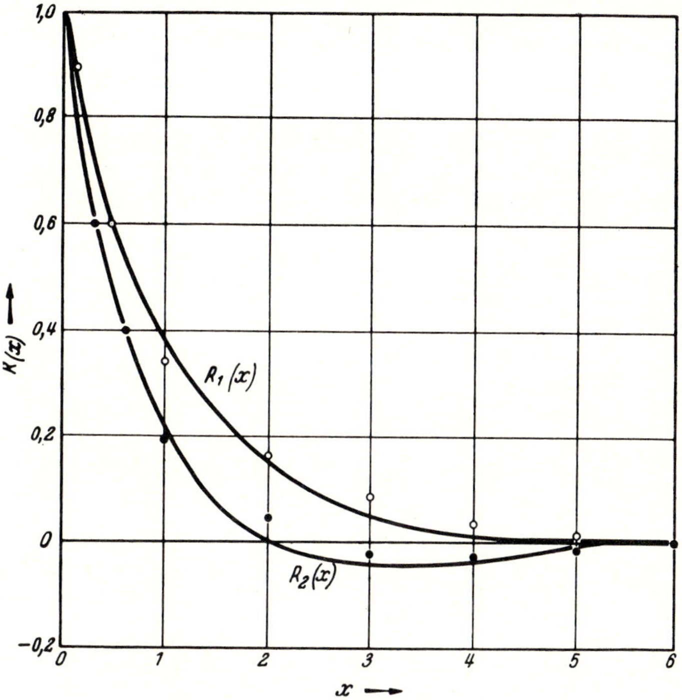

Abb. 3. Die Korrelationsfunktionen.

Dabei ist wieder L_0 mit der Maschenweite des Gitters identifiziert und λ aus dem Spektrum für $U = 1060$ cm/sec berechnet worden. Die experimentellen Werte stimmen tatsächlich mit den theoretischen bei kleineren Werten von x sehr genau überein, eigentlich genauer, als bei der Unbestimmtheit von L_0 erwartet werden konnte. Von $x k_0 \sim 1$ ab wird die Abweichung der experimentellen Punkte von den theoretischen Kurven merklich, was auch nach der Ableitung anzunehmen war. Denn der Verlauf bei größeren x-Werten hängt vom Verhalten des Spektrums in der Nähe von k_0 ab, das durch unsere Formeln prinzipiell nicht erfaßt werden kann. Aber selbst bei größeren x-Werten bleiben die Abweichungen von den theoretischen Kurven klein.

[1] TAYLOR, G. J.: Journ. Aeron. Sci. **4**, Nr. 8, 311 (1937).

Zeitschrift für Physik *124*, 628–657 (1948)

4. *Die Druckschwankungen.*

Beim Studium der Diffusion in einem turbulenten Luftstrom hat Taylor (l. c.) eine Beziehung zwischen der für die Diffusion maßgebenden Korrelationsfunktion und dem quadratischen Mittelwert des Druckgradienten abgeleitet. Wir wollen daher auch die quadratischen Mittelwerte der Druckschwankungen vom Standpunkt der hier beschriebenen Theorie aus untersuchen.

In Anlehnung an Gl. (2) kann man den Druck in eine Fourierreihe entwickeln

$$p = \sum_{\mathfrak{k}} p_{\mathfrak{k}}\, e^{i\mathfrak{k}\mathfrak{r}}, \quad p_{-\mathfrak{k}} = p_{\mathfrak{k}}^{*}, \tag{58}$$

und die hydrodynamische Grundgleichung

$$\dot{\mathfrak{v}} = -(\mathfrak{v}\nabla)\,\mathfrak{v} - \frac{1}{\rho}\nabla p + \frac{\mu}{\rho}\Delta\mathfrak{v}$$

geht über in

$$\dot{\mathfrak{v}}_{\mathfrak{k}} = -\sum_{\mathfrak{k}'} i\,(\mathfrak{v}_{\mathfrak{k}'}\,\mathfrak{k})\,\mathfrak{v}_{\mathfrak{k}-\mathfrak{k}'} - \frac{i\mathfrak{k}}{\rho}\,p_{\mathfrak{k}} - \frac{\mu}{\rho}\,k^2\,\mathfrak{v}_{\mathfrak{k}}. \tag{59}$$

Wegen $(\mathfrak{v}_{\mathfrak{k}}\,\mathfrak{k}) = 0$ folgt daraus

$$\dot{\mathfrak{v}}_{\mathfrak{k}} = -\sum_{\mathfrak{k}'} i\,(\mathfrak{v}_{\mathfrak{k}'}\,\mathfrak{k})\left[\mathfrak{v}_{\mathfrak{k}-\mathfrak{k}'} - \frac{\mathfrak{k}}{k^2}\,(\mathfrak{v}_{\mathfrak{k}-\mathfrak{k}'},\,\mathfrak{k})\right] - \frac{\mu}{\rho}\,k^2\,\mathfrak{v}_k,$$

$$p_{\mathfrak{k}} = -\frac{\rho}{k^2}\sum_{\mathfrak{k}'}(\mathfrak{v}_{\mathfrak{k}'}\,\mathfrak{k})\,(\mathfrak{v}_{\mathfrak{k}-\mathfrak{k}'},\,\mathfrak{k}). \tag{60}$$

Für die quadratischen Mittelwerte von Druck und Druckgradient ergibt sich

$$\overline{p^2} = \sum_{\mathfrak{k}} \left|\overline{p_{\mathfrak{k}}^2}\right| \quad , \quad \overline{\operatorname{grad}^2 p} = \sum_{\mathfrak{k}} k^2 \left|\overline{p_{\mathfrak{k}}^2}\right|. \tag{61}$$

Wir interessieren uns zunächst für diesen letzteren Mittelwert:

$$\overline{\operatorname{grad}^2 p} = \sum_{\mathfrak{k}}\sum_{\mathfrak{k}'}\sum_{\mathfrak{k}''}\frac{\rho^2}{k^2}\,\overline{(\mathfrak{v}_{\mathfrak{k}'}\,\mathfrak{k})\,(\mathfrak{v}_{\mathfrak{k}-\mathfrak{k}'}\,\mathfrak{k})\,(\mathfrak{v}_{\mathfrak{k}''}\,\mathfrak{k})\,(\mathfrak{v}_{\mathfrak{k}-\mathfrak{k}''}\,\mathfrak{k})}. \tag{62}$$

Der Mittelungsstrich bedeutet hier einfach zeitliche Mittelung. Wenn man Mittelwerte der Art (62) ausrechnen will, muß man in irgendeiner Form eine „Unordnungsannahme" über die turbulente Bewegung zugrunde legen. Dabei kann man davon ausgehen, daß die Amplituden $\mathfrak{v}_{\mathfrak{k}}$ im Lauf der Zeit um einen durch (10) bzw. (28) gegebenen Wert schwanken, so daß der zeitliche Mittelwert $\overline{\mathfrak{v}_{\mathfrak{k}}\mathfrak{v}_{-\mathfrak{k}}}$ einfach durch das Spektrum (28) gegeben ist. Die Phasen der $\mathfrak{v}_{\mathfrak{k}}$ aber werden im Lauf der Zeit alle möglichen Werte durchlaufen, alle Werte der Phase werden im Durchschnitt gleich häufig vorkommen. Wenn man die Phasen zu verschiedenen Wellenzahlen als statistisch völlig unabhängig ansehen könnte, so würden bei der Mittelung solcher Produkte von vier Faktoren

$$\overline{\mathfrak{v}_{\mathfrak{k}_1}\mathfrak{v}_{\mathfrak{k}_2}\mathfrak{v}_{\mathfrak{k}_3}\mathfrak{v}_{\mathfrak{k}_4}}$$

nur die Glieder übrigbleiben, bei denen je zwei Wellenzahlen entgegengesetzt gleich sind; also Glieder vom Typus

$$\overline{\mathfrak{v}_{\mathfrak{k}_1}\mathfrak{v}_{-\mathfrak{k}_1}\mathfrak{v}_{\mathfrak{k}_2}\mathfrak{v}_{-\mathfrak{k}_2}},$$

und diese Mittelwerte könnten durch die Produkte der Mittelwerte der Amplitudenquadrate ersetzt werden:

$$\overline{\mathfrak{v}_{\mathfrak{k}_1}\mathfrak{v}_{-\mathfrak{k}_1}\mathfrak{v}_{\mathfrak{k}_2}\mathfrak{v}_{-\mathfrak{k}_2}} = \overline{\mathfrak{v}_{\mathfrak{k}_1}\mathfrak{v}_{-\mathfrak{k}_1}}\;\overline{\mathfrak{v}_{\mathfrak{k}_2}\mathfrak{v}_{-\mathfrak{k}_2}}. \tag{63}$$

In Wirklichkeit werden jedoch statistische Korrelationen zwischen den Phasen zu verschiedenen Wellenzahlen bestehen, da die Wellen sich gegenseitig beeinflussen. Im Abschnitt 5 wird versucht werden, in einem einfachen Fall solche Korrelationen abzuschätzen. Trotz des Vorhandenseins der Korrelationen möchten wir aber annehmen, daß in einer Summe von der Art (62) die Glieder vom Typus (63) den größten Anteil beisteuern, denn ihr Mittelwert ist schon in der ersten Näherung, ohne jede Annahme über das zeitliche Verhalten der Wellen, von Null verschieden, während die anderen Mittelwerte erst durch das feinere Wechselspiel verschiedener Wellen einen von Null verschiedenen Wert erhalten. Wir glauben also, daß man eine brauchbare Näherung erhält, wenn man in (62) nur die Glieder vom Typus (63) berücksichtigt. Dann ergibt sich

$$\overline{\operatorname{grad}^2 p} = 2\rho^2 \sum_{\mathfrak{k}'\mathfrak{k}''} \frac{\overline{(\mathfrak{v}_{\mathfrak{k}'}\mathfrak{k}'')\,(\mathfrak{v}_{-\mathfrak{k}'}\mathfrak{k}'')\,(\mathfrak{v}_{\mathfrak{k}''}\mathfrak{k}')\,(\mathfrak{v}_{-\mathfrak{k}''}\mathfrak{k}')}}{(\mathfrak{k}'-\mathfrak{k}'')^2}. \tag{64}$$

Bei der Mitteilung über die Richtungen der $\mathfrak{v}_{\mathfrak{k}}$ – man nimmt dabei wieder an, daß alle Richtungen $\perp\,\mathfrak{k}$ für $\mathfrak{v}_{\mathfrak{k}}$ gleichwahrscheinlich seien – benützt man zweckmäßig die Relation

$$\overline{(\mathfrak{v}_{\mathfrak{k}}\mathfrak{a})\,(\mathfrak{v}_{-\mathfrak{k}}\mathfrak{b})} = \frac{|\overline{\mathfrak{v}_{\mathfrak{k}}^2}|}{2}\left[\mathfrak{a}\mathfrak{b} - \frac{(\mathfrak{a}\mathfrak{k})\,(\mathfrak{b}\mathfrak{k})}{k^2}\right]. \tag{65}$$

So erhält man

$$\overline{\operatorname{grad}^2 p} = \frac{\rho^2}{2} \sum_{\mathfrak{k}'\mathfrak{k}''} |\overline{v_{\mathfrak{k}'}^2}|\;|\overline{v_{\mathfrak{k}''}^2}|\;\frac{[k'^2 k''^2 - (\mathfrak{k}'\mathfrak{k}'')^2]^2}{(\mathfrak{k}'-\mathfrak{k}'')^2}. \tag{66}$$

Setzt man $(\mathfrak{k}'\,\mathfrak{k}'') = k'\,k''\,\zeta$, so kann man die Integration über ζ ausführen und findet:

$$\overline{\mathrm{grad}^2\,p} = \frac{\rho^2}{2} \sum_{\mathfrak{k}'\mathfrak{k}''} |\overline{v_{\mathfrak{k}'}^2}|\,|\overline{v_{\mathfrak{k}''}^2}|\;k'\,k''\,\psi\left(\frac{k'}{k''}\right), \tag{67}$$

wobei

$$\psi(s) = \psi\left(\frac{1}{s}\right) = \frac{1}{16\,s^3}\left[-1 + \frac{11}{3}s^2 + \frac{11}{3}s^4 - s^6 + \frac{(1-s^2)^4}{2s}\,\lg\frac{1+s}{|1-s|}\right]. \tag{68}$$

Für $0 \le s \ll 1$ gilt näherungsweise:

$$\psi(s) \approx \frac{8s}{15}\left(1 - \frac{3}{7}\,s^2 + \frac{s^4}{21}\right). \tag{69}$$

Verwandelt man die Summen in Integrale und setzt das Spektrum (28) in (67) ein, so folgt schließlich

$$\overline{\mathrm{grad}^2\,p} = \frac{2}{9}\,\rho^2\,v_0^4\,k_0^{\frac{4}{3}} \int\limits_{k_0}^{\infty} \frac{d\,k'}{k'^{\frac{2}{3}}} \int\limits_{k_0}^{\infty} \frac{d\,k''}{k''^{\frac{2}{3}}}\; \frac{\psi\left(\frac{k'}{k''}\right)}{\left[1 + \left(\frac{k'}{k_s}\right)^{\frac{8}{3}}\right]^2 \left[1 + \left(\frac{k''}{k_s}\right)^{\frac{8}{3}}\right]^2}. \tag{70}$$

Man erkennt aus (70), daß die Integrale bei kleinen k-Werten konvergieren, daß man die Integration also ohne erheblichen Fehler von $k = 0$ ab führen kann. Dies zeigt, daß $\overline{\mathrm{grad}^2 p}$ tatsächlich durch das Verhalten des Spektrums bei großen k, d. h. durch die *kleinsten* Turbulenzelemente bestimmt ist. Hätten wir $\overline{p^2}$ berechnet, so hätte sich im Gegenteil ergeben, daß das Integral bei kleinen Werten von k divergiert, daß sein Wert also ganz durch die größten Turbulenzelemente bestimmt wird. Der Wert von $\overline{p^2}$ kann daher überhaupt nicht nach dem hier angewandten Verfahren berechnet werden; denn erstens hat das Spektrum bei kleinen k eine von der Geometrie abhängige Form, und zweitens wäre es bei den größten Turbulenzelementen sicher ganz unberechtigt, nur die Mittelwerte vom Typus (63) zu berücksichtigen, da die Geometrie dem System der größten Wirbel sicher bestimmte Phasenbeziehungen aufprägt.

Aus (70) wird jetzt:

$$\overline{\mathrm{grad}^2\,p} = \frac{2}{9}\,\rho^2\,v_0^4\,k_0^{\frac{4}{3}}\,k_s^{\frac{2}{3}} \int\limits_0^{\infty} \frac{d\zeta}{\zeta^{\frac{2}{3}}} \int\limits_0^{\infty} \frac{d\eta}{\eta^{\frac{2}{3}}}\; \frac{\psi\left(\frac{\zeta}{\eta}\right)}{\left[1 + \zeta^{\frac{8}{3}}\right]^2 \left[1 + \eta^{\frac{8}{3}}\right]^2}. \tag{71}$$

Das Doppelintegral rechts wurde nach einem graphischen Verfahren zu 0,763 abgeschätzt; so folgt schließlich [vgl. (30)]

$$\overline{\mathrm{grad}^2\,p} = 0{,}17\;\rho^2\,v_0^4\,k_0^{\frac{4}{3}}\,k_s^{\frac{2}{3}} \tag{72}$$

$$= 0{,}05\;\rho^2\,v_0^4\,k_0^2\,\sqrt{R_0\,\varkappa}\,. \tag{73}$$

TAYLOR (l. c.) hatte die Vermutung ausgesprochen, daß $\overline{\text{grad}^2 p}$ die gleiche Größenordnung wie $\overline{\rho^2 v_0^2 \left(\frac{\partial v}{\partial x}\right)^2}$ haben sollte, also die Größenordnung $\rho^2 v_0^4 k_0^{2/3} k_s^{4/3}$. Man erkennt nun aus (72), daß $\overline{\text{grad}^2 p}$ erheblich kleiner sein muß, und zwar um so mehr, je größer das Verhältnis $\frac{k_s}{k_0}$ ist. Die von TAYLOR definierte Länge λ_η:

$$\overline{\left(\frac{\partial p}{\partial y}\right)^2} = 2\rho^2 \frac{(\overline{\mathfrak{v}_y^2})^2}{\lambda_\eta^2}, \text{ d. h.} \tag{74}$$

$$\overline{\text{grad}^2 p} = \frac{2}{3} \rho^2 \frac{v_0^4}{\lambda_\eta^2} \tag{75}$$

muß daher für große Reynoldssche Zahlen erheblich größer werden als die Länge λ von Gl. (55). Aus (75), (73) und (56) folgt:

$$\lambda_\eta = \lambda \cdot 0{,}42 \sqrt[4]{R_0 \varkappa}. \tag{76}$$

Dieses Ergebnis paßt allerdings schlecht zum experimentellen Befund; denn TAYLOR gibt bei einem Versuch, den SIMMONS im Anschluß an ähnliche Experimente von SCHUBAUER[1] ausgeführt hatte, $\frac{\lambda_\eta}{\lambda} = 0{,}5$ an, wobei nach den Versuchsbedingungen etwa $\sqrt[4]{R_0 \varkappa} \approx 3{,}9$ angenommen werden muß. Man muß daher die Frage stellen, ob etwa das Ergebnis (72) dadurch gefälscht worden ist, daß nur die Glieder (63) bei der Mittelung berücksichtigt wurden. Man kann aber leicht einsehen, daß zwar vielleicht der Zahlenfaktor in (72) dadurch beeinflußt werden kann, daß aber die Abhängigkeit von k_0 und k_s, d. h. die Abhängigkeit des λ_η von $(R_0 \varkappa)$ nichts mit dieser Vernachlässigung zu tun hat. Denn schon die Gl. (62) zeigt, daß auf der rechten Seite wegen Gl. (10) der Normierungsfaktor $v_0^4 k_0^{4/3}$ auftreten muß. Dieser Faktor wird nach Ausführung der Mittelung ergänzt durch einen Faktor der Dimension $k^{2/3}$, der offenbar höchstens von der Ordnung $k_s^{2/3}$ sein kann – und auch sein muß, denn das zugehörige Integral über k würde eben wie $k^{2/3}$ divergieren, wenn nicht bei $k \sim k_s$ der Abfall des Spektrums mit k^{-7} einsetzen würde. Es bliebe noch die Möglichkeit, daß nur der Zahlenfaktor in (72) durch die alleinige Berücksichtigung der Glieder (63) zu niedrig geschätzt worden ist. Aber man kann sich schwer vorstellen, daß der richtige Ausdruck um mehr als das Zehnfache größer würde – was zur Deutung der Experimente nötig wäre.

Vielleicht läßt sich der Widerspruch in folgender Weise aufklären: Der Hauptbeitrag zu $\overline{\text{grad}^2 p}$ rührt von Wellenzahlen der Ordnung k_s her, also von Turbulenzelementen, deren Durchmesser wenige Milli-

[1] SCHUBAUER: Rep. Nat. Adv. Comm. Aero. Nachr. Nr. 524 (1935).

Zeitschrift für Physik *124*, 628–657 (1948)

meter beträgt. Beim Versuch von Simmons wird der Luftstrom durch einen 20 cm langen, quer zum Windkanal gespannten Heizdraht erwärmt und dann die Verteilung der erwärmten Luft in einem gewissen Abstand hinter dem Heizdraht gemessen. Für die Bestimmung von λ_η sind gerade die kleineren Abstände (5 bis 15 cm) maßgebend. Es liegt nahe anzunehmen, daß der Heizdraht selbst im Luftstrom eine kleine Wirbelstraße und zusätzliche Turbulenz hervorruft, wobei die Turbulenzelemente einige Millimeter groß sein dürften – d. h. der Draht erhöht gerade die Intensität der Turbulenz in dem Spektralgebiet, das auf $\overline{\text{grad}^2\, p}$ den stärksten Einfluß hat. Dabei dürfte die zusätzliche Turbulenz in der unmittelbaren Nachbarschaft des Drahtes viel größer sein als die ursprüngliche Turbulenz des gleichen Wellenlängenbereiches. Aber diese zusätzliche Turbulenz klingt natürlich schnell ab, und es ist wohl schwer abzuschätzen, ob sie allein die Diskrepanz zwischen (76) und dem empirischen λ_η-Wert erklären kann.

5. *Die Energiedissipation bei normaler isotroper Turbulenz.*

Die Untersuchungen des vorausgehenden Abschnitts hängen bereits eng mit einem Grundproblem der statistischen Turbulenztheorie zusammen: mit der Bestimmung der Energiedissipation bei normaler Energieverteilung, d. h. der Bestimmung der Konstante $\varkappa$ in Gl. (12). Bei diesem Problem kann die molekulare Reibung völlig vernachlässigt werden. Die hydrodynamischen Grundgleichungen können also in der Form

$$\dot{\mathfrak{v}} = -(\mathfrak{v}\nabla)\,\mathfrak{v} - \frac{1}{\rho}\,\nabla p\,,\ \nabla\mathfrak{v} = 0$$

vorausgesetzt werden. Ferner soll als Normierungsvolumen ein geeignet herausgegriffenes Teilvolumen der Flüssigkeit gewählt werden, das unter Umständen mit der Flüssigkeit – etwa dem Mittelwert der Geschwindigkeit über das Volumen entsprechend – mitbewegt wird. Wir nehmen also an, das Volumen bewege sich mit der Geschwindigkeit $\mathfrak{u}$. Dann geht (60) über in

$$\dot{\mathfrak{v}}_{\mathfrak{k}} = -i\sum_{\mathfrak{k}'}(\mathfrak{v}_{\mathfrak{k}'}\,\mathfrak{k})\left[\mathfrak{v}_{\mathfrak{k}-\mathfrak{k}'} - \frac{\mathfrak{k}}{k^2}(\mathfrak{v}_{\mathfrak{k}-\mathfrak{k}'}\,,\,\mathfrak{k})\right] + i\,(\mathfrak{u}\,\mathfrak{k})\,\mathfrak{v}_{\mathfrak{k}}\,. \qquad (77)$$

Für die Berechnung der Energiedissipation muß man ermitteln, wie sich die Intensität $|\mathfrak{v}_{\mathfrak{k}}^2|$ einer bestimmten Eigenschwingung (oder vielleicht besser: die Summe solcher Amplitudenquadrate über einen kleinen Spektralbereich Δk: $\sum_{k}^{k+\Delta k} |\mathfrak{v}_{\mathfrak{k}}^2|$) im Lauf der Zeit ändert. Man erkennt aus (77), daß man dafür Zeitmittelwerte über Produkte vom Typus

$$\overline{\mathfrak{v}_{\mathfrak{k}_1}\,\mathfrak{v}_{\mathfrak{k}_2}\,\mathfrak{v}_{\mathfrak{k}_3}} \qquad (78)$$

braucht, wobei $\mathfrak{k}_3 = -\mathfrak{k}_1 - \mathfrak{k}_2$ ist. Wegen der statistisch gleichmäßigen Verteilung der Phasen würden diese Mittelwerte verschwinden, wenn es nicht statistische Korrelationen zwischen den Phasen zu verschiedenen $\mathfrak{k}$ gäbe, die von der gegenseitigen Beeinflussung der verschiedenen Wellen herrühren, wie bereits im Abschnitt 4 auseinandergesetzt wurde. Um diese Korrelationen zu ermitteln, muß man den Einfluß, der auf eine Welle mit gegebenem $\mathfrak{k}$ von Wellen mit anderem $\mathfrak{k}$ ausgeübt wird, irgendwie in den Gleichungen ausdrücken; das kann z. B. dadurch geschehen, daß man eine der drei Amplituden in (78) als Zeitintegral über $\dot{\mathfrak{v}}_{\mathfrak{k}}$ darstellt und $\dot{\mathfrak{v}}_{\mathfrak{k}}$ wieder durch eine Summe über zwei andere $\mathfrak{v}_{\mathfrak{k}}$ nach (77) ausdrückt. Dann erhält man Produkte von je vier Amplituden $\mathfrak{v}_{\mathfrak{k}}$, die z. T. aber zu verschiedenen Zeiten zu nehmen sind. Für solche Produkte gelten die Überlegungen des 4. Abschnittes, nach denen man eine erste Näherung dadurch bekommt, daß man nur Produkte vom Typus (63) berücksichtigt. Natürlich könnte man das Verfahren im Prinzip fortsetzen und die anderen Mittelwerte von vierfachen Produkten dadurch zu berechnen suchen, daß man sie auf sechsfache zurückführt usw. Aber solche Rechnungen würden wohl viel zu kompliziert werden, die höheren Glieder tragen wohl auch weniger bei, und wir werden uns mit dem ersten Schritt begnügen.

Bei diesen Rechnungen wird man offenbar Mittelwerte vom Typus

$$\overline{\mathfrak{v}_{\mathfrak{k}}(t)\,\mathfrak{v}_{-\mathfrak{k}}(t+\tau)}$$

brauchen und wir definieren daher:

$$R_k(t,\tau) = \frac{\sum\limits_{k}^{k+\Delta k}{}' \, \overline{\mathfrak{v}_{\mathfrak{k}}\left(t+\frac{\tau}{2}\right)\mathfrak{v}_{-\mathfrak{k}}\left(t-\frac{\tau}{2}\right)}}{\sum\limits_{k}^{k+\Delta k}{}' \, \overline{\mathfrak{v}_{\mathfrak{k}}\,\mathfrak{v}_{-\mathfrak{k}}}}. \qquad (79)$$

Dabei ist die Summation über einen kleinen Spektralbereich Δk in die Definition von $R_k(t,\tau)$ aufgenommen worden, damit in $R_k(t,\tau)$ nicht die Größe des Normierungsvolumens unmittelbar eingeht und damit die Mittelung über alle Richtungen von $\mathfrak{k}$ gleich vorgenommen wird. Der Spektralbereich Δk ist dabei offenbar so breit zu wählen, daß noch viele Eigenschwingungen des Normierungsvolumens in ihm Platz haben (d. h. $k^2 \Delta k\, V \gg 1$), aber doch sehr klein gegen k selbst. Diese Forderungen sind bei den Turbulenzelementen von der Größenordnung V selbst nicht mehr vereinbar, aber für diese kann man die statistischen Methoden ohnehin nicht anwenden. Das ganze Verfahren ist also nur durchführbar, wenn sich herausstellt, daß die großen Turbulenzelemente zu den zu untersuchenden Mittelwerten praktisch nichts mehr beitragen.

Zeitschrift für Physik *124*, 628–657 (1948)

Um aus den hydrodynamischen Gleichungen Aussagen über die Größen $R_k(t, \tau)$ zu erhalten, liegt es nahe, den folgenden Ausdruck zu untersuchen:

$$\dot{\mathfrak{v}}_{\mathfrak{k}}\left(t+\frac{\tau}{2}\right)\mathfrak{v}_{-\mathfrak{k}}\left(t-\frac{\tau}{2}\right) = -i\sum_{\mathfrak{k}'}\left(\mathfrak{v}_{\mathfrak{k}'}^{t+\frac{\tau}{2}}, \mathfrak{k}\right)\left(\mathfrak{v}_{\mathfrak{k}-\mathfrak{k}'}^{t+\frac{\tau}{2}}\, \mathfrak{v}_{-\mathfrak{k}}^{t-\frac{\tau}{2}}\right) + i\,(\mathfrak{u}\,\mathfrak{k})\,\mathfrak{v}_{\mathfrak{k}}^{t+\frac{\tau}{2}}\,\mathfrak{v}_{-k}^{t-\frac{\tau}{2}}. \tag{80}$$

In diesem Ausdruck kann man $\mathfrak{v}_{\mathfrak{k}'}^{t+\frac{\tau}{2}}$ ersetzen durch ein Zeitintegral über $\dot{\mathfrak{v}}_{\mathfrak{k}'}^{t+\frac{\tau}{2}}$:

$$\mathfrak{v}_{\mathfrak{k}'}^{t+\frac{\tau}{2}} = \int_{-T}^{t+\frac{\tau}{2}} \dot{\mathfrak{v}}_{\mathfrak{k}'}^{t'}\,dt' + \mathfrak{v}_{\mathfrak{k}'}(-T) = \int_0^{t+\frac{\tau}{2}+T} d\tau'\,\dot{\mathfrak{v}}_{\mathfrak{k}'}^{t+\frac{\tau}{2}-\tau'} + \mathfrak{v}_{\mathfrak{k}'}(-T). \tag{81}$$

Wenn man T hinreichend groß wählt, wird die Korrelation zwischen $\mathfrak{v}_{-k'}(t)$ und $\mathfrak{v}_{k'}(T)$ verschwinden; es ist daher zweckmäßig, nach dem Einsetzen in (80) den Grenzübergang $T \to \infty$ vorzunehmen. Wenn man auch noch die Mittelung über die Richtungen vornimmt – dadurch fällt das Glied mit $\mathfrak{u}$ weg –, erhält man aus (80) und (81):

$$\begin{aligned} &\sum_k^{k+\Delta k} \dot{\mathfrak{v}}_k\left(t+\frac{\tau}{2}\right)\mathfrak{v}_{-k}\left(t-\frac{\tau}{2}\right) = \\ &= -\sum_k^{k+\Delta k}\sum_{\mathfrak{k}'}\sum_{\mathfrak{k}''}\int_0^\infty d\tau'\left(\mathfrak{v}_{\mathfrak{k}''}^{t+\frac{\tau}{2}-\tau'}\,\mathfrak{k}'\right)\left[\mathfrak{v}_{\mathfrak{k}'-\mathfrak{k}''}^{t+\frac{\tau}{2}-\tau'}\,\mathfrak{k} - \frac{(\mathfrak{k}\,\mathfrak{k}')\left(\mathfrak{v}_{\mathfrak{k}'-\mathfrak{k}''}^{t+\frac{\tau}{2}-\tau'}\,\mathfrak{k}'\right)}{k'^2}\right]\left(\mathfrak{v}_{\mathfrak{k}-\mathfrak{k}'}^{t+\frac{\tau}{2}}\,\mathfrak{v}_{-\mathfrak{k}}^{t-\frac{\tau}{2}}\right). \end{aligned} \tag{82}$$

Berücksichtigt man weiter, wie im Abschnitt 4, nur die Glieder vom Typus (63), ersetzt ferner $\mathfrak{k} - \mathfrak{k}'$ durch $\mathfrak{k}'$ und integriert über den Kosinus des Winkels zwischen $\mathfrak{k}$ und diesem Vektor, so folgt:

$$\begin{aligned} &\sum_k^{k+\Delta k} \dot{\mathfrak{v}}_{\mathfrak{k}}\left(t+\frac{\tau}{2}\right)\mathfrak{v}_{-\mathfrak{k}}\left(t-\frac{\tau}{2}\right) = \\ &= \sum_k^{k+\Delta k}\int dk'\,k'^2\,\frac{V}{(2\pi)^3}\,\frac{\pi}{16}\int_0^\infty d\tau'\,\left|\overline{\mathfrak{v}_k^2}\right|\,\left|\overline{\overline{\mathfrak{v}_{k'}^2}}\right|\,R_k\left(t-\frac{\tau'}{2}, \tau-\tau'\right)R_{k'}\left(t+\frac{\tau-\tau'}{2}, \tau'\right)\cdot \\ &\cdot k^{-3}\,k'^{-3}\,(k^2-k'^2)\left[2\,k\,k'\left(k^4+k'^4-\frac{2}{3}\,k^2\,k'^2\right)-(k^2+k'^2)\,(k^2-k'^2)\,\lg\frac{k+k'}{|k-k'|}\right]. \end{aligned} \tag{83}$$

Diese Gleichung gibt die Möglichkeit, die zeitlichen Differentialquotienten $\frac{d R_k(t,\tau)}{d t}$ und $\frac{d R_k(t,\tau)}{d \tau}$ durch die R_k selbst auszudrücken, außerdem kann man, wenn die R_k bekannt sind, aus (83) die Energiedissipation berechnen, indem man $\tau = 0$ setzt. Dabei wollen wir annehmen, daß der ganze Turbulenzvorgang entweder stationär ist oder nur sehr langsam abklingt, so daß die Zeiten, in denen sich die Intensität $|\mathfrak{v}_k^2|$ merklich ändert, sehr lang sind verglichen mit den Schwankungsperioden von $\mathfrak{v}_k$. $|\overline{\mathfrak{v}_k^2}|$ bedeutet dann den Mittelwert über eine Zeit, die zwar sehr lang gegen die Schwankungsperioden, aber sehr viel kürzer als die Abklingzeit ist.

Die Gleichung für $\frac{d R_k(t,\tau)}{d t}$ gibt ein Maß für die Schwankungen der Größe $R_k(t, \tau)$ als Funktion von t um ihren zeitlichen Mittelwert:

$$R_k(\tau) = \overline{R_k(t\,,\,\tau)}\,. \tag{84}$$

Man kann vermuten, daß diese Schwankungen klein sind im Gebiet kleiner τ, das durch die kleinen Turbulenzelemente bestimmt ist, und daß sie mit wachsendem τ größer werden; diese Frage wird später noch untersucht werden.

Vor der Durchführung der weiteren Rechnungen muß nun noch festgelegt werden, wie das Teilvolumen V und seine Geschwindigkeit $\mathfrak{u}$ gewählt werden sollen. Man könnte zuerst versuchen, $\mathfrak{u} = 0$ zu setzen und V mit dem Gesamtvolumen zu identifizieren. Dadurch würde man aber ein falsches Bild der Verhältnisse erhalten. Der zeitliche Abfall der Korrelationsfunktion $\overline{R_k(t, \tau)} = R_k(\tau)$ als Funktion von τ ist in diesem Koordinatensystem nämlich durch die größten Turbulenzelemente bestimmt, und daher sehr rasch. Man kann zeigen, daß die Korrelationsfunktion in diesem Koordinatensystem in hinreichender Näherung durch

$$R_k(\tau) = 1 - \frac{k v_0 \tau}{\sqrt{3}} \cdot e^{-\frac{k^2 v_0^2 \tau^2}{12}} \int_0^{\frac{k v_0 \tau}{\sqrt{12}}} dx\, e^{x^2}$$

gegeben ist; auf die Rechnungen, die zu diesem Ausdruck geführt haben, braucht nicht näher eingegangen zu werden, da der Ausdruck später nicht weiter verwendet wird. Die physikalische Deutung des Ausdrucks wird durch die folgende Überlegung gegeben: Die Funktion $R_k(\tau)$ in ihm fällt nach einer Zeit der Ordnung $\frac{2\pi}{k v_0}$ ab; das ist die Zeit, in der etwa gerade ein Wirbel der Wellenlänge $\frac{2\pi}{k}$ infolge der großen Geschwindigkeit in den größten Turbulenzelementen am Beobachtungspunkt vorbeizieht. Daß

Zeitschrift für Physik *124*, 628–657 (1948)

die Korrelationsfunktion nach dieser Zeit absinkt, bedeutet also einfach, daß die Geschwindigkeit in den größten Turbulenzelementen die Größenordnung v_0 hat, aber statistisch um Werte dieser Ordnung schwankt. Mit der Auflösung des Wirbels von der Wellenlänge $\frac{2\pi}{k}$ hat dieser Vorgang nichts zu tun. Vielmehr ist bei Gültigkeit des $F(k) \sim k^{-5/3}$-Gesetzes nach den Ähnlichkeitsbetrachtungen v. WEIZSÄCKERS zu erwarten, daß sich die Auflösung des Wirbels erst nach Zeiten der Ordnung $2\pi v_0^{-1} k^{-\frac{2}{3}} k_0^{-\frac{1}{3}}$ vollzieht. Andererseits hat die Energiedissipation aber mit der Auflösung der Wirbel zu tun und nicht mit der Bewegung im großen. Wenn man die Auflösung der Wirbel in den Gleichungen beschreiben will, muß man das Koordinatensystem also mitbewegen. Man muß dann das Teilvolumen V in den Lineardimensionen etwas, aber nicht sehr viel größer machen als $\frac{2\pi}{k}$ und es nach Maßgabe der mittleren Geschwindigkeit in ihm mitbewegen. Wir wollen versuchsweise annehmen, daß man für jedes k ein ihm entsprechendes Volumen V angeben kann, derart, daß $V k^3$ von k unabhängig wird und daß für die so gewählten mitbewegten Volumenelemente die Korrelationsfunktion $R(t, \tau)$ eine universelle Funktion der Variabeln $v_0\, k^{2/3}\, k_0^{1/3}\, t$ bzw. $v_0\, k^{2/3}\, k_0^{1/3}\, \tau$ ist, wie nach der v. WEIZSÄCKERschen Ähnlichkeitsüberlegung zu erwarten ist. Wir werden zeigen, daß die aus (83) hervorgehenden Beziehungen für $R_k(t, \tau)$ durch diese Annahme tatsächlich befriedigt werden können, wenn die F_k nach dem $k^{-5/3}$-Gesetz verteilt sind.

Setzt man nämlich $R_k(t, \tau) = g(\zeta, \eta)$ wobei

$$\zeta = \frac{v_0}{6}\, k^{2/3}\, k_0^{1/3}\, t \quad , \quad \eta = \frac{v_0}{6}\, k^{2/3}\, k_0^{1/3}\, \tau \quad , \quad \frac{k'}{k} = y\,, \tag{85}$$

so folgt aus (83):

$$\frac{d g(\zeta, \eta)}{d\eta} = \frac{3}{16} \int_0^\infty dy\, f(y) \int_0^\infty d\eta' \tag{86}$$

$$\left[g\left(\zeta - \frac{\eta'}{2}, \eta' - \eta\right) g\left(\left(\zeta + \frac{\eta - \eta'}{2}\right) y^{2/3}, \eta' y^{2/3}\right) - g\left(\zeta - \frac{\eta'}{2}, \eta + \eta'\right) g\left(\left(\zeta - \frac{\eta + \eta'}{2}\right) y^{\frac{2}{3}}, \eta' y^{\frac{2}{3}}\right)\right],$$

$$\frac{d g(\zeta, \eta)}{d\zeta} = \frac{3}{8} \int_0^\infty dy\, f(y) \int_0^\infty d\eta'$$

$$\left[g\left(\zeta - \frac{\eta'}{2}, \eta' - \eta\right) g\left(\left(\zeta + \frac{\eta - \eta'}{2}\right) y^{2/3}, \eta' y^{2/3}\right) - g\left(\zeta - \frac{\eta'}{2}, \eta + \eta'\right) g\left(\left(\zeta - \frac{\eta + \eta'}{2}\right) y^{\frac{2}{3}}, \eta' y^{\frac{2}{3}}\right)\right],$$

wobei

$$f(y) = \frac{1-y^2}{y^{\frac{14}{3}}}\left[2y\left(1-\frac{2}{3}y^2+y^4\right)-(1-y^2)(1-y^4)\lg\frac{1+y}{|1-y|}\right]. \quad (87)$$

Diese Gleichungen enthalten tatsächlich die Konstanten k_0, v_0 nicht mehr. Dies liegt daran, daß das Integral über y sowohl bei kleinen wie bei großen y konvergiert [$f(y)$ verschwindet hinreichend stark sowohl bei $y=0$ wie für $y\to\infty$]; daher kann man das Integral über k' ohne erheblichen Fehler statt von k_0 einfach von 0 ab führen; außerdem zeigt die Konvergenz des Integrals bei großen Werten von k, daß die molekulare Reibung in diesem Problem tatsächlich keine Rolle spielt; das Verhalten des Spektrums im Gebiet der kleinsten Turbulenzelemente ist für die Korrelationsfunktionen $R(t,\tau)$ und die Energiedissipation bei mittleren k-Werten unwichtig.

Ehe wir die numerische Lösung von (86) versuchen, soll die Gl. (83) noch benutzt werden, um in der hier angestrebten Näherung die Energiedissipation zu berechnen. Zu diesem Zweck setzen wir $\tau=0$ und integrieren die Gl. (83) über k zwischen zwei willkürlichen Grenzen K_1 und K_2:

$$\frac{d}{dt}\int_{K_1}^{K_2} 4\pi k^2\,dk\left|\frac{\mathfrak{v}_{\mathfrak{k}}^2}{2}\right| = \quad (88)$$

$$= \frac{\pi^2 V}{4(2\pi)^3}\int_{K_1}^{K_2}\frac{dk}{k}\int_0^\infty\frac{dk'}{k'}\left|\overline{\mathfrak{v}_k^2}\right|\left|\overline{\mathfrak{v}_{k'}^2}\right|\int_0^\infty d\tau'\,\overline{R_k\left(t-\frac{\tau'}{2},\tau'\right)R_{k'}\left(t-\frac{\tau'}{2},\tau'\right)}\,\cdot$$

$$\cdot\,(k^2-k'^2)\left[2kk'\left(k^4+k'^4-\frac{2}{3}k^2k'^2\right)-(k^2+k'^2)(k^2-k'^2)^2\lg\frac{|k-k'|}{k+k'}\right].$$

Der Integrand auf der rechten Seite ist eine antisymmetrische Funktion in k und k'. Wenn man die zeitliche Änderung der Gesamtenergie berechnet, d. h. $K_1=0$, $K_2=\infty$ setzt, so ergibt sich also Null, sofern das Integral rechts überhaupt konvergiert. D. h. die Gesamtenergie ist zeitlich konstant, was auch gefordert werden muß, da die molekulare Reibung nicht berücksichtigt worden ist. Wenn man jedoch die zeitliche Änderung der Energie betrachtet, die in dem zwischen K_1 und K_2 gelegenen Teil des Spektrums enthalten ist, so läßt sich das Integral auf der rechten Seite in der folgenden Weise umformen (der antisymmetrische Integrand sei einfach J genannt):

$$\frac{d}{dt}\int_{K_1}^{K_2}4\pi k^2\,dk\left|\frac{v_k^2}{2}\right| = \int_{K_1}^{K_2}dk\int_0^\infty dk'\,J = \int_{K_1}^{K_2}dk\int_0^{K_1}dk'\,J - \int_{K_1}^{K_2}dk\int_{K_2}^\infty dk'\,(-J). \quad (89)$$

Zeitschrift für Physik *124*, 628–657 (1948)

Im ersten der beiden Integrale rechts ist J, im zweiten $(-J)$ stets positiv. Aus dieser Schreibweise folgt, daß man das erste Integral auffassen darf als die Energie, die pro Zeiteinheit von kleineren Wellenzahlen ($k' < K_1$) her in das Gebiet zwischen K_1 und K_2 einströmt, das zweite Integral als die Energie, die nach größeren Wellenzahlen ($k' > K_2$) abströmt. Setzt man insbesondere $K_1 = 0$ und $K_2 \gg k_0$, so stellt das zweite Integral die gesamte Energiedissipation dar; sie muß sich beim Normalspektrum [$F(k) \sim k^{-5/3}$] als von K_2 unabhängig erweisen. So erhält man aus (83), (86) und (89) für die Energiedissipation den Ausdruck

$$S = \frac{\rho}{24} v_0^3 k_0 \int\limits_{K_2}^{\infty} \frac{dk''}{k''} \int\limits_0^{\frac{K_2}{k''}} dy\, f(y) \int\limits_0^{\eta} d\eta\, \overline{g\left(\zeta - \frac{\eta}{2}, \eta\right) g\left(\left(\zeta - \frac{\eta}{2}\right) y^{\frac{2}{3}}, \eta\, y^{\frac{2}{3}}\right)} = \tag{90}$$

$$= \frac{\rho}{24} v_0^3 h_0 \int\limits_0^1 dy\, (-\lg y)\, f(y) \int\limits_0^{\infty} d\eta\, \overline{g\left(\zeta - \frac{\eta}{2}, \eta\right) g\left(\left(\zeta - \frac{\eta}{2}\right) y^{\frac{2}{3}}, \eta\, y^{\frac{2}{3}}\right)}.$$

Dieser Ausdruck ist tatsächlich von K_2 unabhängig, wie es sein muß. Da die gesamte Energiedissipation nach Gl. (44) auch durch $\rho \varkappa \frac{\sqrt{3}}{8} v_0^3 k_0$ gegeben ist, so folgt:

$$\varkappa = \frac{1}{3\sqrt{3}} \int\limits_0^1 dy\, (-\lg y)\, f(y) \int\limits_0^{\infty} d\eta\, \overline{g\left(\zeta - \frac{\eta}{2}, \eta\right) g\left(\left(\zeta - \frac{\eta}{2}\right) y^{\frac{2}{3}}, \eta\, y^{\frac{2}{3}}\right)}. \tag{91}$$

Aus dieser Gleichung kann $\varkappa$ numerisch berechnet werden, wenn die Funktion $g(\zeta, \eta)$ bekannt ist.

Wir gehen nun zur Behandlung des Gleichungssystems (86) über. Dieses System stellt insofern eine große Vereinfachung gegenüber der Ausgangsgleichung (77) dar, als es keine dimensionsbehafteten Größen mehr enthält und schon aus dem Gleichgewichtsspektrum $k^{-5/3}$ hergeleitet ist. Andererseits enthält auch (86) noch Aussagen über die Schwankungen der $g(\zeta, \eta)$ als Funktion der ζ und ist deswegen wohl zu kompliziert, um strenge Lösungen zuzulassen. Man könnte versuchen, in einer ersten Näherung die Schwankungen ganz zu vernachlässigen und mit den Mittelwerten zu rechnen. Leider stellt sich aber heraus, daß der Beitrag der Schwankungen in gewissen Gebieten groß ist. Man erkennt dies aus der zweiten Gl. (86).

Setzt man nämlich

$$g(\zeta, \eta) = g(\eta) + \Delta g(\zeta, \eta), \quad \text{wobei} \tag{92}$$

$g(\eta)$ den Mittelwert über ζ von $g(\zeta, \eta)$ bedeutet:

$$g(\eta) = \overline{g(\zeta, \eta)}, \tag{93}$$

so folgt aus dem Zeitmittelwert der zweiten Gl. (86):

$$\frac{3}{8} \int_0^\infty dy\, f(y) \int_0^\infty d\eta' \, [g(\eta' - \eta) + g(\eta' + \eta)]\, g\left(\eta' y^{\frac{2}{3}}\right) =$$

$$= -\frac{3}{8} \int_0^\infty dy\, f(y) \int_0^\infty d\eta' \Big[\overline{\Delta g\left(\zeta - \frac{\eta'}{2}, \eta' - \eta\right) \Delta g\left(\left(\zeta + \frac{\eta - \eta'}{2}\right) y^{\frac{2}{3}}, \eta' y^{\frac{2}{3}}\right)}$$

$$+ \overline{\Delta g\left(\zeta - \frac{\eta'}{2}, \eta' + \eta\right) \Delta g\left(\left(\zeta - \frac{\eta + \eta'}{2}\right) y^{\frac{2}{3}}, \eta' y^{\frac{2}{3}}\right)} \Big]. \tag{94}$$

Diese Beziehung zeigt, daß die Schwankungen $\Delta g(\zeta, \eta)$ nicht immer klein sein können. Zwar verschwindet die linke Seite von (94) für $\eta = 0$; dies folgt aus der später noch zu besprechenden Beziehung

$$f\left(\frac{1}{y}\right) = -y^{4/3} f(y) \tag{95}$$

und bedeutet, daß das Spektrum $k^{-5/3}$ tatsächlich im Gleichgewicht ist; aber für größere $|\eta|$ nimmt die linke Seite erhebliche Werte an. Es ist daher auch zweifelhaft, ob man eine ausreichende Näherung bekommt, wenn man beim Übergang von (82) nach (83) nur die Mittelwerte vom Typus (63) berücksichtigt. Es ist mir aber nicht gelungen, hier die Näherung zu verbessern oder mehr als eine ganz grobe Abschätzung der Lösung von (86) zu gewinnen.

Man kann für eine solche Abschätzung vielleicht annehmen, daß für große Werte von η der erste Summand auf der rechten Seite von (94) viel größer ist, als der zweite. Denn im ersten wird das Integral über das Gebiet $\eta' \sim \eta$ hinweggeführt, was wahrscheinlich viel beträgt, während im zweiten für große η die Faktoren Δg im ganzen Integrationsgebiet schon stark abgefallen sind. Man kann also, wenigstens für große η, die Annahme versuchen, daß der zweite Summand auf der rechten Seite von (94) vernachlässigt werden darf. Dann wird in dieser Näherung aus dem Zeitmittelwert der ersten Gl. (86) unter Benutzung von (94):

$$\frac{dg(\eta)}{d\eta} = -\frac{3}{8} \int_0^\infty (fy\, dy) \int_0^\infty d\eta' \, g(\eta + \eta')\, g\left(\eta' y^{\frac{2}{3}}\right). \quad (\text{für } \eta \gg 0) \tag{96}$$

Zeitschrift für Physik *124*, 628–657 (1948)

Diese Gleichung kann man etwa in der Weise ausnutzen, daß man eine plausible Form für $g(\eta)$ annimmt, wobei man den Maßstab in der η-Richtung zuerst offenläßt und ihn nachträglich so bestimmt, daß die Gl. (96) für große η möglichst genau gilt. Damit wird man die Steilheit des Abfalls von $g(\eta)$ bei großen η einigermaßen richtig treffen, und eben diese Steilheit ist für den Wert von $\varkappa$ maßgebend.

Bei der praktischen Durchführung der Rechnung ist es zweckmäßig, an Stelle von y und $f(y)$ neue Variable

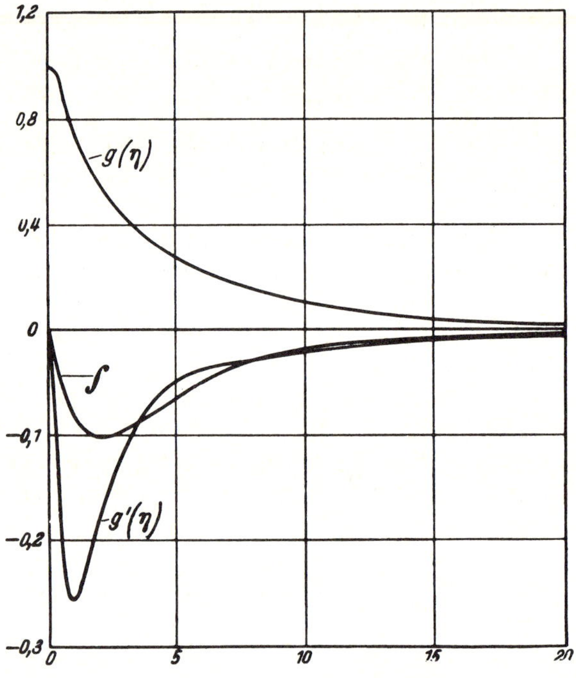

Abb. 4. Die Korrelationsfunktion $g(\eta)$.

$$s = y^{\frac{2}{3}}; \quad \varphi(s)\, ds = f(y)\, dy \tag{97}$$

einzuführen. Dann gilt, wie man aus (87) erkennt [vgl. auch (95)]:

$$\varphi\left(\frac{1}{s}\right) = -s\,\varphi(s). \tag{98}$$

Diese Gleichung beruht auf dem Umstand, daß die Energiedissipation von der Wellenzahl $\frac{1}{\alpha} k$ nach der Wellenzahl k bis auf einen durch die Ähnlichkeitstransformation bedingten Faktor übereinstimmt mit der von k nach αk. Ferner gilt dann in guter Näherung

$$\varphi(s) = \frac{32}{5} s (1 - s^3) \left(1 - \frac{2}{7} s^3\right) \quad \text{für } 0 \leq s \leq 1 \tag{99}$$

und für größere s kann man $\varphi(s)$ durch (98) auf den Bereich $0 \leq s \leq 1$ zurückführen.

Die Abb. 4 gibt eine plausible Kurve für $g(\eta)$, dazu die rechte Seite von (96) als $\int_0^\infty d\eta \ldots$ und $g'(\eta)$. Der Maßstab ist bereits so gewählt, daß die beiden letzten Kurven bei großen η übereinstimmen. Bei kleinen η gibt es dann erhebliche Unterschiede, aber dort kann auch die Gl. (96) nicht mehr richtig sein. Setzt man die so gewonnene Funktion $g(\eta)$ in (91) ein und vernachlässigt die Schwankungen, so ergibt sich für $\varkappa$:

$$\varkappa = \frac{1}{2\sqrt{3}} \int_0^1 ds\,(-\lg s)\,\varphi(s) \int_0^\infty d\eta\, g(\eta)\, g(\eta s) = 0{,}98. \tag{100}$$

Diese grobe Abschätzung gibt also die richtige Größenordnung für $\varkappa$, aber der exakte Wert mag gut um einen Faktor 2 von 0,98 verschieden sein. Die Rechnungen dieses Abschnitts haben also zwar nicht zu einer exakten Berechnung der Konstante $\varkappa$ geführt, wohl aber zu einer qualitativen mathematischen Darstellung der Vorgänge, auf denen die Energiedissipation beruht. Vielleicht wird es möglich sein, durch eine umfassende Diskussion der verschiedenen Versuche von Simmons, Dryden (l. c.), Prandtl[1] und anderen über das Spektrum und über das Abklingen der Turbulenz zu einer recht genauen experimentellen Bestimmung von $\varkappa$ zu gelangen.

[1] Prandtl, L.: Proc. VI. Intern. Congr. f. Appl. Mech. Cambridge (Mass.) 1938, S. 340.

Die Gestalt der Spiralnebel *.

Von

W. Heisenberg und C. F. von Weizsäcker.

Mit 2 Textabbildungen.

(Eingegangen am 13. Dezember 1947).

Es ist in der Astronomie seit längerem diskutiert worden, ob die Spiralstruktur extragalaktischer Nebel einer Verzerrung radialer Materie-

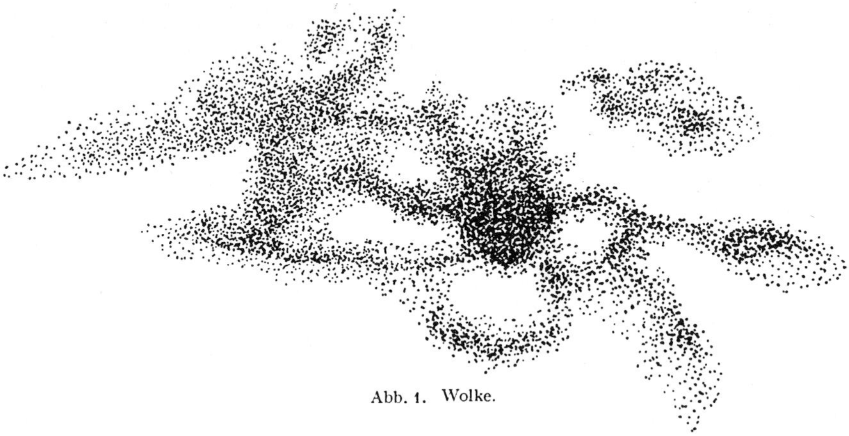

Abb. 1. Wolke.

verteilungen durch nichtstarre Rotation zu verdanken sei[1]. Diese Deutung kann nur als wahrscheinlich gelten, wenn sie zwei Fragen zu beantworten vermag:

1. Wie kommt es zur Bildung radialer, zweistrahliger Materieansammlungen, aus denen sich die zwei Spiralarme entwickeln?
2. Warum wickeln sich die Spiralarme nicht mit der Zeit wie eine Fadenrolle auf, d. h. wodurch werden sie rechtzeitig wieder aufgelöst?

Die zweite Frage ist in der oben zitierten Arbeit des einen von uns behandelt. Die Turbulenz dürfte zur Einstellung eines fluktuierenden Gleichgewichtes der Entstehung und Wiederauflösung der Arme genügen. Die vorliegende Notiz beschäftigt sich mit der ersten Frage.

---

* Prof. W. Lenz zum 60. Geburtstag gewidmet.

[1] Wilczynski, E. Z.: Ap. J. **4**, 97 (1896). — Astron. Z. **20**, 67 (1899). — Brown, E. W.: Observatory **51**, 277 (1928). Vgl. H. Vogt: Die Spiralnebel, S. 149, 1946. Von den gegenwärtigen Verfassern wurde diese Vermutung [C. F. v. Weizsäcker, Z. Astrophys. **24**, 181 (1947)] erneut ausgesprochen; inzwischen wurden uns die älteren Arbeiten bekannt.

Zunächst sei eine Veranschaulichung der Überlegungen gegeben. In der Abb. 1 ist eine gewöhnliche atmosphärische Wolke abgebildet, wie sie an einem Herbstabend vom Himmel abgezeichnet wurde; also ein beliebiges turbulentes Gebilde. Dieses Gebilde wurde rechnerisch um einen Punkt in seinem Inneren so gedreht, wie es dem dritten KEPLERschen Gesetz entspricht, und zwar um einen solchen Betrag, daß das äußerste rechte Ende um etwa 45° verschoben wurde, die inneren Teile entsprechend stärker. Das Ergebnis ist in der Abb. 2 abgebildet; es ist einem Spiralnebel nicht unähnlich. Die Zweizahl der Spiralarme ergibt sich bereits aus der nicht sehr ausgeprägten „Länglichkeit" der Wolke. Die Rotationsverzerrung zieht alle Gestalten in die Länge und erzeugt dadurch die typischen „schmalen" Spiralarme. Auf einem Kreis um den Schwerpunkt gemessen sind die Spiralarme noch immer ebenso „dick" wie die ursprüngliche Wolke. Die gewählte Wolke hatte außer der Länglichkeit gar keine für die entstandene Spiralstruktur wesentlichen Formmerkmale. Das heißt jede beliebige längliche Figur muß durch differentielle Rotation eine zweizählige Spiralstruktur ergeben.

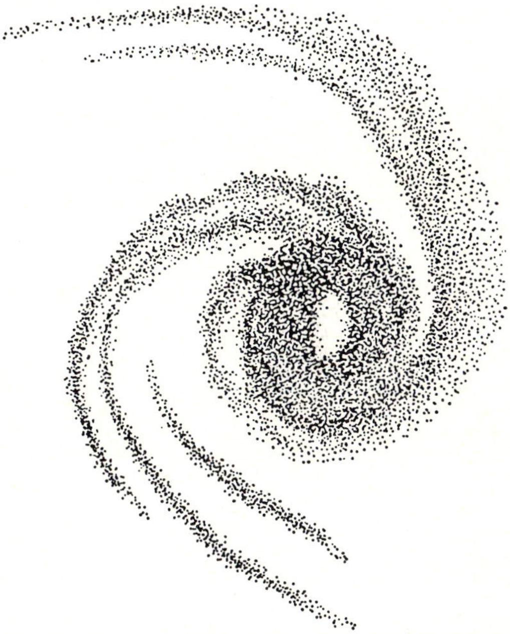

Abb. 2. Durch Rotation transformierte Wolke.

Man kann aber in der Begründung der Zweizahl radialer Arme weitergehen. Im Spiralnebel wirken Gravitation, Zentrifugalkraft und Reibung. Die Gravitation strebt, alle Materie auf möglichst engem Raum zusammenzudrängen. Die Reibung bringt die zufälligen Bewegungen mit der Zeit zum Erlöschen; daher strebt der Nebel dem Zustand tiefster Gravitationsenergie in der Tat zu. Aber die Reibung kann die Erhaltungssätze, insbesondere den Drehimpulssatz nicht aufheben. Daher kann nicht alle Materie des Nebels in den Schwerpunkt sinken. Hingegen ist es mit den Erhaltungssätzen verträglich, daß sich alle die Materie, die einen fest vorgegebenen Abstand vom Schwerpunkt hat, nicht auf einen Kreis verteilt, sondern in einem Punkt dieses Kreises sammelt. Hierdurch wird die Gravitationsenergie noch erniedrigt. Der mit dem Drehimpulssatz verträgliche energetisch tiefste Zustand des ganzen Nebels wird also der sein, in dem sich alle Materie auf einer Kurve (und zwar, wie man leicht sieht, auf einer Geraden) sammelt, die durch den Schwerpunkt geht. Der Schwerpunktsatz fordert die Symmetrie der Verteilung um den Schwerpunkt. Der tiefste Zustand ist also ein „Balken", in dessen Mitte der Schwerpunkt des Systems liegt.

Genau dieses Bild zeigen die „Balkenspiralen" in ihrem Inneren. Daß in ihnen der Balken nicht zur Spirale verzerrt ist, muß bedeuten, daß sie starr rotieren. Wo die Balken außen in Spiralarme übergehen, muß die starre Rotation aufhören. Es ist dynamisch zu erwarten, daß ein Nebel innen starr, außen unstarr rotiert; die bisherigen Beobachtungen[1] deuten in derselben Richtung. Wenn der Übergang zwischen beiden Rotationsweisen sehr unstetig ist, können sich die Spiralarme vielleicht wie ein Ring um den Balken aufwickeln und die bekannte Θ-Figur ergeben. Doch tendieren auch Gravitation und Reibung zur Ringbildung[2].

Die bisherigen Beobachtungen[1] zeigen, daß die sichtbaren Spiralstrukturen der großen Nebel in einem Gebiet liegen, in dem die Rotation der starren noch sehr nahekommt. Man kann daher nicht versuchen, die Form der Spiralarme genauer zu berechnen, solange das Rotationsgesetz nicht sehr viel genauer bekannt ist. Die Sonne, in deren Umgebung die Milchstraße sicher nicht starr rotiert, liegt also wohl schon außerhalb des Teils des Systems, der von außen eine stark hervortretende Spiralstruktur zeigen würde.

Auf zwei empirische Folgen der hier vertretenen Ansicht sei noch hingewiesen:

1. Die Spiralarme müssen mit der konvexen Seite voraus rotieren (höchstens nahe dem Innenrand eines Rings muß davon eine Ausnahme gelten, weil dort die Winkelgeschwindigkeit nach außen zunimmt[2]). Nach einer Arbeit von Hubble, die uns noch nicht zugänglich wurde, scheint dies der Fall zu sein.

2. Da an der Gestaltbildung der Spiralarme die Turbulenz Anteil hat, sollte die Spiralstruktur sich vorzugsweise in der interstellaren Materie und den jungen Sternen (B-Sterne) ausprägen. In der Tat scheint die Spiralstruktur im roten Licht nicht oder schwächer aufzutreten[3].

[1] Babcock, H. W.: Lick Obs. Bull. **19**, 41 (1939). — Mayall, N. V. u. L. H. Aller: Ap. J. **95**, 5 (1942).

[2] Weizsäcker, C. F. v.: Z. Naturforsch. (im Erscheinen).

[3] Seyfert: Ap. J. **100**, 1947 (1944).

REFERENCES

Bamford, C. H. & Dewar, M. J. S. 1948*a* *Trans. Faraday Soc.* (in the Press).
Bamford, C. H. & Dewar, M. J. S. 1948*b* *Proc. Roy. Soc.* A, **192**, 309.
Bolland, J. L. & Gee, G. 1946 *Trans. Faraday Soc.* **42**, 244.
Burnett, G. M. & Melville, H. W. 1947 *Proc. Roy. Soc.* A, **189**, 456.
Chapman, D. L., Briers, F. & Walters, E. 1926 *J. Chem. Soc.* p. 562.
Gregg, R. A. & Mayo, F. R. 1948 *Trans. Faraday Soc.* (in the Press).
Haden, W. L. & Rice, O. K. 1942 *J. Chem. Phys.* **10**, 445.
Leighton, P. A. & Noyes, W. A. 1941 *The photochemistry of gases*, p. 202. New York: Reinhold.
Rice, O. K. 1942 *J. Chem. Phys.* **10**, 440.
Rice, F. O. & Herzfeld, K. F. 1934 *J. Amer. Chem. Soc.* **56**, 284.
Steacie, E. W. R., Darwent, D. de B. & Trost, W. R. 1948 *Trans. Faraday Soc.* (in the Press).
Swain, C. G. & Bartlett, P. D. 1946 *J. Amer. Chem. Soc.* **68**, 2377.

On the theory of statistical and isotropic turbulence

BY W. HEISENBERG, *Göttingen*

(*Communicated by Sir Geoffrey Taylor, F.R.S.—Received* 15 *June* 1948)

The statistical theory of turbulence, initiated by Taylor (1935) and v. Kármán & Howarth (1938), has recently been developed so far that a satisfactory explanation of the spectral distribution of energy among the turbulent eddies can be given. In fact Kolmogoroff (1941 *a, b*) and independently Onsager (1945) and v. Weizsaecker (1948) have introduced a similarity hypothesis, which allows a determination of the spectrum for eddies with large Reynolds numbers, and the author (Heisenberg 1948) has extended these calculations to include those frequency components which have small Reynolds numbers. Since the distribution of energy among the largest eddies must be a geometrical and not a statistical problem, one may say that the statistical part of the spectrum is now well understood. Recently Batchelor & Townsend (1947, 1948 *a, b*) have studied the decay of turbulent motion caused by a mesh grating in a wind tunnel, and the following discussions will apply the statistical theory to this problem.

For the calculations the notation of Heisenberg (1948) will be used. If $\rho F(k)\,dk$ denotes the energy contained between the wave numbers k and $k+dk$, the following equation for the dissipation of energy was given (Heisenberg 1948, equation (13)):

$$S_k = \left\{\mu + \rho\kappa\int_k^\infty \sqrt{\left(\frac{F(k'')}{k''^3}\right)}\,dk''\right\}\int_0^k 2F(k')\,k'^2\,dk'. \tag{1}$$

S_k means the total loss of the energy of that part of the spectrum that is contained between $k = 0$ and k. The physical assumption underlying equation (1) is that for all eddies between 0 and k the action of the smaller eddies can be represented by an additional viscosity, since the way in which these smaller eddies transfer momentum

Proceedings of the Royal Society of London *A 195*, 402–406 (1948)

is similar to the action of ordinary friction (Prandtl 1945). This additional viscosity must depend upon the intensity $F(k)$ of the small eddies, i.e. on the part of the spectrum with large k. If one admits this hypothesis, then the value

$$\eta_k = \rho\kappa \int_k^\infty \sqrt{\left(\frac{F(k'')}{k''^3}\right)} dk'' \tag{2}$$

for the turbulent viscosity follows as the simplest assumption from purely dimensional reasons, since η_k must be the product of density, velocity and length. κ is a numerical constant, which from the experimental data seems to have a value of about 0·8. Equation (2) is of course just a simple mathematical expression of the similarity hypothesis introduced by Kolmogoroff, Onsager and v. Weizsaecker.

In Heisenberg (1948) the shape of $F(k)$ has been discussed for a stationary turbulent motion. It has been shown that for smaller k (i.e. for large Reynolds numbers of the eddies) $F(k) \sim k^{-\frac{5}{3}}$, while for large k (small Reynolds numbers of the eddies) $F(k) \sim k^{-7}$. Both results follow from the underlying similarity hypothesis. In fact, if one denotes by v_k the characteristic velocity of an eddy of diameter π/k, then for medium values of k ($k \gg k_0$, where k_0 characterizes the biggest eddies, but $v_k/k\nu \gg 1$) the product of the turbulent viscosity $\left(\sim \rho \frac{v_k}{k}\right)$ and the average value $\overline{\text{curl}^2 v}$ for this part of the spectrum ($\sim v_k^2 k^2$) must be constant, since S_k means just the total energy flow and must therefore be independent of k; that leads to

$$\frac{v_k}{k} v_k^2 k^2 = \text{const.}, \quad v_k \sim k^{-\frac{1}{3}}, \tag{3}$$

which is equivalent to $F(k) \sim k^{-\frac{5}{3}}$. For very large k, on the other hand, ($v_k/k\nu \ll 1$) the total flow of energy from all larger eddies into these small ones is given by $\sim \rho \frac{v_k}{k} \overline{\text{curl}^2 v}$, where the average $\overline{\text{curl}^2 v}$ can now be taken practically over the whole spectrum and does in this approximation not depend upon k. This energy must equal the energy that is consumed in the small eddies by ordinary viscosity, i.e. $\sim \mu v_k^2 k^2$. Therefore

$$\rho \frac{v_k}{k} \overline{\text{curl}^2 v} \sim \mu v_k^2 k^2, \quad v_k \sim k^{-3}, \tag{4}$$

which is equivalent to $F(k) \sim k^{-7}$.

If the turbulent motion is not maintained by means of external forces, it must decay and the rate of decay can be directly derived from (1), which now can be put into the form ($\nu = \mu/\rho$)

$$\frac{\partial}{\partial t} \int_0^k F(k,t)\, dk = -\left\{\nu + \kappa \int_k^\infty \sqrt{\left(\frac{F(k'',t)}{k''^3}\right)} dk''\right\} \int_0^k 2F(k',t)\, k'^2 dk'. \tag{5}$$

If the initial spectral distribution $F(k, 0)$ is given, then (5) determines the spectrum for any later time. Here again we may look for solutions with simple similarity properties. One sees at once, that a special group of solutions can be given by assuming

$$F(k,t) = \frac{1}{\sqrt{t}} f(k\sqrt{t}). \tag{6}$$

Equation (5) then goes over into

$$\int_0^x f(x)\,dx - \tfrac{1}{2}xf(x) = \left\{\nu + \kappa \int_x^\infty \sqrt{\left(\frac{f(x'')}{x''^3}\right)}\,dx''\right\} \int_0^x 2f(x')\,x'^2\,dx'. \tag{7}$$

The physical meaning of the similarity hypothesis (6) can be most easily expressed by saying that the total spectrum is determined essentially by one length, which we may call π/k_0, the diameter of the 'largest' eddies, and their velocity v_0 only. Then it follows from dimensional reasoning that

$$\left.\begin{aligned} &\frac{d}{dt}\left(\frac{1}{k_0}\right) \sim v_0 \quad \text{and} \quad \frac{d}{dt}\left(\frac{1}{v_0}\right) \sim k_0, \\ \text{which gives at once}\quad & k_0 \sim v_0 \sim 1/\sqrt{t}. \end{aligned}\right\} \tag{8}$$

It is important to note that the similarity (6) is not destroyed by the presence of the viscosity μ, and that the Reynolds number of the turbulent motion stays constant throughout the whole decay, since this number is essentially $\rho v_0/\mu k_0$, $\sim$ const.

The solution of (7) will be discussed first in the case $\nu = 0$, i.e. for very large Reynolds numbers. Then $f(x)$ seems to be uniquely determined except again for a similarity transformation; for if $f = g(x)$ is a solution of (7) for $\nu = 0$, then $f = \alpha^3 g(\alpha x)$ is also a solution.

As essential properties of $f(x)$ one finds from (7):

$$\left.\begin{aligned} &f(x) \sim \text{const.}\,x \quad \text{for very small } x, \\ \text{and}\quad &f(x) \sim \text{const.}\,x^{-\frac{5}{3}} \quad \text{for very large } x. \end{aligned}\right\} \tag{9}$$

The second property is of course just the result of the Kolmogoroff hypothesis. If one puts $\kappa = 0{\cdot}8$, then an approximate solution of (7) is given by

$$f(x) = \frac{x\alpha^4}{\left[1 + \left(\frac{x\alpha}{1{\cdot}1}\right)^4\right]^{\frac{4}{3}}}. \tag{10}$$

This solution is correct for small and large values of x and is a reasonably good approximation for intermediate values; α is an arbitrary constant of integration.

If one characterizes the initial state $t = t_0$ by two constants v_0 and k_0 one gets finally from (6) and (10)

$$F(k,t) = \frac{v_0^2}{k_0^2}\,\frac{k}{\left[1 + \left(\frac{k}{1{\cdot}1}\right)^4\left(\frac{tv_0}{k_0}\right)^4\right]^{\frac{4}{3}}}. \tag{11}$$

The reason for $F(k,t)$ being proportional to k at small values of k is that energy dissipation in the very big eddies is very small, and if there is to be similarity in the sense of equation (6), it can only be by a linear dependence of $F(k,t)$ on k.

In the case of finite but small values of ν the spectrum (11) can only be valid for not too large values of k and will then be continued by the k^{-7} law as discussed in Heisenberg (1948).

In the opposite case of very large values of ν, i.e. very small Reynolds numbers of the total turbulent motion one can again derive an approximate solution of (7).

This solution varies for small values of x as $x\,e^{-2\nu x^2}$ and for large values according to the x^{-7} law. To a reasonable approximation it is given by ($\beta \ll 1$),

$$F(k,t) = \beta^2\nu^2 k\left[e^{-k^2\nu t} + \frac{\kappa\beta}{4}\frac{k^2\nu t}{(1+2^{\frac{1}{2}}k^2\nu t)^3}\right]^2, \tag{12}$$

where $\beta \ll 1$ measures essentially the Reynolds number of the motion.

We now have to turn to the question whether the actual turbulent motion should during decay approach the simple solution (11). This solution describes a state of equilibrium. The smaller eddies are produced by the bigger ones and lose their energy to still smaller ones. The energy is derived always from the largest eddies. One may describe the similarity hypotheses (6) and (8) also by saying that the biggest eddies grow partly larger and partly smaller. When they become smaller they get very quickly into the general Kolmogoroff equilibrium. When they become larger they must have approximately the same angular momentum as they had before, since even if new parts of the fluid are included in the larger eddy, on the average no angular momentum will have been added. So v_0/k_0 for the largest eddies will approximately stay constant, which again leads to (8).

Now the question arises whether the actual turbulent motion will approach this equilibrium. The experiments of Batchelor & Townsend (1948*a*) show that after a short time the decay actually takes place according to (8), the energy decreasing as $1/t$, but after a certain period the decay becomes faster, the Reynolds number of the turbulence decreases, and finally the inertia terms play no appreciable role in the turbulent motion, as has been studied in detail by Batchelor & Townsend (1948*b*) and the energy decreases at $t^{-\frac{5}{2}}$.

The actual turbulent motion will certainly for large k values, i.e. the small eddies, very soon come into a state of equilibrium. The real problem arises from the region of the large eddies. In Heisenberg (1948) it had been assumed that the size of the largest eddies should be approximately constant; this is certainly not true and would never follow from any reasonable spectrum according to (5). But neither is it possible that the real spectrum ultimately approaches (11) for small k values, because the real spectrum must approach zero for very small values of k much faster than (11). Batchelor (1948) has shown that for very small k, $F(k)$ should vary as

$$F(k) \sim k^4 \text{ const.} \tag{13}$$

Usually the real spectrum will, in a certain range of not too small k be well represented by a function like (11). But there will be a region of very small k in which $F(k)$ from the beginning is much smaller than (11); the very biggest eddies can never contain as much energy as indicated by (11). Therefore the linear decay of the energy ($v^2 \sim 1/t$) will take place until finally the maximum of the curve (11) is shifted to values of k which approach this region. Then the biggest eddies no longer contain enough energy to supply the $1/t$ law, the Reynolds number of the turbulent motion drops and finally gets so low that the inertia terms become unimportant, as has been shown by Batchelor & Townsend (1948*b*).

The point at which the $1/t$ law is replaced by some faster decay depends on the initial conditions, especially on the amount of energy contained in Fourier components belonging to wave-lengths that are larger than the mesh width of the grating.

This whole description applies even if the intial Reynolds number of the turbulent motion is not large, since the similarity hypothesis (6) leads to solutions of (5) for any value of the Reynolds number. The essential condition for the occurrence of a $1/t$ law for the decay is the existence of a rather large range of the initial spectrum (or its creation in the very first establishment of turbulent equilibrium) in which $F(k) \sim k$, and the $1/t$ law is limited only by the breakdown of this law $F(k) \sim k$ at very small values of k.

A few remarks may be added with regard to the physical picture of turbulence presented in the recent papers. In the earlier years one thought that turbulence was caused by viscosity. This seemed to be true since without viscosity the liquid could theoretically perform all the classical laminar motions, where the liquid glides along the walls; it is only through the viscosity that rotational motions are produced near the walls. At present we know that it is almost the other way round. The liquid without friction is a system with an infinite number of degrees of freedom. It is extremely improbable that only those few degrees of freedom which a laminar motion represents should be excited. As soon as one puts energy into a liquid without friction, this energy will be distributed among all degrees of freedom, and what finally results is a certain equilibrium distribution, corresponding to the Maxwellian distribution in gases, and represented by the spectrum found by Kolmogoroff, Onsager and v. Weizsaecker. It is the viscosity that reduces the number of degrees of freedom, since it damps very quickly all motions in the very small eddies. Therefore only through viscosity is a laminar motion at all possible. Turbulence is an essentially statistical problem of the same type as one meets in statistical mechanics, since it is the problem of distribution of energy among a very large number of degrees of freedom. Just as in Maxwell theory this problem can be solved without going into any details of the mechanical motion, so it can be solved here by simple considerations of similarity. Equation (1) is just a very simple way of describing these similarity properties.

References

Batchelor G. K. 1948 *Proc. Roy. Soc.* A, **195** (in the Press).
Batchelor G. K. & Townsend A. A. 1947 *Proc. Roy. Soc.* A, **190**, 534.
Batchelor G. K. & Townsend A. A. 1948*a* *Proc. Roy. Soc.* A, **193**, 539.
Batchelor G. K. & Townsend, A. A. 1948*b* *Proc. Roy. Soc.* A, **194**, 527.
Heisenberg, W. 1948 *Z. Phys.* **124**, 628.
von Kármán, T. & Howarth, L. 1938 *Proc. Roy. Soc.* A, **164**, 192.
Kolmogoroff, A. N. 1941*a* *C.R. Acad. Sci. U.R.S.S.* **30**, 301.
Kolmogoroff, A. N. 1941*b* *C.R. Acad. Sci. U.R.S.S.* **32**, 16.
Onsager, L. 1945 *Phys. Rev.* **68**, 286 (abstract only).
Prandtl, L. 1945 *Nachr. Ges. Wiss. Göttingen*, p. 6.
Taylor, G. I. 1935 *Proc. Roy. Soc.* A, **151**, 421.
von Weizsaecker, C. F. 1948 *Z. Phys.* **124**, 614.

so wird nach (9)

$$\frac{d\chi(x,\sigma)}{dx_\nu} = e^{iU(\sigma)} \frac{\partial\varphi(x)}{\partial x_\nu} e^{-iU(\sigma)} - i \int_\sigma d\sigma_\nu' \left[\varphi(x), \frac{\partial U(\sigma)}{\partial\sigma(x')} \right],$$

wo das Flächenintegral nach (16) und (7) verschwindet, und

$$\frac{d^2\chi(x,\sigma)}{dx_\nu^2} = e^{iU(\sigma)} \frac{\partial^2\varphi(x)}{\partial x_\nu^2} e^{-iU(\sigma)} - i \int_\sigma d\sigma_\nu' \left[\frac{\partial\varphi(x)}{\partial x_\nu}, \frac{\partial U(\sigma)}{\partial\sigma(x')} \right],$$

woraus mit Hilfe von (6) und (7) wie früher (vgl. Abschn. 2) folgt:

$$\frac{d^2\chi(x,\sigma)}{dx_\nu^2} = (\mu^2 + \gamma)\,\chi(x,\sigma),$$

entsprechend der Ruhmasse $\sqrt{\mu^2 + \gamma}\,\hbar/c$.

Anstatt den gemäß (15) transformierten Operator χ einzuführen, kann man natürlich auch φ beibehalten und die Schrödinger-Gleichung entsprechend transformieren. Setzen wir nämlich in (10), (11), (12)

$$S(\sigma) = e^{iU(\sigma)},$$

so daß

$$\Psi^*(\sigma)\,\chi(x,\sigma)\,\Psi(\sigma) = \Psi'^*(\sigma)\,\varphi(x)\,\Psi'(\sigma),$$

so wird

$$\boldsymbol{H}'(x,\sigma) = -i\hbar c\, S^{-1}(\sigma) \frac{\partial \dot{S}(\sigma)}{\partial\sigma(x)} = \hbar c \frac{\partial U(\sigma)}{\partial\sigma(x)} + \ldots = 1/2 \cdot \gamma\,\varphi^2(x) + \ldots,$$

und unter Verwendung dieser Hamilton-Funktion (statt $\boldsymbol{H}$) in (5) folgt wie oben

$$\frac{d^2\varphi(x)}{dx_\nu^2} = (\mu^2 + \gamma)\,\varphi(x).$$

Daß diese Betrachtungen auf Teilchen mit Spin übertragen werden können, bedarf wohl keiner Erläuterung.

Bemerkungen zum Turbulenzproblem

Von Werner Heisenberg

Aus dem Max-Planck-Institut für Physik, Göttingen

(Z. Naturforschg. **3a**, 434—437 [1948]; eingegangen am 15. Juni 1948)

Die verschiedenen physikalischen Gesichtspunkte, die einerseits den früheren Stabilitätsuntersuchungen an laminaren Strömungen, andererseits der neueren statistischen Turbulenztheorie zugrunde liegen, werden miteinander verglichen und kritisch besprochen.

In früheren Jahren hat Sommerfeld mit seinen Schülern häufig das Turbulenzproblem diskutiert, das lange Zeit als das ungelöste Grundproblem der neueren Hydrodynamik galt. Inzwischen hat die statistische Theorie der Turbulenz, die vor etwa zehn Jahren von Taylor[1] und v. Kármán[2] begründet worden ist, so entscheidende Fortschritte ermöglicht, daß das Turbulenzproblem in seinem physikalischen Kern wohl als gelöst angesehen werden kann. In den folgenden Zeilen sollen verschiedene physikalische Gesichtspunkte, die im Laufe der Zeit auf das Turbulenzproblem angewandt worden sind, noch einmal kritisch erörtert werden.

Vor etwa vierzig Jahren wurde von Sommerfeld[3] zunächst die Frage behandelt: Woher kommt es, daß gewisse laminare Bewegungen, die bis zu beliebigen Geschwindigkeiten Lösungen der hydrodynamischen Gleichungen sind, von bestimmten Werten der Reynoldsschen Zahl ab instabil werden? Diese Fragestellung ging von der Auffassung aus, daß in einer reibungslosen Flüssigkeit alle die laminaren Bewegungen auftreten könnten, die man aus der klassischen Hydrodynamik kennt, daß aber offenbar die Rei-

[1] G. J. Taylor, Proc. Roy. Soc. [London] Ser. A **151**, 421 [1935].

[2] Th. v. Kármán u. L. Howarth, Proc. Roy. Soc. [London] Ser. A **164**, 192 [1938].

[3] A. Sommerfeld, Int. Math. Kongr. Rom 1908, Vol. III, S. 116.

bung diese Bewegungen unter gewissen Voraussetzungen instabil macht. Die Annahme, daß der Übergang von der laminaren in die turbulente Strömungsform erst durch die Reibung zustande kommt, wurde etwa durch die Helmholtzschen Wirbelsätze begründet. Eine turbulente Bewegung ist ja eine Bewegung, in der überall in der Flüssigkeit größere und kleinere Wirbel verteilt sind. Eine solche Bewegung kann in einer reibungsfreien Flüssigkeit nicht entstehen, wenn nicht schon von Anfang an in weiten Bereichen Wirbelung vorhanden ist.

Die Stabilitätsuntersuchungen der folgenden Jahre ergaben für einige laminare Strömungen Stabilität[4], für bestimmte Klassen von Strömungen aber in der Tat von hinreichend hohen Reynoldsschen Zahlen ab Instabilität[5]. Die Art, wie bei wachsender Reynoldsscher Zahl neue Lösungstypen und damit Instabilitäten der Ausgangslösung entstehen können, ist später von Burgers[6] an einem vereinfachten mathematischen Modell studiert worden.

Die neuere statistische Theorie der Turbulenz geht demgegenüber von einer ganz anderen Fragestellung aus. Sie interessiert sich für die voll ausgebildete Turbulenz. Die Flüssigkeit ist ein mechanisches System von außerordentlich vielen Freiheitsgraden. Da es von vornherein viel wahrscheinlicher ist, daß die Energie irgendwie über alle Freiheitsgrade verteilt ist, als speziell auf eine bestimmte laminare Bewegungsform, wird, ähnlich wie in der statistischen Mechanik, nach den allgemeinen Eigenschaften solcher Verteilungen gefragt werden müssen. Offenbar ist es bei einem System von so vielen Freiheitsgraden von vornherein zwecklos, die feinsten Details des Strömungszustandes untersuchen zu wollen, da der Zustand dauernd wechselt und die Einzelheiten praktisch unwichtig sind. Man muß vielmehr zunächst festlegen, für welche Eigenschaften einer turbulenten Strömung man sich noch interessieren will. Hier wird zweckmäßig die gleiche Frage gestellt wie in der Optik der kontinuierlichen Spektren. Ebenso wie man bei einem kontinuierlichen Spektrum elektromagnetischer Wellen im allgemeinen nur nach der Verteilung der Energie über die verschiedenen Wellenlängen fragt, nicht aber nach den Phasenbeziehungen, so denkt man sich auch die Geschwindigkeit in der turbulenten Strömung als Funktion des Ortes in eine Fourier-Reihe entwickelt und bezeichnet etwa mit $\varrho F(k)dk$ die Energie (ϱ = Massendichte), die zwischen den Wellenzahlen k und $k + dk$ in der Volumeneinheit vorhanden ist. Man fragt also nach der Gestalt der Verteilung $F(k)$, d. h. nach dem „Spektrum" der Turbulenz, verzichtet aber auf die Kenntnis der Phasenbeziehungen, die ja doch dauernd wechseln. Auch die statistischen Korrelationen zwischen den Geschwindigkeiten an zwei verschiedenen Punkten in der Flüssigkeit sind im wesentlichen durch das Spektrum festgelegt.

Ähnlich wie es in der Optik nun eine bestimmte Verteilung gibt, die im thermodynamischen Gleichgewicht stets auftritt, nämlich die Plancksche Strahlungsverteilung, so muß es auch eine bestimmte Gleichgewichtsverteilung der Energie auf die Wirbel verschiedener Größe geben. Eine solche Gleichgewichtsverteilung kann z. B. praktisch dadurch erreicht werden, daß man die Flüssigkeit dauernd umrührt. Zwar wird dann bei den größten Wirbeln keine statistische Verteilung eintreten, da ihre Form und Geschwindigkeit durch die Art des Rührens bestimmt wird. Aber die großen Wirbel erzeugen kleinere oder lösen sich in solche auf, und für kürzere Wellenlängen wird sich schnell ein stationärer Betrieb einstellen, bei dem Energie stets von größeren in kleinere Wirbel transportiert und schließlich durch Reibung in Wärme verwandelt wird.

Das Spektrum dieser Gleichgewichtsverteilung ist im Grenzfall sehr hoher Reynoldsscher Zahlen unabhängig durch Kolmogoroff[7], Onsager[8] und v. Weizsäcker[9] abgeleitet worden; eine Untersuchung von Prandtl[10] enthält ebenfalls die dieser Ableitung zugrunde liegenden

[4] L. Hopf, Ann. Physik **43**, 1 [1914]; F. Noether, Z. angew. Math. Mechan. **6**, 232 [1926]; Th. Sexl, Ann. Physik **83**, 835 [1927]; **84**, 807 [1927]; **87**, 570 [1928].

[5] W. Heisenberg, Ann. Physik **74**, 577 [1924]; W. Tollmien, Nachr. Ges. Wiss. Göttingen math.-physik. Kl. **1929**; N. Schlichting, Ann. Physik **14**, 905 [1932]. Neuerdings ist das Stabilitätsproblem nochmals zusammenfassend behandelt worden von C. C. Lin, Proc. Nat. Acad. Sci. USA. **30**, 316 [1944] und Quart. J. appl. Math. **3**, 117, 218, 277 [1945—1946].

[6] J. M. Burgers, Kon. Akad. Wetensch. Amsterdam, Proc. **17**, 1 [1939].

[7] A. N. Kolmogoroff, C. R. [Doklady] Akad. Sci. URSS. **30**, 301 [1941]; **32**, 16 [1941].

[8] L. Onsager, Physic. Rev. **68**, 286 [1945].

[9] C. F. v. Weizsäcker, Z. Physik **124**, 614 [1942].

[10] H. Prandtl, Nachr. Ges. Wiss. Göttingen math.-physik. Kl. **1945**, S. 6.

Zeitschrift für Naturforschung 3a, 434–437 (1948)

Ähnlichkeitsbetrachtungen; die Fortsetzung des Spektrums in das Gebiet der kleinsten Wirbel (deren Reynolds-Zahlen nicht mehr als groß gelten können) wurde vom Verfasser berechnet[11].

Diese Berechnungen beruhen auf einem Gleichgewichtsansatz, der formal und physikalisch als Analogon zum Boltzmannschen Stoßzahlansatz aufgefaßt werden kann. Man fragt nämlich nach der Energiemenge S_k, die pro Zeiteinheit aus dem Wellenzahlgebiet zwischen 0 und k in das Gebiet größerer Wellenzahlen (also kleinerer Wirbel) transportiert wird. Dieser Transport geschieht durch die Reibung. Dabei tritt jedoch gleichberechtigt neben die gewöhnliche Viskosität der Flüssigkeit noch die turbulente Viskosität, also die Übertragung von Impuls durch die kleineren turbulenten Wirbelbewegungen; denn diese kleinen unregelmäßigen Bewegungen übertragen bei Vorhandensein eines mittleren Geschwindigkeitsgradienten in ähnlicher Weise Impuls wie die molekularen Bewegungen und führen daher ebenso zu einer Art von Zähigkeit. Man kann diese turbulente Zähigkeit der Größenordnung nach gleich $\varrho l_k v_k$ setzen, wobei l_k den Prandtlschen Mischungsweg, v_k die Geschwindigkeit der hierfür in Betracht kommenden Wirbel von der Wellenzahl k darstellt. Drückt man die turbulente Zähigkeit durch das Spektrum $F(k)$ für alle Wellenzahlen zwischen k und ∞ aus, so erhält man aus Dimensionsbetrachtungen als einfachsten Ansatz für die turbulente Zähigkeit:

$$\varkappa \varrho \int_k^\infty \sqrt{\frac{F(k')}{k'^3}}\, dk' , \qquad (1)$$

wobei $\varkappa$ eine dimensionslose Zahl ist, die empirisch etwa den Wert 0,8 hat. Bedeutet μ die molekulare Zähigkeit, so wird schließlich unter Anwendung bekannter Formeln der Hydrodynamik:

$$S_k = \left(\mu + \varkappa \varrho \int_k^\infty \sqrt{\frac{F(k')}{k'^3}}\, dk'\right) \int_0^k 2\, F(k'')\, k''^2\, dk'' . \qquad (2)$$

Beim stationären Betrieb muß S_k für größere k von k unabhängig sein, wenn die Energiezufuhr bei kleinen k-Werten (d. h. durch große Wirbel) erfolgt. Aus dieser Bedingung ergibt sich die Gestalt des Spektrums, nämlich $F(k) \sim k^{-5/3}$ für Wirbel mit großen Reynoldsschen Zahlen, $F(k) \sim k^{-7}$ für Wirbel mit kleinen Reynoldsschen Zahlen, d. h. für das kurzwellige Ende des Spektrums[11]. Auch nichtstationäre Vorgänge, wie z. B. das zeitliche Abklingen einer turbulenten Bewegung, können nach Gl. (2) diskutiert werden, wenn es sich um eine voll ausgebildete turbulente Bewegung handelt. Wenn keinerlei Energie von außen zugeführt wird, so folgt ja aus (2) sofort

$$\frac{d}{dt}\left[\varrho \int_0^k F(k')\, dk'\right] = -\left(\mu + \varkappa \varrho \int_k^\infty \sqrt{\frac{F(k')}{k'^3}}\, dk'\right) \int_0^k 2\, F(k'')\, k''^2\, dk'' . \qquad (3)$$

Aus dieser Gleichung kann für die abklingende turbulente Bewegung die Form des Spektrums und die Art des Abklingens ermittelt werden[12].

Die statistische Theorie der Turbulenz geht also grundsätzlich von der Annahme aus, daß die Bewegung auch einer nicht reibenden Flüssigkeit „im allgemeinen" turbulent sei, denn die nicht reibende Flüssigkeit ist ein System von vielen Freiheitsgraden. Die laminare Bewegung ist ein ebenso unwahrscheinlicher Spezialfall wie die geordnete Bewegung der Moleküle in einem Gas. Erst eine hinreichend große Viskosität begünstigt laminare Bewegungen, da sie bei kleinen Reynoldsschen Zahlen die Wirbelbildung unterdrückt, die Bewegung also wenigstens im Bereich der kleinen Wellenlängen glättet. Während bei den Stabilitätsuntersuchungen der laminaren Bewegungen die Reibung als die Kraft erschien, die diese Bewegungen labil macht und die Turbulenz erzeugt, erscheint jetzt die Reibung als die Kraft, die Wirbelungen unterdrückt und die Bewegungen laminar zu machen sucht.

Der Widerspruch zwischen diesen beiden Behauptungen ist nur scheinbar. Denn im ersten Fall handelt es sich nicht um einen Gleichgewichtszustand, sondern um die Geschwindigkeit, mit der ein nicht stabiler Vorgang in einen Gleichgewichtszustand übergeht (in der Sprache der Chemie um eine Reaktionsgeschwindigkeit), im zweiten Fall wird nach der Gleichgewichtsver-

11 W. Heisenberg, Z. Physik **124**, 628 [1948].
12 W. Heisenberg, Proc. Roy. Soc. [London] Ser. A, im Erscheinen.

teilung selbst gefragt. In der nicht reibenden Flüssigkeit ist zwar die laminare Bewegung ein unwahrscheinlicher Spezialfall, aber die Geschwindigkeit des Übergangs in den turbulenten Gleichgewichtszustand ist zunächst verschwindend klein; sie würde von Null verschieden nur, wenn etwa schon vorher viele kleine Wirbel vorhanden sind, d. h. wenn durch Turbulenz im kleinen eine turbulente Reibung hervorgerufen wird. Aus dieser letzten Überlegung geht hervor, daß es im Grunde nie vorkommen kann, daß ein Strömungszustand, bei dem die Flüssigkeit an den Wänden haftet und der ohne Reibung stabil ist, bei Einführung der Reibung labil wird. Denn die Reibung ließe sich ebensogut durch eine von kleinen Wirbeln hervorgerufene turbulente Reibung ersetzen; nur kann man diese Art der Labilität nicht mit der üblichen Methode der kleinen Schwingungen behandeln. Denn die Zeit, in der eine Schwingung anwächst, hängt dann von der Schwingungsamplitude der kleinen Wirbel ab und würde bei verschwindender Amplitude unendlich groß. Zu ihrer Berechnung müßte man also auch nichtlineare Glieder in den Schwingungen mitberücksichtigen, was freilich die relative mathematische Einfachheit der Methode der kleinen Schwingungen zerstören würde. Man erkennt aus dieser Überlegung, daß die Methode der kleinen Schwingungen bei hydrodynamischen Stabilitätsfragen grundsätzlich nicht ausreicht, was zum Teil die physikalisch etwas widersprechenden Ergebnisse früherer Untersuchungen erklären mag.

Im Gleichgewicht selbst dagegen unterdrückt die Reibung die kleinen Wirbel und glättet die Bewegung; sie verschiebt das Gleichgewicht zu Ungunsten der kleinen turbulenten Bewegungen. Wenn man die voll ausgebildete turbulente Strömung in dieser Weise als Gleichgewichtsvorgang ansieht und etwa mit der Maxwellschen Geschwindigkeitsverteilung vergleicht, so muß man allerdings darauf achten, daß dieser Vergleich in einem Punkte nicht zutreffen kann. Während sich das thermodynamische Gleichgewicht stets von selbst in einem abgeschlossenen System einstellt, wenn man nur hinreichend lange wartet, benötigt die turbulente Gleichgewichtsströmung eine dauernde Energiezufuhr, da ja durch Reibung auch dauernd Energie in Wärme verwandelt und abgeführt wird. Daher zerfällt das Spektrum der turbulenten Bewegung meist in drei physikalisch unterschiedene Wellenlängenbereiche. Die Energie wird im allgemeinen im Gebiet der langen Wellen auf die Flüssigkeit übertragen, etwa durch einen relativ zur Flüssigkeit bewegten Körper; in diesem langwelligen Gebiet herrscht natürlich kein Gleichgewicht, die Form und Häufigkeit der entstehenden Wirbel hängt von dem speziellen Vorgang ab, der die Energie überträgt, die Berechnung der Strömung ist sicher kein rein statistisches Problem. Dann kommt ein Bereich kleinerer Wirbel, die von den größeren erzeugt werden und die ihre Energie wieder in kleinere Wirbel verteilen; in diesem Bereich kann der Idealfall des von Kolmogoroff[7], Onsager[8] und v. Weizsäcker[9] studierten statistischen Gleichgewichts in guter Näherung verwirklicht sein $[F(k) \sim k^{-5/3}]$. Nach kurzen Wellenlängen schließt sich endlich ein dritter Gleichgewichtsbereich an[11], in dem die Energie der Wirbel zum größten Teil durch die Reibungswärme verzehrt wird $[F(k) \sim k^{-7}]$.

In vielen Fällen wird es natürlich auch nicht zur Ausbildung des Gleichgewichts kommen, oft wird sich an den ersten Bereich unmittelbar der dritte anschließen, in dem die Energie in Wärme verwandelt wird, ohne daß sie vorher in erheblichem Ausmaß auf alle Freiheitsgrade verteilt würde. Die Verteilung der Energie auf die anderen Wellenlängenbereiche geschieht ja stets durch die nichtlinearen, die Trägheit darstellenden Glieder der hydrodynamischen Differentialgleichungen; die Geschwindigkeiten müssen also einen gewissen Mindestbetrag erreichen, damit die Trägheitsglieder überhaupt eine erhebliche Rolle spielen und die Energie in andere Wellenlängenbereiche übertragen können. Die mathematische Behandlung der Zwischenfälle turbulenter Bewegung, bei denen es noch nicht zur Ausbildung des richtigen Gleichgewichts kommt, ist natürlich besonders kompliziert, ähnlich wie in der kinetischen Gastheorie. Nur die beiden Grenzfälle der streng geordneten (laminaren) und der völlig ungeordneten (statistisch turbulenten) Strömung können einfach mathematisch beschrieben werden. Daß die turbulente Bewegung die Regel, die laminare die Ausnahme ist, folgt einfach aus der Tatsache, daß eine Flüssigkeit sehr viele Freiheitsgrade besitzt.

Group 2

On Atomic and Molecular Structure (1922–1925)

On Atomic and Molecular Structure (1922–1925)

An Annotation by F. Hund, Göttingen*

In the years from 1925 to 1927 two fundamental insights were established in physics. One was the recognition of the degree to which the concepts of classical physics had to be restricted in the atomic domain because of the indivisibility of the quantum of action h. The second was the observation that, in the case of electrons as indistinguishable particles, Pauli's exclusion principle and Fermi's statistics applies, and that the electrons possess a spin of magnitude $h/4\pi$. These two insights belong to two different levels of the description of physical reality, to quantum mechanics and to the theory of elementary particles. The corresponding phenomena were, however, closely intermeshed in the spectroscopic data. Accordingly, prior to 1925 quantum theory had to struggle with two unresolved difficulties, and it is thus understandable that, in the process, wrong paths were taken.

This early phase of the history of quantum theory, which may be denoted by the title of A. Sommerfeld's book: *Atomic Structure and Spectral Lines* [1], was ushered by N. Bohr in 1913 with his theory of the hydrogen spectrum. The latter already contained the germ of the correspondence principle, according to which the frequencies and intensities of radiation emitted from atoms correspond to the frequencies and intensities derived from a classical model of the atom and become identical in the limit of large quantum numbers. Thus, the merging of atomic dynamics, as yet unknown, into the already familiar classical mechanics became conceivable. With these notions Sommerfeld and Bohr succeeded in explaining the basic pattern of spectral lines. The starting model consisted of an "external electron" moving in the electric field of the "atomic core", which was assumed to be centrally symmetric. To the frequencies $\tau\nu_1 + \nu_2$ of this motion corresponded the transitions of two quantum numbers, n and k, by $\Delta n = \tau$ and $\Delta k = \pm 1$. At that time, $k = 1, 2, 3, \ldots$ counted the terms s, p, d, ... (today, $l = 0, 1, 2, \ldots$). The fact that there were sometimes doublets and sometimes triplets was understood only incompletely. A particular challenge to the theory moreover, was that these terms exhibited an anomalous Zeeman effect.

For a better understanding of Heisenberg's papers on atomic structure, the current interpretation of "normal" multiplets and their Zeeman effects might be mentioned. In the initial model (see Fig. 1), the total angular momentum $L = 0, 1, 2, \ldots$ (in units of $\hbar = h/2\pi$) and the total spin angular momentum S (half integral for an odd, integral for an even number of external or series electrons) combine as vectors to the resultant J. Since S provides twice the amount to

* The English translation was prepared by H. Rechenberg. (Editors)

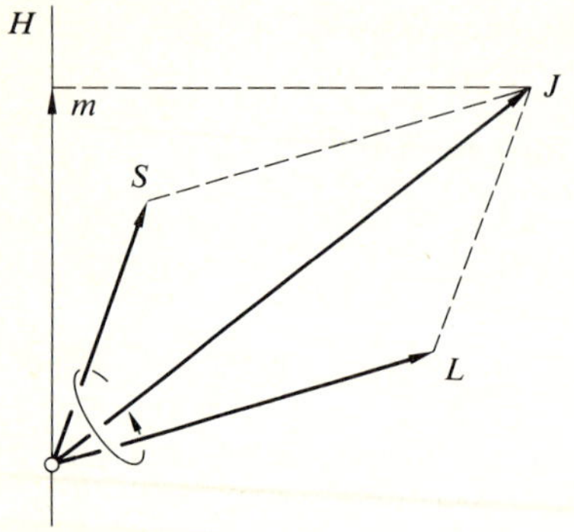

Fig. 1. Anomalous Zeeman Effect

the magnetic moment that the orbital momentum of the moving electron furnishes, the magnetic momentum corresponding to J is characterized by

$$Jg = J + S\cos(S,J) = J + \frac{J^2 + S^2 - L^2}{2J^2}.$$

In a weak external magnetic field H, the angular momentum component of J in the field direction is m; the component of the magnetic momentum is then

$$mg = \cos(H,J)\,[J + S\cos(S,J)] = m\left(1 + \frac{J^2 + S^2 - L^2}{2J^2}\right).$$

The quantum theoretical translation of the squares of the angular momentum thus yields A. Landé's g formula, i.e.,

$$g = 1 + \frac{J(J+1) + S(S+1) - L(L+1)}{2J(J+1)}. \tag{1}$$

The explanation of the anomalous Zeeman effects, therefore, demands first the assumption of the double contribution of the spin (which is a property of the particle electron) and, secondly, the modification of the concept of angular momentum (which is a feature of quantum theory).

The anomalous Zeeman effects occupied Heisenberg in Munich (1920 – 1922) and later in Göttingen (1924) and Copenhagen (1924 – 1925). Between 1922 and 1923, he collaborated with M. Born on other aspects of atomic and molecular theory. Accordingly, we may divide his published papers on these topics into three subgroups and discuss these subgroups systematically in the following sections.

1. With Sommerfeld in Munich (Papers Nos. 1 – 3)

By 1920 good data were available on the anomalous Zeeman effects of doublet and triplet spectra. A. Landé in Frankfurt and W. Heisenberg in Munich had obtained experimental results from E. Back of Tübingen, which both analyzed, deriving from them term schemes. Landé published his results first in 1921 [2]. Heisenberg developed a mechanical model accounting from the results in paper No. 1 (p. 134).

In the case of doublets and weak magnetic fields, the terms derived from the data were given by the formula

$$mg = m\left(1 \pm \frac{1}{2k-1}\right) = m\frac{k-1/2 \pm 1/2}{k-1/2} = m\frac{j}{k-1/2}, \tag{2}$$

with $m = \pm 1/2, \pm 3/2, \ldots, m < j$. In Heisenberg's model, the external electron donated from its angular momentum $k = 1, 2, 3 \ldots$ the portion $1/2$ (in units of $\hbar$) to the core, keeping $k - 1/2$. Only the latter part was subject to spatial quantization, yielding the component m in the direction of the magnetic field. The angular momentum $1/2$ of the core was arranged parallel or antiparallel to the angular momentum of the electron and contributed to the magnetic moment. Thus, it was possible to explain Eq. (2), except for the limitation $m < j$. The special role of the angular momentum $1/2$ resided in the fact that it was not considered in the process of spatial quantization (of the angular momentum). Hence, a wrong (inconsistent) theory gave the right results. If Heisenberg had instead guessed the double magnetism of the core and subjected the total angular momentum j to spatial quantization, he would not have arrived at the right formula (even with a different numbering of the k and j).

In a stronger magnetic field, Heisenberg assumed that the core's momentum takes on the direction of the resultant obtained from the external magnetic field and the (internal) magnetic field of the electron. It is not surprising that he then arrived at W. Voigt's formula, which described the data correctly. This is nearly inevitable, since the limiting cases of strong and weak magnetic fields are fixed.

The noble gas-like Na^+ and Mg^{++} ions could be figured as having zero angular momentum; the Na atom and the Mg^+ ion should obtain core momenta only from the external electron. The Mg core of the Mg atom would then obtain a further angular momentum of magnitude $1/2$ when the second electron is added; hence a singlet system (spectral line) arising from the core momentum $r = 0$ and a triplet system arising from $r = 1$ followed. In this way, the core momentum was composed of parts of the angular momenta of the external electrons. Later in 1925, these parts were instead attributed to the spin of the electron.

In the third edition of *Atomic Structure and Spectral Lines*, completed in the beginning of 1922, Sommerfeld quoted Landé's term values for doublets and triplets, but used for doublets $m = \pm 1, \pm 3, \pm 5, \ldots$ with the enigmatic selection rule $\Delta m = \pm 2$. In an appendix, he presented Heisenberg's model. N. Bohr, on the other hand, considered the latter as "hardly compatible with general principles".

In the papers authored jointly by Heisenberg and Sommerfeld, Bohr's correspondence principle is applied. In paper No. 2 (p. 159), the emission calculated from the classical model provided an approximate determination of atomic emission. Thus, the authors obtained a good estimate for the lifetime of atomic states and the width of spectral lines. In paper No. 3 (p. 165), detailed consequences of the correspondence principle are presented for the selection and polarization rules and the intensities of multiplet lines and their Zeeman components. The problem of half-integral or integral quantum numbers did not enter here because the authors dealt with the limiting case of large quantum numbers. Sommerfeld had previously been somewhat reluctant with respect to the correspondence principle, but in the third (1922) edition of his book he fully acknowledged its value. He and Heisenberg also attended the lectures in Göttingen in June 1922 at which Bohr emphasized the guiding role of the principle. About two months after the lectures, the papers Nos. 2 and 3 were submitted.

2. With Born in Göttingen (Papers Nos. 4–7, 10)

During the Bohr lectures in Göttingen, M. Born had taken notice of Heisenberg and, arranging for him to come to Göttingen in the winter semester of 1922/1923, got harnessed Heisenberg in his program. He was then seeking to adapt the astronomical perturbation theory to atoms. Classical mechanics allowed for certain exceptional cases which, according to quantum conditions, might prove to be the rule in atoms and molecules. Thus, paper No. 4 (p. 189), written entirely in Born's style, treated the case of phase relations which had to be assumed between the motions of electrons.

In paper No. 5 (p. 201), Born and Heisenberg sought to verify whether atomic theory (as formulated at the time), in which certain stationary motions were singled out via quantum conditions from the continuum of classically possible motions, enabled one to calculate the energy states of the helium atom, next to the hydrogen atom the simplest atomic structure. The result was negative: "Either the quantum conditions are wrong, or mechanics is not valid even for stationary states". (See also paper No. 9, p. 278, on this problem.) Later, after Kramers and Heisenberg's work on dispersion theory (see paper No. 2 of Group 3 below), this failure was interpreted in the following way: For a system with one electron, the known theory was sufficient, however, in the case of two interacting electrons in an atom (or, more generally, two interacting atomic systems) the quantum theoretical frequencies must replace the classical ones used in the above calculation. Heisenberg tried to introduce changes into atomic theory, such as half-integral quantum numbers, in order to improve on the helium problem before and after his work on paper No. 5. As he did not achieve full agreement with observation, he avoided publishing anything about these investigations.

Heisenberg finished paper No. 6 (p. 216) with Born during his second Göttingen period (autumn 1923 to spring 1926). In it, the authors presented the methods of molecular theory, applied more or less intuitively up to that time, in a systematic scheme of successive approximations. In his letter to W. Pauli of 7 December 1923, he mocked summations having up to eight indices in the paper [3].

In collaborating with Born, Heisenberg was perhaps more interested in the research carried out for paper No. 7 (p. 247), which dealt with electrons orbiting in the periphery of an atom. The Coulomb field arising from the core thus becomes modified by virtue of the simple fact that the outer electron polarizes the core. It was possible to calculate the core's polarizability by taking the difference between the terms of the atoms under consideration and the corresponding terms of the hydrogen atom. The result was that only when the electron has angular momentum $k = 1/2, 3/2, 5/2, \ldots$ do the calculated values for the polarizability agree with those found from other properties of the atoms. In paper No. 10 (p. 280), Heisenberg showed that, for a larger polarizability of an ion in a molecule, a symmetrical structure, which is stable without polarization, might become unstable. Although the asymmetric linear model of the water molecule that he proposed was later shown to be unstable against buckling, paper No. 10 drew attention to the fact that crystal lattices or molecules containing ions that can be strongly polarized may attain a less regular shape.

3. Multiplets and Zeeman Effects (Papers Nos. 8, 11, 12)

Since 1922, not only had the occurrence of singlet, doublet, and triplet spectra emitted from atoms been familiar phenomena, but also more complex quartet, quintet, etc., spectra. These, too, were organized into terms with the help of three quantum numbers r, k, j. By looking at the Zeeman effects of these complex spectra, Landé found in 1923 his g formula, which he first wrote as [4]:

$$g = 1 + \frac{J^2 - 1/4 + R^2 - K^2}{2(J^2 - 1/4)} = 3/2 + \frac{R^2 - K^2}{2(J^2 - 1/4)}. \qquad (3)$$

It should be emphasized that he counted R, K, J somewhat differently from the r, k, j in earlier use, and also differently from the presently used numbers (see Eq. (1)) S, L, and J. Cautiously he named the quantities R, K, J "apparent" angular momenta. W. Pauli immediately succeeded in interpreting this result in terms of a model [5]. Starting from the limiting case of very strong (external) magnetic fields, in which he put the magnetic quantum number m of the atom equal to $m_k + m_r$, the magnetic energy proportional to $m_k + 2m_r$ (integral number) and considered m_k and m_r as components of the angular momenta k and r, he was able to derive the behavior in weak fields. He also associated the angular momentum r on the "inner system" (core) with the double magnetic moment, obtaining

$$g = 1 + \frac{j^2 + r^2 - k^2}{2j^2} = 3/2 + \frac{r^2 - k^2}{2j^2}. \qquad (4)$$

In this equation, he merely had to replace j^2 by $(j - 1/2)\,(j + 1/2)$, or a differential quotient by a difference quotient, i.e.

$$\frac{\mathrm{d}}{\mathrm{d}j}\left(-\frac{1}{j}\right) = \frac{1}{j^2} \rightarrow \frac{1}{j - 1/2} - \frac{1}{j + 1/2} = \frac{1}{j^2 - 1/4}, \qquad (5)$$

in order to derive from Eq. (4) Landé's Eq. (3). Later, in the beginning of 1925, Pauli attributed the four quantum numbers n, k, m_k, m_r, or n, k, j, m, to the electron and discovered the exclusion principle [6].

Paper No. 8 of A. Landé and Heisenberg (p. 270) does not deal with "normal" multiplets, for which Landé's g formula (3) is satisfied. The rules derived there apply rather to what was later called atoms with $j - j$ coupling.

A reformulation of differentials as differences, similar to Eq. (5), led Heisenberg in paper No. 11 (p. 289) to a generalization of the quantum theory of the anomalous Zeeman effects, which as shown later proved a wrong path. Just as a classical frequency can be written as $(\mathrm{d}E/h\mathrm{d}J)$ and has to be replaced in quantum theory by the difference $E(J+1/2)-E(J-1/2)$, Heisenberg replaced a classical computed energy $E_{\mathrm{cl}}(J)$ by a difference, i.e.,

$$E_{\mathrm{cl}}(J)=\mathrm{d}F/\mathrm{d}J \rightarrow E_{\mathrm{qu}}=F(J+1/2)-F(J-1/2)\,. \tag{6}$$

The reason for this move was that, frequently, one state seemed to be described by two quantum numbers, say $k+1/2$ and $k-1/2$. Later, it was discovered that one "double-valuedness" arose from the as yet undiscovered spin, the other from the quantum theoretical reformulation of the concept of angular momentum. However, Heisenberg expected that the half-integral quantum numbers and the double magnetism of a part of the atomic system emerged from the same root as the change due to relation (6). Paper No. 11 was published only after a very long hesitation. Although he had communicated the basic idea to Pauli in October 1923 [7], Heisenberg only submitted the paper in June 1924. He also used it for his habilitation thesis at the University of Göttingen in July 1924.

Paper No. 12 (p. 306) was written during Heisenberg's stay in Copenhagen. In it he struggled again with "double-valuedness". By then he was particularly aware of a reciprocal double-valuedness in the interaction between electrons and the atomic core. In the simplest cases, this suggested two models, both of which depicted part of the real situation in a complex atom and thereby complemented one another. Here, there is a suggestion of the thoughts of Niels Bohr, which later became fruitful in his idea of complementarity. In paper No. 12, the hints of such ideas were more puzzling than enlightening. Of practical help, however, was Heisenberg's reference to the model of H. N. Russell and F. A. Saunders [8], in which the angular momentum vectors $\boldsymbol{k}$ of the individual (series) electrons combine to a total angular momentum vector $\boldsymbol{k}$ (corresponding to Landé's $\boldsymbol{K}$, later, total orbit momentum). Hence, e.g., a p- and a d-electron yielded singlets and triplets, for which Russell and Saunders proposed the notation P, D, F (today, these are written $l_1=1$, $l_2=2$, $L=1,2,3$; ^{1}P, ^{1}D, ^{1}F, ^{3}P, ^{3}D, ^{3}F). With this model Heisenberg combined Pauli's exclusion principle and managed for example in the case of two p-electrons to organize the states with the same principal quantum number into normal multiplets ^{1}S, ^{3}P, ^{1}D. He thus sketched

l	m_l	m_s	Σm_l	Σm_s	$L\ S$
1, 1	±(1, 1)	+1/2, −1/2	±2	0	2 0 ^{1}D
	0, 0	+1/2, −1/2	0	0	
	±1, 0	±1/2, ±1/2	±1	0	
				±1, 0	1 1 ^{3}D
	+1, −1	±1/2, ±1/2	0	±1, 0	
				0	0 0 ^{1}S

the procedure presented in the table above, for two p-electrons (using today's notation).

Paper No. 12 opened the way for explaining normal multiplets of complex spectra from the states of single electrons.

In the fall of 1925, G. E. Uhlenbeck and S. Goudsmit published their hypothesis of spin $\hbar/2$ derived from Pauli's four quantum numbers associated with each single electron [9]. Their note provided a simple way of completing the model of angular momentum vectors of the atom, as far as it was possible to develop it without strict quantum mechanics (which Heisenberg introduced in his paper No. 3 of the following Group 3). If in the classical atomic model the absolute value of the angular momentum of an electron orbit was constant or nearly constant, it was made equal to an integer multiple of $\hbar$. Both the orbit angular momenta of the electrons and the spin momenta were composed to a resultant. The spin momenta possessed the double magnetic moment. The square of an angular momentum, J^2, had to be replaced by $J(J+1)$. This part of the prescription was clarified in publications in early 1926: The replacement $J(J+1)$ followed from the commutation relation of angular momenta (see papers No. 4 in Group 3 and No. 1 in Group 4); the apparently strong spin-spin coupling did not arise from a spin-spin interaction at all, but rather from differences in the Coulomb energy of the orbital parts (of the interaction energy), whose symmetry is connected with the total spin (see papers No. 5 of Group 3 and No. 2 of Group 4). Finally, the strange double magnetism (mentioned in Sect. 1 above) was explained by P. A. D. Dirac's relativistic electron equation [10].

References

1 A. Sommerfeld: *Atombau und Spektrallinien* (Vieweg, Braunschweig 1919). Second edition 1920, third edition 1922, fourth edition 1924

2 A. Landé: Über den anomalen Zeemaneffekt. Z. Phys. **5**, 231–241 (1921)

3 W. Pauli: *Wissenschaftlicher Briefwechsel / Scientific Correspondence, Band/Volume I: 1919–1929,* ed. by A. Hermann, K. von Meyenn, V. F. Weisskopf (Springer, New York, Heidelberg, Berlin 1979) p. 132

4 A. Landé: Termstruktur und Zeemaneffekt der Multiplets. Z. Phys. **15**, 189–205 (1923)

5 W. Pauli: Über die Gesetzmäßigkeiten des anomalen Zeemaneffekts. Z. Phys. **16**, 155–164 (1923)

6 W. Pauli: Über den Zusammenhang des Abschlusses der Elektronengruppen im Atom mit der Komplexstruktur der Spektren. Z. Phys. **31**, 765–783 (1925)

7 See Heisenberg's letters, dated 9 October 1923, in *W. Pauli: Wissenschaftlicher Briefwechsel,* pp. 125–127

8 H. N. Russel, F. A. Saunders: New regularities in the spectra of alkaline earths. Astrophys. J. **61**, 38–61 (1925)

9 G. E. Uhlenbeck, S. Goudsmit: Ersetzung der Hypothese vom unmechanischen Zwang durch eine Forderung bezüglich des inneren Verhaltens jedes einzelnen Elektrons. Die Naturwissenschaften **13**, 953–954 (1925)

10 P. A. M. Dirac: The quantum theory of the electron I. Proc. Roy. Soc. London **A 117**, 610–624 (1928)

Zeitschrift für Physik 8, 273–297 (1922)

Zur Quantentheorie der Linienstruktur und der anomalen Zeemaneffekte.

Von **Werner Heisenberg** in München.

Mit einer Abbildung. (Eingegangen am 17. Dezember 1921.)

Die große Regelmäßigkeit und Gesetzmäßigkeit der Aufspaltungsbilder im anomalen Zeemaneffekt ist in letzter Zeit mehrfach Gegenstand eingehender Untersuchungen geworden. Zunächst hat Back[1]), dem man ja auch das empirische Material wesentlich verdankt, den einheitlichen Aufbau der Zeemantypen klargelegt. Dann hat insbesondere Landé[2]) in Weiterführung des quantentheoretischen Standpunktes, wie er im „magnetooptischen Zerlegungssatz" von Sommerfeld[3]) eingenommen wurde und unter Verwendung von dessen „inneren Quantenzahlen", die magnetischen Aufspaltungen für kleine Felder in einfache quantentheoretische Formeln zusammengefaßt, die sich auch extrapolatorisch gut bewährt haben. In Teil II[4]) versucht Landé bereits, seinen Formeln gewisse modellmäßige Vorstellungen unterzulegen. In allerletzter Zeit haben Paschen und Back[5]) merkwürdige magnetische Eigenschaften der zusammengesetzten Dubletts und Tripletts aufgedeckt, die mit dem behandelten Problem in innigem Zusammenhang zu stehen scheinen. Ein weiterer Ausbau des empirischen Materials kann noch dadurch erfolgen, daß man die Voigtsche Theorie der D-Linienaufspaltung bzw. deren von Sommerfeld[6]) herrührende Vereinfachung in die Sprache der Quantentheorie übersetzt. Bei den Dubletts können wir dann den ganzen Übergang von kleinen magnetischen Feldern (anomaler Zeemaneffekt) bis zum Paschen-Backeffekt genau verfolgen. — Von einer Theorie aller dieser Erscheinungen müssen wir fordern, daß sie 1. die Linienstruktur aller Elemente qualitativ und, soweit dies ohne Kenntnis der Einzelheiten der Modelle möglich ist, quantitativ erklärt, und daß sie 2. das gesamte bisherige empirische Material über anomale Zeemaneffekte, also auch den Übergang zum Paschen-Backeffekt, darstellt. Bevor wir zur Entwicklung einer Theorie übergehen, soll das empirische Material

[1]) E. Back, Die Naturwissenschaften **9**, 199, 1921.
[2]) A. Landé, ZS. f. Phys. **5**, 231, 1921.
[3]) A. Sommerfeld, Ann. d. Phys. **63**, 221, 1920.
[4]) A. Landé, ZS. f. Phys. **7**, 398, 1921.
[5]) Zeemanjubiläumsheft, Physica **1**, 261, 1921.
[6]) A. Sommerfeld, Gött. Nachr. 1914, März; ZS. f. Phys. **8**, 257—272, 1922.

noch einmal kurz zusammengefaßt werden; für die Begründung und Prüfung verweisen wir auf die genannten Abhandlungen oder auf die im Erscheinen begriffene Neuauflage des Buches[1]) von Sommerfeld.

I. Dublettstruktur.

Ist $\varDelta \nu_n$ die normale magnetische Aufspaltung, gemessen durch $\varDelta \nu_n = \frac{1}{4\pi}\frac{e}{mc}H$, und $\varDelta \nu$ der ursprüngliche Linienabstand im Dublett bzw. $\varDelta \nu \,.\, h$ die Energiedifferenz der beiden Dubletterme, und endlich v das Verhältnis: $v = \frac{\varDelta \nu}{\varDelta \nu_n}$, so ist nach Landé für $v \sim \infty$ die Zusatzenergie eines Energieniveaus im Magnetfeld, dessen azimutale Quantenzahl n, dessen „innere" Quantenzahl[2]) n_J sei, gegeben durch

$$\varDelta E = \frac{n_J \,.\, m^*}{n - \frac{1}{2}} \cdot \varDelta \nu_n \,.\, h, \tag{1}$$

wo m^* eine magnetische Quantenzahl ist, die die Werte $\pm\left(\frac{1}{2}, \frac{3}{2}, \frac{5}{2} \ldots\right)$ annehmen kann[3]), mit der Nebenbedingung $|m^*| < n_J$. Wir können diese Formel auch so schreiben:

$$\varDelta E = m^*\left(1 \pm \frac{1}{2n-1}\right) \varDelta \nu_n \,.\, h, \tag{2}$$

wobei das obere Vorzeichen für das obere Dublettniveau ($n_J = n$), das untere für das untere Niveau ($n_J = n - 1$) gilt. Ferner besagt die viel allgemeinere Voigtsche Theorie[4]), in ihrer Extrapolation, folgendes: Ist $\overline{E}$ das Mittel der Energien der beiden Dubletterme, so ist die Gesamtenergie gegeben durch:

$$E = \overline{E} + \varDelta \nu_n \,.\, h \left[m^* \pm \frac{1}{2}\sqrt{v^2 + 2\frac{m^*}{n^*}v + 1}\right]. \tag{3}$$

[1]) A. Sommerfeld, Atombau und Spektrallinien, 3. Aufl., Braunschweig 1922, Kapitel 6.

[2]) Die Sommerfeldsche Bezeichnung n_i ändern wir ab in n_J, um anzudeuten, daß bei den Dubletts, wie später entwickelt werden soll und wie auch schon von Landé vermutet wurde, n_J die Quantenzahl des Gesamtimpulses des Atoms darstellt. Bei den Tripletts ist der Zusammenhang zwischen n_J und dem gesamten Drehimpuls jedoch komplizierter.

[3]) Diese „Halbzahligkeit" deuten wir durch das Symbol $m^* = m - \frac{1}{2}$ an. Ebenso sei $n^* = n - \frac{1}{2}$.

[4]) Ihre Übereinstimmung mit der Erfahrung bei Li, $\lambda = 6708$ zeigt Kent, Astrophys. Journ. **40**, 343, 1914.

Diese Formel enthält natürlich (1) und (2) in sich. Für kleine Felder ist nämlich v sehr groß; entwickeln wir also nach Potenzen von $\frac{1}{v}$, so ergibt sich:

$$E = \overline{E} + \varDelta \nu_n . h \left[m^* \pm \frac{v}{2} \left(1 + \frac{m^*}{n^*} \frac{1}{v} \right) \right]$$

$$= \overline{E} \pm \frac{\varDelta \nu}{2} \cdot h + \varDelta \nu_n . h . m^* \left(1 \pm \frac{1}{2n-1} \right).$$

Für große Felder, also $v \sim 0$, gibt die Formel (3) den Paschen-Backeffekt:

$$E = \overline{E} + \varDelta \nu_n . h . \left(m^* \pm \frac{1}{2} \right);$$

$m^* \pm {}^1/_2$ ist nämlich eine ganze Zahl. Durch die Formel (3) wird also das ganze empirische Material der Dubletts beschrieben.

II. Triplettstruktur.

Hier gilt für kleine Felder nach Landé die folgende Formel für die Zusatzenergie im Magnetfeld:

$$\left.\begin{aligned} \varDelta E &= m \left(1 + \frac{1}{n} \right) \varDelta \nu_n . h && \text{für } n_J = n; \\ \varDelta E &= m \left(1 - \frac{1}{n} + \frac{1}{n-1} \right) \varDelta \nu_n . h && \text{für } n_J = n - 1; \\ \varDelta E &= m \left(1 - \frac{1}{n-1} \right) \varDelta \nu_n . h && \text{für } n_J = n - 2. \end{aligned}\right\} \quad (4)$$

Die magnetische Quantenzahl m soll jetzt eine ganze Zahl sein mit der Nebenbedingung

$$|m| \leqq n_J.$$

Für starke Felder haben wir auch hier den Paschen-Backeffekt. Ein Analogon zur Voigtschen Theorie der D-Linien für den Übergang von schwachen zu starken Feldern existiert bei den Tripletts nicht.

Nachdem wir so alle bisherigen Messungen und Beobachtungen in ein einfaches Quantenschema geordnet haben, soll die Frage nach der modellmäßigen Deutung aufgenommen werden.

§ 1. Allgemeine Annahme über Dublettatome. Im periodischen System der Elemente ist für die Linienstruktur eine Eigenschaft außerordentlich charakteristisch: Dublettstruktur ist stets mit ungerader Valenz verbunden, Triplettstruktur mit gerader Valenz. Dieser Satz ist schon von Rydberg [1]) ausgesprochen worden, exakter

[1]) A. Rydberg, Astrophys. Journ. 6, 239, 1897.

begründet wurde er von Sommerfeld[1]). Die einfachste Deutung dieser Tatsache würde in der Annahme liegen, daß bei den Dubletts ein, bei den Tripletts zwei äußere Elektronen durch losere Kopplung eine gewisse ausgezeichnete Rolle spielen.

Wir beschreiben also ein „Dublettatom", speziell z. B. ein Alkali, in folgender Weise:

Es besteht aus einem um den Kern gruppierten, stärker zusammengekoppelten Komplex von Elektronen, den wir in Zukunft den „Atomrumpf" nennen wollen, und einem äußeren Elektron, dem „Serienelektron" oder „Valenzelektron". Im Normalzustand, der durch den Term $1s$ gegeben ist, hat das ganze Atom den Drehimpuls $1 \cdot \frac{h}{2\pi}$. Den Faktor $\frac{h}{2\pi}$ wollen wir in Zukunft der einfacheren Ausdrucksweise halber weglassen und sagen: Das ganze Atom hat im Normalzustand den Impuls 1. Dies entspricht der bisher üblichen Annahme über den s-Term. Nun fragt sich: Können wir über den Impuls des äußeren Elektrons allein eine Aussage machen? Zunächst ist klar, daß dieser Impuls keineswegs konstant sein wird; im Verlauf der Bewegung wird vielmehr Impulsmoment vom Rumpf auf das Elektron und umgekehrt übergehen. Wir wollen jetzt die für alles folgende grundlegende Annahme machen, daß im zeitlichen Mittel die Verteilung des Impulses zwischen beiden eine gleiche, also der mittlere Impuls des äußeren Elektrons wie der des Rumpfes $^1/_2$ sei. Diese Annahme ist identisch mit der folgenden: Das äußere Elektron ist hinsichtlich seines Impulses analog gequantelt, wie ein (relativistisches) Wasserstoffatom, jedoch mit der Quantenzahl $^1/_2$. Damit wir die letztere Form der Annahme aus der ersten ableiten, bemerken wir:

Um die stationären Bahnen des äußeren Elektrons festzulegen, benutzt man Winkelvariable[2]), deren Einführung, wie sich aus der Tatsache der Serienspektren folgern läßt, derjenigen bei Wasserstoff nachzubilden ist. Durch die erste Quantenbedingung wird dann bekanntlich die allgemeine Serienform festgelegt:

$$E = -\frac{R \, . \, h}{m^2},$$

wobei R die Rydbergzahl und m die in den Serien auftretende Gesamtquantenzahl ist. Dieses m hat natürlich nichts mit der magnetischen Quantenzahl der Einleitung zu tun.

[1]) Ann. d. Phys. l. c. oder „Atombau" 3, Kap. 6, § 6; 467.

[2]) Vgl. etwa Kramers, ZS. f. Phys. 3, 199, 1920.

Die zweite, uns interessierende, Quantenbedingung liefert das Auseinandertreten der Balmerserie in die verschiedenen (s, p, d usw.) Serien. Sie lautet:

$$\int_0^{2\pi} p\, d\beta = n^* . h.$$

Hierin bedeutet p das Impulsmoment des Elektrons, β die ihm konjugierte Winkelvariable, nämlich den Winkel, den die große Achse der Bahn mit einer Koordinatenachse (bei Polarkoordinaten $\varphi = 0$) einschließt. n^* sei die „Impulsquantenzahl" des äußeren Elektrons. Auf der linken Seite steht offenbar das 2π-fache des zeitlichen Mittelwertes von p, wenn dieser Mittelwert über einen vollen Perihelumlauf erstreckt wird. Da innerhalb eines Perihelumlaufs das Elektron oft in seiner Bahn kreist, so wird sich dieser Mittelwert $\bar{p}$ in hinreichender Näherung nicht von dem oben definierten, der gleich $^1/_2$ gesetzt wurde, unterscheiden.

Wir schließen also aus $\bar{p} = {}^1/_2$ in der Tat $n^* = {}^1/_2$ für die Grundbahn (1 s), ebenso für jede Bahn ($m\,s$).

Wir nehmen jetzt an, daß der Atomrumpf seinen mittleren Impuls $^1/_2$ auch in den angeregten Zuständen behalten soll. Für das äußere Elektron bedenke man, daß die in den Serienformeln auftretende Quantenzahl m stets ganzzahlig ist und daß m als Summe der Impulsquantenzahl und der radialen Quantenzahl betrachtet werden kann. Daraus folgt, daß die radiale Quantenzahl halbzahlig, d. h. mit den Werten ($^1/_2$, $^3/_2 \ldots$) fortschreitet und daß ihr Minimalwert $^1/_2$, nicht 0 ist[1]). Der Minimalwert von m wird dementsprechend $n^* + {}^1/_2$, wenn wir das eben geschilderte Verhalten der radialen Quantenzahl auch für die anderen Serienterme fordern. Nun ist der Minimalwert von m identisch mit der bisher üblichen konventionellen azimutalen Quantenzahl n. Wir erhalten so:

$$n^* = n - {}^1/_2. \tag{5}$$

Es ergibt sich also das folgende Atommodell: In der Mitte ein „Rumpf" mit dem mittleren Impuls $^1/_2$, außen ein Elektron mit dem mittleren Impuls $n - {}^1/_2$. Jetzt entsteht die Frage: Wodurch ist die relative Lage dieser beiden Drehimpulse $n - {}^1/_2$ und $^1/_2$ festgelegt? Hier ist nun zu bedenken, daß zwischen diesen beiden Impulsen eine magnetische Wechselwirkung besteht, die sich, für ein spezielles Atom-

[1]) Schrödinger (ZS. f. Phys. 4, 347, 1921) macht es wahrscheinlich, daß in den s-Termen der Minimalwert von m gleich 2, der der radialen Quantenzahl also gleich $^3/_2$ ist. In den anderen Serientermen haben wir aber den Minimalwert $^1/_2$ als gesichert zu betrachten.

modell, genau berechnen läßt. Sei ϑ der Winkel zwischen den beiden Impulsachsen, so ist die magnetische Wechselwirkungsenergie

$$\varDelta E \text{ prop. } H_i \cos\vartheta.$$

H_i sei das Magnetfeld, welches das Valenzelektron in der Nähe des Rumpfes erzeugt.

Der Proportionalitätsfaktor wird (vgl. Einleitung): Impuls des Rumpfes $\times \frac{e}{2mc}$.

Also beträgt die Wechselwirkungsenergie

$$\varDelta E = \frac{1}{2} \cdot \frac{h}{2\pi} \frac{e}{2mc} H_i \cos\vartheta. \tag{6}$$

Der Winkel ϑ ist hierbei jedoch nicht beliebig, sondern nach Sommerfeld[1]) durch die sogenannte räumliche Quantelung festgelegt. Das Prinzip der räumlichen Quantelung besagt folgendes: Außer dem Impulsvektor ist auch seine Projektion auf die Achse eines etwaigen magnetischen Feldes, also die Komponente des Impulses um die Feldachse, bestimmt durch feste Quantenzahlen.

Nach unserer Annahme (5)

$$n^* = n - {}^1/_2$$

schreitet der mögliche Impuls des Elektrons halbzahlig ($^1/_2$, $^3/_2 \ldots$) fort. Wir müssen dasselbe auch von der Projektion seines Impulsvektors auf die Feldachse (also hier auf die Achse des vom Rumpf herrührenden inneren Magnetfeldes) fordern. Diese Notwendigkeit ergibt sich schon daraus, daß wir jedenfalls die (bei Abwesenheit äußerer Felder!) mechanisch natürlichste Stellung $\vartheta = 0$, $\cos\vartheta = 1$ in den Bereich der Möglichkeit zu ziehen haben.

Bei dieser Annahme folgt für die mögliche relative Lage des Atomrumpfes und des Valenzelektrons

$$n^* \cos\vartheta = m_i^*. \tag{7}$$

Für die Quantelung im äußeren Felde wenden wir gleich die analoge Gleichung

$$n^* \cos\vartheta_1 = m^* \tag{7a}$$

an, deren Bedeutung wir in § 2, S. 297 und S. 300, Gleichung (17) näher erläutern werden.

Hier bedeutet m_i^* eine „innere" magnetische Quantenzahl. Diese kann, wie später auseinandergesetzt werden soll, im Falle der Dublettatome nur die beiden Werte $+n^*$ und $-n^*$ annehmen; also ist

$$\cos\vartheta = \pm 1.$$

[1]) A. Sommerfeld, Ann. d. Phys. **51**, 1, 1916, § 7.

Wir erhalten daher für unser Atommodell zwei fundamentale Stellungen:

1. Die Richtung der beiden Impulsvektoren $n - {}^1/_2$ und ${}^1/_2$ ist gleich;
2. sie ist genau entgegengesetzt.

Der Gesamtimpuls des ganzen Atoms ist im ersten Falle

$$n - {}^1/_2 + {}^1/_2 = n,$$

im zweiten Falle

$$n - {}^1/_2 - {}^1/_2 = n - 1.$$

Diese beiden Möglichkeiten liefern, wie wir behaupten, die beiden verschiedenen Dublettenergieterme[1]). Die Bedeutung der inneren Quantenzahl n_J ergibt sich dann von selbst; sie stellt den Gesamtimpuls des Atoms dar, während n die Summe der absoluten Beträge der beiden Impulse — Elektron—Rumpf — angibt. Dem entspricht es, daß ein Dublettniveau mit der azimutalen Quantenzahl n stets in zwei Niveaus mit den inneren Quantenzahlen $n_J = n$ und $n_J = n - 1$ zerfällt. Die Energiedifferenz dieser beiden Dublettniveaus folgt aus (6) und (7) zu[2]):

$$h \,.\, \Delta\nu = \frac{1}{2} \frac{h}{2\pi} \frac{e}{2mc} H_i - \left(-\frac{1}{2}\right) \frac{h}{2\pi} \frac{e}{2mc} \cdot H_i,$$

also

$$\Delta\nu = \frac{1}{4\pi} \frac{e}{mc} \cdot H_i. \tag{8}$$

Hierin bedeutet $\Delta\nu$ den in Wellenzahlen (cm^{-1}) gemessenen Abstand der beiden Linien.

Damit haben wir eine Hypothese über das Wesen der Dublettstruktur eingeführt; wir wollen sehen, wie sie sich dem Problem der anomalen Zeemaneffekte gegenüber bewährt.

§ 2. Verhalten der Dublettatome im äußeren Magnetfelde. H sei die Stärke eines äußeren magnetischen Feldes. Wir führen, wie in der Einleitung, die Größe

$$\Delta\nu_n = \frac{1}{4\pi} \frac{e}{mc} \cdot H$$

ein und setzen

$$v = \frac{\Delta\nu}{\Delta\nu_n} = \frac{H_i}{H}.$$

[1]) Es wäre hier die Frage aufzuwerfen, weshalb der s-Term nicht doppelt ist. Wenn wir mit Schrödinger annehmen müssen, daß das s-Elektron in den Rumpf eindringt, so ist eine Auszeichnung des gleichsinnigen Drehimpulses sehr plausibel. Jedenfalls müßte die ungleichsinnige zweite Bahnmöglichkeit eine ganz andere Energie — wohl eine sehr viel kleinere — ergeben. Die Beobachtung scheint aber ein in Strenge einheitliches s-Niveau zu verlangen.

[2]) H_i hat für beide Niveaus dieselbe Größe, da diese nur vom äußeren Elektron abhängt, das in beiden Fällen wesentlich dieselbe Bahn beschreibt.

Um zunächst die Stellung der Bahn des äußeren Elektrons gegen das Feld zu definieren, führen wir ein räumliches Polarkoordinatensystem ein, dessen Polarachse dem Feldvektor H parallel geht, und beschreiben die Lage der mittleren Bahnebene (von der Bahnebene schlechtweg zu sprechen, hätte keinen Sinn, da die Bahn im allgemeinen nicht eben sein wird) durch den Winkelabstand[1]) γ_1 der Knotenlinie von der Meridianebene $\varphi = 0$ und durch den mittleren Impuls $\bar{q}$ des Elektrons um die Feldachse. Der Winkel ϑ_1, den die Impulsachse des äußeren Elektrons mit der Feldachse einschließt, ist dann festgelegt durch $\cos\vartheta_1 = \frac{\bar{q}}{\bar{p}}$ ($\bar{p}$ der mittlere Impuls des äußeren Elektrons). Die Koordinaten γ_1, $\bar{q}$ sind als konjugierte Variable eines ungestörten Wasserstoffatoms hinreichend bekannt. Analog können wir die Lage des Atomrumpfes durch den Winkel ϑ_2 beschreiben, den sein Impulsvektor mit dem Feldvektor H einschließt (der dem q entsprechende Drehimpuls $\bar{q}_2$ des Rumpfes um die Feldachse interessiert uns nicht weiter), und durch den Winkel γ_2, den wir als Winkel deuten zwischen $\varphi = 0$ und der Schnittlinie einer zum Drehimpulsvektor des Rumpfes orthogonalen Ebene und der φ-Ebene. Der Winkel ϑ_{12} zwischen den beiden Vektoren des mittleren Impulses $n^* = n - \frac{1}{2}$ und $^1/_2$ ist dann elementargeometrisch gegeben als

$$\cos\vartheta_{12} = \cos\vartheta_1 \cos\vartheta_2 + \sin\vartheta_1 \sin\vartheta_2 \cos(\gamma_1 - \gamma_2).$$

Dazu merken wir an:

$$-\frac{\partial}{\partial\vartheta_2}\cos\vartheta_{12} = \cos\vartheta_1 \sin\vartheta_2 - \sin\vartheta_1 \cos\vartheta_2 \cos(\gamma_1 - \gamma_2).$$

Jetzt sind wir in der Lage, den durch das innere und äußere Feld verursachten Energiezuwachs $\varDelta E$ als Funktion der Koordinaten zu schreiben. $\varDelta E$ spielt die Rolle einer Störungsenergie (daß man Dublettstruktur und Zeemaneffekt als „kleine Störung“ auffassen kann, ist ohne weiteres klar) und ist deshalb als ein zeitlicher Mittelwert definiert, wobei die zeitliche Mittelung über die längste Periode des ungestörten Systems zu erstrecken ist[2]). Diese Mittelung ist der Grund dafür, daß, trotz der Inkonstanz des Impulses des Valenzelektrons einerseits, des Rumpfes andererseits, in die Energie der Zeemaneffekte nur die azimutale Quantenzahl eingeht. Wir erhalten so, wie in (6)

$$\varDelta E = \varDelta\nu_n \cdot h\left[n^* \cos\vartheta_1 + \frac{1}{2}\left(\cos\vartheta_2 + v\cos\vartheta_{12}\right)\right]. \tag{9}$$

[1]) Der Index 1 bezieht sich auf das äußere Elektron, der Index 2 auf den Rumpf.

[2]) Bohr, On the Quantum theory of line spectra. Part II. S. 41 u. 87.

Hierbei rührt der erste Term vom Einfluß des äußeren Feldes auf das Serienelektron her; er könnte durch $\Delta\nu_n . 2\pi^-$ ersetzt werden. Der zweite Term bedeutet den Einfluß des äußeren Feldes auf den Rumpf, der dritte endlich vertritt die Wechselwirkungsenergie.

Einschaltungsweise wollen wir jetzt, bevor wir Ansatz (9) allgemein auswerten, ΔE für den Grenzfall $v \sim \infty$ für beliebige Werte von m_i^* festlegen und den Beweis nachholen, daß m_i^* nur die beiden Werte $n^* = n - {}^1/_2$; $-n^* = -(n - {}^1/_2)$ annehmen kann. Zu diesem Zweck stellen wir ein allgemeines Prinzip auf, das für eine beliebige Anzahl (λ) möglicher Werte von m_i^* Gültigkeit haben soll.

Sind $\Delta E_1, \Delta E_2 \ldots \Delta E_\lambda$ die zu den möglichen Werten von m_i^*, nämlich $m_{i1}^*, m_{i2}^* \ldots m_{i\lambda}^*$, gehörigen Zusatzenergien, so muß gelten:

$$\sum_{k=1}^{\lambda} \Delta E_k = \lambda . m^* . \Delta\nu . h + \text{const.} \tag{10}$$

Hierin bedeutet m^* die unten näher zu bestimmende magnetische Quantenzahl, von der wir hier nur die Eigenschaft voraussetzen, daß der Übergang $m^* \longrightarrow m^*$ π-Komponenten, der Übergang $m^* \longrightarrow (m^* + 1)$ σ-Komponenten erzeugt. Die Konstante enthält auch die Terme der Wechselwirkungsenergie, die von m^* unabhängig sind. Ob m^*, wie hier geschrieben, halbzahlig oder ob es ganzzahlig ist, ist gleichgültig.

Dieses Prinzip kann etwa folgendermaßen begründet werden: Die Überlegungen, die Rubinowicz[1]) über die Ausstrahlung von Impulsmoment anstellt, würden, auf das einzelne Atom angewendet, stets den normalen Zeemaneffekt liefern. In der Tat verlangt Rubinowicz für eine σ-Komponente stets die Ausstrahlung des Drehimpulses $\frac{h}{2\pi}$ um die Feldachse; da aber ΔE für $v \sim \infty$ proportional dem Impuls um die Feldachse ist, so bedeutet dies normale Aufspaltung der Linie. Wir können also, um nicht in Widerspruch mit der Erfahrung zu kommen, das Rubinowiczsche Prinzip nur für die Gesamtheit der Atome[2]) gelten lassen. Der in senkrechter Polarisation ausgestrahlte Gesamtimpuls muß hiernach bei N Atomen $N \cdot \frac{h}{2\pi}$ betragen. Nehmen wir noch an, daß unsere λ verschiedenen Atome gleichwahrscheinlich sind, so folgt daraus der obige Satz.

Nun wollen wir ΔE $(v \sim \infty)$ als Funktion von m_i^* und m^* anschreiben. Wir verlangen, daß die stationären Zustände des Atoms

[1]) Rubinowicz, Phys. ZS. **19**, 441, 465, 1918.

[2]) Damit stellen wir uns in bewußten Gegensatz zur klassischen Ausstrahlung. Die Folgerungen aus diesem physikalisch höchst wichtigen Ergebnis mögen der unten noch zu erwähnenden, späteren Arbeit vorbehalten bleiben.

im Felde H, wie in Gl. (7), dadurch gegeben seien, daß die Projektion des mittleren Impulses des äußeren Elektrons auf die Feldachse H gleich $m^* \cdot \frac{h}{2\pi}$ sei. [Für $m^* = \pm n^*$ stimmt diese Forderung mit (7a) überein]. Es wird sich später ergeben, daß dann für m^* das Auswahlprinzip Gültigkeit hat. Der Gesamtimpuls des Atoms sei J, sein Winkel gegen die Feldachse Θ. Dann gilt, wegen $\cos\vartheta = \frac{m_i^*}{n^*}$ [vgl. (7)]:

$$J = \sqrt{n^{*2} + 2 \cdot {}^1\!/_2 \cdot n^* \cos\vartheta + ({}^1\!/_2)^2} = \sqrt{n^{*2} + m_i^* + ({}^1\!/_2)^2}.$$

Der mittlere Impuls $\overline{\overline{p}}$ des äußeren Elektrons entsteht jetzt durch Mittelbildung auch über die durch H_i hervorgerufenen Bewegungen und stimmt überein mit der Projektion von $\overline{p} = n^* \cdot \frac{h}{2\pi}$ auf die Richtung von J. Denn vermöge der „Larmorpräzession“ von H_i (die wegen $H \ll H_i$ noch nicht merklich gestört wird) führen die Vektoren n^* und ${}^1\!/_2$ Präzessionen um J aus. $\overline{\overline{p}}$ wird also geometrisch:

$$\overline{\overline{p}} = \frac{J^2 + n^{*2} - ({}^1\!/_2)^2}{2J} \cdot \frac{h}{2\pi}.$$

Daraus folgt dann, weil $\Delta E = (J . \cos\Theta + {}^1\!/_2\, v \cos\vartheta)\, \Delta v_n . h$ und $\cos\Theta = \frac{m^* \cdot \frac{h}{2\pi}}{\overline{\overline{p}}}$:

$$\left.\begin{aligned} \Delta E &= \Delta v_n . h \left[\frac{2(n^{*2} + m_i^* + ({}^1\!/_2)^2)}{2n^{*2} + m_i^*} m^* + \frac{v}{2} \cdot \frac{m_i^*}{n^*}\right] \\ &= \Delta v_n . h \left[\left(1 + \frac{m_i^* + {}^1\!/_2}{2n^{*2} + m_i^*}\right) m^* + \frac{v}{2} \cdot \frac{m_i^*}{n^*}\right]. \end{aligned}\right\} \tag{11}$$

Gleichung (10) verlangt jetzt zur Bestimmung von λ und m_i^*:

$$\sum_{k=1}^{\lambda} \left(1 + \frac{m_{ik}^* + {}^1\!/_2}{2n^{*2} + m_{ik}^*}\right) = \lambda; \quad \text{d. h.} \sum_{k=1}^{\lambda} \frac{m_{ik}^* + {}^1\!/_2}{2n^{*2} + m_{ik}^*} = 0.$$

Denkt man sich die m_{ik}^* als Funktion von n^* geschrieben, so findet man, daß diese Gleichung durch die Werte $\lambda = 2$, $m_{i1}^* = n^*$, $m_{i2}^* = -n^*$ und nur durch diese befriedigt werden kann. Der Wert $m_i^* = -{}^1\!/_2$ ist auszuschließen, weil er für den Atomrumpf zu einer Bewegung führen würde, die derjenigen entspricht, die z. B. ein Wasserstoffatom im äußeren Magnetfelde ausführt, wenn sein Impuls um die Feldachse 0 ist. Dieser Bewegungszustand ist aber nach Bohr verboten. Die Gleichung (10) scheidet also aus der Schar der möglichen Werte von m_i^* gerade zwei „brauchbare“ aus.

Setzen wir die beiden, nun gesicherten Größen $m_i^* = n^*, \; -n^*$ in Gleichung (11) ein, so ergeben sich sofort die Landéschen Formeln des anomalen Zeemaneffektes ($H \ll H_i$):

$$\Delta E_1 = \Delta \nu_n \,.\, h \left[m^* \left(1 + \frac{1}{2n-1}\right) + \frac{v}{2}\right];$$

$$\Delta E_2 = \Delta \nu_n \,.\, h \left[m^* \left(1 - \frac{1}{2n-1}\right) - \frac{v}{2}\right].$$

Wir gehen jetzt wieder zum allgemeinen Fall (beliebiges v) über. In Gleichung (9) soll ϑ_1 später festgelegt werden, über ϑ_2 wollen wir zunächst verfügen. Dies geschieht einfach durch die Forderung, daß die Richtung des Drehimpulsvektors des Atomrumpfes stets mit der Richtung der Resultante von H und H_i zusammenfallen solle. Der damit charakterisierte Übergang von kleinem zu großem Felde H stellt einen adiabatischen Prozeß dar (der für $H \ll H_i$ natürlich $\cos \vartheta_{12} = +1$ liefert). Man erkennt dies daraus, daß dann das Phasenintegral des Rumpfes um die Feldrichtung, d. i. die Projektion seines mittleren Impulses auf die Resultante, tatsächlich invariant ist.

Die Einstellung in die Resultante definieren wir am bequemsten nach dem Prinzip der virtuellen Verrückungen so, daß wir setzen $\frac{\partial \Delta E}{\partial \vartheta_2} = \frac{\partial \Delta E}{\partial \gamma_2} = 0$ unter konstant gehaltenem ϑ_1. Bezeichnen wir also den vom Rumpf herrührenden Teil der Zusatzenergie mit X, so haben wir

$$X = \cos \vartheta_2 + v \cos \vartheta_{12}, \tag{12}$$

und die Bedingungen

$$-\frac{\partial X}{\partial \vartheta_2} = 0 = \sin \vartheta_2 - v \frac{\partial}{\partial \vartheta_2} \cos \vartheta_{12} \tag{13}$$

und

$$\frac{\partial X}{\partial \gamma_2} = 0, \quad \text{d. h.} \quad \sin(\gamma_1 - \gamma_2) = 0. \tag{14}$$

Die drei Impulsvektoren liegen stets in einer Ebene. Multiplizieren wir jetzt (12) mit $\cos \vartheta_2$ und (13) mit $\sin \vartheta_2$, ein anderes Mal (12) mit $\cos \vartheta_{12}$ und (13) mit $-\frac{\partial}{\partial \vartheta_2} \cos \vartheta_{12}$, so ergibt sich durch Addieren:

$$X \cos \vartheta_2 = 1 + v \cos \vartheta_1$$

und

$$X \cos \vartheta_{12} = \cos \vartheta_1 + v \left[\cos^2 \vartheta_1 + \sin^2 \vartheta_1 \cos^2 (\gamma_1 - \gamma_2)\right] = v + \cos \vartheta_1.$$

Multiplizieren wir (12) mit X und setzen die Werte von $X \cos \vartheta_2$ und $X \cos \vartheta_{12}$ ein, so folgt:

$$X^2 = 1 + 2 v \cos \vartheta_1 + v^2. \tag{15}$$

Gehen wir damit wiederum in (11) ein und ersetzen $\cos\vartheta_1$ durch $\frac{\overline{q}}{\overline{p}} = \frac{2\pi\overline{q}}{n^*.h}$, so folgt:

$$\varDelta E = \varDelta\nu_n . h\left[\frac{2\pi\overline{q}}{h} \pm \frac{1}{2}\sqrt{1 + 2\frac{2\pi\overline{q}}{n^*.h}v + v^2}\right]. \qquad (16)$$

In dieser Energieformel ist charakteristischerweise γ_1 herausgefallen; γ_1 ist also für die Störungsrechnung eine zyklische Koordinate. Daraus folgt ein Doppeltes[1]):

1. daß wir auch für beliebige Felder

$$\int_0^{2\pi} \overline{q}\,d\gamma = 2\pi\overline{q} = m^*.h \qquad (17)$$

durch die hier einzuführende Quantenzahl m^* festzulegen haben (vgl. 7a) und

2. daß für m^* ein Auswahlprinzip der bekannten Art gilt, nämlich: $m^* \longrightarrow m^*$ gibt π-Komponenten, $m^* \longrightarrow (m^* \pm 1)$ gibt σ-Komponenten; daß die für m^* in Betracht kommenden Werte, wie schon in der Bezeichnungsweise angedeutet, halbzahlig werden, nämlich: $\pm(^1/_2, {}^3/_2, {}^5/_2 \cdots)$, wurde schon in Gleichung (7) begründet.

Setzen wir (17) in (16) ein, so ergibt sich

$$\varDelta E = \varDelta\nu_n . h\left[m^* \pm \frac{1}{2}\sqrt{1 + 2\frac{m^*}{n^*}v + v^2}\right]; \qquad (18)$$

wie in der Einleitung (3) hervorgehoben, gibt diese Gleichung die anomalen Zeemaneffekte der Dubletts erschöpfend wieder. [Zu (18) müssen wir hinzufügen, daß die Bedingung gilt $|m^*| < n_y$. Dies wird in einer sogleich zu erwähnenden späteren Arbeit begründet werden.] Wir sind also, durch Anwendung quantentheoretischer Störungsrechnungen auf unser hypothetisches Dublettatom, zu derselben Formel gekommen, die aus der Voigtschen Theorie extrapoliert war, und haben damit den Anschluß an sämtliche Beobachtungen erreicht, die über die anomalen Zeemaneffekte von Dublettsystemen bis jetzt gemacht worden sind. Eine spätere gemeinsam mit Herrn Prof. Sommerfeld auszuführende Arbeit wird zeigen, daß sich, als Folge dieser Ableitung, auch die Intensitätsfrage mittels des Korrespondenzprinzips befriedigend lösen läßt. Insbesondere wird das Verschwinden einiger Komponenten im Paschen-Backeffekt und die Durchbrechung des Auswahlprinzips von n_J durch Magnetfelder[2]) eine Begründung erheischen. In der dargestellten Betrachtungsweise hat übrigens der

[1]) Vgl. Bohr, l. c. Part. II, S. 54—60.

[2]) Paschen und Back, l. c. Zeemanjubiläumsschrift, Physica 1, 261, 1921.

Zeitschrift für Physik 8, 273–297 (1922)

Paschen-Backeffekt eine einfache, anschauliche Bedeutung. Während die Bahn des äußeren Elektrons von schwachen bis zu starken Feldern stets dieselbe Lage (7a) oder (17) gegen das äußere Feld hat, wird der Atomrumpf durch das stärker werdende äußere Feld allmählich gedreht, bis sich sein Impuls im totalen Paschen-Backeffekt parallel zum Feldvektor einstellt. — Im totalen Paschen-Backeffekt werden ferner im allgemeinen die resultierenden Linien des Lorentztripletts schmale Mehrfachlinien von der Größenordnung der ursprünglichen Dubletts bleiben, in derselben Weise, wie dies z. B. beim Starkeffekt des Wasserstoffs der Fall ist. Neueste Beobachtungen von Paschen scheinen bei Li, $\lambda = 6708$, auch dies Resultat der Theorie quantitativ zu bestätigen, doch sind die Untersuchungen darüber nicht abgeschlossen.

Eine genaue Prüfung hat die Theorie bei mittleren Magnetfeldern nur im Falle des D-Linientypus erfahren [1]). Für den Beginn der „magnetischen Beeinflussung" liegen außerdem jetzt Messungen von Paschen (l. c.) an den verbotenen Linien $d_1 p_2$ für verschiedene Magnetfelder vor. Paschen beobachtet dort eine gewisse „Verschiebung ϑ", die die Abweichung der Linien vom gewöhnlichen anomalen Zeemaneffekt mißt. Diese Verschiebungen lassen sich theoretisch aus (18) berechnen mit den Daten $\Delta \nu$ und H. Wir stellen z. B. die empirischen Werte bei Al mit den aus Gleichung (18) berechneten zusammen: Es ist dort $\Delta \nu_{3d} = 1{,}32\,\text{cm}^{-1} = 0{,}126$ Å.

Tabelle 1. $\lambda = 3802{,}1$ Å; $\gamma = 2p_2 - 3d_1$; $H = 39\,000$ Gauß.
(ϑ in Å-Einheiten.)

Aufspaltung in $\Delta\lambda_{\text{normal}}$	$\vartheta_{\text{beobachtet}}$	$\vartheta_{\text{berechnet}}$	$\vartheta_{\text{beob.}} - \vartheta_{\text{ber.}}$
σ $+ 22/15$	$-0{,}054$	$-0{,}060$	$+0{,}006$
σ $+ 14/15$	—	$-0{,}050$	—
π $+ 4/15$	—	$-0{,}050$	—
π $- 4/15$	$-0{,}042$	$-0{,}039$	$-0{,}003$
σ $- 14/15$	$-0{,}035$	$-0{,}039$	$+0{,}004$
σ $- 22/15$	$-0{,}023$	$-0{,}023$	$0{,}000$

Die Übereinstimmung ist vollständig, schon wegen des wechselnden Vorzeichens der Abweichung.

§ 3. **Quantitative Folgerungen über den Dublettabstand im periodischen System.** Einfache Rechnungen über den Dublettabstand lassen sich offenbar nur dann mit einiger Sicherheit aus-

[1]) Kent, l. c.

führen, wenn wir den Rumpf in erster Näherung als unendlich klein voraussetzen können. Für den p-Term ist das nur bei Lithium 3 der Fall, für den d-Term wohl noch bei Natrium 11, viel weniger gut schon bei Kalium 19. Die Dimensionen des Rumpfes, d. h. die Ionenradien, nehmen nämlich mit der Ordnungszahl stark zu[1]. Ehe wir nun den Gang mit der Ordnungszahl näher betrachten, wollen wir den Dublettabstand bei Lithium ($2p$) unter möglichst einfachen Annahmen ausrechnen. Wir nehmen an, daß die Bahn des äußeren Elektrons eben sei und daß der Dublettabstand nur durch magnetische Wechselwirkung bedingt werde. (Prinzipiell ist dies letztere nämlich bei den vorangehenden Betrachtungen gar nicht vorausgesetzt — vorausgesetzt ist nur eine Wechselwirkung, die [vgl. (9)] mit $\cos\vartheta_{12}$ geht.) Wir haben also eine Bahn des äußeren Elektrons mit den zwei Achsen[2]: Halbe große Achse $a_1 = a \cdot 2^2 = 4a$, halbe kleine Achse $a_2 = a\, 2 \cdot {}^3/_2 = 3a$. a sei der Wasserstoffradius. Die effektive Kernladung ist für das äußere Elektron 1.

Das Magnetfeld ist in der Nähe des Rumpfes nach Biot-Savart gegeben durch

$$\mathfrak{H} = \frac{e}{c}\frac{[\mathfrak{v}\mathfrak{r}]}{r^3} = \frac{e}{mc}\cdot\frac{m[\mathfrak{v}\mathfrak{r}]}{r^3}. \tag{19}$$

r bedeutet den Kernabstand, $m[\mathfrak{v}\mathfrak{r}]$ stellt das Impulsmoment des Elektrons dar und ist als solches $\frac{h}{2\pi}\cdot\frac{3}{2}\cdot$ Da uns der Mittelwert von H_i interessiert, müssen wir noch $\overline{\frac{1}{r^3}}$ berechnen. Durch Einführung der exzentrischen Anomalie u und der Exzentrizität ε erhalten wir ohne weiteres:

$$\overline{\frac{1}{r^3}} = \frac{1}{2\pi}\int_0^{2\pi}\frac{du}{a_1^3(1+\varepsilon\cos u)^3} = \frac{1}{a_1^3(1-\varepsilon^2)^{3/2}} = \frac{1}{a_2^3}. \tag{20}$$

Aus (19) und (20) folgt:

$$\overline{H_i} = \frac{h}{2\pi}\frac{e}{mc}\frac{1}{a^3 \cdot 2^3 \cdot ({}^3/_2)^2}. \tag{19 a}$$

Die Aufspaltung wird nach (8)

$$\Delta\nu = \overline{H_i}\cdot\frac{1}{4\pi}\frac{e}{mc}.$$

[1]) Vgl. Grimm, ZS. f. phys. Chem. 98, 353, 1921.

[2]) Siehe z. B. Sommerfeld, Atombau, 3. Aufl., Kap. 4, S. 290. Man beachte, daß an Stelle von n und n' hier halbe Zahlen, bei Li ($2p$) nämlich ${}^3/_2$ bzw. ${}^1/_2$, zu treten haben.

Das ergibt mit den bekannten Zahlenwerten

$$h = 6{,}55 \,.\, 10^{-27}; \quad \frac{e}{m.c} = 1{,}77 \,.\, 10^{7} \quad \text{und} \quad a = 0{,}532 \,.\, 10^{-8}:$$

$$\underline{\varDelta \nu_{2p} = 0{,}32\,\text{cm}^{-1}.}$$

Der empirische Wert ist nach Kent $0{,}34\,\text{cm}^{-1}$, die Übereinstimmung ist, in Anbetracht des groben Näherungsverfahrens bei der Rechnung, sehr gut. Aus (19a) entnehmen wir auch sofort das Gesetz, mit dem $\varDelta \nu$ je in den ersten Seriengliedern mit der azimutalen Quantenzahl abnimmt:

$$\varDelta \nu \sim \frac{1}{n^3 \,.\, (n - {}^1\!/_2)^2}.$$

Der Abstand $\varDelta \nu_{3d}$ wird also, wenn nicht andere Unterschiede sich geltend machen:

$$\varDelta \nu_{3d} = \frac{2^3 \,.\, ({}^3\!/_2)^2}{3^3 \,.\, ({}^5\!/_2)^2} \,.\, \varDelta \nu_{2p} = 0{,}11 \,.\, \varDelta \nu_{2p}.$$

(Größere Abweichungen von diesem Gesetz zeigen sich naturgemäß da, wo der Rumpf zwar für den d-Term, nicht aber für den p-Term als unendlich klein zu gelten hat; dann wird $\varDelta \nu_{3d} < 0{,}11\, \varDelta \nu_{2p}$ werden.) Er würde sich also bei Lithium zu $0{,}035\,\text{cm}^{-1}$ ergeben. Zu dieser Größenordnung stimmt gut, daß Kent in der zweiten Nebenserie $0{,}34\,\text{cm}^{-1}$, in der ersten Nebenserie aber nur $0{,}327\,\text{cm}^{-1}$ für $\varDelta \nu_{2p}$ erhält, ein Unterschied, den Sommerfeld auf die Multiplizität des d-Terms zurückführt. In der Tat würden wir in der ersten Nebenserie ein Liniengebilde vom Typus der zusammengesetzten Dubletts erwarten, dessen Schwerpunktsabstand sich zu etwa 0,32 ergäbe. Doch kann man auf die hier in Betracht kommenden letzten Dezimalen sicher nicht allzu viel Wert legen.

Für das Verhalten der Dublettabstände mit wachsender Ordnungszahl kommen drei Ursachen in Betracht:

1. Hat der Rumpf nicht mehr als unendlich klein zu gelten, so wird die für das äußere Elektron wirksame Kernladung nicht mehr genau 1, sondern größer, sagen wir gleich Z_{eff} sein. Allgemein geht H_i mit $\frac{1}{r^3}$ (19). Da aber wiederum r mit $\frac{1}{Z_{eff}}$ geht, so folgt, daß H_i mit Z_{eff}^3 proportional wird.

2. Wenn wir etwa das äußere Elektron und den Rumpf durch zwei Elektronenbahnen vom Radius r_1 und r_2 repräsentieren, so wächst die magnetische Wechselwirkung in dem Maße, wie $r_1 = r_2$ wird, gegen unendlich. Je näher also die Bahn des äußeren Elektrons der des Rumpfes kommt, um so größer wird die Wechselwirkungsenergie.

3. Wenn Elektronenbahn und Rumpf sich sehr nahe kommen, wird man wohl annehmen müssen, daß auch elektrische Größen in den Bahnunterschieden eine Rolle spielen. Man könnte sogar die Frage aufwerfen, ob das nicht auch bei den leichten Atomen (z. B. Li) der Fall ist und ob diese elektrischen Unterschiede nicht von höherer Größenordnung sind als die magnetischen. Das möchten wir nicht annehmen, denn schon die Tatsache, daß sich die Serienspektren im wesentlichen durch modifizierte Kernfelder beschreiben lassen, weist auf eine recht große Symmetrie des Rumpfes hin, auf eine Symmetrie nicht im geometrischen, sondern im physikalischen Sinne.

Die angeführten Gründe machen ein allgemeines Anwachsen der Dublettdifferenz $\Delta\nu$ mit der Ordnungszahl Z verständlich. Das quantitative Gesetz dieses Anwachsens (ungefähr mit Z^2) läßt sich daraus jedoch nicht begründen; dazu sind offenbar speziellere Vorstellungen über den Schalenbau notwendig.

§ 4. Allgemeine Annahme über Triplettatome. Wie schon zu Anfang von § 2 hervorgehoben wurde, müssen wir dem Triplettatom konsequenterweise zwei Valenzelektronen zuschreiben. Wir fügen also zu unserem Dublettatom noch ein weiteres s-Elektron mit dem Impuls 1, verteilen diesen Impuls 1 gleichmäßig auf Rumpf und Elektron und erhalten so für die Grundbahn des Triplettatoms zwei fundamentale Möglichkeiten:

1. Der Atomrumpf hat den mittleren Impuls 0, das eine Valenzelektron $1/2$, das andere $-1/2$. Der Gesamtimpuls des Atoms ist 0, das Atom ist also diamagnetisch. Wegen des Gesamtimpulses 0 wird diese Konfiguration besonders stabil sein. Sie liefert den Grundterm $1S$ der Einfachlinien. Wie sich aus Schlüssen, die denen bei den Dubletts genau nachgebildet sind, ergibt, verallgemeinert sich diese These im angeregten Zustand dahin, daß das eine Valenzelektron den mittleren Impuls $n - 1/2$, das andere $-1/2$ erhält. Die Verteilung $n - 1/2$ und $+1/2$ wollen wir aus demselben Grunde ausschließen, dem wir die Einfachheit des s-Terms bei den Dubletts entnommen haben; später wird noch von dieser Möglichkeit zu sprechen sein. Der Gesamtimpuls des Atoms ist also $n - 1$, in Übereinstimmung mit der inneren Quantenzahl n_J, die Landé[1]) den Einfachlinien zuordnet. Der Grund, warum wir es hier mit wirklichen Einfachlinien zu tun haben, ist ersichtlich der, daß der Atomrumpf den mittleren Impuls 0 hat. In der Tat haben wir schon bei den Dubletts erkannt, daß ein

[1]) A. Landé, Über das Serienspektrum des Ne und Hg. Phys. ZS. **22**, 417, 1921.

Zeitschrift für Physik 8, 273–297 (1922)

magnetisches Moment des Rumpfes die hinreichende und notwendige Bedingung für Linienstruktur und anomalen Zeemaneffekt darstellt. Wir sehen also, daß ein Triplettatom stets auch ein Einfachliniensystem enthalten muß, wie dies ja auch bei den betreffenden Elementen stets beobachtet wurde.

2. Die zweite Möglichkeit für den s-Term ist folgende: Der Atomrumpf hat den mittleren Impuls $+1$, das innere Valenzelektron $+{}^1/_2$, das äußere ebenfalls $+{}^1/_2$. Der Gesamtimpuls ist hier 2. Dies entspricht durchaus den Termaufspaltungen, die Landé am Zeemaneffekt des s-Terms feststellt. Im angeregten Zustande würde dies Atom, das eigentliche Triplettatom, dann folgendermaßen aussehen: Ein Rumpf mit dem mittleren Drehimpuls 1, ein inneres Valenzelektron mit dem Impuls ${}^1/_2$, ein äußeres mit $n - {}^1/_2$. Die gegenseitige Stellung dieser drei Impulsvektoren soll wieder durch magnetische Kopplung, genau wie im Falle der Dublettatome, festgelegt werden. Zunächst konstatieren wir, daß der s-Term einfach sein wird. Nach Schrödinger werden nämlich die beiden s-Elektronen in den Rumpf eindringen. Also wird die gleichsinnige Drehrichtung ausgezeichnet, außerdem die gegenseitige Kopplung viel zu stark sein, um andere Stellungen als die Gleichrichtung aller drei Impulse zuzulassen. Dieser Schluß gilt aber vom inneren Valenzelektron allgemein, auch für die angeregten Zustände. Bis auf Größen, die von der Größenordnung

$$\delta = \frac{\text{Kopplung des äußeren Elektrons mit dem Rumpf}}{\text{Kopplung des inneren Elektrons mit dem Rumpf}}$$

sind, können wir deshalb sagen, daß der mittlere Impuls von Rumpf + innerem Elektron ${}^3/_2$ wird.

Die gegenseitige Lage der drei Impulse wird auch hier wieder durch räumliche Quantelung (und zwar diesmal beider Elektronen gegen den Rumpf und sein Feld) festgelegt sein. Wir beziehen den Index 1 auf das äußere, den Index 2 auf das innere Elektron, den Index 3 auf den Rumpf. Dann heißt die der Gleichung (7) entsprechende Quantenbedingung:

$$ {}^1/_2 \cos\vartheta_{23} + n^* \cos\vartheta_{13} = m_i. \tag{21}$$

Hierbei wird m_i ganzzahlig sein, weil auch $n^* + {}^1/_2$ eine ganze Zahl ist. Da die relative Lage vom Rumpf zum inneren Elektron jedoch schon durch $\vartheta_2 = \vartheta_3$, $\cos\vartheta_{23} = 1$ definiert ist[1]), liefert uns Gleichung (21) den Winkel $\vartheta_{13} = \vartheta_{12}$ zwischen den beiden Impulsen ${}^3/_2$ und $n - {}^1/_2$:

$$n^* \cos\vartheta_{12} = m_i^*.$$

[1]) Dies gilt bis auf die Größen von der Ordnung δ, die wir prinzipiell vernachlässigen wollen, was bei leichteren Atomen wohl eine hinreichende Näherung bedeutet.

Hier muß, wie später gezeigt werden wird, m_i^* die drei Werte

$$m_i^* = n^*, \quad 0^*, \quad -n^* \left[= n - \frac{1}{2}, \quad -\frac{1}{2}, \quad -\left(n - \frac{1}{2}\right)\right]$$

annehmen. Für die Tripletts sind also drei charakteristische Fälle möglich: Die Gleichrichtung beider Impulse, die entgegengesetzt gleiche Richtung und die „gekreuzte" Stellung von Rumpf zu Elektron und umgekehrt. Jetzt soll noch der Abstand des obersten und untersten Triplettniveaus $\Delta\nu_{13}$ in Wellenzahlen festgelegt werden. Bedeutet H_i wieder das innere Wechselwirkungsfeld, so wird

$$\Delta\nu_{13} = \left[\left(+\frac{3}{2}\right)H_i - \left(-\frac{3}{2}\right)H_i\right]\frac{1}{4\pi}\frac{e}{mc} = \frac{3}{4\pi}\frac{e}{mc}H_i. \tag{22}$$

Die Einstellung des Rumpfes (vgl. § 2, S. 299) bei beliebigem v können wir für das obere und untere Triplettniveau wieder formal durch die Bedingung festlegen:

$$\frac{\partial}{\partial\vartheta_2}\Delta E = 0; \quad \frac{\partial}{\partial\gamma_2}\Delta E = 0$$

bei konstant gehaltenem ϑ_1.

§ 5. Verhalten der Triplettatome im Magnetfeld. Sehr einfach läßt sich die Frage des Zeemaneffektes bei der ersten Möglichkeit, dem Einfachliniensystem, beantworten. Die der Gl. (21) entsprechende[1]) Bedingung der räumlichen Quantelung im Felde H lautet:

$$\frac{1}{2}\cos\vartheta_2 + n^*\cos\vartheta_1 = m. \tag{23}$$

Schreiben wir die Störungsenergie auf, so wird, wie in Gl. (11)

$$\left.\begin{aligned} \Delta E &= \Delta\nu_n . h\left[n^*\cos\vartheta_1 + \frac{1}{2}\cos\vartheta_2 + \frac{1}{2}v\cos\vartheta_{12}\right] \\ &= \Delta\nu_n . h\left[m + \frac{v}{2}\cos\vartheta_{12}\right]. \end{aligned}\right\} \tag{24}$$

$\cos\vartheta_{12}$ wird, wie aus den Annahmen von § 4 folgt, für $v \sim \infty$ gleich -1. Für kleines Feld H haben wir also bereits normalen Zeemaneffekt, was sich durch Wachsen von H nicht ändern kann. Wir schreiben also an:

$$\Delta E = \Delta\nu_n . h\left(m - \frac{v}{2}\right). \tag{25}$$

Der normale Zeemaneffekt des Einfachliniensystems war ja auch, wegen des Rumpfimpulses 0, von vornherein klar.

[1]) In (21) selbst handelte es sich um das Wechselwirkungsfeld H_i. Analog der Gleichung (7a) muß (23) für $\cos\vartheta_{12} \neq \pm 1$ etwas allgemeiner (vgl. S. 308) gefaßt werden.

Zeitschrift für Physik *8*, 273–297 (1922)

In ganz derselben Weise schreiben wir jetzt für die zweite Möglichkeit, die eigentlichen Tripletterme, die Störungsenergie auf. Sie wird

$$\Delta E = \Delta\nu_n . h\left[n^* \cos\vartheta_1 + \frac{3}{2}\cos\vartheta_2 + \frac{3}{2} v \cos(\vartheta_1 - \vartheta_2)\right]. \quad (26)$$

Hier ist die Bedingung $\partial\Delta E/\partial\gamma_2 = 0$, $\vartheta_{12} = \vartheta_1 - \vartheta_2$ bereits berücksichtigt. Wie bei den Dubletts wollen wir $\cos\vartheta_2 + v\cos(\vartheta_1 - \vartheta_2)$ mit X bezeichnen. Dabei beschränken wir uns zunächst auf das oberste und unterste Triplettniveau und legen ΔE gleich für beliebige Werte von v fest; dann haben wir zur Bestimmung von ΔE vier Gleichungen:

$$\Delta E = \Delta\nu_n . h\left[n^* \cos\vartheta_1 + \frac{3}{2} X\right] \quad (27)$$

$$X = \cos\vartheta_2 + v\cos(\vartheta_1 - \vartheta_2) \quad (28)$$

$$\frac{\partial X}{\partial\vartheta_2} = 0 = \sin\vartheta_2 + v\sin(\vartheta_2 - \vartheta_1) \quad (29)$$

und dazu die Quantenbedingung (23).

Aus (28) und (29) folgt wieder, wie in § 2:

$$X^2 = 1 + 2v\cos\vartheta_1 + v^2. \quad (30)$$

Aus (29) ferner entnehmen wir:

$$0 = \sin\vartheta_2 . (1 + v\cos\vartheta_1) - v\sin\vartheta_1\cos\vartheta_2.$$

Daraus folgt trigonometrisch

$$\cos\vartheta_2 = \frac{1 + v\cos\vartheta_1}{\sqrt{1 + 2v\cos\vartheta_1 + v^2}} = \frac{1 + v\cos\vartheta_1}{X}. \quad (31\,\mathrm{a})$$

Setzen wir hierin $\cos\vartheta_2$ aus (23) ein, so ergibt sich

$$2(m - n^*\cos\vartheta_1) = \frac{1 + v\cos\vartheta_1}{X}$$

und

$$\cos\vartheta_1 = \frac{2mX - 1}{2n^*X + v}. \quad (31\,\mathrm{b})$$

Aus (30) und (31 b) erhalten wir

$$X^2 - 1 - v^2 = 2v\frac{2Xm - 1}{2Xn^* + v},$$

oder

$$X^3 + \frac{v}{2n^*}X^2 - X\left(1 + \frac{2m}{n^*}v + v^2\right) - \frac{v^3 - v}{2n^*} = 0. \quad (32)$$

Wir sind also zu einer Gleichung dritten Grades für X gekommen, von deren Wurzeln aber nur zwei physikalische Bedeutung haben können; für die dritte Wurzel ergeben sich imaginäre Winkel.

Für das oberste und unterste Triplettniveau schließen wir von X auf ΔE durch die Gleichung [vgl. (27), (31 b)]

$$\Delta E = \Delta v \,.\, h \,.\left[n^* \frac{2\,Xm - 1}{2\,Xn^* + v} + \frac{3}{2} X\right]. \tag{33}$$

Gl. (32) hat für kleines magnetisches Feld H ($v \sim \infty$) die zwei für uns in Betracht kommenden Wurzeln

$$X = v + \frac{m}{n}; \quad X = -v - \frac{m}{n-1}.$$

Kürzen wir $\frac{\Delta E}{\Delta v_n \,.\, h}$ durch $\Delta E_1'$, $\Delta E_2'$, $\Delta E_3'$ für das erste, zweite, dritte Triplettniveau ab, so folgt daraus für $v \sim \infty$

$$\Delta E_1' = \frac{3}{2} v + m\left(1 + \frac{1}{n}\right); \quad \Delta E_3' = -\frac{3}{2} v + m\left(1 - \frac{1}{n-1}\right). \tag{34}$$

Zur Diskussion dieses Ergebnisses soll erst später übergegangen werden.

Zuerst müssen wir noch den Beweis dafür nachholen, daß m_i^* nur die drei Werte n^*, 0^*, $-n^*$ annehmen kann; zu diesem Zweck, werden wir analog der Beweisführung in § 2, ΔE für $v \sim \infty$ und beliebiges m_i^* ausrechnen.

Es wird

$$\Delta E = \left[J \cos \Theta + \frac{3}{2} v \cdot \frac{m_i^*}{n^*}\right] \Delta v_n \,.\, h.$$

J, der Gesamtimpuls, ist wegen $\cos \vartheta_{12} = \frac{m_i^*}{n^*}$ gegeben durch

$$J = \sqrt{n^{*2} + 2 \cos \vartheta_{12} \cdot \frac{3}{2} \cdot n^* + \left(\frac{3}{2}\right)^2} = \sqrt{n^{*2} + 3\,m_i^* + \left(\frac{3}{2}\right)^2}.$$

$\cos \Theta$ (wo Θ der Winkel des Gesamtimpulses gegen die Feldachse) wird entsprechend (23), gleich $\frac{m}{\overline{p}_{12}}$, wo $\overline{p}_{12}$ den Mittelwert der Impulse n^* und $\frac{1}{2}$, also deren Projektion auf die Richtung von J darstellt. Die vektorielle Summe von n^* und $\frac{1}{2}$ ergibt sich zu $\sqrt{n^{*2} + \left(\frac{1}{2}\right)^2 + m_i^*}$, deren Projektion auf J wird

$$\overline{p}_{12} = \frac{n^{*2} + \left(\frac{1}{2}\right)^2 + m_i^* + J^2 - 1}{2\,J}.$$

Daraus folgt für ΔE:

$$\Delta E = \Delta \nu_n . h \left[\frac{2\left(n^{*2} + \left(\frac{3}{2}\right)^2 + 3\, m_i^*\right)}{2\, n^{*2} + \frac{3}{2} + 4\, m_i^*} m + \frac{3}{2} v \frac{m_i^*}{n^*}\right]$$

$$= \Delta \nu_n . h \left[\left(1 + \frac{2\left(m_i^* + \frac{3}{2}\right)}{2\, n^{*2} + 4\, m_i^* + \frac{3}{2}}\right) m + \frac{3}{2} v \frac{m_i^*}{n^*}\right].$$

Gleichung (10) liefert jetzt als Bedingung für die Größen m_i^*:

$$\sum_{k=1}^{\lambda} \frac{2\left(m_{ik}^* + \frac{3}{2}\right)}{2\, n^{*2} + 4\, m_{ik}^* + \frac{3}{2}} = 0.$$

Denkt man sich wieder m_i^* als Funktion von n geschrieben, so findet man, daß diese Bedingung durch das Wertsystem $\lambda = 3$, $m_{i1}^* = n^*$, $m_{i2}^* = 0^*$, $m_{i3}^* = -n^*$ und nur durch dieses befriedigt wird. Wie wir also im Falle des einen Valenzelektrons zwangsläufig auf ein Dublett, so werden wir im Falle der zwei Valenzelektronen zwangsläufig auf ein Triplett geführt. Der Fall $m_i^* = -\frac{3}{2}$ ist für $n > 2$ aus demselben Grunde auszuschließen, aus dem wir $m_i^* = -\frac{1}{2}$ bei den Dubletts als unmöglich erklärten.

Setzen wir jetzt die gefundenen Werte von m_i^* in ΔE für $v \sim \infty$ ein, so ergeben sich die Landéschen Formeln (für $H \leqslant H_i$), die für ΔE_1 und ΔE_3 schon auf andere Weise hergeleitet wurden:

$$\left.\begin{aligned} \Delta E_1 &= \Delta \nu_n . h \left[m\left(1 + \frac{1}{n}\right) + \frac{3}{2} v\right] \\ \Delta E_2 &= \Delta \nu_n . h \left[m\left(1 - \frac{1}{n} + \frac{1}{n-1}\right) + \frac{3}{2} \frac{0^*}{n^*} v\right] \\ \Delta E_3 &= \Delta \nu_n . h \left[m\left(1 - \frac{1}{n-1}\right) - \frac{3}{2} v\right] \end{aligned}\right\} \quad (35)$$

Gleichung (10) hat hier die spezielle Form:

$$\Delta E_1 + \Delta E_2 + \Delta E_3 = \Delta \nu_n . h . \left(3\, m + \frac{3}{2} v \cdot \frac{0^*}{n^*}\right). \quad (36)$$

Sie muß, ihrer Ableitung entsprechend, für beliebige Werte von v gelten und liefert daher in Verbindung mit (33) ΔE_2 auch für mittlere und starke Felder ohne weitere Rechnung. Daß die hierdurch gegebene Energie des mittleren Triplettniveaus wirklich einem adiabatischen Übergang aus (35) entspricht, folgt daraus, daß dies einerseits für $\varDelta E_1$ und $\varDelta E_3$ entsprechend ihrer Ableitung, andererseits für die rechte Seite von Gl. (36) der Fall ist.

Wir gehen nun zur Diskussion der Resultate über. Ein Vergleich von (34) und (35) mit (4) der Einleitung zeigt, daß die entwickelte Theorie die anomalen Zeemaneffekte der Triplettsysteme erschöpfend wiederzugeben vermag (Über die Bedingung $|m| \leqq n_J$ vgl. S. 284). — Für $v = 0$ liefert (32) und (36) den Paschen-Backeffekt: Es wird

$$\varDelta E_1' = m + 1; \quad \varDelta E_2' = m; \quad \varDelta E_3' = m - 1.$$

Die einzelnen magnetischen Niveaus haben sich sehr charakteristisch übereinandergeschoben. Die Übereinstimmung dieses Resultats mit der Erfahrung beim partiellen Paschen-Backeffekt von Mg ($p_i d$) ist näher ausgeführt von Sommerfeld[1]). Die Figur stellt

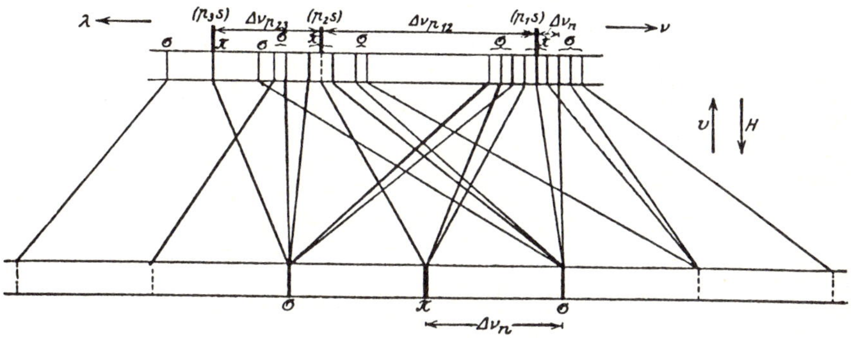

den Übergang der Linien im (sp_i) Triplett vom anomalen Zeemaneffekt bis zum Paschen-Backeffekt dar; die entsprechende Figur für die Dublettstruktur teilt Sommerfeld in dieser Zeitschrift mit. Ein quantitativer Vergleich der theoretischen Ergebnisse für mittlere Magnetfelder mit der Erfahrung ist bis jetzt unmöglich. Das einzige vorliegende Material scheinen die Messungen von Paschen und Back[2]) am Sauerstofftriplett $\lambda = 3947$ zu sein. Diese stehen mit dem abgeleiteten Ergebnis sehr gut im Einklang. Die Gleichungen (35) und (36) fordern z. B. für die endgültige mittlere π-Komponente bei starken Feldern, daß sie in Strenge (d. h. bis auf Größen von der Ordnung v^2, nicht nur bis auf Größen der Ordnung v) einfach sei, daß jedoch die beiden σ-Komponenten Mehrfachlinien mit Abständen von der Ordnung der ursprünglichen Triplettabstände bleiben müssen. Beides, sowie die theoretische Lage der π-Komponente ($\varDelta E_\pi = \frac{v}{2}$

[1]) Sommerfeld, Atombau, 3. Aufl., Kap. 6, S. 495.

[2]) Paschen u. Back, Ann. d. Phys. 40, 960, 1913.

für $v \ll 1$) findet in den Aufnahmen volle Bestätigung. Für mittlere Magnetfelder läßt sich wegen der Verwaschenheit der Linien nur qualitativ eine Übereinstimmung zwischen Theorie und Erfahrung feststellen. Eine weitere Möglichkeit der quantitativen Prüfung besteht wieder bei den an den verbotenen Linien ($p_2 d_1$ usw.) gemessenen „Verschiebungen ϑ" (vgl. S. 301). Diese Verschiebungen zeigen deutlich den zu erwartenden Gang. Wegen der Vernachlässigung der Größen von der Ordnung δ wird jedoch die direkte Berechnung von ϑ unsicher.

Für die Genauigkeit unserer Formeln für mittlere Magnetfelder müssen wir noch eine allgemeine, auch für die Dubletts gültige Bemerkung machen. Die Rechnungen wurden unter wesentlich idealisierenden Voraussetzungen durchgeführt, z. B. Homogenität des Feldes H_i in der Nähe des Rumpfes, Fehlen elektrischer Wirkungen. Die damit begangene Ungenauigkeit ist für die Grenzfälle $v \ll 1$ und $v \gg 1$ offenbar gleichgültig, spielt aber für $v \backsim 1$ eine gewisse Rolle. Es ist sehr wahrscheinlich, daß z. B. die kleinen Diskrepanzen, die Kent bei Li, $\lambda = 6708$ im Gebiet der mittleren Felder zwischen der Voigtschen Theorie und der Erfahrung feststellt, auf die genannten Einflüsse zurückzuführen sind. Die theoretischen Formeln bedeuten bis jetzt zweifellos nur eine erste Näherung. Erst wenn uns die Einzelheiten des Modells bekannt sind, ist auch hier die völlig exakte Rechnung möglich.

§ 6. Quantitative Folgerungen über den Triplettabstand. Als wichtigstes Resultat können wir aus den Formeln (34) und (35) das Verhältnis der Triplettabstände $\frac{\Delta \nu^{12}}{\Delta \nu^{23}}$ (ohne Magnetfeld) ableiten. Es wird

$$\frac{\Delta \nu^{12}}{\Delta \nu^{23}} = \frac{1 - \frac{0^*}{n^*}}{1 + \frac{0^*}{n^*}} = \frac{1 + \frac{1}{2n^*}}{1 - \frac{1}{2n^*}} = \frac{n}{n-1}.$$

Diese Gleichung sagt aus:

Für den p-Term ist $\frac{\Delta \nu_p^{12}}{\Delta \nu_p^{23}} = 2$, für den d-Term ist $\frac{\Delta \nu_d^{12}}{\Delta \nu_d^{23}} = 1{,}5$.

In der Spektroskopie ist dieses Verhältnis längst als charakteristisch für den p- und d-Term bekannt; es unterscheidet die magnetischen Tripletts bezeichnend von den relativistischen. Wie eine Tabelle [1]) zeigt, wird es für die leichteren Elemente stets bis auf einige Prozent genau so beobachtet, wie es die Theorie fordert, nur bei schweren Atomen zeigen sich systematische Abweichungen. Auch

[1]) Vgl. Sommerfeld, l. c. Atombau, 3. Aufl., Kap. 6, S. 455.

dies ist sehr verständlich, denn dort ist eben die Vernachlässigung der Größe δ nicht mehr statthaft. Auch die Messungen von **Saunders**[1]) am b-Term von Barium passen sich dem theoretisch zu erwartenden Verhältnis $\frac{\Delta \nu_b^{12}}{\Delta \nu_b^{23}} = 1{,}33$ gut an.

Über die absolute Größe des Triplettabstandes lassen sich insbesondere bei den Elementen der zweiten Spalte des periodischen Systems Angaben machen. Die Bahn des äußeren Valenzelektrons wird nämlich in roher Annäherung für ein Alkali und das folgende Erdalkali wesentlich dieselbe sein. Also kann man ungefähr setzen $H_{i\,\text{Alkali}} \backsim H_{i\,\text{Erdalkali}}$ Größere Abweichungen werden sich da zeigen, wo im Dublettatom das Atominnere noch als sehr klein vorausgesetzt werden kann, beim Triplettatom (wegen des inneren Valenzelektrons) aber nicht mehr.

Aus (22) und (8) schließen wir dann, daß der Abstand des obersten vom untersten Triplettniveau im Erdalkali etwa dreimal so groß ist wie der entsprechende Dublettabstand im vorausgehenden Alkali. Auch diese Beziehung ist hinreichend gut erfüllt. Im übrigen gelten für das Verhalten der Triplettabstände mit wachsender Ordnungszahl dieselben Überlegungen, wie wir sie bei den Dublettsystemen angedeutet haben.

Abschließend sei zu den Triplettsystemen noch bemerkt, daß bei ihnen die theoretischen Schwierigkeiten in mancher Beziehung erheblich größer sind als bei den Dublettsystemen. Die „innere Quantenzahl“ hat nicht mehr die einfache Bedeutung des Gesamtimpulses (dies trifft nur für die Einfachlinie und das jedesmal unterste Triplettniveau zu), sondern muß eher als eine Art von Numerierung der drei möglichen Rumpfstellungen aufgefaßt werden. Die Auswahlregel für n_J kann daher bei den Triplettsystemen nur unter dem Gesichtspunkte des Korrespondenzprinzips verstanden werden. Auch die Intensitätsfrage kann dann, wie bei den Dubletts schon hervorgehoben wurde, auf diesem Wege befriedigend gelöst werden.

Endlich mag noch die Frage der sogenannten „uneigentlichen Dubletts“ kurz berührt werden, wie sie z. B. bei Orthohelium beobachtet sind. Es liegt nahe, sie als Dubletts von „Einfachlinien“ mit den Möglichkeiten der Impulsstellung des äußeren und inneren Valenzelektrons (vgl. S. 288) ($n - 1/2$ und $-1/2$) und ($n - 1/2$, $+ 1/2$) aufzufassen, denn gerade der Grund, der für das Ausfallen der letzteren Möglichkeit geltend gemacht wurde, trifft bei He wegen des fehlenden Rumpfes nicht zu. Der normale Zemanneffekt dieser Dubletts, wie er ja beobachtet ist, versteht sich dann von selbst.

[1]) **Saunders**, Astrophys. Journ. **51**, 23, 1920.

Zur einfacheren Übersicht wollen wir noch einmal die Deutungen und Umdeutungen der verwendeten Quantenzahlen zusammenstellen. Die in den Serienformeln im Nenner auftretende Quantenzahl bleibt ganzzahlig und ist als Summe der radialen und der Impulsquantenzahl für das äußere Elektron aufzufassen. Letztere stimmt nicht überein mit der bisher üblichen azimutalen Quantenzahl n, sondern sie ist gleich $n - {}^1/_2$. Die radiale Quantenzahl schreitet mit den Werten (${}^1/_2$, ${}^3/_2$ ...) fort. Die „innere" Quantenzahl n_J, die für die innermagnetisch verschiedenen Niveaus verschieden ist, stimmt bei den Dublettsystemen überein mit dem Gesamtimpuls des Atoms; bei diesen kann man dann n als Summe der Absolutwerte des Rumpfimpulses und des dem äußeren Elektron zugehörigen Drehimpulses deuten. Bei den Tripletts ist der Zusammenhang zwischen n_J und dem Gesamtimpuls komplizierter. Festgelegt wird die Dublett- bzw. Triplettstruktur durch die innere magnetische Quantenzahl m_i^* bzw. m_i. Sie bedeutet die Projektion des mittleren Impulses der Valenzelektronen auf die Achse des vom Rumpf herrührenden Magnetfeldes. Die magnetische Quantenzahl m^* bzw. m ist gegeben durch die Projektion des mittleren Impulses der Valenzelektronen auf die Feldachse von H.

Zusammenfassung. In § 1 und § 4 werden über die Struktur von Dublett- und Triplettatomen einfache Annahmen gemacht, die die Linienmultiplizitäten durch innere Magnetfelder erklären. In § 2 und § 5 wird das Verhalten dieser Atome im äußeren Magnetfelde untersucht. Die Untersuchung liefert die Gesamtheit der anomalen Zeemaneffekte in Übereinstimmung mit der Erfahrung und stellt den Übergang zum Paschen-Backeffekt formelmäßig dar. Als wesentlicher physikalischer Grund für die Möglichkeit anomaler Zeemaneffekte wird ein Widerspruch gegen die klassische Ausstrahlung (Kugelwelle) festgestellt. In § 3 und § 6 werden quantitative Schlüsse auf die Dublett- und Triplettabstände gezogen. Die Aufspaltung der Lithiumlinie $\lambda = 6708$ wird berechnet in guter Übereinstimmung mit der Erfahrung; auch die Abnahme der Aufspaltung in der Reihe (s, p, d ...) wird formelmäßig festgelegt. Für das Verhältnis der Triplettabstände $\frac{\Delta\nu^{12}}{\Delta\nu^{23}}$ wird eine einfache Formel hergeleitet, die das charakteristische Verhalten dieses Quotienten in der Reihe der Elemente: 2 : 1 beim p-Term, 3 : 2 beim d-Term usw. wiedergibt.

Am Schluß möchte ich meinem hochverehrten Lehrer, Herrn Prof. Sommerfeld, meinen Dank aussprechen für die Anregung zu dieser Arbeit und für vielfache Förderung.

München, Institut für theoretische Physik.

Eine Bemerkung über relativistische Röntgendubletts und Linienschärfe.

Von **A. Sommerfeld** und **W. Heisenberg** in München.

(Eingegangen am 3. August 1922.)

In den neueren Überlegungen Bohrs spielt der Strahlungswiderstand eine prinzipielle Rolle. Da die Ergebnisse der klassischen Theorie durch das Korrespondenzprinzip für die Quantentheorie nutzbar gemacht werden und da der Strahlungswiderstand ein notwendiges Glied der klassischen Theorie bildet, so sollte er auch für die quantentheoretische Behandlung von Bedeutung sein, obwohl er dort vernachlässigt werden muß: er bestimmt nach Bohr die Unschärfe, die den quantentheoretischen Methoden notwendig anhaftet. Eine Quantenrechnung, welche Glieder berücksichtigt, die kleiner sind als der vernachlässigte Energieverlust durch klassische Ausstrahlung, ist nach Bohr prinzipiell unzulässig. Zu solchen Gliedern zählt man, wie es scheint, die höheren Relativitätskorrektionen, die in der Theorie der Röntgendubletts bei den schweren Atomen auftreten. Es soll demgegenüber gezeigt werden, daß die Berücksichtigung dieser Korrektionen mit dem Bohrschen Gesichtspunkte von der Rolle des Strahlungswiderstandes vollkommen verträglich ist. Ferner soll besprochen werden, was sich aus dem gleichen Gesichtspunkte für die Abklingungsversuche von W. Wien[1]) beim Leuchten der Kanalstrahlen schließen läßt.

§ 1. Die Größenordnung der verschiedenen Relativitätskorrektionen im Vergleich mit dem Strahlungswiderstande. Vom empirischen Standpunkte aus kann kein Zweifel sein über die Notwendigkeit der höheren Relativitätskorrektionen bei den L-Dubletts der schweren Elemente. (Bei den M- und N-Dubletts sowie bei den leichteren Elementen sind diese Korrektionen ohnehin so klein, daß sie numerisch belanglos werden.) Man vgl. z. B. Tab. 51 in dem Buche[2]) „Atombau und Spektrallinien". Hier ergibt sich der in der ganzen Reihe der Elemente konstante Wert 3,5 der Abschirmungszahl nur dann, wenn man z. B. bei Uran außer dem Hauptgliede mit α^2 $\left(\alpha = \text{Konstante der Feinstruktur} = \frac{2\pi e^2}{ch} = 7{,}29 \, . \, 10^{-3}\right)$ noch die Glieder mit α^4 und α^6 berücksichtigt. Wollte man bei dem

[1]) Ann. d. Phys., Mitteilung I, **60**, 597, 1919 u. Mitteilung II, **66**, 229, 1921.
[2]) 3. Aufl., Braunschweig 1922, S. 611.

Hauptgliede stehen bleiben, so ergäbe sich für Uran statt $s = +3{,}5$ der unsinnige Wert $s = -3{,}7$.

Überschlagen wir nun theoretisch die sukzessiven Relativitätskorrektionen im Verhältnis zum Strahlungswiderstande. Jene entstehen aus dem Trägheitsgliede, das wir für transversale Beschleunigung — wir denken etwa an eine Kreisbahn — anschreiben:

$$\frac{m_0}{\sqrt{1-\beta^2}}\,\dot{\mathfrak{v}} = m_0\,\dot{\mathfrak{v}}\left(1+\frac{1}{2}\beta^2+\frac{3}{8}\beta^4+\cdots\right), \tag{1}$$

dieser ist für eine Kreisbahn von der Periode $\frac{2\pi}{\omega}$ umgerechnet:

$$\frac{2}{3}\frac{e^2}{c^3}\,\ddot{\mathfrak{v}} = \frac{2}{3}\frac{e^2}{c^3}\,i\,\omega\,\dot{\mathfrak{v}}. \tag{2}$$

Jedem Gliede β^{2s} von (1) entspricht in der schließlichen relativistischen Dublettformel je das Glied mit α^{2s}. Um zu sehen, welche Glieder in (1) größer sind als (2), wie viele Glieder der Entwickelung in (1) also im Sinne Bohrs beizubehalten sind, bilden wir das Verhältnis von (2) gegen das Glied mit β^{2s} von (1), unter Weglassung unwesentlicher Zahlenfaktoren:

$$\left.\begin{aligned}\frac{e^2\omega}{m_0 c^3\beta^{2s}} &= \frac{e^2\omega^3}{\omega^2 m_0 c^3\beta^{2s}} = \frac{e^2 v^3}{a^3\omega^2 c^3 m_0\beta^{2s}} = \frac{e^2\beta^{3-2s}}{a^3\omega^2 m_0} = \frac{\beta^{3-2s}}{Z}\\ &= \frac{\alpha}{n}\left(\frac{\alpha Z}{n}\right)^{2-2s}.\end{aligned}\right\} \tag{3}$$

Hier entsteht die zweite Form aus der ersten durch Erweiterung von Zähler und Nenner mit ω^2, die dritte durch die Substitution $\omega = \frac{v}{a}$, wo a den Bahnradius bedeutet; die vierte berücksichtigt $\frac{v}{c} = \beta$; die fünfte benutzt die Gleichung der Zentrifugalkraft:

$$m\,a^3\omega^2 = e^2 Z, \tag{4}$$

wo Z die effektive Kernladungszahl für den als wasserstoff- und kreisähnlich berechneten Bahntypus bedeutet; schließlich folgt die letzte Form, wenn man berücksichtigt:

$$\beta = \frac{\alpha Z}{n}, \tag{5}$$

welche Beziehung in bekannter Weise aus (4) und der Gleichung des Impulsmomentes

$$m\,a^2\omega = \frac{nh}{2\pi} \tag{6}$$

durch Division beider hervorgeht.

Das in (3) berechnete Verhältnis muß nun $\leqq 1$ sein, wenn das Glied α^{2s} noch beizubehalten ist, d. h. es muß sein:

$$\left(\frac{\alpha Z}{n}\right)^{2s-2} \geqq \frac{\alpha}{n}. \tag{7}$$

Die folgende Tabelle gibt die linke Seite von (7) für $n = 2$ (L-Dublett) und für verschiedene Werte von s und Z an. Wir haben diejenige Zahl in jeder Reihe eingeklammert, welche bereits kleiner als $\frac{\alpha}{2} = 3{,}6 \,.\, 10^{-3}$ ist; das zugehörige s gibt diejenige Relativitätskorrektion α^{2s} an, welche dementsprechend wegen des Strahlungswiderstandes bereits als nichtssagend fortzulassen ist. Man sieht: Bei Wasserstoff und den leichtesten Elementen hat man nur das erste Relativitätsglied $\frac{R\alpha^2 Z^4}{2^4}$ zu berücksichtigen. Oberhalb Ca tritt das zweite Glied hinzu. Bei Uran, $Z = 92 - 3{,}5$, ist es berechtigt, auch das dritte Glied mitzurechnen. Die hierdurch bestimmten Grenzen decken sich vollständig mit den Genauigkeitsgrenzen, in denen die seitherige Berechnung des L-Dubletts ausgeführt wurde.

$s =$	1	2	3	4
$Z = 10$	1	$(1{,}3 \,.\, 10^{-3})$	—	—
20	1	$5{,}3 \,.\, 10^{-3}$	$(2{,}8 \,.\, 10^{-5})$	—
50	1	$3{,}3 \,.\, 10^{-2}$	$(1{,}1 \,.\, 10^{-3})$	—
90	1	$1{,}1 \,.\, 10^{-1}$	$1{,}2 \,.\, 10^{-2}$	$(1{,}3 \,.\, 10^{-3})$

Wir ziehen daraus den Schluß:

Der von Bohr ins Auge gefaßten korrespondenzmäßigen Verwertung des Strahlungswiderstandes steht von seiten der Relativitätskorrektionen der Röntgendubletts nichts im Wege.

§ 2. Abklingungsdauer und Linienbreite als Folge des Strahlungswiderstandes. Das in (3) berechnete Verhältnis gibt für $s = 0$ den Strahlungswiderstand geteilt durch das erste Trägheitsglied, welches hinreichend genau den gesamten Trägheitswiderstand darstellt. Indem wir das Verhältnis der Widerstände gleich dem der entsprechenden Energiebeträge setzen[1]), schreiben wir (3):

$$\frac{U}{W} = \frac{\alpha^3 Z^2}{n^3}. \tag{8}$$

[1]) Hierin liegt eine gewisse Willkür. Wenn wir die Energie U direkt berechnen, z. B. als die während eines Umlaufes ausgestrahlte Energie, so tritt in (8) rechter Hand der Faktor $8\pi/3$ hinzu. Durch unsere korrespondenzmäßige Normierung in Gleichung (13) würde dieser Faktor nachträglich wieder fortgeschafft werden. Statt 4/15 würde es dann in (13) heißen $1/10\pi$.

Hier bedeutet U die Unschärfe, die durch den vernachlässigten Strahlungswiderstand in die Energiebestimmung hineinkommt, W ist die quantenmäßig bestimmte Energie der Bahn, also für ein wasserstoffähnliches Atom:

$$W = \frac{h R Z^2}{n^2} \tag{9}$$

(R = Rydbergkonstante). Aus (8) und (9) folgt:

$$\frac{U}{h} = \frac{\alpha^3 R Z^4}{n^5}. \tag{10}$$

Es kommt jetzt darauf an, aus dem Korrespondenzprinzip auf den Zusammenhang zu schließen zwischen den Unschärfen U_1, U_2 der beiden Termenergien und der Breite $\Delta\nu$ der Linie, die sich aus den beiden Termen zusammensetzt. Es zeigt sich, daß $h\Delta\nu$ proportional gesetzt werden muß mit $U_1 - U_2$ (nicht etwa, wie man vom Begriff der Energieschwankung ausgehend lieber annehmen würde, proportional mit $\sqrt{U_1^2 + U_2^2}$), und daß der Proportionalitätsfaktor von der Größenordnung 1 wird.

Um das Korrespondenzprinzip mit Sicherheit anwenden zu können, gehen wir in den Quantenzahlen (n_2, n_1) von Anfangs- und Endterm zur Grenze $n_2 = n_1 + 1 = n = \infty$ über und erhalten aus (10):

$$\frac{1}{h}(U_1 - U_2) = \alpha^3 R Z^4 \left(\frac{1}{n_1^5} - \frac{1}{n_2^5}\right) = \alpha^3 R Z^4 \frac{5}{n^6}. \tag{11}$$

Andererseits ist die klassisch gerechnete Linienbreite, soweit sie lediglich vom Strahlungswiderstande herrührt:

$$\Delta\nu_{kl} = \frac{4\pi}{3} \frac{e^2}{c^3 m} \nu^2;$$

für wasserstoffähnliche Atome ergibt sie bei entsprechendem Grenzübergang:

$$\Delta\nu_{kl} = \frac{4\pi}{3} \frac{e^2}{c^3 m} R^2 Z^4 \left(\frac{1}{n_1^2} - \frac{1}{n_2^2}\right)^2 = \frac{4\pi}{3} \frac{e^2}{c^3 m} R^2 Z^4 \left(\frac{2}{n^3}\right)^2. \tag{12}$$

Die Abhängigkeit von n, Z und den Elektronenkonstanten ist, wie man nach der Bedeutung von α und R leicht bestätigt, in (11) und (12) die gleiche. Um den Zahlenfaktor zu vergleichen, bilden wir den Quotienten von (12) und (11):

$$\frac{4\pi}{3} \frac{e^2}{c^3 m} \frac{4R}{5\alpha^3} = \frac{4}{15}. \tag{13}$$

Mit diesem Faktor haben wir (11) zu multiplizieren, um $\frac{1}{h}(U_1 - U_2)$ zu normieren, d. h. an den klassisch berechneten Wert von $\Delta\nu_{kl}$ anzupassen. Den so normierten Betrag nennen wir die quantentheoretische Unschärfe:

$$\Delta\nu_{qu} = \frac{4}{15}\alpha^3 R Z^4\left(\frac{1}{n_1{}^5} - \frac{1}{n_2{}^5}\right) = \frac{16\pi}{15}\frac{e^2}{c^3 m} R^2 Z^4\left(\frac{1}{n_1{}^5} - \frac{1}{n_2{}^5}\right). \quad (14)$$

Wir wollen nun angeben, wie sich diese Formeln ändern, wenn wir von Kreis- zu Ellipsenbahnen übergehen. Indem wir mit Bohr n für die Hauptquantenzahl (Summe von azimutaler und radialer Quantenzahl) beibehalten, bezeichnen wir die azimutale Quantenzahl mit k. Wir berechnen U für eine Ellipsenbahn als die während eines Umlaufes ausgestrahlte Energie, wobei wir uns vorbehalten, den dabei auftretenden Zahlenfaktor (vgl. die Anmerkung zu S. 395) korrespondenzmäßig zu normieren. Wir erhalten dann statt Gleichung (10):

$$\frac{U}{h} = \frac{\alpha^3 R Z^4}{k^5} f, \quad f = 1 + \frac{\varepsilon^2}{2} = \frac{1}{2}\left[3 - \left(\frac{k}{n}\right)^2\right]. \quad (10\,a)$$

k^5 tritt also jetzt an die Stelle von n^5; der Faktor f, der für Kreisbahnen ($k = n$) gleich 1 wird, trägt der Exzentrizität ε der Ellipsenbahn Rechnung. Dementsprechend ergibt sich statt Gleichung (14):

$$\Delta\nu_{qu} = \frac{16\pi}{15}\frac{e^2}{c^3 m} R^2 Z^4\left(\frac{f_1}{k_1{}^5} - \frac{f_2}{k_2{}^5}\right). \quad (14\,a)$$

f_2 und f_1 sind die Werte von f in der Anfangs- und Endbahn.

Wir gehen jetzt auf die Linien der Balmerserie ($Z = 1$) und die Wienschen Kanalstrahlmessungen ihrer Abklingung ein. Indem wir jedesmal an die stärkste Linie der Feinstruktur denken, setzen wir $k_1 = 2$, $k_2 = 3$, $f_1 = 1$ und bzw. für H_α, H_β, H_γ

$$f_2 = 1 \text{ bzw.} = \frac{1}{2}\left[3 - \left(\frac{3}{4}\right)^2\right] = 1{,}22 \text{ bzw.} = \frac{1}{2}\left[3 - \left(\frac{3}{5}\right)^2\right] = 1{,}32,$$

also

$$\frac{f_1}{k_1{}^5} - \frac{f_2}{k_2{}^5} = \frac{1}{2^5}(1 - 0{,}13) \text{ für } H_\alpha,$$
$$= \frac{1}{2^5}(1 - 0{,}16) \;\;\text{„}\;\; H_\beta,$$
$$= \frac{1}{2^5}(1 - 0{,}17) \;\;\text{„}\;\; H_\gamma.$$

Aus Gleichung (14a) folgt daraufhin, daß unser $\Delta\nu_{qu}$ merklich konstant, d. h. für alle Balmerlinien dasselbe wird. Gerade dies hat Wien beobachtet.

Wien vergleicht seinen Beobachtungswert mit dem klassisch gerechneten Werte von H_α. Wir bilden deshalb aus (14 a) und (12) für $k_1 = n_1 = 2$, $k_2 = n_2 = 3$, $f_1 = f_2 = 1$:

$$\frac{\Delta\nu_{qu}}{\Delta\nu_{kl}} = \frac{4}{5}\,\frac{\frac{1}{2^5}-\frac{1}{3^5}}{\left(\frac{1}{2^2}-\frac{1}{3^2}\right)^2} = 1{,}125.$$

Ganz ähnlich findet Wien als Verhältnis der beobachteten zur klassisch berechneten Abklingungszeit in Mitteilung I: 1,25, in Mitteilung II: 0,81.

Wien hat auch Beobachtungen an O-Linien mitgeteilt, zum Teil Bogen-, zum Teil Funkenlinien. Da indessen (außer bei $\lambda = 395\,\mu\mu$) ihre Seriendarstellung unbekannt ist, so lassen sich einstweilen von hieraus keine theoretischen Schlüsse ziehen. Auch über die Abklingungszeit von Bandenlinien können wir nichts aussagen. Sehr interessant wird es sein, unsere Gleichung (14 a) an Messungen serientheoretisch wohl definierter Spektrallinien von verschiedener azimutaler Quantenzahl zu prüfen; die hierzu erforderliche Erweiterung der Gleichung (14 a) auf wasserstoffunähnliche Atome soll mitgeteilt werden, wenn solche Messungen verfügbar sein werden.

Im Anschluß an die Überlegungen bei der Balmerserie wird man schon jetzt sagen können, daß innerhalb einer Serie die Abklingungsdauer für alle Linien als gleich zu erwarten ist, da sie wesentlich nur von der Quantenzahl k, nicht von n abhängt.

Wir wollen nicht behaupten, daß unsere recht formale Betrachtung der physikalischen Bedeutung der Wienschen Abklingungsversuche gerecht wird oder daß sie uns über den Dämpfungsprozeß aufklärt. In der Tat liegt das nicht in der Absicht des Korrespondenzprinzipes, welches ja auf jedes modellmäßige Verständnis verzichtet. Man kann es auch für unbefriedigend ansehen, daß in unserer Betrachtung an Stelle der kausal bestimmten Abklingung das viel unbestimmtere Merkmal der Linienverbreiterung gesetzt wird, das nur als Folgeerscheinung mit jenem Prozeß verknüpft ist, und zwar nur durch die klassische Theorie. Immerhin scheint unsere Betrachtung nützliche Schlüsse auf den Abklingungsprozeß zu gestatten. Durch ihre Übereinstimmung mit der Erfahrung bestätigt sie umgekehrt die Grundlage unserer Überlegungen, nämlich die von Bohr ins Auge gefaßte korrespondenzmäßige Verwertung des Strahlungswiderstandes.

Die Intensität der Mehrfachlinien und ihrer Zeemankomponenten.

Von **A. Sommerfeld** und **W. Heisenberg** in München.

Mit zwei Abbildungen. (Eingegangen am 26. August 1922.)

Das Korrespondenzprinzip gestattet, wie bekannt, aus dem kinematischen Charakter der Atombahnen auf die Intensität der Spektrallinien zu schließen, die dem Wechsel der Atombahnen entspringen. Der kinematische Charakter einer Atombahn ist qualitativ bekannt, wenn die sie charakterisierenden Quantenzahlen festgestellt sind. Die letztere Aufgabe darf heutzutage für große Gruppen von Spektrallinien als gelöst gelten, nicht nur für Einfachlinien, sondern auch für Mehrfachlinien (Dubletts, Tripletts, allgemein „Multipletts") und ihre magnetischen Aufspaltungen (die anomalen Zeemankomponenten).

Für die Einfachterme genügen zwei Quantenzahlen, die Hauptquantenzahl n (wir benutzen Bohrs Bezeichnungen), welche die Laufzahl des Serienterms bestimmt, und die azimutale Quantenzahl k, welche die verschiedenen Serienterme unterscheidet ($k = 1$: s-Term, $k = 2$: p-Term, $k = 3$: d-Term, $k = 4$: b-Term). Alle Bemühungen der Spektroskopiker um die Serienordnung der Linien zielten, schon vor der Quantentheorie der Serienspektra, unbewußter Weise auf die Feststellung dieser beiden Quantenzahlen. Für die Mehrfachterme kommt die innere Quantenzahl j hinzu, welche die natürlichen Aufspaltungen eines Terms, z. B. die drei Niveaus eines Tripletts, voneinander unterscheidet. Auch die Bestimmung dieser inneren Quantenzahl ist heutzutage für viele Klassen von Mehrfachtermen gesichert. Die durch ein äußeres Magnetfeld hervorgerufenen Zeemanaufspaltungen schließlich werden durch eine vierte, die magnetische Quantenzahl m unterschieden; und die formale Theorie, die Landé für die Zeemanterme aufgestellt hat, bestimmt die Werte von m innerhalb des einzelnen Zeemanschen Zerlegungsbildes.

Die kinematische Bedeutung dieser Quantenzahlen ist die folgende: n charakterisiert (mittels des in n enthaltenen Radialquantums) die Bahnform des äußeren Elektrons („Serienelektrons"), wobei $n = k$, kurz gesagt, eine Kreisbahn, $n > k$ eine Ellipsenbahn bedeutet, von um so größerer Exzentrizität, je größer $n - k$ ist. Der Quantenzahl k andererseits ist kinematisch ein Umlauf des Perihels der Bahnellipse in der Ebene derselben zugeordnet. Die Normale zu dieser Ebene, um welche das Perihel fortschreitend gedreht wird, nennen wir die k-Achse. Wegen des Perihelumlaufes nimmt die Bahnkurve die Form einer Rosette an. Die Quantenzahl j bedeutet das gesamte Impulsmoment des jeweiligen Atomzustandes. Die Achse desselben nennen wir j-Achse. Sie ist im allgemeinen von der k-Achse verschieden. Um die j-Achse führt die Bahnebene eine Drehung oder Präzession („Innenpräzession") aus, entsprechend der Bedeutung von j als Impulsmoment des Atoms. Schließlich führt das ganze so bewegte System noch eine Drehung um die Kraftlinienrichtung des äußeren Magnetfeldes, die wir auch m-Achse nennen, aus. Diese „magnetische Präzession" ist durch die Quantenzahl m bestimmt und besteht im einfachsten Falle aus einer Larmorpräzession.

Die Umlaufzahlen dieser vier Bewegungen, pro Zeiteinheit gerechnet, nennen wir bzw.

$$\nu_n, \quad \nu_k, \quad \nu_j, \quad \nu_m,$$

die zugehörigen Winkelgeschwindigkeiten, d. h. die mit 2π multiplizierten Umlaufzahlen, heißen bzw.

$$\mathfrak{o}_n, \quad \mathfrak{o}_k, \quad \mathfrak{o}_j, \quad \mathfrak{o}_m.$$

Sie sind in dieser Folge der Größe nach geordnet, d. h. es ist stets $\mathfrak{o}_n \gg \mathfrak{o}_k$, $\mathfrak{o}_k \gg \mathfrak{o}_j$ und im allgemeinen auch $\mathfrak{o}_j \gg \mathfrak{o}_m$. Wenn letztere Bedingung bei ausnehmend starken Feldern verletzt wird, treten Besonderheiten (Paschen-Back-Effekte) auf, auf die wir in § 3 zurückkommen werden. Bei unserer eigentlichen Behandlung der Zeemanintensitäten in § 6 und 7 werden wir uns ausdrücklich auf kleine Felder beschränken, so daß hier $\mathfrak{o}_m \ll \mathfrak{o}_j$ gelten wird. Die im allgemeinen bestehende Größenfolge der $\mathfrak{o}$ bringt es mit sich, daß wir jede dieser Bewegungen ohne Rücksicht auf die folgenden, im Verhältnis unendlich langsamen Bewegungen betrachten dürfen, entsprechend dem Verfahren einer sukzessiven Störungsrechnung.

Nebenbei bemerkt lassen sich nach allgemeinen Regeln der Mechanik die Umlaufzahlen ν aus dem Ausdruck der Bahnenergie durch Differentiation nach der betreffenden Quantenzahl berechnen. Da nach dem Gesagten in vielen Fällen die Termenergie als Funktion der Quantenzahlen empirisch bekannt ist, können die ν, also auch, wegen $\mathfrak{o} = 2\pi\nu$, die $\mathfrak{o}$ als bekannt angesehen werden. Doch werden wir ihre genaueren Werte im folgenden nicht nötig haben.

Auch über den modellmäßigen Ursprung der Atombahnen brauchen wir uns keine genaueren Vorstellungen zu bilden; es genügt für das Folgende die vorangehende allgemeine kinematische Beschreibung. Wir werden Dublett- und Triplettsysteme (allgemeiner: geradzahlige und ungeradzahlige Spektren) nach demselben Schema behandeln können, während die genaue Modelltheorie sehr verschiedene Voraussetzungen für beide machen muß. Da Intensitätsfragen wesentlich qualitative Fragen sind, wird es auf quantitative Einzelheiten vielfach nicht ankommen. Z. B. ist es nicht von Belang, ob wir das Impulsmoment des Serienspektrums direkt gleich der azimutalen Quantenzahl k oder gleich $k^* = k - 1/2$ setzen, wie es im Falle der Dublettsysteme das von Heisenberg vorgeschlagene magnetooptische Modell verlangt.

§ 1. Allgemeines über die Bewegung und ihre Fourierdarstellung. Streng genommen würde es sich bei der im klassischen Sinne gerechneten Ausstrahlung eines Atoms um das gesamte elektrische Moment, d. h. um die Koordinatensumme über alle Elektronen des Atoms handeln. Wenn wir trotzdem nur von den Koordinaten eines Elektrons, des „Serienelektrons", sprechen, so liegt das daran, daß wir nur Beobachtungen im sichtbaren Gebiet im Auge haben. Die Bewegung der anderen, dem Atominnern angehörenden und daher steif gebundenen Elektronen, stehen zu den Röntgenlinien in derselben korrespondenzmäßigen Beziehung wie die Bewegung des Serienelektrons mit den sichtbaren Linien. Es scheint daher erlaubt, bei Beschränkung auf das sichtbare Gebiet die Bewegung der inneren Elektronen zu übergehen.

Die Bewegung des Serienelektrons betrachten wir in erster Näherung als eben. In die Bahnebene legen wir die rechtwinkligen Koordinatenachsen ξ, η. Wäre die Bahn rein periodisch, z. B. elliptisch, so hätten wir bei komplexer Zusammenfassung von ξ und η:

$$\xi + i\eta = \sum_{s=-\infty}^{s=+\infty} a_s e^{i s \mathfrak{o}_n t}. \tag{1}$$

Die ganze Zahl s vertritt korrespondenzmäßig die Quantensprünge Δn. Die Koeffizienten a_s der Fourierreihe sind komplexe Konstante.

In Wirklichkeit ist die Bahn bei allgemeiner Atomstruktur nicht periodisch, sondern mit einer Periheldrehung $\mathfrak{o}_k$ versehen. Wir dürfen voraussetzen, daß die von der Elektronenhülle herrührenden störenden Kräfte klein gegen die Kernanziehung sind und folgern daraus (vgl. die Einleitung)

$$\mathfrak{o}_k \ll \mathfrak{o}_n.$$

Für die gestörte Bewegung ist ebenso wie für die ursprüngliche periodische das Azimut eine zyklische Koordinate, in der demnach der Flächensatz gilt. Unter der besonderen Voraussetzung, daß die störenden Kräfte als Zentralkräfte angesetzt werden können, läßt sich dies sofort verifizieren. Aber auch abgesehen von dieser speziellen Annahme ist die

Zyklizität des Azimuts gesichert, schon allein dadurch, daß es überhaupt eine azimutale Quantenzahl k gibt; die Berechtigung zu ihrer Einführung folgt gerade aus der Zyklizität des Azimuts. Gleichzeitig folgt daraus, daß die Periheldrehung gleichförmig erfolgt.

Demnach tritt an die Stelle von (1)

$$\xi + i\eta = e^{i\nu_k t} \sum a_s e^{is\nu_n t}. \tag{2}$$

Wir gehen von dem $\xi\eta$-System zu einem xyz-System über. Die z-Achse sei die „j-Achse", ϑ sei der Winkel zwischen ihr und der Normalen zur Bahnebene, unserer „k-Achse", vgl. die Einleitung. Die y-Achse wollen wir mit der η-Achse zusammenfallen lassen. Die Transformationsformeln lauten dann:

$$x = \xi \cos\vartheta, \quad y = \eta, \quad z = \xi \sin\vartheta. \tag{3}$$

Daraus folgt

$$x + iy = \xi \cos\vartheta + i\eta. \tag{3a}$$

Nun liegt die Bahnebene, die $\xi\eta$-Ebene, im System der xyz nicht fest, sondern wird um die z-Achse mit der Winkelgeschwindigkeit ν_j umgedreht („Innenpräzession"). (3a) geht dadurch über in

$$x + iy = e^{i\nu_j t}(\xi \cos\vartheta + i\eta). \tag{3b}$$

Wir nehmen an, vgl. die Einleitung, daß

$$\nu_j \ll \nu_k.$$

Dies bedeutet, daß die Abweichung der wirklichen Störung von der eines Zentralfeldes klein ist, und es spiegelt sich darin wieder, daß die Feinstruktur der Linien (z. B. die Dublett- oder Triplettstruktur) in Wellenzahlen gerechnet klein ist gegen die Wellenzahldifferenz der Linien verschiedener Serien.

Da j das Impulsmoment des ganzen Atoms bedeutet (vgl. die Einleitung), gilt beim kräftefreien Atom der Flächensatz um die j-Achse in Strenge; er rechtfertigt seinerseits die Einführung der Quantenzahl j. Die Drehung um die j-Achse erfolgt daher gleichförmig, wie Gleichung (3b) zum Ausdruck bringt. Befindet sich andererseits das Atom in einem äußeren Kraftfelde, z. B. einem Magnetfelde, so gilt das Gesagte in Annäherung so lange, als das äußere Feld hinreichend schwach ist gegenüber den störenden Kräften, welche die Innenpräzession bewirken.

Wir berechnen aus (2), indem wir den reellen und imaginären Teil bilden:

$$\left.\begin{aligned} \xi &= \frac{1}{2}\left(\sum a_s e^{i(s\nu_n + \nu_k)t} + \sum \bar{a}_s e^{-i(s\nu_n + \nu_k)t}\right), \\ \eta &= \frac{1}{2i}\left(\sum a_s e^{i(s\nu_n + \nu_k)t} - \sum \bar{a}_s e^{-i(s\nu_n + \nu_k)t}\right). \end{aligned}\right\} \tag{4}$$

$\bar{a}_s$ bedeutet dabei den konjugierten Wert zu a_s. Wir setzen (4) in (3b) und (3) ein und erhalten zunächst:

$$\left.\begin{aligned} x + iy &= e^{i\nu_j t}\left\{\frac{\cos\vartheta + 1}{2}\sum a_s e^{i(s\nu_n + \nu_k)t}\right. \\ &\quad \left. + \frac{\cos\vartheta - 1}{2}\sum \bar{a}_s e^{-i(s\nu_n + \nu_k)t}\right\}, \\ z &= \frac{\sin\vartheta}{2}\left\{\sum a_s e^{i(s\nu_n + \nu_k)t} + \sum \bar{a}_s e^{-i(s\nu_n + \nu_k)t}\right\}. \end{aligned}\right\} \quad (5)$$

Diese Formeln lassen sich bequemer als Doppelsumme schreiben: die eine Summation erstreckt sich über s von $-\infty$ bis $+\infty$, wobei wir benutzen

$$\sum \bar{a}_s e^{-is\nu_n t} = \sum \bar{a}_{-s} e^{+is\nu_n t};$$

die andere Summation betrifft den Summationsindex r in

$$e^{ir\nu_k t}$$

und erstreckt sich nur auf die zwei Werte $r = \pm 1$. Statt (5) ergibt sich so

$$\left.\begin{aligned} x + iy &= \sum_{s=-\infty}^{+\infty} \sum_{r=-1}^{+1}{}' c_{s,r} e^{i(s\nu_n + r\nu_k + \nu_j)t}, \\ z &= \sum_{s=-\infty}^{+\infty} \sum_{r=-1}^{+1}{}' d_{s,r} e^{i(s\nu_n + r\nu_k)t}. \end{aligned}\right\} \quad (6)$$

Das Zeichen $'$ bei der zweiten Summe deutet an, daß der Wert $r = 0$ auszuschließen ist. Die neu eingeführten Koeffizienten c und d haben, wie der Vergleich von (6) und (5) ergibt, die folgende Bedeutung:

$$\left.\begin{aligned} c_{s,1} &= \frac{1 + \cos\vartheta}{2} a_s, & c_{s,-1} &= -\frac{1 - \cos\vartheta}{2}\bar{a}_{-s}, \\ d_{s,1} &= \frac{\sin\vartheta}{2} a_s, & d_{s,-1} &= \frac{\sin\vartheta}{2}\bar{a}_{-s}. \end{aligned}\right\} \quad (7)$$

Wir gehen zu einem dritten Koordinatensystem über, das wir XYZ nennen. Die Z-Achse sei die „m-Achse", d. h. die Richtung der magnetischen Kraftlinien. Θ sei der Winkel zwischen ihr und der j-Achse. Die Y-Achse lassen wir mit der y-Achse zusammenfallen. Die Transformationsformeln lauten jetzt:

$$\left.\begin{aligned} X &= x\cos\Theta - z\sin\Theta, \quad Y = y, \\ Z &= x\sin\Theta + z\cos\Theta. \end{aligned}\right\} \quad (8)$$

Man hat also zunächst

$$X + iY = x\cos\Theta + iy - z\sin\Theta. \quad (8\text{a})$$

Wir beachten nun die „magnetische Präzession", der zufolge das ganze Atom, insbesondere die j-Achse, mit der gleichförmigen

Winkelgeschwindigkeit $\mathfrak{o}_m$ um die Z-Achse gedreht wird. Daß die Geschwindigkeit gleichförmig ist, folgt aus der Homogenität des Magnetfeldes. Wir ersetzen deshalb (8a) durch

$$X + iY = e^{i\mathfrak{o}_m t}(x\cos\Theta + iy - z\sin\Theta). \tag{8b}$$

Aus der ersten Gleichung (6) folgt, wenn wir den reellen und imaginären Teil bilden:

$$x = \frac{1}{2}\sum_s\sum_r{}'\left(c_{s,r}\, e^{i(s\mathfrak{o}_n + r\mathfrak{o}_k + \mathfrak{o}_j)t} + \bar{c}_{s,r}\, e^{-i(s\mathfrak{o}_n + r\mathfrak{o}_k + \mathfrak{o}_j)t}\right),$$

$$y = \frac{1}{2i}\sum_s\sum_r{}'\left(c_{s,r}\, e^{i(s\mathfrak{o}_n + r\mathfrak{o}_k + \mathfrak{o}_j)t} - \bar{c}_{s,r}\, e^{-i(s\mathfrak{o}_n + r\mathfrak{o}_k + \mathfrak{o}_j)t}\right).$$

Tragen wir dies und den Wert von z aus (6) in (8b) und (8) ein, so entsteht:

$$X + iY = e^{i\mathfrak{o}_m t}\left\{\frac{1+\cos\Theta}{2}\sum_s\sum_r{}' c_{s,r}\, e^{i(s\mathfrak{o}_n + r\mathfrak{o}_k + \mathfrak{o}_j)t}\right.$$
$$\left. - \frac{1-\cos\Theta}{2}\sum_s\sum_r{}' \bar{c}_{s,r}\, e^{-i(s\mathfrak{o}_n + r\mathfrak{o}_k + \mathfrak{o}_j)t} - \sin\Theta\sum_s\sum_r{}' d_{s,r}\, e^{i(s\mathfrak{o}_n + r\mathfrak{o}_k)t}\right\}$$

$$Z = \frac{\sin\Theta}{2}\sum_s\sum_r{}'\left\{c_{s,r}\, e^{i(s\mathfrak{o}_n + r\mathfrak{o}_k + \mathfrak{o}_j)t} + \bar{c}_{s,r}\, e^{-i(s\mathfrak{o}_n + r\mathfrak{o}_k + \mathfrak{o}_j)t}\right\}$$
$$+ \cos\Theta\sum_s\sum_r{}' d_{s,r}\, e^{i(s\mathfrak{o}_n + r\mathfrak{o}_k)t}.$$

Wir vereinfachen die Schreibweise, indem wir dreifache Summen einführen, nämlich außer den Summationen nach s und r eine solche nach q. Wir fassen nämlich

$$e^{+i\mathfrak{o}_j t},\quad 1,\quad e^{-i\mathfrak{o}_j t}$$

zusammen in die Form

$$e^{iq\mathfrak{o}_j t},\qquad q = +1, 0, -1.$$

Daraufhin können wir schreiben

$$\left.\begin{aligned} X + iY &= \sum_{-\infty}^{+\infty}{}_s \sum_{-1}^{+1}{}'_r \sum_{-1}^{+1}{}_q C_{s,r,q}\, e^{i(s\mathfrak{o}_n + r\mathfrak{o}_k + q\mathfrak{o}_j + \mathfrak{o}_m)t} \\ Z &= \sum_{-\infty}^{+\infty}{}_s \sum_{-1}^{+1}{}'_r \sum_{-1}^{+1}{}_q D_{s,r,q}\, e^{i(s\mathfrak{o}_n + r\mathfrak{o}_k + q\mathfrak{o}_j)t} \end{aligned}\right\} \tag{9}$$

Die hier eingeführten Koeffizienten C und D haben die folgende Bedeutung:

$$\left.\begin{aligned} &C_{s,r,1} = \frac{1+\cos\Theta}{2}c_{s,r}, \quad C_{s,r,0} = -\sin\Theta\, d_{s,r}, \\ &\qquad\qquad C_{s,r,-1} = -\frac{1-\cos\Theta}{2}\bar{c}_{-s,-r} \\ &D_{s,r,1} = \frac{\sin\Theta}{2}c_{s,r}, \quad D_{s,r,0} = \cos\Theta\, d_{s,r}, \quad D_{s,r,-1} = \frac{\sin\Theta}{2}\bar{c}_{-s,-r}. \end{aligned}\right\} \tag{10}$$

Die Gleichungen (6) und (7) beantworten in einem allgemeinen Sinne das Intensitätsproblem für die feldfreien Linienstrukturen. In gleicher Weise enthalten die Gleichungen (9) und (10) die Aussagen, die man über die Intensität in magnetooptischen Zerlegungsbildern machen kann.

§ 2. Die Auswahlregeln für die Quantenzahlen k, j, m. Es ist wohlbekannt, wie Bohr aus dem Ansatz (2) heraus auf das Auswahlprinzip für die azimutale Quantenzahl k schließt. Während hinsichtlich der „n-Bewegung" beliebige Oberschwingungen, d. h. Glieder vom Exponenten $is\omega_n t$ in (2) auftreten, ist hinsichtlich der „k-Bewegung" nur die Grundschwingung vom Exponenten $i\omega_k t$ vertreten. Korrespondenzmäßig können daher beliebige Sprünge $\Delta n = s$ der Hauptquantenzahl auftreten, aber die mit ihnen verbundenen Spünge der azimutalen Quantenzahl sind auf die Werte $\Delta k = \pm 1$ eingeschränkt. Dabei ist der Sprung $\Delta k = -1$ insofern durch (2) angezeigt, als hier das Vorzeichen von i auf der rechten und linken Seite umgekehrt werden kann.

Auf die Gruppen (pp') und (dd') der Erdalkalien und analoge Gruppen in den komplizierten Spektren, die eine Ausnahme von diesem Auswahlprinzip zu bilden scheinen, kommen wir in § 4 zurück.

Genau so schließen wir aus (6) auf das Auswahlprinzip für die innere Quantenzahl j. Hinsichtlich der j-Bewegung kommen in $x + iy$ bzw. bei Vertauschung des Vorzeichens von i in $x - iy$ die Exponenten $\pm i\omega_j t$ vor, zugleich kommt in z die j-Bewegung überhaupt nicht vor. Dasselbe sagen wir uns, wenn wir den in (9) eingeführten Summationsindex q auf die Werte $+1$, 0, -1 einschränkten. Aus dem Korrespondenzprinzip folgt nun die gleiche Einschränkung für den Quantensprung Δj: er ist nur der Werte ± 1 und 0 fähig. Die Auswahlregel der inneren Quantenzahl, insbesondere auch ihre Abweichung von derjenigen für die azimutale Quantenzahl, ursprünglich rein empirisch aus den Linienstrukturen der Nebenserien erschlossen, wird auf diese Weise eine kinematische Selbstverständlichkeit. Wir könnten auch eine Polarisationsregel hinzufügen (die Komponente $\Delta j = 0$ schwingt parallel, die Komponenten $\Delta j = \pm 1$ senkrecht zur j-Achse), aber diese wäre bei der räumlich undefinierten Lage der j-Achse im feldfreien Falle unkontrollierbar.

Sehr lehrreich ist es, daß die Durchbrechung des Auswahlprinzips der inneren Quantenzahl, nämlich die Realisierung [1]) verbotener Linien in starken Magnetfeldern durch Paschen und Back, auf Grund

[1]) Zeemanheft der Physica 1, 261, 1921.

unserer Korrespondenzformel (6) hätte vorhergesehen werden können. Wie schon bei der Aufstellung dieser Gleichung im vorigen Paragraphen bemerkt wurde, wird ein Magnetfeld, welches nicht mehr schwach ist gegen die Kräfte, die die Innenpräzession hervorrufen, die Gleichförmigkeit dieser Präzession stören. Dann müssen aber in (6) außer dem Exponenten $i\mathfrak{o}_j t$ die allgemeinen Zusatzglieder $iq\mathfrak{o}_j t$ auftreten mit $q > 1$. Korrespondenzmäßig werden dann auch die Übergänge $\Delta j > 1$ möglich, d. h. das Auswahlprinzip der j wird durchbrochen. Bemerkenswert ist in dieser Hinsicht auch die Beobachtung von Paschen und Back, daß der Übergang $\Delta j = 2$ leichter, nämlich bei kleineren Magnetfeldern zu realisieren ist, als der Übergang $\Delta j = 3$.

Bohr hat diesen Sachverhalt bereits als Zusatz [1]) zu einem Vortrage vom Jahre 1921 angedeutet. Er hat später auch den Grund dafür erkannt, warum gerade ein magnetisches, nicht ein elektrisches Feld dazu befähigt ist, die verbotenen Linien hervorzulocken. Das elektrische Zentrum, d. h. der mittlere Angriffspunkt der elektrischen Kraft während der Rosettenbewegung des Elektrons in einer Bahnebene, fällt nämlich mit dem Kern zusammen. Der Hebelarm der elektrischen Kraft ist also im Mittel gleich Null; das elektrische Feld übt also kein Drehmoment aus und ändert die Lage der Bahnebene nicht ab — umgekehrt wie das Magnetfeld, welches in erster Linie auf eine Präzession der Bahn hinwirkt.

Wir kommen schließlich zu dem Auswahlprinzip für die magnetische Quantenzahl m, welches in den Gleichungen (9) enthalten ist. Da die Frequenz $\mathfrak{o}_m$ hier in gleicher Weise vorkommt, wie die Frequenz $\mathfrak{o}_j$ in (6), lautet auch das Auswahlprinzip für m formal ebenso wie das für j: Es sind nur die Quantensprünge $\Delta m = \pm 1$ und 0 zulässig. Auch diese Regel war zunächst empirisch postuliert in der Landéschen Theorie der anomalen Zeemaneffekte; auch sie wird nunmehr im Sinne des Korrespondenzprinzips eine kinematische Selbstverständlichkeit.

In Hinsicht auf die Polarisationsverhältnisse unterscheiden sich aber die Sprünge der magnetischen Quantenzahl m von denjenigen des inneren Quantums j: Die zu $\Delta m = 0$ gehörenden Komponenten sind nach der Z-Achse, d. h. der Richtung der Kraftlinien polarisiert, die zu $\Delta m = \pm 1$ sind im Quereffekt senkrecht dazu, im Längseffekt zirkular polarisiert. Diese Polarisationen sind in der Beobachtung wohl nachweisbar, da die Kraftlinienrichtung im Raume festliegt.

[1]) Vgl. ZS. f. Phys. 9, 27, 1922.

Es gibt noch einen anderen Unterschied zwischen dem Auswahlprinzip der m und der j: Jenes bleibt auch bei stärksten Magnetfeldern strenge gültig, dieses wird, wie wir sahen, schließlich durchbrochen. In der Tat begründeten wir die Gleichförmigkeit der magnetischen Präzession, aus der das Auswahlprinzip der m folgt, lediglich mit der Homogenität des Magnetfeldes [vgl. den Übergang von Gleichung (8a) zu (8b)], und diese wird durch Verstärkung des Feldes nicht beeinträchtigt. Wollten wir auch das magnetische Auswahlprinzip durchbrechen, so müßten wir, umgekehrt wie bei den j, nicht magnetische, sondern elektrische Felder, z. B. senkrecht gegen die Richtung des homogenen Magnetfeldes, anlegen. Diese würden die magnetische Präzession ungleichförmig machen und daher Übergänge $\Delta m > 1$ realisieren. Andeutungen solcher Übergänge hat man, wie bekannt, in Aufnahmen der Balmerlinien[1]) von Paschen und Back aus dem Jahre 1912 bemerkt, wo überzählige magnetische Komponenten sichtbar sind.

§ 3. Die Landéschen Zusatzverbote $0 \longrightarrow 0$ für die innere und die magnetische Quantenzahl. Während im allgemeinen der Übergang $\Delta j = 0$ nach dem Auswahlprinzip der inneren Quantenzahl möglich ist, ist er verboten, wenn $j = 0$ ist. Dies ist die erste der Landéschen Zusatzregeln. Die zweite lautet: Der Übergang $0 \longrightarrow 0$ in der magnetischen Quantenzahl ist verboten, wenn gleichzeitig $\Delta j = 0$ ist (ohne daß hierbei j selbst gleich Null zu sein braucht).

Beide Zusatzverbote sind empirisch gesichert, insbesondere das erste durch den Ausfall der Linie Sp_3 im Spektrum der Erdalkalien und der analogen Elemente. Wir zeigen, daß sie nach dem Korrespondenzprinzip in der Tat zu erwarten sind.

Im ersten Falle ist der Gesamtimpuls j des Atoms sowohl im Anfangs- wie im Endzustande Null. Die Lage der j-Achse ist also völlig unbestimmt. Daher wird auch die Intensität der Linie, in welche, wie wir im nächsten Paragraphen noch näher ausführen werden, der Winkel ϑ zwischen der j- und k-Achse eingeht, unbestimmt. Ebenso die Polarisation des zwar nicht zu beobachtenden, aber physikalisch zu fordernden Einzelvorganges. Wir können mit Landé sagen: Wegen $\Delta j = 0$ müßte das emittierte Licht parallel zur j-Achse polarisiert sein. Diese Achse hat aber keine physikalische Bedeutung. Also wird auch die Lichtemission physikalisch bedeutunglos.

[1]) Ann. d. Phys. 39, Tafel VIII, Fig. 4.

Was das zweite Zusatzverbot betrifft, so folgt dieses als Sonderfall aus unserer allgemeinen Intensitätsberechnung der Zeemankomponenten in § 6; vgl. auch § 7 im Anschluß an Gleichung (20). Es ist daher eine notwendige Folge des Korrespondenzprinzips.

Weshalb wir schon hier besonders auf diese Verbote hinweisen, hat folgenden Grund: Das Verbot $0 \longrightarrow 0$ in der inneren Quantenzahl und seine korrespondenzmäßige Deutung zeigt zweifelsfrei, daß die innere Quantenzahl, wie zuerst Landé begründet hat, das gesamte mechanische Impulsmoment des Atoms bedeutet. Wenn man andererseits nach der Methode von Stern und Gerlach das magnetische Moment in einem Zustande feststellt, dessen j bekannt ist, so kann man prüfen, ob der klassische Zusammenhang zwischen magnetischem und mechanischem Moment zu recht besteht. Diese Prüfung ist für die gegenwärtige Entwicklung der Atomphysik entscheidend.

Ein anderer Grund, weshalb das Zusatzverbot $0 \longrightarrow 0$ für die innere Quantenzahl fundamental ist, besteht darin, daß es gestattet, die absoluten Werte der j zu normieren, die sonst nur bis auf eine additive Größe bestimmt sind. Dies läßt sich z. B. am Spektrum von Chrom zeigen [1]).

§ 4. **Die Intensität der natürlichen, magnetisch nicht zerlegten Linienstrukturen.** Allgemein sind die Intensitäten nach dem Korrespondenzprinzip als Fourierkoeffizienten bestimmt [2]). In einfachster Form treten solche in Gleichung (2) unter der Bezeichnung a_s auf. Sie geben ein Maß für die Stärke der aufeinander folgenden Linien einer Serie. Bei den Bahnen, die nicht in das Atominnere eindringen und die daher als voll ausgebildete, angenäherte Ellipsen verlaufen, könnte man die a_s ähnlich berechnen, wie es Bessel bei den Planetenbahnen getan hat (Ursprung der Besselschen Funktionen!) und wie es Kramers in seiner Dissertation für die Wasserstoffellipsen durchgeführt hat. Diese Koeffizienten interessieren uns hier indessen nicht, da sie (wegen der Anregungsbedingungen) nicht gut mit der Erfahrung vergleichbar werden. Unser Problem betrifft vielmehr die relativen Intensitäten innerhalb eines Liniengebildes, welches bei gegebenem $\Delta n = s$ und gegebenem Δk den verschiedenen Werten

[1]) Gieseler, Diss. Tübingen; A. Sommerfeld, Ann. d. Phys. 1923 (im Erscheinen).

[2]) Dabei tritt die bekannte, für das Korrespondenzprinzip charakteristische Schwierigkeit auf: Soll man die Fourierreihe des Anfangs- oder Endtermes betrachten oder die einer geeignet gemittelten Zwischenbahn? Wir brauchen uns hierüber nicht zu entscheiden, da unsere Schlüsse mehr qualitativer Art und für Anfangs-, End- oder Zwischenbahn wesentlich dieselben sind.

von Δj und j entspricht[1]). Die hierbei in Betracht kommenden Fourierkoeffizienten sind durch die $c_{s,r}$ und $d_{s,r}$ in Gleichung (6) und (7) gegeben. Die a_s, welche in letzteren als Faktoren auftreten, fallen für die Frage der relativen Intensitäten heraus.

Wir betrachten an erster Stelle ein Gebilde, welches dem gewöhnlichen Auswahlprinzip der k genügt und setzen $\Delta k = +1$ voraus. Wegen des Auswahlprinzips für j haben wir drei Fälle zu unterscheiden:

$$\Delta k = +1, \quad \begin{cases} \Delta j = +1, \\ \Delta j = 0, \\ \Delta j = -1. \end{cases}$$

Bei den Dubletts fällt der letzte Übergang fort; bei den zusammengesetzten Tripletts und erst recht bei Multipletts der komplizierten Spektren treten einzelne Übergänge mehrfach auf, entsprechend verschiedenen Werten von j (bei gleichem Δj).

Die den beiden ersten Übergängen korrespondenzmäßig zuzuordnenden Fourierkoeffizienten können aus den Gleichungen (6) und (7) direkt übernommen werden. Man hat $r = \Delta k = +1$ zu nehmen und $s = \Delta n$ beliebig zu lassen, indem man für $\Delta j = +1$ die erste, für $\Delta j = 0$ die zweite Gleichung (6) benutzt. Die fraglichen Koeffizienten lauten daher nach (7)

$$\left.\begin{aligned} \Delta j = +1:\ c_{s,1} &= \frac{1+\cos\vartheta}{2}\,a_s = \cos^2\frac{\vartheta}{2}\,a_s, \\ \Delta j = 0:\quad d_{s,1} &= \frac{\sin\vartheta}{2}\,a_s = \sin\frac{\vartheta}{2}\cos\frac{\vartheta}{2}\,a_s, \end{aligned}\right\} \tag{11}$$

Für $\Delta j = -1$ hat man den Koeffizienten des Fouriergliedes

$$e^{i(s\nu_n + \nu_k - \nu_j)t}$$

aufzusuchen. Dieses Glied tritt in der ersten Gleichung (6) auf, wenn man $+i$ mit $-i$, $+s$ mit $-s$ vertauscht und $r = -1$ setzt. Der zugehörige Koeffizient ist daher mit Rücksicht auf den Wert von $c_{s,-1}$ in (7)

$$\bar{c}_{-s,-1} = -\frac{1-\cos\vartheta}{2}\,a_s.$$

Somit ergibt sich in Vervollständigung der Gleichung (11)

$$\Delta j = -1:\ \bar{c}_{-s,-1} = -\frac{1-\cos\vartheta}{2}\,a_s = -\sin^2\frac{\vartheta}{2}\,a_s.$$

[1]) Wir verabreden, unter Δ die Differenz: Anfangs- minus Endwert der betr. Quantenzahl: zu verstehen, also die Abnahme einer Quantenzahl mit positivem Vorzeichen zu rechnen. Der Hauptserie (Übergang $p \longrightarrow s$) entspricht daher $\Delta k = +1$, der II. N. S. (Übergang von $s \longrightarrow p$) $\Delta k = -1$.

Um die diesen drei Schwingungen entsprechenden Intensitäten zu berechnen — wir nennen sie J_{+1}, J_0 und J_{-1} —, nehmen wir eine beliebige Beobachtungsrichtung w und ergänzen sie durch zwei dazu senkrechte Achsen u, v. Wir haben also rechtwinklige, beliebig gegeneinander gedrehte Achsenkreuze xyz und uvw. Um z. B. J_{+1} zu berechnen, nehmen wir bei unserer ersten Schwingung die Komponenten von x und y nach u und v — die Komponenten nach w kommen für die Ausstrahlung nicht in Frage —, bilden

$$(x_u + y_u)^2 + (x_v + y_v)^2$$

und mitteln über alle Lagen von u, v, w. Wir finden leicht

$$J_{+1} = \frac{2}{3}|c_{s,1}|^2 \quad = \frac{2}{3}\cos^4\frac{\vartheta}{2}|a_s|^2,$$

ebenso

$$J_{-1} = \frac{2}{3}|\bar{c}_{-s,-1}|^2 = \frac{2}{3}\sin^4\frac{\vartheta}{2}|a_s|^2.$$

Bei der Berechnung von J_0 haben wir zu bedenken, daß in der zweiten Gleichung (6) die beiden konjugierten Glieder

$$d_{s,1}e^{i(s\nu_n+\nu_k)t} + d_{-s,-1}e^{-i(s\nu_n+\nu_k)t}$$

die gleiche Frequenz haben und zusammenzufassen sind; ferner daß hier noch eine Mittelbildung nach der Zeit hinzukommt. Schließlich wird

$$J_0 = \frac{1}{3}|2\,d_{s,1}|^2 = \frac{4}{3}\cos^2\frac{\vartheta}{2}\sin^2\frac{\vartheta}{2}|a_s|^2.$$

Die relativen Intensitäten sind also gegeben durch die Proportion

$$J_{+1}:J_0:J_{-1} = \cos^4\frac{\vartheta}{2} : 2\sin^2\frac{\vartheta}{2}\cos^2\frac{\vartheta}{2} : \sin^4\frac{\vartheta}{2}. \tag{12}$$

Man beachte wohl, daß der Winkel ϑ in den drei Gliedern unserer Proportion je ein anderer ist. Auch wenn es sich um Übergänge handelt, die vom gleichen Anfangsniveau ausgehen, sind die Endniveaus, zu denen sie führen, verschieden. Als Winkel ϑ kommen dann die Winkel zwischen j und k-Achse in drei voneinander verschiedenen Zwischenzuständen in Frage (vgl. den Anfang dieses Paragraphen). Ebenso, wenn es sich um Übergänge handelt, die zu demselben Endniveau führen.

Wir betrachten an zweiter Stelle den Übergang $\Delta k = -1$ mit den drei Unterfällen

$$\Delta k = -1 \begin{cases} \Delta j = -1, \\ \Delta j = 0, \\ \Delta j = +1. \end{cases}$$

Als Fourierkoeffizienten für die beiden ersten entnimmt man aus (6), wenn man $r = 1$ setzt und $+i$, $+s$ mit $-i$, $-s$ vertauscht:

$$\bar{c}_{-s,1} = \frac{1+\cos\vartheta}{2}\,\bar{a}_{-s},\quad \bar{d}_{-s,1} = \frac{\sin\vartheta}{2}\,\bar{a}_{-s}.$$

Der dritte Fourierkoeffizient ergibt sich aus (6), wenn man $r = -1$ und $-s$ statt s setzt:

$$c_{-s,-1} = -\frac{1-\cos\vartheta}{2}\,\bar{a}_{-s}.$$

Somit die gemittelten Intensitäten:

$$J_{-1}:J_0:J_{+1} = \cos^4\frac{\vartheta}{2} : 2\sin^2\frac{\vartheta}{2}\cos^2\frac{\vartheta}{2} : \sin^4\frac{\vartheta}{2}. \qquad (13)$$

Man beachte, daß die Reihenfolge jetzt gegenüber (12) umgekehrt ist.

An dritter Stelle betrachten wir den Übergang $\Delta k = 0$, der im Widerspruch mit dem Auswahlprinzip für die azimutale Quantenzahl steht. Er tritt auf bei den Kombinationen (pp') und (dd') im Ca- und Sr-Spektrum, die Herr R. Götze formal nach der Methode der inneren Quantenzahl gedeutet hat und deren modellmäßige Auffassung, wie wir gesprächsweise erfahren haben, von Herrn J. Franck im Anschluß an die neuen Bohrschen Atommodelle vertieft worden ist. Auch im Spektrum des Mn und Cr gibt es analoge, aber viel kompliziertere Strukturen.

Offenbar unterscheiden sich zwei solche Terme wie np, np' bei gleicher azimutaler Quantenzahl und annähernd gleicher Bahn des äußeren Serienelektrons durch die Konstitution des Atomrumpfes. Wir müssen also hier außer dem Serienelektron mindestens ein atominneres Elektron mit in Betracht ziehen und seine Bewegung in der Fourierentwicklung zum Ausdruck bringen. In der Gleichung (2) treten deshalb Zusatzglieder auf, welche nicht ohne einige Willkür hingeschrieben werden können. Da die Bahnebene des inneren Elektrons von der des äußeren verschieden sein wird, kommt zu der Gleichung für $\xi + i\eta$ (jetzt als $\Sigma\xi + i\Sigma\eta$ zu schreiben) eine zur $\xi\eta$-Ebene senkrechte ζ-Komponente der Bewegung hinzu, die zum Teil von der Bahn des inneren, zum Teil von der Koppelung zwischen innerem und äußerem Elektron herrührt.

Gewisse Überlegungen (Mittelbildung über die Rosette des äußeren Elektrons, Unterscheidung zwischen den Schwingungen, die dem inneren Elektron als solchem zugehören und nach Abtrennung des äußeren Elektrons zum Funkenspektrum führen, und den Schwingungen, die nur durch die Mitwirkung des äußeren Elektrons ermöglicht werden) führen dazu, anzunehmen, daß die Hauptglieder der Entwicklung die von o_k freien Koppelungsglieder in der ζ-Komponente sind. Ihr Fourierkoeffizient sei b. Indem wir die Transformation (3), jetzt mit Rücksicht auf die ζ-Komponente, ausführen, erhalten wir als Koeffizienten der Hauptglieder in der nunmehrigen Entwicklung (6) für $x + iy$ bzw. z:

$$\left.\begin{aligned} \Delta j &= \pm 1 : c = b \cos \vartheta, \\ \Delta j &= 0 \quad : d = b \sin \vartheta. \end{aligned}\right\} \tag{14}$$

Hieraus folgt für die Intensitäten J_0, J_{+1}, J_{-1} der drei Übergänge

$$\Delta k = 0 \begin{cases} \Delta j = 0 \\ \Delta j = +1 \\ \Delta j = -1 \end{cases}$$

die Proportion

$$J_0 : J_{+1} : J_{-1} = 2 \cos^2 \vartheta : \sin^2 \vartheta : \sin^2 \vartheta. \tag{15}$$

Die Reihenfolge haben wir abermals gegenüber (12) und (13) abgeändert, indem wir immer an erster Stelle die (unter der noch zu diskutierenden Annahme eines kleinen Winkels ϑ) größte Intensität geschrieben haben. Wir bringen so die folgende Intensitätsregel zum Ausdruck:

Diejenigen Übergänge in der inneren Quantenzahl sind die stärksten, die mit dem Übergange in der azimutalen Quantenzahl parallel gehen (gleichsinnige Übergänge); die Intensität wird um so schwächer, je mehr der Sinn des einen Überganges von dem des anderen abweicht (minder oder mehr ungleichsinnige Übergänge).

Diese Regel wurde bereits früher aus den empirischen Intensitätsdaten abgelesen[1]). Daß sie aus den vorstehenden Formeln folgt, soll im nächsten Paragraphen erörtert werden.

§ 5. Prüfung der Intensitätsregel an der Erfahrung. Unsere Intensitätsformeln (12), (13) und (15) ergeben eine deutliche Größenabstufung, wenn der Winkel ϑ klein ist. (Richtiger gesagt: „die Winkel ϑ", da jeder Übergang seinen eigenen Winkel ϑ hat.) In diesem Falle gehen die rechten Seiten von (12), (13) und (15) über in:

$$1 : \frac{1}{2} \vartheta^2 : \frac{1}{16} \vartheta^4 \quad \text{bzw.} \quad 1 : \frac{1}{2} \vartheta^2 : \frac{1}{2} \vartheta^2. \tag{16}$$

[1]) Atombau und Spektrallinien, 3. Aufl., S. 447.

Es fragt sich also, wann der Winkel ϑ, d. h. die Neigung der j- gegen die k-Achse klein sein wird. Nun setzt sich das gesamte Impulsmoment j zusammen (und zwar im allgemeinen vektoriell) aus dem Impulsmoment des unangeregten Atoms (j_0) und dem Impulsmoment der Anregung (j_1). Letzteres ist im wesentlichen[1]) gleich der azimutalen Quantenzahl k. Wenn also k groß ist gegen j_0, so bildet k den Hauptbestandteil von j und die j-Achse fällt merklich mit der k-Achse zusammen.

I. Wir erwarten also, daß unsere Intensitätsregel in den Termen von großem k, dem d-, b, ... -Term, ausgeprägt zur Geltung kommen wird, daß sie dagegen bei kleinem k weniger hervortreten wird. Wir werden sogleich sehen, wie vollständig die Erfahrung diesen Schluß bestätigt.

Der Winkel ϑ wird aber auch klein, wenn zwar nicht $k \gg j_0$ ist, aber j_0 und j_1 (bzw. k) sich algebraisch, nicht vektoriell addieren. Dieses findet statt im Hauptniveau jedes Termes, im Niveau der größten inneren Quantenzahl, welche zugleich im allgemeinen — wenigstens bei den gewöhnlichen Dublett- und Triplettsystemen — das oberste Niveau, dasjenige des größten Termwertes darstellt.

II. Wir erwarten also, daß unsere Intensitätsregel bei den verschiedenen Termniveaus verschieden stark zur Geltung kommen wird, am ausgesprochensten bei dem Niveau des größten j, und sich bei kleinerem j verwischen wird. Auch diese Erwartung wird durch die Erfahrung bestätigt.

a) $\Delta k = \pm 1$, Tripletts. Wir prüfen unseren Satz (I) zunächst bei den Triplettsystemen. Für die Kombinationen (pd) der I. N. S. von Ca, z. B. $2p - 4d$, findet man in Kaysers Handbuch die folgenden Intensitätsangaben (10 stark, 0 schwach):

$$J_{+1} : J_0 : J_{-1} = 10 : 8 : 4.$$

Es handelt sich hier um die wohlbekannte Struktur des zusammengesetzten Tripletts. Der Übergang $\Delta j = +1$ tritt dreimal auf,

---

[1]) Wenn der unangeregte Atomzustand eine s-Bahn ($k = 1$) ist, hätte man $j_1 = k - 1$ zu setzen. Dies hat in der Tat bei den Triplettsystemen (allgemeiner den „ungeradzahligen Systemen") zu geschehen. Dagegen scheint es geboten, bei den Dublettsystemen (allgemein den „geradzahligen Systemen") $j_1 = k^* = k - {}^1\!/_2$ zu nehmen. Vgl. hierzu die S. 140 zitierte Arbeit von Sommerfeld in den Ann. d. Phys. Offenbar sind diese Unterscheidungen für unsere mehr qualitativen Intensitätsbetrachtungen belanglos.

nämlich zwischen den Termen[1]) $d_3 \longrightarrow p_2$, $d_2 \longrightarrow p_1$, $d_1 \longrightarrow p_0$. Dies sind die „Hauptlinien" der Struktur und sämtlich von der Intensität 10. Der Übergang $\Delta j = 0$ kommt zweimal vor, nämlich bei $d_2 \longrightarrow p_2$ und $d_1 \longrightarrow p_1$. Dies sind „die Satelliten erster Ordnung" und beide von der schwächeren Intensität 8. Der Übergang $\Delta j = -1$ ist der „Satellit zweiter Ordnung" $d_1 \longrightarrow p_2$ und hat die geringste Intensität. Da in der I. N. S. $\Delta k = +1$ ist, sind die Hauptlinien gleichsinnige, die Satelliten entgegensinnige Übergänge, wobei der Satellit zweiter Ordnung in bezug auf den Änderungssinn von k und j stärker abweicht als die Satelliten erster Ordnung. Durch unsere Intensitätsregel wird also erstmalig der Gegensatz — Satellit und Hauptlinie — theoretisch gefaßt.

Von den Satelliten in der Bergmannserie der Tripletts stehen uns Intensitätsmessungen leider nicht zur Verfügung. Wir erwarten nach (I), daß hier der Unterschied zwischen Hauptlinien und Satelliten noch stärker ausgeprägt sein wird als in der I. N. S., weil die azimutale Quantenzahl hier in Anfangs- und Endterm je um 1 höher ist als dort.

Wir gehen zur H. S. und II. N. S., d. h. zu den Kombinationen (sp) zurück. Hier ist zunächst zu bemerken, daß die Intensitätsabstufung in der H. S. zwischen den Triplettlinien $p_2 \longrightarrow s$, $p_1 \longrightarrow s$ und $p_0 \longrightarrow s$, $\Delta k = +1$, dieselbe ist wie in der II. N. S. zwischen den (umgekehrt gelegenen) Linien $s \longrightarrow p_2$, $s \longrightarrow p_1$, $s \longrightarrow p_0$, $\Delta k = -1$. Dies ist eine unmittelbare Folge unserer Regel von den gleichsinnigen und ungleichsinnigen Übergängen, nach der es nicht auf die Richtung des Überganges in j, sondern nur auf ihr Verhältnis zur Richtung des Überganges in k ankommt. Wir verweisen auch auf die Gleichung (12) (H. S., $\Delta k = +1$) und (13) (II. N. S., $\Delta k = -1$), welche beide dieselben Intensitäten in der entsprechend umgekehrten Zuordnung zu den Quantensprüngen Δj aufweisen.

Als Zahlenwerte finden wir bei Kayser z. B. für Ca $2p - 1s$ (II. N. S.)

$$J_{-1} : J_0 : J_{+1} = 10:10:8.$$

[1]) Wir bezeichnen die Terme hier so, daß wir die innere Quantenzahl als Index beifügen, und möchten diese Bezeichnungsweise, obwohl sie von der üblichen abweicht, allgemein empfehlen. Bei den sonst mit d_1, d_2, d_3 bezeichneten Termen kehrt sich dann die Indizierung gerade um in d_3, d_2, d_1; statt der bisherigen p_1, p_2, p_3 hat man zu schreiben p_2, p_1, p_0. In der neuen Bezeichnung gehören die Linien D_1, D_2 zu den Termen p_1, p_2, nicht wie bisher zu bzw. p_2, p_1. Die früher rein konventionellen Indizes bekommen bei diesem Verfahren eine wichtige physikalische Bedeutung und lassen unmittelbar die möglichen Kombinationen übersehen.

Voigt sieht beim Zeemaneffekt von Sauerstoff 3:3:2 als typischen Wert an. Diese ausgesprochen geringe Intensitätsabnahme entspricht in Anbetracht des kleinen Anfangs- und Endwertes von k ganz der Erwartung I).

b) $\Delta k = \pm 1$, Dubletts. Von den Dubletts der Alkalien wählen wir als am weitesten getrennt das Cs. Die Kayserschen Angaben für die H.S. und II. N.S. schwanken untereinander und je nach der Gliednummer (vgl. auch die Untersuchungen von Füchtbauer und Roschdestwensky). Für die I. N. S. finden wir z. B. folgende Angaben:

$$J_1 : J_0 = 9 : 6 \quad 2p - 5d$$
$$8 : 2 \quad 2p - 6d, \quad 2p - 7d.$$

Am charakteristischsten ist die Angabe von K. W. Meissner[1]) über die Bergmann-Satelliten von Cs:

$$J_1 : J_0 = 10 : 2 \quad 3d - 5b, \quad 3d - 6b.$$

Hier ebenso wie in den vorhergehenden Proportionen bezieht sich die erste Zahl auf die beiden Hauptlinien $b_4 \longrightarrow d_3$ und $b_3 \longrightarrow d_2$ des zusammengesetzten Dubletts, die zweite auf den einzigen vorhandenen Satelliten $b_3 \longrightarrow d_2$. Ergänzen wir nach dem Vorbilde von $D_2 : D_1$ als entsprechendes Intensitätsverhältnis für die Hauptserie etwa 2 : 1, so haben wir in der Skala

$$\underbrace{2:1}_{\text{H.S.}} \qquad \underbrace{4:1}_{\text{I.N.S.}} \qquad \underbrace{5:1}_{\text{B.S.}}$$

einen Maßstab für die Verkleinerung des Winkels ϑ und für die Gültigkeit unseres Satzes I).

c) $\Delta k = \pm 1$, Multipletts. Den Satz II) können wir nur an den „Multipletts" der verwickelten Spektren von Cr, Mn, prüfen, weil nur hier die nötige Zahl vergleichbarer und voll entwickelter Termniveaus vorhanden ist, während z. B. bei den zusammengesetzten Tripletts nur ein Triplett voll, die beiden anderen abgebrochen sind. Die Zuordnung der Linien und die Intensitätsangaben für Cr und Mn entnehmen wir einer Arbeit von Catalan[2]), die Deutung und Bezeichnung[3]) der Linien entspricht der S. 140 zitierten Arbeit von Sommerfeld.

[1]) Ann. d. Phys. **65**, 378, 1921.

[2]) Series and other regularities in the spectrum of manganese. Trans. R. Soc. London (im Erscheinen).

[3]) Mit dem Unterschied, daß die Indizierung der Terme dort in der bisher üblichen Weise, hier nach den inneren Quantenzahlen (vgl. die Anmerkung zu S. 145) erfolgt ist.

Von Cr benutzen wir ein Multiplett vom Charakter (PD) bei $\lambda = 525$ bis $541\,\mu\mu$ und eines vom Charakter (DB) bei $\lambda = 388$ bis $394\,\mu\mu$. Die Anordnung läßt den Aufbau der Liniengebilde nach dem Auswahlprinzip der inneren Quantenzahlen erkennen.

	D_4		D_3		D_2		D_1		D_0
P_3	8	:	6	:	4				
P_2			7	:	7	:	6		
P_1					5	:	6	:	6

	B_5		B_4		B_3		B_2		B_1
D_4	40	:	20	:	6				
D_3			25	:	20	:	6		
D_2					20	:	20	:	6
D_1							20	:	15
D_0									12

Am ausgeprägtesten ist der Intensitätsabfall $J_{+1}:J_0:J_{-1}$, wie wir es nach Satz II erwarten, in der obersten Zeile des (DB)-Schemas; er schwächt sich von hieraus in den unteren Zeilen schrittweise ab. Auch in der obersten Zeile von (PD) ist der Intensitätsabfall deutlich, wenn auch, vgl. Satz I, nicht so stark wie in (DB); in den unteren Zeilen von (PD) verwischt er sich bis zur Umkehr.

Wir fügen entsprechende Schemata für Mn hinzu. Hier sind die inneren Quantenzahlen nicht absolut, sondern nur relativ bekannt. Wir bezeichnen daher die Indizes mit $x, x+1$ usw. Die gegensätzliche Bezeichnung p, d; D, B ist durch das Vorhandensein von Interkombinationslinien zwischen beiden Termsystemen, ähnlich wie bei den Erdalkalien, begründet.

	d_{x+3}		d_{x+2}		d_{x+1}		d_x		d_{x-1}
p_{x+2}	10	:	9	:	7				
p_{x+1}			$8^1/_2$	:	8	:	$7^1/_2$		
p_x					$7^1/_2$	:	8	:	8

	B_{x+3}		B_{x+2}		B_{x+1}		B_x		B_{x-1}		B_{x-2}		B_{x-3}
D_{x+2}	10	:	3	:	1								
D_{x+1}			9	:	6	:	2						
D_x					8	:	5	:	2				
D_{x-1}							5	:	6	:	2		
D_{x-2}									4	:	4	:	2

Sehr bezeichnend ist wieder die starke Ausprägung des Intensitätsabfalles in der obersten Zeile von *(DB)* und seine graduelle Abschwächung nach unten hin. Dasselbe zeigt (pd) in schwächerem Maße, wie es nach Satz I sein soll.

d) $\Delta k = 0$, Erdalkalien, Chrom, Mangan. Wir kommen jetzt zu den Kombinationen (pp'), (dd'), $\Delta k = 0$, der Erdalkalien. Gleichsinnig sind hier die Übergänge $\Delta j = 0$, die in der Diagonalreihe unserer Schemata stehen; die Übergänge $\Delta j = \pm 1$ sind in gleichem Maße ungleichsinnig. Für letztere erwarten wir unter sich gleiche, aber schwächere, für erstere stärkere Intensitäten. Der Intensitätsunterschied zwischen ersteren und letzteren soll nach Satz I stärker ausgeprägt sein bei (dd') als bei (pp') und ferner nach Satz II deutlicher sein in den oberen Zeilen als in den unteren. Alles dies bestätigen in der Tat die beiden folgenden, aus der Arbeit von R. Götze[1]) entnommenen Schemata:

	p_2	p_1	p_0
p'_2	20	15	
p'_1	15	15	15
p'_0		15	—

	d_3	d_2	d_1
d'_3	20	8	
d'_2	8	15	8
d'_1		8	10

Die entsprechenden Schemata von Cr und Mn bei $\lambda = 388$ bis $394\,\mu\mu$ bzw. $\lambda = 402$ bis $483\,\mu\mu$ sind:

	D_4	D_3	D_2	D_1	D_0
D'_4	12	8			
D'_3	10	10	10		
D'_2		10	6	8	
D'_1			9	5	7
D'_0				6	—

	D_{x+2}	D_{x+1}	D_x	D_{x-1}	D_{x-2}
D'_{x+2}	10	6			
D'_{x+1}	8	8	6		
D'_x		5	4	6	
D'_{x-1}			4	2	6
D'_{x-2}				2	2

Auch hier bemerken wir: Betonung der Übergänge $\Delta j = 0$ gegenüber $\Delta j = \pm 1$, wenigstens in den oberen Zeilen, ungefähre Gleichheit von $\Delta j = +1$ und $\Delta j = -1$. Das Verbot des Überganges $0 \longrightarrow 0$, das bei Cr zur absoluten Festlegung der j gedient hat, ist in dem Schema für Cr und im Schema (pp') der Erdalkalien durch das Zeichen — angedeutet.

[1]) Ann. d. Phys. **66**, 285, 1921. (pp') ist bei Ca die Fraunhofersche Gruppe G, $\lambda = 428$ bis $432\,\mu\mu$, bei Sr die Gruppe $\lambda = 427$ bis 488. (dd') ist bei Ca, Sr und Ba je einmal mit gleicher Verteilung der Intensitäten beobachtet. Man beachte auch, daß die durch — angedeutete Linie $0 \longrightarrow 0$ bei (pp'), ebenso bei (DD') ausfällt.

§ 6. **Die Intensität der Zeeman-Komponenten.** Die Intensitäten der Zeemankomponenten sind nach dem Korrespondenzprinzip durch die Koeffizienten $C_{s,r,q}$ und $D_{s,r,q}$ der Gleichung (9) gegeben, die die Bewegung des Serienelektrons im XYZ-System darstellen (Z-Achse = m-Achse = Richtung der magnetischen Kraftlinien). Während wir die analogen Überlegungen des § 4 nach den Fällen $\Delta k = +1, -1, 0$ anordneten, haben wir jetzt die Fälle $\Delta j = +1, -1, 0$ zu unterscheiden; während früher die Unterfälle zu jedem Falle sich auf die Werte von Δj bezogen, betreffen sie jetzt die Änderungen Δm der magnetischen Quantenzahl.

Wir betrachten also zuerst

$$\Delta j = +1 \begin{cases} \Delta m = +1, \\ \Delta m = 0, \\ \Delta m = -1. \end{cases}$$

Zu $\Delta j = +1$, $\Delta m = +1$ gehört bei übrigens beliebigen Übergängen in den Quantenzahlen n und k nach der ersten Gleichung (9) der Fourierkoeffizient $C_{s,r,+1}$, zu $\Delta j = +1$, $\Delta m = 0$ nach der zweiten Gleichung (9) der Koeffizient $D_{s,r,+1}$. Im Falle $\Delta j = +1$, $\Delta m = -1$ haben wir in der ersten Gleichung (9) $+i$, $+s$, $+r$ zu vertauschen mit $-i$, $-s$, $-r$ und dementsprechend $q = -1$ zu setzen. Das für diesen Fall benötigte Fourierglied lautet dann

$$\overline{C}_{-s,-r,-1}\, e^{i(s\,\nu_n + r\,\nu_k + \nu_j - \nu_m)t}.$$

Für die drei hiernach in Betracht kommenden Fourierkoeffizienten entnimmt man aus den Gleichungen (10)

$$C_{s,r,+1} = \frac{1+\cos\Theta}{2}\, c_{s,r}, \quad D_{s,r,+1} = \frac{\sin\Theta}{2}\, c_{s,r},$$

$$\overline{C}_{-s,-r,-1} = -\frac{1-\cos\Theta}{2}\, c_{s,r}.$$

Die Intensitäten bezeichnen wir je nach den Werten von Δm mit J_{+1}, J_0, J_{-1}; wir haben sie gleich den Quadraten der absoluten Beträge der drei vorstehenden Koeffizienten zu setzen, unter Hinzufügung des Faktors 4 bei J_0 (die räumliche Mittelung fällt jetzt fort, da die Lage der Z-Achse im Atom physikalisch bestimmt ist; die zeitliche Mittelung bleibt bestehen). Somit

$$J_{+1} : J_0 : J_{-1} = \tfrac{1}{4}(1+\cos\Theta)^2 : \sin^2\Theta : \tfrac{1}{4}(1-\cos\Theta)^2 \qquad (17)$$

Wir haben hier nicht, wie in der analogen Gleichung (12), den halben Winkel eingeführt, weil dies für das folgende unbequem wäre.

Wir betrachten an zweiter Stelle (man bemerke die abgeänderte Reihenfolge der Unterfälle):

$$\Delta j = -1 \begin{cases} \Delta m = -1, \\ \Delta m = 0, \\ \Delta m = +1. \end{cases}$$

Der Fourierkoeffizient im ersten Unterfalle ist nach (9) $\bar{C}_{-s,-r,+1}$, im zweiten $D_{s,r,-1}$, im dritten $C_{s,r,-1}$. Nach (10) ist

$$\bar{C}_{-s,-r,+1} = \frac{1+\cos\Theta}{2}\,\bar{c}_{-s,-r}, \quad D_{s,r,-1} = \frac{\sin\Theta}{2}\,\bar{c}_{-s,-r},$$

$$C_{s,r,-1} = -\frac{1-\cos\Theta}{2}\,\bar{c}_{-s,-r}.$$

Die Intensitäten verhalten sich also

$$J_{-1} : J_0 : J_{+1} = \tfrac{1}{4}(1+\cos\Theta)^2 : \sin^2\Theta : \tfrac{1}{4}(1-\cos\Theta)^2. \qquad (18)$$

Es bleiben noch die Fälle

$$\Delta j = 0 \begin{cases} \Delta m = 0, \\ \Delta m = +1, \\ \Delta m = -1. \end{cases}$$

Die in Frage kommenden Fourierkoeffizienten sind jetzt nach (9) und (10):

$$D_{s,r,0} = \cos\Theta\, d_{s,r}, \quad C_{s,r,0} = -\sin\Theta\, d_{s,r},$$

$$\bar{C}_{-s,-r,0} = -\sin\Theta\, \bar{d}_{-s,-r}.$$

Da in Gleichung (16) z reell ist, müssen die Koeffizienten d konjugiert imaginär sein, also $d_{-s,-r} = \bar{d}_{s,r}$, oder auch $\bar{d}_{-s,-r} = d_{s,r}$. Wir erhalten von da aus für die Intensitäten:

$$J_0 : J_{+1} : J_{-1} = \cos^2\Theta : \tfrac{1}{4}\sin^2\Theta : \tfrac{1}{4}\sin^2\Theta. \qquad (19)$$

Wie schon in § 4 für den Winkel ϑ hervorgehoben, hat auch Θ in jedem Gliede dieser und der vorangehenden Proportionen je einen anderen, für den betreffenden Übergang eigens zu wählenden Wert.

§ 7. Graphische Darstellung der Intensität der Zeemankomponenten und Vergleich mit der Erfahrung. Zunächst müssen wir uns über die Wahl des Winkels Θ schlüssig werden, bei der wir uns nicht wie bei ϑ in § 5 auf kleine Werte beschränken dürfen. Θ bedeutet, wie wir wissen, den Winkel zwischen der m- und der j-Achse. Hätten wir mit den einfachen Verhältnissen des Wasserstoffmodells und seiner räumlichen Quantelung zu tun, so hätten wir zu nehmen

$$\cos\Theta = \frac{m}{j}. \qquad (20)$$

Diesem Ansatz liegt die Auffassung zugrunde, daß die magnetische Quantenzahl m die Projektion des Gesamtimpulses j auf die Richtung der magnetischen Kraftlinien bedeutet, eine Auffassung, die allgemein durch die empirisch bestätigte Bedingung $|m| \leqq j$ gestützt wird (vgl. die Landésche Theorie der anomalen Zeemaneffekte). Wieweit Gleichung (20) allgemein zutrifft, ließe sich nur aus den Einzelheiten der Atommodelle entnehmen. Im folgenden werden wir uns im wesentlichen an Gleichung (20) halten. Nur bei kleinen Werten von j, wo die Berechnung von Θ besonders empfindlich wird, werden wir Gleichung (20) im Sinne der von Heisenberg vorgeschlagenen magnetooptischen Modelle korrigieren.

Einschaltungsweise wollen wir hier das Zusatzverbot für die magnetische Quantenzahl aus § 3 begründen. Es besagte, daß der Übergang $0 \longrightarrow 0$ in m verboten ist, wenn $\Delta j = 0$. Nun sehen wir aus (20), daß beim Übergange $0 \longrightarrow 0$ $\cos\Theta = 0, \sin\Theta = 1$ wird, und zwar sowohl für die Anfangs- wie für die Endbahn, also auch für eine beliebige Zwischenbahn. Aus Gleichung (19) folgt aber dann für $\Delta j = 0$:

$$J_0 : J_{\pm 1} = 0 : \tfrac{1}{4};$$

also gilt in der Tat $J_0 = 0$, was zu beweisen war.

Die folgenden Figuren sind so gewonnen, daß Θ bei jedem Übergang für die Anfangs- und Endbahn einzeln in der angegebenen Weise bestimmt und in die Intensitätsformeln (17), (18), (19) des vorigen Paragraphen eingesetzt wurde. Zum Schluß wurde aus den so erhaltenen Intensitäten das arithmetische Mittel gebildet und in die Figur als Länge eingetragen. Die Art der Polarisation ist dadurch angedeutet, daß die σ-Komponenten (senkrecht zu den Kraftlinien polarisiert) ausgezogen, die π-Komponenten (parallel dazu polarisiert) punktiert wurden.

Die Größe der normalen Aufspaltung $\Delta\nu_n$ ist in der ersten Figur jeder Gruppe eingetragen, so daß für jede Komponente ihre Aufspaltung in Teilen von $\Delta\nu_n$ beurteilt werden kann. Die σ-Komponenten entstehen nach der Polarisationsregel aus den Übergängen $\Delta m = \pm 1$, und zwar die auf der kurzwelligen bzw. langwelligen Seite gelegenen aus $\Delta m = +1$ bzw. $\Delta m = -1$; die π-Komponenten entsprechen $\Delta m = 0$ und liegen ebenfalls symmetrisch zur Mitte des Zerlegungsbildes. Die Anfangswerte von m, von denen aus die Sprünge Δm erfolgen, sind in der Figur am oberen Ende der Intensitätslinien angegeben. Daß diese Werte bei den Dubletts halbzahlig sind, entspricht der Landéschen Numerierung der magnetischen Niveaus und somit auch den empirischen Zerlegungsbildern. Modellvorstellungen kommen dabei nicht in Frage. Die Zahlen am unteren Ende der Intensitätslinien bedeuten die von Herrn Back ge-

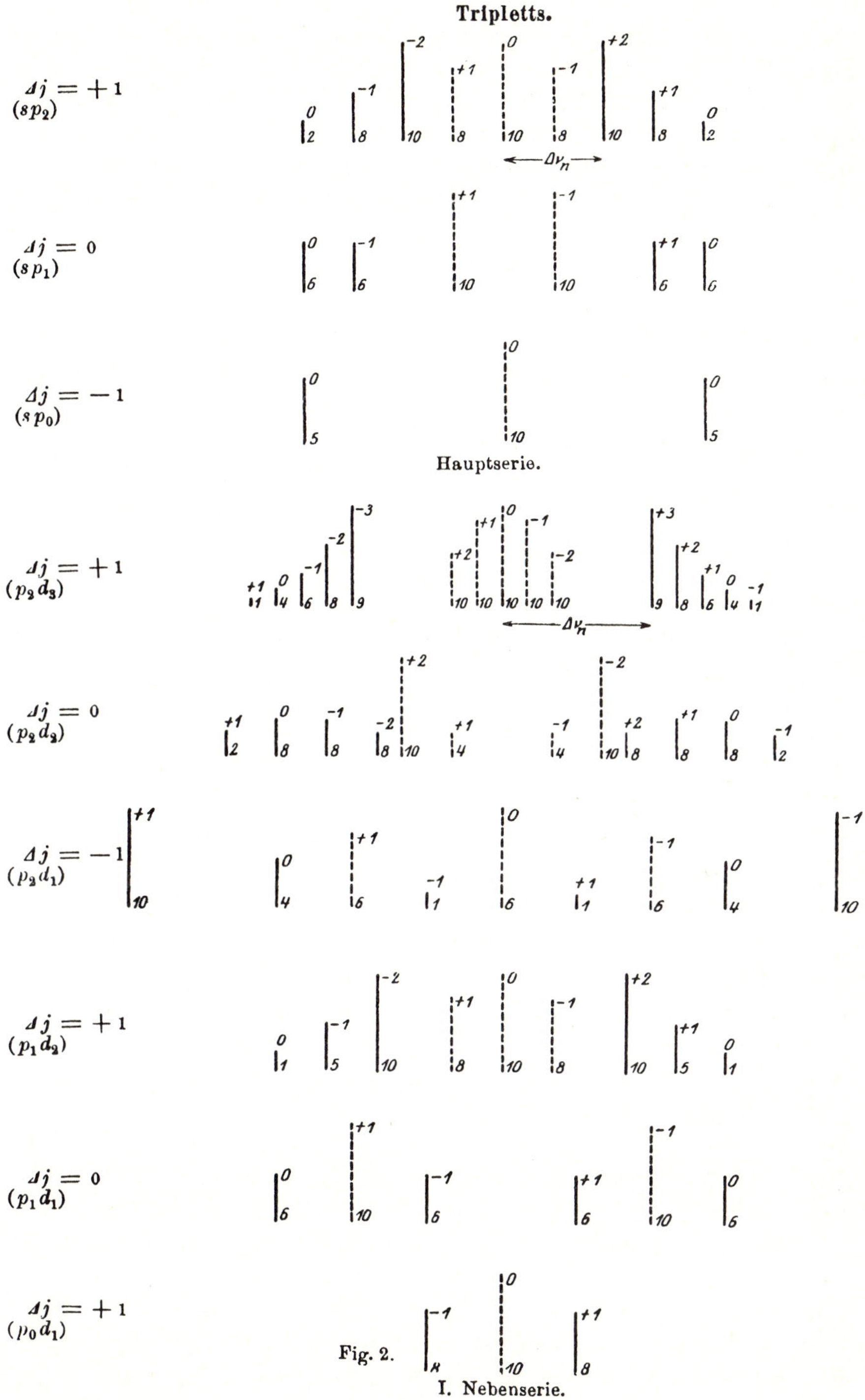
Tripletts.
Δj = +1 (sp₂)
Δj = 0 (sp₁)
Δj = −1 (sp₀)
Hauptserie.
Δj = +1 (p₂d₃)
Δj = 0 (p₂d₂)
Δj = −1 (p₂d₁)
Δj = +1 (p₁d₂)
Δj = 0 (p₁d₁)
Δj = +1 (p₀d₁)
I. Nebenserie.

Fig. 2.

schätzten und uns freundlichst mitgeteilten empirischen Intensitäten seiner Aufnahmen. Ihre Übereinstimmung mit den theoretischen, durch die Linienlänge dargestellten Intensitäten ist im allgemeinen sehr gut.

Bemerkenswert ist, daß sowohl empirisch wie theoretisch die Gesamtintensität der σ-Komponenten im allgemeinen größer ist als die

Dubletts.

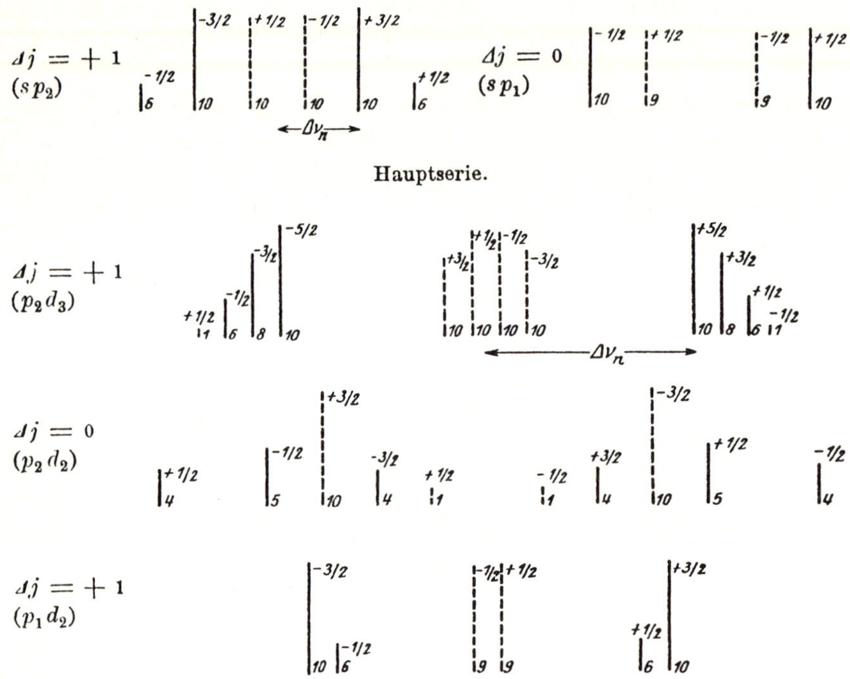

I. Nebenserie.

Fig. 1.

der π-Komponenten. Diese merkwürdige Folge des Korrespondenzprinzips wird schon in der Dissertation von Kramers bei den anomalen Zeemaneffekten bemerkt. Wir bestätigen hier auf Grund der Backschen Angaben, daß sie, wie auch Bohr neuerdings vermutet, reell sein dürfte.

Im ganzen dürfen wir sagen, daß sich das Korrespondenzprinzip in seiner Anwendung auf die hier studierten Intensitätsfragen, sowohl bei den spontanen, feldfreien Termaufspaltungen als auch bei den eigentlichen Zeemanaufspaltungen vorzüglich bewährt hat. Der im allgemeinen qualitative Charakter dieser Fragen brachte es mit sich, daß wir keine näheren Vorstellungen über den Atombau nötig hatten, wodurch die Sicherheit unserer Schlüsse verstärkt wird.

Über Phasenbeziehungen bei den Bohrschen Modellen von Atomen und Molekeln.

Von **M. Born** und **W. Heisenberg** in Göttingen.

(Eingegangen am 16. Januar 1923.)

Bei Gelegenheit einer Untersuchung der quantentheoretisch möglichen Modelle der Wasserstoffmolekel [1]) hat sich herausgestellt, daß bei konsequenter Anwendung der Quantenbedingungen zwischen den Bewegungen der Elektronen zweier Atome, die in Wechselwirkung treten, bestimmte Phasenbeziehungen bestehen müssen.

Wir haben diese Frage weiter verfolgt und dabei gefunden, daß solche Phasenbeziehungen auch schon zwischen den Bewegungen der Elektronen ein und desselben Atoms auftreten müssen. Für den Fall des Orthoheliums hat Bohr zur Erklärung des Dubletts eine solche Möglichkeit erörtert.

Wir glauben jetzt auf Grund unserer Rechnungen die Vermutung aussprechen zu dürfen, daß bei jedem Atom im Normalzustand das ganze System der Elektronenbahnen exakt in Phase (d. h. rein periodisch) ist. Hier wollen wir aber die Verhältnisse nur an einfachen Fällen studieren; an diesen kann man die Reduktion der unabhängigen Perioden verfolgen und zugleich sehen, daß dabei keineswegs eine Verminderung der Zahl der unabhängig zu quantelnden Phasenintegrale eintritt. Die mathematische Methode, die wir benutzen, ist eine Weiterführung der von Poincaré und Lindstedt [2]) angegebenen, die Born und Pauli [3]) der Quantentheorie angepaßt haben; es hat sich nämlich gezeigt, daß dieses Verfahren für die hier betrachteten Probleme nicht völlig ausreicht, sondern einer Ergänzung bedarf. Wir haben diese durch Anwendung der Bohlinschen Methode [4]) geben können, die Poincaré im zweiten Bande seiner Mécanique Céleste (Kap. XIX, S. 443) ausführlich diskutiert.

Am Schluß besprechen wir einige prinzipielle Schwierigkeiten der Quantentheorie, die bei der Frage nach der Entstehung der Quantenzustände aus den unverbundenen Teilchen auftreten.

§ 1. Für die Anwendung der Störungstheorie auf die Astronomie und auf die Atommodelle besteht ein wichtiger Unterschied:

---

[1]) M. Born, Die Naturwissenschaften **10**, 677, 1922.

[2]) S. H. Poincaré, Méthodes nouvelles de la mécanique céleste (Paris 1893), Bd. II, Kap. 9 ff.

[3]) M. Born u. W. Pauli jun., ZS. f. Phys. **10**, 137, 1922.

[4]) K. Bohlin, Über eine neue Annäherungsmethode in der Störungstheorie. Bihang till K. Svenska Vet. Akad. Handlingar **14**, Afd. I, Nr. 5. Stockholm 1888.

Bei der Bewegung der Himmelskörper kann der Fall als seltene und merkwürdige Ausnahme betrachtet werden, daß die (unter Vernachlässigung der Störungen berechneten) Umlaufszeiten zweier Gestirne nahezu kommensurabel sind; und daß dies exakt der Fall ist, kann als unendlich unwahrscheinlich gelten.

Bei den Bewegungen der Elektronen im Atom aber sind solche Kommensurabilitäten die Regel. Denn die Quantenbedingungen fordern geradezu für die ungestörten Bahnen exakt rationale Verhältnisse der Umlaufszeiten.

Dieser Unterschied verlangt nun auch eine wesentlich verschiedene Behandlungsweise. Ehe wir diese allgemein diskutieren, mögen die Verhältnisse an einem einfachen Beispiel erläutert werden.

Wir denken uns auf derselben Achse zwei gleiche Rotatoren vom Trägheitsmoment A, deren Stellungen durch die Winkel φ_1, φ_2 bestimmt seien. Sie mögen eine Wechselwirkung aufeinander ausüben, deren Drehmoment mit $\sin(\varphi_1 - \varphi_2)$ proportional sei. Sind p_1, p_2 die Impulsmomente, so sind

$$w_1 = \frac{\varphi_1}{2\pi}, \quad J_1 = 2\pi p_1,$$

$$w_2 = \frac{\varphi_2}{2\pi}, \quad J_2 = 2\pi p_2$$

kanonisch konjugierte Variable, und die Energiefunktion lautet:

$$H = \frac{J_1^2}{8\pi^2 A} + \frac{J_2^2}{8\pi^2 A} + \lambda\,[1 - \cos(\varphi_1 - \varphi_2)]. \tag{1}$$

Durch die kanonische Transformation

$$2\pi w = \varphi = \frac{\varphi_1 + \varphi_2}{2}, \qquad J = J_1 + J_2,$$

$$2\pi w' = \varphi' = \frac{\varphi_1 - \varphi_2}{2}, \qquad J' = J_1 - J_2,$$

geht H über in

$$H = \frac{J^2 + J'^2}{16\pi^2 A} + \lambda\,(1 - \cos 2\varphi'). \tag{2}$$

φ ist also zyklische Variable, und man hat $J = hn$ zu setzen. Ist W die Energiekonstante, so wird das zweite Wirkungsintegral

$$\left.\begin{aligned} J^* &= \oint J'\,dw' = \frac{1}{2\pi}\oint \sqrt{16\pi^2 A[W - \lambda(1-\cos 2\varphi')] - J^2}\,d\varphi' \\ &= \oint \frac{\sqrt{8\lambda A}}{\varkappa}\cdot\sqrt{1-\varkappa^2\sin^2\varphi'}\,d\varphi' = 8\cdot\frac{\sqrt{2\lambda A}}{\varkappa}\cdot E(\varkappa), \\ \varkappa^2 &= \frac{2\lambda}{W - \dfrac{J^2}{16\pi^2 A}}, \end{aligned}\right\} \tag{3}$$

wo E für $\varkappa < 1$ das vollständige elliptische Integral zweiter Gattung bedeutet.

Wir betrachten nun zunächst die „ungestörte“ Bewegung ($\lambda = 0$); deren Energie wird:

$$W_0 = \frac{J_1^2 + J_2^2}{8\pi^2 A} = \frac{J^2 + J'^2}{16\pi^2 A}.$$

Nun sind zwei Fälle zu unterscheiden:

I. In der ungestörten Bewegung sind die Frequenzen, also auch die Einzelimpulsmomente verschieden: $J_1 \neq J_2$, $J' \neq 0$. Dann ist $W_0 - \frac{J^2}{16\pi A}$ von Null verschieden; $\varkappa$ verschwindet daher mit λ und man kann für hinreichend kleine λ die Entwicklung $E(\varkappa) = \frac{\pi}{2}\left(1 - \frac{\varkappa^2}{4} + \cdots\right)$ benutzen. Man erhält:

$$W = \frac{J^2}{16\pi^2 A} + \frac{J^{*2}}{16\pi^2 A}\left(1 + \lambda\frac{16\pi^2 A}{J^{*2}} + \cdots\right).$$

J^* stimmt bis auf Größen der Ordnung λ mit J_0' des ungestörten Systems überein. Verlangt man noch dazu, daß bei adiabatischem Nullwerden des Parameters λ das gestörte System in das ungestörte übergehen solle, so wird $J^* = J_0'$.

II. In der ungestörten Bewegung sind die Frequenzen, also auch die Einzelimpulsmomente gleich: $J_1 = J_2$, $J' = 0$. Dann liegt ein besonderer Fall der Entartung vor. Im ungestörten System wird $W_0 = \frac{J^2}{16\pi^2 A}$.

Über die Größenordnung von $\varkappa$ läßt sich zunächst nichts aussagen, J^* wird von der Ordnung $\frac{\sqrt{\lambda}}{\varkappa}$.

Verlangt man, daß sich auch W nach Potenzen von λ entwickeln lasse: $W = W_0 + \lambda W_1 + \cdots$, so wird $\varkappa$ von der Größenordnung 1, J^* von der Ordnung $\sqrt{\lambda}$. Verlangt man noch dazu, daß bei adiabatischem Nullwerden von λ das gestörte System wieder in das ungestörte übergehen solle, so wird $J^* = 0$.

Bei dem hier gedachten adiabatischen Prozeß ist jedoch zu beachten, daß wir eventuell an eine Unstetigkeitsstelle der Bewegung kommen können, wenn wir λ langsam abnehmen lassen, nämlich dort, wo die Libration von w' in Oszillation umschlägt ($\varkappa^2 = 1$). Es ist aber leicht einzusehen, daß wir dann für $\lambda = 0$ nicht zur ungestörten Bewegung $J' = 0$ zurückkehren. Der einzige Wert von J^* vielmehr, für den wir an diese Unstetigkeitsstelle niemals, d. h. erst bei $\lambda = 0$

gelangen, ist $J^* = 0$. Für diesen Wert aber haben wir für $\lambda = 0$ auch $J' = 0$, d. h. die ungestörte Bewegung.

Wenn J^* quantenmäßig festgelegt werden muß, so ist aus dem Vorhergehenden klar, daß in diesem Falle II $J^* = 0 . h$ zu setzen ist. Und dies Resultat bliebe im allgemeinen auch bestehen, wenn das gestörte System nicht adiabatisch aus dem ungestörten hervorginge, denn J^* ist, wie oben gezeigt, von der Ordnung $\sqrt{\lambda}$, kann also nicht $1 . h$, $2 . h \ldots$ sein, wenn anders die Entwicklung nach Potenzen von λ einen Sinn haben soll.

Aus $J^* = 0$ folgt nach (3), daß $\varphi' =$ const. ist; also wird auch $J' =$ const. $= 0$, und aus den Bewegungsgleichungen ergibt sich: $\frac{\partial H}{\partial \varphi'} = 0$.

Als Lösungen dieser Gleichung erhält man nach (2) $\varphi' = 0$, also $\varphi_1 = \varphi_2$.

Das bedeutet vollständige Phasengleichheit beider Rotatoren.

Dieses Resultat ist, wie wir sahen, eine Folge der Anwendung der Quantenregeln. Würde es sich um einen grobmechanischen Vorgang handeln, bei dem keine Quantenbedingungen gelten, so wäre eine solche Phasenbeziehung unendlich unwahrscheinlich; der Winkel φ' würde Schwingungen (Librationen) ausführen, deren Energie eventuell sehr klein sein kann. Die Quantentheorie schließt aber solche kleine Schwingungsenergien aus, und daraus entspringen jene Phasenbeziehungen.

§ 2. Um den betrachteten Fall sinngemäß zu verallgemeinern, werden wir den Begriff der Entartung eines Systems schärfer zu fassen haben, als es üblich ist.

Wir betrachten ein System, dessen Energiefunktion in der Form

$$H = H_0 + H_1 \lambda + H_2 \lambda^2 + \cdots$$

darstellbar sei, wo die Energie H_0 des ungestörten Systems als Funktion der Wirkungsintegrale $J_1^0, J_2^0, \ldots J_f^0$ bekannt sein möge. Wir unterscheiden nun zwei Fälle der Entartung:

1. Eigentliche Entartung, wenn H_0 nur von einem Teile der $J_1^0, \ldots J_f^0$, etwa von $J_1^0, \ldots J_\varrho^0$ abhängt, $J_{\varrho+1}^0, \ldots J_f^0$ aber darin nicht vorkommen. Dann sind die Frequenzen

$$\nu_k^0 = \frac{\partial H_0}{\partial J_k^0}$$

identisch Null für $k = \varrho + 1, \ldots f$. Im ungestörten System dürfen jedenfalls die Größen $J_{\varrho+1}^0, \ldots J_f^0$ nicht quantenmäßig festgelegt werden.

2. **Zufällige Entartung**, wenn die in H_0 vorkommenden Größen $J_1^0, .., J_\varrho^0$ zufällig solche Werte haben, daß von den Frequenzen $\nu_1^0, \ldots \nu_\varrho^0$ einige Null sind; es seien etwa $\nu_1^0, \ldots \nu_\sigma^0$ von Null verschieden, aber $\nu_{\sigma+1}^0 = \ldots \nu_\varrho^0 = 0$.

Dieser Fall tritt auch immer dann ein, wenn zwischen den $\nu_1^0, \ldots \nu_\varrho^0$ eine oder mehrere Kommensurabilitäten bestehen; denn dann ist es durch lineare Transformation der Variabeln J_k^0, w_k^0 stets möglich, einer bzw. mehreren der neuen Winkelvariabeln die mittlere Bewegung (Frequenz) Null zu erteilen.

Der letzte Fall ist offenbar der unseres Beispiels in § 1. Wir erkennen an diesem deutlich, was wir dabei zu erwarten haben:

Auch die Wirkungsintegrale $J_{\sigma+1}, \ldots J_\varrho$ der zufällig entarteten Freiheitsgrade müssen quantenmäßig festgelegt werden. Wenn aber die Energie des gestörten Systems nur wenig von der des ungestörten abweicht, also etwa als Potenzreihe der Form

$$W = W_0 + \lambda W_1 + \cdots$$

angesetzt wird, werden die $J_{\sigma+1}^*, \ldots J_\varrho^*$ gleich Null; das aber hat zur Folge, daß die zugehörigen Winkelvariabeln konstant sind, die entsprechenden Frequenzen also verschwinden.

Da bei den **Bohrschen** Atommodellen die Umlaufsfrequenzen der ungestörten Bahnen gerade auf Grund der Quantenbedingungen kommensurabel sind, haben wir hier regelmäßig diesen Fall der „zufälligen" Entartung vor uns. Wir werden daher erwarten, daß die Umläufe der verschiedenen Elektronen in einfachen, konstanten Phasenbeziehungen stehen.

Um das zu zeigen, ist es nötig, auch hier die Bewegungen in Umläufe mit überlagerter Libration zu zerlegen, wie bei unserem Beispiel. Das soll im folgenden zunächst für den Fall durchgeführt werden, daß nur **eine** zufällig entartete Koordinate vorhanden ist und eigentlich entartete ganz fehlen.

§ 3. Zum Beweise unserer Behauptung benutzen wir die Methoden der Störungstheorie. Das von **Born** und **Pauli** auf die Probleme der Quantentheorie angewandte Verfahren ist jedoch nicht ohne weiteres auf den hier behandelten Fall „zufälliger" Entartung anwendbar. Um das zu sehen, wiederholen wir den Grundgedanken dieser Methode:

Die **Hamiltonsche** Funktion sei entwickelbar nach Potenzen eines kleinen Paramenters λ; als Argumente seien die konjugierten Winkel- und Wirkungsvariabeln des ungestörten Systems gewählt:

$$H(J_k^0, w_k^0) = H_0(J_k^0) + \lambda H_1(J_k^0, w_k^0) + \lambda^2 H_2(J_k^0, w_k^0) + \cdots.$$

Durch eine kanonische Transformation soll zu den Winkel- und Wirkungsvariabeln J_k, w_k des gestörten Systems übergegangen werden; es wird also gesetzt:

$$S = S(J_k, w_k^0); \; J_k^0 = \frac{\partial S}{\partial w_k^0}, \; w_k = \frac{\partial S}{\partial J_k}.$$

Dann ist S als Lösung der Hamilton-Jacobischen Differentialgleichung

$$H\left(\frac{\partial S}{\partial w_k^0}, w_k^0\right) = W.$$

zu bestimmen. Hierzu wird S als Potenzreihe von λ angesetzt, ebenso die Energiekonstante W;

$$S = S_0 + \lambda S_1 + \lambda^2 S_2 + \cdots, \quad W = W_0 + \lambda W_1 + \lambda^2 W_2 + \cdots.$$

Dann wird

$$S_0 = \sum_k J_k w_k^0, \; S_1, S_2, \ldots \text{ rein periodisch in den } w_k^0.$$

Man erhält für $S_1, S_2, \ldots$ durch Einsetzen einfache Rekursionsformeln.

In dieser Form ist das Verfahren nur anwendbar, wenn keinerlei Entartung vorliegt, also keine der Frequenzen $\nu_k^0 = \frac{\partial H_0}{\partial J_k^0}$ weder identisch, noch für bestimmte Werte der J_k verschwindet.

Born und Pauli haben überdies noch den Fall der eigentlichen Entartung behandelt, indem sie die Theorie der säkularen Störungen in geeigneter Weise anwandten.

Dagegen ist das Verfahren im Falle zufälliger Entartung nicht ohne weiteres brauchbar. Dann existiert nämlich im allgemeinen gar keine Entwicklung von S nach Potenzen von λ, sondern nach Potenzen von $\sqrt{\lambda}$. Hierauf beruht die Bohlinsche Methode, deren Grundgedanken Charlier[1]) durch ein Analogon aus der Algebra klar macht:

Es sei J als Funktion von w gegeben durch die algebraische Gleichung

$$F(J, w) = F_0(J) + \lambda F_1(J, w) - W = 0,$$

und es sei

$$F_0(J_0) - W = 0.$$

Dann läßt sich im allgemeinen $J(w)$ nach Potenzen von λ entwickeln:

$$J = J_0 + \lambda J_1(w) + \cdots.$$

Dies ist nicht mehr möglich, wenn

$$\left(\frac{\partial F_0}{\partial J}\right)_{J = J_0} = 0$$

[1]) C. L. Charlier, Mechanik des Himmels, Bd. 2, S. 466.

ist; dann aber läßt sich J durch eine nach Potenzen von $\sqrt{\lambda}$ fortschreitende Reihe darstellen.

Analog wird man bei dem Störungsproblem mit zufälliger Entartung versuchen müssen, S nach Potenzen von $\sqrt{\lambda}$ zu entwickeln, und es zeigt sich, daß dies in der Tat möglich ist. Für die vollständige Durchführung der Methode verweisen wir auf die ausführliche Darstellung durch Poincaré. Wir wollen hier nur die ersten Glieder der Entwicklung herstellen und die quantentheoretischen Folgerungen ziehen.

§ 4. Wir schreiben die Hamiltonsche Funktion (unter Änderung der Indizierung) so:

$$H = H_0(J_k) + \lambda H_2(J_k, w_k) + \lambda^2 H_4(J_k, w_k) + \cdots \tag{4}$$

und nehmen an, daß

$$\left.\begin{aligned} \nu_k^0 &= \frac{\partial H_0}{\partial J_k} \neq 0 \ (k = 1, 2, \cdots f-1), \quad \text{aber} \\ \nu_f^0 &= \frac{\partial H_0}{\partial J_f} = 0 \end{aligned}\right\} \tag{5}$$

sei für das bestimmte Wertsystem der J_k, von dem wir ausgehen. Dabei werde zur Vereinfachung immer w_k, J_k statt w_k^0, J_k^0 geschrieben.

Wir schließen vorläufig das Vorkommen eigentlicher Entartung aus und beschränken uns auf einen zufällig entarteten Freiheitsgrad.

Nun machen wir den Ansatz

$$S = S_0 + \sqrt{\lambda} S_1 + \lambda S_2 + \lambda\sqrt{\lambda} S_3 + \cdots \left\{\begin{aligned} &S_0 = \sum_k J_k w_k, \\ &S_1, S_2, \ldots \text{ periodisch in den } w_k, \end{aligned}\right\} \tag{6}$$

und behaupten, daß man dadurch zu einer Lösung des mechanischen Problems gelangen kann, bei der die Energie nach ganzen Potenzen von λ entwickelbar ist:

$$W = W_0 + \lambda W_2 + \lambda^2 W_4 + \cdots. \tag{7}$$

Setzt man nämlich $J = \frac{\partial S}{\partial w_k}$ in H ein, so erhält man folgende Rekursionsformeln:

$$\left.\begin{aligned} &\sum_k \frac{\partial H_0}{\partial J_k} \frac{\partial S_1}{\partial w_k} = 0 \\ &\sum_k \frac{\partial H_0}{\partial J_k} \frac{\partial S_2}{\partial w_k} + \frac{1}{2!} \sum_{kl} \frac{\partial^2 H_0}{\partial J_k \partial J_l} \frac{\partial S_1}{\partial w_k} \frac{\partial S_1}{\partial w_l} + H_2 = W_2 \\ &\cdots\cdots\cdots\cdots\cdots\cdots \end{aligned}\right\} \tag{8}$$

Aus der ersten Gleichung (8) und der Periodizitätsforderung folgt sofort, daß

$$\frac{\partial S_1}{\partial w_k} = 0 \text{ für } k = 1, 2, \ldots f-1; \tag{9}$$

$\frac{\partial S_1}{\partial w_f}$ bleibt zunächst unbestimmt und muß aus der zweiten Gleichung (8) ermittelt werden. Zu diesem Zwecke bilden wir den Mittelwert dieser zweiten Gleichung (8) über die $w_1, w_2, \ldots w_{f-1}$ und setzen:

$$\overline{H}_2 = \int_0^1\int_0^1 \ldots \int_0^1 H_2 \, d w_1 \, d w_2 \ldots d w_{f-1}; \tag{10}$$

$\overline{H}_2$ hängt dann nur noch von w_f ab. Wegen (9) erhält man nun

$$\frac{1}{2!}\frac{\partial^2 H_0}{\partial J_f^2}\left(\frac{\partial S_1}{\partial w_f}\right)^2 + \overline{H}_2(w_f) = W_2. \tag{11}$$

Diese Gleichung hat den Charakter einer Hamilton-Jacobischen Differentialgleichung für einen Freiheitsgrad und liefert S_1 durch eine Quadratur:

$$S_1 = \int \frac{\partial S_1}{\partial w_f} d w_f = \int \sqrt{\frac{W_2 - \overline{H}_2(w_f)}{\frac{1}{2!}\frac{\partial^2 H_0}{\partial J_f^2}}}\, d w_f. \tag{12}$$

Die dabei auftretende Integrationskonstante muß durch eine Quantenbedingung festgelegt werden. Da nun $\oint \frac{\partial S_0}{d w_f} d w_f = J_f$ von vornherein als Multiplum von h anzunehmen ist, so folgt nach (6) als Quantenbedingung für S_1:

$$\sqrt{\lambda} \oint \frac{\partial S_1}{\partial w_f} d w_f = n h. \tag{13}$$

Hier kann aber nach Schlüssen, die denen in § 1 genau nachgebildet sind, n nur den Wert Null haben, sofern der Ansatz (7) für die Energie des gestörten Systems zu recht besteht. Es muß also

$$\oint \sqrt{W_2 - \overline{H}_2(w_f)}\, d w_f = 0 \tag{14}$$

sein; das Integral ist um einen unendlich kleinen Verzweigungsschnitt der komplexen w_f-Ebene zu erstrecken, d. h. die beiden in Betracht kommenden Wurzeln der Gleichung $W_2 - \overline{H}_2(w_f) = 0$ müssen zusammenfallen. Daher muß $w_f =$ konst. sein und den Gleichungen

$$\overline{H}_2(w_f) = W_2, \quad \frac{\partial \overline{H}_2}{\partial w_f} = 0 \tag{15}$$

genügen. Hieraus sieht man, daß zwischen den Bewegungen der Elektronen eine **strenge Phasenbeziehung** herrscht. Die zweite

Gleichung (15) wird im allgemeinen mehrere Wurzeln haben; doch werden nur diejenigen von ihnen, die zu einem Minimum von $\overline{H_2} : \frac{\partial^2 H_0}{\partial J_f^2}$ gehören, einer stabilen Lösung entsprechen. Die erste Gleichung (15) liefert die zu diesen Werten von w_f gehörigen Energiekonstanten W_2. Hat $H_2 : \frac{\partial^2 H_0}{\partial J_f^2}$ mehrere Minima, so haben wir vielleicht mit Mehrfachtermen zu rechnen.

In der Astronomie ist natürlich im allgemeinen das Wirkungsintegral (13) nicht gleich Null; vielmehr werden Librationen der Variabeln w_f um das durch (15) bestimmte Librationszentrum stattfinden. Die Periode dieser Librationen kommt zu den $(f - 1)$ Perioden der nicht entarteten Freiheitsgrade hinzu, so daß im ganzen wieder f Perioden vorhanden sind. In der Atomtheorie aber können auf Grund der Quantenbedingungen diese Librationen nicht stattfinden; die entsprechende Periode ist Null und die Gesamtzahl der Perioden reduziert sich auf $f - 1$. Diese speziellen Bewegungstypen stehen im engsten Zusammenhang mit den von der Astronomie behandelten periodischen Lösungen des Drei- oder Mehr-Körperproblems [1]). Die Anzahl der zu quantelnden Phasenintegrale bleibt unter allen Umständen gleich der Anzahl der Freiheitsgrade f [2]).

§ 5. Die Fortsetzung der Methode zu den höheren Näherungen macht nicht mehr die geringsten Schwierigkeiten. Man findet das Verfahren bei Poincaré, l. c., für den allgemeinen Fall der Librationen dargestellt. Hier vereinfacht es sich noch wesentlich auf Grund der Quantenbedingung (14), indem es gerade in den von Born und Pauli l. c. behandelten Fall übergeht: aus der Entwicklung von S nach Potenzen von $\sqrt{\lambda}$ fallen alle ungeraden Glieder fort, und es bleibt eine Potenzreihe nach λ selbst. Für das erste Glied ist das schon aus den obigen Formeln abzulesen; nach (11) und (14) wird $\frac{\partial S_1}{\partial w_f} = 0$, und man kann daher mit Rücksicht auf (9) $S_1 = 0$ setzen. Den allgemeinen Beweis, daß auch S_3, S_5, ... verschwinden, wollen wir hier nicht durchführen, obwohl er nicht schwierig ist. Ebensowenig wollen wir den allgemeinen Fall behandeln, bei welchem die zufällige Entartung mehrmals vorkommt ($\nu^0_{\sigma+1}, \ldots \nu^0_{\varrho} = 0$) und wo außerdem eine mehrfache wirkliche Entartung vorliegt ($\nu^0_{\varrho+1}, \ldots \nu^0_f = 0$). Eine voll-

---

[1]) Vgl. etwa C. L. Charlier, l. c., Bd. 2, S. 187 u. f.

[2]) Dieses Resultat ist nicht in Übereinstimmung mit der Meinung von Herrn A. Smekal (ZS. f. Phys. 11, 294, 1922), daß die Anzahl der Quantenzahlen kleiner als die Anzahl der Freiheitsgrade sei.

ständige Behandlung dieses Falles mit Berücksichtigung der Librationen wäre wohl auch nach dem heutigen Stande der astronomischen Methoden nicht möglich; doch vereinfacht sich das Problem in der Atomtheorie durch den Fortfall der Librationen auf Grund der zu (14) analogen Quantenbedingungen beträchtlich. Wir wollen den Gang der Rechnung noch an einem Falle illustrieren, nämlich für einen zufällig und mehrere eigentlich entartete Freiheitsgrade.

Wir betrachten zunächst die zufällig ausgeartete Variable, etwa w_ϱ, als Konstante und integrieren die Hamilton-Jacobische Differentialgleichung

$$\left.\begin{array}{l}\overline{H}_2(J_1,\dots J_\varrho;\ J_{\varrho+1},\dots J_f,\ w_{\varrho+1},\dots w_f;\ w_\varrho) = W_2 \\ \quad(\text{gemittelt über } w_1,\dots w_{\varrho-1})\end{array}\right\} \tag{16}$$

für die Variabeln $w_{\varrho+1},\dots w_f$. Wir nehmen also an, daß es eine kanonische Transformation

$$\begin{pmatrix} w_{\varrho+1},\dots w_f \\ J_{\varrho+1},\dots J_f\end{pmatrix} \longrightarrow \begin{pmatrix} w'_{\varrho+1},\dots w'_f \\ J'_{\varrho+1},\dots J'_f\end{pmatrix}$$

gäbe, durch die $\overline{H}_2$ von den $w'_{\varrho+1},\dots w'_f$ unabhängig wird:

$$\overline{H}_2\ (J_1,\dots J_\varrho;\ J'_{\varrho+1},\dots J'_f;\ w_\varrho) = W_2. \tag{17}$$

Dann bestimmen wir eine Konstante w^0_ϱ aus der Gleichung

$$\frac{\partial \overline{H}_2}{\partial w_\varrho} = 0. \tag{18}$$

Nachträglich müssen wir aber, um der Allgemeinheit sicher zu sein, die Beschränkung aufheben, daß w_ϱ konstant sei, und setzen daher $w_\varrho = w^0_\varrho + w'_\varrho$, wo w'_ϱ als klein gegen w^0_ϱ angenommen werde. Dann schreiben wir die Hamiltonsche Funktion in der Form

$$H = H^*_0 + \lambda H^*_2 + \lambda^2 H_4 + \cdots, \tag{19}$$

wo

$$\left.\begin{array}{l} H^*_0(J_1,\dots J_\varrho;\ J'_{\varrho+1},\dots J'_f) = H_0(J_0,\dots J_\varrho) \\ \qquad + \lambda \overline{H}_2(J_1,\dots J'_\varrho;\ J'_{\varrho+1},\dots J'_f;\ w^0_\varrho), \\ H^*_2(J_1,\dots J_\varrho;\ J'_{\varrho+1},\dots J'_f,\ w_1,\dots w_{\varrho-1},\ w'_\varrho, w'_{\varrho+1},\dots w_f) \\ \qquad = H_2(J_1,\dots J'_f,\ w_1,\dots w'_f) - \overline{H}_2(J_1,\dots J'_f;\ w^0_\varrho) \end{array}\right| \tag{20}$$

gesetzt ist. Da H^*_0 alle $J_1,\dots J'_f$ enthält, ist das Problem auf das ursprüngliche (keine wirkliche Entartung) zurückgeführt; aus der Quantenbedingung (14) und der Gleichung (18) folgt, daß w'_ϱ exakt gleich Null ist.

§ 6. Indem wir dieses Resultat verallgemeinern, möchten wir vermutungsweise folgenden Satz aussprechen:

Um die Hamilton-Jacobische Differentialgleichung im Falle zufälliger Entartung (Indizes $\sigma + 1,\dots \varrho$) und eigentlicher Entartung

(Indizes $\varrho + 1, \ldots f$) unter Berücksichtigung der Quantenbedingungen zu integrieren, betrachte man die $J_{\sigma+1}, \ldots J_\varrho$, $w_{\sigma+1}, \ldots w_\varrho$ zunächst als konstante Parameter und bestimme die Wirkungsfunktion S für die übrigen Freiheitsgrade ($1, \ldots \sigma$, $\varrho + 1, \ldots f$) nach dem Born-Paulischen Verfahren durch Entwicklung nach Potenzen von λ (wobei die $J_{\varrho+1}, \ldots J_f$ durch die säkularen Störungen festgelegt und durch $J'_{\varrho-1}, \ldots J'_f$ ersetzt werden; die Striche werden im folgenden wieder weggelassen). Das Endresultat dieser Integration hat dann die Form

$$W = W(J_k) + \lambda W_2(J_k, w_{\sigma+1}, \ldots w_\varrho) + \lambda^2 W_4(J_k, w_{\sigma+1}, \ldots w_\varrho) + \cdots \quad (21)$$

Dann bestimme man die Werte von $w_{\sigma+1}, \ldots w_\varrho$ durch die Gleichungen

$$\frac{\partial W_2}{\partial w_{\sigma+1}} = 0, \cdots \frac{\partial W_2}{\partial w_\varrho} = 0. \quad (22)$$

Damit wird W eine Funktion aller $J_1, \ldots J_f$, die sämtlich gleich ganzen Vielfachen von h zu setzen sind[1]).

Wenn sich dieser Satz als richtig erweist, bedeutet er eine große Vereinfachung für die theoretische Bestimmung von Atommodellen, die Bohr mit so großem Erfolg durch Heranziehung der empirischen Kenntnisse über Spektren in Angriff genommen hat. Die Stabilitätsuntersuchung solcher dynamischer Modelle gewinnt dann eine große Ähnlichkeit mit der entsprechenden Frage bei statischen Modellen, da es wesentlich auf die durch (22) bestimmten Minima der Größe W_2 als Funktion der konstanten Phasenwinkel $w_{\sigma+1}, \ldots w_\varrho$ ankommt.

Die ausführliche Diskussion dieser Fragen mag der späteren Anwendung auf ein bestimmtes Problem der Störungstheorie vorbehalten bleiben.

§ 7. Zum Schluß wollen wir noch die Frage kurz berühren, wie die nach diesen Gesetzen aufgebauten Systeme aus ihren Bestandteilen entstehen können. Es erheben sich dabei dieselben Schwierigkeiten, die Einstein und Ehrenfest[2]) bei Gelegenheit der Versuche von Stern und Gerlach[3]) erörtert haben. Betrachten wir der Einfachheit halber das in § 1 besprochene Modell und denken uns etwa den Para-

---

[1]) Die so gewonnenen Lösungen sind etwas allgemeiner als die von Poincaré untersuchte Klasse periodischer Lösungen; diese ist dadurch gekennzeichnet, daß die mittleren Bewegungen der säkularen Störungen Null sind, daß also neben (22) noch die Beziehungen gelten:

$$\frac{\partial W_2}{\partial J_{\varrho+1}} = 0, \cdots \frac{\partial W_2}{\partial J_f} = 0.$$

[2]) A. Einstein und P. Ehrenfest, ZS. f. Phys. **11**, 31, 1922.

[3]) O. Stern und W. Gerlach, ZS. f. Phys. **9**, 349, 1922.

meter λ so von der Entfernung der beiden Rotatoren abhängig, daß er bei unendlichem Abstand verschwindet und bei Annäherung wächst, so werden die beiden (gleich schnell rotierenden) Körper zunächst für hinreichend lose Koppelung ($\lambda = 0$) keine feste Phaseneinstellung gegeneinander haben, weil dann die überall vorhandenen, unregelmäßigen Felder die Koppelungskräfte überwiegen werden. Wenn sie sich nun einander nähern, so müssen sie irgendwie in die Stellung $\varphi' = 0$ gelangen; erfolgt die Annäherung wirklich unendlich langsam (adiabatisch), so kann man sich klar machen, daß die Einstellung nach den Gesetzen der klassischen Mechanik wirklich eintritt, sofern man nur zuläßt, daß das Quantenintegral (3) nicht exakt Null, sondern beliebig klein (etwa kleiner als Größen von der Ordnung des klassisch berechneten Strahlungswiderstandes) ist. Dann wird nämlich bei ganz schwacher Koppelung eine äußerst langsame Libration eintreten, die adiabatisch bei wachsender Koppelung in äußerst schnelle Schwingungen mit äußerst kleiner Amplitude um den Einstellungswinkel der Phasen übergeht.

Bei den meisten wirklichen Prozessen wird, wie bei Stern-Gerlach, der Übergang nicht so langsam sein, daß man ihn als adiabatisch behandeln darf. Dann wird im allgemeinen Energie frei werden. Was wird aus dieser Energie?

Die möglichen Deutungen, die Einstein und Ehrenfest für den Fall der Stern-Gerlachschen Versuche diskutiert haben, lassen sich auch hier anwenden und führen zu denselben Schwierigkeiten. Der Ausgleich der Energien durch Strahlung scheint uns aus den dort angegebenen Gründen unmöglich. Wir neigen dazu, eine Verletzung der mechanischen Gleichungen anzunehmen; die Einstellungen (der Atomachsen bzw. Phasen) werden auch bei nichtadiabatischen Übergängen ohne Kompensation erfolgen. Der Energie- und vielleicht auch der Impulssatz werden für den Einzelvorgang nicht gelten.

Natürlich wird man annehmen müssen, daß der Energiesatz wenigstens statistisch richtig bleibt. Dem entspricht, daß bei den Versuchen von Stern und Gerlach sich je die Hälfte der magnetischen Atome mit ihrer Achse parallel und antiparallel zur Feldrichtung einstellt. Ob dieser Ausweg auch bei den Phaseneinstellungen der Atome möglich ist, muß erst die nähere Untersuchung lehren.

Die Elektronenbahnen im angeregten Heliumatom.

Von **M. Born** in Göttingen und **W. Heisenberg** in München.

Mit neun Abbildungen. (Eingegangen am 11. Mai 1923.)

Einleitung. Kramers[1]) hat kürzlich gezeigt, daß das Bohr-Kemblesche Modell des normalen Heliumatoms zu einem falschen Wert für die Ionisierungsspannung führt. Ferner hat Bohr[2]) für die angeregten Zustände des Orthoheliums ein ebenes Modell untersucht, bei dem die Elektronen im gleichen Sinne umlaufen, und hat gefunden, daß der berechnete p-Term roh, die übrigen Terme (d, b) gar nicht mit den empirischen übereinstimmen. Wir haben uns nun die Aufgabe gestellt, alle möglichen Bahntypen des angeregten Heliumatoms systematisch aufzusuchen, die quantentheoretisch zulässigen Lösungen auszusondern und die Energiewerte zu berechnen, um festzustellen, ob nicht Bahnen vorhanden sind, die die empirischen Terme richtig liefern[3]). Das Ergebnis unserer Rechnung ist negativ: man gelangt durch konsequente Anwendung der bekannten Quantenregeln zu keiner Erklärung des Heliumspektrums[4]).

§ 1. Entwicklung der Störungsfunktion. Die mathematische Aufgabe besteht in der Lösung des Dreikörperproblems für den Fall, daß das eine Elektron in erheblich größerer Entfernung um den Kern kreist, als das andere. Man gewinnt durch diesen Grenzfall zwar nicht die s-Terme, wohl aber die p-, d-, b-Terme des He (natürlich auch die entsprechenden Terme von Li^+, Be^{++} usw.). Wir werden zeigen, daß die Methoden der Störungstheorie[5]) für dieses Problem ausreichen; und zwar vereinfacht sich das Dreikörperproblem unter den geschilderten Voraussetzungen auf zwei Weisen:

Erstens ist es für die Berechnung der Bahnstörungen des äußeren Elektrons in erster Näherung erlaubt, über die Bewegung des inneren

---

[1]) H. A. Kramers, ZS. f. Phys. **13**, 312, 1923.

[2]) N. Bohr, Göttinger Vorträge 1922 (unpubliziert).

[3]) Den ersten Versuch einer solchen systematischen Berechnung des Heliumspektrums hat A. Landé (Phys. ZS. **20**, 228, 1919) gemacht; doch hat er nicht die säkularen Bewegungen von den kurzperiodischen gesondert und ist darum nicht zu einer vollständigen Übersicht über die Bewegungstypen gelangt. Ferner hat P. Epstein (Naturw. **6**, 230, 1918; s. insbesondere § 18, S. 252) mitgeteilt, daß er bei einem Versuche der Berechnung des Heliumspektrums zu keinem positiven Ergebnis gelangt ist.

[4]) Wir wollen nicht versäumen, an dieser Stelle Herrn W. Pauli für wertvolle Ratschläge unseren Dank auszusprechen.

[5]) Siehe etwa M. Born und W. Pauli, ZS. f. Phys. **10**, 137, 1922.

Elektrons zu mitteln; zweitens wird man konsequent die Störungsfunktion nach negativen Potenzen der Entfernung des äußeren Elektrons vom Kern oder, was dasselbe ist, nach negativen Potenzen der azimutalen Quantenzahl k_1 des äußeren Elektrons entwickeln.

Die Hamiltonsche Funktion des allgemeinen Dreikörperproblems vom He-Typus lautet:

$$H = \frac{1}{2m}\left(p_{r_1}^2 + \frac{1}{r_1^2}\left[\frac{1}{\sin^2\vartheta_1}p_{\varphi_1}^2 + p_{\vartheta_1}^2\right]\right) + \frac{1}{2m}\left(p_{r_2}^2 + \frac{1}{r_2^2}\left[\frac{1}{\sin^2\vartheta_2}p_{\varphi_2}^2 + p_{\vartheta_2}^2\right]\right)$$
$$-\frac{e^2Z}{r_1} - \frac{e^2Z}{r_2} + \frac{e^2}{\sqrt{r_1^2 + r_2^2 - 2r_1r_2[\cos\vartheta_1\cos\vartheta_2 + \sin\vartheta_1\sin\vartheta_2\cos(\varphi_1-\varphi_2)]}} \quad (1)$$

Hierin bedeuten r_1, ϑ_1, φ_1 die Polarkoordinaten des äußeren, r_2, ϑ_2, φ_2 die des inneren Elektrons; Z sei die Kernladung.

Da über die Bewegung des inneren Elektrons gemittelt werden soll, so ist

$$\frac{1}{2m}\left(p_{r_2}^2 + \frac{1}{r_2^2}\left[\frac{1}{\sin^2\vartheta_2}p_{\varphi_2}^2 + p_{\vartheta_2}^2\right]\right) - \frac{e^2Z}{r_2} = W_0 Z^2$$

eine additive Konstante, die man weglassen kann; dabei bedeutet W_0 die Energie des unerregten Wasserstoffatoms, und es ist hier wie im folgenden angenommen, daß die Hauptquantenzahl des inneren Elektrons $n_2 = 1$ ist (was beim He sicher erfüllt ist).

Wir betrachten also die Energie des inneren Elektrons als die Hamiltonsche Funktion des „ungestörten" Systems und den Rest der Energie als Störungsfunktion; diese ist über die ungestörte Bewegung zu mitteln und als neue Hamiltonsche partielle Differentialgleichung für die übrigbleibenden Freiheitsgrade zu behandeln.

Diese Freiheitsgrade lassen sich noch durch Benutzung der elementaren Integrale (Impulssatz) reduzieren.

Man lege die Polarachse in die Richtung des Gesamtimpulses P (der invariablen Geraden). Dann steht die Knotenlinie (Schnittebene der beiden ungestörten Ellipsen) auf der Polarachse senkrecht; ihr Winkelabstand von einer festen, durch die Polarachse gehenden Ebene ist eine zyklische Koordinate des Systems, zu der der Gesamtimpuls P kanonisch konjugiert ist. Als weitere Koordinaten benutzen wir die Periheldistanz ψ_2 des inneren Elektrons von der Knotenlinie, zu der $p_{\psi_2} = \sqrt{\frac{1}{\sin^2\vartheta_2}p_{\varphi_2}^2 + p_{\vartheta_2}^2}$ kanonisch konjugiert ist; den Winkelabstand ψ_1 des äußeren Elektrons vom Knoten, zu dem $p_{\psi_1} = \sqrt{\frac{1}{\sin^2\vartheta_1}p_{\varphi_1}^2 + p_{\vartheta_1}^2}$ konjugiert ist; endlich den Radiusvektor des äußeren Elektrons r_1 mit dem konjugierten radialen Impuls p_{r_1}. Es ist praktisch, statt der Impulse die entsprechenden Quantenzahlen einzuführen; wir setzen also

$$2\pi p_{\psi_1} = k_1 . h; \quad 2\pi p_{\psi_2} = k_2 . h; \quad 2\pi P = j . h.$$

Dann lauten die konjugierten Variablenpaare:

$$\frac{\psi_1}{2\pi} \longleftrightarrow k_1 . h,$$

$$\frac{\psi_2}{2\pi} \longleftrightarrow k_2 . h,$$

$$r_1 \longleftrightarrow p_{r_1}.$$

Die zu P konjugierte zyklische Variable braucht nicht eingeführt zu werden.

Wenn wir jetzt in H (wobei wir die additive Konstante $W_0 . Z^2$ weggelassen haben) den Wechselwirkungsterm in eine Kugelfunktionenreihe entwickeln und mit dem Gliede r_1^{-3} abbrechen, so lautet die Energiefunktion unseres Problems:

$$H = \frac{1}{2m}\left(p_{r_1}^2 + \frac{1}{r_1^2} p_{\psi_1}^2\right) - \frac{e^2(Z-1)}{r_1} + \frac{\Delta_1}{r_1^2} + \frac{\Delta_2}{r_1^3}, \tag{2}$$

wobei

$$\left.\begin{aligned} \Delta_1 &= - e^2 \frac{a_0}{Z}\frac{3}{2}\sqrt{1-k_2^2}\left(\cos\psi_1 \cos\psi_2 + \sin\psi_1 \sin\psi_2 \frac{j^2-k_1^2-k_2^2}{2k_1k_2}\right) \\ \Delta_2 &= \quad e^2\left(\frac{a_0}{Z}\right)^2 \cdot \frac{1}{2}\left\{\frac{1}{2} - \frac{3}{2}\sin^2\psi_1\left[1-\left(\frac{j^2-k_1^2-k_2^2}{2k_1k_2}\right)^2\right]\right. \\ &\quad + 3(1-k_2^2)\left[2\left(\cos\psi_1\cos\psi_2 + \sin\psi_1\sin\psi_2\frac{j^2-k_1^2-k_2^2}{2k_1k_2}\right)\right. \\ &\quad \left.\left. - \frac{1}{2}\left(\cos\psi_1\sin\psi_2 - \sin\psi_1\cos\psi_2\frac{j^2-k_1^2-k_2^2}{2k_1k_2}\right)^2 - \frac{1}{2}\right]\right\}. \end{aligned}\right\} \tag{2a}$$

Hierin bedeutet a_0 den Radius des normalen Wasserstoffatoms. Die Berücksichtigung von Gliedern höherer Ordnung als r_1^{-3} hat prinzipiell keinen Sinn, wenn man über die Bewegung des inneren Elektrons mittelt. So weit aber haben wir die Grenze der von uns erstrebten Genauigkeit gesteckt.

§ 2. Anwendung der Störungstheorie. Die Hamiltonsche Funktion enthält noch drei Freiheitsgrade und ist zunächst nicht separierbar. Da sie aber in Glieder verschiedener Größenordnung zerfällt, so können wir sie wieder störungstheoretisch behandeln. Als „ungestörtes Problem" betrachten wir das, was übrig bleibt, wenn man das letzte Glied $\frac{\Delta_2}{r_1^3}$ wegstreicht; nachträglich wird der Beitrag dieses Gliedes berechnet, indem sein Mittelwert über die ungestörte

Bewegung gebildet wird. Die resultierende „ungestörte" Hamiltonsche Gleichung ist jetzt einmal separierbar und liefert:

$$p_{\psi_1}^2 + 2m\Delta_1 = \text{const} = \alpha^2, \tag{3}$$

$$\frac{1}{2m}\left(p_{r_1}^2 + \frac{\alpha^2}{r_1^2}\right) - \frac{e^2(Z-1)}{r_1} = H. \tag{4}$$

Die Gleichung (3) enthält nur noch zwei Freiheitsgrade; aber sie läßt sich nicht separieren. Wir müssen daher zum dritten Male die Störungstheorie in Anwendung bringen, und dies ist möglich, da $p_{\psi_1}^2$ wieder groß gegen $2m\Delta_1$ ist.

Wir wollen die Größenordnung der verschiedenen vorkommenden Glieder genauer ins Auge fassen.

Bekanntlich [1]) hat der „Parameter" (halbe Sehne durch den Brennpunkt senkrecht zur großen Achse) einer Ellipse mit den Quantenzahlen n_1, k_1 den Wert $\frac{1}{2} a_0 k_1^2$. Nehmen wir dies als Maß von r_1 und bezeichnen „der Größenordnung nach gleich" mit $\sim$, so ist also

$$r_1 \sim a_0 k_1^2.$$

Bedeutet ferner W_0 die Energie des normalen Wasserstoffatoms, so ist nach (4)

$$H \sim \frac{W_0}{k_1^2},$$

ferner nach (2a)

$$\frac{\Delta_1}{r_1^2} \sim \frac{W_0}{k_1^4}, \quad \frac{\Delta_2}{r_1^3} \sim \frac{W_0}{k_1^6},$$

und wegen $2\pi p_{\psi_1} = k_1 . h$, $\frac{h^2}{4\pi^2 m a_0} = e^2$:

$$\frac{p_{\psi_1}^2}{2mr_1^2} \sim \frac{W_0}{k_1^2}.$$

Dividiert man also die Glieder von (3) durch r_1^2 [so wie sie in (4) vorkommen], so sieht man, daß die Glieder niederster Ordnung $\sim \frac{W_0}{k_1^2}$, die Störungsglieder erster Ordnung $\sim \frac{W_0}{k_1^4}$, die zweiter Ordnung $\sim \frac{W_0}{k_1^6}$ werden. Da nun $\frac{\Delta_2}{r_1^3}$ ebenfalls $\sim \frac{W_0}{k_1^6}$ ist, so muß man dieses Glied notwendigerweise zu den Störungsgliedern zweiter Ordnung von Gleichung (3) hinzunehmen.

Die Störungen dritter Ordnung haben im Rahmen unserer Genauigkeit keinen Sinn mehr.

Es ist ein besonders günstiger Umstand, daß gerade bis zu der angestrebten Genauigkeit $\sim \frac{W_0}{k_1^6}$ die Energie des ganzen Systems

[1]) Siehe etwa N. Bohr, Drei Aufsätze über Spektren und Atombau. (Friedr. Vieweg u. Sohn Akt.-Ges., Braunschweig 1922); 3. Aufsatz, S. 78.

durch eine Formel vom Rydbergschen Typus darstellbar ist; d. h. wir behaupten:

$$H = W = \frac{W_0 (Z-1)^2}{(n_1 + \delta)^2}. \tag{5}$$

Hierin soll n_1 die Hauptquantenzahl des äußeren Elektrons bedeuten, δ die Rydbergkorrektion, die von den anderen Quantenzahlen j, k_1, k_2 abhängt.

Wenn man das Glied $\frac{\Delta_2}{r_1^3}$ vernachlässigt, ist diese Behauptung ohne weiteres aus der Form der Differentialgleichung (4) abzulesen. Aber auch der Beitrag von $\frac{\Delta_2}{r_1^3}$ läßt sich dieser Formel durch geeignete Wahl von δ einfügen; denn es zeigt sich, daß der über die Bewegungen beider Elektronen gemittelte Wert von $\frac{\Delta_2}{r_1^3}$ hinsichtlich seiner Abhängigkeit von n_1 die Gestalt $\frac{\Delta_2'}{n_1^3}$ hat, wo Δ_2' nicht mehr von n_1 abhängt.

Der Gang der Rechnung ist also jetzt folgender: Wir integrieren (3) für die kanonischen Variablenpaare (ψ_1, p_{ψ_1}), (ψ_2, p_{ψ_2}) nach den Methoden der Störungstheorie. Dabei wird die Bewegung von ψ_1 schon durch die ungestörte Energie festgelegt; die „säkularen" Bewegungen von ψ_2 ergeben sich aus der Störungsfunktion, und zwar nicht der erster, sondern der zweiter Ordnung. Es zeigt sich nämlich, daß der Mittelwert der Störungsfunktion erster Ordnung identisch verschwindet. Zu den Gliedern zweiter Ordnung der Störungsfunktion ist aber der Beitrag von $\frac{\Delta_2}{r_1^3}$ hinzuzufügen; dieser ist also für die säkularen Bewegungen von ψ_2 wesentlich mitbestimmend.

Ein besonderes Verfahren muß eingeschlagen werden, wenn die Exzentrizität des inneren Elektrons sehr klein ist[1]) ($k_2 \approx 1$). Es zeigt sich aber, daß die schließliche Energieformel genau dieselbe ist, wie für endliche Exzentrizitäten, so daß wir uns auf die Durchrechnung dieses Falles beschränken können.

§ 3. Durchrechnung des Störungsproblems. Wir setzen in (3)

$$(2\pi\alpha)^2 = h^2 . W' \tag{6}$$

und erhalten:

$$W' = k_1^2 - \frac{3}{Z}\sqrt{1-k_2^2}\cdot\left\{\cos\psi_1\cos\psi_2 + \sin\psi_1\sin\psi_2\frac{j^2-k_1^2-k_2^2}{2k_1k_2}\right\} \tag{7}$$

[1]) Siehe H. Poincaré, Mec. cél. II., Kap. XII, S. 74.

Nach dem Schema der Born-Paulischen Methode[1]) hat man nun $W' = H_0' + H_1'$ anzusetzen; dabei ist die ungestörte Energie

$$H_0' = k_1^2 = W_0'$$

und die Störungsfunktion:

$$H_1' = -\frac{3}{Z}\sqrt{1-k_2^2}\cdot\left\{\cos\psi_1\cos\psi_2 + \sin\psi_1\sin\psi_2\frac{j^2-k_1^2-k_2^2}{2k_1k_2}\right\}.$$

Als die zu k_1, k_2 konjugierten Winkelvariabeln sind die Größen

$$w_1 = \frac{h}{2\pi}\psi_1, \quad w_2 = \frac{h}{2\pi}\psi_2$$

anzusehen.

Für die Wirkungsfunktion F erhält man die Näherungsgleichungen[2]):

$$\frac{\partial H_0'}{\partial k_1}\cdot\frac{\partial F_1}{\partial w_1} + H_1' = W_1'; \quad W_1' = \overline{H_1'} = 0;$$

$$\frac{1}{2}\frac{\partial^2 H_0'}{\partial k_1^2}\cdot\left(\frac{\partial F_1}{\partial w_1}\right)^2 + \frac{\partial H_1'}{\partial k_1}\cdot\frac{\partial F_1}{\partial w_1} + \frac{\partial H_1'}{\partial k_2}\cdot\frac{\partial F_1}{\partial w_2} + \frac{\partial H_0'}{\partial k_1}\cdot\frac{\partial F_2}{\partial w_1} + H_2' = W_2';$$

$$W_2' = \overline{\frac{1}{2}\frac{\partial^2 H_0'}{\partial k_1^2}\cdot\left(\frac{\partial F_1}{\partial w_1}\right)^2} + \overline{\frac{\partial H_1'}{\partial k_1}\cdot\frac{\partial F_1}{\partial w_1}} + \overline{\frac{\partial H_1'}{\partial k_2}\cdot\frac{\partial F_1}{\partial w_2}}.$$

Eine elementare Rechnung liefert:

$$W_2' = \frac{9}{2Z^2k_1^2}\left\{\frac{1-j^2+k_1^2}{4} + (1-k_2^2)\sin^2\psi_2\,.\,V\right\}, \tag{8}$$

wobei

$$V = \frac{1}{2}\frac{j^2-k_1^2-k_2^2}{2k_1k_2}\cdot\frac{j^2+k_1^2-k_2^2}{2k_1k_2} - \frac{1}{2k_2}\frac{j^2-k_1^2+k_2^2}{2k_2} - \frac{1}{4}\left[1-\left(\frac{j^2-k_1^2-k_2^2}{2k_1k_2}\right)^2\right] \tag{8a}$$

eingeführt ist.

Wir setzen nun $j = k_1 + p$, entwickeln W_2' nach Potenzen von k_1^{-1} und brechen die Reihe mit k_1^{-3} ab; dann erhalten wir:

$$W_2' = \frac{9}{4Z^2k_1}\left(-p + \frac{1-p^2}{2k_1}\right) + \frac{27}{8Z^2k_1^2}(1-k_2^2)\left(\frac{p^2}{k_2^2}-1\right)\sin^2\psi_2. \tag{9}$$

Diese Größe liefert direkt einen Beitrag δ_1 zur Rydbergschen Korrektion δ. Ein weiterer Beitrag δ_2 zu δ rührt nach dem oben Gesagten von $\frac{\Delta_2}{r_1^3}$ her; und zwar wird man diese Größe nicht nur über die ungestörte Bewegung des inneren, sondern auch des äußeren Elektrons zu mitteln haben. Es zeigt sich, daß dabei die Perihellänge des äußeren Elektrons von selbst herausfällt, so daß der

[1]) Will man diese Methode ganz formal anwenden, so muß man die Störungsfunktion nach fallenden Potenzen von k_1 (nicht von k_1^2) entwickeln. Das hier eingeschlagene Verfahren ist einfacher, liefert aber nur die Glieder bis zur Ordnung k_1^{-2} einschließlich genau.

[2]) Siehe M. Born u. W. Pauli, l. c.

Doppelmittelwert $\overline{\overline{\frac{\Delta_2}{r_1^3}}}$ von denselben Variablen p, k_1, k_2, ψ_2 abhängt, wie W_2'. Man findet nach längerer, elementarer Rechnung

$$\overline{\overline{\frac{\Delta_2}{r_1^3}}} = -W_0 \frac{(Z-1)^3}{Z^2} \frac{1}{k_1^3 (n_1+\delta_1)^3} \left\{ -\frac{1}{4} + \frac{3}{2}(1-k_2^2) \right.$$

$$\left. + \frac{3}{4} k_2^2 \left(\frac{j^2-k_1^2-k_2^2}{2 k_1 k_2}\right)^2 - \frac{15}{4} \sin^2\psi_2 (1-k_2^2) \left[1 - \left(\frac{j^2-k_1^2-k_2^2}{2 k_1 k_2}\right)^2\right] \right\}, \quad (10)$$

oder mit $j = k_1 + p$ bis auf Glieder von höherer als vierter Ordnung in k_1^{-1}:

$$\overline{\overline{\frac{\Delta_2}{r_1^3}}} = -W_0 \frac{(Z-1)^3}{Z^2} \cdot \frac{1}{4 k_1^3 (n_1+\delta_1)^3} \Gamma, \quad (11)$$

wo

$$\Gamma = 5 + 3p^2\left(1 + \frac{p}{k_1}\right) - 3k_2^2\left(2 + \frac{p}{k_1}\right)$$

$$- 15 \sin^2\psi_2 (1-k_2^2)\left(1 - \frac{p^2}{k_2^2}\right)\left(1 + \frac{p}{k_1}\right). \quad (11\,a)$$

Nun folgt aus (4)

$$H = \frac{W_0 (Z-1)^2}{(n_r + \sqrt{W'})^2} = \frac{W_0 (Z-1)^2}{(n_1+\delta)^2},$$

wo $n_r = n_1 - k_1$ die radiale Quantenzahl ist; also

$$\delta_1 = \sqrt{W'} - k_1 = \sqrt{k_1^2 + W_2'} - k_1 = \frac{W_2'}{2k_1} + \cdots \quad (12)$$

Fügt man das Glied (11) hinzu, so wird

$$H = \frac{W_0 (Z-1)^2}{(n_1+\delta_1)^2}\left\{1 - \frac{Z-1}{Z^2} \frac{1}{4 k_1^3 (n_1+\delta_1)} \cdot \Gamma\right\} = \frac{W_0 (Z-1)^2}{(n_1+\delta_1+\delta_2)^2},$$

also

$$\delta_2 = \frac{1}{8} \frac{Z-1}{Z^2} \frac{1}{k_1^3} \cdot \Gamma. \quad (13)$$

Nunmehr kann man die gesamte Rydbergsche Korrektion zusammenfügen [bis auf Glieder der Ordnung k_1^{-3} einschließlich; vgl. Gleichung (9)]:

$$\delta = \delta_1 + \delta_2 = \frac{9}{8 k_1^2 Z^2}\left(-p + \frac{1-p^2}{2k_1}\right) + \frac{Z-1}{8 k_1^3 Z^2} \cdot (5 + 3p^2 - 6k_2^2)$$

$$- \frac{15}{8} \sin^2\psi_2 \cdot \frac{Z-1}{k_1^3 Z^2} (1-k_2^2)\left(1 - \frac{p^2}{k_2^2}\right). \quad (14)$$

Führt man die Abkürzung γ ein vermöge der Relation:

$$\delta - \frac{9}{8 k_1^2 Z^2}\left(-p + \frac{1-p^2}{2k_1}\right) - \frac{Z-1}{8 Z^2 k_1^3}\{5 + 3p^2\} = -\gamma \cdot \frac{3(Z-1)}{4 Z^2 . k_1^3}, \quad (15)$$

so bekommt man folgende Gleichung für ψ_2, k_2:

$$\gamma = k_2^2 + \frac{5}{2} \sin^2\psi_2 (1-k_2^2)\left(1 - \frac{p^2}{k_2^2}\right), \quad (16)$$

woraus folgt:

$$k_2^2 = \frac{\gamma - \frac{5}{2}(1+p^2)\sin^2\psi_2}{2(1-\frac{5}{2}\sin^2\psi_2)}$$

$$\pm \frac{1}{2(1-\frac{5}{2}\sin^2\psi_2)}\sqrt{[\gamma-\frac{5}{2}(1+p^2)\sin^2\psi_2]^2+10\sin^2\psi_2\, p^2(1-\frac{5}{2}\sin^2\psi_2)}. \quad (17)$$

Durch die Diskussion dieser Gleichung müssen wir alle vom formalen Standpunkt der Störungstheorie aus möglichen Bewegungstypen des angeregten He gewinnen.

§ 4. Die Bewegungstypen. Wir erhalten die verschiedenen möglichen Bewegungen offenbar dadurch, daß wir die reellen Zweige der Kurve $k_2 = f(\psi_2)$, oder, was dasselbe ist, die reellen und positiven Zweige der Kurve $k_2^2 = \varphi(\psi_2)$ [Gleichung (17)] aufsuchen. Und zwar muß diese Untersuchung für alle reellen Werte von γ und für alle Werte von p zwischen $+1$ und -1 durchgeführt werden.

1. (Fig. 1). Für hinreichend große Werte von γ, nämlich

$$\gamma > \frac{5}{2}(1+p^2) + p\sqrt{15},$$

ist k_2^2 für alle Werte von ψ_2 reell. Der eine Kurvenast (a), der dem negativen Vorzeichen der Quadratwurzel entspricht, verläuft überall stetig und liefert ein negatives k_2^2. Der dem positiven Vorzeichen

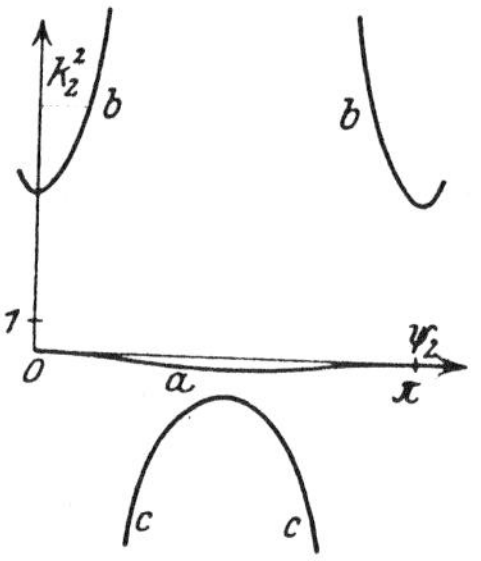

Fig. 1. ($p^2 = \frac{1}{5}$, $\gamma = 5$).

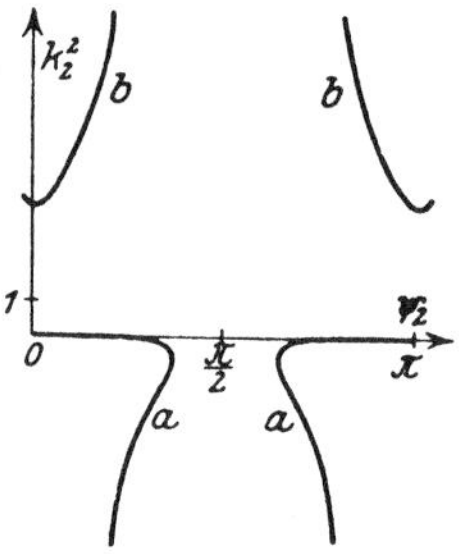

Fig. 2. ($p^2 = \frac{1}{5}$, $\gamma = 4$).

entsprechende Ast zerfällt in zwei Teile. Der eine (b), zu dem Wertebereich $\sin^2\psi_2 < \frac{2}{5}$ gehörig, ist positiv und geht bei $\sin^2\psi_2 = \frac{2}{5}$ gegen $+\infty$; der andere (c), $\sin^2\psi_2 > \frac{2}{5}$, ist negativ und geht bei $\sin^2\psi_2 = \frac{2}{5}$ gegen $-\infty$.

2. (Fig. 2). Für

$$\frac{5}{2}(1+p^2) + p\sqrt{15} > \gamma > \frac{5}{2}(1+p^2) - p\sqrt{15}$$

wird k_2^2 in einem gewissen Gebiet um $\sin^2\psi_2 = 1$ herum imaginär. Die Kurve besteht nur noch aus zwei Ästen. Ist $\gamma > 1 + p^2$, so ist der eine (b), $\sin^2\psi_2 < \frac{2}{5}$, wie bei 1. positiv und geht bei $\sin^2\psi_2 = \frac{2}{5}$ gegen $+\infty$, der andere (a) ist negativ, überschreitet zunächst den Punkt $\sin^2\psi_2 = \frac{2}{5}$ stetig, kehrt dann bei einem gewissen Werte

$1 > \sin^2 \psi_2 > \frac{2}{5}$ um und geht nunmehr bei $\sin^2 \psi_2 = \frac{2}{5}$ gegen $-\infty$. Ist $\gamma < 1 + p^2$, so vertauschen die beiden Kurvenäste ihre Rollen, d. h. der negative geht sofort bei $\sin^2 \psi_2 = \frac{2}{5}$ gegen $-\infty$, der positive überschreitet diesen Punkt zuerst stetig, kehrt um und geht dann gegen $+\infty$.

3. (Fig. 3). Ist

$$\tfrac{5}{2}(1 + p^2) - p\sqrt{15} > \gamma > 1,$$

so bleiben die Verhältnisse von 2. wesentlich bestehen, nur wird k_2^2 auch in einem kleinen Bereich um $\sin^2 \psi_2 = 1$ herum reell (und positiv), die Kurve erhält wieder einen dritten Zweig (c), und zwar ist dieser dritte Zweig — das ist für die Bewegung wesentlich — geschlossen.

4. (Fig. 4). Für $\gamma < 1$ wird k_2^2 wieder für alle Werte von ψ_2 reell. Der dem positiven Vorzeichen der Quadratwurzel entsprechende Ast (c) verläuft überall stetig und liefert ein positives k_2^2. Die anderen beiden Zweige gehören zum negativen Vorzeichen der Quadratwurzel; der eine (a), $\sin^2 \psi_2 < \frac{2}{5}$, ist negativ und geht für $\sin^2 \psi_2 = \frac{2}{5}$ gegen $-\infty$, der andere (b), $\sin^2 \psi_2 > \frac{2}{5}$, ist positiv und geht für

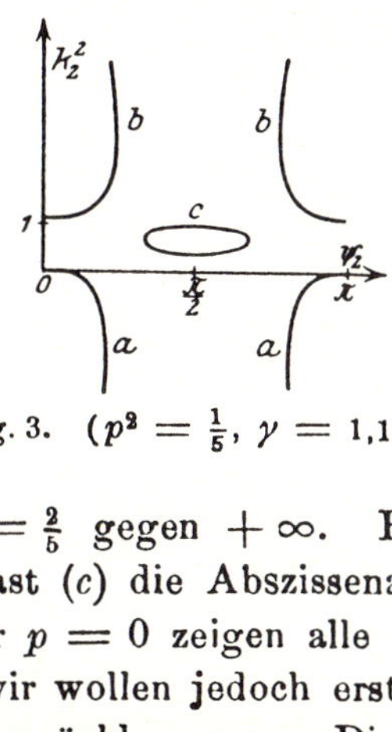

Fig. 3. ($p^2 = \frac{1}{5}$, $\gamma = 1{,}1$).

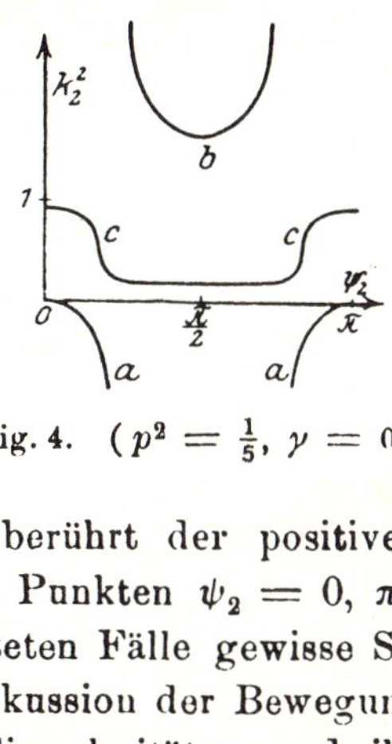

Fig. 4. ($p^2 = \frac{1}{5}$, $\gamma = 0{,}9$).

$\sin^2 \psi_2 = \frac{2}{5}$ gegen $+\infty$. Für $\gamma \leqq 0$ berührt der positive stetige Kurvenast (c) die Abszissenachse in den Punkten $\psi_2 = 0, \pi$.

Für $p = 0$ zeigen alle vier betrachteten Fälle gewisse Singularitäten; wir wollen jedoch erst bei der Diskussion der Bewegung selbst darauf zurückkommen. Die Art der Singularitäten und ihre Entstehung wird durch Fig. 5 und 6 dargestellt.

Wenn wir nach den bedingt periodischen Lösungen der Bewegungsgleichungen fragen, so müssen wir von den Kurvenästen $k_2 = f(\psi_2)$ diejenigen heraussuchen, bei denen $k_2 (p \leqq k_2 \leqq 1)$ eine stetige periodische Funktion von ψ_2 ist, oder wir müssen die geschlossenen Kurvenzüge wählen. Die ersteren geben zu Rotationsbewegung, die letzteren zu Libration Anlaß. Aus 3. und 4. sehen wir, daß beides möglich ist.

I. Für $\gamma \leqq 1$ haben wir reine Rotationsbewegung, das Perihel des inneren Elektrons läuft immer im gleichen Sinne um.

Die Bewegung kann dabei sowohl eben (Fig. 7), wie räumlich (Fig. 8) sein. Der ebene Fall ist durch

$$k_1 + k_2 = j, \qquad k_2 = \text{const} = p$$

festgelegt. Die komplanare Bewegung ist also dadurch ausgezeichnet, daß die Impulsmomente beider Elektronen in erster Näherung konstant, d. h. keinen säkularen Veränderungen unterworfen sind. Die

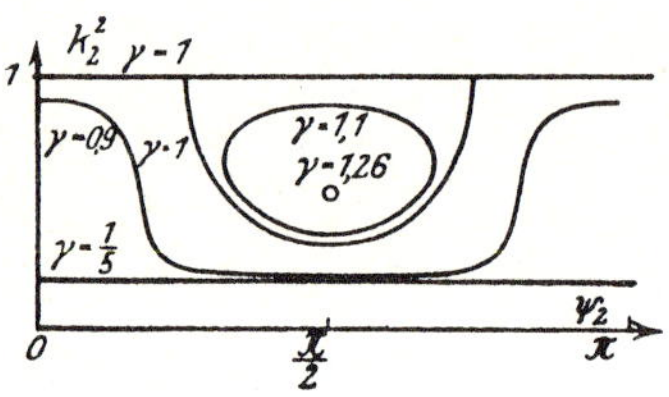

Fig. 5. $(p^2 = 0)$.

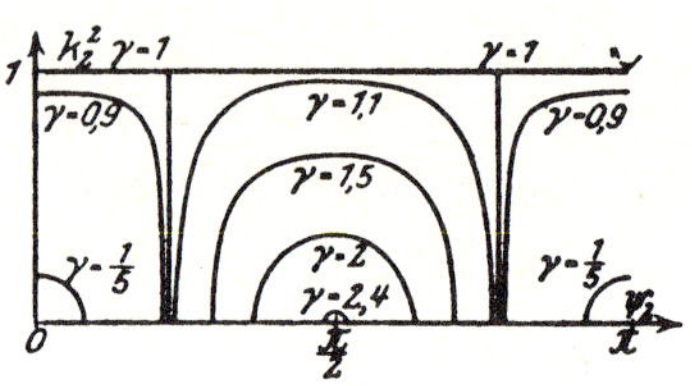

Fig. 6. $(p^2 = \frac{1}{5})$.

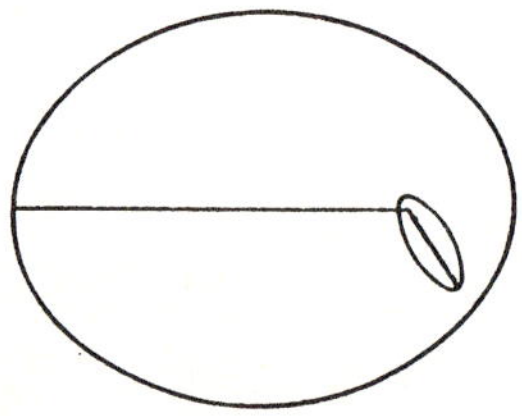

Fig. 7.

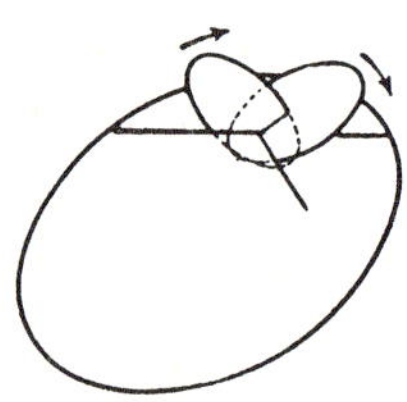

Fig. 8.

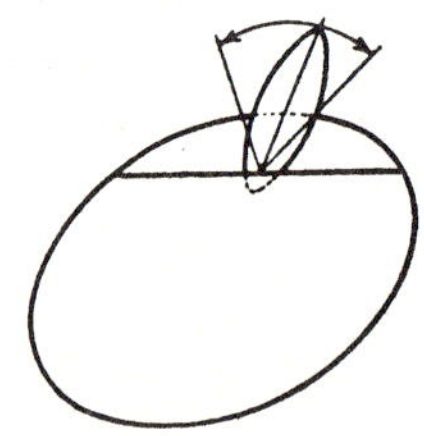

Fig. 9.

Periheldrehung sowohl des inneren wie des äußeren Elektrons ist gleichförmig, und zwar, wie man durch Differentiation der Energieformel findet, entgegengesetzt gerichtet.

Im allgemeinen ist aber die Bewegung räumlich, k_2 ist säkular veränderlich; wegen der Konstanz von j und k_1 ist der Winkel zwischen beiden Bahnebenen ebenfalls säkular veränderlich (vgl. Fig. 8). Die Periheldrehung des inneren Elektrons erfolgt nicht mehr gleichförmig.

Ist speziell $\gamma = 1$, so wird $k_2 = 1$; das innere Elektron beschreibt eine Kreisbahn, die gegen die Bahnebene des äußeren Elektrons beliebig geneigt sein kann.

Aus $k_2 \geqq p$ folgt ferner für den Bewegungstypus I: $\gamma \geqq p^2$.

II. Für $1 < \gamma < \frac{5}{2}(1 + p^2) - p\sqrt{15}$ beschreibt das Perihel des inneren Elektrons eine Librationsbewegung (Fig. 9), es pendelt zwischen zwei festen Grenzen um den Wert $\sin^2 \psi_2 = 1$, also $\psi_2 = \frac{\pi}{2}$ oder $\psi_2 = \frac{3\pi}{2}$ herum. Hierbei kehrt sich das Impulsmoment des inneren Elektrons nicht um, die Exzentrizität seiner Bahn bleibt stets endlich. Im speziellen kann der Librationsbereich auf das Librationszentrum zusammenschrumpfen; dann steht das Perihel des inneren Elektrons bei $\psi = \frac{\pi}{2}\left(\frac{3\pi}{2}\right)$ fest, k_2 hat den konstanten Wert

$$k_2 = \sqrt{p\sqrt{\tfrac{5}{3}}} = \sqrt{p}\,.\,1{,}14.$$

Aus der Bedingung $k_2 \leqq 1$ entnehmen wir, daß die Librationsbewegung II nur möglich ist, wenn

$$p \leqq \sqrt{\tfrac{3}{5}} = 0{,}775.$$

III. Für $p = 0$ bleibt der Bewegungstypus I nur dann bestehen, wenn $\gamma = 1$ ist; für kleinere Werte von γ entsteht wieder eine Librationsbewegung, und zwar um den Punkt $\psi_2 = 0\,(\pi)$ als Zentrum, für größere γ um $\psi_2 = \frac{\pi}{2}\left(\frac{3\pi}{2}\right)$ als Zentrum (vgl. Fig. 6). Das Impulsmoment des inneren Elektrons nimmt dabei immer wieder den Wert Null an (der einer Pendelbahn entspricht), und wenn die Libration langsam erfolgt, so wird das innere Elektron dem Kern sehr nahe kommen. Der spezielle Bewegungstypus II, $k_2 =$ const. wird für $p = 0$ insofern singulär, als k_2 den konstanten Wert Null annimmt. Es muß aber betont werden, daß sich der Impuls des inneren Elektrons von k_2 noch um Größen der Ordnung k_1^{-1} unterscheidet; also folgt aus $k_2 = 0$ zunächst nur, daß der Impuls des inneren Elektrons klein wird von der Ordnung k_1^{-1}.

Hiermit haben wir alle möglichen bedingt periodischen Bewegungstypen des angeregten He aufgefunden, soweit sie von dem Formalismus der Störungstheorie aus erfaßt werden können.

Der Bewegungstypus I wurde von Bohr für $p = 1$ zur Erklärung des Orthoheliumspektrums, für $p = 0$ zur Deutung des Parheliumspektrums versuchsweise herangezogen. Auf den Typus II hat für $p = 0$ Bohr in seinen Göttinger Vorträgen hingewiesen, ihn jedoch wegen $k_2 \to 0$ für $k_1 \to \infty$ nicht zur Deutung physikalischer Erscheinungen benutzt.

Bei unseren bisherigen Überlegungen haben wir stillschweigend die Voraussetzung gemacht, daß die benutzten Reihen der Störungstheorie wirklich konvergieren. Dies ist aber im allgemeinen nicht der Fall. Nach dem Brunsschen Theorem liegen vielmehr die Konvergenz- und Divergenzstellen der Reihen überall dicht. Wir müssen also erwarten, daß die von uns gefundenen Bewegungstypen wirklich alle existieren, daß es aber in jeder noch so nahen Umgebung einer Lösung unendlich viele andere, nicht bedingt periodische Lösungen der Bewegungsgleichungen gibt, die von unseren Formeln nicht geliefert werden.

Dieser Sachverhalt führt auf große Schwierigkeiten, wenn wir das Korrespondenzprinzip auf das He-Modell anwenden wollen.

Wenn wir fordern, daß nur solche Terme kombinieren sollen, die durch eine kontinuierliche Mannigfaltigkeit bedingt periodischer Lösungen verbunden werden können, so wäre eine ganz neue Untersuchung des Problems nach solchen kontinuierlichen harmonischen Lösungen nötig. Abgesehen davon, daß die Existenz derartiger Mannigfaltigkeiten fraglich ist, gibt es eine Reihe physikalischer Gründe, die gegen diese Auffassung des Korrespondenzprinzips sprechen. Man kann aber fordern, daß nur Terme kombinieren sollen, die dann kontinuierlich ineinander übergeführt werden können, wenn über die Bewegung des inneren Elektrons gemittelt wird; daß also wirklich harmonische Lösungssysteme der Hamiltonschen Gleichung (2) zur Kombination zugelassen werden. Für eine solche Auffassung scheinen nach Bohr gewichtige physikalische Gründe zu sprechen. Endlich könnte man zum dritten die ganze unendlich dichte Mannigfaltigkeit harmonischer Lösungen als kontinuierlich im Sinne des Korrespondenzprinzips auffassen.

In diesem letzten Falle wäre es konsequent, auch die Stabilität der gefundenen Bewegungstypen zu untersuchen in der Weise, wie dies sonst bei bedingt periodischen Systemen üblich ist. Man muß dann fordern, daß eine Bahn, die durch kleine Variation der Anfangswerte von k_2 und ψ_2 aus der zu untersuchenden Bahn hervorgeht, dieser in ihrem ganzen Verlauf unendlich benachbart bleibt. Es zeigt sich, daß der Bewegungstypus I im allgemeinen stabil ist. Doch ist die spezielle Lösung $k_2 = 1$, $\gamma = 1$ nur für

$$p > \sqrt{\tfrac{3}{5}} = 0{,}775$$

stabil, für kleinere p labil. Typus II ist im allgemeinen stabil, ebenso die spezielle Lösung $k_2 = \text{const} = 1{,}14 \cdot \sqrt{p}$.

Eine Entscheidung zugunsten irgend einer der drei geschilderten Betrachtungsweisen soll hier nicht versucht werden, besonders, da das quantitative Ergebnis unserer Rechnungen zeigen wird, daß die gewöhnliche Mechanik nicht zur Beschreibung der Mehrkörperprobleme der Atomphysik ausreicht. Wir kehren also wieder zu den gefundenen Bewegungstypen zurück und wählen aus allen möglichen harmonischen Lösungen die quantentheoretisch ausgezeichneten aus.

§ 5. **Die Quantenbahnen.** Die Auswahl der Quantenbahnen geschieht durch die Forderung, daß

$$k_2^* = \frac{1}{2\pi} \oint k_2 \, d\psi \tag{18}$$

eine ganze Zahl sein soll.

Man erkennt leicht, daß $k_2^* = 1$, also auch $k_2 = \text{const} = 1$, was in erster Näherung Kreisbahn bedeutet, die einzigen Lösungen liefert, bei denen k_2^* eine ganze Zahl ist und bei denen das innere Elektron nicht dem Kern sehr nahe kommt (d. h. $k_2 \neq 0$). Auch der spezielle Bewegungstypus II: $k_2 = \text{const} = \sqrt{p} \cdot 1{,}14$ liefert $k_2^* = 0$. Doch wird er wegen der Ganzzahligkeit von p, die hier nur $p = 0$ übrig läßt, physikalisch unbrauchbar.

Für die Quantenlösungen erhalten wir also die Energieformel

$$\delta = \frac{9}{8 k_1^2 Z^2}\left(-p + \frac{1-p^2}{2 k_1}\right) + \frac{Z-1}{8 k_1^3 Z^2} \cdot (3 p^2 - 1). \tag{19}$$

Da ferner $p = j - k_1$ nur die Werte $+1$, 0, -1 annehmen kann, erhalten wir drei Klassen von Quantenbahnen:

I. $p = 1$: Komplanare Bahnen von gleichem Umlaufssinn,

$$\delta = -\frac{1}{4 Z^2 k_1^2}\left(\frac{9}{2} - \frac{Z-1}{k_1}\right);$$

II. $p = 0$: Nahezu aufeinander senkrechte Bahnen,

$$\delta = \frac{1}{8 Z^2 k_1^3}\left(\frac{11}{2} - Z\right);$$

III. $p = -1$: Komplanare Bahnen mit entgegengesetztem Umlaufssinn:

$$\delta = \frac{1}{4 Z^2 k_1^2}\left(\frac{9}{2} + \frac{Z-1}{k_1}\right).$$

Hierbei ist zu betonen, daß die Bahnen II, $p = 0$ labil sind. Insbesondere erhält man für $Z = 2$:

$$\text{I. } p = 1: \quad \delta = -\frac{1}{32 k_1^2}\left(9 - \frac{2}{k_1}\right).$$

$$\text{II. } p = 0: \quad \delta = +\frac{Z}{64 k_1^3}.$$

$$\text{III. } p = -1: \delta = +\frac{1}{32 k_1^2}\left(9 + \frac{2}{k_1}\right).$$

Wir stellen nun die Werte von δ für $k_1 = 2, 3, 4$ in einer Tabelle zusammen:

Tabelle 1.

	$k_1 = 2$	$k_1 = 3$	$k_1 = 4$
$p = 1$	$-0{,}0625$	$-0{,}0289$	$-0{,}0166$
$p = 0$	$+0{,}0137$	$+0{,}0040$	$+0{,}0017$
$p = -1$	$+0{,}0781$	$+0{,}0336$	$+0{,}0185$

Dieser theoretischen Tabelle stellen wir eine empirische gegenüber[1]):

Tabelle 2.

	$k_1 = 2$	$k_1 = 3$	$k_1 = 4$
Orthohelium . . .	— 0,067	— 0,0036	— 0,00103
Parhelium	+ 0,0103	— 0,0025	— 0,0014?

Ein Vergleich beider Tabellen lehrt, daß das Resultat unserer Untersuchung völlig negativ ist.

Zwar ließe sich für $k_1 = 2$ eine rohe Übereinstimmung beider Tabellen erreichen, wenn man für Orthohelium $p = 1$, für Parhelium $p = 0$ setzt, wie dies Bohr[2]) versucht hat. Aber in den höheren Termen kann von Übereinstimmung keine Rede mehr sein.

Um der Vollständigkeit der Untersuchung sicher zu sein, müssen wir noch den Einfluß der Relativität auf die Bewegung abschätzen. Die Relativitätskorrektion des äußeren Elektrons ergibt einfach zu dem Gliede Δ_2 einen kleinen Zusatz, der sich in δ als kleine additive Konstante äußert; sie kann die Bewegung des inneren Elektrons (ψ_2, k_2) nicht beeinflussen. Die Relativitätskorrektion des inneren Elektrons ergibt zu δ einen Zusatz von der Form $C \cdot \frac{n_1^3}{k_2}$ und kann daher für große Werte von n_2 merklich werden und den Bewegungstypus abändern, um so mehr, je kleiner k_2 im Verlauf der Bewegung wird. Gerade aber die quantentheoretisch ausgezeichnete Lösung $k_2 = 1$ bleibt unverändert bestehen. Erst wenn der Einfluß der Relativität auf die vom äußeren Elektron herrührenden periodischen Störungen von k_2 merklich wird, kann die Relativitätskorrektion den Bewegungstypus abändern. Die dann entstehende Bewegung läßt sich aber wieder leicht behandeln, weil man die Wirkung des äußeren Elektrons als schwaches homogenes elektrisches Feld auffassen kann, das am inneren Elektron einen Dipol induziert. Dieser induzierte Dipol ist von Kramers[3]) berechnet worden. Eine Abschätzung der Größenordnung zeigt, daß erst für $n_1 > 100$ eine solche „relativistische" Bewegung eintritt.

[1]) Da die Terme einer Serie in Wirklichkeit nicht genau einer Rydbergschen Formel genügen, sind die Zahlen der Tabelle als Mittelwerte anzusehen. — Der Wert für δ im b-Term ($k_1 = 4$) von Parhe muß als sehr unsicher betrachtet werden, da er durch Mittelung aus zwei stark verschiedenen Werten — 0,00085; — 0,00204) erhalten ist.

[2]) Die Formel für $p = 1$ wurde bereits früher von Bohr abgeleitet und auf das Orthohelium angewandt.

[3]) H. A. Kramers, ZS. f. Phys. 3, 199, 1920.

Wir kommen also zu dem Resultat, daß eine konsequente quantentheoretische Durchrechnung des Heliumproblems zu falschen Werten für die Energieterme führt [1]). Es gibt aus dieser Schwierigkeit offenbar nur zwei Auswege: Entweder sind die Quantenbedingungen falsch, d. h. die Forderung, daß k_1, k_2^* und j ganzzahlig sein sollen, besteht nicht zu Recht (die Ganzzahligkeit von n_1 und n_2 ist wohl empirisch bewiesen), oder die Bewegung der Elektronen genügt auch in den stationären Zuständen nicht mehr den mechanischen Gleichungen. Beide Auffassungen führen zu fundamentalen Schwierigkeiten für das Verständnis der bisher gesicherten quantentheoretischen Ergebnisse.

Jedenfalls wird es nötig sein, neue Hypothesen einzuführen, und dieser Aufgabe wird die Kenntnis der Bewegungstypen und die Energieformel wohl als Vorbereitung dienen können.

---

[1]) Anmerkung bei der Korrektur: J. H. van Vleck teilt in Phys. Rev. **21**, 372, 1923 mit, daß er bei einer Berechnung der Energiestufen des Heliumatoms keine Übereinstimmung mit der Erfahrung gefunden hat.

1924. № 9.

ANNALEN DER PHYSIK.

VIERTE FOLGE. BAND 74.

1. *Zur Quantentheorie der Molekeln;*

von M. Born und W. Heisenberg.

Einleitung.

In der Anwendung der Quantentheorie auf die Molekeln kann man folgende Entwicklungsstufen unterscheiden: Das erste Stadium[1]) ersetzt die zweiatomige Molekel durch das Hantelmodell, das als einfacher „Rotator" behandelt wird. Mehratomige Molekeln werden in entsprechender Weise als starre „Kreisel" angesehen.[2]) Dieser Standpunkt erlaubt es, die einfachsten Gesetze der Bandenspektren und der spezifischen Wärme mehratomiger Gase zu erklären. Das nächste Stadium[1]) läßt die Annahme starrer Verbindungen zwischen den Atomen fallen und berücksichtigt die Kernschwingungen, zunächst als harmonische Schwingungen; dabei ergeben sich nach Sponer[3]) und Kratzer[4]) Zusammenhänge zwischen den einzelnen Banden eines Bandensystems. Im dritten Stadium[5]) wird die Wechselwirkung zwischen Rotation und Kernschwingung in Rechnung gezogen; zugleich muß beachtet werden, daß die Schwingungen nicht streng harmonisch sind. Hierdurch erklären sich die feineren Gesetze der Linienabstände einer Bande und der Aufeinanderfolge der Banden. Bei diesen Untersuchungen werden wesentlich nur Kernbewegungen behandelt, während von den Elektronen, die um

1) K. Schwarzschild, Berl. Akad. Ber. 1916. S. 548; N. Bjerrum, Nernst-Festschrift 1912. S. 90.

2) T. Heurlinger, Phys. Zeitschr. **20.** S. 188. 1919; Zeitschr. f. Phys. **1.** S. 82. 1920; W. Lenz, Verh. d. Deutsch. phys. Ges. **21.** S. 632. 1919.

3) H. Sponer, Diss. Göttingen 1920; Jahrb. d. Phil. Fak. 1921. S. 153.

4) A. Kratzer, Zeitschr. f. Phys. **3.** S. 289. 1920. **16.** S. 353. 1923; Phys. Zeitschr. **22.** S. 552. 1921; Ann. d. Phys. **67.** S. 127. 1922; **71.** S. 72. 1913; Sitzber. Bayr. Ak. 1922. S. 107; Naturw. **11.** S. 577. 1913.

5) A. Kratzer, a. a. O.; ferner M. Born u. E. Hückel, Phys. Zeitschr. **24.** S. 1. 1923.

die Kerne herumlaufen und sie zusammenhalten, kaum die Rede ist; die Quantensprünge der Elektronenbahnen werden formal berücksichtigt, indem die Energie der Elektronen additiv hinzugefügt wird und die Konstanten der Energie der Kernbewegung (Trägheitsmomente, Schwingungszahlen) als theoretisch nicht bestimmbare Funktionen der Elektronenquantenzahlen betrachtet werden. Das vierte Stadium der Theorie[1]) zieht wenigstens eine Eigenschaft der Elektronenbewegung heran, nämlich die Existenz eines Drehimpulses der Elektronen von gleicher Größenordnung, wie der Drehimpuls der Kerndrehung; Kratzer, Pauli und Kramers haben gezeigt, daß hierdurch die Kreiselbewegung wesentlich beeinflußt wird, und Kratzer hat versucht, die bei manchen Banden auftretenden Dublettlinien durch die Wirkung des Elektronenimpulses zu erklären.

Das fünfte und letzte Stadium der Theorie macht mit der Auffassung der Molekel als eines Gebäudes von Kernen und Elektronen Ernst. Für zweiatomige Molekeln hat Kratzer in seinen letzten Arbeiten diesen Standpunkt eingenommen. Wir wollen im folgenden versuchen, ganz allgemein die Aufgabe zu lösen: Wie bewegt sich eine Molekel als mechanisches System von Kernen und Elektronen, und wieweit gibt die gewöhnliche Bandentheorie von dieser Bewegung Rechenschaft?

Wir werden als Hauptmerkmal der Molekel die Tatsache betrachten, daß die Kernmasse sehr groß gegen die Elektronenmasse ist. Dann werden wir unter der von der Quantentheorie geforderten Annahme, daß alle Impulse und Drehimpulse von gleicher Größenordnung sind, den Nachweis führen, daß die oben aufgezählten Stadien der Bandentheorie im großen und ganzen der natürlichen Reihenfolge der sukzessiven Näherungen entsprechen, auf die man bei Entwicklung der Energie nach Potenzen der Quadratwurzel aus dem Verhältnis Elektronenmasse zu Kernmasse geführt wird. Als mathematisches Hilfsmittel verwenden wir dabei die Störungstheorie[2]); doch

1) A. Kratzer, a. a. O.; ferner H. A. Kramers, Zeitschr. f. Phys. **13**. S. 343. 1923; H. A. Kramers u. W. Pauli jun., Zeitschr. f. Phys. **13**. S. 351. 1923.

2) Vgl. P. Epstein, Zeitschr. f. Phys. **8**. S. 211. 305. 1922; **9**. S. 92. 1922; M. Born u. W. Pauli, Zeitschr. f. Phys. **10**. S. 137. 1922; M. Born u. W. Heisenberg, Zeitschr. f. Phys. **14**. S. 44. 1923; L. Nordheim, Zeitschr. f. Phys. **17**. S. 316. 1923; **21**. S. 242. 1924.

ist der vorliegende Fall dadurch ausgezeichnet, daß das übliche Verfahren zur Bestimmung der säkularen Störungen versagt, weil der Mittelwert der Störungsfunktion 1. Ordnung identisch verschwindet. Wir schicken daher der eigentlichen Molekeltheorie (Kap. II) eine Untersuchung über die in Betracht kommenden Sätze der Störungstheorie voraus (Kap. I).

Kapitel I. Hilfssätze der Störungstheorie.

Das ungestörte System sei bedingt periodisch und in Strenge durch Separation der Variabeln integriert. Dabei sollen von den f-Freiheitsgraden $f - s$ entartet sein; s heißt nach Bohr der Periodizitätsgrad des Systems. Wir setzen fest,

daß die Indices $k, l, \ldots$ von 1 bis f,
$\alpha, \beta, \gamma, \ldots$ „ 1 „ s,
$\varrho, \sigma, \tau, \ldots$ „ $s+1, \ldots f$

laufen sollen.

Man kann es immer so einrichten, daß die Hamiltonsche Funktion des ungestörten Systems nur von den Wirkungsvariabeln j_α^0 ($\alpha = 1, 2, \ldots s$) abhängt:

$$H_0 = H_0(j_1^0, \ldots j_s^0). \tag{1}$$

Dann sind die Freiheitsgrade $s+1$, $s+2$, $\ldots f$ entartet, d. h. die Winkelvariabeln w_ϱ^0 ($\varrho = s+1, \ldots f$) des ungestörten Systems sind konstant.

Die Störungsfunktion sei nach einem Parameter λ entwickelbar; die gesamte Hamiltonsche Funktion sei:

$$H = H_0 + \lambda H_1 + \lambda^2 H_2 + \ldots \tag{2}$$

Hier hängen $H_1, H_2, \ldots$ von allen Variabeln j_k^0, w_k^0 ($k = 1, \ldots f$) ab.

Das gewöhnliche Verfahren der Störungstheorie besteht nun darin, daß man zunächst die Entartung der j_ϱ^0, w_ϱ^0 beseitigt durch die Berechnung der säkularen Störungen aus der Differentialgleichung

$$\overline{H}_1 = \mathsf{E}_1;$$

dabei bedeutet der Strich Mittelbildung über die $w_1^0, w_2^0, \ldots w_s^0$.

Wir wollen nun aber annehmen, daß dieser Mittelwert identisch in j_ϱ^0, w_ϱ^0 verschwindet:

$$\overline{H}_1 = \mathsf{E}_1 = 0. \tag{3}$$

Damit ist die gewöhnliche Methode zur Bestimmung der säkularen Störungen hinfällig und es entsteht die Aufgabe, für diesen Fall das Bewegungsproblem zu lösen.

Es liegt nahe, die säkularen Bewegungen der j_ϱ^0, w_ϱ^0 jetzt mit Hilfe von $\overline{H_2}$ zu bestimmen; doch geht das nicht ohne weiteres, weil ja H_1 nicht identisch in $w_a{}^0$, $j_a{}^0$ verschwindet und in den folgenden Näherungen immer wieder vorkommt. Es ist dann keineswegs erlaubt, die mit Hilfe von $\overline{H_2}$ bestimmten, zu j_ϱ^0, w_ϱ^0 gehörigen transformierten Winkel- und Wirkungsvariabeln j_ϱ, w_ϱ nachträglich in H_1 einzuführen und damit weiter zu rechnen.

Der erste Schritt wird vielmehr darin bestehen, daß man H_1 gänzlich fortschafft.

Hierzu dient eine kanonische Substitution

$$(w_k{}^0 \, j_k{}^0) \rightarrow (w_k \, j_k) \qquad (k = 1, 2, \ldots f)$$

mit der erzeugenden Funktion:

$$F = \sum_k w_k{}^0 \, j_k + \lambda F_1 (w_1{}^0, w_2{}^0, \ldots j_1, j_2, \ldots). \tag{4}$$

Sie wird also durch die Gleichungen

$$(4) \quad \begin{cases} \text{a)} \quad w_k = \dfrac{\partial F}{\partial j_k} = w_k{}^0 + \lambda \dfrac{\partial F_1}{\partial j_k}, \\ \text{b)} \quad j_k{}^0 = \dfrac{\partial F}{\partial w_k{}^0} = j_k + \lambda \dfrac{\partial F_1}{\partial w_k{}^0} \end{cases}$$

vermittelt.

Wir ersetzen zunächst unter Beibehaltung von $w_k{}^0$ in jedem Gliede von H (2) die $j_k{}^0$ durch die j_k nach (4b); dann ordnen wir wieder nach Potenzen von λ und erhalten:

$$H = H_0{}^* + \lambda H_1{}^* + \lambda^2 H_2{}^* + \ldots, \quad \text{wo} \tag{5}$$

$$(5) \quad \begin{cases} \text{a)} \quad H_0{}^* = H_0 (j_1, j_2, \ldots j_s), \\ \text{b)} \quad H_1{}^* = \displaystyle\sum_{\alpha=1}^{s} \frac{\partial H_0}{\partial j_\alpha} \frac{\partial F_1}{\partial w_\alpha{}^0} + H_1, \\ \text{c)} \quad H_2{}^* = \dfrac{1}{2} \displaystyle\sum_{\alpha,\beta=1}^{s} \frac{\partial^2 H_0}{\partial j_\alpha \partial j_\beta} \frac{\partial F_1}{\partial w_\alpha{}^0} \frac{\partial F_1}{\partial w_\beta{}^0} + \sum_{k=1}^{f} \frac{\partial H_1}{\partial j_k} \frac{\partial F_1}{\partial w_k{}^0} + H_2, \\ \text{d)} \quad H_3{}^* = \dfrac{1}{3} \displaystyle\sum_{\alpha,\beta,\gamma=1}^{s} \frac{\partial^3 H_0}{\partial j_\alpha \partial j_\beta \partial j_\gamma} \frac{\partial F_1}{\partial w_\alpha{}^0} \frac{\partial F_1}{\partial w_\beta{}^0} \frac{\partial F_1}{\partial w_\gamma{}^0} \\ \qquad + \dfrac{1}{2} \displaystyle\sum_{k,l=1}^{f} \frac{\partial^2 H_1}{\partial j_k \partial j_l} \frac{\partial F_1}{\partial w_k{}^0} \frac{\partial F_1}{\partial w_l{}^0} + \sum_{k=1}^{f} \frac{\partial H_2}{\partial j_k} \frac{\partial F_1}{\partial w_k{}^0} + H_3, \\ \quad - - - - - - - - - - - - . \end{cases}$$

Da nun nach der Voraussetzung $\overline{H_1} = 0$ ist, kann man F_1 als periodische Funktion der $w_a{}^0$ (die übrigens noch die j_a, $w_\varrho{}^0$, j_ϱ enthält) aus der Differentialgleichung

$$(6) \qquad H_1{}^* = \sum_{a=1}^{s} \frac{\partial H_0}{\partial j_a} \frac{\partial F_1}{\partial w_a{}^0} + H_1 = 0$$

bestimmen (wobei eine additive Funktion der $w_\varrho{}^0$, j_ϱ, j_a unbestimmt bleibt). Ist das geschehen, so wird

$$(5a) \qquad H = H_0{}^* + \lambda^2 H_2{}^* + \lambda^3 H_3{}^* + \ldots,$$

wo $H_2{}^*$, $H_3{}^*$, ... aus (5c, d, ...) eindeutig bestimmte Funktionen der $w_k{}^0$, j_k sind.

Da nun diese Variablen nicht kanonisch konjugiert sind, so wird man noch statt der $w_k{}^0$ die w_k mit Hilfe von (4a) einführen müssen. Löst man diese Gleichungen nach den $w_k{}^0$ auf, so erhält man leicht folgende Reihe:

$$(7) \qquad w_k{}^0 = w_k - \lambda \frac{\partial F_1}{\partial j_k} + \lambda^2 \sum_l \frac{\partial^2 F_1}{\partial j_k \partial w_l} \frac{\partial F_1}{\partial j_l} + \ldots..$$

Diese muß man in (5a) einsetzen und wieder nach λ entwickeln; dann wird:

$$(8) \qquad H = \mathsf{H}_0 + \lambda^2 \mathsf{H}_2 + \lambda^3 \mathsf{H}_3 + \ldots.,$$

wo

$$(8) \quad \begin{cases} \text{a)} & \mathsf{H}_0 = H_0(j_1, j_2, \ldots j_s), \\ \text{b)} & \mathsf{H}_2 = H_2{}^*(w_1, w_2, \ldots w_f, j_1, j_2, \ldots j_f) \\ \text{c)} & \mathsf{H}_3 = H_3{}^* - \sum_k \frac{\partial H_2{}^*}{\partial w_k} \frac{\partial F_1}{\partial j_k}, \\ \text{d)} & \mathsf{H}_4 = H_4{}^* - \sum_k \frac{\partial H_3{}^*}{\partial w_k} \frac{\partial F_1}{\partial j_k} + \frac{1}{2} \sum_{kl} \left\{ \frac{\partial^2 H_2{}^*}{\partial w_k \partial w_l} \frac{\partial F_1}{\partial j_k} \frac{\partial F_1}{\partial j_l} \right. \\ & \left. \qquad + \frac{\partial H_2{}^*}{\partial w_k} \frac{\partial^2 F_1}{\partial j_k \partial w_l} \frac{\partial F_1}{\partial j_l} + \frac{\partial H_2{}^*}{\partial w_l} \frac{\partial^2 F_1}{\partial j_l \partial w_k} \frac{\partial F_1}{\partial j_k} \right\} \\ & - - - - - - - - - \end{cases}$$

In (8) sind alle Koeffizienten H_0, H_2, H_3, ... Funktionen der konjugierten Größen j_1, w_1; j_2, w_2; ... j_f, w_f (H_0 nur der j_1, j_2, ... j_s).

Die entarteten Variabeln w_ϱ, j_ϱ ($\varrho = s + 1, \ldots f$) können jetzt nach der Methode der säkularen Störungen mittels

$$\overline{\mathsf{H}_2} = \mathsf{E}_2$$

festgelegt werden, wo der Strich Mittelbildung über die w_a bedeutet. Wir setzen voraus, daß diese Gleichung durch

Separation der Variabeln lösbar ist; dann können wir eine kanonische Transformation

$$(w_\varrho, j_\varrho) \rightarrow (W_\varrho J_\varrho)$$

angeben von der Art, daß die w_ϱ, j_ϱ periodisch sind in den W_ϱ (mit der Periode 1) und daß $\overline{\mathsf{H}}_2$ in eine Funktion der J_ϱ übergeht (die natürlich auch die j_α enthält):

$$\overline{\mathsf{H}}_2 = \mathsf{E}_2 (j_1, \ldots j_s, J_{s+1}, \ldots J_f). \tag{9}$$

Im folgenden soll Mittelbildung über die w_α durch *einen* Strich, Mittelbildung über die w_α und W_ϱ durch *zwei* Striche angedeutet werden.

Wir denken uns jetzt in $\mathsf{H}_2, \mathsf{H}_3, \ldots$ überall die W_ϱ, J_ϱ statt der w_ϱ, j_ϱ eingeführt.

Sodann bilden wir eine kanonische Substitution

$$(j_\alpha w_\alpha J_\varrho W_\varrho) \rightarrow (j_\alpha^* w_\alpha^* J_\varrho^* W_\varrho^*)$$

mittels der erzeugenden Funktion

$$\left\{ \begin{aligned} F = \sum_{\alpha=1}^{s} j_\alpha^* w_\alpha + \sum_{\varrho=s+1}^{f} J_\varrho^* W_\varrho + \lambda F_1 (j_\alpha^*, w_\alpha, J_\varrho^*, W_\varrho) \\ + \lambda^2 F_2 (j_\alpha^*, w_\alpha, J_\varrho^*, W_\varrho) + \ldots, \end{aligned} \right. \tag{10}$$

also:

$$j_\alpha = \frac{\partial F}{\partial w_\alpha}, \quad w_\alpha^* = \frac{\partial F}{\partial j_\alpha^*}; \quad J_\varrho = \frac{\partial F}{\partial W_\varrho}, \quad W_\varrho^* = \frac{\partial F}{\partial J_\varrho^*}. \tag{10a}$$

Dann erhält man folgende Reihe von Näherungsgleichungen:

$$\sum_\alpha \frac{\partial \mathsf{H}_0}{\partial j_\alpha^*} \frac{\partial F_1}{\partial w_\alpha} = 0, \tag{11a}$$

daraus $F_1 = G_1 (W_\varrho, J_\varrho^*, j_\alpha^*)$, wo G_1 eine zunächst willkürliche Funktion der angegebenen Argumente bedeutet; weiter:

$$\left\{ \begin{aligned} &\text{b)} \quad \sum_\alpha \frac{\partial \mathsf{H}_0}{\partial j_\alpha^*} \frac{\partial F_2}{\partial w_\alpha} + \mathsf{H}_2 = \mathsf{E}_2, \\ &\text{c)} \quad \sum_\alpha \frac{\partial \mathsf{H}_0}{\partial j_\alpha^*} \frac{\partial F_3}{\partial w_\alpha} + \sum_\varrho \frac{\partial \mathsf{H}_2}{\partial J_\varrho^*} \frac{\partial F_1}{\partial W_\varrho} + \mathsf{H}_3 = \mathsf{E}_3, \\ &\text{d)} \quad \sum_\alpha \frac{\partial \mathsf{H}_0}{\partial j_\alpha^*} \frac{\partial F_4}{\partial w_\alpha} + \frac{1}{2} \sum_{\alpha,\beta} \frac{\partial^2 \mathsf{H}_0}{\partial j_\alpha^* \partial j_\beta^*} \frac{\partial F_2}{\partial w_\alpha} \frac{\partial F_2}{\partial w_\beta} + \sum_\alpha \frac{\partial \mathsf{H}_2}{\partial j_\alpha^*} \frac{\partial F_2}{\partial w_\alpha} \\ &\qquad + \sum_\varrho \frac{\partial \mathsf{H}_3}{\partial J_\varrho^*} \frac{\partial F_2}{\partial W_\varrho} + \frac{1}{2} \sum_{\varrho\sigma} \frac{\partial \mathsf{H}_2}{\partial J_\varrho^* \partial J_\sigma^*} \frac{\partial F_1}{\partial W_\varrho} \frac{\partial F_1}{\partial W_\sigma} \\ &\qquad + \sum_\varrho \frac{\partial \mathsf{H}_3}{\partial J_\varrho^*} \frac{\partial F_1}{\partial W_\varrho} + \mathsf{H}_4 = \mathsf{E}_4, \\ &\text{— — — — — — — — — — —} \end{aligned} \right. \tag{11}$$

Die Mittelung von (11b) über die w_a führt auf (9) zurück. Daher läßt sich (11b) integrieren und liefert

$$F_2 = F_2^0(w_a, j_a^*; W_\varrho, J_\varrho^*) + G_2(W_\varrho, J_\varrho^*, j_a^*)\,, \tag{12b}$$

wo F_2^0 eine bestimmte, in w_a, W_ϱ periodische Funktion, G_2 wieder eine willkürliche Funktion ist.

Mittelung von (11c) über die w_a gibt die Gleichung

$$\sum_\varrho \frac{\partial \mathsf{E}_2}{\partial J_\varrho^*}\frac{\partial G_1}{\partial W_\varrho} + \overline{\mathsf{H}}_3 = \mathsf{E}_3 \tag{13a}$$

zur Bestimmung von G_1. Durch Mittelung über die W_ϱ folgt zunächst

$$\overline{\overline{\mathsf{H}}}_3 = \mathsf{E}_3; \tag{13b}$$

jetzt läßt sich die Gleichung (13a) integrieren, wodurch G_1 festgelegt ist. Nunmehr läßt sich auch (11c) integrieren und liefert

$$F_3 = F_3^0(w_a, j_a^*, W_\varrho, J_\varrho^*) + G_3(W_\varrho, J_\varrho^*, j_a^*); \tag{12c}$$

wo F_3^0 bestimmt, G_3 willkürlich ist.

Mittelung von (11d) über die w_a gibt die Gleichung

$$\sum_\varrho \frac{\partial \mathsf{E}_2}{\partial J_\varrho^*}\frac{\partial G_2}{\partial W_\varrho} + \overline{\Phi}_4 = \mathsf{E}_4\,, \tag{13b}$$

wo zur Abkürzung

$$\left\{\begin{aligned}\overline{\Phi}_4 = {} & \tfrac{1}{2}\sum_{a,\beta}\frac{\partial^2 \mathsf{H}_0}{\partial j_a^*\,\partial j_\beta^*}\,\overline{\frac{\partial F_2^0}{\partial w_a}\frac{\partial F_2^0}{\partial w_\beta}} + \tfrac{1}{2}\sum_{\varrho,\sigma}\frac{\partial^2 \mathsf{E}_2}{\partial J_\varrho^*\,\partial J_\sigma^*}\frac{\partial G_1}{\partial W_\varrho}\frac{\partial G_1}{\partial W_\sigma}\\ & + \sum_\varrho \frac{\partial \overline{\mathsf{H}}_3}{\partial J_\varrho^*}\frac{\partial G_1}{\partial W_\varrho} + \overline{\mathsf{H}}_4\end{aligned}\right. \tag{14}$$

gesetzt ist. Somit wird

$$\overline{\overline{\Phi}}_4 = \mathsf{E}_4 \tag{15}$$

und F_4 bekommt die Form

$$F_4 = F_4^0(w_a, j_a^*; W_\varrho, J_\varrho^*) + G_4(W_\varrho, J_\varrho^*, j_a^*)\,,$$

wo F_4^0 bestimmt, G_4 willkürlich ist.

In derselben Weise läßt sich offenbar das Verfahren beliebig fortsetzen. G_3 wird durch die fünfte, G_4 durch die sechste Näherung festgelegt, usf.

Für die Energie erhält man eine Reihe der Form:

$$H = \mathsf{E} = \mathsf{E}_0 + \lambda^2 \mathsf{E}_2 + \lambda^3 \mathsf{E}_3 + \ldots,$$

wo $\mathsf{E}_0 = H_0$ von $j_1^*, \ldots j_s^*$, $\mathsf{E}_2, \mathsf{E}_3, \ldots$ von $j_1^*, \ldots j_s^*$, $J_{s+1}^*, \ldots J_f^*$ abhängen.

Das Verfahren versagt, wenn der Mittelwert $\overline{\mathsf{H}}_2$ identisch in w_ϱ, j_ϱ $(\varrho = s+1, \ldots f)$ verschwindet. Dann muß man das ganze Verfahren noch einmal wiederholen: Zunächst muß H_2 durch eine geeignete kanonische Transformation gänzlich aus der Reihe (8) beseitigt werden, so daß diese die Form

$$H = \mathsf{H}_0 + \lambda^3 \mathsf{H}_3 + \lambda^4 \mathsf{H}_4 + \ldots$$

annimmt; sodann mache man eine zu der oben gebrauchten (10) ganz analoge Entwicklung der kanonischen Transformation und gelangt so zu einer Entwicklung der Energie von der Form

$$H = \mathsf{E} = \mathsf{E}_0 + \lambda^3 \mathsf{E}_3 + \ldots,$$

ausgenommen den Fall, daß auch $\overline{\mathsf{H}}_3$ wieder verschwindet. Dann muß das Verfahren nochmals wiederholt werden usw.

Kapitel II. Mechanik der Molekeln.

§ 1. Bezeichnungen und Definitionen.

Eine Molekel ist ein aus Kernen und Elektronen bestehendes Massensystem, das durch die elektrostatischen Kräfte zwischen diesen zusammengehalten wird.

Die Massen M_k der Kerne sind groß gegen die der Elektronen; wir setzen $M_k = \frac{1}{\lambda^2} m_k$, wo λ ein Parameter von der Größenordnung $\sqrt{\mu / M_0}$ ist, unter μ die Masse des Elektrons, unter M_0 die des Wasserstoffkerns verstanden. Dann sind die m_k von derselben Größenordnung wie μ.

Die Aufgabe, die im folgenden behandelt wird, besteht darin, alle aus dieser Größenordnungsbeziehung folgenden Eigenschaften der Molekel abzuleiten.

Wir bezeichnen alle auf die Elektronen bezüglichen Größen mit kleinen Buchstaben, die auf die Kerne bezüglichen mit großen, also:

	Elektronen	Kerne
Anzahl . . .	n	N
Masse . . .	μ	$M_\varrho = m_\varrho / \lambda^2$
Ortsvektor . .	$\mathfrak{r}_\alpha$	$\mathfrak{R}_\varrho$
Geschwindigkeit	$\mathfrak{u}_\alpha = \dot{\mathfrak{r}}_\alpha$	$\mathfrak{U}_\varrho = \dot{\mathfrak{R}}_\varrho$.

Ferner treffen wir über die *Bezeichnung der Indizes* ein für allemal die Festsetzung, daß für

Elektronen	**Kerne**
$\alpha, \beta, \gamma \ldots$	$\varrho, \sigma, \tau, \ldots$

gebraucht werden soll.

Da eine Molekel in einem stationären Zustande ein stabiles Gebilde sein soll, muß es gewisse, miteinander starr verbundene Punkte $\mathfrak{R}_\varrho^0$ geben, von denen die Kerne sich während der Bewegung der Molekel niemals weit entfernen. Wir werden daher in Analogie zur Theorie des Oszillators (2atomige Molekel) für die Bewegungen der Kerne den Ansatz

$$(1) \qquad \mathfrak{R}_\varrho = \mathfrak{R}_\varrho^0 + \lambda \mathfrak{R}_\varrho^{(1)}$$

machen, wobei die $\mathfrak{R}_\varrho^{(1)}$ von derselben Größenordnung wie die $\mathfrak{R}_\varrho^0$ sein sollen.

Wir setzen im folgenden stets voraus, daß die Punkte $\mathfrak{R}_\varrho^0$ nicht sämtlich auf einer Geraden liegen; damit schließen wir insbesondere den Fall zweiatomiger Molekeln von der Betrachtung aus. Die Behandlung des wichtigen Sonderfalls geradliniger Molekeln soll einer besonderen Mitteilung vorbehalten bleiben. Die zu untersuchenden Bewegungen schränken wir weiter durch die Forderung ein, daß sämtliche Impulse und Impulsmomente von gleicher Größenordnung, und zwar proportional der nullten Potenz von λ, sein sollen. Dies ist notwendig, um die Quantentheorie anwenden zu können, nach der diese Größen bis auf Zahlenfaktoren gleich der Planckschen Konstanten h gesetzt werden.

Wir führen nun ein mit den Punkten $\mathfrak{R}_\varrho^0$ fest verbundenes Koordinatensystem Σ ein, und bezeichnen für

	Elektronen	**Kerne**
Koordinaten	$x_\alpha, y_\alpha, z_\alpha$	$X_\varrho, Y_\varrho, Z_\varrho$
Geschwindigkeitskomponenten	$\mathfrak{u}_{\alpha x} = \dot{x}_\alpha, \ldots$	$\mathfrak{U}_{\varrho x} = \dot{X}_\varrho, \ldots$

Dieses System Σ wird sich im allgemeinen im Raume bewegen.

Daneben gebrauchen wir ein im Raume festes Koordinatensystem $\overline{\Sigma}$ und kennzeichnen alle auf dieses bezogenen Größen durch Überstreichen. Die gegenseitige Lage der beiden Systeme Σ, $\overline{\Sigma}$ läßt sich durch sechs Parameter beschreiben. Drei davon sind die Koordinaten des Nullpunkts von Σ bezüglich $\overline{\Sigma}$; die Translationsgeschwindigkeit von Σ gegen $\overline{\Sigma}$ werde mit $\mathfrak{v}$ bezeichnet. Die drei anderen Parameter bestimmen die Orientierung (z. B. die Eulerschen Winkel); ohne uns auf eine bestimmte Wahl festzulegen, bezeichnen wir sie mit $\varphi_1, \varphi_2, \varphi_3$. Der Vektor der instantanen Winkelgeschwindigkeit von Σ gegen $\overline{\Sigma}$ sei $\mathfrak{d}$. Seine Komponenten bezüglich Σ sind lineare Funktionen der Ableitungen der φ_k nach der Zeit[1]):

(2a) $$\mathfrak{d}_x = \sum_{k=1}^{3} a_{xk}\,\dot{\varphi}_k\,.$$

Die Auflösung dieser Gleichungen sei

(2b) $$\dot{\varphi}_k = \sum_x b_{kx}\,\mathfrak{d}_x\,.$$

Die relativen Geschwindigkeiten der Elektronen und der Kerne bezüglich Σ, $\mathfrak{u}_\alpha$ und $\mathfrak{U}_\varrho$, hängen mit den absoluten Geschwindigkeiten $\overline{\mathfrak{u}}_\alpha$, $\overline{\mathfrak{U}}_\varrho$ so zusammen:

(3) $$\begin{cases} \text{a)} & \overline{\mathfrak{u}}_\alpha = \mathfrak{u}_\alpha + \mathfrak{v} + [\mathfrak{d}\,\mathfrak{r}_\alpha]\,, \\ \text{b)} & \overline{\mathfrak{U}}_\varrho = \mathfrak{U}_\varrho + \mathfrak{v} + [\mathfrak{d}\,\mathfrak{R}_\varrho]\,. \end{cases}$$

Die kinetische Energie der Elektronen ist

(4a) $$T^{(E)} = \frac{\mu}{2}\sum_\alpha \overline{\mathfrak{u}}_\alpha^{\,2}$$

und die der Kerne

(4b) $$T^{(K)} = \tfrac{1}{2}\sum_\varrho M_\varrho \overline{\mathfrak{U}}_\varrho^{\,2} = \frac{1}{\lambda^2}\,\tfrac{1}{2}\sum_\varrho m_\varrho \overline{\mathfrak{U}}_\varrho^{\,2}\,.$$

Die potentielle Energie ist

(5) $$U = \frac{e^2}{2}\left\{\sum_{\alpha\beta}{}' \frac{1}{|\mathfrak{r}_\alpha - \mathfrak{r}_\beta|} + \sum_{\varrho\sigma}{}' \frac{Z_\varrho Z_\sigma}{|\mathfrak{R}_\varrho - \mathfrak{R}_\sigma|} - 2\sum_{\alpha\varrho}\frac{Z_\varrho}{|\mathfrak{r}_\alpha - \mathfrak{R}_\varrho|}\right\},$$

wo Z_ϱ die Ordnungszahl des ϱten Kernes ist.

§ 2. **Überzählige Koordinaten und Nebenbedingungen.**

Da sich nach unseren Annahmen die Kerne nur wenig aus den starr verbundenen Lagen $\mathfrak{R}_\varrho^0$ entfernen, liegt es nahe, die sechs Bestimmungsstücke dieses starren Körpers neben

1) Wir gebrauchen hier und im folgenden die Zeichen x, y, z auch als Indizes, die die drei Koordinatenrichtungen durchlaufen; so schreiben wir z. B. div $\mathfrak{A} = \sum_x \mathfrak{A}_x$ usw. Vgl. M. Born, Atomtheorie des festen Zustandes (Leipzig 1923).

den relativen Koordinaten der Kerne und Elektronen als gleichberechtigte Unbekannte einzuführen; dann muß man natürlich sechs Nebenbedingungen aufstellen, damit die Anzahl der Freiheitsgrade gewahrt bleibt. Diese Nebenbedingungen wird man so wählen, daß das bewegte Koordinatensystem Σ mit den darin ruhenden Punkten $\mathfrak{R}_\varrho^0$ sich der wirklichen Kernbewegung möglichst gut anschmiegt.

Die relativen Lagen der Punkte $\mathfrak{R}_\varrho^0$ sind zunächst als willkürlich, aber fest gewählt zu denken; nachträglich wird sich dann zeigen, daß es nur eine endliche Anzahl von Lagen gibt, bei denen die Kernbewegung den Charakter kleiner Schwingungen um diese Lagen hat, wie es unser Ansatz (1) fordert. Die Orientierung des Koordinatensystems Σ gegen die $\mathfrak{R}_\varrho^0$ ist zunächst ebenfalls willkürlich; es ist bequem, sie so zu wählen, daß der Nullpunkt in den Schwerpunkt und die Koordinatenachsen in die Hauptträgheitsachsen der in $\mathfrak{R}_\varrho^0$ ruhend gedachten Kerne fallen. Dann gilt:

$$(6)\begin{cases} \text{a)} \quad \sum_\varrho m_\varrho \mathfrak{R}_\varrho^0 = 0\,, \\ \text{b)} \sum_\varrho m_\varrho Y_\varrho^0 Z_\varrho^0 = 0\,, \quad \sum_\varrho m_\varrho Z_\varrho^0 X_\varrho^0 = 0\,, \quad \sum_\varrho m_\varrho X_\varrho^0 Y_\varrho^0 = 0\,. \end{cases}$$

Nachdem so das Koordinatensystem Σ gegen die $\mathfrak{R}_\varrho^0$ festgelegt ist, kann man die genannten sechs Nebenbedingungen wählen, die seine Bewegung an die der Kerne anschließen. Die Wahl der Bedingungen ist natürlich bis zu einem gewissen Grade willkürlich und wird nur durch die Frage der Zweckmäßigkeit bestimmt. Wir haben gefunden, daß folgende Wahl der Nebenbedingungen vorteilhaft ist:

$$(7)\quad\begin{cases} \text{a)} \quad \sum_\varrho m_\varrho \mathfrak{R}_\varrho^{(1)} = 0\,, \\ \text{b)} \quad \sum_\varrho m_\varrho [\mathfrak{R}_\varrho^0 \mathfrak{R}_\varrho^{(1)}] = 0\,. \end{cases}$$

Die erste bedeutet offenbar, daß der Schwerpunkt dauernd im Nullpunkt von Σ bleiben soll. Die Bedeutung der zweiten Bedingung erhellt, wenn man sie nach der Zeit differenziert; nach (1) ist $\mathfrak{U}_\varrho = \lambda\, \dot{\mathfrak{R}}_\varrho^{(1)}$, und man hat daher:

$$(8)\quad\begin{cases} \text{a)} \quad \sum_\varrho m_\varrho \mathfrak{U}_\varrho = 0\,, \\ \text{b)} \quad \sum_\varrho m_\varrho [\mathfrak{R}_\varrho^0 \mathfrak{U}_\varrho] = 0\,. \end{cases}$$

Die linke Seite der zweiten Gleichung ist das Impulsmoment der Kerne relativ zu Σ in den Lagen $\mathfrak{R}_\varrho^0$; wir fordern also das Verschwinden des Drehmoments der Kerne, wenn sie mit irgendwelchen Geschwindigkeiten die Ruhelagen $\mathfrak{R}_\varrho^0$ passieren.[1])

Man hat nun die Wahl, ob man die Nebenbedingungen im der ursprünglichen Form (7) oder der differenzierten Form (8) bei der Aufstellung der Bewegungsgleichungen benützen will. Es scheint vorteilhaft, die letztere Form (8) zu wählen, weil sich dann die Hamilton-Jacobische Integrationsmethode in besonders einfacher Weise anwenden läßt, wie wir sogleich sehen werden.

Die Lagrangesche Funktion (Integrand des Hamiltonschen Prinzips) bei fehlenden Nebenbedingungen würde lauten:

$$L = T^{(E)} + T^{(K)} - U\,. \tag{9}$$

Hierzu hat man nun die mit Multiplikatoren versehenen Nebenbedingungen hinzuzufügen; wir können diese sechs Multiplikatoren als Komponenten zweier Vektoren auffassen, die wir mit $\mathfrak{A}$, $\mathfrak{B}$ bezeichnen. Dann ist die erweiterte Lagrangesche Funktion:

$$L^* = L + \mathfrak{A} \sum_\varrho m_\varrho \mathfrak{U}_\varrho + \mathfrak{B} \sum_\varrho m_\varrho [\mathfrak{R}_\varrho^0 \mathfrak{U}_\varrho]\,. \tag{10}$$

Hier kann man nunmehr als gleichberechtigte, voneinander unabhängige Unbekannte die Koordinaten der Elektronen und Kerne, die sechs Bestimmungsstücke des Koordinatensystems Σ und die sechs Komponenten von $\mathfrak{A}$, $\mathfrak{B}$ ansehen.

§ 3. Impulse und Hamiltonsche Funktion.

Wir bilden jetzt die zu den soeben aufgezählten Variabeln gehörigen Impulse; indem man (3), (4a), (4b) berücksichtigt, erhält man:

---

1) Es liegt nahe, statt (8b) das Verschwinden des relativen Impulsmoments in jedem Augenblick zu fordern:

$$\sum_\varrho m_\varrho [\mathfrak{R}_\varrho \mathfrak{U}_\varrho] = 0\,.$$

Diese Bedingung ist aber nicht linear in den Verrückungen und ihren zeitlichen Ableitungen; aus diesem Grunde erweist sie sich später als unbequem.

$$(11)\quad\begin{cases} \text{a)}\ \mathfrak{p}_{\alpha x} = \dfrac{\partial L^*}{\partial \mathfrak{u}_{\alpha x}} = \mu\, \bar{\mathfrak{u}}_{\alpha x}, \\ \text{b)}\ \mathfrak{P}_{\varrho x} = \dfrac{\partial L^*}{\partial \mathfrak{U}_{\varrho x}} = m_\varrho \left(\dfrac{1}{\lambda^2} \bar{\mathfrak{U}}_{\varrho x} + \mathfrak{A}_x + [\mathfrak{B}\,\mathfrak{R}_\varrho^0]_x\right), \\ \text{c)}\ \mathfrak{B}_x = \dfrac{\partial L^*}{\partial \mathfrak{v}_x} = \mu \sum\limits_\alpha \bar{\mathfrak{u}}_{\alpha x} + \dfrac{1}{\lambda^2} \sum\limits_\varrho m_\varrho \bar{\mathfrak{U}}_{\varrho x}, \\ \text{d)}\ \Phi_k = \dfrac{\partial L^*}{\partial \dot{\varphi}_k} = \sum\limits_x \dfrac{\partial L^*}{\partial \mathfrak{d}_x} a_{xk} = \sum\limits_x a_{xk} \mathfrak{D}_x, \\ \qquad \text{wo} \\ \text{e)}\ \mathfrak{D}_x = \dfrac{\partial L^*}{\partial \mathfrak{d}_x} = \mu \sum\limits_\alpha [\mathfrak{r}_\alpha \bar{\mathfrak{u}}_\alpha]_x + \dfrac{1}{\lambda^2} \sum\limits_\varrho m_\varrho [\mathfrak{R}_\varrho, \bar{\mathfrak{U}}_\varrho]_x. \end{cases}$$

$\mathfrak{B}$ ist offenbar der gesamte Impuls, $\mathfrak{D}$ der gesamte Drehimpuls der Molekel. Die Hamiltonsche Funktion ist folgendermaßen definiert:

$$(12)\quad H = \sum_\alpha (\mathfrak{u}_\alpha \mathfrak{p}_\alpha) + \sum_\varrho (\mathfrak{U}_\varrho \mathfrak{P}_\varrho) + (\mathfrak{v}\,\mathfrak{B}) + \sum_{k=1}^{3} \dot{\varphi}_k \Phi_k - L^*.$$

Da nun die von den Nebenbedingungen herrührenden Glieder von L^* linear in den $\mathfrak{U}_\varrho$ sind, fallen sie aus H heraus und man erhält einfach:

$$(12\text{a})\qquad H = T^{(E)} + T^{(K)} + U.$$

Die Hamiltonsche Funktion ist also mit der Energie identisch; dieser vorteilhafte Umstand beruht darauf, daß wir die Nebenbedingungen in der differenzierten Form (8) und nicht in der ursprünglichen Form (7) benutzt haben.

Wir berechnen nun aus (11a) und (11b) $\bar{\mathfrak{u}}_\alpha$ und $\bar{\mathfrak{U}}_\varrho$ und setzen die gefundenen Werte in (11c) und (11e) ein; dann erhalten wir mit Rücksicht auf (6a) und (7a):

$$(13)\quad\begin{cases} \text{a)}\ \mathfrak{B} = \mathfrak{p} + \mathfrak{P} - m\,\mathfrak{A}, \\ \text{b)}\ \mathfrak{D} = \mathfrak{g} + \mathfrak{G} - \sum\limits_\varrho m_\varrho [\mathfrak{R}_\varrho [\mathfrak{B}\,\mathfrak{R}_\varrho^0]]. \end{cases}$$

Dabei bedeuten

$$(14)\qquad \mathfrak{p} = \sum_\alpha \mathfrak{p}_\alpha, \qquad \mathfrak{g} = \sum_\alpha [\mathfrak{r}_\alpha\, \mathfrak{p}_\alpha]$$

den Impuls und Drehimpuls der Elektronen,

$$(15)\qquad \mathfrak{P} = \sum_\varrho \mathfrak{P}_\varrho, \qquad \mathfrak{G} = \sum_\varrho [\mathfrak{R}_\varrho\, \mathfrak{P}_\varrho]$$

den Impuls und Drehimpuls der Kerne; wegen der Nebenbedingungen ist nicht einfach $\mathfrak{B} = \mathfrak{p} + \mathfrak{P}$ und $\mathfrak{D} = \mathfrak{g} + \mathfrak{G}$, sondern es treten Zusatzglieder auf, die die Multiplikatoren $\mathfrak{A}$, $\mathfrak{B}$ enthalten. Mit Hilfe dieser Gleichungen (13) kann man nun $\mathfrak{A}$ und $\mathfrak{B}$ aus dem Problem eliminieren. Die erste liefert

$$(16) \qquad m\mathfrak{A} = \mathfrak{p} + \mathfrak{P} - \mathfrak{B}.$$

Die Lösung der zweiten nach $\mathfrak{B}$ läßt sich durch Entwicklung in eine Potenzreihe von λ gewinnen. Wir schreiben (13b):

$$(17) \quad \begin{cases} \sum_\varrho m_\varrho [\mathfrak{R}_\varrho{}^0[\mathfrak{B}\mathfrak{R}_\varrho{}^0]] + \lambda \sum_\varrho m_\varrho [\mathfrak{R}_\varrho{}^{(1)}[\mathfrak{B}\mathfrak{R}_\varrho{}^0]] \\ \qquad = \mathfrak{g} + \mathfrak{G}^0 + \lambda \mathfrak{G}^{(1)} - \mathfrak{D}, \end{cases}$$

wo

$$(18) \qquad \mathfrak{G}^0 = \sum_\varrho [\mathfrak{R}_\varrho{}^0 \mathfrak{P}_\varrho], \qquad \mathfrak{G}^{(1)} = \sum_\varrho [\mathfrak{R}_\varrho{}^{(1)} \mathfrak{P}_\varrho]$$

gesetzt ist.

Die Summen linker Hand haben eine einfache Bedeutung; sie stehen nämlich mit den Trägheits- und Deviationsmomenten der Kerne

$$(19) \quad \begin{cases} A_{xx} = \sum_\varrho m_\varrho (Y_\varrho{}^2 + Z_\varrho{}^2), \dots \\ A_{yz} = -\sum_\varrho m_\varrho Y_\varrho Z_\varrho, \dots \end{cases}$$

im Zusammenhang. Setzt man hierin (1) ein, so erhält man

$$(20) \quad \begin{cases} A_{xx} = \overset{0}{A}_{xx} + \lambda \overset{(1)}{A}_{xx} + \lambda^2 \overset{(2)}{A}_{xx}, \dots \\ A_{yz} = \overset{0}{A}_{yz} + \lambda \overset{(1)}{A}_{yz} + \lambda^2 \overset{(2)}{A}_{yz}, \dots, \end{cases}$$

wo mit Rücksicht auf (6b), (7b):

$$(21) \quad \begin{cases} \overset{0}{A}_{xx} = \overset{0}{A}_x = \sum_\varrho m_\varrho (Y_\varrho^{02} + Z_\varrho^{02}), \dots \\ \overset{0}{A}_{yz} = 0, \end{cases}$$

$$(22) \quad \begin{cases} \overset{(1)}{A}_{xx} = 2\sum_\varrho m_\varrho (Y_\varrho{}^0 Y_\varrho{}^{(1)} + Z_\varrho{}^0 Z_\varrho{}^{(1)}), \dots \\ \overset{(1)}{A}_{yz} = -2\sum_\varrho m_\varrho Y_\varrho{}^0 Z_\varrho{}^{(1)} = -2\sum_\varrho m_\varrho Y_\varrho{}^{(1)} Z_\varrho{}^0, \dots \end{cases}$$

$$(23) \quad \begin{cases} \overset{(2)}{A}_{xx} = \sum_\varrho m_\varrho (Y_\varrho^{(1)2} + Z_\varrho^{(1)2}), \dots \\ \overset{(2)}{A}_{yz} = -\sum_\varrho m_\varrho Y_\varrho^{(1)} Z_\varrho^{(1)}, \dots \end{cases}$$

Wegen unserer Annahme, daß die Punkte $\mathfrak{R}_\varrho{}^0$ nicht sämtlich auf einer Geraden liegen, sind die Hauptträgheitsmomente nullter Näherung $\overset{0}{A}_x$, $\overset{0}{A}_y$, $\overset{0}{A}_z$ sicher von Null verschieden.

Man sieht nun leicht, daß die Gleichung (17) folgendermaßen geschrieben werden kann:

$$A_x{}^0 \mathfrak{B}_x + \lambda \tfrac{1}{2} \sum_y A_{xy}^{(1)} \mathfrak{B}_y = \mathfrak{g}_x + \mathfrak{G}_x{}^0 - \mathfrak{D}_x + \lambda \mathfrak{G}_x{}^{(1)}. \tag{24}$$

Wir setzen die Lösung an in der Form:

$$\mathfrak{B} = \mathfrak{B}^0 + \lambda \mathfrak{B}^{(1)} + \lambda^2 \mathfrak{B}^{(2)} + \dots; \tag{25}$$

dann erhält man:

$$\tag{26} \left\{ \begin{aligned} &\text{a)}\ \mathfrak{B}_x{}^0 = \frac{1}{A_x{}^0}(\mathfrak{g}_x + \mathfrak{G}_x^0 - \mathfrak{D}_x), \\ &\text{b)}\ \mathfrak{B}_x{}^{(1)} = \frac{1}{A_x{}^0}\left(\mathfrak{G}_x^{(1)} - \tfrac{1}{2}\sum_y A_{xy}^{(1)} \mathfrak{B}_y^0\right), \\ &\text{c)}\ \mathfrak{B}_x{}^{(2)} = -\frac{1}{2A_x{}^0}\sum_y A_{xy}^{(1)} \mathfrak{B}_y^{(1)} \\ &\qquad = -\frac{1}{2A_x{}^0}\left(\sum_y \frac{A_{xy}^{(1)}}{A_y{}^0}\mathfrak{G}_y{}^{(1)} - \tfrac{1}{2}\sum_{yz}\frac{A_{xy}^{(1)}A_{yz}^{(1)}}{A_y{}^0}\mathfrak{B}_z^0\right) \\ &\qquad - - - - - - - - - - - - - - - \end{aligned} \right.$$

Wir führen jetzt die Impulse in die kinetische Energie ein. Es wird nach (4a), (4b) und (11):

$$\tag{27} \left\{ \begin{aligned} &\text{a)}\ T^{(\mathrm{E})} = \frac{1}{2\mu}\sum_a \mathfrak{p}_a{}^2, \\ &\text{b)}\ T^{(\mathrm{K})} = \lambda^2 \Big\{ \tfrac{1}{2}\sum_\varrho \frac{1}{m_\varrho}\mathfrak{P}_\varrho{}^2 + \tfrac{1}{2} m \mathfrak{A}^2 + \tfrac{1}{2}\sum_x A_x^0 \mathfrak{B}_x^2 \\ &\qquad\qquad - \mathfrak{A}\mathfrak{P} - \mathfrak{B}\mathfrak{G}^0 \Big\}. \end{aligned} \right.$$

Hier werden für $\mathfrak{A}$, $\mathfrak{B}$ die Werte (16), (25) eingesetzt:

$$\tag{27c} \left\{ \begin{aligned} T^{(\mathrm{K})} = \lambda^2 \Big\{ &\tfrac{1}{2}\sum_\varrho \frac{1}{m_\varrho}\mathfrak{P}_\varrho{}^2 + \frac{1}{2m}\left((\mathfrak{B} - \mathfrak{p})^2 - \mathfrak{P}^2\right) \\ &+ \tfrac{1}{2}\sum_x \frac{1}{A_x{}^0}\left((\mathfrak{D}_x - \mathfrak{g}_x)^2 - \mathfrak{G}_x^{0\,2}\right) \Big\} \\ - \lambda^3 \Big\{ &\sum_x \frac{1}{A_x{}^0}\mathfrak{G}_x{}^{(1)}(\mathfrak{D}_x - \mathfrak{g}_x) \\ &+ \tfrac{1}{2}\sum_{xy}\frac{A_{xy}^{(1)}}{A_x{}^0 A_y{}^0}(\mathfrak{D}_x - \mathfrak{g}_x)(\mathfrak{D}_y - \mathfrak{g}_y - \mathfrak{G}_y{}^0) \Big\} \\ + \lambda^4 \Big\{ &\tfrac{1}{2}\sum_x \frac{1}{A_x{}^0}\mathfrak{G}_x{}^{(1)\,2} + \sum_{xy}\frac{A_{xy}^{(1)}}{A_x{}^0 A_y{}^0}\mathfrak{G}_x{}^{(1)}(\mathfrak{D}_y - \mathfrak{g}_y - \tfrac{1}{2}\mathfrak{G}_y{}^0) \\ &+ \tfrac{3}{8}\sum_{xyz}\frac{A_{xz}^{(1)}A_{zy}^{(1)}}{A_x{}^0 A_y{}^0 A_z{}^0}(\mathfrak{D}_x - \mathfrak{g}_x - \mathfrak{G}_x{}^0)(\mathfrak{D}_y - \mathfrak{g}_y - \tfrac{1}{3}\mathfrak{G}_y{}^0) \Big\} \\ &+ \dots \end{aligned} \right.$$

Wir entwickeln auch die potentielle Energie nach λ:

$$U = U_0 + \lambda U_1 + \lambda^2 U_2 + \ldots \tag{28}$$

Hier ist U_0 die Energie der Elektronen bei ruhenden Kernen, und es gilt ferner:

$$(28)\quad \begin{cases} \text{a)}\ U_1 = \sum_{\varrho} \sum_{x} \left(\frac{\partial U}{\partial X_{\varrho}} \right)_0 X_{\varrho}^{(1)}, \\ \text{b)}\ U_2 = \frac{1}{2} \sum_{\varrho\sigma} \sum_{xy} \left(\frac{\partial U}{\partial X_{\varrho}\, \partial Y_{\sigma}} \right)_0 X_{\varrho}^{(1)}\, Y_{\sigma}^{(1)}, \\ - - - - - - - - - - \end{cases}$$

Die gesamte Energie (Hamiltonsche Funktion) wird daher:

$$H = H_0 + \lambda H_1 + \lambda^2 H_2 + \ldots, \tag{29}$$

wo:

$$(29)\quad \begin{cases} \text{a)}\ H_0 = \frac{1}{2\mu} \sum_{\alpha} \mathfrak{p}_{\alpha}{}^2 + U_0, \\ \text{b)}\ H_1 = U_1, \\ \text{c)}\ H_2 = \frac{1}{2} \sum_{\varrho} \frac{1}{m_{\varrho}} \mathfrak{P}_{\varrho}{}^2 + \frac{1}{2m} [(\mathfrak{B} - \mathfrak{p})^2 - \mathfrak{P}^2] \\ \qquad + \frac{1}{2} \sum_{x} \frac{1}{A_x{}^0} [(\mathfrak{D}_x - \mathfrak{g}_x)^2 - \mathfrak{G}_x^{0\,2}] + U_2, \\ \text{d)}\ H_3 = - \sum_{x} \frac{1}{A_x{}^0} \mathfrak{G}_x{}^{(1)} (\mathfrak{D}_x - \mathfrak{g}_x) \\ \qquad - \frac{1}{2} \sum_{xy} \frac{A_{xy}^{(1)}}{A_x{}^0 A_y{}^0} (\mathfrak{D}_x - \mathfrak{g}_x)(\mathfrak{D}_y - \mathfrak{g}_y - \mathfrak{G}_y{}^0) + U_3, \\ \text{e)}\ H_4 = \frac{1}{2} \sum_{x} \frac{1}{A_x{}^0} \mathfrak{G}_x{}^{(1)\,2} + \sum_{xy} \frac{A_{xy}^{(1)}}{A_x{}^0 A_y{}^0} \mathfrak{G}_x{}^{(1)} (\mathfrak{D}_y - \mathfrak{g}_y - \tfrac{1}{2} \mathfrak{G}_y{}^0) \\ \qquad + \frac{3}{8} \sum_{xyz} \frac{A_{xz}^{(1)} A_{zy}^{(1)}}{A_x{}^0 A_y{}^0 A_z{}^0} (\mathfrak{D}_x - \mathfrak{g}_x - \mathfrak{G}_x{}^0)(\mathfrak{D}_y - \mathfrak{g}_y - \tfrac{1}{3} \mathfrak{G}_y{}^0) + U_4, \\ - - - - - - - - - - - - - - - \end{cases}$$

Hierin kann man noch

$$\mathfrak{B} = 0$$

setzen, indem man von einer Translation der ganzen Molekel absieht. Der gesamte Drehimpuls $\mathfrak{D}$ ist ein im Raume (d. h. bezüglich $\overline{\Sigma}$, nicht bezüglich Σ) fester Vektor.

Auf diese Funktion H kann man nun die Hamilton-Jacobische Methode anwenden. Dabei sind als konjugierte Variabelnpaare anzusehen:

$$(x_\alpha, \mathfrak{p}_{\alpha x}), \ldots (\lambda X_\varrho^{(1)}, \mathfrak{P}_{\varrho x}); \ldots (\varphi_k, \Phi_k),$$

wobei Φ_k nach (11d) mit $\mathfrak{D}$ zusammenhängt.

§ 4. Elektronenbewegung bei ruhenden Kernen und Kerngleichgewicht.

Da H nach Potenzen von λ entwickelt vorliegt, wird man die Methoden der Störungstheorie anwenden. Dabei muß man voraussetzen, daß das ungestörte Problem mit der Hamiltonschen Gleichung

$$H_0 = \frac{1}{2\mu} \sum_\alpha \mathfrak{p}_\alpha^2 + U_0 = \mathsf{E}_0$$

lösbar ist. Hier ist U_0 eine Funktion der zunächst willkürlich gewählten Koordinaten der Kerne X_ϱ^0, Y_ϱ^0, Z_ϱ^0 und der Elektronenkoordinaten x_α, y_α, z_α. Es handelt sich also um das Problem, die Elektronenbewegung bei willkürlich gewählten Kernlagen $\mathfrak{R}_k^0$ zu bestimmen. Wir nehmen an, daß diese Aufgabe gelöst sei (wie es z. B. für das positive Wasserstoff-Molekel-Ion H_2^+ von Pauli[1]) und Niessen[2]) ausgeführt ist); es sei also H_0 für beliebige $X_\varrho^0, \ldots$ als Funktion gewisser Wirkungsintegrale $j_1, j_2, \ldots j_s$ gefunden:

$$H_0 = \mathsf{E}_0 (j_1, j_2, \ldots j_s; \quad X_\varrho^0, Y_\varrho^0, Z_\varrho^0)\,. \tag{30}$$

Unter den $j_1, j_2, \ldots$ wird das Impulsmoment der Elektronen, d. h. der Betrag des Vektors

$$\mathfrak{g} = \sum_\alpha [\mathfrak{r}_\alpha \mathfrak{p}_\alpha]$$

im allgemeinen *nicht* vorkommen; denn der Impulssatz gilt nur für frei drehbare Systeme, während hier die Kerne festgehalten sind.

Die zu den $j_1, j_2, \ldots$ konjugierten Winkelvariabeln seien $w_1, w_2, \ldots$; wir denken uns diese Variabeln w_α, j_α in alle Glieder der Störungsfunktion statt der Koordinaten und Impulse der Elektronen eingeführt.

1) V. Pauli jun., Ann. d. Phys. 68. S. 177. 1922.

2) F. F. Niessen, Zur Quantentheorie des Wasserstoffmolekülions. Diss. Utrecht 1922.

Die Funktion H_0 enthält die Impulse $\mathfrak{P}_{\varrho x}, \ldots \Phi_k$ nicht; sie ist also bezüglich der Koordinaten $X_\varrho^{(1)}, \ldots \varphi_k$ entartet. Diese werden säkulare Störungen erleiden, und man wird versuchen, ob man diese aus der Hamiltonschen Gleichung

$$\overline{H_1} = \overline{U_1} = \mathsf{E}_1 \tag{31}$$

bestimmen kann, wo der Strich Mittelbildung über $w_1, w_2, \ldots$ bedeutet; wir werden sogleich sehen, daß dies nicht möglich ist. $\overline{U_1}$ ist nämlich eine lineare Funktion der $X_\varrho^{(1)}, \ldots,$ deren Koeffizienten Funktionen der $j_1, j_2, \ldots$ und der $X_\varrho^0, \ldots$ sind. Damit nun unser Ansatz, daß das System der Kerne kleine Schwingungen um die Lagen $\mathfrak{R}_\varrho^0$ machen soll, durchführbar ist, müssen wir erreichen, daß im Ausdruck der Energie E als Funktion der Verrückungen $X_\varrho^{(1)}, \ldots$ keine linearen Glieder auftreten. Wir müssen daher die Kernlagen so wählen, daß

$$\overline{\left(\frac{\partial U}{\partial X_\varrho}\right)_0} = 0; \quad \overline{\left(\frac{\partial U}{\partial Y_\varrho}\right)_0} = 0; \quad \overline{\left(\frac{\partial U}{\partial Z_\varrho}\right)_0} = 0. \tag{32}$$

Dies sind die bekannten Gleichgewichtsbedingungen für die Kerne, wie sie z. B. Pauli und Niessen (a. a. O.) beim H_2^+-Modell angewendet haben. Die Differentiation nach den X_ϱ in $(\partial U/\partial X_\varrho)_0$ ist dabei, entsprechend der Ableitung von (32), bei konstant gehaltenem $x_a, y_a, z_a, \mathfrak{p}_{ax} \ldots$ zu verstehen. Pauli beweist (a. a. O.) auch den Satz, daß

$$\overline{\left(\frac{\partial U}{\partial X_\varrho}\right)_{x, \mathfrak{p}_x}} = \left(\frac{\partial H_0 (j_k, X_\varrho, Y_\varrho, Z_\varrho)}{\partial X_\varrho}\right)_{j_a} \tag{32a}$$

— — — — — — —

(Hierin ist durch die Indizes x, $\mathfrak{p}_x$ bzw, j_a angedeutet, welche Größen bei der Differation nach X_ϱ konstant zu halten sind.) Der Beweis beruht wesentlich auf den Grundlagen des Adiabatensatzes. Es ist nämlich

$$\left(\frac{\partial U}{\partial X_\varrho}\right)_{x, \mathfrak{p}_x} = \left(\frac{\partial H_0}{\partial X_\varrho}\right)_{x, \mathfrak{p}_x} = \left(\frac{\partial H_0}{\partial X_\varrho}\right)_{j_a} + \sum_a \frac{\partial H_0}{\partial j_a} \cdot \left(\frac{\partial j_a}{\partial X_\varrho}\right)_{x, \mathfrak{p}_x}. \tag{33}$$

Die Größe $\partial j_a/\partial X_\varrho$ ist periodisch in den w_a; es gilt[1])

$$\frac{\partial j_a}{\partial X_\varrho} = - \frac{\partial}{\partial w_a} \frac{\partial S^*}{\partial X_\varrho}, \tag{33a}$$

1) Vgl. z. B. A. Sommerfeld, Atombau und Spektrallinien. 3. Aufl. S. 719.

wo $S^* = S - \sum_\alpha j_\alpha w_\alpha$ und S die Wirkungsfunktion des (ungestörten) Problems ist. Der zeitliche Mittelwert von $\partial j_\alpha / \partial X_\varrho$ ist also Null, womit (32a) bewiesen ist.

Die Kernkoordinaten sind aus (32) als Funktionen der Elektronenintegrale $j_1, j_2, \ldots$ zu bestimmen. Im allgemeinen wird es zu jeder Elektronenkonfiguration mehrere Lösungen von (32) geben; brauchbar sind schließlich nur solche, die stabilem Gleichgewicht entsprechen, d. h. für die die Fortsetzung des Näherungsverfahrens nur reelle Eigenschwingungen der Kerne ergibt.

Wir schließen, wie oben gesagt, überdies alle solchen Gleichgewichtslagen aus, bei denen die Kerne auf einer Geraden liegen, da dieser Fall eine gesonderte Behandlung erfordert.

Wegen (32) wird nun identisch in den $X_\varrho^{(1)}, \ldots$:

(31a) $$\overline{H}_1 = \overline{U}_1 = \mathsf{E}_1 = 0\,.$$

Wir haben daher den besonderen Fall der Störungstheorie vor uns, bei dem die säkularen Bewegungen der entarteten Variabeln nicht aus dem Mittelwert der Störungsfunktion 1. Ordnung bestimmt werden können, weil dieser Mittelwert identisch verschwindet. Das Verfahren, welches dann einzuschlagen ist, haben wir im Kap. I auseinandergesetzt; wir benutzen hier die dort erhaltenen Resultate.

In unserem Falle besteht noch die besondere Vereinfachung, daß die Impulse $\mathfrak{P}_{\varrho x}, \ldots$ der entarteten Variabeln in H_1 gar nicht vorkommen. Wir schaffen zunächst H_1 ganz aus der Störungsfunktion heraus mit Hilfe der kanonischen Transformation, Kap. I (4), deren erzeugende Funktion der Gleichung Kap. I (6) genügt; diese lautet hier

(34) $$\sum_\alpha \frac{\partial H_0}{\partial j_\alpha} \frac{\partial F_1}{\partial w_\alpha} + \sum_\varrho \sum_x \left(\frac{\partial U}{\partial X_\varrho}\right)_0 X_\varrho^{(1)} = 0\,,$$

und ihre Lösung hat die Form:

(35) $$F_1 = \sum_\varrho \sum_x \mathfrak{F}_{\varrho x} X_\varrho^{(1)} = \sum_\varrho (\mathfrak{F}_\varrho \mathfrak{R}_\varrho^{(1)})\,.$$

Die hierdurch definierten Vektoren $\mathfrak{F}_\varrho$ erfüllen die Bedingungen

(35a) $$\sum_\varrho \mathfrak{F}_\varrho = 0\,, \qquad \sum_\varrho [\mathfrak{R}_\varrho^0 \mathfrak{F}_\varrho] = 0\,.$$

Setzt man nämlich in

$$U_1 = \sum_\varrho \sum_x \left(\frac{\partial U}{\partial X_o}\right)_0 X_o^{(1)}$$

für die Verrückungen $\mathfrak{R}_\varrho^{(1)}$ eine Translation oder infinitesimale Rotation

$$\mathfrak{R}_o^{(1)} = \mathfrak{A}\,, \quad \text{bzw.} \quad \mathfrak{R}_\varrho^{(1)} = [\mathfrak{d}\,\mathfrak{R}_\varrho^0]\,,$$

wobei man die j_α, w_α ungeändert läßt, so muß U_1 verschwinden; dann folgt aus (34), daß auch F_1 für diese Verrückungen der Kerne identisch in den $X_\varrho^{(1)}, \ldots$ gleich Null sein muß, und daraus ergibt sich die Behauptung.

Die Funktion (35) entspricht der in Kap. I (4) ebenso bezeichneten; durch die kanonische Transformation, Kap. I (4a, b), werden die w_α, j_α, $X_\varrho^{(1)}$, $\mathfrak{P}_{\varrho x}, \ldots$ in neue Größen transformiert, die wir der Einfachheit halber mit denselben Buchstaben bezeichnen wollen. Da F_1 die $\mathfrak{P}_{\varrho x}$ gar nicht enthält und in den $X_\varrho^{(1)}, \ldots$ linear ist, so gehen bei der Transformation die $X_o^{(1)}, \ldots$ in sich über und die $\mathfrak{P}_{\varrho x}, \ldots$ vermehren sich um $-\lambda\,\mathfrak{F}_{\varrho x}, \ldots$ Wegen (35a) bleiben dabei

(36) $$\mathfrak{P} = \sum_\varrho \mathfrak{P}_\varrho \quad \text{und} \quad \mathfrak{G}^0 = \sum_\varrho [\mathfrak{R}_\varrho^0\,\mathfrak{P}_\varrho] \quad \text{ungeändert,}$$

und

(36a) $$\mathfrak{G}^{(1)} = \sum_\varrho [\mathfrak{R}_o^{(1)}\,\mathfrak{P}_\varrho] \quad \text{geht über in} \quad \mathfrak{G}^{(1)} - \lambda\,\mathfrak{Q}^{(1)}\,,$$

wo

(36b) $$\mathfrak{Q}^{(1)} = \sum_\varrho [\mathfrak{R}_o^{(1)}\,\mathfrak{F}_\varrho]$$

gesetzt ist.

Nun können wir die in Kap. I (5), (8) mitgeteilten Formeln zur Umrechnung von H in die neuen Variabeln benutzen. Wir begnügen uns damit, das Resultat der etwas langwierigen Rechnung mitzuteilen: es lautet:

(37) $$H = \mathsf{H}_0 + \lambda^2 \mathsf{H}_2 + \lambda^3 \mathsf{H}_3 + \lambda^4 \mathsf{H}_4 + \ldots,$$

wo

$$
(37)\quad\left\{\begin{aligned}
&\text{a)}\ \mathsf{H}_0 = H_0\,,\\
&\text{b)}\ \mathsf{H}_2 = \tfrac{1}{2}\sum_{\varrho}\frac{1}{m_\varrho}\mathfrak{P}_\varrho^{\,2} + \frac{1}{2m}(\mathfrak{p}^2-\mathfrak{P}^2)\\
&\qquad + \tfrac{1}{2}\sum_{x}\frac{1}{A_x^0}\left[(\mathfrak{D}_x-\mathfrak{g}_x)^2-\mathfrak{G}_x^{0\,2}\right] + \tfrac{1}{2}\sum_{\varrho\sigma}\sum_{xy}\left[2\,{}^{xy}_{\varrho\sigma}\right]X_\varrho^{(1)}Y_\sigma^{(1)},\\
&\text{c)}\ \mathsf{H}_3 = \sum_{\varrho}\sum_{x}\left[3\,{}^{x}_{\varrho}\right]\mathfrak{P}_{\varrho x} + \sum_{\varrho}\sum_{x}X_\varrho^{(1)}\Big\{\left[\left[3\,{}^{x}_{\varrho}\right]\right] + \sum_{y}\left[3\,{}^{xy}_{\varrho}\right]\mathfrak{D}_\varrho\\
&\qquad + \sum_{yz}\left[3\,{}^{xyz}_{\varrho}\right]\mathfrak{D}_y\mathfrak{D}_z + \sum_{\sigma}\sum_{y}\left[3\,{}^{xy}_{\varrho\sigma}\right]\mathfrak{P}_{\sigma y} + \sum_{\sigma}\sum_{yz}\left[3\,{}^{xyz}_{\varrho\sigma}\right]\mathfrak{D}_y\mathfrak{P}_{\varrho z}\Big\}\\
&\qquad\qquad + \sum_{\varrho\sigma\tau}\sum_{xyz}\left[3\,{}^{xyz}_{\varrho\sigma\tau}\right]X_\varrho^{(1)}Y_\sigma^{(1)}Z_\tau^{(1)},\\
&\text{d)}\ \mathsf{H}_4 = [4] + \sum_{\varrho}\sum_{x}X_\varrho^{(1)}\Big\{\left[4\,{}^{x}_{\varrho}\right] + \sum_{y}\left[4\,{}^{xy}_{\varrho}\right]\mathfrak{D}_y\\
&\qquad + \sum_{y}\sum_{\sigma}\left[4\,{}^{xy}_{\varrho\sigma}\right]\mathfrak{P}_{\sigma y}\Big\} + \sum_{\varrho\sigma}\sum_{xy}X_\varrho^{(1)}Y_\sigma^{(1)}\Big\{\left[\left[4\,{}^{zy}_{\varrho\sigma}\right]\right]\\
&\qquad + \sum_{z}\left[4\,{}^{xyz}_{\varrho\sigma}\right]\mathfrak{D}_z + \sum_{\bar{x}\bar{y}}\left[4\,{}^{x\bar{x}y\bar{y}}_{\varrho\sigma}\right]\mathfrak{D}_{\bar{x}}\mathfrak{D}_{\bar{y}}\\
&\qquad + \sum_{z}\sum_{\tau}\left[4\,{}^{xyz}_{\varrho\sigma\tau}\right]\mathfrak{P}_{\tau z} + \sum_{\tau}\sum_{\bar{x}\bar{y}}\left[4\,{}^{x\bar{x}y\bar{y}}_{\varrho\sigma\tau}\right]\mathfrak{D}_{\bar{x}}\mathfrak{P}_{\tau\bar{y}}\\
&\qquad\qquad + \sum_{\bar{\varrho}\bar{\sigma}}\sum_{\bar{x}\bar{y}}\left\|4\,{}^{x\bar{x}y\bar{y}}_{\varrho\bar{\varrho}\sigma\bar{\sigma}}\right\|\mathfrak{P}_{\bar{\varrho}\bar{x}}\mathfrak{P}_{\bar{\sigma}\bar{y}}\Big\}\\
&\qquad + \sum_{\varrho\sigma\bar{\varrho}\bar{\sigma}}\sum_{xy\bar{x}\bar{y}}\left[4\,{}^{x\bar{x}y\bar{y}}_{\varrho\bar{\varrho}\sigma\bar{\sigma}}\right]X_\varrho^{(1)}\bar{X}_{\bar{\varrho}}Y_\sigma^{(1)}\bar{Y}_{\bar{\sigma}}^{(1)}.
\end{aligned}\right.
$$

Dabei sind die Klammersymbole Größen, die nur von der Bewegung der Elektronen abhängen; ihr Bildungsgesetz ist im allgemeinen so kompliziert, daß es keinen Sinn hat, es mitzuteilen. Nur das erste Klammersymbol (in H_2) wollen wir angeben:

$$
(38)\quad\left\{\begin{aligned}
\left[2\,{}^{xy}_{\varrho\sigma}\right] &= \left(\frac{\partial^2 U}{\partial X_\varrho\,\partial Y_\sigma}\right)_0 + \sum_{\alpha\beta}\frac{\partial^2\mathsf{H}_0}{\partial j_\alpha\,\partial j_\beta}\,\frac{\partial\mathfrak{F}_{\varrho x}}{\partial w_\alpha}\,\frac{\partial\mathfrak{F}_{\sigma y}}{\partial w_\beta}\\
&\quad + 2\sum_{\alpha}\left\{\frac{\partial}{\partial j_\alpha}\left(\frac{\partial U}{\partial X_\varrho}\right)_0\cdot\frac{\partial\mathfrak{F}_{\sigma y}}{\partial w_\alpha} + \frac{\partial}{\partial j_\alpha}\left(\frac{\partial U}{\partial Y_\sigma}\right)_0\cdot\frac{\partial\mathfrak{F}_{\varrho x}}{\partial w_\alpha}\right\}.
\end{aligned}\right.
$$

Wir merken ferner noch an, daß die beiden Glieder von H_3, die wir hier

$$
(39\,\text{a})\qquad \sum_{\varrho}\sum_{x}X_\varrho^{(1)}\Big\{\sum_{\sigma}\sum_{y}\left[3\,{}^{xy}_{\varrho\sigma}\right]\mathfrak{P}_{y\sigma} + \sum_{\sigma}\sum_{yz}\left[3\,{}^{xyz}_{\varrho\sigma}\right]\mathfrak{D}_y\mathfrak{P}_{z\sigma}\Big\}
$$

geschrieben haben, aus den in H_3 (29d) auftretenden Termen

$$\text{(39b)} \qquad \frac{1}{2}\sum_{xy}\frac{A_{xy}^{(1)}}{A_x^0 A_y^0}(\mathfrak{D}_x - \mathfrak{g}_x)\mathfrak{G}_y^0 - \sum_x \frac{1}{A_x^0}\mathfrak{G}_x^{(1)}(\mathfrak{D}_x - \mathfrak{g}_x)$$

entstanden sind; hiervon werden wir nachher Gebrauch machen.

§ 5. Kreiselbewegung und Schwingungen der Elektronen als säkulare Störungen.

Man kann jetzt die säkularen Bewegungen der Kerne aus

$$\text{(40)} \qquad \left\{\begin{aligned} \overline{\mathsf{H}}_2 = \frac{\overline{\mathfrak{p}^2}}{2m} + \frac{1}{2}\sum_x \frac{1}{A_x^0}\left(\overline{\mathfrak{g}_x^2} - \overline{\mathfrak{g}}_x^2\right) + \frac{1}{2}\sum_\varrho \frac{1}{m_\varrho}\mathfrak{P}_\varrho^2 - \frac{1}{2m}\mathfrak{P}^2 \\ + \frac{1}{2}\sum_{\varrho\sigma}\sum_{xy}\overline{\left[2^{xy}_{\varrho\sigma}\right]} X_\varrho^{(1)} Y_\sigma^{(1)} \\ + \frac{1}{2}\sum_x \frac{1}{A_x^0}\left[(\mathfrak{D}_x - \overline{\mathfrak{g}}_x)^2 - \mathfrak{G}_x^{0\,2}\right] = \mathsf{E}_2 \end{aligned}\right.$$

bestimmen, wo die Striche Mittelbildung über die Elektronenbewegungen bedeuten.

Diese Hamiltonsche Differentialgleichung läßt sich durch Separation der Variabeln lösen; sie zerfällt nämlich ohne weiteres in zwei Teile, deren erster den Kernschwingungen, deren zweiter den Kreiselbewegungen entspricht.

Wir betrachten zunächst den Anteil der Schwingungen, nämlich die Terme

$$\text{(41)} \qquad \left\{\begin{aligned} \frac{1}{2}\sum_\varrho \frac{1}{m_\varrho}\mathfrak{P}_\varrho^2 + \frac{1}{2}\sum_{\varrho\sigma}\sum_{xy}\overline{\left[2^{xy}_{\varrho\sigma}\right]} X_\varrho^{(1)} Y_\sigma^{(1)} \\ - \frac{1}{2m}\mathfrak{P}^2 - \frac{1}{2}\sum_x \frac{1}{A_x^0}\mathfrak{G}_x^{0\,2} = \mathsf{E}_2^{(s)}. \end{aligned}\right.$$

Als quasielastische Kräfte treten hier die Mittelwerte der Klammersymbole $[2^{xy}_{\varrho\sigma}]$ auf. Wir wollen zeigen, daß diese mit den von Pauli und Niessen a. a. O. zur Berechnung der Kernschwingung benutzten Größen

$$\left(\frac{\partial^2 H_0}{\partial X_\varrho \, \partial Y_\sigma}\right)_{j_\alpha}$$

identisch sind.

Zunächst bemerken wir, daß aus (32), (32a), (33), (33a) und (35) folgt

$$\text{(33b)} \qquad \mathfrak{F}_{\varrho x} = \frac{\partial S^*}{\partial X_\varrho}; \quad \frac{\partial \mathfrak{F}_{\varrho x}}{\partial w_\alpha} = -\frac{\partial j_\alpha}{\partial X_\varrho}.$$

Nun ist analog wie in (33)

$$\left(\frac{\partial^2 U}{\partial X_\varrho \partial Y_\sigma}\right)_{x,\mathfrak{p}_x} = \left(\frac{\partial^2 H_0}{\partial X_\varrho \partial Y_\sigma}\right)_{j_\alpha} + 2\sum_\alpha \left\{\left(\frac{\partial^2 H_0}{\partial X_\varrho \partial j_\alpha}\right)\frac{\partial j_\alpha}{\partial Y_\sigma} + \left(\frac{\partial^2 H_0}{\partial Y_\sigma \partial j_\alpha}\right)\frac{\partial j_\alpha}{\partial X_\varrho}\right\} + \sum_{\alpha,\beta}\frac{\partial^2 H_0}{\partial j_\alpha \partial j_\beta}\frac{\partial j_\alpha}{\partial X_\varrho}\frac{\partial j_\beta}{\partial Y_\sigma} + \sum_\alpha \frac{\partial H_0}{\partial j_\alpha}\frac{\partial^2 j_\alpha}{\partial X_\varrho \partial Y_\sigma}.$$

Hieraus ergibt sich durch Einsetzen von $\partial j_\alpha / \partial X_\varrho$ aus (33b) und $\left(\frac{\partial H}{\partial X_\varrho}\right)_{j_\alpha}$ aus (32a) und durch Mittelbildung (wobei $\overline{\frac{\partial^2 j_\alpha}{\partial X_\varrho \partial Y_\sigma}}$ nach (33a) gleich Null ist),

$$(38\text{a})\quad \begin{cases} \left(\frac{\partial^2 H_0}{\partial X_\varrho \partial Y_\sigma}\right)_{j_\alpha} = \overline{\left(\frac{\partial^2 U}{\partial X_\varrho \partial Y_\sigma}\right)_{x,\mathfrak{p}_x}} + 2\sum_\alpha \left\{\overline{\frac{\partial}{\partial j_\alpha}\left(\frac{\partial U}{\partial X_\varrho}\right)_{x,\mathfrak{p}_x} \cdot \frac{\partial \mathfrak{F}_{\sigma y}}{\partial w_\alpha}} \right. \\ \left. + \overline{\frac{\partial}{\partial j_\alpha}\left(\frac{\partial U}{\partial Y_\sigma}\right)_{x,\mathfrak{p}_x}\frac{\partial \mathfrak{F}_{\varrho x}}{\partial w_\alpha}}\right\} + \sum_{\alpha,\beta}\overline{\frac{\partial^2 H_0}{\partial j_\alpha \partial j_\beta}\frac{\partial \mathfrak{F}_{\varrho x}}{\partial w_\alpha}\frac{\partial \mathfrak{F}_{\sigma y}}{\partial w_\beta}} = \left[2\,{}^{xy}_{\varrho\sigma}\right]. \end{cases}$$

Die quasielastischen Kräfte sind also, wie zu erwarten war, durch die zweiten Differentialquotienten der Energie nach den Kernkoordinaten bei konstant gehaltenen Wirkungsintegralen gegeben.

Die Energiefunktion $\mathsf{E}_2^{(s)}$ läßt sich separieren, indem man die Normalkoordinaten einführt. Um diese zu finden, schreiben wir die Schwingungsgleichungen hin:

$$(42)\qquad m\ddot{X}_\varrho^{(1)} + \sum_\sigma \sum_y \overline{\left[2\,{}^{xy}_{\varrho\sigma}\right]}\, Y_\sigma^{(1)} = 0,$$

und suchen sie durch den Ansatz

$$X_\varrho^{(1)} \text{ proportional } e^{i\omega t}$$

zu lösen. Dann ergeben sich die linearen Gleichungen:

$$(42\text{a})\qquad \omega^2 m X_\varrho^{(1)} - \sum_\sigma \sum_y \overline{\left[2\,{}^{xy}_{\varrho\sigma}\right]}\, Y_\sigma^{(1)} = 0.$$

Damit sie lösbar sind, muß die Determinante verschwinden; das gibt eine Gleichung $3N$ten Grades für ω^2, die $3N$ reelle nicht negative Wurzeln

$$\omega_1^{0\,2},\quad \omega_2^{0\,2},\quad \cdots\quad \omega_{3N}^{0\,2}$$

hat. Von diesen verschwinden im allgemeinen sechs, die den Translationen

$$\mathfrak{R}_\varrho^{(1)} = \mathfrak{A}$$

und den Rotationen (Winkelgeschwindigkeit $\mathfrak{d}$)

$$\mathfrak{R}_\varrho^{(1)} = [\mathfrak{d}\,\mathfrak{R}_\varrho^0]$$

entsprechen; denn bei diesen Bewegungen bleiben die relativen Lagen der Kerne ungeändert, die quasielastischen Kräfte verschwinden also.

Die einzige Ausnahme ist wieder der Fall, daß alle Kernruhelagen $\mathfrak{R}_\varrho^0$ auf derselben geraden Linie sind; dann verschwinden nur fünf Frequenzen, da die Rotation um die Kernverbindungslinie ausfällt. Diesen Fall werden wir besonders behandeln.

Wir numerieren die ω_j^0 so, daß die sechs Nullfrequenzen die letzten sind $(3N-5, \ldots 3N)$.

Schreibt man jede Frequenz so oft, als ihre Vielfachheit beträgt, dann läßt sich zu einer jeden ω_τ^0 eine Lösung

$$X_\varrho^{(1)} = \mathfrak{A}_{\varrho x}^{(\tau)}, \ldots \text{ (vektoriell } \mathfrak{R}_\varrho^{(1)} = \mathfrak{A}_\varrho^{(\tau)};\ \tau = 1, 2, \ldots 3N)$$

so zuordnen, daß folgende Orthogonalitäts- und Normierungsbedingungen gelten:

$$(43) \qquad \sum_\varrho m_\varrho \mathfrak{A}_\varrho^{(\tau)} \mathfrak{A}_\varrho^{(\tau')} = \begin{cases} 0 \text{ für } \tau \neq \tau', \\ 1 \text{ für } \tau = \tau'. \end{cases}$$

Wir hatten oben das Koordinatensystem so gewählt, daß der Nullpunkt in den Schwerpunkt und die Koordinatenachsen in die Hauptträgheitsachsen des Systems der im Gleichgewicht ruhenden Kerne fallen, so daß also

$$\sum_\varrho m_\varrho \mathfrak{R}_\varrho^0 = 0, \qquad \sum_\varrho m_\varrho Y_\varrho^0 Z_\varrho^0 = 0, \ldots$$

gilt. Mit Rücksicht hierauf lassen sich die zu den sechs Nullfrequenzen gehörigen Lösungen, die wir „uneigentliche Eigenschwingungen" nennen wollen, durch die Einheitsvektoren parallel zu den Koordinatenachsen $\mathfrak{i}_1$, $\mathfrak{i}_2$, $\mathfrak{i}_3$ so ausdrücken:

$$(44a) \begin{cases} \omega_{3N-5}^0 = 0, & \omega_{3N-4}^0 = 0, & \omega_{3N-3}^0 = 0, \\ \mathfrak{A}_\varrho^{3N-5} = \dfrac{\mathfrak{i}_1}{\sqrt{m}}, & \mathfrak{A}_\varrho^{3N-4} = \dfrac{\mathfrak{i}_2}{\sqrt{m}}, & \mathfrak{A}_\varrho^{3N-3} = \dfrac{\mathfrak{i}_3}{\sqrt{m}}; \end{cases}$$

$$(44\text{b})\begin{cases} \omega^0_{3N-2} = 0\,, & \omega^0_{3N-1} = 0\,, & \omega^0_{3N} = 0\,, \\ \mathfrak{A}_\varrho^{3N-2} = \dfrac{[\mathfrak{i}_1\,\mathfrak{R}_\varrho{}^0]}{\sqrt{A_x{}^0}}\,, & \mathfrak{A}_\varrho^{3N-1} = \dfrac{[\mathfrak{i}_2\,\mathfrak{R}_\varrho{}^0]}{\sqrt{A_y{}^0}}\,, & \mathfrak{A}_\varrho^{3N} = \dfrac{[\mathfrak{i}_3\,\mathfrak{R}_\varrho{}^0]}{\sqrt{A_z{}^0}}\,. \end{cases}$$

Die „eigentlichen" Eigenschwingungen, an Zahl $3N-6$, genügen nunmehr folgenden Orthogonalitätsrelationen, die man aus (43) durch Benutzung von (44a, b) erhält:

$$(45)\begin{cases} \text{a)}\ \sum\limits_\varrho m_\varrho\,\mathfrak{A}_\varrho^{(\tau)} = 0\,, \qquad \sum\limits_\varrho m_\varrho\,[\mathfrak{R}_\varrho^0\,\mathfrak{A}_\varrho^{(\tau)}] = 0\,, \\ \text{b)}\ \sum\limits_\varrho m_\varrho\,\mathfrak{A}_\varrho^{(\tau)}\,\mathfrak{A}_\varrho^{(\sigma)} = \begin{cases} 0 & \text{für } \tau \neq \sigma\,, \\ 1 & \text{„} \quad \tau = \sigma \end{cases} (\tau, \sigma = 1, 2, \ldots 3N-6). \end{cases}$$

Die Normalkoordinaten Q_τ werden durch die lineare Transformation

$$(46\text{a})\qquad Q_\tau = \sum_\varrho m_\varrho\,\mathfrak{A}_\varrho^{(\tau)}\,\mathfrak{R}_\varrho^{(1)} \qquad \text{oder} \qquad \mathfrak{R}_\varrho^{(1)} = \sum_\tau \mathfrak{A}_\varrho^{(\tau)}\,Q_\tau$$

definiert; die zugehörigen Impulse entsprechend durch

$$(46\text{b})\qquad P_\tau = \sum_\varrho \mathfrak{A}_\varrho^{(\tau)}\,\mathfrak{P}_\varrho\,, \qquad \text{oder} \qquad \frac{1}{m_\varrho}\,\mathfrak{P}_\varrho = \sum_\tau \mathfrak{A}_\varrho^{(\tau)}\,P_\tau\,.$$

Die zu den Nullfrequenzen gehörigen Normalkoordinaten und -impulse haben eine einfache Bedeutung; man findet aus (44a, b) und (46a, b):

$$(47\text{a})\begin{cases} Q_{3N-5} = \dfrac{1}{\sqrt{m}}\sum\limits_\varrho m_\varrho\,X_\varrho^{(1)}\,, & Q_{3N-2} = \dfrac{1}{\sqrt{A_x{}^0}}\sum\limits_\varrho m_\varrho\,[\mathfrak{R}_\varrho^0\,\mathfrak{R}_\varrho^{(1)}]_x\,, \\ Q_{3N-4} = \dfrac{1}{\sqrt{m}}\sum\limits_\varrho m_\varrho\,Y_\varrho^{(1)}\,, & Q_{3N-1} = \dfrac{1}{\sqrt{A_y{}^0}}\sum\limits_\varrho m_\varrho\,[\mathfrak{R}_\varrho^0\,\mathfrak{R}_\varrho^{(1)}]_y\,, \\ Q_{3N-3} = \dfrac{1}{\sqrt{m}}\sum\limits_\varrho m_\varrho\,Z_\varrho^{(1)}\,, & Q_{3N} \;\;= \dfrac{1}{\sqrt{A_z{}^0}}\sum\limits_\varrho m_\varrho\,[\mathfrak{R}_\varrho^0\,\mathfrak{R}_\varrho^{(1)}]_z\,; \end{cases}$$

ferner:

$$(47\text{b})\begin{cases} P_{2N-5} = \dfrac{1}{\sqrt{m}}\,\mathfrak{P}_x\,, & P_{3N-4} = \dfrac{1}{\sqrt{m}}\,\mathfrak{P}_y\,, & P_{3N-3} = \dfrac{1}{\sqrt{m}}\,\mathfrak{P}_z\,, \\ P_{3N-2} = \dfrac{1}{\sqrt{A_x{}^0}}\,\mathfrak{G}_x{}^0\,, & P_{3N-1} = \dfrac{1}{\sqrt{A_y{}^0}}\,\mathfrak{G}_y{}^0\,, & P_{3N} = \dfrac{1}{\sqrt{A_y{}^0}}\,\mathfrak{G}_z{}^0\,. \end{cases}$$

Wegen der Nebenbedingungen (7a, b) erhält man:

$$(48)\qquad Q_{3N-5} = 0\,, \qquad \ldots\ldots\ Q_{3N} = 0\,.$$

Denkt man sich die Q_τ, P_τ statt der $X_\varrho^{(1)}$, $\mathfrak{P}_\varrho$ als kanonische Variable eingeführt, so haben die Bewegungsgleichungen

$$\dot{Q}_\tau = \frac{\partial H}{\partial P_\tau}, \qquad \dot{P}_\tau = -\frac{\partial H}{\partial Q_\tau}$$

die Lösungen (48); daraus folgt aber, daß H von den P_{3N-5}, ... P_{3N} unabhängig sein muß.

In der Tat sieht man zunächst, daß sich die gemittelte Energie $\mathsf{E}_2^{(s)}$ (41) auf

(49) $$\mathsf{E}_2^{(s)} = \tfrac{1}{2}\sum_\tau (P_\tau^2 + \omega_\tau^{0\,2} Q_\tau^2)$$

reduziert, wo die Summe nur über die *eigentlichen* Eigenfrequenzen zu erstrecken ist. Führt man ferner die P_τ, Q_τ in alle Glieder H_2, H_3, ... der Störungsfunktion ein, so müssen sich mit Rücksicht auf (47b) die P_{3N-5}, ... P_{3N} von selbst herausheben. Man kann das z. B. leicht an den beiden, in (39b) angeschriebenen Termen von H_3 nachrechnen. Jedenfalls ist es erlaubt, alle auf die uneigentlichen Eigenfrequenzen ($\tau = 3N-5$, ... $3N$) bezüglichen Terme und gleichzeitig die Vektoren $\mathfrak{P}$ und $\mathfrak{G}^0$ zu streichen.

In (49) sind die Variabeln separiert; man kann die Lösung des Schwingungsproblems durch die kanonische Transformation

(50) $$\begin{cases} Q_\tau = \dfrac{1}{\pi}\sqrt{\dfrac{J_\tau}{2\nu_\tau^0}}\cos 2\pi W_\tau, \\ P_\tau = -\sqrt{2 J_\tau \nu_\tau^0}\sin 2\pi W_\tau \end{cases} \qquad \left(\nu_\tau^0 = \frac{\omega_\tau^0}{2\pi}\right)$$

gewinnen, wo die J_τ, W_τ die Wirkungs- und Winkelvariabeln der Kernschwingungen sind. Dann wird

(51) $$\mathsf{E}_2^{(s)} = \sum_\tau J_\tau \nu_\tau^0 .$$

Wir wenden uns jetzt zum Kreiselanteil von E_2, der, ausführlich geschrieben, so lautet:

(52) $$\mathsf{E}_2^{(r)} = \frac{1}{2}\left(\frac{(\mathfrak{D}_x - \bar{\mathfrak{g}}_x)^2}{A_x^0} + \frac{(\mathfrak{D}_y - \bar{\mathfrak{g}}_y)^2}{A_y^0} + \frac{(\mathfrak{D}_z - \bar{\mathfrak{g}}_z)^2}{A_z^0}\right).$$

Es handelt sich also um keinen gewöhnlichen Kreisel, sondern um einen „Kreiselkörper mit eingebautem Schwungrad", wie ihn Kramers und Pauli[1]) kürzlich behandelt haben.

1) H. A. Kramers, Zeitschr. f. Physik **13**. S. 343. 1923. H. A. Kramers und W. Pauli, Zeitschr. f. Physik **13**. S. 351. 1923. Unsere Formel (52) ist mit der Formel (19) der ersten und Formel (1) der zweiten der zitierten Arbeiten identisch.

Diese Forscher haben gezeigt, wie die Quantenbedingungen für ein solches Gebilde aufgestellt werden können; wir brauchen daher auf diese Frage nicht näher einzugehen und übernehmen nur das Resultat. Man kann zwei Paare von Wirkungs- und Winkelvariabeln der Rotation J, W und J', W' einführen so, daß $\mathsf{E}_2^{(r)}$ eine Funktion von J, J' wird (die Kreiselfunktion K):

$$\mathsf{E}_2^{(r)} = \mathsf{K}(J, J'). \tag{53}$$

Dabei bedeutet $J = |\mathfrak{D}|$ das Gesamtimpulsmoment der Molekel; die zugehörige Winkelvariable W ist in Strenge zyklisch (kommt also auch in den höheren Näherungen der Energiefunktion $\mathsf{H}_3, \mathsf{H}_4, \ldots$ nicht vor).

Die gesamte Energie der säkularen Bewegungen wird nun:

$$\mathsf{E}_2 = \mathsf{E}_2^0 + \mathsf{E}_2^{(s)} + \mathsf{E}_2^{(r)} = \mathsf{E}_2^0 + \sum_\tau \nu_\tau^0 J_\tau + \mathsf{K}(JJ'), \tag{54}$$

wo zur Abkürzung

$$\mathsf{E}_2^0 = \frac{\overline{\mathfrak{p}^2}}{2m} + \frac{1}{2}\sum_x \frac{1}{A_x^0}\left(\overline{\mathfrak{g}_x^2} - \overline{\mathfrak{g}_x}^2\right) \tag{54a}$$

gesetzt ist.

§ 6. Höhere Näherungen.

Will man zu höheren Näherungen übergehen, so muß man die Variabeln J_τ, W_τ $(\tau = 1, \ldots 3N-6)$ und J, W, J', W' in $\mathsf{H}_3, \mathsf{H}_4, \ldots$ einführen; was die Abhängigkeit von den W_τ betrifft, so werden $\mathsf{H}_3, \mathsf{H}_4, \ldots$ Polynome 3., 4., ... Grades in $\cos 2\pi W_\tau$ und $\sin 2\pi W_\tau$. Man hat nun nach dem in Kap. I gegebenen Verfahren vorzugehen.

Zunächst kann man zeigen, daß der über die Elektronen- und Kernbewegung genommene Mittelwert

$$\overline{\overline{\mathsf{H}}}_3 = \mathsf{E}_3 = 0 \tag{55}$$

ist. Denn betrachten wir den Ausdruck (37c) von H_3, so sehen wir, daß er außer Gliedern ungerader Ordnung in $X_\varrho^{(1)}, \ldots$ und $\mathfrak{P}_{\varrho x}, \ldots$ nur solche gerader Ordnung enthält, in denen Produkte $X_\varrho^{(1)}\,\mathfrak{P}_{\sigma y}$ vorkommen. Nun hängen nach (46a, b) die $X_\varrho^{(1)}$ nur von den Q_τ, die $\mathfrak{P}_{\varrho x}$ nur von den P_τ ab; Q_τ und P_τ haben nach (50) die Phasendifferenz $\pi/2$. Folglich verschwinden bei der Mittelung über die W_τ sämtliche Glieder von H_3. Um nun E_4 berechnen zu können, muß man nach dem Rezept des Kap. I zunächst die Funktion F_2 durch die Differentialgleichung (11b) Kap. I bestimmen; diese Funktion drückt die Rückwirkung der Kernschwingungen und Kreiselbewegungen auf die Elektronenbahnen aus; man sieht leicht, daß der Anteil F_2^0 von F_2, der nach (12b) Kap. I durch die Differentialgleichung eindeutig bestimmt ist, ein Polymon 2. Grades von $\sqrt{J_\tau} \cos 2\pi W_\tau$ wird.

Der nächste Schritt besteht darin, daß man den Mittelwert $\overline{\mathsf{H}_3}$ über die Elektronenbewegung bildet und dann G_1 aus der Differentialgleichung (13a) Kap. I berechnet, in der wegen (55) $\mathsf{E}_3 = 0$ zu setzen ist. G_1 wird vom 3. Grade in $\sqrt{J_\tau} \cos 2\pi W_\tau$ und $\sqrt{J_\tau} \sin 2\pi W_\tau$.

Nunmehr kann man $\overline{\Phi_4}$ nach (14) Kap. I bilden; man erhält ein Polynom 4. Grades in $\sqrt{J_\tau} \cos 2\pi W_\tau$ und $\sqrt{J_\tau} \sin 2\pi W_\tau$. Bildet man schließlich nach (15) Kap. I den Mittelwert von $\overline{\Phi_4}$ über die Kernbewegungen

$$\overline{\overline{\Phi_4}} = \mathsf{E}_4,$$

so bekommt man ein Polynom der $\sqrt{J_\tau}$, das Glieder nullter, zweiter und vierter Ordnung in $\sqrt{J_\tau}$ enthält; die Koeffizienten der beiden ersten Arten sind außer von den Elektroneninte- gralen $j_1, j_2, \ldots$ noch von den Kreiselintegralen J, J' abhängig, die Glieder vierter Ordnung nur von den $j_1, j_2, \ldots$ Wir können daher schreiben:

$$\mathsf{E}_4 = \mathsf{E}_0^{(1)}(J, J') + \sum_\tau J_\tau \nu_\tau^{(1)}(J, J') + \tfrac{1}{2} \sum_{\tau\sigma} \nu_{\tau\sigma}^0 J_\tau J_\sigma. \tag{56}$$

Bricht man das Näherungsverfahren an dieser Stelle ab, so erhält man für die gesamte Energie der Molekel folgende Darstellung:

$$(57)\left\{\begin{aligned} \mathsf{E} = {} & [\mathsf{E}_0 + \lambda^4 \mathsf{E}_0^{(1)}(J, J')] + \lambda^2 \{\mathsf{K}(J J') \\ & + \sum_\tau [\nu_\tau^0 + \lambda^2 \nu_\tau^{(1)}(J, J')] J_\tau\} + \lambda^4 \cdot \tfrac{1}{2} \sum_{\tau\sigma} \nu_{\tau\sigma}^0 J_\tau J_\sigma. \end{aligned}\right.$$

Die Formel (57) hat bezüglich der Abhängigkeit von den J_τ genau dieselbe Form, wie die Energie eines Resonatorensystems bei Berücksichtigung der Glieder bis zur 4. Ordnung in der potentiellen Energie.[1]) Die Koeffizienten sind hier aber nicht Konstanten, sondern Funktionen der Quanten-

1) Vgl. M. Born u. E. Brody, Zeitschr. f. Physik **6.** S. 140. 1921; vgl. auch M. Born u. E. Hückel, Phys. Zeitschr. **24.** S. 1. 1923.

zahlen, die zu den Elektronenbewegungen und den Kreiselungen gehören. Die Abhängigkeit von den Elektronen-Quantenzahlen läßt sich natürlich überhaupt nur festlegen, wenn ein bestimmtes Molekelmodell gegeben ist. Die Abhängigkeit von den Quantenzahlen der Kreiselbewegungen läßt sich im Prinzip vollständig berechnen, doch sind im allgemeinen Falle die Rechnungen viel zu verwickelt, als daß ihre Durchführung beim jetzigen Stande der Beobachtungen über Bandenspektren lohnend erschiene. Wesentliche Vereinfachungen treten nur in dem Falle ein, daß die Gleichgewichtslagen aller Kerne auf einer Geraden angeordnet sind; wir werden auf diesen Fall, der in der vorliegenden Arbeit ausdrücklich ausgeschlossen worden ist, später zurückkommen. Hier möchten wir nur so viel sagen, daß dann neue Verwicklungen durch das Auftreten von Entartungen entstehen. Dies gilt schon für 2atomige Molekeln, wenn man mit Kratzer[1]), Kramers und Pauli[2]) annimmt, daß der mittlere Drehimpuls der Elektronen $\bar{\mathfrak{g}}$ nicht in die Kernverbindungslinie fällt; das ist in der Tat nur möglich, wenn die zyklische Koordinate der Elektronenbewegung (Azimut um die Kernachse) entweder eigentlich oder zufällig entartet ist[3]), ein Umstand, der von den genannten Autoren nicht beachtet worden ist. Bei drei- und mehratomigen Molekeln kommt eine weitere Entartung hinzu, die die auf der Kernachse senkrechten Schwingungen der Kerne betrifft. Dieser Fall soll im Hinblick auf die Möglichkeit, daß er bei Molekeln wie H_2O oder CO_2 verwirklicht ist, ausführlich untersucht werden.

Zusammenfassung.

1. Eine Molekel wird als ein aus Kernen und Elektronen bestehendes mechanisches System betrachtet, das im dynamischen Gleichgewicht ist; aus dieser allgemeinen Vorstellung heraus soll die Theorie der Bandenspektren systematisch entwickelt werden.

2. Die mathematische Durchführung dieser Aufgabe beruht darauf, daß die Energiefunktion nach Potenzen von λ

1) A. Kratzer, Ann. d. Phys. **71.** S. 72. 1923.

2) a. a. O.

3) M. Born u. W. Heisenberg, Zeitschr. f. Physik **14.** S. 44. 1923

entwickelt wird, wo λ^2 von der Größenordnung des Verhältnisses Elektronenmasse zu Kernmasse ist.

3. Gesucht werden solche Lösungen des Bewegungsproblems, bei denen die Kerne sich von starr verbundenen Gleichgewichtslagen nur um Beträge der Größenordnung λ entfernen. Von den Gleichgewichtslagen der Kerne wird angenommen, daß nicht alle in gerader Linie liegen; dieser entartete Fall soll später behandelt werden. Der Quantentheorie wird dadurch Rechnung getragen, daß für die gesuchten Bewegungen alle Impulse und Drehimpulse für $\lambda = 0$ endlich bleiben sollen.

4. Zur mathematischen Formulierung der unter 3. genannten Forderungen wird ein bewegtes Koordinatensystem eingeführt, in dem die Gleichgewichtslagen der Kerne ruhen. Die sechs Bestimmungsstücke dieses Bezugssystems werden als überzählige Koordinaten behandelt; zur Aufrechterhaltung der Zahl der Freiheitsgrade dienen sechs Nebenbedingungen, die ausdrücken, daß das bewegte Koordinatensystem sich der wirklichen Kernbewegung möglichst anschmiegt.

5. Aus der Hamiltonschen Funktion dieses Problems lassen sich die den Nebenbedingungen entsprechenden Lagrangeschen Multiplikatoren eliminieren.

6. Das mechanische Problem läßt sich dann mit den Methoden der Störungstheorie behandeln. Das ungestörte System wird von den Elektronen bei festgehaltenen Kernen gebildet. Die Koordinaten der Kerne spielen die Rolle entarteter Variabeln.

7. Die Störungsfunktion 1. Ordnung hat infolge der Forderung, daß Gleichgewichtslagen der Kerne existieren sollen, einen verschwindenden Mittelwert. Daher versagt das gewöhnliche Verfahren zur Festlegung der säkularen Störungen der Kernkoordinaten. Die hier einzuschlagende Methode ist in abstracto im Kap. I auseinandergesetzt; sie läuft darauf hinaus, daß man durch eine geeignete kanonische Substitution die Störungsfunktion 1. Ordnung zum identischen Verschwinden bringt.

8. Die Störungsfunktion 2. Ordnung bestimmt zugleich die Kernschwingungen und Kreiselbewegungen als säkulare Störungen. Die Separation der Variabeln für die Schwingungen läßt sich in bekannter Weise durch Einführung von Normal-

koordinaten erreichen. Die Kreiselglieder stimmen mit der Energie eines „Kreiselkörpers mit eingebautem Schwungrad" überein, die Kramers und Pauli untersucht haben; sie lassen sich nach den von diesen Forschern angegebenen Methoden durch Separation der Variabeln behandeln.

9. Die Störungsfunktion 3. Ordnung liefert keinen Beitrag zur Energie; die 4. Ordnung einen solchen, der von den Quantenzahlen der Kernschwingungen quadratisch, von denen der Elektronenbewegung und der Rotation in sehr komplizierter Weise abhängt.

10. Der in dieser Abhandlung ausgeschlossene Fall, daß die Kerne im Gleichgewicht auf einer geraden Linie liegen, führt auf neue Entartungen; und zwar ist im Falle 2atomiger Molekeln, deren Drehimpuls nicht in die Kernachse fällt, die zu ihm konjugierte zyklische Variable entartet, bei mehratomigen Molekeln sind es überdies die Kernschwingungen senkrecht zur Achse.

(Eingegangen 21. Dezember 1923.)

Über den Einfluß der Deformierbarkeit der Ionen auf optische und chemische Konstanten. I[1]).

Von **M. Born** und **W. Heisenberg** in Göttingen.

Mit einer Abbildung. (Eingegangen am 22. März 1924.)

Unter der Deformierbarkeit α eines Atoms oder Ions verstehen wir den Betrag des elektrischen Momentes, den es im Felde 1 annimmt. — Der Wert von α läßt sich aus den Rydberg- und Ritz-Korrektionen der Serienspektra ableiten. Man erhält dabei verschiedene Werte, je nachdem man die azimutale Quantenzahl k ganz- oder halbzahlig annimmt. Die so erhaltenen α-Werte genügen mit beträchtlicher Genauigkeit dem Gesetze $\alpha = C Z_{\text{eff.}}^{-3}$, wo die zur Bestimmung von $Z_{\text{eff.}}$ benutzte Abschirmungszahl ungefähr mit der für die Röntgenterme gültigen übereinstimmt. Die Einordnung der aus der Atomrefraktion bestimmten α-Werte der Edelgase in dieses Gesetz führt dazu, daß k halbzahlig gewählt werden muß. Unter Heranziehung der Refraktion der Salze läßt sich dann eine vollständige Tabelle der α-Werte aller edelgasähnlichen Ionen aufstellen; mit Hilfe dieser Zahlen wird die Sublimationswärme der Alkalihaloide in guter Übereinstimmung mit der Beobachtung berechnet; die wichtigsten Eigenschaften der Salzmolekel lassen sich dann numerisch vorhersagen.

Einleitung. Während die Quantentheorie für die Aufklärung des Atombaues sehr viel geleistet hat, hat sie zur Frage des Molekelbaues bis jetzt nur wenig beitragen können. Daher war der Gedanke von Kossel besonders fruchtbar, die Betrachtung zunächst auf solche Molekelbindungen zu beschränken, bei denen die Quantentheorie eine möglichst geringe Rolle spielt, nämlich auf die streng polaren Bindungen, die in der Hauptsache durch die elektrostatischen Anziehungen der Ionen zustande kommen. Kossel konnte eine große Menge qualitativer Resultate über die chemischen Kräfte auf diesem Wege gewinnen; für quantitative Überlegungen mußte man sich allerdings auf die einfachsten symmetrischen Gitter beschränken und auch dann nur solche Eigenschaften der Gitter ins Auge fassen, bei denen keine asymmetrische Beanspruchung und daher keine einseitige Deformation der Ionen möglich ist.

Es zeigte sich bald, daß viele mechanische, elektrische und optische Eigenschaften der Kristalle mit dem einfachen Ansatz, bei dem die Ionen als elektrische Punktladungen angesehen werden, nicht erfaßt

[1]) Der wesentliche Inhalt dieser Arbeit und von Teilen der folgenden wurde auf der Sitzung des Gauvereins Niedersachsen der Deutschen Physikalischen Gesellschaft in Braunschweig am 9. Februar vorgetragen.

werden können [z. B. der Unterschied der Elastizitätskonstanten $c_{12} - c_{44}$ bei gewissen regulären Kristallen, die piezo-elektrische Konstante, der Zusammenhang zwischen Dielektrizitätskonstante und Reststrahl-Wellenlänge usw.] [1]); diese Diskrepanzen wurden auf Deformationen der Ionen zurückgeführt, und es konnte gezeigt werden, daß sie sich dadurch in der Tat formal beheben lassen [2]). Kossel selbst bemerkte, daß auch die chemische Valenztätigkeit mit der Deformation der Ionen im Zusammenhang steht, wobei er als Maß der Deformation die Lichtbrechung ansah [3]). In ganz anderer Richtung hat Debye die Deformierbarkeit der Atome zur Erklärung einer universellen Eigenschaft der Substanzen herangezogen, nämlich der van der Waalsschen Kohäsion der Gase [4]); es gelang ihm, quantitative Beziehungen zwischen der auch von ihm als Maß der Deformation betrachteten Refraktionskonstanten und der van der Waalsschen Kohäsionskonstanten abzuleiten. Qualitative Betrachtungen über den Zusammenhang von chemischer Konstitution und Drehungsvermögen spielen in der physikalischen und chemischen Literatur seit langer Zeit eine große Rolle. So hat z. B. Heydweiller [5]) seine ausführlichen Untersuchungen über das Brechungsvermögen elektrolytischer Lösungen unter gleichen Gesichtspunkten ausgeführt. Neuerdings ist das ganze Material besonders von Wasastjerna [6]) und von Fajans und Joos [7]) durch-

[1]) Siehe M. Born u. E. Bormann, Ann. d. Phys. **62**, 218, 1920. Siehe insbesondere S. 224.

[2]) M. Born, Phys. ZS. **19**, 539, 1918. Siehe insbesondere § 3.

[3]) W. Kossel, Sitzungsber. d. Bayr. Akad., math.-naturw. Kl., 1917, S. 127. Auf Grund der zitierten Arbeiten von Kossel und Born ist wohl eine Äußerung von Fajans und Joos in der nachher zu erwähnenden Abhandlung nicht ganz zutreffend, die dahin lautet, jene Arbeiten seien als ein Versuch anzusehen, die Bindungsart in Molekeln und Kristallen salzartiger Verbindungen auf eine rein elektrostatische Anziehung starrer Ionen zurückzuführen. Daß von den für Ionendeformation empfindlichen chemischen und physikalischen Eigenschaften zunächst abgesehen wurde, beruhte auf der Erkenntnis, daß in energetischer Hinsicht die Coulombschen Kräfte jedenfalls als erste Annäherung zu gelten haben. Eine Berücksichtigung der auf den Deformationen beruhenden nächsten Näherung konnte erst erfolgen, wenn die Resultate der ersten Approximation hinreichend gesichert schienen und die Methoden zur Fortsetzung des Verfahrens ausgebildet waren.

[4]) P. Debye, Phys. ZS. **21**, 178, 1920.

[5]) A. Heydweiller, Ann. d. Phys. **41**, 499, 1913; **48**, 681, 1915; **49**, 653, 1916; Verh. d. D. Phys. Ges. **16**, 722, 1914.

[6]) J. A. Wasastjerna, Comm. Fenn. **1**, 7, 1913, Nr. 37. Da uns die Originalarbeit nicht zugänglich war, haben wir das Referat von Kauffmann in den Phys. Ber. **5**, 226, 1924 benutzt.

[7]) K. Fajans, Die Naturwissenschaften **10**, 165, 1923; K. Fajans und G. Joos, Molrefraktion von Ionen und Molekülen im Lichte der Atomstruktur, ZS. f. Phys. **23**, 1—46, 1924. Die Autoren hatten die Freundlichkeit, uns Einblick in die Korrekturen zu geben.

gearbeitet worden; diese Autoren haben Tabellen für die Refraktionskonstanten der einzelnen Ionen aufgestellt. Fajans und Joos haben besonders das Ziel verfolgt, den Zusammenhang der Veränderlichkeit dieser Konstanten in verschiedenen Verbindungen des betrachteten Ions mit der Art der chemischen Bindung aufzuklären.

Die vorliegende Arbeit stellt sich die Aufgabe, den Begriff der Ionendeformation mit der Quantentheorie der Atome in Verbindung zu bringen und die so gewonnenen Zahlen dazu zu verwerten, quantitative Schlüsse auf die physikalischen und chemischen Eigenschaften der polar gebauten Substanzen zu ziehen.

Unter all den komplizierten Deformationen, die ein Ion unter der Einwirkung seiner Umgebung erleiden kann, ist die einfachste diejenige, welche unter der Einwirkung eines homogenen elektrischen Feldes zustande kommt. Als Maß derselben betrachten wir die „Deformierbarkeit" α, nämlich das Verhältnis des durch ein homogenes elektrisches Feld $\mathfrak{E}$ erzeugten Momentes $\mathfrak{p}$ zur Größe des Feldes: $\mathfrak{p} = \alpha \mathfrak{E}$. Wir werden im wesentlichen nur solche Eigenschaften der Substanzen ins Auge fassen, bei denen α als eine dem Ion eigentümliche Konstante aufgefaßt werden kann (Deformationseffekte erster Ordnung). Ihnen gegenüber kann man einen großen Teil der Untersuchungen von Fajans und Joos dadurch kennzeichnen, daß sie die Effekte zweiter Ordnung studieren, die mit einer Änderung von α verbunden sind (z. B. den Übergang eines Ions aus der Lösung in das feste Salz oder den Gaszustand, die Verschiedenheit des α-Wertes des Ions in verschiedenen Verbindungen).

Für die Bestimmung der α-Werte gehen wir von einer neuen Methode aus, die den Zusammenhang zwischen der Deformierbarkeit der Ionen mit den Konstanten der Serienspektren benutzt. Als wichtiges Nebenresultat ergibt sich dabei ein neuer Beleg für die von der Theorie der Zeemaneffekte geforderte Halbzahligkeit der azimutalen Quantenzahl k. Unter Heranziehung der Refraktionen der Edelgase und der Salze stellen wir eine Tabelle der α-Werte der edelgasähnlichen Ionen auf, bei der sich ein einfaches Gesetz für die Abhängigkeit von der Kernladungszahl ergibt. Als Beispiel für die Anwendbarkeit dieser Tabellen berechnen wir die Sublimationswärme zweiatomiger Salze und alle wichtigen Eigenschaften der Salzdampfmolekeln. In folgenden Arbeiten sollen kompliziertere molekulare Gebilde in ähnlicher Weise untersucht werden.

§ 1. **Abhängigkeit der Rydbergkorrektion von der Deformierbarkeit des Atomrumpfes.** Die Terme der Serienspektren lassen sich bekanntlich in ihrer Abhängigkeit von der Laufzahl (Hauptquantenzahl) durch die Annahme eines zentralsymmetrischen Kraftfeldes mit beträchtlicher Genauigkeit darstellen [1]). Hierbei sind nach Schrödinger [2]) und Bohr [3]) zwei Arten von Bahnen des Leuchtelektrons zu unterscheiden, solche, die ganz außerhalb des Atomrumpfes verlaufen (Bahnen erster Art), und solche, die in den Rumpf eindringen (Bahnen zweiter Art oder Tauchbahnen). Für die Bahnen erster Art kann man das Zentralfeld nach Potenzen von $1/r$ entwickeln [4]); im folgenden werden wir nur mit diesem Fall zu tun haben.

Wir wollen nur solche Spektren betrachten, bei denen der Rumpf edelgasartig ist; also z. B. Spektren von Na, Mg^+, Al^{++}, Si^{+++} und die analogen Reihen.

Wir diskutieren jetzt die wichtigsten Möglichkeiten für die Abweichungen des einen solchen Rumpf umgebenden Feldes vom Coulombschen Feld.

Zunächst weiß man aus der Gittertheorie, daß die Entwicklung des Potentials eines edelgasähnlichen Rumpfes wegen seiner Würfelsymmetrie mit dem Gliede r^{-5} anfängt; niedrigere Potenzen können in merklichem Betrage nicht vorkommen, soweit die Rückwirkung des Leuchtelektrons auf den Rumpf vernachlässigt wird. Sobald man diese aber in Rechnung setzt, erhält man, wie leicht zu sehen, ein Glied mit r^{-4}; dieses hat Bohr bereits zur Abschätzung der Größenordnung der Rydbergkorrektion bei Lithium benutzt. Man darf nämlich annehmen, daß der Rumpf in einem äußeren homogenen, elektrischen Felde $\mathfrak{E}$ ein elektrisches Moment $p = \alpha |\mathfrak{E}|$ bekommt; diese Konstante α hängt in einfacher Weise mit der Dielektrizitätskonstante zusammen. Wenn nun das Leuchtelektron hinreichend weit vom Rumpf entfernt ist, kann man das von ihm erregte Feld $|\mathfrak{E} = \frac{e}{r^2}$ im Bereich des Rumpfes als homogen betrachten. Man erhält also ein Moment $p = \frac{\alpha e}{r^2}$. Dieses Moment übt wieder eine Anziehungskraft auf das Leuchtelektron im Betrage von $\frac{2pe}{r^3} = \frac{2\alpha e^2}{r^5}$ aus; ihr Potential ist $-\frac{\alpha e^2}{2r^4}$.

---

[1]) S. E. Fues, ZS. f. Phys. **11**, 364, 1922 und **12**, 1, 1922.
[2]) E. Schrödinger, ebenda **4**, 347, 1921.
[3]) N. Bohr, Ann. d. Phys. **71**, 228, 1923.
[4]) A. Sommerfeld, Atombau und Spektrallinien, 3. Aufl., 1922, Zusatz 18, S. 721.

Wir wollen nun diese beiden niedersten Glieder

$$\frac{b_4}{r^4} + \frac{b_5}{r^5} \tag{1}$$

ins Auge fassen und zunächst die ihnen entsprechende Serienformel ausrechnen. Nach der Theorie der säkularen Störungen erzeugen diese beiden Glieder eine Zusatzenergie zur Energie der Keplerbewegung, die man durch Mittelung über die ungestörte Bewegung erhält:

$$b_4 \overline{\frac{1}{r^4}} + b_5 \overline{\frac{1}{r^5}}\,.$$

Ist Z die effektive Kernladung (Ladung des Rumpfes: $\mathrm{Na}|Z = 1$, $\mathrm{Mg}^+|Z = 2$, $\mathrm{Al}^{++}|Z = 3$ usw.), a_H der Radius der einquantigen Wasserstoffbahn, n die Hauptquantenzahl und k die azimutale Quantenzahl, so erhält man in bekannter Weise:

$$\left.\begin{aligned} \overline{\frac{1}{r^4}} &= \frac{3 - \frac{k^2}{n^2}}{2\,a_H^4 n^3 k^5} Z^4, \\ \overline{\frac{1}{r^5}} &= \frac{5 - \frac{3\,k^2}{n^2}}{2\,a_H^5 n^3 k^7} Z^5. \end{aligned}\right\} \tag{2}$$

Die Energie W ist also bestimmt durch

$$W = -\frac{RhZ^2}{n^2} + b_4 \frac{3 - \frac{k^2}{n^2}}{2\,a_H^4 n^3 k^5} Z^4 + b_5 \frac{5 - \frac{3\,k^2}{n^2}}{2\,a_H^5 n^3 k^7} Z^5.$$

Wir schreiben das in der Rydberg-Ritzschen Form:

$$-\frac{W}{Rh} = \frac{Z^2}{\left(n + \delta_1 + \frac{\delta_2}{n^2}\right)^2}, \tag{3}$$

wo

$$\left.\begin{aligned} \delta_1 &= \frac{3\,Z^2 b_4}{4\,Rhk^5 a_H^4} + \frac{5\,Z^3 b_5}{4\,Rhk^7 a_H^5}, \\ \delta_2 &= -\frac{Z^2 b_4}{4\,Rhk^3 a_H^4} - \frac{3\,Z^3 b_5}{4\,Rhk^5 a_H^5} \end{aligned}\right| \tag{4}$$

sind.

Wir wollen nun die Größenordnung von b_5 aus den durch die Gittertheorie bekannten Dimensionen der Ionen abschätzen. Das Potential eines neutralen Elektronenwürfels, dessen Mittelpunkt (achtfach geladener Kern) im Koordinaten-Nullpunkt liegt und dessen Kanten den Koordinatenachsen parallel laufen, lautet in erster Näherung[1]):

$$\frac{14}{9} e a^4 \frac{5(x^4 + y^4 + z^4) - 3 r^4}{r^9},$$

wo a die halbe Diagonale des Würfels ist. Der Mittelwert dieses Ausdrucks über alle Lagen des Aufpunkts auf einer um den Nullpunkt geschlagenen Kugel ist Null. Nun weiß man aus den Zeemaneffekten, daß der Rumpf einen Drehimpuls hat, also mit relativ hoher Frequenz rotiert. Es wird sich daher der Einfluß der elektrostatischen Kräfte des Elektronenwürfels zum großen Teile wegmitteln. Um den maximalen Einfluß abzuschätzen, betrachten wir Lagen des Aufpunktes in einer der auf den Koordinatenachsen senkrechten Symmetrieebenen (z. B. in der xy-Ebene; $x = r \cos\varphi$, $y = r \sin\varphi$), wo das Potential den Ausdruck

$$\frac{14}{9} \frac{a^4 e}{r^5} (10 \cos^4\varphi - 3)$$

hat, und nehmen an, daß die Rotation des Rumpfes um die z-Achse erfolgt. Dann ist die mittlere Energie auf das äußere Elektron (Ladung $-e$):

$$-\frac{7}{6} \frac{a^4}{r^5} e^2,$$

also hat man

$$b_5 = -\frac{7}{6} a^4 e^2.$$

Die entsprechenden Ausdrücke von δ_1 und δ_2 sind:

$$\delta_1' = -\frac{35}{12} \frac{Z^3}{k^7} \left(\frac{a}{a_H}\right)^4,$$

$$\delta_2' = +\frac{7}{4} \frac{Z^3}{k^5} \left(\frac{a}{a_H}\right)^4.$$

Für die Größe a benutzen wir die Ergebnisse von Fajans, Grimm und Herzfeld[2]), die a aus den Dimensionen und Energien der Salzgitter bestimmt haben. Für Na^+ geben diese an: $a = 0{,}517 \cdot 10^{-8}$ cm, also mit $Z = 1$

$$\delta_1' = -\frac{2{,}63}{k^7}, \quad \delta_2' = \frac{1{,}57}{k^5}.$$

---

[1]) M. Born, Verh. d. D. Phys. Ges. 20, 230, 1918.

[2]) K. Fajans und H. Grimm, ZS. f. Phys. 2, 299, 1920; K. Fajans und K. F. Herzfeld, ebenda 2, 309, 1920; siehe auch A. Sommerfeld, Atombau und Spektrallinien, 3. Aufl., 2. Kap., § 7, S. 132.

Schon bei den d-Termen [$k = 3$ bzw.[1] $\frac{5}{2}$], erst recht bei den f-Termen ($k = 4$ bzw. $\frac{7}{2}$) und den höheren sind diese Werte δ'_1, δ'_2 klein gegen die beobachteten.

Da es sich überdies wohl um obere Grenzen handelt, so werden wir versuchsweise die Glieder mit b_5 bei den weit außen verlaufenden Bahnen weglassen. Wieweit dies berechtigt ist, kann man an dem empirischen Material selbst kontrollieren. Denn nach unseren Formeln muß bei Weglassung des b_5-Gliedes

$$\frac{\delta_2}{\delta_1} = -\frac{k^2}{3} \tag{5}$$

sein; nur wenn das annähernd der Fall ist, halten wir die Bestimmung von δ_1 und δ_2 aus dem vom induzierten Dipol herrührenden Gliede mit b_4 für berechtigt.

Wir haben oben die potentielle Energie der Dipolwirkung auf das äußere Elektron zu $-\frac{\alpha e^2}{2 r^4}$ berechnet, also ist

$$b_4 = -\frac{\alpha e^2}{2}. \tag{6}$$

Wir legen demnach den folgenden Rechnungen die aus (4) und (6) entstehenden Formeln

$$\delta_1 = -\frac{3 Z^2 e^2 \alpha}{8 R h a_H^4 k^5}, \quad \delta_2 = \frac{Z^2 e^2 \alpha}{8 R h a_H^4 k^3}$$

zugrunde; wir vereinfachen diese noch, indem wir

$$\frac{e^2}{2 a_H} = Rh$$

einsetzen. Dann wird:

$$\delta_1 = -\frac{3 Z^2 \alpha}{4 a_H^3 k^5}, \quad \delta_2 = \frac{Z^2 \alpha}{4 a_H^3 k^3}. \tag{7}$$

Wir stellen jetzt die für die folgenden Rechnungen benutzten spektroskopischen Daten zusammen:

Tabelle 1.

		Li	Na	Mg^+	Al^{++}	Si^{+++}	K	Ca^+	Rb	Sr^+	Cs	Ba^+
p	$-\delta_1$ . .	0,049	T	T	T	T	T	T	T	T	T	T
	δ_2 . .	0,031										
d	$-\delta_1$ . .	—	0,015	0,045	0,070	0,089	0,25	T	0,35	T	0,48 T?	T
	δ_2 . .	—	0,036	0,10	0,18	0,21	0,80		0,99		0,08	
f	$-\delta_1$ . .	—	0,0020	0,0044	0,0055	0,0066	0,009	0,027	0,036	0,055	0,032	T
	δ_2 . .	—	0,0064	0,016	0,020	0,032	0,035	0,16	0,35	0,26	0,16	

[1]) Auf die Frage der Ganz- oder Halbzahligkeit von k kommen wir unten zurück.

Die Zahlen sind im allgemeinen den Tabellen von Paschen und Götze und denen von Fowler entnommen. Die Berechnung der Korrektionen δ_1, δ_2 aus den dort angegebenen Termwerten erweist sich als sehr unsicher, sobald die Werte etwa 0,01 unterschreiten. Bei den *d*-Termen von Li z. B. zeigen die bei Paschen-Götze angegebenen Termgrößen einen so unvernünftigen Gang, daß sich daraus überhaupt keine δ_1, δ_2-Werte ableiten lassen [1]). Bei den *f*-Termen von Na, Mg^+, Al^{++}, Si^{+++} sind die Korrektionen der Arbeit von Fowler [2]) entnommen und wir möchten annehmen, daß sie trotz ihrer Kleinheit recht sicher sind.

Allgemein kann man vorsichtig die Genauigkeit aller in der Tabelle 1 angegebenen Korrektionen auf etwa 15 Proz. schätzen.

Das Zeichen T in der Tabelle 1 bedeutet „Tauchbahn".

Wenn man aus den Angaben der Tabelle 1 die α-Werte berechnen will, so muß man sich entscheiden, ob man die azimutale Quantenzahl k ganzzahlig annimmt, wie es bisher am gebräuchlichsten war, oder halbzahlig, wofür die Zeemaneffekte sprechen. Je nach dieser Entscheidung erhält man ein verschiedenes Verhältnis δ_2/δ_1 nach Formel (5). Die folgende Tabelle zeigt die theoretischen und die empirischen Werte dieses Quotienten $-\frac{\delta_2}{\delta_1}$:

Tabelle 2.

$-\delta_2/\delta_1$	$k^2/3$ k ganzz.	$k^2/3$ k halbz.	Li	Na	Mg^+	Al^{++}	Si^{+++}	K	Ca^+	Rb	Sr^+	Cs	Ba^+
p . . .	1,83	0,75	0,63	T	T	T	T	T	T	T	T	T	T
d . . .	3	2,08	—	2,4	2,2	2,6	2,4	3,2	T	2,8	T	0,06 T?	T
f . . .	5,33	4,08	—	3,2	3,6	3,6	4,8	3,9	5,9 ?	9,8 ?	4,7	5,0	T

Man sieht, daß, soweit es bei der Ungenauigkeit der Daten erwartet werden kann, das Verhältnis δ_2/δ_1 im allgemeinen richtig herauskommt; unsere Grundannahme: nahezu alleinige Wirksamkeit des Dipolpotentials, wird dadurch gerechtfertigt; die herausfallenden Werte haben wir durch Fragezeichen gekennzeichnet, was nur bedeuten soll, daß sie sich mit dem Dipolpotential $\sim r^{-4}$ nicht darstellen lassen.

[1]) So ist z. B. der Li-Term $8d$ auf $1/_{100}\,\mathrm{cm}^{-1}$ angegeben, während der Vergleich mit dem Wasserstoffspektrum zeigt, daß schon die 10000 mal größere Stelle ($100\,\mathrm{cm}^{-1}$) falsch sein muß.

[2]) A. Fowler, Proc. Roy. Soc. (A) **103**, 413, 1923.

Eine Entscheidung für ganz- oder halbzahlige k ist aus der Tabelle 2 wohl kaum zu treffen. Wir rechnen im folgenden zunächst mit beiden Annahmen. Man erhält so:

Tabelle 3.

$\alpha \cdot 10^{24}$	Li^+	Na^+	Mg^{++}	Al^{+++}	Si^{++++}	K^+	Ca^{++}	Rb^+	Sr^{++}	Cs^+	Ba^{++}
k ganzz. . .	0,314	0,405	0,23	0,125	0,083	1,68	?	?	2,74	6,48 ?	?
k halbz. . .	0,075	0,21	0,12	0,065	0,043	0,87	?	?	1,42	3,36 ?	?

Die Zahlen sind aus den f-Termen berechnet, mit Ausnahme von Li, wo der p-Term benutzt ist; ferner sind die δ_1-Werte zugrunde gelegt worden, weil diese wohl genauer als die δ_2-Werte sind. Man sieht, daß vom Na^+ aufwärts die ganzzahlige Wahl von k etwa doppelt so große Werte für α liefert als die halbzahlige.

§ 2. Deformierbarkeit und Refraktionskonstante der Edelgase und der Salze. Das in Tabelle 1 zusammengestellte Material erweitern wir nun durch die Deformationskonstanten der Edelgase, die man aus ihrem Brechungsvermögen berechnen kann. Ist n der Grenzwert des Brechungsindex für unendlich lange Wellen, N die Anzahl der Atome in der Volumeneinheit, so gilt die Lorentz-Lorenzsche Formel

$$\alpha = \frac{3}{4\pi N}\frac{n^2-1}{n^2+2}. \tag{8}$$

Bei einatomigen Gasen ist n^2 mit der Dielektrizitätskonstante identisch. Wir wollen aber nicht die elektrisch gemessene Dielektrizitätskonstante benutzen, sondern ziehen es vor, aus den optischen Messungen des Brechungsindex n_λ den Grenzwert n für unendlich lange Wellen nach der Formel

$$n_\lambda = n + \frac{a}{\lambda^2} \tag{8a}$$

zu extrapolieren. Nach den Angaben von Landolt-Börnstein erhält man:

Tabelle 4.

	He	Ne	A	Kr	X
$\alpha \cdot 10^{24}$ . .	0,202	0,392	1,629	2,46	4,00

Diese Zahlen kann man benutzen, um zwischen der halbzahligen oder ganzzahligen Wahl von k zu entscheiden. Die Deformierbarkeit α muß natürlich innerhalb jeder Periode mit wachsender Kernladungszahl abnehmen; tatsächlich sieht man das auch deutlich in der Reihe Na^+, Mg^{++}, Al^{+++}, Si^{++++} der Tabelle 3. Würde man nun k ganzzahlig wählen, so würden die Edelgase sich nicht richtig an die Ionen anschließen; z. B. hätte man für k ganzzahlig

	He	Li^+	Ne	Na^+	A	K^+
$\alpha \cdot 10^{24}$	0,202	0,314	0,392	0,405	1,629	1,68,

also stets die verkehrte Reihenfolge. Dagegen wird für k halbzahlig

	He	Li^+	Ne	Na^+	A	K^+
$\alpha \cdot 10^{24}$	0,202	0,075	0,392	0,21	1,629	0,87

die Reihenfolge und auch der Abfall der Größe richtig.

Wir erhalten also das wichtige Resultat:

Aus den Deformationskonstanten der Edelgase und der edelgasähnlichen Ionen folgt die Halbzahligkeit der Quantenzahl k[1]).

Weiteres Material gewinnen wir mit Hilfe der Refraktion der Salze. Diese setzt sich aus zwei Anteilen zusammen; der erste beruht auf der Verschiebung der Ionen im Gitter als ganze, der zweite auf der Deformation der einzelnen Ionen. Der erste Anteil liefert den Anstieg des Brechungsexponenten beim Durchgang durch die ultrarote Eigenfrequenz des Gitters (Reststrahl) und läßt sich leicht abtrennen; um den zweiten Anteil (Ionendeformation) rein zu bekommen, hat man einfach den Brechungsindex n vom Sichtbaren bis ins kurzwellige Ultrarot zu nehmen bis zu der Stelle, wo sich das erneute Absinken, verursacht durch die Reststrahlfrequenz, bemerkbar macht, und ihn durch eine einfache Dispersionsformel, wie (8a), darzustellen. Die Größe n bestimmt die Summe der Deformationskonstanten der einzelnen Ionen $\Sigma\alpha$ des Salzes wieder durch die Formel[2]) (8), wobei man N durch die Gitterkonstante δ ausdrücken kann.

Wir entnehmen die δ-Werte der Zusammenstellung in dem Buche von Ewald[3]), die Brechungsindizes der Alkalihaloide einer

---

[1]) Dieser Schluß hängt offenbar wesentlich von der Möglichkeit ab, die Quantenzahl k in Analogie zur Theorie der mehrfach periodischen Systeme mit dem mechanischen Impulsmoment des Elektrons zu identifizieren und die Mechanik im Sinne der Gleichungen (2) anzuwenden. Ob sich die wirkliche Quantenzahl später einmal als ganz herausstellen wird, kann hierdurch nicht entschieden werden.

[2]) Die Gültigkeit der Formel (8) bei Kristallen ist auf kubische Gitter beschränkt, wo der Einfluß der Nachbarionen auf ein Ion durch die „Lorentzsche Kraft" $\frac{4\pi}{3}\mathfrak{P}$ dargestellt ist (s. M. Born, Atomtheorie des festen Zustandes, § 43, S. 773. Leipzig 1923).

[3]) P. P. Ewald, Kristalle und Röntgenstrahlen. Berlin 1923.

Arbeit von Spangenberg [1]), der sie für die drei Linien $H_\alpha (\lambda = 656)$, $Na D (\lambda = 589)$, $H_\beta (\lambda = 486)$ gemessen hat. Die Extrapolation auf $n (\lambda = \infty)$ haben wir graphisch ausgeführt (in einigen zweifelhaften Fällen zur Kontrolle mit der Methode der kleinsten Quadrate).

Tabelle 5.

$\delta \,.\, 10^8$	F	Cl	Br	J
Li	4,02	5,14	5,49	6,06 [2])
Na	4,63	5,63	5,97	6,48
K	5,38	6,26	6,59	7,08
Rb	3,66 [3])	6,57	6,90	7,33
Cs	6,02	4,10 [3])	4,30 [3])	4,56 [3])

Tabelle 6.

n	F	Cl	Br	J
Li	1,3846	1,6360	1,742	1,883
Na	1,3198	1,5255	1,608	1,718
K	1,3547	1,4725	1,534	1,625
Rb	1,3897 [3])	1,476	1,528	1,607
Cs	1,4683	1,620 [3])	1,669 [3])	1,739 [3])

Hieraus berechnen wir folgende $\Sigma \alpha$-Werte:

Tabelle 7.

$\Sigma \alpha \,.\, 10^{24}$	F	Cl	Br	J
Li	0,923	2,92	3,98	6,10
Na	1,180	3,26	4,37	6,41
K	2,02	4,12	5,32	7,48
Rb	2,78	4,78	6,04	8,11
Cs	3,62	5,79	7,10	9,15

Aus den ersten drei Horizontalreihen dieser Tabelle finden wir durch Abziehen der aus Tabelle 3 entnommenen α-Werte von Li, Na, K je drei α-Werte für die Halogenionen:

Tabelle 8.

$\alpha \,.\, 10^{24}$	F^-	Cl^-	Br^-	J^-
aus dem Li-Salz . .	0,848	2,84	3,90	6,02
„ „ Na- „ . .	0,97	3,05	4,16	6,20
„ „ K- „ . .	1,15	3,25	4,45	6,61
Mittel	0,99	3,05	4,17	6,28

[1]) K. Spangenberg, ZS. f. Kristallogr. **57**, 517, 1923.

[2]) Eine zweite Angabe von Davey ($\delta = 7{,}074 \,.\, 10^{-8}$) stimmt nicht mit dem aus der Dichte (4,061) berechneten Werte ($\delta = 6{,}01 \,.\, 10^{-8}$).

[3]) Dieses Gitter ist vom CsCl-Typ.

Die Zahlen jeder Spalte zeigen einen systematischen Gang, der darauf zurückzuführen ist, daß die Ionen in den verschiedenen Gittern nicht vollständig gleich sind, sondern selbst eine allseitige Kompression oder Dilatation erfahren. Es ist dies einer der Effekte, die Fajans und Joos in ihrer zitierten Arbeit genauer untersuchen, die aber für den Zweck der vorliegenden Arbeit als „Effekte zweiter Ordnung" unberücksichtigt bleiben sollen. Von diesem Standpunkte aus bilden wir die Mittelwerte jeder Spalte.

Die so gefundenen Mittel für Halogenionen ziehen wir von den Zahlen der Tabelle 7 zeilenweise ab und erhalten dadurch α-Werte für die Alkaliionen, von denen die ersten drei zur Kontrolle unserer früheren Zahlen dienen, während die letzten beiden für Rb^+, Cs^+ neu sind.

Tabelle 9.

$\alpha \cdot 10^{24}$	Aus dem F-Salz	Aus dem Cl-Salz	Aus dem Br-Salz	Aus dem J-Salz	Mittel	Aus dem Spektrum
Li^+	— 0,07	— 0,13	— 0,19	— 0,18	— 0,14	0,075
Na^+	0,19	0,21	0,20	0,13	0,18	0,21
K^+	1,03	1,07	1,15	1,20	1,11	0,87
Rb^+	1,79	1,73	1,87	1,83	1,81	—
Cs^+	2,63	2,74	2,93	2,87	2,79	(3,36?)

Daß der α-Wert für Li hier negativ herauskommt, beruht auf seiner Kleinheit im Verhältnis zu den gesamten α-Werten der Salze und auf dem soeben erwähnten Gang in den Spalten von Tabelle 8.

Wir stellen die Resultate noch einmal zusammen [1]).

Tabelle 10.

		He	Li^+			
$\alpha \cdot 10^{24}$ =		0,202	0,075			
	F^-	Ne	Na^+	Mg^{++}	Al^{+++}	Si^{++++}
$\alpha \cdot 10^{24}$ =	0,99	0,392	0,21	0,12	0,065	0,043
	Cl^-	A	K^+			
$\alpha \cdot 10^{24}$ =	3,05	1,629	0,87			
	Br^-	Kr	Rb^+	Sr^{++}		
$\alpha \cdot 10^{24}$ =	4,17	2,46	1,81	1,42		
	J^-	X	Cs^+			
$\alpha \cdot 10^{24}$ =	6,28	4,00	2,79			

[1]) In Anbetracht der Vernachlässigung jener mehrfach genannten „Effekte zweiter Ordnung" begnügen wir uns mit einer Genauigkeit von etwa 10 Proz. Bei den Metallen ziehen wir die spektroskopischen α-Werte vor, soweit sie vorhanden sind. Das große Material über Refraktion von Ionen in Lösungen, das besonders von Heydweiller und seinen Schülern gesammelt worden ist, haben wir unberücksichtigt gelassen. Man findet es ausführlich diskutiert in der zitierten Arbeit von Fajans und Joos.

§ 3. Abhängigkeit der Deformation von der Kernladung. Die Zahlen der Tabelle 10 zeigen einen sehr regelmäßigen Gang; man erkennt die Zunahme der Festigkeit der Elektronenhülle mit wachsender Kernladung, und man sieht auch deutlich die Abnahme der Festigkeit in den höheren Perioden.

Über die Änderung der Festigkeit innerhalb einer Periode, d. h. die Abhängigkeit des α von der Atomnummer Z, gibt es kein streng gültiges Gesetz. Für den einfachen Fall des Bohrschen Heliummodells hat Kramers[1]) die Polarisierbarkeit berechnet und gefunden, daß α proportional der dritten Potenz des Radius der deformierbaren Elektronenschale, also etwa umgekehrt proportional der dritten Potenz der effektiven Kernladungszahl $Z_{\text{eff.}}$ ist. Dieses Gesetz

$$\alpha = C \cdot Z_{\text{eff.}}^{-3} \tag{9}$$

dürfte weitgehend unabhängig von dem speziellen Modell sein. Dazu führt z. B. folgende grobe Dimensionsbetrachtung: α hat die Dimension eines Volumens, wird also proportional a^3 sein, wenn a der Radius der äußeren Schale ist, auf deren Deformation es hauptsächlich ankommt; a ist aber umgekehrt proportional zu $Z_{\text{eff.}}$.

Man kann nun zeigen, daß dieses Gesetz tatsächlich recht genau zutrifft, wenn man eine vernünftige Wahl der effektiven Kernladung trifft. Bei He und Li^+ sind nur die beiden $K(1_1)$-Elektronen vorhanden, die den deformierbaren Rumpf bilden. Man wird daher erwarten, daß ungefähr für He $Z_{\text{eff.}} = 2$, für Li $Z_{\text{eff.}} = 3$ zu setzen ist.

In den höheren Perioden hängt die Deformierbarkeit hauptsächlich von den am weitesten nach außen dringenden Elektronenbahnen ab; eine genaue theoretische Berechnung der Abschirmungszahl ist natürlich unausführbar. Einen Anhaltspunkt hat man in den Abschirmungszahlen, die man zur Darstellung der Röntgenterme (Ablösungsarbeit eines Elektrons aus der Edelgaskonfiguration) gebraucht. Für die zweite Periode läßt sich diese Zahl aus den Angaben von Bohr und Coster[2]) relativ genau berechnen. Die folgende Tabelle enthält die von ihnen angegebenen Werte für die Quadratwurzeln aus den (in Einheiten der Rydbergzahl) gemessenen Termen für die ersten Glieder der L-Reihe, ergänzt durch den von Hertz[3]) bestimmten Wert für Ne:

Tabelle 11.

	Ne	Na	Mg	Al	Si	P	S	Cl
$Z =$	10	11	12	13	14	15	16	17
$\sqrt{\frac{T}{R}} =$	1,26	—	1,87	2,28	—	3,15	3,44	3,84

[1]) H. A. Kramers, ZS. f. Phys. **13**, 312, 1923.
[2]) N. Bohr und D. Coster, ebenda **12**, 342, 1923.
[3]) G. Hertz, ebenda **18**, 307, 1923.

Sucht man nun diese Terme in der Form

$$\frac{T}{R} = A(Z-\gamma)^2$$

darzustellen, so findet man mit der Methode der kleinsten Quadrate $\gamma = 6{,}8$.

Wir haben gefunden, daß zur Darstellung der α-Werte die Abschirmungszahl $\gamma = 6$ am besten paßt.

Für die folgenden Perioden sind die Röntgenterme zur quantitativen Bestimmung von γ nicht zu brauchen; doch haben wir uns an der bekannten Figur von Bohr und Coster überzeugt, daß die von uns zur Darstellung der α-Werte gewählten Abschirmungszahlen auch für die Röntgenterme ungefähr richtig sind. Wir wählen:

	γ		γ
K-Schale	0	N-Schale	28
L- „	6	O- „	46
M- „	13		

Damit erhält man folgende Tabelle:

Tabelle 12.

			He	Li^+	Be^{++}	B^{+++}		
$Z_{\text{eff.}}$			2	3	4	5		
$10^{24}\,\alpha$			0,202	0,075				Mittel
$C = \alpha Z^3_{\text{eff.}}$			1,616	2,02				von C:
$10^{24}\,\alpha_{\text{ber.}}$			0,228	0,067	0,028	0,0145		1,82
	O^{--}	F^-	Ne	Na^+	Mg^{++}	Al^{+++}	Si^{++++}	
$Z_{\text{eff.}}$	2	3	4	5	6	7	8	
$10^{24}\,\alpha$		0,99	0,392	0,21	0,12	0,065	0,043	Mittel
$C = \alpha Z^3_{\text{eff.}}$		26,8	25,1	26,2	25,9	22,3	22,0	von C:
$10^{24}\,\alpha_{\text{ber.}}$	(3,1)	0,91	0,386	0,197	0,114	0,072	0,048	24,7
	S^{--}	Cl^-	A	K^+	Ca^{++}	Sc^{+++}	Ti^{++++}	
$Z_{\text{eff.}}$	3	4	5	6	7	8	9	
$10^{24}\,\alpha$		3,05	1,63	0,87				Mittel
$C = \alpha Z^3_{\text{eff.}}$		195	204	188				von C:
$10^{24}\,\alpha_{\text{ber.}}$	(7,25)	3,06	1,57	0,91	0,57	0,38	0,27	196
	Se^{--}	Br^-	Kr	Rb^+	Sr^{++}	Y^{+++}	Zr^{++++}	
$Z_{\text{eff.}}$	6	7	8	9	10	11	12	
$10^{24}\,\alpha$		4,17	2,46	1,81	1,42			Mittel
$C = \alpha Z^3_{\text{eff.}}$		1430	1260	1320	1420			von C:
$10^{24}\,\alpha_{\text{ber.}}$	6,4	4,04	2,70	1,90	1,38	1,04	0,80	1380
	Te^{--}	J^-	X	Cs^+	Ba^{++}	La^{+++}	Ce^{++++}	
$Z_{\text{eff.}}$	6	7	8	9	10	11	12	
$10^{24}\,\alpha$		6,28	4,00	2,79				Mittel
$C = \alpha Z^3_{\text{eff.}}$		2150	2050	2035				von C:
$10^{24}\,\alpha_{\text{ber.}}$	9,6	6,05	4,05	2,85	2,08	1,56	1,20	2080

Hierin sind aus den vorher bestimmten α-Werten (Tab. 10) die Koeffizienten $C = \alpha Z^3_{\text{eff.}}$ berechnet, dann aus diesen die Mittel gebildet und mit deren Hilfe die α-Werte zurückberechnet und für die benachbarten Ionen extrapoliert. Die Extrapolation nach wachsender Kernladung scheint uns recht sicher, die nach abnehmender aber bedenklich, vor allem bei O^{--} und S^{--}, deren α-Werte sicherlich in Wirklichkeit erheblich kleiner sind (siehe unten). Fig. 1 veranschaulicht die Tabelle 12; die Abhängigkeit zwischen $Z_{\text{eff.}}$ und α nach Formel (9) wird dabei in logarithmischer Skala durch gerade Linien mit der Neigung $tg\,\varphi = -3$ dargestellt. Wir haben unsere Zahlen mit denen von Wasastjerna [1]), sowie denen von Fajans und Joos [2]) verglichen, die zu diesem Zwecke mit $\frac{4\pi N}{3}$ (N Loschmidtsche Zahl pro Mol) zu dividieren sind. Wir fanden dabei, daß keine besonders auffälligen Abweichungen vorliegen; die Unterschiede sind zum Teil zweifellos reell und beruhen auf dem verschiedenen Ursprung

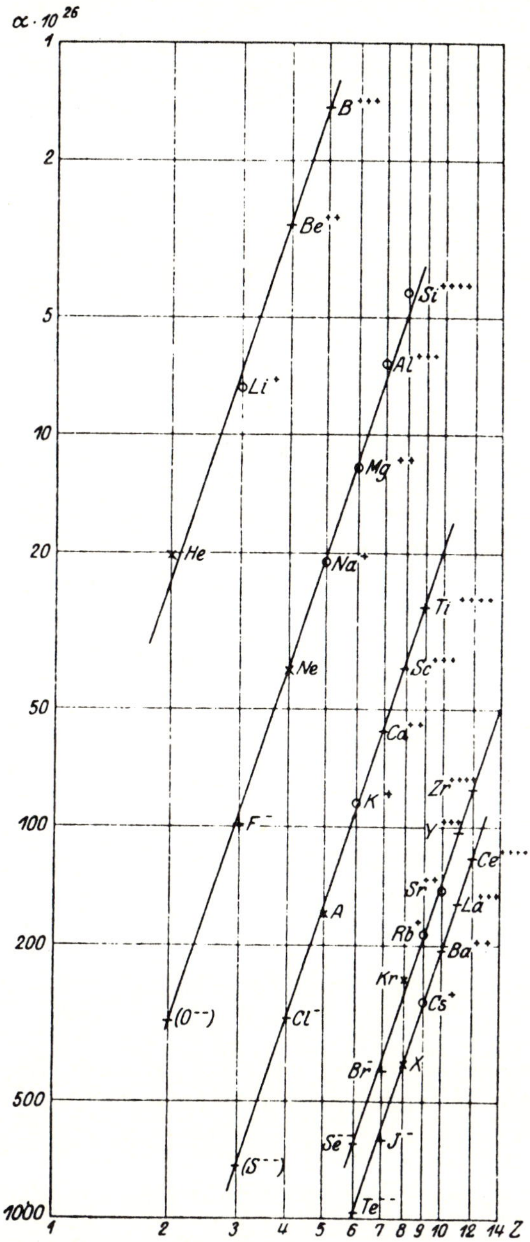

Fig. 1.

[1]) J. A. Wasastjerna, l. c.

[2]) K. Fajans u. G. Joos, l. c.

der Zahlen, vor allem darauf, daß die zitierten Forscher die Refraktion der Lösungen besonders stark benutzten. Ein Teil des Unterschiedes kommt auch daher, daß wir auf unendlich lange Wellen extrapoliert haben, während Fajans und Joos die Natrium-D-Linie zugrunde legen. Für die von uns betrachteten „Effekte erster Ordnung“ spielen die Unterschiede keine wesentliche Rolle.

Als Probe auf den extrapolierten Ca^{++}-Wert berechnen wir die Refraktion des Flußspats CaF_2 und vergleichen sie mit der Beobachtung [1]):

$$\alpha_{\text{ber.}} = 0{,}57 + 2 \times 0{,}99 = 2{,}55,$$
$$\alpha_{\text{beob.}} = 2{,}50.$$

Die Halogenwasserstoffe kann man zur Bestimmung der α-Werte für die Halogenionen nicht benutzen; denn bei ihnen dringt der Wasserstoffkern in die Elektronenhülle des Anions ein und verändert ihre Festigkeit. Man kann aber schließen, daß der α-Wert des Halogenwasserstoffs zwischen dem α-Wert des Halogenions und dem des benachbarten Edelgases liegen muß; denn denkt man sich den Abstand des H^+-Kerns vom Halogenkern verändert, so muß bei sehr großem Abstand der α-Wert des Halogenions gelten, beim Abstand Null der des Edelgases, bei mittleren Abständen also ein Zwischenwert. In der Tat ist das auch der Fall, wie folgende Daten zeigen.

	HCl	HBr	HJ
$n =$	1,000435	1,000594	1,000881.

Man erhält:

Tabelle 13.

	$\alpha \cdot 10^{24}$		$\alpha \cdot 10^{24}$		$\alpha \cdot 10^{24}$
A	1,63	Kr	2,46	X	4,00
HCl . . .	2,56	HBr . . .	3,50	HJ	5,69
Cl^-	3,05	Br^- . . .	4,17	J^-	6,28

Offenbar kann man durch eine analoge Überlegung auch die Kleinheit der Refraktion der Li-Salze verstehen, wie sie in der Anomalie der ersten Zeile von Tabelle 9 zum Ausdruck kommt; sie beruht auf der großen Annäherung des sehr kleinen Li^+-Ions an die großen Halogenionen.

§ 4. Die Sublimationswärmen der Salze. Wir wollen die α-Werte zur Berechnung der Sublimationswärme der Salze verwerten. Während in den regulär kristallisierenden festen Salzen die Ionen so

[1]) Eine Diskussion der α-Werte mit Hilfe der Refraktion komplizierterer doppelbrechender Kristalle scheint uns nur im Zusammenhang mit einer Theorie der Doppelbrechung selbst nach der Ewaldschen Methode fruchtbar zu sein. Wir behalten uns vor, darauf zurückzukommen.

regelmäßig im Gitter gelagert sind, daß sie keine einseitige Deformation erfahren, werden in der Dampfmolekel die Ionen aufeinander gerichtete Kräfte ausüben, die zu starken Deformationen führen. Von diesen hängt also auch die Arbeit V ab, die bei der Bildung der Molekel aus den Ionen frei wird; V ist offenbar gleich der Differenz aus der Gitterenergie U und der Sublimationswärme S beim absoluten Nullpunkt, wie sie mit Hilfe des Dampfdruckes direkt gemessen wird:

$$V = U - S. \tag{10}$$

Die quasielastische Energie, die bei Erregung eines Dipols p entsteht, ist bekanntlich $\frac{1}{2\alpha} p^2$. Die potentielle Energie φ zweier Ionen $(+e, \alpha_1)$ und $(-e, \alpha_2)$ im Abstand r besteht aus der Coulombschen Anziehung der Gesamtladungen $-\frac{e^2}{r}$, einer Abstoßung $\frac{b}{r^9}$, die aus der Gittertheorie bekannt ist, den Anziehungen der Ladungen auf die induzierten Dipole $-\frac{ep_1}{r^2} - \frac{ep_2}{r^2}$, der quasielastischen Energie der Dipole $\frac{p_1^2}{2\alpha_1} + \frac{p_2^2}{2\alpha_2}$, schließlich der Wirkung der Dipole aufeinander $-\frac{2p_1p_2}{r^3}$; also im ganzen

$$\varphi = -\frac{e^2}{r} + \frac{b e^2}{r^9} - \frac{p_1 e}{r^2} - \frac{p_2 e}{r^2} + \frac{p_1^2}{2\alpha_1} + \frac{p_2^2}{2\alpha_2} - \frac{2 p_1 p_2}{r^3}. \tag{11}$$

Die Gleichgewichtsbedingungen lauten:

$$\frac{\partial \varphi}{\partial p_1} = 0; \quad \frac{\partial \varphi}{\partial p_2} = 0; \quad \frac{\partial \varphi}{\partial r} = 0.$$

Die beiden ersten dieser Gleichungen heißen ausführlich:

$$p_1 - \frac{2\alpha_1}{r^3} p_2 = \frac{e\alpha_1}{r^2},$$

$$p_2 - \frac{2\alpha_2}{r^3} p_1 = \frac{e\alpha_2}{r^2}.$$

Durch suksessive Näherung $\left(\text{Entwicklung nach } \frac{\alpha}{r^3}\right)$ bekommt man:

$$\left.\begin{aligned} p_1 &= \frac{e\alpha_1}{r^2} + \frac{2\alpha_1\alpha_2 e}{r^5} + \cdots \\ p_2 &= \frac{e\alpha_2}{r^2} + \frac{2\alpha_1\alpha_2 e}{r^5} + \cdots \end{aligned}\right\} \tag{12}$$

Dies setzen wir in φ ein und erhalten:

$$\varphi = -\frac{e^2}{r} + \frac{b e^2}{r^9} - \frac{(\alpha_1 + \alpha_2) e^2}{2 r^4} - \frac{2 e^2 \alpha_1 \alpha_2}{r^7}. \tag{13}$$

Jetzt bilden wir die dritte Gleichgewichtsbedingung:

$$\frac{\partial \varphi}{\partial r} = \frac{e^2}{r^2}\left(1 - \frac{9b}{r^8} + \frac{2(\alpha_1 + \alpha_2)}{r^3} + \frac{14\alpha_1\alpha_2}{r^6}\right) = 0. \tag{14}$$

Um die Energie φ bis auf Glieder 2. Ordnung zu bekommen, genügt es, r aus Gleichung (14) bis auf Glieder 1. Ordnung zu bestimmen:

$$\left.\begin{aligned} & r = r_0 + r_1; \quad 1 - \frac{9b}{r_0^8} = 0; \quad r_1 \cdot \frac{72b}{r_0^9} + \frac{2(\alpha_1 + \alpha_2)}{r_0^3} = 0; \\ & \text{also} \\ & r = r_0 - \frac{(\alpha_1 + \alpha_2) r_0^6}{36b}, \quad \text{wo} \quad r_0 = \sqrt[8]{9b} \end{aligned}\right\} \tag{15}$$

ist. Setzt man dies in φ ein, so erhält man nach leichten Umformungen

$$\varphi = -\frac{e^2}{r_0}\left[\frac{8}{9} + \frac{\alpha_1 + \alpha_2}{2r_0^3} + \left(\frac{\alpha_1 + \alpha_2}{2r_0^3}\right)^2 + \frac{2\alpha_1\alpha_2}{r_0^6}\right]. \tag{16}$$

Nach der Gittertheorie ist

$$b = r_g^8 \cdot \frac{a}{9 \,.\, S(9)}; \tag{17}$$

dabei bedeutet $r_g = \frac{\delta}{2}$ die halbe Gitterkonstante, a die Madelungsche Zahl $\left(-\frac{a e^2}{r_g}\right.$ ist die elektrostatische Energie des Gitters pro Molekel$\left.\right)$, endlich $S(9) = S'(9) + \beta S''(9)$, wo $S'(9)$ und $S''(9)$ die Gittersummen der Abstoßungskräfte ungleichartiger bzw. gleichartiger Ionen sind. Nach den Erfahrungen [1]) über die Elastizitätskonstanten von NaCl und KCl muß man $\beta = -0{,}5$ setzen. Die Zahlenwerte sind:

$$\left.\begin{aligned} &\text{NaCl-Gitter:}\ a = 1{,}748; \quad S(9) = 5{,}75, \\ &\text{CsCl-Gitter:}\ a = 1{,}018; \quad S(9) = 0{,}0514. \end{aligned}\right\} \tag{18}$$

Den Ausdruck $\alpha_1 + \alpha_2$ kann man mit Hilfe von (8) direkt durch den Brechungsindex n des Kristalls darstellen; denn ist $z = N \,.\, \delta^3 = 8 N r_g^3$ die Zahl der Molekeln in der Zelle, so wird nach (8)

$$\alpha_1 + \alpha_2 = \frac{6 r_g^3}{\pi z} \frac{n^2 - 1}{n^2 + 2}.$$

Für das NaCl-Gitter ist $z = 4$, für das CsCl-Gitter $z = 1$. Aus $-\varphi$ erhält man die Bildungswärme V der Molekel aus den Ionen in kcal (pro Mol) durch Multiplikation mit $2{,}388 \,.\, 10^{-11} \,.\, N$ ($N = 6{,}06 \,.\, 10^{23}$ Loschmidtsche Zahl pro Mol). Man findet:

$$\left.\begin{aligned} V = \frac{588 \,.\, 10^{-8}}{\delta} \sqrt[8]{\frac{S(9)}{a}} \Bigg\{ & 1 + \frac{27}{8\pi z} \frac{n^2 - 1}{n^2 + 2} \left(\frac{S(9)}{a}\right)^{3/8} \\ & + \frac{9}{8}\left[\frac{3}{\pi z} \frac{n^2 - 1}{n^2 + 2} \left(\frac{S(9)}{a}\right)^{3/8}\right]^2 + \frac{27\alpha_1\alpha_2}{\delta^6}\left(\frac{S(9)}{a}\right)^{3/4}\Bigg\}. \end{aligned}\right\} \tag{19}$$

[1]) M. Born, Atomtheorie des festen Zustandes, S. 740. Leipzig 1923.

Mit dieser Formel ist die folgende Tabelle berechnet. Darin sind in der ersten Spalte diejenigen Werte V_0 der Bildungswärme eingetragen, die bei undeformierbaren Ionen gelten würden; in der zweiten Spalte die Werte des Faktors, dessen Abweichung von 1 den Einfluß der Deformierbarkeit ausdrückt. In der dritten und vierten Kolonne stehen die Werte der Gitterenergie [1]) U und der Sublimationswärme [2]) S. Die beiden letzten Spalten enthalten schließlich die Größen $U - S$ und $V_{\text{ber.}}$, die einander gleich sein sollen.

Tabelle 14.

	V_0	$\frac{V_{\text{ber.}}}{V_0}$	U	S	$U - S$	$V_{\text{ber.}}$
Na F	147	1,097	222	56,6	165	161
Na Cl	121	1,150	182	44,3	138	139
Na Br	114	1,171	172	38,6	133	133
Na J	105	1,196	158	37,0	121	126
K F	126	1,115	191	41,9	149	140
K Cl	108	1,145	164	40,5	123	124
K Br	103	1,161	156	38,2	118	120
K J	96	1,175	145	37,2	108	113
Rb F	110	1,101	163	40,0	123	122
Rb Cl	103	1,154	156	37,8	118	119
Rb Br	99	1,168	149	37,0	112	115
Rb J	93	1,190	140	37,0	103	110
Cs F	113	1,150	170	34,7	135	129
Cs Cl	99	1,164	146	37,4	109	115
Cs Br	94	1,173	139	36,7	102	110
Cs J	89	1,188	131	36,6	94	105

Die Tabelle zeigt, daß der theoretische Wert von $V = U - S$ im allgemeinen nur wenige Prozent vom beobachteten abweicht; nur in der Reihe der Cs-Salze treten größere Unstimmigkeiten auf. Ein Vergleich zwischen der ersten und den beiden letzten Spalten der Tabelle lehrt, daß die Polarisation wesentlich ist für die gute Übereinstimmung. 10 bis 20 Proz. der Bildungswärme rühren vom Einfluß der Ionendeformation her [3]).

[1]) Die Gitterenergien haben wir mit den δ-Werten der Tabelle 5 neu gerechnet nach den Formeln

$$U = \frac{8}{9} \cdot N \cdot \frac{a e^2}{r_g} = \begin{cases} 1026 \cdot 10^{-8} \cdot \frac{1}{\delta} \text{ kcal für NaCl}, \\ 597 \cdot 10^{-8} \frac{1}{\delta} \text{ kcal für CsCl}. \end{cases}$$

[2]) H. v. Wartenberg und Ph. Albrecht, ZS. f. Elektrochem. **27**, 162, 1921 und H. v. Wartenberg und H. Schulz, ebenda **27**, 568, 1921.

[3]) Die Zahlen der Tabelle 14 bedeuten natürlich auch eine direkte Bestätigung der Gitterenergien. Verbindet man damit die Knippingschen Werte für die Trennungsarbeiten der Halogenwasserstoffe in Ionen, so folgt daraus, wie an

Dies Ergebnis berechtigt uns, auch die übrigen Daten der Salzmolekeln aus (11) zu berechnen. Die folgende Tabelle enthält den Kernabstand r nach Gleichung (15), das aus diesem berechnete Trägheitsmoment A und das elektrische Moment p der Molekel, berechnet nach der Formel

$$p = er - (p_1 + p_2) = er\left(1 - \frac{\alpha_1 + \alpha_2}{r_0^3} + \cdots\right). \tag{20}$$

Bei der Berechnung von r, A und p haben wir uns auf Glieder 1. Ordnung in $\frac{\alpha}{r^3}$ beschränkt, da die Glieder 2. Ordnung nur wenige Prozent ausmachen.

Tabelle 15.

	$r \cdot 10^8$	$A \cdot 10^{38}$	$p \cdot 10^{17}$		$r \cdot 10^8$	$A \cdot 10^{38}$	$p \cdot 10^{17}$
NaF	1,92	0,63	0,92	RbF	2,56	1,68	1,22
NaCl	2,29	1,21	1,09	RbCl	2,68	2,96	1,28
NaBr	2,41	1,72	1,15	RbBr	2,80	5,34	1,34
NaJ	2,59	2,16	1,24	RbJ	2,96	7,38	1,41
KF	2,22	1,04	1,06	CsF	2,46	1,69	1,17
KCl	2,56	2,01	1,22	CsCl	2,82	3,69	1,34
KBr	2,68	3,10	1,28	CsBr	2,94	7,14	1,40
KJ	2,86	4,08	1,36	CsJ	3,11	10,38	1,48

Empirisches Material zur Prüfung dieser Daten liegt noch nicht vor.

§ 5. **Schwingungen binärer Molekeln.** Die Rechnungen des vorhergehenden Abschnitts geben uns auch die Möglichkeit, die Schwingungen der Salzmolekeln in allen Einzelheiten zu studieren. Die quasielastische Kraft, durch welche die Kerne an ihre Ruhelage gebunden sind, ist bekanntlich durch die zweite Ableitung des Potentials φ gegeben. Die Frequenz ω der Kernschwingung steht also mit φ'' in der Beziehung

$$\mu\omega^2 = \varphi'', \quad \text{wo} \quad \frac{1}{\mu} = \frac{1}{m_1} + \frac{1}{m_2}. \tag{21}$$

Dabei bedeuten m_1 und m_2 die Massen der beiden Ionen. Aus (14) und (15) ergibt sich φ'' bis auf Glieder erster Ordnung einschließlich:

$$\varphi'' = -\frac{2e^2}{r^3}\left(1 - \frac{45\,b}{r^8} + \frac{5(\alpha_1 + \alpha_2)}{r^3}\right) = \frac{8e^2}{r_0^3}\left(1 + \frac{2(\alpha_1 + \alpha_2)}{r_0^3}\right). \tag{22}$$

anderer Stelle gezeigt worden ist (M. Born, Atomtheorie des festen Zustandes **39**, 751, 1923), für die Dissoziationswärme des Wasserstoffs 80 kcal (mit einem Fehler von etwa ± 5 kcal). Es scheint uns dies vorläufig die zuverlässigste Bestimmung dieser wichtigen thermochemischen Größe zu sein.

Setzt man noch $\lambda = \frac{2\pi c}{\omega}$, so erhält man die (ultrarote) Wellenlänge der Kernschwingungen:

$$\lambda = 2\pi c \sqrt{\frac{\mu r_0^3}{8 e^2}} \left(1 - \frac{\alpha_1 + \alpha_2}{r_0^3}\right). \tag{23}$$

Die folgende Tabelle enthält die nach (23) aus den Daten des § 4 berechneten Werte von λ in 10^{-4} cm (μ-Einheiten).

Tabelle 16.

$\lambda \cdot 10^4$	F	Cl	Br	J
Na	13,9	19,5	23,4	26,1
K	19,0	27,2	33,7	39,3
Rb	26,1	33,8	45,6	53,3
Cs	24,5	38,2	53,0	65,5

Ebenso, wie die Wellenlänge bzw. die Frequenz der Kernschwingung, lassen sich aus den Formeln des § 4 die übrigen Konstanten der Kratzerschen Bandentheorie der zweiatomigen Molekel berechnen[1]). Doch fehlt für alle diese Größen bisher jede Möglichkeit einer Prüfung an der Erfahrung.

Eine solche Prüfungsmöglichkeit besteht nur bei den Halogenwasserstoffen, auf die unsere Überlegungen in der bisherigen Form nicht anwendbar sind; wir wollen aber versuchen, durch rohe Annäherungen auch auf die Halogenwasserstoffe die bisherigen Formeln anzuwenden.

Aus den bekannten Kernabständen der Halogenwasserstoffe ist zu schließen, daß der H-Kern in die Elektronenhülle des Halogenions eindringt. Repräsentiert man die äußerste Schale des edelgasähnlichen Ions durch eine mit homogener Flächenladung belegte Kugel, so schließt man, daß beim Eindringen des H-Kerns in die Schale der für den H-Kern in Betracht kommende α-Wert unstetig auf einen bestimmten, zunächst unbekannten kleineren Wert — wir nennen ihn α_{eff} — springt.

Ferner ist nach allgemeinen Überlegungen über würfelähnliche Atommodelle anzunehmen, daß die abstoßende Kraft zwischen H+ und Halogenion proportional mit r^{-5} (statt r^{-9} bei Wirkung zweier Würfel aufeinander) angesetzt werden muß. Bezeichnet man den

[1]) A. Kratzer, Ann. d. Phys. **67**, 127, 1922; M. Born u. E. Hückel, Phys. ZS. **24**, 1, 1923.

Radius der äußeren Schale mit r_s, so ergibt sich also folgender Ansatz für φ (vgl. § 1, S. 391):

$$\left.\begin{aligned}\varphi &= -\frac{e^2}{r} + \frac{b e^2}{r^5} + 2 e^2 \alpha \int_{\infty}^{r_s} \frac{d r}{r^5} + 2 e^2 \alpha_{\text{eff}} \int_{r_s}^{r} \frac{d r}{r^5} \\ &= -\frac{e^2}{r} + \frac{b e^2}{r^5} - \frac{e^2 \alpha_{\text{eff}}}{2 r^4} - \frac{e^2 (\alpha - \alpha_{\text{eff}})}{2 r_s^4}.\end{aligned}\right\} \tag{24}$$

Die Gleichgewichtsbedingung lautet:

$$1 - \frac{5 b}{r^4} + \frac{2 \alpha_{\text{eff}}}{r^3} = 0, \tag{25}$$

ferner gilt:

$$\varphi'' = -\frac{2 e^2}{r^3}\left(1 - \frac{3 . 5 b}{r^4} + \frac{5 \alpha_{\text{eff}}}{r^3}\right), \quad = +\frac{2 e^2}{r^3}\left(2 + \frac{\alpha_{\text{eff}}}{r^3}\right)$$

und nach (21)

$$\lambda = 2 \pi c \sqrt{\frac{\mu r^3}{4 e^2 \left(1 + \frac{\alpha_{\text{eff}}}{2 r^3}\right)}}. \tag{26}$$

Die Größe von r muß man, da b und α_{eff} zunächst unbekannt sind, aus dem gemessenen Trägheitsmoment entnehmen. Ebenso kann man α_{eff} direkt aus dem gemessenen elektrischen Moment erhalten, da nach (20)

$$p = e r \left(1 - \frac{\alpha_{\text{eff}}}{r^3}\right).$$

Dies Moment p ist nur bei HCl untersucht[1]): $p = 2{,}15 . 10^{-18}$; die r-Werte entnehmen wir der Arbeit von Kratzer (l. c.):

	HF	HCl	HBr
$10^8 . r =$	0,92	1,265	1,407

Daraus ergibt sich für HCl: $\alpha_{\text{eff}} = 1{,}30 . 10^{-24}$, ein Wert, der etwa $^2/_5$ des α-Wertes von Cl^- ($3{,}05 . 10^{-24}$) beträgt. Nehmen wir näherungsweise an, daß das Verhältnis $\frac{\alpha_{\text{eff}}}{\alpha}$ für alle Halogenionen dasselbe ist, also

	HF	HCl	HBr	HJ
$10^{24} . \alpha_{\text{eff}}$	0,42	1,30	1,78	2,68

so ergeben sich für λ die folgenden Werte, die wir mit den aus der Bandentheorie genau bekannten vergleichen:

	HF	HCl	HBr
$\lambda_{\text{ber.}} \cdot 10^4$.	2,48	3,42	3,85
$\lambda_{\text{Bande}} \cdot 10^4$.	1,93	3,09	4,50

[1]) H. Falkenhagen, Phys. ZS. **23**, 87, 1922, s. bes. S. 93.

Man gelangt also in die richtige Größenordnung; eine genaue Übereinstimmung ist natürlich bei der Roheit des Ansatzes nicht zu verlangen.

Auch die Dissoziationswärme der Halogenwasserstoffe kann man aus (24) der ungefähren Größe nach ableiten. Nimmt man nämlich an, daß der H-Kern nur wenig in die äußere Schale eindringt, so ist $r \sim r_s$ und man erhält für φ:

$$\varphi = -\frac{e^2}{r}\left(1 - \frac{b}{r^4} + \frac{\alpha}{2r^3}\right) = -\frac{e^2}{r}\left(\frac{4}{5} - \frac{2}{5}\frac{\alpha_{\text{eff}}}{r^3} + \frac{\alpha}{2r^3}\right). \quad (27)$$

Die Zahlenwerte für die Dissoziationsarbeit D lauten daher:

	HF	HCl	HBr
$D_{\text{ber.}}$ . . .	436	339	272
$D_{\text{beob.}}$ . . .	—	313	299

Naturgemäß ist die Übereinstimmung hier erheblich schlechter, als in Tabelle 14 für die Bildungswärme der Salzmolekeln, da in Wirklichkeit der Aufbau der Halogenwasserstoffmolekel durch das Eindringen des H-Kerns außerordentlich kompliziert wird.

Bisher haben wir uns in den Anwendungen der Kenntnis der Ionendeformierbarkeiten auf zweiatomige Molekeln beschränkt. In einer folgenden Arbeit sollen die Untersuchungen auf kompliziertere Molekeln und Ionengebilde ausgedehnt werden. Vorläufige rohe Rechnungen zeigten, daß sich z. B. die von Schaefer[1]) gemessenen Eigenschwingungen der CO_3-Gruppe auf Grund von Ansätzen, die denen des § 4 analog sind, mit guter Näherung berechnen lassen. In den Eigenschwingungen von ClO_3, BrO_3, JO_3, SO_4 usw.[2]) liegt ein großes empirisches Material zur Prüfung der Theorie vor. Ferner ergeben sich für den Aufbau der Molekel vom Typus H_2O, CO_2 usw. neue Gesichtspunkte durch die Berücksichtigung der Polarisierbarkeit der Ionen. Schließlich ist zu erwarten, daß sich in der Theorie der Kristallgitter manche schon in der Einleitung erwähnte Diskrepanzen, die bis jetzt noch zwischen Theorie und Beobachtung bestanden, z. B. bei der Piezoelektrizität, durch die Berücksichtigung der Deformierbarkeiten werden erklären lassen.

1) Cl. Schaefer u. M. Schubert, Ann. d. Phys. (4) 50, 283, 1916.
2) Dieselben, l. c., u. ZS. f. Phys. 7, 309, 1921.

Termstruktur der Multipletts höherer Stufe.

Von **A. Landé** in Tübingen und **W. Heisenberg** in Göttingen.

(Eingegangen am 18. Mai 1924.)

Im Anschluß an die Struktur des Neonspektrums wird die Struktur der Multipletts höherer Stufe abgeleitet auf Grund eines Verzweigungsprinzips.

Im folgenden soll die Multiplettstruktur, d. h. die Termvielfachheit und die Ordnung der „inneren“ Quantenzahlen bei denjenigen Komplextermen besprochen werden, welche nicht dem früher behandelten System[1]) der Singulett-, Dublett- usw. Terme angehören. Obwohl das Neonspektrum das einzige bisher bekannte solche System darstellt, kann man doch die dort gültigen Gesetze leicht verallgemeinern. Als modellmäßiges Charakteristikum für jene früheren Multipletts, die wir als solche erster Stufe bezeichnen wollen, wurde angesehen, daß zum Drehimpuls R des Atomrumpfes nur Elektronen der niedrigsten azimutalen Quantenzahl ($k = 1$ in Sommerfelds, $K = {}^1/_2$ in unserer Normierung) beitragen, indem die im Rumpf enthaltenen Elektronen als s-Terme gebunden wurden. Sind im Atomrumpf auch n_k-Elektronen mit $k > 1$ vorhanden, so müssen wir annehmen, daß die Termzustände des Atoms immer noch zur ersten Stufe gehören, wenn jene Impulsbeiträge $k > 1$ sich nach Größe und Richtung gegenseitig fortheben, d. h. impulslos abgeschlossene Schalen bilden (z. B. geben K und Na denselben Spektraltypus wie Li, weil die 2_1-Bahnen sich dort in ihren Impulsen neutralisieren, ebenso die 2_2-, die 3_1- und die 3_2-Bahnen). Trägt dagegen zum Rumpfimpuls außer n_1-Bahnen auch eine n_k-Bahn mit $k > 1$ bei, indem eines der impulsbeitragenden Rumpfelektronen als p-Term oder d-Term usw. gebunden wurde, so gibt das Atom Multipletts zweiter Stufe. Tragen zwei n_k-Bahnen mit $k > 1$ zum Rumpfimpuls bei, indem etwa zwei Elektronen als p-Terme oder auch eines als p-Term und eines als d-Term usw. gebunden wurde, derart, daß die Impulse sich nicht fortheben, so gibt das Atom Multipletts dritter Stufe usw.; die verschiedenen Stufen bauen sich sukzessive aufeinander auf.

Ein Beispiel für ein Komplexsystem zweiter Stufe gibt das von Paschen analysierte Neonspektrum[2]). Dieses läßt sich zwar nicht

[1]) A. Landé, Termstruktur und Zeemaneffekt der Multipletts. ZS. f. Phys. **15**, 189, 1923 und **19**, 112, 1923.

[2]) F. Paschen, Das Spektrum des Neons. Ann. d. Phys. **60**, 405, 1915 und **63**, 201, 1920.

bezüglich seiner Intervallverhältnisse und seines Zeemaneffektes, wohl aber bezüglich seiner Struktur (d. h. Vielfachheit und „innere" Quantenzahlen J) zerlegt denken in ein Singulettsystem, zwei Triplettsysteme und ein Quintettsystem[1]). Und zwar baut sich das Singulettsystem und das eine Triplettsystem über dem Dublett-$\mathfrak{p}_2$-Term des Neonions auf, das andere Triplett- und das Quintettsystem über dem Dublett-$\mathfrak{p}_1$-Term des Neonions[2]); dies ist angedeutet durch das in Tabelle 1 reproduzierte Schema der inneren Quantenzahlen J beim Neon:

Tabelle 1.

Neonion		$\mathfrak{p}_2$-Term $J = 1$ ↙	↘	$\mathfrak{p}_1$-Term $J = 2$ ↙	↘
Neonatom:	s	$1/2$	$3/2$	$3/2$	$5/2$
Terme 2. Stufe .	p	$3/2$	$1/2$ $3/2$ $5/2$	$1/2$ $3/2$ $5/2$	$3/2$ $5/2$ $7/2$
	d	$5/2$	$3/2$ $5/2$ $7/2$	$3/2$ $5/2$ $7/2$	$1/2$ $3/2$ $5/2$ $7/2$ $9/2$

Dem Ne^+-Term $J = 1$ entspringen (durch „Verzweigung" von J in $J \pm 1/2$) zwei Ne-s-Terme mit $J = 1/2$ bzw. $3/2$, und dem Ne^+-Term $J = 2$ entspringen zwei Ne-s-Terme mit $J = 3/2$ und $J = 5/2$. Diese vier s-Terme werden dann zur Grundlage der vier ineinandergreifenden Termsysteme des Neonatoms.

Es muß besonders betont werden, daß diese Einteilung des Neonspektrums in ein Singulett-Triplett- und ein Triplett-Quintett-Termsystem nicht etwa bedeuten soll, daß auch die Intervall- und Intensitätsverhältnisse und der Zeemaneffekt hier dieselben Gesetze befolgen wie bei den entsprechenden Multipletts erster Stufe; vielmehr soll diese Einteilung nur eine übersichtliche Ordnung der vorkommenden J-Werte erleichtern. In Wirklichkeit bilden die vier s-Terme zusammen eine enge Einheit und ebenso bilden die zehn p-Terme zusammen eine enge Einheit usw., wie besonders bei ihren zurzeit von Back untersuchten Zeemaneffekten hervortritt. Diese engen Einheiten werden nur dadurch gegliedert, daß die von Ne^+-$\mathfrak{p}_2$ abgezweigte Termgruppe nach Paschen um rund 780 Wellenzahlen gegenüber der von Ne^+-$\mathfrak{p}_1$ abgezweigten Termgruppe verschoben ist. Diese Gruppeneinteilung gegeneinander verschobener Terme ist in Tabelle 1 (und allgemein in Tabelle 2) durch senkrechte Trennungsstriche angedeutet.

Das Schema der J-Werte von Tabelle 1 läßt sich nun hypothetisch erweitern zu dem allgemeinen Schema der Multipletterme

[1]) A. Landé, Zur Struktur des Neonspektrums. ZS. f. Phys. **17**, 292, 1923.
[2]) W. Grotrian, Das L-Dublett des Neons. ZS. f. Phys. **8**, 116, 1922.

zweiter Stufe mit den in Tabelle 2 angegebenen J-Werten[1]). Die Normierung des J schließt sich der schon bei den Multipletts erster Stufe bewährten an; die Bezeichnungsweise der Spektroskopie behält diese J-Werte bei, soweit sie ganzzahlig sind, schreibt aber statt der halbzahligen J-Werte um $1/2$ kleinere Werte als „innere" Quantenzahlen j.

Bei den Multipletts erster Stufe war als praktische Termbezeichnung das Symbol n_{kj}^{r} eingeführt worden (n als Hauptquantenzahl, j als ganze innere Quantenzahl, $k = 1, 2, 3\ldots$ bei den $s, p, d\ldots$-Termen, $r = 1, 2, 3\ldots$ bei Singulett-, Dublett-, Triplett...-Termen). Bei den Komplextermen zweiter Stufe werden wir mehrere Werte r als obere Indizes anfügen n_{kj}^{rrr}, also etwa einen Neonterm, weil er dem Singulett-Triplett-Quintett-System angehört, mit n_{kj}^{1335} bezeichnen. Soll noch besonders angedeutet werden, daß der betreffende Term speziell zu der von $Ne^{+} - \mathfrak{p}_1$ abgezweigten Triplett-Quintettgruppe gehört, so wird durch Überstreichen geschrieben $n_{kj}^{13\overline{35}}$. Da es dann immer noch zwei solche Terme gibt, müssen diese vorläufig noch durch angehängte Buchstaben a und b unterschieden werden, wie es folgende Gegenüberstellung der systematischen Bezeichnung und der von Paschen beim Neon gebrauchten erläutert.

Tabelle 3.

s_3	s_2	s_4	s_5	p_1	p_2	p_5	p_4	p_3	p_7	p_{10}	p_6	p_8	p_9
$n_{10}^{\overline{1}335}$	$n_{11}^{\overline{1}33\overline{5}}$	$n_{11}^{13\overline{35}}$	$n_{12}^{13\overline{35}}$	$n_{20}^{\overline{1}335}$	$n_{21a}^{\overline{1}335}$	$n_{21b}^{\overline{1}335}$	$n_{22}^{\overline{1}335}$	$n_{20}^{13\overline{35}}$	$n_{21a}^{13\overline{35}}$	$n_{21b}^{13\overline{35}}$	$n_{22a}^{13\overline{35}}$	$n_{22b}^{13\overline{35}}$	$n_{23}^{13\overline{35}}$

s_1'	s_1''	s_1''''	s_1'''	d_6	d_2	d_5	d_1''	d_3	d_1'	d_4	d_4'
$n_{31}^{\overline{1}335}$	$n_{32a}^{\overline{1}335}$	$n_{32b}^{\overline{1}335}$	$n_{33}^{\overline{1}335}$	$n_{30}^{13\overline{35}}$	$n_{31a}^{13\overline{35}}$	$n_{31b}^{13\overline{35}}$	$n_{32a}^{13\overline{35}}$	$n_{32b}^{13\overline{35}}$	$n_{33a}^{13\overline{35}}$	$n_{33b}^{13\overline{35}}$	$n_{34}^{13\overline{35}}$

Diese Bezeichnungsweise für die Terme zweiter Stufe ist eindeutig, wenn man noch verabredet, mit a den kleineren, mit b den größeren der beiden Terme gleicher übriger Indizes zu benennen.

Das Neonspektrum bildet bisher das einzige bekannte Beispiel eines Multiplettsystems zweiter Stufe. Trotzdem kann sein in Tabelle 1 gegebenes Strukturschema mit gutem Grund zu dem Schema der Tabelle 2 erweitert werden; denn wir wenden dabei das gleiche Verzweigungsprinzip an, das sich schon bei den Multipletts

[1]) Bei der Verzweigung von $J = 1/2$ in 0 und 1 hat nur der letztere Zweig reale Bedeutung, da Terme mit $J = 0$ durch die allgemeine Strukturregel: $|R - K| + 1/2 \leq J \leq |R + K| - 1/2$ ausgeschlossen sind.

Tabelle 2.

Singulett-p-Term 1. Stufe	n_{21}^{1}	$3/2$ → 1		$3/2$ → 2		
Dublett-Quartett-System 2. Stufe	n_{kj}^{24}	1 2 2 3 3 4		1 2 3 1 2 3 4 2 3 4 5		
Dublett-p-Terme 1. Stufe	n_{2j}^{2}	1 → $1/2$	1 → $3/2$	2 → $3/2$	2 → $5/2$	
Singulett-Triplett-Quintett-System 2. Stufe	n_{kj}^{1335}	$3/2$ $5/2$ $7/2$	$1/2$ $3/2$ $5/2$ $3/2$ $5/2$ $7/2$ $5/2$ $7/2$ $9/2$	$1/2$ $3/2$ $5/2$ $3/2$ $5/2$ $7/2$ $5/2$ $7/2$ $9/2$	$3/2$ $5/2$ $7/2$ $1/2$ $3/2$ $5/2$ $7/2$ $9/2$ $3/2$ $5/2$ $7/2$ $9/2$ $11/2$	
Triplett-p-Terme 1. Stufe	n_{2j}^{3}	$1/2$ → 1	$3/2$ → 1	$3/2$ → 2	$5/2$ → 2	$5/2$ → 3
Dublett-Quartett-Sextett-System 2. Stufe	n_{kj}^{22446}	1 2 2 3 3 4	1 2 2 3 3 4	1 2 3 1 2 3 4 2 3 4 5	1 2 3 1 2 3 4 2 3 4 5	2 3 4 1 2 3 4 5 1 2 3 4 5 6

Tabelle 2 (Fortsetzung).

Singulett-*d*-Term 1. Stufe	n^1_{32}	5/2 → 2	5/2 → 3
		1 2 3	2 3 4
Quartett-Sextett-System 2. Stufe	n^{46}_{kj}	1 2 3 4	1 2 3 4 5
		2 3 4 5	1 2 3 4 5 6

Dublett-*d*-Terme 1. Stufe	n^2_{3j}	2 → 3/2	2 → 5/2	3 → 5/2	3 → 7/2
		1/2 3/2 5/2	3/2 5/2 7/2	3/2 5/2 7/2	5/2 7/2 9/2
Triplett-Quintett-Septett-System 2. Stufe	n^{3557}_{kj}	3/2 5/2 7/2	1/2 3/2 5/2 7/2 9/2	1/2 3/2 5/2 7/2 9/2	3/2 5/2 7/2 9/2 11/2
		5/2 7/2 9/2	3/2 5/2 7/2 9/2 11/2	3/2 5/2 7/2 9/2 11/2	1/2 3/2 5/2 7/2 9/2 11/2 13/2

Triplett-*d*-Terme 1. Stufe	n^3_{3j}	3/2 → 1	3/2 → 2	5/2 → 2	5/2 → 3	7/2 → 3	7/2 → 4
		1 2	1 2 3	1 2 3	2 3 4	2 3 4	3 4 5
Dublett-Quartett-Sextett-Oktett-System 2. Stufe	n^{244668}_{kj}	2 3	1 2 3 4	1 2 3 4	1 2 3 4 5	1 2 3 4 5	2 3 4 5 6
		3 4	2 3 4 5	2 3 4 5	1 2 3 4 5 6	1 2 3 4 5 6	1 2 3 4 5 6 7

erster Stufe bewährt. Hat etwa ein Ion (Mg^+) einen Dublett-$\mathfrak{s}$-Term als Grundzustand mit $J = 1$, so hat das zugehörige Atom (Mg) zwei s-Terme mit $J = {}^1/_2$ und $J = {}^3/_2$ und zeigt demnach ein Singulett- und ein Triplettsystem erster Stufe.

Wir verallgemeinern diese Verzweigungsregel wie folgt: Wenn sich das Atom auf Zuständen des Ions aufbaut, die durch die J-Werte $J_1, J_2, \ldots J_n$ charakterisiert sind, so nehmen wir an, daß das Atom $2n$ s-Terme besitzt mit den J-Werten $J_1 \pm {}^1/_2, \ldots J_n \pm {}^1/_2$, und daß das Atom dementsprechend $2n$ Multiplettsysteme zeigt von der Vielfachheit $2J = 2(J_1 \pm {}^1/_2)$, $2(J_2 \pm {}^1/_2)$, ... usw.[1]) Dieser Verzweigungssatz erscheint als Erweiterung und Präzisierung des Rydbergschen Wechselsatzes.

Charakteristisch für die Verzweigung ist noch, daß die beiden Multiplettsysteme, die sich auf einem Zustand des Ions aufbauen, offenbar eine besondere Zusammengehörigkeit zeigen. Sie besitzen nämlich die gleiche Seriengrenze, da ja bei völliger Ionisation ein bestimmter Term des Ions übrigbleibt[2]).

Die Anwendung dieses Prinzips auf die Terme n^r_{kj} 1. Stufe mit den von ihnen abgezweigten ($\rightarrow$) Termsystemen 1. und 2. Stufe zeigt die hypothetische Tabelle 4.

Tabelle 4.

$n^1_{10} \rightarrow n^2_{kj}$	$n^2_{11} \rightarrow n^1_{kj}\ n^3_{kj}$	$n^3_{11} \rightarrow n^2_{kj}\ n^4_{kj}$	$n^4_{12} \rightarrow n^3_{kj}\ n^5_{kj}$	...
$n^1_{21} \rightarrow n^{24}_{kj}$	$n^2_{2j} \rightarrow n^{1335}_{kj}$	$n^3_{2j} \rightarrow n^{22446}_{kj}$	$n^4_{2j} \rightarrow n^{133557}_{kj}$	...
$n^1_{32} \rightarrow n^{46}_{kj}$	$n^2_{3j} \rightarrow n^{3557}_{kj}$	$n^3_{3j} \rightarrow n^{244668}_{kj}$	$n^4_{3j} \rightarrow n^{13355779}_{kj}$	...
$n^1_{43} \rightarrow n^{68}_{kj}$	$n^2_{4j} \rightarrow n^{5779}_{kj}$	$n^8_{4j} \rightarrow n^{4668810}_{kj}$	$n^4_{4j} \rightarrow n^{355779911}_{kj}$	...
......	....	....	....	...

Zu beachten ist noch, daß die Terme des Ions, über denen sich die Terme des Atoms aufbauen, nicht identisch zu sein brauchen mit

[1]) Nur wenn J_1 selbst gleich ${}^1/_2$ ist, findet bei ihm eine Verzweigung nicht in $J_1 \pm {}^1/_2 = 0$ und 1, sondern nur in $J_1 + {}^1/_2 = 1$ statt, so daß das Atom dann nur $(2n - 1)$ s-Terme besitzt, auf denen sich $2n - 1$ Multiplettsysteme aufbauen (vgl. S. 281, Anm. 1).

[2]) Als Beispiel diene das Cd, dessen Singuletterme und Tripletterme bei hoher Laufzahl derselben Grenze zustreben (denn sie sind verzweigt aus dem Dublett $\mathfrak{s}$-Term des Cd-Ions). Als zweites Beispiel diene das Ne, bei dem die vom Dublett $\mathfrak{p}_1$-Term des Ne-Ions abgezweigten Terme $n^{\overline{1335}}$ einer gemeinsamen Grenze zustreben, die vom Dublett $\mathfrak{p}_2$-Term abgezweigten Terme $n^{\overline{1335}}$ einer anderen gemeinsamen Grenze zustreben, die gegen erstere um rund 780 Wellenzahlen verschoben ist.

With A. Landé in Zeitschrift für Physik *25*, 279–286 (1924)

den Grundtermen des Ions. Vielmehr können auch im Ion, indem es zum Atomrumpf wird, noch beträchtliche Quantensprünge seiner Elektronen vorkommen; man darf also das Aufbauprinzip nicht allzu eng auffassen.

Ebenso wie sich die Multipletts zweiter Stufe durch Verzweigung auf den höheren Termen erster Stufe aufbauen, so gewinnt man durch naturgemäße Verallgemeinerung von Tabelle 2 von der zweiten Stufe ausgehend durch Verzweigung die Komplexterme dritter Stufe. Ist z. B. das zu vorletzt gebundene Elektron als Term n_{kj}^{1335} (p-Term des Neonspektrums) gebunden, so wird jeder p-Term zweiter Stufe sich durch Hinzukommen des zuletzt gebundenen Elektrons in zwei s-Terme dritter Stufe verzweigen mit den folgenden J-Werten (Tabelle 5):

Tabelle 5.

n_{2j}^{1335}-p-Terme 2. Stufe $J =$	$^3/_2$	$^1/_2$	$^3/_2$	$^5/_2$	$^1/_2$	$^3/_2$	$^5/_2$	$^3/_2$	$^5/_2$	$^7/_2$
	↙ ↘	↓	↙ ↘	↙ ↘	↓	↙ ↘	↙ ↘	↙ ↘	↙ ↘	↙ ↘
s-Terme 3. Stufe $J =$	1 2	1	1 2	2 3	1	1 2	2 3	1 2	2 3	3 4

Man hat also hier 18 s-Terme dritter Stufe zu erwarten, unter ihnen 6 s-Terme mit $J = 1 = {}^2/_2$, 7 s-Terme mit $J = {}^4/_2$, 4 s-Terme mit $J = {}^6/_2$ und 1 s-Term mit $J = {}^8/_2$; demnach soll das Spektrum dritter Stufe hier aus 6 Dublettsystemen, 7 Quartett-, 4 Sextett- und 1 Oktettsystem bestehen, deren Terme sich jedoch bezüglich ihrer Intervalle und ihres Zeemaneffektes ganz anders verhalten werden als bei den betreffenden Multipletts erster Stufe. Man sieht, daß hier eine große Fülle von ineinandergreifenden Termsystemen zu erwarten ist, von der aus man durch Verzweigung weiter zu Komplextermen höherer Stufen übergehen kann. Zum Glück werden aber wohl Komplextermsysteme höherer Stufen nur selten vorkommen, weil mehrere n_k-Rumpfelektronen mit $k > 1$ ihre Impulse nach Größe und Richtung leicht aufheben können, indem sie zu abgeschlossenen Schalen zusammentreten. Immerhin darf man Spektra höherer Stufen besonders dort suchen, wo etwa nach Bohrs System mehrere n_k-Bahnen mit $k > 1$ in unabgeschlossenen Schalen vorliegen sollen.

Von besonderem Interesse wird es sein, die Differenzen der gegeneinander verschobenen Termgruppen (z. B. $A = 780$ bei Ne) zu verfolgen und zu sehen, ob etwa bei den über Triplettermen erster Stufe aufgebauten Termen zweiter Stufe sich in der Tat drei Gruppen aussondern lassen, deren Verschiebungen A und B mit den

Intervallen der zugrunde liegenden drei Tripletterme erster Stufe übereinstimmen.

Zur modellmäßigen Bedeutung des Verzweigens möge hier nur weniges bemerkt werden. Hat etwa das Neonion (vgl. Tabelle 1) die Gesamtimpulsquantenzahlen $J = 1$ (als $\mathfrak{p}_2$-Term) bzw. $J = 2$ (als $\mathfrak{p}_1$-Term), so erwartet man eigentlich nach dem Bohrschen Aufbauprinzip, daß der Rumpf des Neonatoms, der ja nichts anderes ist als das Neonion, jetzt die Rumpfquantenzahlen $R = 1$ bzw. $R = 2$ zeigen wird. Statt dessen besitzt er die Rumpfquantenzahlen $R = 1 \pm 1/2$ bzw. $R = 2 \pm 1/2$ [wie man aus Tabelle 1 sieht; denn R ist[1]) stets identisch mit dem J-Wert des s-Terms, hier also gleich $1/2$, $3/2$, $3/2$ und $5/2$]. Die Verzweigung bedeutet also, daß das Ion, wenn es durch Einfangung eines äußeren Elektrons zum Rumpf eines Atoms wird, scheinbar entgegen dem Aufbauprinzip zwei neue Zustände mit um $\pm 1/2$ geändertem Impuls annimmt. Die nähere Diskussion dieses auf Grund der Bohrschen Theorie aus der spektroskopischen Erfahrung abzulesenden Tatbestandes führt aber zu so prinzipiellen Schwierigkeiten[2]), daß gerade dieser formal so einfach darstellbare Impulsverzweigungsprozeß einer der Haupteinwände gegen die Anwendbarkeit der für bedingt-periodische Bewegungen gültigen Quantenregeln auf die gekoppelten Systeme geworden ist[3]) und weiterhin zu einem einfachen Modifikationsversuch[4]) der bisherigen Quantenregeln gedient hat.

Die Zeemaneffekte bei den höheren Multiplettstufen werden demnächst im Anschluß an eine von Back unternommene magnetische Untersuchung des Neonspektrums besprochen, wobei sich das Verzweigungsprinzip in Verbindung mit dem Permanenzprinzip der g-Summen ebenfalls bewährt.

---

1) Wegen der allgemeinen Strukturregel $|R - K| + 1/2 \leq J \leq |R + K| - 1/2$.

2) N. Bohr, Kayserheft der Ann. d. Phys. **71**, 228, 1923.

3) Vgl. z. B. A. Landé, Phys. ZS. **24**, 441, 1923 (Vortrag in Bonn).

4) W. Heisenberg, ZS. f. Phys. (erscheint demnächst).

Zeitschrift für Physik 25, 175–176 (1924)

Bemerkung zu einer Arbeit von F. v. Wisniewski: „Zur Theorie des Heliums".

Von W. Heisenberg in Göttingen.

(Eingegangen am 21. Mai 1924.)

In einer vor kurzem erschienenen Abhandlung[1]) versucht Wiśniewski das Problem des angeregten Heliums auf neuartige Weise zu behandeln, und er behauptet, entgegen den Untersuchungen von P. Epstein[2]), A. Landé[3]), N. Bohr[4]), v. Vleck[5]), M. Born und W. Heisenberg[6]), O. Halpern[7]) bei Anwendung der bisherigen Prinzipien der Mechanik und Quantentheorie zu den richtigen Termwerten für das Parheliumspektrum zu kommen.

Es ist von vornherein klar, daß der Behandlungsweise Wiśniewskis, da er zu anderen Resultaten kommt als die bisherigen Autoren, eine unmechanische Annahme zugrunde liegen muß, wenn nicht zufällig die übereinstimmenden Resultate der bisherigen Heliumarbeiten alle auf demselben Fehler beruhen sollten. Die unmechanische Annahme besteht bei Wiśniewski darin, daß zwar die Wirkung des inneren Elektrons auf das äußere, nicht aber die Rückwirkung des äußeren auf das innere berücksichtigt wird. Dies hat aber unter anderem zur Folge, daß der Gesamtimpuls des Atoms nicht konstant ist.

Wenn es auch sicher ist, daß nur eine Abänderung der mechanischen oder quantentheoretischen Prinzipien zu einer Beschreibung des Heliumspektrums führen kann, so kann doch dieser Versuch Wiśniewskis kaum als gelungen bezeichnet werden, da die Konstanz des Gesamtimpulses durch die Auswahlregeln der Sommerfeldschen inneren Quantenzahl als bewiesen gelten muß, da ferner die Rechnungen Wiśniewskis keinen Anhaltspunkt geben für das Auftreten und Nichtkombinieren zweier Spektra Par- und Orthohelium und da schließlich die Wahl der Quantenzahlen, sowie die Zuordnung derselben zu Ortho- und Parhelium sehr willkürlich erscheint.

---

1) F. v. Wiśniewski, Phys. ZS. **25**, 135, 1924.
2) P. Epstein, Naturw. **6**, 230, 1918.
3) A. Landé, Phys. ZS. **5**, 114, 1920.
4) N. Bohr, Göttinger Vorträge (nicht veröffentlicht).
5) v. Vleck, Phys. Rev. **21**, 372, 1923.
6) M. Born u. W. Heisenberg, ZS. f. Phys. **16**, 229, 1923.
7) O. Halpern, ebenda **18**, 344, 1923.

Es möge die Gelegenheit benutzt werden, einen Rechenfehler zu berichtigen, welcher in der Arbeit von Born und Heisenberg (l. c.) stehengeblieben ist. In der rechten Seite der Gleichung (14) dieser Arbeit ist das zweite Glied der rechten Seite von Gleichung (9) vergessen worden, so daß an Stelle des Faktors $-\frac{15}{8}$ im letzten Glied von (14) der Faktor $-\frac{1}{8}\left[15+\frac{27}{2(Z-1)}\right]$ zu treten hat. Dies führt keine qualitativen, wohl aber quantitative Änderungen in den Zahlen der numerischen Diskussion § 4 herbei. Hinsichtlich der Quantenbahnen § 5 bleiben jedoch alle quantitativen und qualitativen Resultate ungeändert bestehen.

Göttingen, Institut für theoretische Physik.

Zeitschrift für Physik *26*, 196–204 (1924)

Über den Einfluß der Deformierbarkeit der Ionen auf optische und chemische Konstanten. II[1]).

Stabilität und Bildungswärme dreiatomiger Molekeln und Ionen.

Von **W. Heisenberg** in Göttingen.

Mit drei Abbildungen. (Eingegangen am 4. Juni 1924.)

Es werden im Anschluß an eine im Teil I versuchte Theorie der binären Molekeln diejenigen dreiatomigen Molekeln studiert, bei denen alle drei Atome bzw. Ionen auf einer Geraden liegen. Als Anwendung wird auf einen Nernstschen Einwand gegen die Kosselsche Theorie eingegangen.

Einleitung. Im folgenden soll im Anschluß an eine in Teil I versuchte Theorie der binären Molekeln vom Typus NaCl das Problem derjenigen polaren, dreiatomigen Molekeln behandelt werden, bei denen alle drei Atome auf einer Geraden liegen. Ob diese letztere Annahme für Molekel, wie H_2O, CO_2, zutrifft, soll nicht diskutiert werden; es sprechen bekanntlich manche Gründe dagegen. Wir werden aber zeigen, daß die Existenz eines elektrischen Moments nicht als Argument für ein Dreiecksmodell dieser Molekeln aufgefaßt werden darf, sondern daß dieses elektrische Moment im allgemeinen hervorgerufen sein kann durch eine charakteristische Unsymmetrie, die durch die Deformierbarkeiten der Ionen verursacht wird. Ferner soll im Anschluß an die hierbei durchzuführenden Rechnungen auf den bekannten Einwand von Nernst[2]) gegen die Kosselsche Theorie eingegangen werden, der behauptet, daß neben Molekeln der Form H^+Cl^- nach Kossel stets auch eine beträchtliche Anzahl von Ionen der Form $H^+Cl^-H^+$ bzw. $Cl^-H^+Cl^-$ zu erwarten seien, was der Beobachtung widerspräche. Die entsprechende Rechnung unter Berücksichtigung der Deformierbarkeiten wollen wir am Beispiel des Na^+Cl^- durchführen [HCl verhält sich (vgl. Teil I) theoretisch komplizierter, weil der H-Kern in Cl^- eindringt] und zeigen, daß nur eine verschwindend kleine Menge solcher dreiatomiger Ionen $Na^+Cl^-Na^+$ bzw. $Cl^-Na^+Cl^-$ zu erwarten ist.

§ 1. Behandlung des Stabilitätsproblems. Wir nehmen (vgl. Fig. 1) an, daß eine dreiatomige Molekel gegeben sei, bestehend aus einem Z_0-fach geladenen Ion der Masse M_0 und zwei Z_1-fach geladenen Ionen der Masse M_1. Ist die ganze Molekel neutral, so muß

1) M. Born und W. Heisenberg, ZS. f. Phys. **23**, 388, 1924.

2) W. Nernst, Theoretische Chemie, S. 438. 8. bis 10. Aufl. Stuttgart 1921.

natürlich gelten $Z_0 + 2Z_1 = 0$. Die Abstände der Ionen seien r_1 bzw. r_2. Als Kräfte kommen in Betracht:

1. Die elektrostatischen Kräfte.
2. Eine Abstoßung von M_0 auf M_1 der Form $\frac{b e^2}{r^n}$.
3. Die Wirkung der Deformierbarkeiten.

Der Einfachheit halber nehmen wir im folgenden an, daß nur das Ion M_0 merklich deformierbar sei (Deformierbarkeit α vgl. Teil I). Die

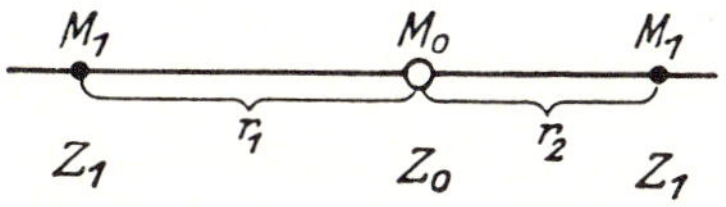

Fig. 1.

Verallgemeinerung auf Deformierbarkeit aller Ionen liefert prinzipiell nichts Neues — auch der Grenzfall, daß allein M_1 deformierbar sei, verhält sich genau, wie der hier behandelte —, würde aber die Rechnung außerordentlich komplizieren.

Bezeichnen wir mit $\mathfrak{p}$ das elektrische Moment von M_0, so wird die gesamte Energie der Molekel [vgl. Teil I, S. 404, Gl. (11)]

$$\varphi = Z_0 Z_1 e^2\left(\frac{1}{r_1} + \frac{1}{r_2}\right) + \frac{Z_1^2 e^2}{r_1 + r_2} + b e^2\left(\frac{1}{r_1^n} + \frac{1}{r_2^n}\right)$$
$$+ \mathfrak{p} e Z_1\left(\frac{1}{r_1^2} - \frac{1}{r_2^2}\right) + \frac{\mathfrak{p}^2}{2\alpha}. \tag{1}$$

Die Gleichgewichtsbedingungen lauten:

$$\frac{\partial \varphi}{\partial \mathfrak{p}} = 0; \quad \frac{\partial \varphi}{\partial r_1} = 0; \quad \frac{\partial \varphi}{\partial r_2} = 0.$$

Die erste liefert

$$\mathfrak{p} = -Z_1 \alpha e\left(\frac{1}{r_1^2} - \frac{1}{r_2^2}\right),$$

also

$$\varphi = Z_0 Z_1 e^2\left(\frac{1}{r_1} + \frac{1}{r_2}\right) + \frac{Z_1^2 e^2}{r_1 + r_2} + b e^2\left(\frac{1}{r_1^n} + \frac{1}{r_2^n}\right) - \frac{\alpha Z_1^2}{2} e^2\left(\frac{1}{r_1^2} - \frac{1}{r_2^2}\right)^2. \tag{2}$$

Wenn wir dimensionslose Größen einführen:

$$r_1 = u \cdot \sqrt[n-1]{n b}; \quad r_2 = v \cdot \sqrt[n-1]{n b}; \quad \frac{\alpha}{(n b)^{\frac{3}{n-1}}} = \beta, \tag{3}$$

so lauten die beiden anderen Gleichgewichtsbedingungen:

$$Z_0 Z_1 \frac{1}{u^2} + Z_1^2 \frac{1}{(u+v)^2} + \frac{1}{u^{n+1}} - Z_1^2 \beta\left(\frac{1}{u^2} - \frac{1}{v^2}\right)\frac{2}{u^3} = 0. \tag{4a}$$

$$Z_0 Z_1 \frac{1}{v^2} + Z_1^2 \frac{1}{(u+v)^2} + \frac{1}{v^{n+1}} + Z_1^2 \beta\left(\frac{1}{u^2} - \frac{1}{v^2}\right)\frac{2}{v^3} = 0. \tag{4b}$$

Betrachten wir nun etwa, um zu quantitativen Resultaten zu kommen, den Fall, wo M_1 die Ladung $+1$ und M_0 die Ladung -2 besitzt:

$$Z_0 = -2; \quad Z_1 = 1; \text{ ferner sei (vgl. Teil I) } n = 9.$$

Wir stellen den Verlauf von u als Funktion von v bzw. umgekehrt graphisch dar (Fig. 2 und 3). Im Fall $\beta = 0{,}5$ schneiden sich die

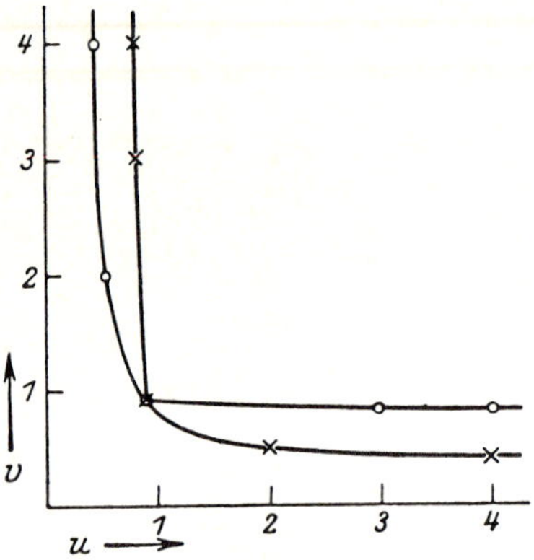

Fig. 2 ($\beta = 0{,}5$).

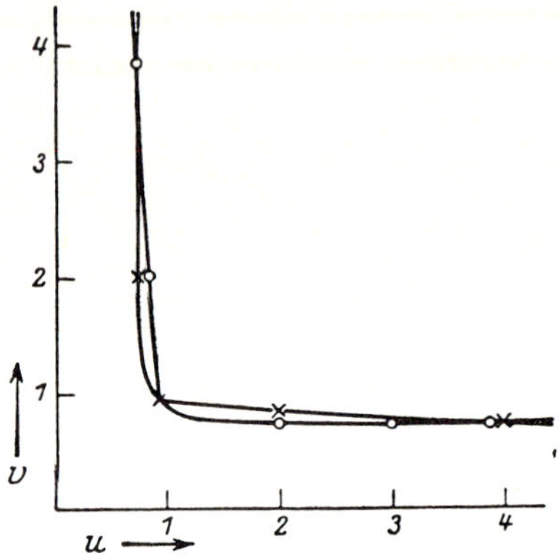

Fig. 3 ($\beta = 1{,}5$).

beiden Kurven (die natürlich symmetrisch um die Gerade $u = v$ liegen) nur in einem Punkt $u = v$. Dieser Schnittpunkt muß immer vorhanden sein. Er bedeutet physikalisch, daß die beiden Ionen M_1 symmetrisch um M_0 herum liegen; es ist klar, daß diese Stellung bei geeigneter Wahl von $r_1 = r_2$ einen möglichen Gleichgewichtszustand darstellt.

Weit interessanter ist dagegen der Fall $\beta = 1{,}5$. Hier schneiden sich die Kurven (4a) und (4b) in drei Punkten. Es gibt also offenbar drei mögliche Gleichgewichtszustände. Neben dem mittleren, der dieselbe Bedeutung wie bei $\beta = 0{,}5$ hat, gibt es zwei gleichwertige unsymmetrische Gleichgewichtslagen, bei denen das eine Ion M_1 dem M_0-Ion erheblich näher ist, als das andere. Der physikalische Grund der Existenz einer solchen unsymmetrischen Gleichgewichtslage beruht auf der Polarisierbarkeit der Ionen. Das dem Ion M_0 nähere M_1-Ion polarisiert M_0 so stark, daß die Anziehungskraft auf das andere Ion M_1 erheblich verringert wird, daß also bei festgehaltenem ersten M_1-Ion die Gleichgewichtslage des zweiten nach außen rückt. Ist die Polarisierbarkeit von M_0 hinreichend groß, so rückt sie so weit nach außen, daß nunmehr auch die Rückwirkung des zweiten Ions auf M_0 und das erste Ion M_1 die Unsymmetrie der Gleichgewichtslage nicht mehr aufhebt.

Der Grenzfall zwischen dem Typus von Fig. 2 ($\beta = 0{,}5$) und Fig. 3 ($\beta = 1{,}5$) ist offenbar dadurch gegeben, daß im Punkte $u = v$

$$\frac{dv}{du} = -1$$

für beide Kurven (4a) und (4b) gilt. Denn dann berühren sich die beiden Kurven im Punkt $u = v$. Rechnen wir den zugehörigen Wert von β aus:

$$\frac{dv}{du} = -1 \quad \text{oder} \quad \frac{\partial \psi}{\partial v} = \frac{\partial \psi}{\partial u}, \tag{5}$$

wenn $\psi(u, v) = 0$ etwa die Gleichung (4a) repräsentiert; wir setzen also

$$\psi = -2 + \frac{1}{\left(1 + \frac{v}{u}\right)^2} + \frac{1}{u^8} - \frac{2\beta}{u}\left(\frac{1}{u^2} - \frac{1}{v^2}\right). \tag{6}$$

Wenn die Gleichung $\frac{dv}{du} = -1$ für $u = v$ erfüllt ist für (4a), so ist sie es von selbst aus Symmetriegründen für (4b). Wir erhalten also

$$\frac{\partial \psi}{\partial u} = -\frac{8}{u^9} + \frac{2\frac{v}{u^2}}{\left(1 + \frac{v}{u}\right)^3} + \frac{2\beta}{u^2}\left(\frac{1}{u^2} - \frac{1}{v^2}\right) + \frac{2\beta}{u} \cdot \frac{2}{u^3},$$

$$\left(\frac{\partial \psi}{\partial u}\right)_{u=v} = -\frac{8}{u^9} + \frac{1}{4u} + \frac{4\beta}{u^4};$$

daraus wegen $\psi_{u=v} = 0$, $\frac{1}{u^8} = \frac{7}{4}$:

$$\left(\frac{\partial \psi}{\partial u}\right)_{u=v} = -\frac{14}{u} + \frac{1}{4u} + \frac{4\beta}{u \cdot \left(\frac{4}{7}\right)^{\frac{3}{8}}},$$

$$\left(\frac{\partial \psi}{\partial v}\right)_{u=v} = \left(-\frac{2\frac{1}{u}}{\left(1 + \frac{v}{u}\right)^3} - \frac{2\beta}{u}\frac{2}{v^3}\right)_{u=v} = -\frac{1}{4u} - \frac{4\beta}{u \cdot \left(\frac{4}{7}\right)^{\frac{3}{8}}}.$$

Also folgt aus $\frac{\partial \psi}{\partial u} = \frac{\partial \psi}{\partial v}$:

$$+14 = 2\left(\frac{1}{4} + \frac{4\beta}{\left(\frac{4}{7}\right)^{\frac{3}{8}}}\right) \tag{7}$$

$$\beta = \frac{6{,}75}{4}\left(\frac{4}{7}\right)^{\frac{3}{8}} = 1{,}37. \tag{8}$$

Zeitschrift für Physik *26*, 196–204 (1924)

Für $\beta < 1{,}37$ haben wir eine, für $\beta > 1{,}37$ drei Gleichgewichtslagen.

Wir müssen nun noch untersuchen, welche Gleichgewichtslage stabil und welche instabil ist. Betrachten wir also, um die Stabilität der Mittellage $u = v$ zu untersuchen, Schwingungen von M_0 gegen die in unverändertem gegenseitigen Abstand gehaltenen Ionen M_1; wir setzen also $r_1 = r + x$; $r_2 = r - x$; $\frac{1}{\mu} = \frac{1}{M_0} + \frac{1}{2M_1}$.

Dann wird

$$\mu \ddot{x} + x\left(\frac{d^2 \varphi(r+x,\, r-x)}{d x^2}\right)_{x=0} = 0,$$

oder wegen $\psi_{u=v} = 0$:

$$\mu \omega^2 = \left(\frac{d^2 \varphi}{d x^2}\right)_{x=0} = 2\left(\frac{\partial^2 \varphi}{\partial r_1^2} - \frac{\partial^2 \varphi}{\partial r_1 \partial r_2}\right)_{x=0}$$

$$= -\frac{2 e^2}{u^2 (9\,\mathrm{b})^{\frac{3}{8}}}\left(\frac{\partial \psi}{\partial u} - \frac{\partial \psi}{\partial v}\right)_{u=v}. \qquad (9)$$

Daraus folgt: Für $\beta < 1{,}37$, d. h. solange nur eine Gleichgewichtslage existiert, ist diese stabil. Sobald die zwei anderen Gleichgewichtslagen auftreten, wird die mittlere labil, die äußeren sind stabil. Dies Ergebnis war auch physikalisch vorauszusehen. Denn wenn man das dritte Ion vom Unendlichen kommend an die anderen heranbringt, so gewinnt man zunächst Arbeit wegen der Überschußladung der beiden ersten Ionen. Die erste auftretende Gleichgewichtslage muß also einem Minimum der Energie entsprechen.

Wir sind schließlich zu dem Resultat gekommen: Molekeln vom Typus $M_1^+ M_0^{--} M_1^+$ werden nur dann symmetrisch um M_0 gebaut sein, wenn die Polarisierbarkeit α von M_0 hinreichend klein ist $\left[\text{nach (3) und (8) muß sein: } \alpha < 1{,}37 \,.\, (9\,\mathrm{b})^{\frac{3}{8}}\right]$. Übersteigt α einen gewissen Grenzwert, so wird die symmetrische Lage der Ionen labil, die Molekel schlägt von selbst in eine unsymmetrische Lage $r_1 \neq r_2$ um. Dieser Schluß bleibt bestehen, auch wenn man die Polarisierbarkeit von M_1 mitberücksichtigt. Wir möchten in der geschilderten Unsymmetrie also den eigentlichen Grund für das Auftreten von elektrischen Momenten bei Molekeln, wie H_2O, CO_2 usw. sehen. Selbst wenn aus anderen Gründen auf eine Dreiecksform der betrachteten Molekel zu schließen sein sollte, so würde bei hinreichend großen Polarisierbarkeiten (und die müssen wir wohl immer bei Sauerstoffverbindungen annehmen) die berechnete Unsymmetrie bestehen bleiben. Wir möchten also für sehr wahrscheinlich halten, daß etwa im H_2O die beiden H-Atome nicht gleich stark gebunden sind. Dieser Schluß bleibt auch wohl dann richtig, wenn die Annahme reiner Polarität für die betrachteten Molekeln nur noch eine sehr schlechte Annäherung bedeutet, denn der Einfluß der Deformierbarkeiten stellt gerade den ersten Schritt zu einem Übergang von der rein polaren zur nichtpolaren Bindung dar.

§ 2. Die Ionen $Na^+Cl^-Na^+$ und $Cl^-Na^+Cl^-$. Alle Schlüsse des vorigen Abschnitts lassen sich auch auf die jetzt zu behandelnden Ionen $Na^+Cl^-Na^+$ und $Cl^-Na^+Cl^-$ ausdehnen. Die Stabilitätsuntersuchung liefert nur insofern etwas anderes, wie in § 1, als oberhalb der Stabilitätsgrenze die äußeren Gleichgewichtslagen mit wachsendem β ins Unendliche rücken. Oberhalb einer zweiten Grenze existiert dann überhaupt keine Gleichgewichtslage mehr, d. h. etwa die Ionen $Na^+Cl^-Na^+$ zerfallen spontan in Na^+ und Na^+Cl^-. Wenn daher die Polarisierbarkeit über dieser zweiten Grenze läge, so existierten Gebilde, wie die von Nernst postulierten, überhaupt nicht; und es ist die Frage, ob nicht bei HCl, wo die relative Polarisierbarkeit offenbar am größten ist, dieser Fall eintritt. Aber da HCl einer sauberen theoretischen Berechnung nicht zugänglich ist, wollen wir NaCl untersuchen. Da läßt sich nun leicht feststellen, daß bei NaCl die Polarisierbarkeit unter beiden Stabilitätsgrenzen liegt, d. h. es existieren Ionen der von Nernst postulierten Form und sie sind symmetrisch gebaut.

Der Nernstsche Einwand gegen die Kosselsche Theorie beruht im wesentlichen darauf, daß nach der einfachen Kosselschen Annahme starrer Molekeln die Wärmetönung des Prozesses

$$3\,Na^+Cl^- \rightleftarrows Na^+Cl^-Na^+ + Cl^-Na^+Cl^-$$

gleich Null ist. In der Tat beträgt die elektrostatische Energie von Na^+Cl^- zunächst $-\frac{e^2}{r}$, die von $Na^+Cl^-Na^+$ bzw. $Cl^-Na^+Cl^-$ je $-\frac{3\,e^2}{2\,r}$, also ist die Wärmetönung

$$-\frac{3\,e^2}{r} + 2\cdot\frac{3}{2}\frac{e^2}{r} = 0.$$

Wir wollen diese Wärmetönung unter Berücksichtigung der Deformierbarkeiten ausrechnen. Bedeute α jetzt die Polarisierbarkeit des äußeren Ions (das innere ist aus Symmetriegründen nicht deformiert), so wird (bis auf quadratische Glieder in α):

$$\varphi = -\frac{3\,e^2}{2\,r} + \frac{2\,b\,e^2}{r^9} - \frac{\alpha}{2}\cdot\frac{9}{8}\frac{e^2}{r^4}\,. \tag{10}$$

Zeitschrift für Physik 26, 196–204 (1924)

Die Gleichgewichtsbedingung liefert:

$$\frac{\partial \varphi}{\partial r} = \frac{3e^2}{2r^2} - \frac{2.9be^2}{r^{10}} + \frac{4\alpha.9}{2.8}\frac{e^2}{r^5} = 0.$$

Sei $r = r_0 + \alpha r_1$, so wird $r_0^8 = 12b$ und in erster Näherung (bis auf Glieder mit α^2):

$$\varphi = -\frac{3e^2}{2r_0} + \frac{2be^2}{r_0.12.b} - \frac{\alpha}{2}\cdot\frac{9}{8}\cdot\frac{e^2}{r_0\cdot(12b)^{\frac{3}{8}}}$$

$$\varphi = -\frac{e^2}{r_0}\left(+\frac{4}{3} + \frac{\alpha}{2}\cdot\frac{9}{8.(12b)^{\frac{3}{8}}}\right). \tag{11}$$

In Teil I, Formel (15), S. 405 ist die Größe $\sqrt[8]{9b}$ als r_0 eingeführt. Nennen wir sie jetzt r', also $r' = \sqrt[8]{9b}$, so wird in dieser Einheit

$$\varphi = -\frac{e^2}{r'}\cdot\sqrt[8]{\frac{3}{4}}\left[\frac{4}{3} + \frac{9\alpha}{16r'^3}\cdot\left(\frac{3}{4}\right)^{\frac{3}{8}}\right]. \tag{12}$$

Für NaCl ist (vgl. Teil I) $r' = 2{,}438.10^{-8}$. Es wird demnach wegen $\alpha_{Na^+} = 0{,}21.10^{-24}$, $\alpha_{Cl^-} = 3{,}05.10^{-24}$:

$$\left.\begin{aligned}\varphi_{Na^+Cl^-Na^+} &= -\frac{e^2}{r'}\cdot 0{,}965\left(1{,}333 + 0{,}505\cdot\frac{0{,}21}{14{,}45}\right) = -\frac{e^2}{r'}\cdot 1{,}29\\ \varphi_{Cl^-Na^+Cl^-} &= -\frac{e^2}{r'}\cdot 0{,}965\left(1{,}333 + 0{,}505\cdot\frac{3{,}05}{14{,}45}\right) = -\frac{e^2}{r'}\cdot 1{,}39\end{aligned}\right\} \tag{13}$$

Um aus (13) die Bildungswärme in kcal zu berechnen, entnehmen wir aus Tabelle 14 von Teil I, daß V_0, welches einem Potential $-\frac{e^2}{r'}\frac{8}{9}$ entspricht, 121 kcal beträgt. Also erhalten wir

$$V_{Na^+Cl^-Na^+} = 121\cdot\tfrac{9}{8}\cdot 1{,}29 = 176 \text{ kcal}$$
$$V_{Cl^-Na^+Cl^-} = 121\cdot\tfrac{9}{8}\cdot 1{,}39 = 189 \text{ „}$$

Die Wärmetönung des Prozesses

$$3\,Na^+Cl^- \rightleftarrows Na^+Cl^-Na^+ + Cl^-Na^+Cl^-$$

beträgt also (wegen $V_{Na^+Cl^-} = 139$, vgl. Teil I, Tab. 14)

$$3.139 - (176 + 189) = 52 \text{ kcal}. \tag{14}$$

Man sieht ohne weiteres, daß eine so große Wärmetönung die Bildung der von Nernst (für den Fall starrer Ionen mit Recht) postulierten dreiatomigen Ionen fast völlig unterdrückt.

Wir wollen aber die Rechnung hier doch kurz andeuten und zu diesem Zweck die chemischen Konstanten von Na^+Cl^-, $Na^+Cl^-Na^+$ und $Cl^-Na^+Cl^-$ quantentheoretisch nach Stern und Fowler ausrechnen.

Wir nehmen an, daß wir uns in einem Temperaturbereich befänden, in welchem die Rotationen völlig, die Kernschwingungen jedoch noch sehr wenig angeregt sind (etwa $T < 1000$). Man kann leicht ausrechnen, daß die Berücksichtigung der Librationen bei höheren Temperaturen das Gleichgewicht noch stark zuungunsten der Bildung von $Na^+Cl^-Na^+$ verschiebt.

Nach Stern und Fowler weicht die chemische Konstante eines Rotators mit dem Trägheitsmoment A von der eines einatomigen Gases um das Glied

$$\log \frac{8\pi^2 k A}{h^2}$$

ab.

Bei symmetrischen Molekeln wie $Na^+Cl^-Na^+$, die eine Vertauschung zulassen, ist der Bruch unter dem Logarithmus durch 2 zu dividieren.

Zur Berechnung der Glieder brauchen wir die Trägheitsmomente von Na^+Cl^- (A_0), $Na^+Cl^-Na^+$ (A_1) und $Cl^-Na^+Cl^-$ (A_2). Setzen wir näherungsweise für die letzteren Ionen $r = r_0$, so wird

$$A_1 = 2\,r_0^2 \,.\, M_{\mathrm{Na}} = 4{,}8 \,.\, 10^{-38}\,\mathrm{g\,cm^2}$$
$$A_2 = 2\,r_0^2 \,.\, M_{\mathrm{Cl}} = 7{,}4 \,.\, 10^{-38}\,\mathrm{g\,cm^2}.$$

Nach Teil I, Tabelle 15 ist ferner

$$A_0 = 1{,}21 \,.\, 10^{-38}\,\mathrm{g\,cm^2}.$$

Die chemische Konstante C wird hiernach

$$C = -1{,}6 + \frac{3}{2}\log M + \log \frac{8\pi^2 k A}{\sigma h^2}$$

wo $\sigma = 2$ für symmetrische Molekeln), also

$$\left.\begin{aligned} C_{\mathrm{Na^+Cl^-}} &= -1{,}6 + 2{,}66 + 0{,}48 = 1{,}54 \\ C_{\mathrm{Na^+Cl^-Na^+}} &= -1{,}6 + 2{,}87 + 0{,}78 = 2{,}05 \\ C_{\mathrm{Cl^-Na^+Cl^-}} &= -1{,}6 + 2{,}96 + 0{,}97 = 2{,}33. \end{aligned}\right\} \qquad (15)$$

Sei nun δ die Anzahl der in Ionen vom Typus $Na^+Cl^-Na^+$ bzw. $Cl^-Na^+Cl^-$ verwandelten NaCl-Molekel im Verhältnis zu ihrer Gesamtzahl. Dann gibt es pro Mol

$(1 - \delta)$ Molekeln NaCl,

$\frac{\delta}{3}$ Ionen $Na^+Cl^-Na^+$,

$\frac{\delta}{3}$ Ionen $Cl^-Na^+Cl^-$.

Zeitschrift für Physik 26, 196–204 (1924)

Die Gleichgewichtskonstante der betrachteten Reaktion ist daher

$$K_p = P\,\frac{27\,.\,(1-\delta)^3}{(3-\delta)\,\delta^2}, \tag{16}$$

wo P den Gesamtdruck angibt. Sei $P = 1$ Atm. und δ (wie sich später herausstellt) klein, so wird

$$K_p = \frac{9}{\delta^2}.$$

Nach allgemeinen Sätzen ist dann [vgl. (14) und (15)]:

$$\log K_p = \frac{52\,000}{4{,}571\,T} + \frac{3}{2}\log T + 3\,.\,1{,}54 - 2{,}05 - 2{,}33$$

$$= \frac{52\,000}{4{,}571\,T} + \frac{3}{2}\log T + 0{,}24. \tag{17}$$

Daraus

$$2\log\delta = -\left[\frac{52\,000}{4{,}571\,T} + \frac{3}{2}\log T - 0{,}71\right]$$

$$= -\left[\frac{113\,80}{T} + \frac{3}{2}\log T - 0{,}71\right]. \tag{18}$$

Für $T = 1000^0$ abs. gibt dies

$$\delta = 2{,}6\,.\,10^{-8}. \tag{19}$$

Bei Berücksichtigung der Deformationen trifft also jedenfalls für Steinsalz der Einwand gegen die Kosselsche Theorie, daß sich $Na^+Cl^-Na^+$-Ionen in merklicher Menge bilden müßten, in keiner Weise zu. Zweifellos liegen die Verhältnisse bei HCl nicht wesentlich anders, als bei NaCl, wahrscheinlich sind sie einer Bildung von $H^+Cl^-H^+$ bzw. $Cl^-H^+Cl^-$ noch erheblich ungünstiger.

Wir schließen mit der Bemerkung, daß, wie sich in den Rechnungen von Teil I und II gezeigt hat, die Berücksichtigung der Ionendeformierbarkeit eine, wenn auch näherungsweise, Beschreibung sämtlicher chemischer Eigenschaften polarer Molekeln, insbesondere der Bildungswärme und der chemischen Konstante, gestattet. Wieweit diese Tatsache empirisch verwertbar sein wird, d. h. für welche Bindungen die Annahme polaren Charakters mit Berücksichtigung der Deformierbarkeit eine vernünftige Näherung darstellt, muß sich durch die Erfahrung entscheiden.

Göttingen, Institut für theoretische Physik.

Zeitschrift für Physik 26, 291–307 (1924)

Über eine Abänderung der formalen Regeln der Quantentheorie beim Problem der anomalen Zeemaneffekte.

Von **W. Heisenberg** in Göttingen.

Mit einer Abbildung. (Eingegangen am 13. Juni 1924.)

In der Arbeit wird der Versuch gemacht, durch eine einfache Erweiterung der quantentheoretischen Regeln zu einer formalen Beschreibung eines Teiles derjenigen Erscheinungen zu gelangen, die bei der Kopplung mehrerer Elektronen auftreten. Insbesondere wird gezeigt, daß die Widersprüche gegen das Aufbauprinzip hinsichtlich der statistischen Gewichte, sowie die „Permanenzgesetze" der Zeemaneffekte sich kaum anders deuten lassen, als durch die Annahme, daß die Kopplungsenergie nicht durch eine, sondern durch ein Paar von Quantenzahlen definiert ist.

Einleitung. Seitdem das empirische Material [1]) bei den anomalen Zeemaneffekten durch Landé [2]) entsprechend den bisherigen quantentheoretischen Prinzipien systematisch geordnet und in Formeln gebracht wurde, stellte es sich immer klarer heraus, daß eine Erklärung der Erscheinungen des anomalen Zeemaneffekts tiefgreifende Änderungen in unseren quantentheoretischen Vorstellungen mit sich bringen müsse [3]). Besonders eindrucksvoll zeigt sich dies im Versagen des Aufbauprinzips hinsichtlich der statistischen Gewichte von Atomrest und Elektron [4]).

Obwohl wir von einer wirklichen Aufklärung der beim Zeemaneffekt auftretenden grundsätzlichen Schwierigkeiten weit entfernt sind, scheint es doch, als ob sich ein ausgesprochener Fortschritt in formaler Hinsicht erzielen ließe, wenn man die Regeln der Quantentheorie in einfacher Weise abändert. Wir wollen zeigen, daß wir durch eine solche Abänderung zu einer Beschreibung nicht nur des Aufbauprinzips, sondern auch der Landéschen g-Werte, sowie der für die anomalen Zeemaneffekte so charakteristischen Permanenzgesetze [5]) gelangen können [6]).

---

[1]) E. Back, Ann. d. Phys. **70**, 333, 1923; ZS. f. Phys. **15**, 206, 1923.

[2]) A. Landé, ZS. f. Phys. **5**, 231, 1921; Phys. ZS. **22**, 417, 1921; ZS. f. Phys. **7**, 398, 1921; **11**, 353, 1922; **15**, 189, 1923 (im folgenden als Landé I zitiert); **16**, 391, 1923; **17**, 292, 1923; **19**, 112, 1923 (im folgenden als Landé II zitiert).

[3]) A. Landé, Phys. ZS. **24**, 441, 1923.

[4]) N. Bohr, Ann. d. Phys. **71**, 228, 1923.

[5]) W. Pauli, ZS. f. Phys. **16**, 155, 1923; Landé II und W. Heisenberg, ZS. f. Phys. **8**, 273, 1922.

[6]) Ich möchte nicht versäumen, an dieser Stelle Herrn Prof. Bohr für die vielfache Förderung, die ich aus seinen Ratschlägen erfahren habe, den herzlichsten Dank auszusprechen.

§ 1. Formulierung des neuen Quantenprinzips. Zu der in Rede stehenden Erweiterung der Quantenregeln wird man unmittelbar durch eine Betrachtung der Landéschen g-Formel [1]) geführt. Wir benutzen die Bezeichnungen und Numerierungen von Landé: $\dot{R}$ = Impuls des Atomrests, K = Impuls des äußeren Elektrons, J = Gesamtimpuls des Atoms. Dann ist

$$g = 1 + \frac{(J+\frac{1}{2})(J-\frac{1}{2}) + R^2 - K^2}{(J+\frac{1}{2})(J-\frac{1}{2})}. \tag{1}$$

Nach dem Aufbauprinzip ist ferner zu erwarten, daß sich der Wert J als Impuls des Atomrests im Atom mit nächsthöherer Elektronenzahl wiederfindet. Es zeigt sich aber, daß sich im Spektrum des nächsten Atoms zwei Multiplettsysteme mit dem R-Impuls $J+\frac{1}{2}$ und $J-\frac{1}{2}$ finden. So kommt man fast zwangläufig dazu, einem Wert der Kopplungsenergie nicht die Impulsquantenzahl J, sondern die beiden Zahlen $(J+\frac{1}{2}, J-\frac{1}{2})$ zuzuordnen, denn sowohl in der g-Formel, wie in den R-Impulsen des nächsthöheren Elements tritt statt J immer das Paar $(J+\frac{1}{2}, J-\frac{1}{2})$ auf. Indem wir alle Betrachtungen über den physikalischen Sinn des folgenden Formalismus auf später verschieben, schreiben wir die neue Quantenregel hin:

Ein bestimmter Wert der von der Kopplung zwischen Elektron und Atomrest herrührenden Energie ist nicht, wie bisher angenommen, mit einem Wert J der Gesamtimpulsquantenzahl, sondern mit zwei solchen Werten ($J+\frac{1}{2}$, $J-\frac{1}{2}$ in unserer Numerierung) verknüpft. Er wird berechnet, indem man in der Formel (wir definieren eine Integralinvariante F als $F = \int H\,dJ$)

$$H = \frac{\partial F}{\partial J} \tag{2}$$

den Differentialquotienten der rechten Seite durch die Differenz $\Delta F = F(J+\frac{1}{2}) - F(J-\frac{1}{2})$ ersetzt. Die so erhaltene Energie der quantenmechanischen Kopplung (sie heiße H_{qu} im Gegensatz zu H_{kl}) genügt also der Gleichung

$$H_{qu} = F(J+\tfrac{1}{2}) - F(J-\tfrac{1}{2}) = \Delta F = \int_{-\frac{1}{2}}^{+\frac{1}{2}} H_{kl}\,dJ. \tag{3}$$

Um die formale Analogie dieser Regel zur Frequenzbedingung hervorzuheben, schreiben wir die letztere hier an:

$$\nu_{qu} = \Delta H = \int \nu_{kl}\,dJ_k. \tag{4}$$

[1]) Landé I.

Aus der Form (3) der neuen Quantenregel erkennt man einerseits, daß im Grenzfall hoher Quantenzahlen H_{qu} und H_{kl} genähert gleich werden, andererseits sieht man, daß H_{qu} formal einfach durch Mittelung von H_{kl} erhalten werden kann.

Wir wollen jetzt zeigen, daß wir auf Grund des von Landé[1]) und Pauli[2]) behandelten Ersatzmodells zu den richtigen g-Werten und dem Permanenzgesetz gelangen.

§ 2. Das Ersatzmodell. Im folgenden bedeute k den Impuls des äußeren Elektrons, r den des Atomrests, m die Komponente des Gesamtimpulses in der Feldrichtung (äußeres Feld $= H$), p_r und p_k die Komponenten von r bzw. k in der Feldrichtung, j den Gesamtimpuls, φ_r, φ_k, ψ die zu p_r, p_k und j gehörigen Winkel. Es bedeuten also φ_r und φ_k die Azimute von r und k um die Feldrichtung, ψ den Winkel zwischen der (r-k-j)-Ebene und der (jH)-Ebene. Mit Pauli (l. c.) nehmen wir für unser Modell an, daß der Atomrest doppelten Magnetismus besitze. Die großen Buchstaben R, K, J gebrauchen wir für die zu r, k und j gehörigen Quantenzahlen. Die Stärke des äußeren Magnetfeldes werde durch die Larmorfrequenz $\mathfrak{o}$ charakterisiert. Bei allen Impulsen lassen wir der Einfachheit halber einen Faktor $h/2\pi$ weg, damit sie von der Dimension einer reinen Zahl werden.

Das durch das Ersatzmodell gegebene mechanische Problem ist bereits ausführlich von Pauli (l. c.) behandelt worden. Wir wiederholen im folgenden nur die für uns wichtigen Einzelheiten der Rechnung.

Zur mathematischen Behandlung kann man zwei Sorten von Variabeln benutzen:

I. r, k, m, j und ψ

oder II. r, k, m, p_k und $(\varphi_k - \varphi_r)$

bzw. r, k, m, p_r und $(\varphi_r - \varphi_k)$.

In beiden Fällen haben wir ein Problem von einem Freiheitsgrad mit den konjugierten Variabeln I. j und ψ oder II. p_k und $(\varphi_k - \varphi_r)$ bzw. p_r und $(\varphi_r - \varphi_k)$. Die Variabelnreihe I. ist dem Falle eines schwachen äußeren Feldes, die Reihe II. dem Falle mittlerer und starker Felder mehr angemessen.

Die Gesamtenergie des Atoms im äußeren Felde setzt sich aus drei Teilen zusammen: der Wirkung von H auf das äußere Elektron $k \cdot \cos(kH)\,\mathfrak{o} \cdot h$, auf den Atomrest $2 \cdot r \cdot \cos(rH)\,\mathfrak{o} \cdot h$ und der Wechselwirkung zwischen Elektron und Atomrest. Die letztere Wechselwirkung

---

[1]) Landé II.

[2]) W. Pauli jr., ZS. f. Phys. **20**, 371, 1924, im folgenden als Pauli, l. c. zitiert.

muß bekanntlich angenähert dem Kosinus zwischen r und k proportional gesetzt werden, wenn man die richtigen sogenannten „Intervallproportionen“ [1]) erhalten will. Wir werden jedoch diese Annahme zunächst noch nicht machen, um zu zeigen, daß die g-Werte, sowie das Permanenzgesetz, von dieser Annahme unabhängig sind.

Die Hamiltonsche Funktion H des zugrundegelegten Modells lautet also in den Variabeln I:

$$\frac{1}{h}\cdot H = \mathfrak{o}\cdot\left\{m\left(2\frac{j^2+r^2-k^2}{2j^2}+\frac{j^2+k^2-r^2}{2j^2}\right)\right.$$

$$\left.+\frac{\cos\psi}{2j^2}\sqrt{(j^2-m^2)\left[(k+r)^2-j^2\right]\left[j^2-(k-r)^2\right]}\right\}+f\left(\frac{j^2-k^2-r^2}{2kr}\right) \quad (5)$$

Es ist nämlich

$$\cos(kj)=\frac{j^2+k^2-r^2}{2kj},\quad \cos(rj)=\frac{j^2+r^2-k^2}{2rj},$$

$$\cos(kr)=\frac{j^2-k^2-r^2}{2kr},\quad \cos(jH)=\frac{m}{j}\quad\text{und}$$

$$\cos(kH)=\cos(jH)\cos(jk)-\sin(jH)\sin(jk)\cos\psi$$

$$\cos(rH)=\cos(jH)\cos(jr)-\sin(jH)\sin(jr)\cos\psi.$$

In den Variabeln II wird

$$\cos(rH)=\frac{p_r}{r},\quad \cos(kH)=\frac{p_k}{k},$$

$$\cos(rk)=\cos(rH)\cos(kH)+\sin(rH)\sin(kH)\cos(\varphi_r-\varphi_k),$$

also

$$\frac{1}{h}H=(2p_r+p_k)\,\mathfrak{o}\,.\,h+f\left(\frac{p_r}{r}\cdot\frac{p}{k}+\frac{\sqrt{r^2-p_r^2}}{r}\frac{\sqrt{k^2-p_k^2}}{k}\cos(\varphi_r-\varphi_k)\right)$$

$$=(p_r+m)\,\mathfrak{o}h+f\left(\frac{p_r}{r}\frac{m-p_r}{k}+\frac{\sqrt{r^2-p_r^2}}{r}\frac{\sqrt{k^2-(m-p_r)^2}}{k}\cos\chi\right), \quad (6)$$

wobei $\chi=\varphi_r-\varphi_k$ zu p_r kanonisch konjugiert ist.

Führen wir jetzt die Phasenintegrale ein:

$$\frac{1}{2\pi}\oint j\,d\psi=J,\qquad \frac{1}{2\pi}\oint p_r\,d\chi=P_r,$$

so ergibt sich zunächst durch einfache störungstheoretische Betrachtungen [2]) aus (5) für $\mathfrak{o}\ll f$ (schwaches Magnetfeld):

$$H=f\left(\frac{J^2-K^2-R^2}{2KR}\right)+\left[1+\frac{J^2+R^2-K^2}{2J^2}\right]M\,\mathfrak{o}h+\mathfrak{o}^2\ldots \quad (7)$$

[1]) Vgl. Landé I und W. Heisenberg, l. c.

[2]) Siehe z. B. N. Bohr, On the Quantum theory of Line-spectra, part II.

und aus (6) für $\mathfrak{o} \gg f$ (starkes Magnetfeld):

$$H = (P_r + M)\,\mathfrak{o}\,.\,h$$

$$+ \frac{1}{2\pi}\int\limits_0^{2\pi} d\chi \cdot f\left(\frac{P_r}{R}\frac{M - P_r}{K} + \frac{\sqrt{(R^2 - P_r^2)[K^2 - (M - P_r)^2]}}{R\,.\,K}\cos\chi\right). \quad (8)$$

Die beiden Phasenintegrale J und P_r sind naturgemäß zur Festlegung der stationären Zustände völlig gleichwertig, nach allgemeinen mechanischen Sätzen können sie sich nur um ganzzahlige Vielfache von M, R und K unterscheiden.

Auf die Frage der Zuordnung der einzelnen Werte J zu denen von P_r wollen wir hier nicht eingehen, da diese Frage eben den Gegenstand der oben zitierten Arbeit von Pauli bildet.

Um das Rechenschema von § 1, Gleichung (3) anzuwenden, bilden wir jetzt die Funktion

$$F = \int H\,d\,J \text{ bzw.} = \int H\,d\,P_r.$$

Es wird für $\mathfrak{o} \ll f$ nach (7)

$$F = \int f\left(\frac{J^2 - K^2 - R^2}{2\,K\,.\,R}\right) d\,J + \left(J + \frac{J^2 + K^2 - R^2}{2\,J}\right) M\,\mathfrak{o}\,h, \quad (9)$$

und nach (8) für $\mathfrak{o} \gg f$

$$F = \left(M P_r + \frac{P_r^2}{2}\right)\mathfrak{o}\,.\,h$$

$$+ \frac{1}{2\pi}\int d\,P_r \int\limits_0^{2\pi} d\chi f\left(\frac{P_r}{R}\frac{M - P_r}{K} + \frac{\sqrt{(R^2 - P_r^2)[K^2 - (M - P_r)]}}{R\,.\,K}\cos\chi\right). \quad (10)$$

Wird speziell die Wechselwirkung zwischen Elektron und Atomrest proportional mit dem Kosinus zwischen R und K angesetzt, so vereinfachen sich die Formeln (9) und (10):

$$f\left(\frac{J^2 - K^2 - R^2}{2\,K R}\right) = \omega \cdot \frac{J^2 - K^2 - R^2}{2\,K R} \cdot h;$$

für $\mathfrak{o} \ll \omega$:

$$F = \frac{\frac{1}{3} J^3 - J(K^2 + R^2)}{2\,K R}\,\omega\,.\,h + \left(J + \frac{J^2 + K^2 - R^2}{2\,J}\right) M\,\mathfrak{o}\,.\,h \quad (11)$$

und für $\mathfrak{o} \gg \omega$:

$$F = \left(M P_r + \frac{P_r^2}{2}\right)\mathfrak{o}\,.\,h + \frac{P_r^2\left(M - \frac{2}{3} P_r\right)}{2\,R K}\,\omega\,.\,h. \quad (12)$$

Zeitschrift für Physik *26*, 291–307 (1924)

§ 3. **Quantentheoretische Deutung nach § 1 und Vergleich mit den empirischen Formeln.** Ein Vergleich von (1) und (7) lehrt, daß wir nach Formel (3) zu den richtigen g-Werten kommen, wenn wir für K, R, M die Landésche Numerierung beibehalten — also

$$K = \frac{1}{2}, \frac{3}{2}, \frac{5}{2}, \frac{7}{2} \dots$$

für s-, p-, d-, b-... Term;

$$R = \frac{1}{2}, \frac{2}{2}, \frac{3}{2}, \frac{4}{2} \dots$$

für Singlett-, Dublett-, Triplett-, Quartett-... Systeme. Der Quantenzahl J aber müssen wir statt des Landéschen Wertes J_0 die zwei Werte $J_0 + \frac{1}{2}$, $J_0 - \frac{1}{2}$ zuschreiben.

Z. B. hätten wir im Falle der Dubletts $R = 1$:

	s-Term		p_1-,		p_2-Term	d_1-,		d_2-Term
R	1			1			1	
K	$\frac{1}{2}$			$\frac{3}{2}$			$\frac{5}{2}$	
J	$\frac{3}{2}$	$\frac{1}{2}$	$\frac{5}{2}$	$\frac{3}{2}$	$\frac{1}{2}$	$\frac{7}{2}$	$\frac{5}{2}$	$\frac{3}{2}$

Die so entstehende Normierung des J und M hat offenbar große formale Vorzüge vor der bisher notwendigen: Die vektorielle Zusammensetzung von K und R zu J liefert bekanntlich die Beziehung:

$$|K - R| \leqq J \leqq |K + R|. \tag{13}$$

In dieser Beziehung mußten bisher die den Gleichheitszeichen entsprechenden Parallelstellungen ausgeschlossen werden, um zur richtigen Anzahl der Niveaus zu gelangen, was entweder durch geeignete Numerierung von J (so, daß $J_{\max} = |K + R - \frac{1}{2}|$ war) oder durch Zusatzverbote erreicht wurde. In der Sommerfeldschen[1]) Numerierung traten diese Schwierigkeiten nicht auf, doch entzieht sich die Sommerfeldsche Numerierung, da sie den Impuls des Grundzustandes mit dem der Anregung zusammensetzt, in dieser Form der modellmäßigen Deutung.

Analoge Schwierigkeiten herrschten bei der Normierung von M insofern, als die Parallelstellung $M = J$ in der bisher üblichen Nume-

[1]) A. Sommerfeld, Ann. d. Phys. **73**, 32, 1923; Phys. ZS. **24**. 360, 1923.

rierung ausgeschlossen war. Bei der hier vorgeschlagenen Erweiterung der Quantenprinzipien ergibt sich von selbst, daß J sämtliche Werte

$$K+R,\ K+R-1\ \ldots\ |K-R|,$$

und ebenso M sämtliche Werte

$$J,\ J-1,\ \ldots\ -J$$

annehmen kann, wie es zunächst zu erwarten war.

Die Anzahl der möglichen Energieniveaus ist nämlich nach § 1 um Eins kleiner als die Anzahl der möglichen J-Werte. Daraus ergibt sich dann die richtige Anzahl der Aufspaltungsterme.

An einem Beispiel soll dieser Sachverhalt klargelegt werden:

Wir betrachten den p-Term eines Triplettsystems, bei welchem das Kosinusgesetz, also Gleichung (11) und (12) gilt.

Es ist dann $R=\frac{3}{2},\ K=\frac{3}{2}\ (\mathfrak{o}\ll\omega)$:

		$F(J,M)$	H_{qu}	$\Delta\nu$ für $\mathfrak{o}=0$
p_1	$J=3$	$-\omega h+\frac{9}{2}M\mathfrak{o}h$	$\frac{11}{27}\omega h+\frac{3}{2}M\mathfrak{o}h$	
	$J=2$	$-\frac{38}{27}\omega h+3M\mathfrak{o}h$		$\frac{24}{27}\omega h$
p_2	$J=2$, $J=1$		$-\frac{13}{27}\omega h+\frac{3}{2}M\mathfrak{o}h$	
	$J=1$	$-\frac{25}{27}\omega h+\frac{3}{2}M\mathfrak{o}h$		$\frac{12}{27}\omega h$
p_3	$J=0$	$0\cdot\omega h+0\cdot M\mathfrak{o}h$	$-\frac{25}{27}\omega h+\frac{3}{2}M\mathfrak{o}h$	

wobei die zugehörigen möglichen M-Werte aus folgendem Schema hervorvorgehen:

$$\begin{array}{lll ccccccc}
 & J=3 & M= & -3 & -2 & -1 & 0 & +1 & +2 & +3\\
p_1 & \updownarrow & & & \updownarrow & \updownarrow & \updownarrow & \updownarrow & \updownarrow & \\
 & J=2 & M= & & -2 & -1 & 0 & +1 & +2 & \\
p_2 & \updownarrow & & & & \updownarrow & \updownarrow & \updownarrow & & \\
 & J=1 & M= & & & -1 & 0 & +1 & & \\
p_3 & \updownarrow & & & & & \updownarrow & & & \\
 & J=0 & M= & & & & 0 & & &
\end{array}$$

Man erkennt hieraus: Wir erhalten drei p-Terme, wie es sein muß, wobei M im oberen bis 2, im zweiten bis 1 und im untersten „bis 0“ laufen kann. Auch das Verhältnis der $\Delta\nu$ ergibt sich bei Annahme des „Kosinusgesetzes“ richtig zu 2 : 1. Aus (11) folgt die Landésche Erweiterung des Gesetzes der Intervallproportionen auf beliebige Multipletts natürlich ganz allgemein.

Zeitschrift für Physik *26*, 291–307 (1924)

Auch im Falle starker Felder weicht die von uns zu wählende Normierung der P_r von der bisherigen ab. Es ist auch hier die Parallelstellung $R \parallel H$ erlaubt, P_r nimmt die Werte an:

$$R,\ R-1,\ \ldots -(R-1),\ -R.$$

§ 4. Die Permanenzprinzipien. Der Inhalt der sogenannten Summationssätze läßt sich in folgende Behauptung zusammenfassen:

Die Summe $\sum_{J_{\min}}^{J_{\max}} {}^J H_{qu}$ ist linear in $\mathfrak{o}$ und ω.

Wir werden zeigen, daß dies unmittelbar aus der Quantenregel von § 1 folgt.

Nach (2) ist nämlich:

$$\sum_{J_{\min}}^{J_{\max}} {}^J H_{qu} = \int_{J_{\min}}^{J_{\max}} H_{kl}\, dJ = \int_{J_{\min}}^{J_{\max}} \int_0^1 H_{kl}\, dJ\, dw. \tag{14}$$

Hierin sei w die zu J konjugierten Winkelvariablen. Wird ferner $w_0 = \frac{\psi}{2\pi}$ gesetzt, so ist nach allgemeinen mechanischen Gesetzen $dJ\,dw = dj\,dw_0$. Aus diesem Grunde läßt sich das Integral (14) verwandeln in ein Integral in der j, w_0-Ebene, erstreckt über das dem Rechteck: $J_{\max} \geqq J \geqq J_{\min}$, $1 \geqq w \geqq 0$ entsprechende Gebiet[1]). Wir müssen also untersuchen, auf welches Gebiet in der j, w_0-Ebene sich das genannte Rechteck abbildet. Die besonderen Grenzen dieses Rechtecks verursachen, und das ist der wesentliche Punkt des zu führenden Beweises, daß sich das Gebiet: $J_{\max} \geqq J \geqq J_{\min}$, $1 \geqq w \geqq 0$ genau wieder auf das Rechteck $j_{\max} \geqq j \geqq j_{\min}$, $1 \geqq w_0 \geqq 0$ abbildet. Aus (5) folgt nämlich, daß $j = j_{\max} = \text{const}$ und $j = j_{\min} = \text{const}$ Lösungen der Bewegungsgleichungen sind, da für $j = j_{\max} = k + r$ und $j = j_{\min} = |k - r|$ bzw. $= |m|$ der Faktor von $\cos\psi$ in (5) verschwindet, so daß

$$\frac{dj}{dt} = -\frac{\partial H}{\partial \psi} = 0$$

wird. Also bilden sich die Grenzen $J = J_{\max}$ und $J = J_{\min}$ auf die Geraden $j = j_{\max}$ und $j = j_{\min}$ ab. Daß sich aber dann auch das ganze Rechteck $J_{\max} \geqq J \geqq J_{\min}$, $1 \geqq w \geqq 0$ auf das ganze Rechteck $j_{\max} \geqq j \geqq j_{\min}$, $1 \geqq w_0 \geqq 0$ abbildet, folgt einfach daraus, daß für beliebige Felder jeder Punkt des letzteren Rechtecks einen reellen Wert

[1]) Diese Bemerkung verdanke ich Herrn Prof. Born, dem ich hierfür, sowie für viele andere wertvolle Ratschläge zu großem Dank verpflichtet bin.

der Energie, also auch eine mögliche Bewegung liefert. Wir haben aus (5) und (14)

$$\sum_{J_{\min}}^{J_{\max}} J\, H_{qu} = \int_{J_{\min}}^{J_{\max}} \int_0^1 H_{kl}\, d\, J\, d\, w = \int_{j_{\min}}^{j_{\max}} \int_0^1 H_{kl}\, d\, j\, d\, w_0$$

$$= m\, \mathfrak{o}\, h \int_{j_{\min}}^{j_{\max}} d\, j \left(1 + \frac{j^2 - k^2 + r^2}{2\, j^2}\right) + h\,.\int_{j_{\min}}^{j_{\max}} d\, j\, f\left(\frac{j^2 - k^2 - r^2}{2\, k\, r}\right) \quad (15)$$

Die Summe (14) ist tatsächlich linear in $\mathfrak{o}$ und ω. Der Grund der Summationssätze ist nicht, wie vom Verfasser früher l. c. angenommen, physikalischer, sondern mathematischer Natur. Und zwar beruht er wesentlich auf dem Umstand, daß die Grenzwerte von J bzw. M oder P_r den „Parallelstellungen" entsprechen. Dies zeigt besonders deutlich, daß ein künstliches Ausschließen der Parallelstellungen der physikalischen Natur des Problems widersprechen würde.

Mit dem Beweis der Summationssätze sind wir im vorhergehenden zu einer formal sehr einfachen und vollständigen Beschreibung derjenigen anomalen Zeemaneffekte gekommen, die bei schwachen Feldern durch die Landéschen Formeln charakterisiert sind. Ein empirischer Fortschritt ist nur insofern erzielt worden, als es auf Grund von (3), (6) und (10) möglich ist, auch für mittlere Magnetfelder die Linienaufspaltungen auszurechnen.

§ 5. Die physikalische Bedeutung des Formalismus von § 1. So sehr die Regel von § 1 in formaler Beziehung das Problem der Zeemaneffekte vereinfacht, so ungewiß scheint ihre physikalische Deutung. Daß nach § 1 ein Wert der Kopplungsenergie mit zwei Quantenzahlen J, und eine Quantenzahl J in gewisser Weise formal mit zwei Werten der Kopplungsenergie verknüpft ist, scheint zunächst ein Zug der Theorie, der den bisherigen Vorstellungen besonders fremdartig gegenübersteht. Aber vielleicht kann die formale Analogie von (3) zur Frequenzbedingung doch später einen physikalischen Gesichtspunkt zur Deutung von § 1 insofern liefern, als ja das Problem der Kopplung zwischen Elektronen mit der Theorie der Strahlung aufs engste zusammenhängt und wir also darauf gefaßt sein müssen, auch bei der Kopplung auf Wirkungen zu stoßen, die mit der Möglichkeit eines Übergangs von einem stationären Zustand zum anderen verknüpft sind. Umgekehrt kann aber die Regel von § 1, die in ihrer jetzigen Form durchaus den Problemen des anomalen Zeemaneffekts angepaßt scheint, vielleicht die Richtung angeben, in

Zeitschrift für Physik *26*, 291–307 (1924)

welcher allgemein die Abänderung der bisherigen Mechanik und die Schaffung einer Quantenmechanik der Systeme mit mehreren Elektronen zu suchen ist.

Der provisorische Charakter der vorhergehenden Rechnungen geht besonders deutlich daraus hervor, daß wir ein von der gewöhnlichen Mechanik abweichendes Ersatzmodell benutzen mußten. In einer endgültigen Theorie der Zeemaneffekte müßten die unmechanischen Eigenschaften unseres Ersatzmodells ebenso, wie etwa die g-Formel in § 3, als Ausfluß der Quantenmechanik erscheinen, die durch die Erhaltung der Stabilität auch bei Kopplung von Elektronen gefordert wird. Zum Erreichen dieses Ziels ist noch kein Schritt getan. Aber man kann vermuten, daß die grundsätzlichen Schwierigkeiten, die der Deutung der halben Quantenzahlen, der relativistischen Dublettformel und des doppelten Magnetismus entgegenstehen, aus der gleichen Wurzel stammen, wie die durch die g-Formel nach § 1 versuchte Abänderung der Quantenmechanik.

Hierfür sind deutliche Anzeichen vorhanden: Z. B. wird die Ganzzahligkeit der Aufspaltung beim Paschen-Backeffekt (bei ganzzahligem M) dadurch formal hervorgebracht, daß auf die halbzahligen Werte von P_r das Schema von § 1 angewendet wird. Die theoretisch so unbefriedigende Halbzahligkeit von K ferner erhält eine neue Beleuchtung durch die von Landé[1]) untersuchte Tatsache, daß etwa die Niveaus p_1 und p_2, die ihrem ganzen magnetischen Verhalten nach mit $K = \frac{3}{2}$ zu normieren sind, auch im Sichtbaren (wie im Röntgengebiet) ein relativistisches Dublett zu bilden scheinen, das den Werten $K = 2$ und 1 entspricht.

Für den doppelten Magnetismus scheint es bemerkenswert, daß nach § 3 die Anlagerung eines Elektrons mit $K = \frac{1}{2}$ den Magnetismusfaktor 2 nicht ändert. Denn alle s-Terme besitzen wieder, wie der Atomrest, den g-Wert 2.

Die physikalische Deutung der Regel von § 1 spielt auch eine wesentliche Rolle bei der Betrachtung des Aufbauprinzips. Eine exakte Begründung des merkwürdigen Verhaltens der statistischen Gewichte (vgl. Einl.) und der Einstellungsmöglichkeiten bei der Kopplung muß dadurch gegeben werden, daß die Regel von § 1 auf Systeme mit mehreren Elektronen sinngemäß erweitert und dann die Anzahl der Einstellungsmöglichkeiten untersucht wird. Dies soll für starke Magnetfelder in § 6 versucht werden. Zunächst können wir trotzdem ein Schema anschreiben, das rein formal zeigt, wie das Aufbauprinzip in naturgemäßem Zusammen-

[1]) A. Landé, l. c., ZS. f. Phys. **16**, 391, 1923 und Naturw. **12**, 332, 1924.

hang steht mit der Regel von § 1. Obwohl wir auch hier auf eine physikalische Deutung zunächst verzichten müssen, so zeigt das folgende Schema doch, wie die Regel von § 1 fast zwangläufig zu denjenigen Aufbaugesetzen führt, die kürzlich Landé und der Verfasser in dieser Zeitschrift angegeben haben.

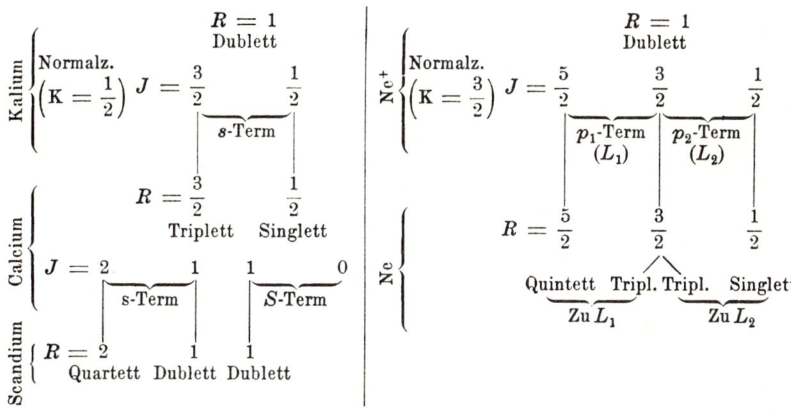

Der „Verzweigungssatz", von dem in der oben angeführten Arbeit die Rede war, ist direkt äquivalent mit der Regel von § 1. Aber eine eigentliche Begründung dieses Aufbauprinzips, die zugleich das Verhalten der statistischen Gewichte direkt als Wirkung der Quantenmechanik deutet, ist nur möglich durch Betrachtung von Systemen mit mehreren äußeren Elektronen.

§ 6. Mehrere äußere Elektronen. Eine sinngemäße Erweiterung des Prinzips § 1 auf den Fall mehrerer Elektronen ist uns bis jetzt nur in dem einfachsten Falle sehr großer Magnetfelder, d. h. also sehr schwacher Kopplung gelungen. In § 2 haben wir bei schwacher Kopplung das Atom beschrieben durch die Größen r, k, m, und p_k, wobei p_k die Projektion von k auf H bedeutet. Beim Vorhandensein mehrerer äußerer Elektronen beschreiben wir es naturgemäß durch die Größen:

$$r,\ k_1,\ k_2 \ldots k_n,\ m \quad \text{und} \quad p_{k_1},\ p_{k_2} \ldots p_{k_n};$$

hierin sollen $k_1, k_2 \ldots k_n$ die Drehimpulse der Elektronen, $p_{k_1}, \ldots p_{k_n}$ deren Projektionen auf H bedeuten. Die quantentheoretische Energie H_{qu} entsteht jetzt — wie früher durch Mittlung über p_k — durch „gleichzeitige" Mittlung über $P_{k_1}, P_{k_2}, \ldots P_{k_n}$. „Gleichzeitig" soll hier bedeuten, daß wir im $P_{k_1}, \ldots P_{k_n}$-Raum mitteln über eine Gerade, die

Zeitschrift für Physik 26, 291–307 (1924)

von einem Punkt etwa mit den Koordinaten $P^0_{k_1}, P^0_{k_2} \dots P^0_{k_n}$ zu einem anderen, z. B. $P^0_{k_1} + 1, P^0_{k_2} + 1 \dots P^0_{k_n} + 1$ gezogen ist, daß wir also über irgend eine Raumdiagonale der n-dimensionalen Würfel $P_{k_1} \dots P_{k_n}$, $P_{k_1} + 1, P_{k_2} \dots P_n; \dots P_{k_1} + 1, P_{k_2} + 1, \dots P_{k_n} + 1$ mitteln.

Es ist leicht zu zeigen, daß man durch diese Erweiterung des Prinzips von § 1 nicht nur zu der vom Aufbauprinzip geforderten richtigen Anzahl der Terme, sondern auch zur richtigen Aufspaltung der Terme bei starken Magnetfeldern kommt. Nehmen wir als Beispiel etwa den Fall, bei welchem aus einem Atom mit Quartett-s-Term als Grundzustand durch Anlagerung eines Elektrons ein Atom mit Quintett- und Triplettsystem entsteht, und betrachten wir die d-Terme dieser Systeme. Wir haben einen Atomrumpf mit $R = 2$ und zwei äußere Elektronen $K_1 = \frac{1}{2}$, $K_2 = \frac{5}{2}$. Die klassische Energie bei hohen Feldern ist nach (6) ($f = 0$) gegeben durch

$$H_{kl} = (2M - P_{k_1} - P_{k_2})\,\mathfrak{o}\,.\,h. \tag{16}$$

Die quantentheoretische Energie wird also:

$$H_{qu} = (2M - \overline{P}_{k_1} - \overline{P}_{k_2})\,\mathfrak{o}\,.\,h. \tag{17}$$

Zeichnen wir uns in der $P_{k_1}P_{k_2}$-Ebene die Grenzen des in Betracht kommenden Gebietes auf: P_{k_1} muß zwischen $+\frac{1}{2}$ und $-\frac{1}{2}$ liegen, P_{k_2}

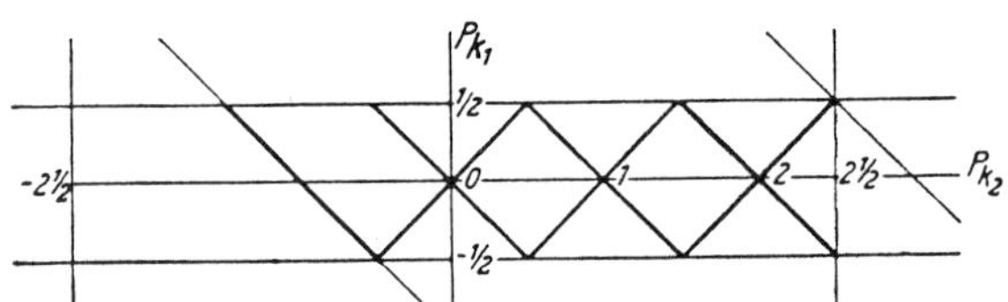

Fig. 1. $\left(K_1 = \dfrac{1}{2},\ K_2 = \dfrac{5}{2},\ M = 1\right)$.

zwischen $+\frac{5}{2}$ und $-\frac{5}{2}$, außerdem bestehen wegen $R = 2$ für ein bestimmtes M die Grenzen $P_{k_1} + P_{k_2} = M + 2$ und $P_{k_1} + P_{k_2} = M - 2$. In Fig. 1 sind die vorkommenden Grenzen für $M = 1$ gezeichnet. Die im Gebiet möglichen Diagonalen sind in der Figur stark ausgezogen. Für $M = 1$ erhalten wir also vier Terme des Quintettsystems durch Diagonalen, die von links oben nach rechts unten laufen, dargestellt. Die ihnen entsprechenden Aufspaltungen sind nach (17):

$$\frac{1}{\mathfrak{o}\,.\,h} H_{qu} = 3,\ 2,\ 1,\ 0.$$

Ferner haben wir drei Terme des Triplettsystems, dargestellt durch Diagonalen, die von rechts oben nach links unten laufen. Ihre Aufspaltungen sind

$$\frac{1}{o \cdot h} H_{qu} = 2,\ 1,\ 0.$$

Führen wir die analogen Rechnungen für andere Werte von M aus, etwa dadurch, daß wir in Fig. 1 die schrägen Grenzen parallel um das dem neuen Wert von M entsprechende Stück verschieben, so finden wir:

		$M = 0$	1	2	3	4
$\frac{1}{oH}$	Quintett	2, 1, 0, — 1, — 2	3, 2, 1, 0	4, 3, 2	5, 4	6
	Triplett	1, 0, — 1	2, 1, 0	3, 2	4	—

Diese Werte stimmen völlig überein mit denen, die Pauli (l. c.) für die Aufspaltung bei starken Feldern findet.

Die Erweiterung des Prinzips § 1 auf mehrere Elektronen macht also offenbar für starke Felder keine Schwierigkeiten. Trotzdem kann diese Erweiterung in der hier angeführten Form nicht ohne weiteres als befriedigend angesehen werden, da sie zunächst keine Erklärung für die Gültigkeit der Summationsprinzipien und für die Aufspaltungen bei kleinen Feldern liefert. Es scheint aber sehr natürlich, daß eine Aufklärung dieser letztgenannten Punkte nur in engem Zusammenhang mit einer Lösung der übrigen grundsätzlichen Schwierigkeiten der anomalen Zeemaneffekte zu suchen ist, und daß hierbei auch die physikalische Deutung von § 1 eine wesentliche Rolle spielt. Wir haben die obige Erweiterung des Prinzips § 1 auseinandergesetzt, um zu zeigen, in welcher Weise das Aufbauprinzip in direktem Zusammenhang mit der Zuordnung von zwei Quantenzahlen zu einer Kopplungsenergie steht, und um außerdem einen Formalismus anzugeben, der gestattet — was empirisch von Bedeutung sein dürfte —, für alle Arten von Multipletts die Aufspaltung bei starken Feldern zu berechnen. Hinsichtlich des Aufbauprinzips kommen wir so auch wieder gerade zu denjenigen Klassen von Multipletts, die von Landé und dem Verfasser kürzlich behandelt wurden.

§ 7. Neonspektrum. Als direkte Anwendung der im letzten Abschnitt gegebenen Gesichtspunkte kann man das von Paschen[1]) analysierte und von Landé[2]) theoretisch geordnete Neonspektrum ansehen. Der Atomrest des angeregten Neons besitzt die Zustände L_1 und L_2[3]),

[1]) Paschen, Ann. d. Phys. **60**, 405, 1919; **63**, 201, 1920.
[2]) Landé, l. c., ZS. f. Phys. **17**, 292, 1923.
[3]) Grotrian, ZS. f. Phys. **8**, 116, 1922.

Zeitschrift für Physik *26*, 291–307 (1924)

die nach Landé[1]) mit p_1 und p_2 zu identifizieren sind. Das Aufbauprinzip liefert also folgendes Schema:

$$J_{\text{Atomrest}} = \underbrace{\overbrace{\frac{1}{2} \qquad \frac{3}{2}}^{}}_{L_2} \underbrace{\qquad \frac{5}{2}}_{L_1}$$

	L_2	L_1
Multiplettklasse des Atoms:	Singlett-Triplett	Triplett-Quintett

Alle durch dies Schema angedeuteten Terme und ihre Aufspaltungen für starke Felder werden durch die Erweiterung des Prinzips von § 1 wiedergegeben.

Besonders einfach gestalten sich die Verhältnisse bei den s-Termen von Neon. Für die Zeemaneffekte ist es nämlich offenbar gleichgültig, ob sich das Elektron mit $k = \frac{3}{2}$ innen befindet und das mit $k = \frac{1}{2}$ außen oder umgekehrt. Die s-Terme des Neons entsprechen also genau den p-Termen eines Erdalkalis, da bei den letzteren das innere Elektron $k = \frac{1}{2}$ und das äußere $k = \frac{3}{2}$ besitzt. Wir müssen daher annehmen, daß nicht nur für starke, sondern auch für schwache Felder die vier s-Terme des Neons magnetisch sich genau wie die vier p-Terme etwa von Mg verhalten, daß sie also z. B. dieselben g-Werte besitzen. Dies ist tatsächlich auch beobachtet. Hinsichtlich der Intervallproportionen müssen aber Mg (p-Terme) und Ne (s-Terme) sich verschieden verhalten, da die Kopplungsverhältnisse durchaus andere sind. Stellen wir schematisch zum Vergleich die s-Terme von Ne und die p-Terme von Mg untereinander:

		L_1		L_2	
Neon	Termsystem	Quintett	Triplett	Triplett	Singlett
	Term	s_5	s_4	s_2	s_3
	g-Wert	$\frac{3}{2}$	$\frac{3}{2}$	1	$\frac{0}{0}$
	M_{max}	2	1	1	0

		Triplett			Singlett
Mg	Termsystem				
	Term	p_1	p_2	p_3	P
	g-Wert	$\frac{3}{2}$	$\frac{3}{2}$	$\frac{0}{0}$	1
	M_{max}	2	1	0	1

Auch bei den höheren (p-, d-...) Termen von Ne ist von einer wirklichen Trennung in Quintett-Triplett- und Triplett-Singlettsystem,

[1]) Landé, l. c., ZS. f. Phys. **16**, 391, 1923.

wie sie etwa bei Mg oder bei Cr, Mn, Fe auftritt, nicht die Rede; die obige Einteilung gibt nur das Schema der J-Werte in einfacher Weise wieder. Dieses Verhalten entspricht ganz dem, was theoretisch an sich immer erwartet werden muß. Die Gültigkeit des „Kosinusgesetzes" ist eine für die „Multipletts erster Stufe" (inneres Elektron in einer s-Bahn) charakteristische Ausnahme, deren Begründung bis jetzt noch aussteht. Einen möglichen Anhaltspunkt für das scheinbare Verschwinden der Wechselwirkung, die in erster Linie aus großen Quadrupolgliedern bestehen müßte, können wir aber vielleicht aus dem Prinzip § 1 und § 6 entnehmen. Wird von zwei äußeren Elektronen das innere in einer s-Bahn $k_1 = \frac{1}{2}$ angelagert, so sagt das Mittlungsschema von § 6 aus, daß der Vektor k_1 über sämtliche ihm zur Verfügung stehenden Lagen $-1 \leqq \cos(k_1 H) \leqq +1$ gemittelt wird. Dies hat zur Folge, daß hinsichtlich seiner Wechselwirkung mit dem äußeren Elektron das innere Elektron wie ein kugelsymmetrisches Potential wirkt. Man kann an dem Mittlungsschema von Fig. 1 leicht zeigen, daß dieses kugelsymmetrische Ersatzpotential für gleiche Diagonalenrichtungen das gleiche ist, daß sich also die große Wechselwirkung zwischen äußerem und innerem Elektron nur in der starken Verschiedenheit der Termgröße von Triplett- und Quintettsystem zeigt, und daß die kleine Aufspaltungsweite im einzelnen Multiplettsystem von der Wegmittlung der Wechselwirkung herrührt.

Läuft das innere Elektron nicht in einer s-Bahn, so können wir das Kosinusgesetz nicht mehr erwarten, wir müssen annehmen, daß alle Terme in Abständen derselben Größenordnung durcheinanderliegen. Wir können dann auch keineswegs fordern, daß die g-Werte rational und durch die Multiplettstruktur eindeutig bestimmt sind, sondern wir werden vermuten, daß in die g-Werte das Kopplungsverhältnis, etwa z. B. für p-Terme charakterisiert durch $\frac{L_1 - L_2}{p_i - p_k}$, eingeht. Daraus würde folgen, daß auch die Prestonsche Regel, nach der die g-Werte von der Hauptquantenzahl unabhängig sind, hier versagt.

Dies alles gilt schon, wenn man nur die Kopplung zwischen den Impulsvektoren, also die Kopplung zwischen den über ihre ganze Rosettenbahn gemittelten Elektronen betrachtet. Dazu kommt aber beim Neon, daß die Kopplung zwischen den Elektronen offenbar schon so groß ist, daß auch die periodischen gegenseitigen Störungen der Elektronen nicht zu vernachlässigen sind, wie das Auswahlprinzip (vgl. § 8)

Zeitschrift für Physik 26, 291–307 (1924)

zeigt. Diese Art von Kopplung wird eventuell schon die Rationalität der g-Werte auch bei den s-Termen stören. Allgemein werden wir also auch bei den Multipletts erster Stufe annehmen, daß die g-Werte nur mit der Genauigkeit rational sind, mit welcher die durch Kopplung hervorgerufenen Bahnstörungen zu vernachlässigen sind.

§ 8. Die Auswahlprinzipien. Durch die Regeln § 1 und § 6 scheint uns ein formaler Fortschritt auch insofern erzielt, als sich dort herausstellt, daß z. B. bei den Erdalkalien Singlett- und Triplettsystem beide mit derselben dem Atomrest angehörigen Kopplungsenergie verknüpft sind. Dies steht offenbar unmittelbar mit der Tatsache im Zusammenhang, daß die beiden Multiplettsysteme die gleiche Seriengrenze zeigen, und daß Interkombinationslinien zwischen Triplett- und Singlettsystem in derselben Stärke auftreten, wie die Linien eines Multiplettsystems.

Es scheint daher auch begreiflich, daß empirisch einige Tatsachen dafür sprechen [z. B. bei Mn [1]) und Fe [2])], daß häufig je zwei Multiplettsysteme bevorzugt kombinieren, daß aber Interkombinationen zwischen nicht zusammengehörigen Systemen ausfallen. Eine endgültige Beantwortung der hier sich darbietenden Fragen kann natürlich auch erst erfolgen, wenn der quantentheoretische Kopplungsmechanismus geklärt ist.

Die korrespondenzmäßige Berechnung der Intensitäten [3]) wird durch die Regeln von § 1 und § 6 nicht beeinflußt.

Schlußbemerkungen. Den Inhalt der vorliegenden Arbeit wollen wir dahin zusammenfassen, daß es sich als möglich erwiesen hat, durch eine einfache Abänderung der formalen Quantenregeln die Gesetze der Multipletts und Zeemaneffekte einfacher und einheitlicher als bisher zu beschreiben. Besonders das Permanenzgesetz und die Aufbauregeln scheinen sich kaum anders deuten zu lassen, als durch die Annahme, daß ein Wert der Kopplungsenergie mit zwei Quantenzahlen verknüpft ist. Was die physikalische Bedeutung des angegebenen Formalismus betrifft, so kann man vermuten, daß die Regel § 1 in Zusammenhang steht mit der Tatsache, daß auch die quantenmechanischen Wechsel-

[1]) E. Back, l. c.

[2]) O. Laporte, ZS. f. Phys. **23**, 135, 1924.

[3]) Vgl. A. Landé, l. c., ZS. f. Phys. **5**, 231, 1922 (§ 5) und A. Sommerfeld und W. Heisenberg, ZS. f. Phys. **11**, 131, 1922.

wirkungen zwischen Atomen und Strahlungsfeldern, die ja zu den Kopplungswirkungen völlig analog sind, notwendig mit der Möglichkeit eines Übergangs von einem stationären Zustand zum anderen verknüpft sind. In diesem Sinne darf man wohl annehmen, daß Formalismen, die denen von § 1 und § 6 ähnlich sind, eine über die Zeemaneffektprobleme hinausgehende allgemeine Rolle in der Quantenmechanik der Kopplungswirkungen werden spielen können.

Göttingen, Institut für theor. Physik.

Zeitschrift für Physik 32, 841–860 (1925)

Zur Quantentheorie der Multiplettstruktur und der anomalen Zeemaneffekte.

Von **W. Heisenberg**, z. Z. in Kopenhagen.

(Eingegangen am 10. April 1925.)

In der Arbeit wird versucht, durch eine nähere Analyse der mechanisch nicht beschreibbaren Zweideutigkeit, welche die Wechselwirkung zwischen Atomrest und Elektron kennzeichnet, zu einer Vereinigung der verschiedenen diese Wechselwirkung beschreibenden Formalismen zu kommen. Die hierdurch gewonnenen Gesichtspunkte gestatten eine Interpretation mancher empirisch gefundener Gesetzmäßigkeiten der Multipletts höherer Stufe, die in den bisherigen Formalismen nicht enthalten waren. Schließlich werden die für gleichzeitigen Sprung mehrerer Elektronen gültigen Auswahlregeln korrespondenzmäßig untersucht.

Obwohl die Quantentheorie der mehrfach periodischen Systeme bei der Interpretation der Gesetzmäßigkeiten des Atombaues zweifellos große Erfolge erzielt hat, hat es sich doch herausgestellt, daß man sich bereits wesentlich außerhalb der Grenzen der Anwendbarkeit dieser Theorie befindet, wenn man den Aufbau der Atome durch sukzessive Bindung der Elektronen untersucht. Eben die Analyse der Multiplettstruktur und des anomalen Zeemaneffekts hat gezeigt, daß die Kopplung der Elektronen im Atom einen mechanisch unbeschreibbaren Zwang enthält [1]). Dieser bringt mit sich, daß trotz der eindeutigen Bestimmtheit des ganzen Atoms in seinen stationären Zuständen die Beschreibung des Verhaltens der einzelnen Elektronen eine gewisse Zweideutigkeit erfordert. Das Ziel der vorliegenden Arbeit ist eine nähere Analyse jener charakteristischen Zweideutigkeit, auf Grund derer wir versuchen wollen, zu einer Vereinigung der verschiedenen scheinbar einander widersprechenden Formalismen zu kommen, die in letzter Zeit vorgeschlagen wurden und bei der Beschreibung der empirischen Tatsachen teilweise große Erfolge erzielt haben.

§ 1. Grundannahmen. Wir legen unseren Untersuchungen folgende Hypothese zugrunde: Stehen ein Elektron und ein Atomrest (um Entartungen zu vermeiden, nehmen wir die Anwesenheit eines äußeren Feldes an) miteinander in Wechselwirkung, so besitzt die Energie dieser Wechselwirkung eine reziproke Zweideutigkeit, derart, daß zu bestimmt gegebenen stationären Zuständen des Atomrestes und des äußeren Elektrons

[1]) Siehe z. B. N. Bohr, Ann. d. Phys. **71**, 228, 1923, oder A. Landé, Phys. ZS. **24**, 441, 1923.

stets zwei Energiewerte der Wechselwirkung und daher auch zwei stationäre Zustände des Gesamtatoms gehören, und daß umgekehrt zu einem Werte der Wechselwirkungsenergie stets zwei Systeme stationärer Zustände von Elektron und Atomrest gehören.

Um diese Hypothese zu verwerten, ist man beim jetzigen Stande der Quantentheorie darauf angewiesen, symbolische modellmäßige Bilder zu brauchen, die mehr oder weniger dem mechanischen Verhalten von Elektronen in der klassischen Theorie nachgebildet sind. Nach dem Wesen der Sache bringt aber die obige Hypothese in der Wahl der modellmäßigen Bilder einen charakteristischen Dualismus mit sich: Man kann die durch den quantentheoretischen Zwang der Wechselwirkung hervorgerufene Zweideutigkeit modellmäßig interpretieren, indem man entweder annimmt:

I. Das Elektron wirkt auf den Atomrest durch einen unmechanischen Zwang derart, daß sich der stationäre Zustand des Atomrestes scheinbar verdoppelt;

oder aber, indem man annimmt:

II. Der Atomrest wirkt auf das Elektron durch einen unmechanischen Zwang derart, daß sich der stationäre Zustand des Elektrons scheinbar verdoppelt.

Diese beiden Bilder, die, wie gezeigt werden soll, mit der „Neigungstheorie“ [1]) der Multipletts einerseits und der „relativistischen“ Theorie [2]) andererseits in Verbindung gebracht werden können, stehen offenbar gleichberechtigt nebeneinander und die Forderung der Eindeutigkeit der stationären Zustände des ganzen Atoms bringt es mit sich, daß sie sich in ihren Ergebnissen niemals widersprechen können. Vielmehr werden sich die beiden Bilder eben wegen jener Eindeutigkeit in ihren Aussagen so ergänzen müssen, daß die Größen, die in dem einen Schema unbestimmt bleiben, im anderen bestimmt werden und umgekehrt; die beiden Schemata zusammen werden sozusagen ein konvergentes Rechenverfahren zur Bestimmung der Eigenschaften der stationären Zustände des Atoms bilden. Betrachtet man die Wechselwirkung mehrerer Elektronen mit einem Atomrest, so wird es sogar eine ganze Reihe gleichwertiger Bilder geben, die erst zusammen eine eindeutige Beschreibung der stationären Zustände des Atoms ermöglichen.

---

[1]) Siehe z. B. A. Landé, ZS. f. Phys. **5**, 231, 1921; W. Heisenberg, ebenda **8**, 273, 1922; W. Pauli, ebenda **16**, 155, 1923; W. Heisenberg, ebenda **26**, 291, 1924.

[2]) R. A. Millikan und J. S. Bowen, Phys. Rev. **24**, 209 u. 223, 1924; A. Landé, ZS. f. Phys. **24**, 88 u. **25**, 46, 1924; W. Pauli, ebenda **31**, 765, 1925.

Zeitschrift für Physik *32*, 841–860 (1925)

Die zugrunde gelegte Hypothese einer reziproken Zweideutigkeit gibt also einen in den bisherigen quantentheoretischen Vorstellungen nicht vorhandenen hohen Grad von Freiheit in der Benutzung modellmäßiger Bilder, der es uns möglich zu machen scheint, zu einer beim gegenwärtigen Stand der Theorie sinngemäßen formalen Beschreibung der empirischen Tatsachen in der Komplexstruktur komplizierter Spektren zu gelangen.

§ 2. Dublettatome. Die unseren Rechnungen zugrunde liegenden allgemeinen Modellvorstellungen sind bei den Dublettatomen folgende:

Die „quasimechanische" Bahn des Elektrons ist durch das Nichtvorhandensein eines Energieaustausches zwischen Atomrest und Elektron definiert und führt deshalb hinsichtlich der Abhängigkeit der Energie des Elektrons von der Hauptquantenzahl zu einer Rydberg-Ritzschen Formel. Die erwähnte Zweideutigkeit soll sich nicht auf die Hauptquantenzahl, sondern auf die Impulsquantenzahlen (k und m_k für das äußere Elektron, r und m_r für den Atomrest) erstrecken. Wenn wir von den Werten der Wechselwirkungsenergie absehen, so ist das äußere Elektron durch ganze Quantenzahlen nach der Art des Bohrschen Wasserstoffmodells gegeben. Der edelgasähnliche Atomrest ist im äußeren Felde nur einer Stellung ($m_r = 0$) fähig, außerdem ist er völlig isotrop[1]). Falls der Gesamtimpuls eines Atoms einen bestimmt definierten Wert hat — und dies möchten wir wegen der Forderung der eindeutigen Beschreibbarkeit der stationären Zustände annehmen —, so scheint es die einfachste Annahme, daß der Gesamtimpuls durch die Zahl N der Einstellungsmöglichkeiten im äußeren Felde gemäß $j = \frac{N-1}{2}$ definiert sei[2]), daß also der Gesamtimpuls des Edelgasatoms Null betrage. Von den beiden möglichen modellmäßigen Bildern gibt das Schema I, in welchem das Elektron als eindeutig und der Atomrest als zweideutig aufgefaßt wird, insofern ein zutreffenderes mechanisches Bild des wirklichen Sachverhalts als Schema II, als es die Auswahlregeln, die für die Struktur des Spektrums charakteristisch sind, zu beschreiben gestattet. Wenn wir aber nach den quantitativen Energiewerten der Terme fragen, so besteht einer der charakteristischen Züge der Dublettseriensysteme in jener von der Theorie der mehrfach periodischen Systeme aus völlig unverständlichen

[1]) W. Heisenberg, ZS. f. Phys. **31**, 617, 1925.

[2]) Die dabei sich ergebenden Werte stimmen mit den von Sommerfeld (Ann. d. Phys. **73**, 209, 1924) für die inneren Quantenzahlen angenommenen j-Werten überein.

Trennung der relativistischen von den Abschirmungsdubletts. Eine Beschreibung dieser empirischen Tatsache dürfte nur durch die Einführung einer Zweideutigkeit, wie die oben angenommene, ermöglicht werden, und kann deshalb hier innerhalb des Schemas II, in welchem das Elektron als zweideutig und der Atomrest als eindeutig erscheint, gegeben werden. Die aus dem erwähnten quasimechanischen Modell folgende Energie des Elektrons $E(n, k)$ kann in zwei Teile geteilt werden, die Wechselwirkungsenergie $E_W(n, k)$ und die Eigenenergie $E_E(n, k)$. Bei Nichtanwesenheit äußerer Felder ist in der letzteren Energie nur der relativistische Teil der Energie von k abhängig. Die Eigenenergie $E_E(n, k)$ wird, wie im Wasserstoffatom, durch die ganzzahligen Werte von k beschrieben. Zu jedem Werte der Eigenenergie gehören jedoch zwei Wechselwirkungsenergien $E_W(n, k)$, die durch die Werte $k \pm {}^1/_2$ gegeben sind. Allgemein seien die Eigenenergien und die Wechselwirkungsenergien nach folgendem Schema einander zugeordnet und zu Termwerten verbunden:

Eigenenergie $E_E(n,k)$: $k =$			1				2				3				4	
Wechsel-wirkungsen. $E_W(n,k)$: $k =$	$^1/_2$	s		p_2	$^3/_2$	p_1		d_2	$^5/_2$	d_1		f_2	$^7/_2$	f_1		$^9/_2$ ...

Zu jeder Wechselwirkungsenergie gehören also auch umgekehrt zwei Werte der Eigenenergie, und diese in der Grundhypothese geforderte Reziprozität erscheint hier wesentlich als Folge der halbzahligen Wahl des für die Wechselwirkungsenergie maßgebenden k-Wertes.

Diese Darstellung gibt offenbar eine vollständige Beschreibung der Termwerte eines Dublettsystems, wenn wir von der Beeinflussung der Terme durch äußere Felder absehen. In der Tat bilden die beiden, zu einem bestimmten Werte von E_W gehörigen Terme ein relativistisches Dublett. Unsere Formulierung geht aber über die von Landé gegebene an einigen Punkten noch hinaus. Es kann vorkommen, daß etwa die durch $k = {}^5/_2$ charakterisierte quasimechanische Bahn nicht in den Atomrest eindringt, wohl aber die durch $k = 2$ dargestellte Bahn. Dann gehört zu einem Werte der Wechselwirkungsenergie, der einer nichteindringenden Bahn entspricht, ein relativistisches Dublett von einer Größe, die auf eindringende Bahnen schließen läßt. Ein solcher Fall ist vielleicht in den d-Termen von Rubidium oder Cäsium realisiert.

Die Darstellung der Dublettspektren durch das Schema II ist aber ihrem Wesen nach einseitig und, wie erwähnt, zwingt uns auch eine wesentliche Seite der experimentellen Resultate, nämlich die Auswahlprinzipien, zur Betrachtung des Schemas I (in welchem das äußere Elektron als eindeutig und der Impuls des Atomrestes als zweideutig aufgefaßt

Zeitschrift für Physik 32, 841–860 (1925)

wird). In der Tat kann das Korrespondenzprinzip nur auf mechanisch eindeutige, durch Quantenzahlen eindeutig festgelegte periodische Bewegungen angewandt werden. Wenn es sich um Sprünge des einen Elektrons handelt, können daher Auswahlregeln nur aus dem Schema I abgeleitet werden. Aus den empirischen Auswahlregeln erfahren wir dann, daß dasjenige Dublett, das innerhalb des Schemas II als Unterschied zwischen zwei Eigenenergien des Elektrons aufgefaßt wurde, in Schema I durch verschiedene Neigungen des Atomrestes gegen das Elektron entsteht. Dieses Schema I muß dann auch die richtige Anzahl von Stellungen des Atoms in einem äußeren Felde wiedergeben, da dies für die Anwendbarkeit des Korrespondenzprinzips eine notwendige Bedingung ist. Bekanntlich kann das Modell des Schemas I folgendermaßen beschrieben werden: Das äußere Elektron läuft in einer durch das halbzahlige k charakterisierten Bahn. Der Atomrest hat die Impulswerte $r = (0), 1$ (wo 0 keine physikalische Bedeutung haben kann), der formal eingehende „Gesamtimpuls" wird entweder ganzzahlig (nach Landé „J") oder durch ein Paar von halbzahligen Quantenzahlen[1] ($J + 1/2$, $J - 1/2$) bestimmt. Es muß aber hervorgehoben werden, daß der mechanisch definierte Gesamtimpuls des Atoms im Gegensatz zum formalen j-Wert des Schemas I eindeutig festgelegt ist durch die Anzahl der Einstellungen im äußeren Felde.

Die Bedingung dafür, daß das Schema I wirklich eine konsequente Umkehrung des Schemas II im Sinne der Hypothese von S. 841 ist, besteht offenbar darin, daß alle Resultate, die sowohl aus I, wie auch aus II hergeleitet werden können, identisch sind.

Bisher besteht Identität beider Schemata, wenn man von der Dublettstruktur absieht, hinsichtlich der Wechselwirkungsenergie, die in beiden Fällen durch den halbzahligen k-Wert gegeben ist. Über den Dublettabstand kann das Schema I nichts aussagen, da die Richtungsabhängigkeit der Wechselwirkungsenergie, ebenso, wie der Impuls des Atomrestes selbst, erst durch quantentheoretischen Zwang entsteht.

Betrachten wir jetzt das Verhalten des Atoms in einem äußeren Magnetfeld. Da die magnetische Anomalie nur durch die Einführung der Duplizität entstehen kann, muß in Schema I der Atomrest, in Schema II das äußere Elektron als Träger der magnetischen Anomalie erscheinen. Wir nehmen zunächst an, daß in Schema I die Richtungsabhängigkeit der Wechselwirkung zwischen Atomrest und Elektron verschwindend klein gegenüber den Wirkungen des Magnetfeldes sei, ebenso

[1]) W. Heisenberg, l. c., ZS. f. Phys. **26**, 291, 1924.

in Schema II, daß der relativistische Teil der Eigenenergie zu vernachlässigen sei gegen den magnetischen Teil der Eigenenergie.

Dann schließen wir aus Schema II, welches wir auf m_k ebenso anwenden können, wie oben auf k, daß die magnetische Energie, als von der Wechselwirkung unabhängige Eigenenergie, durch die ganzzahligen Werte von m_k, die für die Wechselwirkung (und daher auch für die Größe des mechanischen Impulsmomentes) maßgebenden Werte m_k aber halbzahlig gegeben seien:

Magnetische Eigenenergie: $m_k = -3 \quad -2 \quad -1 \quad 0 \quad 1 \quad 2 \quad 3 \ldots$

Wechselwirkungsenergie: $m_k = -5/2 \quad -3/2 \quad -1/2 \quad 1/2 \quad 3/2 \quad 5/2 \ldots$

Dieses Schema gibt bekanntlich die magnetischen Energiewerte im Paschen-Backeffekt wieder. Wir sehen daraus zunächst, daß wir, um Identität der Resultate beider Schemata zu erzielen, innerhalb des Schemas I der (im Gegensatz zum formalen j-Wert dieses Schemas eindeutig mechanisch definierten) Projektion m des Gesamtimpulses in die Richtung eines äußeren Feldes halbzahlige Werte erteilen müssen. Ferner ist in Schema I die magnetische Anomalie des Atomrestes zunächst unbestimmt, da sowohl das mechanische als auch das magnetische Moment des Atomrestes erst durch Zwang entstehen. Wir lernen dann aus den Energiewerten des Schemas II, daß man dem Atomrest den anomalen Magnetismusfaktor 2 zuschreiben muß, wenn die beiden Schemata dieselben Resultate ergeben sollen.

Gehen wir nun vom Falle starker Magnetfelder zu dem mittlerer und schwacher über, so können wir aus dem Schema II über diesen Übergang nichts erfahren. In Schema I dagegen erscheint der betrachtete Übergang als ein quasimechanisches Problem. Bekanntlich ist dieser Übergang durch das Kosinusgesetz und die Summationsgesetze charakterisiert. Das Kosinusgesetz kann mechanisch nicht begründet werden, es erscheint als einfachste Möglichkeit für die Richtungsabhängigkeit der durch unmechanischen Zwang in Schema I entstehenden Wechselwirkungsenergie; eine quantentheoretische Vereinfachung ähnlicher Art werden wir beim Problem der pp'-Linien kennenlernen. Die Summationsregeln dagegen können bekanntlich durch eine einfache Darstellung der quantentheoretischen Energie durch ein Integral über die entsprechende klassische Energie gewonnen werden. Es mag aber darauf hingewiesen werden, daß das Summationsprinzip nicht durch eine solche Darstellbarkeit der Energie durch ein Integral über eine quasimechanische Energie bedingt ist, sondern zweifellos eine noch allgemeinere Gültigkeit (z. B. bei

Zeitschrift für Physik 32, 841–860 (1925)

mehreren äußeren Elektronen) besitzt. Der eigentliche Grund der Summationsgesetze wird vielmehr derselbe sein, der auch zu den Ornstein-Burgerschen Intensitätsregeln und zur spektroskopischen Stabilität Anlaß gibt[1]). Immer dort, wo in der klassischen Theorie einfache Integralsätze bestehen, scheinen in der Quantentheorie die korrespondenzmäßig entsprechenden Summensätze zu gelten. Die den magnetischen Summationssätzen mechanisch entsprechende Integralformel ist in der oben erwähnten Arbeit des Verfassers abgeleitet, und man kann leicht die Existenz dieser Integralformel auch beim Vorhandensein mehrerer äußerer Elektronen beweisen. Ein ähnlicher Summensatz ist auch der von Landé[2]) angegebene Schwerpunktsatz.

Diesem Schwerpunktsatz korrespondiert klassisch ein einfacher Satz über Kugelfunktionen. Stellt man nämlich die Energie des Atoms E als Funktion von r, k, j dar, so wird sie im allgemeinen entwickelbar sein nach Kugelfunktionen von $\cos\vartheta$, wo $\cos\vartheta = \cos(k, r) = \frac{j^2 - k^2 - r^2}{2kr}$; also

$$E = \sum_{n=0}^{\infty} f_n(k, r) \cdot P_n(\cos\vartheta).$$

In der klassischen Theorie wird man daher den „Schwerpunkt des Multipletts", d. h. den von j unabhängigen Teil der Energie $E_S = f_0(k, r)$ finden können nach der Formel

$$E_S = \frac{\int_{-1}^{+1} E\, d(\cos\vartheta)}{\int_{-1}^{+1} d(\cos\vartheta)} = \frac{\int_{j_{\min}}^{j_{\max}} E j\, dj}{\int_{j_{\min}}^{j_{\max}} j\, dj};$$

die entsprechende quantentheoritische Summenformel lautet offenbar:

$$E_S^{qu} = \frac{\sum_{j_{\min}}^{j_{\max}} E^{qu} \cdot j}{\sum_{j_{\min}}^{j_{\max}} j}.$$

Dies ist eben die Aussage von Landés „Schwerpunktsatz".

Das Schema I gibt also eine einfache Darstellung für den Übergang von starken zu schwachen Magnetfeldern, die g-Werte bei schwachen

[1]) W. Heisenberg, l. c., ZS. f. Phys. **31**, 617, 1925.

[2]) E. Back und A. Landé, Zeemaneffekte und Multiplettstruktur der Spektrallinien, (Berlin 1925), S. 58 u. f.

Feldern erscheinen in Schema I wesentlich als einfache Folge der Summationsregeln.

Andererseits ist es interessant hervorzuheben, daß sich für die g-Werte innerhalb des Schemas II eine besonders einfache Interpretation geben läßt, die sich eng an die formale Deutung der Dubletts anschließt. Innerhalb des Schemas II kann man nämlich für die magnetischen Wirkungen das äußere Elektron allein verantwortlich machen, da hier der Atomrest den Impuls 0 hat. Dann scheint es konsequent, das magnetische Moment des Elektrons als eine von der Wechselwirkung unabhängige Eigenschaft des Elektrons (man kann sie prüfen, ohne die Bahn des Elektrons zu stören) an den ganzzahligen k-Wert zu knüpfen, das mechanische Moment des Elektrons jedoch, als durch die Wechselwirkung definiert (weil nur durch Wechselwirkung prüfbar), an den halbzahligen k-Wert zu knüpfen. Der g-Wert erscheint also hier als Quotient aus dem zu dem betreffenden Term gehörigen ganzzahligen und dem halbzahligen k-Wert:

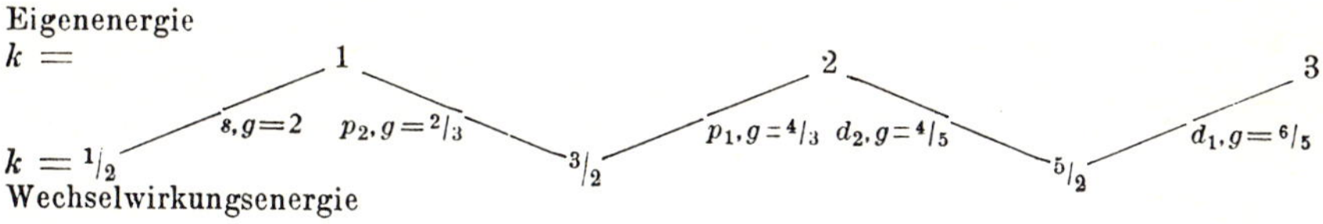

Insbesondere ergibt sich der Magnetismusfaktor 2 des s-Terms hier als einfache Folge der Halbzahligkeit des mechanischen k-Wertes.

§ 3. Atome mit zwei äußeren Elektronen. Betrachten wir ein Erdalkaliatom und nehmen wir zunächst an, daß die Relativitätswirkung in Schema II sowie die entsprechende Wechselwirkung in Schema I verschwindet; dann scheint das Problem, die Wechselwirkung zwischen den Elektronen auszurechnen, eindeutig und mechanisch einfach durchführbar. Die beiden Elektronen mit den (halbzahligen) Impulsen k_1, k_2 werden räumlich gegeneinander gequantelt werden müssen, so daß der ihnen entsprechende Gesamtimpuls i die Werte

$$k_1 + k_2 \geqq i \geqq |k_1 - k_2|$$

annehmen kann (i ist wegen der Halbzahligkeit der beiden k hier ganzzahlig). Befindet sich z. B. das eine der beiden Elektronen in einer s-Bahn ($k_2 = {}^1/_2$), so hat i die zwei Werte $k_1 \pm {}^1/_2$. Dieses Resultat gibt offenbar naturgemäß Rechenschaft von der Tatsache, daß die gewöhnlichen Serienspektren der Erdalkalien in zwei Systeme ($s, p, d, f\ldots$;

Zeitschrift für Physik 32, 841–860 (1925)

$S, P, D, F \ldots$) zerfallen, welche die gleiche Seriengrenze besitzen. Befindet sich das innere Elektron z. B. in einer p-Bahn ($k = {}^3/_2$), so erwarten wir vier Seriensysteme (mit gleicher Grenze). Die Wirkung der Zweideutigkeit kommt erst in den Multipletts und ihren magnetischen Aufspaltungen zutage.

Betrachten wir nun genauer, wie sich die Wirkung der Zweideutigkeit in den Schemata I und II beschreiben läßt.

Die einfachste empirische Regel, die auch für mehrere äußere Elektronen gültig bleibt, ist die von Landé und dem Verfasser[1]) aufgestellte Verzweigungsregel über die Anzahl und Werte der Quantenzahl j im Gesamtatom. In Schema I erscheint sie (vgl. die mehrfach zitierte Arbeit des Verfassers) als Folge der Zweideutigkeit des Impulses des dublettähnlichen Atomrestes, in Schema II (vgl. die oben zitierte Arbeit von Pauli) erscheint sie als Folge der Zweideutigkeit des Elektrons. Innerhalb des Schemas II haben Stoner[2]) und Pauli[3]) auch noch die empirischen Gesetzmäßigkeiten beim Aufbau der Schalen und beim Auftreten äquivalenter Elektronen einfach dargestellt durch die Forderung, daß nie zwei Elektronen vorkommen können, die dieselben Werte der Quantenzahlen (n, $k_{\text{Eig.}}$, $k_{\text{Wechs.}}$, m_k) besitzen.

Keines der beiden Schemata gibt jedoch Rechenschaft von der Tatsache, daß wir es bei den auftretenden Multipletts mit Tripletts und Singuletts zu tun haben. Denn auch in Schema I, das im Falle der Multipletts erster Stufe auf Tripletts und Singuletts führt, hat diese Einteilung zunächst keine einfache Bedeutung für die quantitativen Termgrößen, da der dublettähnliche Atomrest nicht isotrop ist und da deswegen eine Beschreibung der Tripletts in Analogie zu der Beschreibung der Dubletts in § 2 in Schema I nicht möglich ist.

Wir können aber die Freiheit, die unsere Grundhypothese in der Verteilung der Zweideutigkeit auf die Teilsysteme des Atoms gewährt, noch weiter nützen und die Duplizität beider Elektronen zusammen auf den edelgasähnlichen Atomrest werfen. Dann erhalten wir an Stelle der 2-Werte (0), 1 die Werte ${}^1/_2$, ${}^3/_2$ und damit die Möglichkeit, das allgemeine Auftreten von Singuletts und Tripletts zu beschreiben. Das so erreichte dritte Schema[4]) stellt sich als naturgemäß verallgemeinerte Anwendung unserer

[1]) A. Landé und W. Heisenberg, ZS. f. Phys. **25**, 279, 1924.

[2]) E. C. Stoner, Phil. Mag. **48**, 719, 1924.

[3]) W. Pauli, l. c., ZS. f. Phys. **31**, 765, 1925.

[4]) Inzwischen haben schon Russell und Saunders (Astr. Journ. **61**, 38, 1925) in einer eben erschienenen reichhaltigen Arbeit über die Erdalkalispektren auf

Grundannahme gleichberechtigt neben die beiden anderen und führt auf das folgende quasimechanische Modell. Der Atomrest besitzt die Impulse $r = 1/2, 3/2$, die Richtungsabhängigkeit seiner Wechselwirkung mit den Elektronen ist durch das näherungsweise gültige Kosinusgesetz charakterisiert, der Absolutwert dieser Wechselwirkung ist in Schema III unbestimmt und tritt in II als relativistische Wirkung auf. Die beiden äußeren Elektronen setzen sich zu einem Gesamtimpuls zusammen, der sich wiederum mit r zum Gesamtimpuls j des Atoms zusammensetzt. Ähnlich wie nun im Falle der Dubletts in das Schema I als k des äußeren Elektrons nicht das (ganzzahlige) k der Eigenenergie, sondern das k der Wechselwirkungsenergie einging, so führt das Schema III nur dann zu einer Beschreibung der empirischen Tatsachen, wenn in ihm als Resultante der beiden Elektronenimpulse nicht i selbst, sondern $i \pm 1/2$ angenommen wird; wir bezeichnen diesen Impuls, der im folgenden eine wichtige Rolle spielt, mit l[1]. l und i stehen also wieder in der schon mehrfach bekannten Beziehung:

$$\begin{array}{lccccccccc} i = & 1 & & 2 & & 3 & & 4 & & 5 \\ & \diagdown & & \diagup\diagdown & & \diagup\diagdown & & \diagup\diagdown & & \diagup \\ l = & & 3/2 & & 5/2 & & 7/2 & & 9/2 & \end{array} .$$

Das Schema III sagt also aus: Solange die Wechselwirkung zwischen r und l klein ist gegenüber der zwischen k_1 und k_2 (also vom Schema II aus: wenn die Relativität klein ist gegenüber der Wechselwirkung der Elektronen), so haben wir immer eine Einteilung in Singulett- und Triplettsysteme, gleichviel, welches die k-Werte der Elektronen sind. Nehmen wir als Beispiel: $k_1 = 5/2$, $k_2 = 3/2$. Dann haben wir die i-Werte:

$$i = 4 \quad 3 \quad 2 \quad 1.$$

Die zugehörigen l-Werte sind $7/2$, $5/2$, $3/2$.

$$\begin{array}{lccccccc} i = & 4 & & 3 & & 2 & & 1 \\ & \diagdown & & \diagup\diagdown & & \diagup\diagdown & & \diagup \\ l = & & 7/2 & & 5/2 & & 3/2 & \end{array} .$$

In Schema III hat r die Werte $1/2$, $3/2$; also kommen wir zu drei Tripletts und drei Singuletts mit den l-Werten $3/2$, $5/2$, $7/2$. Wie sich die Singuletts und Tripletts zu den i-Werten zuordnen, ist nicht ein-

---

die Zweckmäßigkeit eines dem Schema III entsprechenden mechanischen Modells für die Deutung mehrerer der von ihnen gemachten wichtigen experimentellen Befunde hingewiesen.

[1]) Diese Bezeichnung ist leider der von Russell und Saunders gerade entgegengesetzt (Vertauschung von l und k); ich möchte aber die hier gewählte Bezeichnung nicht verändern, damit man das Symbol n_k für eine Elektronenbahn beibehalten kann.

Zeitschrift für Physik 32, 841–860 (1925)

deutig. Zu einem i-Wert gehören stets zwei l-Werte und zwei Werte von r; wir erhalten die zwei Möglichkeiten:

$$\begin{array}{ll|}
i = 4 & \begin{array}{l} j = 3 \\ l = 7/2 \end{array} \\ \hline
i = 3 & \begin{array}{l} j = \underbrace{4 \quad 3 \quad 2}_{} \quad 2 \\ l = \quad 7/2 \qquad 5/2 \end{array} \\ \hline
i = 2 & \begin{array}{l} j = \underbrace{3 \quad 2 \quad 1}_{} \quad 1 \\ l = \quad 5/2 \qquad 3/2 \end{array} \\ \hline
i = 1 & \begin{array}{l} j = \underbrace{2 \quad 1 \quad 0}_{} \\ l = \quad 3/2 \end{array}
\end{array}
\qquad \text{oder} \qquad
\begin{array}{ll}
j = 4 & \begin{array}{l} j = \underbrace{4 \quad 3 \quad 2}_{} \\ l = \quad 7/2 \end{array} \\ \hline
j = 3 & \begin{array}{l} j = \underbrace{3 \quad 2 \quad 1}_{} \quad 3 \\ l = \quad 5/2 \qquad 7/2 \end{array} \\ \hline
i = 2 & \begin{array}{l} j = \underbrace{2 \quad 1 \quad 0}_{} \quad 2 \\ l = \quad 3/2 \qquad 5/2 \end{array} \\ \hline
i = 1 & \begin{array}{l} j = 1 \\ l = 3/2 \end{array}
\end{array}$$

Eine Entscheidung zwischen den beiden Möglichkeiten kann zunächst nicht getroffen werden.

Bisher war es nach dem Vorhergehenden möglich, das Schema III so aufzubauen, daß das Verzweigungsprinzip in Schema III als Folge der zweifachen Duplizität des Atomrestes erscheint; in der Tat erhalten wir dieselben j-Werte des Gesamtimpulses, die auch das Verzweigungsprinzip des Schemas I direkt liefert:

$$j = 4 \quad 3 \quad 2 \quad 1 \quad 0 \qquad 3 \quad 2 \quad 1 \qquad 3 \quad 2 \quad 1 \qquad 2.$$

Gleichzeitig sehen wir jedoch, daß das Schema III über I und II hinausgehend eine Beschreibung des Auftretens von Singuletts und Tripletts im Spektrum ermöglicht.

Die Bedingung dafür, daß die Schemata I, II, III als verschiedene Seiten desselben physikalischen Problems im Sinne der Grundhypothese S. 841 aufgefaßt werden können, ist wieder die, daß die Schemata immer dort, wo bestimmte Resultate aus mehreren von ihnen hergeleitet werden können, identische Resultate geben.

Betrachten wir nun das Atom in einem starken äußeren Magnetfeld (groß gegenüber der Relativitätswirkung in Schema II bzw. der entsprechenden Wechselwirkung in I und III). Der empirische Sachverhalt[1]) erscheint hier in Schema II wie bei den Dubletts (vgl. S. 846) als einfache Folge der Zweideutigkeit von m_k[2]) (während bei den Dubletts zu einem Werte von m die magnetischen Aufspaltungen $m \pm 1/2$ gehörten, haben wir bei den Tripletts im allgemeinen die Aufspaltungen m, $m \pm 1$); in Schema III, in welchem die magnetische Anomalie ganz auf den edelgasähnlichen Atomrest fällt, ergeben sich wieder die Aufspaltungen im

[1]) A. Landé, Ann. d. Phys. **76**, 273, 1925.

[2]) Die genaue Durchführung vgl. W. Pauli, l. c.

Paschen-Backeffekt in Übereinstimmung mit Schema II, wenn man dem Atomrest doppelten Magnetismus zuschreibt.

Der Übergang von starken zu schwachen Magnetfeldern ist einfach nur in Schema III beschreibbar und wird hier wieder durch die Summationsgesetze und das Kosinusgesetz charakterisiert. Als Bedingung für eine so einfache Beschreibbarkeit (in die r und l eingehen, wie früher r und k bei den Dubletts) ergibt sich jedoch aus den quasimechanischen Eigenschaften des Modells des Schemas III, daß die Wechselwirkung zwischen r und l klein sein muß gegen die zwischen k_1 und k_2, oder, in der Sprache des Schemas II, daß die Relativität klein ist gegen die Wechselwirkung der Elektronen. Ist dies der Fall, so haben wir offenbar bei schwachen Feldern immer die Landéschen g-Werte, wobei aber an Stelle von k in die g-Formel der Wert von l eingeht. Also

$$g = 1 + \frac{j(j+1) - l^2 + r^2}{2j(j+1)}.$$

Bei den Multipletts erster Stufe ($k_2 = {}^1/_2$) stimmen l und k überein. Als Beispiel für Multipletts höherer Stufe sei wieder der Fall $k_1 = {}^5/_2$ $k_2 = {}^3/_2$ (vgl. S. 851) erwähnt. Wir erhalten dort für die g-Werte:

$i = 4$	$j =$	3			
	$g =$	1			
	$l =$	$^7/_2$			
$i = 3$	$j =$	4	3	2	2
	$g =$	$^5/_4$	$^{13}/_{12}$	$^2/_3$	1
	$l =$		$^7/_2$		$^5/_2$
$i = 2$	$j =$	3	2	1	1
	$g =$	$^4/_3$	$^7/_6$	$^1/_2$	1
	$l =$		$^5/_2$		$^3/_2$
$i = 1$	$j =$	2	1	0	
	$g =$	$^3/_2$	$^3/_2$	$^0/_0$	
	$l =$		$^3/_2$		

Die g-Werte genügen natürlich auch der allgemeineren, von Landé für den Fall mehrerer Elektronen aufgestellten Summenregel. Es muß aber hervorgehoben werden, daß die Bedingung für das Auftreten der Multipletts mit den Landéschen g-Werten eben die Kleinheit der „Relativität" gegenüber der Wechselwirkung ist. Ist diese Bedingung nicht erfüllt, so ist eine einfache Einteilung der Terme in Multipletts überhaupt nicht möglich, die g-Werte werden im allgemeinen als irrational

Zeitschrift für Physik 32, 841–860 (1925)

zu erwarten sein[1]); dies scheint z. B. der Fall im Spektrum des Neon[2]). Außerdem besteht offenbar ein fundamentaler Unterschied zwischen den Multipletts erster Stufe und denjenigen Multipletts höherer Stufe, die sich bei kleiner Relativität so wie Multipletts erster Stufe verhalten — wir wollen sie normale Multipletts nennen —, darin, daß für die Multipletts höherer Stufe die Prestonsche Regel nicht gelten wird. In der Tat wird in einer Serie mit wachsender Hauptquantenzahl die Wechselwirkungsenergie immer kleiner und schließlich von derselben Größenordnung oder gar kleiner als die „Relativitätsgröße" des inneren (nicht in einer s-Bahn befindlichen) Elektrons. Daher werden auch mit wachsender Hauptquantenzahl die g-Werte sowie die Intervallproportionen immer mehr von ihren normalen Werten abweichen[3]).

Fragen wir nun weiter nach den Multiplettaufspaltungen ohne Feld, so können wir, wie bei den Dubletts in Schema I, hier in Schema III nur die relativen Werte der Aufspaltung der „Intervalle" erhalten. Man erhält diese Werte, wenn man, wie bei den Dubletts, die Annahme des Kosinusgesetzes macht. Dann folgt aus dem quasimechanischen Modell, das dem Schema III zugrunde liegt, daß die Wechselwirkungsenergie von j, k_1, k_2, r, l abhängt nach der Formel:

$$\cos(l, r)\,[c_1 \cos(k_1, l) + c_2 \cos(k_2, l)],$$

wo c_1 und c_2 zwei bzw. zu den beiden äußeren Elektronen gehörige Konstanten sind, die innerhalb III nicht bestimmbar sind. Die quantentheoretische Größe der Wechselwirkungsenergie erhält man dann wegen der Summationssätze durch Integration der quasiklassischen Energie über j und (wegen des Bestehens eines anologen Summationssatzes für l) über l (von $l - {}^1/_2$ bis $l + {}^1/_2$). Wir erhalten so die Formel: Die Aufspaltung ohne Feld beträgt, von einer additiven, von j unabhängigen Konstante abgesehen:

$$j(j+1)\left[c_1 \frac{k_1^2 - k_2^2 + l^2 - {}^1/_4}{4\,k_1 r\,(l^2 - {}^1/_4)} + c_2 \frac{k_2^2 - k_1^2 + l^2 - {}^1/_4}{4\,k_2 r\,(l^2 - {}^1/_4)}\right]. \qquad (1)$$

---

[1]) Dagegen werden wir in den anderen Grenzfällen, z. B. etwa „Relativität groß gegen Wechselwirkung" wohl wieder rationale g-Werte zu erwarten haben, die allerdings kaum durch die Landéschen Formeln dargestellt werden dürften.

[2]) E. Back (Ann. d. Phys. **76**, 317, 1925) macht zwar darauf aufmerksam, daß innerhalb der empirischen Fehler die g-Werte rational dargestellt werden können; die benutzte Darstellung weicht aber wesentlich von der bei anderen Spektren geltenden ab.

[3]) Hierfür sind in der von Russell und Saunders, l. c., beobachteten pp'-Serie bei Ca deutliche Anzeichen vorhanden.

Was die absoluten Werte der Intervalle anlangt, so können sie, wie bei den Dubletts, nur aus Schema II entnommen werden. Aber auch in Schema II ist eine einfache Bestimmung dieser Intervalle nicht ohne weiteres möglich. Denn das Problem der zwei Elektronen ist in gewisser Weise analog dem Problem eines Elektrons in einem starken äußeren Felde, und auch in einem starken äußeren Magnetfeld war es ja bei den Dubletts nicht möglich, die der Relativitätsgröße entsprechenden, vom Felde unabhängigen Teile der Aufspaltungswerte in Schema II zu beschreiben. Ein besonders einfacher Fall für das Problem der zwei Elektronen tritt jedoch ein, wenn ein Elektron sich in einer s-Bahn befindet; das ihm entsprechende Glied in Formel (1) fällt dann fort, das betreffende s-Elektron trägt naturgemäß zur Relativitätsaufspaltung nicht bei. Landé hat in diesem Falle die naheliegende Annahme versucht, daß die Differenz zwischen den äußersten Termen des Multipletts ($p_1 p_3$, $d_1 d_3 \ldots$) in Schema II als relativistisches Dublett gedeutet werden kann und hat dadurch Übereinstimmung mit der Erfahrung erzielt. Eine Aufklärung dieser Annahme scheint wohl nur möglich, wenn man auch den Fall: Relativität groß gegen die Wechselwirkungsenergie und die Übergänge zwischen den verschiedenen Fällen näher studiert hat. Macht man die Landésche Annahme, so erhält man eine Bestimmung der Konstanten c_1 und c_2. Bezeichnet man das relativistische Dublett der Elektronen mit a_1 bzw. a_2, so wird die Aufspaltung beim Triplett ($r = {}^3/_2$):

$$j(j+1)\left[a_1 \frac{k_1^2 - k_2^2 + l^2 - {}^1/_4}{8\,k_1 (l^2 - {}^1/_4)} + a_2 \frac{k_2^2 - k_1^2 + l^2 - {}^1/_4}{8\,k_2 (l^2 - {}^1/_4)}\right].$$

Schließlich ist es noch von Interesse, zu untersuchen, in welcher Weise die Multipletts höherer Stufe im Falle des Vorhandenseins mehrerer

l	Zwei äquivalente s-Elektronen		Zwei äquivalente p-Elektronen [1]		Zwei äquivalente d-Elektronen	
	Tripletts	Singuletts	Tripletts	Singuletts	Tripletts	Singuletts
$^1/_2$		0 $g = {}^0/_0$		0 $g = {}^0/_0$		0 $g = {}^0/_0$
$^3/_2$			2 1 0 $g = {}^3/_2\ {}^3/_2\ {}^0/_0$		2 1 0 $g = {}^3/_2\ {}^3/_2\ {}^0/_0$	
$^5/_2$				2 $g = 1$		2 $g = 1$
$^7/_2$					4 3 2 $g = {}^5/_4\ {}^{13}/_{12}\ {}^2/_3$	
$^9/_2$						4 $g = 1$

[1]) Das Vorhandensein des p-Tripletts bei zwei äquivalenten p-Bahnen scheint empirisch bestätigt bei Sauerstoff; hier haben wir zwar in Wirklichkeit eine noch

äquivalenter Elektronen vereinfacht werden. Pauli hat, l. c., eine einfache Regel innerhalb des Schemas II angegeben (vgl. S. 849), die gestattet, bei äquivalenten Bahnen gewisse j-Werte auszuschließen und die magnetischen Aufspaltungen bei starken Feldern für die übrigbleibenden j-Werte auszurechnen. Diese Kenntnis der magnetischen Aufspaltung bei starken Feldern gibt dann eindeutig auch die Bedeutung der übrigbleibenden j-Werte innerhalb des Schemas III und daraus die Multiplettstruktur und die Zeemaneffekte bei schwachen Feldern und kleiner Relativität. In unserem Falle zweier äquivalenter Elektronen folgt aus der Paulischen Regel die vorstehende Tabelle der vorhandenen j-Werte.

§ 4. Mehr als zwei äußere Elektronen. Gehen wir nun vom Falle zweier äußerer Elektronen zu dem mehrerer Valenzelektronen über, so lassen sich die Ergebnisse, die durch Zusammenwirken der drei Schemata I, II, III entstehen, leicht angeben. Die möglichen j-Werte und die Werte der magnetischen Aufspaltung bei starken Feldern folgen aus dem Verzweigungsprinzip des Schemas I ebenso, wie aus den beiden anderen Schemata. Die Seriensysteme werden durch einfache räumliche Quantelung der halbzahligen k-Werte gewonnen. Die Multipletts werden, solange die Relativität, d. h. also die Multiplettaufspaltung klein ist gegen die Wechselwirkung, durch die Landéschen Formeln charakterisiert, sind also „normale“ Multipletts. Die zu diesen Multipletts gehörigen r-Werte entstehen durch stete „Verdopplung“ bei Anlagerung eines neuen Elektrons, so daß wir für normale Multipletts zu dem schon von Laporte[1]) empirisch gefundenen Schema kommen.

	Zahl der Valenzelektronen				
	1	2	3	4	5
Mögliche normale Multipletts . . .	Dubletts	Singuletts Tripletts	Dubletts Quartetts	Singuletts Tripletts Quintetts	Dubletts Quartetts Sextetts

größere Anzahl Valenzelektronen an das vorausgehende Edelgas angelagert. Aber es fehlen eben zwei p-Elektronen zur nächstfolgenden Edelgaskonfiguration; wegen der im Paulischen Schema enthaltenen Reziprozität der Ausbildung und Abschließung der Gruppen dürfte die betreffende Konfiguration daher äquivalent sein mit der im Text betrachteten.

[1]) O. Laporte, Naturw. **11**, 779, 1923.

Der Höchstwert des Rumpfimpulses r ist durch die Zahl N der nicht in einer abgeschlossenen Schale befindlichen Elektronen gegeben durch

$$r_{\max} = \frac{N+1}{2}.$$

Der für diese normalen Multipletts in die g-Formel an Stelle des Landéschen K eingehende Wert ist durch die oben definierte Quantenzahl l gegeben und hat mit dem mechanischen k-Wert einer Elektronenbahn nichts zu tun. Dies scheint die Erklärung für die auffallende Tatsache zu sein, daß in den komplizierten Spektren der Eisengruppe sehr tiefe Terme mit großen Aufspaltungen auftreten, die einem sehr hohen Werte von l entsprechen. Man kann ausrechnen, daß Elektronenbahnen, deren k durch diesen hohen Wert gegeben wäre, überhaupt nicht in den Atomrest eindringen könnten, außerdem würden bekanntlich die aus den Multipletts bisher irrtümlich gefolgerten k-Werte der Systematik des periodischen Systems widersprechen. In Wirklichkeit ist eben der Wert von l von dem wirklichen k-Wert der Elektronenbahn ganz verschieden. Nur wenn es sich um echte Multipletts erster Stufe (d. h. solche, bei denen im Atomrest nur s-Bahnen zum Impuls beitragen) handelt, stimmen l und der k-Wert des äußeren Elektrons überein. Aus der Größe und der Aufspaltung der Terme mit hohem l in der Eisengruppe dürfte man also direkt schließen können, daß wir es hier nicht mit echten Multipletts erster Stufe zu tun haben, sondern mit „normalen Multipletts" (höherer Stufe).

§ 5. Anwendung des Korrespondenzprinzips. Die Anwendung des Korrespondenzprinzips zur Ableitung von Auswahlregeln und Intensitätsgesetzen ist nur möglich bei Benutzung eindeutiger mechanischer Modelle. Wenn es sich daher um die Berechnung der Intensitäten von Linien, die Sprüngen eines Elektrons entsprechen, handelt, so werden wir die Fourierentwicklung der Bewegung des Elektrons in demjenigen Schema zu untersuchen haben, in welchem das äußere Elektron eindeutig beschrieben wird, also in Schema I bzw. III. Man kommt dann zu den bekannten einfachen Auswahlregeln und Intensitätsschätzungen [1]), die in neuester Zeit durch die Anwendung von Summationsregeln zu exakten

[1]) A. Landé, ZS. f. Phys. **5**, 231, 1921, § 5; A. Sommerfeld und W. Heisenberg, ebenda **11**, 131, 1922.

Zeitschrift für Physik 32, 841–860 (1925)

Gesetzen[1]) ausgebaut wurden. Insbesondere darf der Impuls k des Elektrons nur nach $k \pm 1$ springen, während für j die Sprünge

$$j \begin{array}{l} \nearrow j+1 \\ \rightarrow j \\ \searrow j-1 \end{array}$$

erlaubt sind.

Um die Intensitäten von Linien zu schätzen, die Sprüngen von mehr als einem Elektron entsprechen, müssen wir die entsprechenden mechanischen Kopplungsschwingungen untersuchen. Betrachten wir zunächst ein verwandtes, einfacheres Problem: Auf das Elektron, dessen Bewegung durch eine Fourierreihe mit den Frequenzen $\omega_{n,k,j}$ dargestellt sein möge, falle äußeres, linear polarisiertes Licht von der Frequenz ν. Kramers und der Verfasser[2]) haben untersucht, welche Kombinationsschwingungen zwischen ν und $\omega_{n,k,j}$ auftreten, und haben gefunden, daß alle Kombinationsschwingungen Frequenzen von der Form $\nu + \omega_{n,k,j} - \omega'_{n,k,j}$ besitzen, wo $\omega_{n,k,j}$ und $\omega'_{n,k,j}$ irgendwelche Frequenzen der ungestörten Bewegung bedeuten. Ihre Amplituden sind durch Produkte aus gewissen Differentialquotienten der zu $\omega_{n,k,j}$ bzw. $\omega'_{n,k,j}$ gehörigen Amplituden der ungestörten Bewegung gegeben. In der ungestörten Bewegung treten nun offenbar in der Fourierreihe für das elektrische Moment des Elektrons alle Oberschwingungen von ω_n, aber von ω_k nur die Grundschwingung $\pm \omega_k$ auf. Daher haben alle Kombinationsschwingungen die Form

$$\nu + \omega_{n,j} \quad \text{oder} \quad \nu + \omega_{n,j} \pm 2\,\omega_k.$$

Die Kombinationsschwingungen entstehen also, wenn wir dieses Resultat gemäß der Quantentheorie deuten, durch Sprünge von k um 0 oder um 2, Sprünge um 1 und um mehr als 2 kommen (bei den Kombinationsschwingungen) nicht vor. Da dieses Resultat eine große Ähnlichkeit mit dem empirischen Befund bei den pp'-Linien aufweist, wollen wir untersuchen, unter welchen Bedingungen es sich auf die Kombinationsschwingungen zweier oder mehrerer Elektronen überträgt.

Denken wir uns zunächst die vorkommenden Kombinationsschwingungen berechnet, die dem periodischen Teil der Wechselwirkungsenergie als Ansatz für die Störungsfunktion entsprechen. Die Fourierreihe dieser

[1]) S. Ornstein und K. Burger, ZS. f. Phys. **24**, 41 u. **29**, 241, 1924; S. Goudsmit und R. de L. Kronig, Naturw. **13**, 90, 1925; H. Hönl, ZS. f. Phys. **31**, 340, 1925; R. de L. Kronig, ebenda **31**, 885, 1925. A. Sommerfeld und H. Hönl, Sitzber. d. preuß. Akad. d. Wiss. 1925, S. 141.

[2]) H. A. Kramers und W. Heisenberg, ZS. f. Phys. **31**, 681, 1925.

Wechselwirkung wird die Frequenzen aller Elektronen enthalten und wenn, wie dies in der klassischen Mechanik der Fall wäre, die Elektronenbahnen sich gegenseitig ineinander verschlingen und stark stören, so hätten wir mit allen denkbaren Kombinationsschwingungen zu rechnen. Dies Bild der klassischen Mechanik trifft aber jedenfalls für die Quantentheorie nicht zu; denn schon die Tatsache, daß die Bewegung der Elektronen „periodisch" ist, läßt auf weitgehende unmechanische Vereinfachungen schließen. Betrachten wir deshalb speziell die Bewegung eines bestimmten Elektrons; unter gewissen Bedingungen wird sich die obengenannte Störungsfunktion der Wechselwirkung entwickeln lassen nach Kugelfunktionen vom Abstand des Elektrons; das erste Glied dieser Entwicklung lautet offenbar $(\mathfrak{p}\,.\,\mathfrak{A})$, wo $\mathfrak{p}$ den Vektor des elektrischen Moments der ungestörten Elektronenbewegung und $\mathfrak{A}$ den Vektor der elektrischen Feldstärke der anderen Elektronen am Orte des Kernes bedeutet.

Die quantentheoretische Vereinfachung scheint nun darin zu bestehen, daß in der Quantenmechanik, unabhängig davon, ob in der klassischen Mechanik die Bedingungen für die Entwickelbarkeit erfüllt sind, und unabhängig von der Größe der höheren Glieder der betreffenden Kugelfunktionenreihe, das genannte erste Glied der Entwicklung alle anderen Glieder weit überwiegt. Diese Annahme hat eine gewisse Ähnlichkeit mit dem Kosinusgesetz, das ja auch aussagt, daß in der Wechselwirkung das erste Glied einer Kugelfunktionenentwicklung weit überwiegt. Eine Begründung für diese quantentheoretische Vereinfachung läßt sich vom Standpunkt der Mechanik aus nicht geben.

Machen wir also die Annahme, daß diese Vereinfachung wirklich eintritt; dann erhalten wir in dieser Näherung dieselben Ergebnisse, wie bei der Kopplungswirkung zwischen Licht und einem Atom. Betrachtet man speziell die Störung eines bestimmten Elektrons, so findet man, daß sie zu Kopplungsschwingungen führt, bei denen für das betrachtete Elektron k um 0 oder ± 2, für irgend ein anderes Elektron k um ± 1 springt (da nur die Grundfrequenz $\pm \omega_k$ der anderen Elektronen und nicht ihre Multipla in der Fourierentwicklung von $\mathfrak{A}$ vorkommen); mehr als zwei Elektronen können gleichzeitig nicht springen. Da es gleichgültig ist, welches Elektron herausgegriffen wird, so gilt dieses Resultat allgemein.

Nach dieser störungstheoretischen Betrachtung kehren wir wieder zu den Spektra der Elemente mit mehreren Valenzelektronen zurück. Man wird nach Schema III fordern, daß die eben abgeleitete Auswahl-

Zeitschrift für Physik *32*, 841–860 (1925)

regel für die mechanischen k-Werte der Elektronen gilt [1]): Wenn zwei Elektronen gleichzeitig springen, so ändert sich das k des einen Elektrons um ± 1, das k des anderen um 0 oder ± 2.

Ferner wird für den wirklichen Gesamtimpuls j des Atoms als einer eindeutigen mechanischen Größe die Auswahlregel

$$j \begin{matrix} \nearrow j+1 \\ \rightarrow j \\ \searrow j-1 \end{matrix}$$

gelten müssen.

Endlich wird man aus Schema III schließen, daß im Falle normaler Multipletts (Relativität klein gegen Wechselwirkung) l, das sich ja in Schema III wie j verhält, derselben Auswahlregel gehorcht, wie j, nämlich [2])

$$l \begin{matrix} \nearrow l+1 \\ \rightarrow l \\ \searrow l-1 \end{matrix} .$$

---

1) Ebensowenig wie das Kosinusgesetz wird man allerdings diese Auswahlregel als exaktes Gesetz ansprechen dürfen. Bei den komplizierteren Spektren müssen wir auf Abweichungen vorbereitet sein.

2) Anm.: Laporte (ZS. f. Phys. **23**, 135, 1924) hat noch eine empirische Regel aufgestellt, nach der jedes Spektrum in zwei Teilsysteme (die „gestrichenen" und die „ungestrichenen" Terme) geteilt werden kann, von denen jedes unter sich nach der Regel $\Delta l = \pm 1$ kombiniert, während bei Kombinationen zwischen gestrichenen und ungestrichenen Termen $\Delta l = 0$ ist. [Jedenfalls muß im letzteren Falle oft auch $\Delta l = \pm 2$ zugelassen werden (vgl. die Messungen von Russell und Saunders, l. c.; Laporte, ZS. f. Phys. **26**, 1, 1924)]. Diese Laportesche Regel ist eine einfache Folge des Auswahlprinzips für k, wenn man zunächst zuläßt, daß allgemein l bei Kombination zwischen gestrichenen und ungestrichenen Termen um beliebige gerade, bei Kombinationen innerhalb der Systeme um beliebige ungerade Zahlen springen kann. Denn man braucht als ungestrichenes System nur alle Terme zu wählen, für welche

$$l + \sum_1^N k_i \equiv \frac{N+1}{2} \pmod 2$$

(die Summe über i bedeutet Summe über alle Elektronen, ihre Anzahl sei N), als gestrichenes alle, für die

$$l + \sum_1^N k_i \equiv \frac{N-1}{2} \pmod 2.$$

Dann folgt die oben erwähnte Auswahlregel für l daraus, daß $\sum_1^N k_i$ stets nur um ungerade Zahlen springen kann. Der Einteilung in gestrichene und ungestrichene Terme dürfte also keine physikalische Bedeutung zukommen, wenn sie auch empirisch für die Entwirrung komplizierter Spektren wesentlich ist.

Doch wird man, wenn die Relativität nicht mehr klein ist, auf Abweichungen von dieser Regel (ebenso etwa wie auf Abweichungen von der Intervallregel) gefaßt sein müssen [1].

Damit ist die Reihe der mechanisch eindeutigen Größen erschöpft. Empirisch scheint noch eine Auswahlregel für r zu bestehen

$$r \begin{array}{l} \nearrow r+1 \\ \rightarrow r \\ \searrow r-1 \end{array} .$$

Eine Ableitung für dieses Gesetz kann man zunächst korrespondenzmäßig nicht geben.

Schlußbemerkungen. Es dürfte kaum nötig sein, den vorläufigen und in vieler Beziehung unbefriedigenden Charakter der Formulierung des Multiplettproblems, der in der vorliegenden Arbeit versucht ist, zu betonen. Obwohl die quantentheoretischen Gesetze der Wechselwirkung der Elektronen im Atom sich zweifellos durch große Einfachheit auszeichnen, scheint es zurzeit jedoch keinen anderen Weg zur Deutung dieser Gesetze zu geben, als den über modellmäßige Bilder symbolischer Bedeutung, bei denen diese Einfachheit kaum genügend zum Ausdruck kommt.

Der amerikanischen Stiftung International Education Board, die meinen Aufenthalt in Kopenhagen ermöglichte, sei an dieser Stelle der herzlichste Dank ausgesprochen.

Kopenhagen, Univers. Inst. for teoretisk Fysik, 4. April 1925.

---

[1] Die hier auftretenden Intensitätsfragen werden ausführlich behandelt in einer im Erscheinen begriffenen Arbeit von R. d. L. Kronig.

Group 3

Quantum Mechanics (1925–1927)

Quantum Mechanics (1925–1927)

An Annotation
by B. L. van der Waerden, Zurich, and H. Rechenberg, Munich

Between 1925 and 1927, Heisenberg participated prominently in the discovery, formulation and application of quantum mechanics. In this period he published 11 original scientific papers, three of which we shall mention here only briefly (they have been allocated to the next Group 4 of papers dealing with applications of quantum mechanics), while another one has been placed into the previous Group 2 (paper No. 12, p. 306). We shall now concentrate on the remaining seven papers, which played a crucial role in creating and establishing quantum mechanics and its physical interpretation. Occasionally, we shall also refer to the three reviews of the theory which Heisenberg wrote or coauthored at the same time and which have been included in *Volume B* of the *Collected Works*.

1. "Sharpening" the Principle of Correspondence

Heisenberg spent the winter of 1924–1925 at Copenhagen, working with Niels Bohr and Hendrik Kramers. In the papers emerging from this collaboration, one on fluorescence (No. 1, p. 344) and the other (No. 2, p. 354) written jointly with Kramers), the underlying ideas were (a) Bohr's principle of correspondence stating that quantum laws must be formulated such that for large quantum numbers they yield agreement with classical laws, and (b) the idea of "virtual oscillators", which requires that, in calculating the frequency, intensity, and polarization of light absorbed and emitted by an atom, we may describe the atom by a set of (harmonic) oscillators having the (observed) frequencies of the absorbed and emitted light.

Ever since 1922, Heisenberg had felt that "something in the atom must vibrate with the right frequency". The idea of virtual oscillators expressing this feeling was implicit already in an earlier (1921) paper of Rudolf Ladenburg on the number of dispersion electrons in atom [1]. In their well-known paper of 1924, Bohr, Kramers and John Slater then formulated it explicitly as follows:

> The correspondence principle has led to comparing the reaction of an atom on a field of radiation with the reaction on such a field which, according to the classical theory of electrodynamics, should be expected from a set of "virtual" harmonic oscillators with frequencies equal to those determined by the Eq. (1) [i.e., $h\nu = E_1 - E_2$, with E_1 and E_2 denoting the energy values of the atom in the initial and final state and h Planck's constant] for the various possible transitions between stationary states [2].

In Heisenberg's paper (No. 1, p. 344), the correspondence principle is mentioned already in the first sentence, and the idea of virtual oscillators in the second sentence. The author used both basic assumptions to derive a series of remarkable results on the polarization of radiation emitted by complex atoms in the presence of external electric or magnetic fields. Even more effectively, Kramers and Heisenberg applied the same assumptions in their joint paper (No. 2, p. 354), as they stated in the introduction:

> Two clues for the investigation were provided: one, by the empirical applicability of the formula for [the electric (scattering) moment] obtained from classical dispersion theory, if the atom is imagined to contain classical oscillators whose natural frequencies coincide with the frequencies of the absorption lines; the other by the correspondence principle [3].

The authors first considered that part of the scattered light having the same frequency ν as the incident radiation. They derived a classical formula for the scattering moment, i.e., formula (26) of their paper; and next, they wrote down a quantum theoretical formula (27) containing difference instead of differential quotients, which coincides asymptotically with Eq. (26) in the limit of large quantum numbers. The quantities $\mathfrak{A}_a$ and $\mathfrak{A}_e$ occurring in Eq. (27) are, as the authors stated, "the amplitudes characteristic for the absorption and emission transitions", i.e., they are the amplitudes of virtual oscillators, multiplied by the charge of the electron $(-e)$ in order to represent the electrical moments determining the intensities of the radiation.

With respect to the distribution of work between the two authors, Heisenberg once told van der Waerden that Kramers was responsible for the composition of the joint paper, while Heisenberg provided in particular the final form of Eqs. (41) and (42); these equations, the most valuable result of the work, determine the dispersive moments $\mathfrak{M}(\nu+\nu^*)$ and $\mathfrak{M}(\nu-\nu^*)$, i.e., the intensity of the incoherently scattered radiation by atoms and molecules (later discovered experimentally in 1928 by C. V. Raman and named after him).

For the further development of Heisenberg's ideas both papers turned out to be of decisive importance, as they did not merely provide applications of well-known principles, but rather employed a new method, which the author called in letters and publications "sharpening the correspondence principle" [4]. The main step in this method was the following: in determining quantum theoretical formulae from classical ones describing atomic systems one has to replace the Fourier coefficients of the latter by quantum theoretical amplitudes, namely either $\mathfrak{A}_a$ or $\mathfrak{A}_e$.

At about the time that Heisenberg and Kramers began to collaborate in Copenhagen, i.e. in October 1924, John Hasbrouck Van Vleck of Minnesota published a paper in *Physical Review*, entitled "The Absorption of Radiation by Multiply Periodic Orbits", in which he also clarified and extended the principle of correspondence [5]. Van Vleck's idea may be expressed this way: If we want to estimate the absorption of radiation by atoms and molecules with the help of the principle of correspondence, we have to compare the absorption, computed classically, with the difference between absorption and induced emission, computed from Einstein's formulae of 1916 [6]. In the limit of higher quantum numbers, this difference must become equal to the classical absorption [7].

2. "Fabricating Quantum Mechanics"

In the spring and early summer of 1925, Heisenberg was "fabricating quantum mechanics", as he wrote in a letter to Pauli dated June 21, 1925 [8]. In particular he reflected about the amplitudes of the virtual oscillators associated with atomic systems. From his letters to Ralph Kronig and Wolfgang Pauli written in June 1925 we see how the law of multiplication of quantum mechanical quantities – that is, of the amplitudes of "virtual oscillators" – was beginning to take shape in Heisenberg's mind. This law, in which $a(n, m)$ denoted the amplitude of a virtual oscillator corresponding to a transition from a level n to a level m, was later recognized as representing matrix multiplication.

For example, in the letter to Kronig, dated June 5 [9], Heisenberg considered the square of a Fourier series containing only one term, i.e.,

$$b_2 \cdot e^{2i\omega t} = (a_1 \cdot e^{i\omega t})^2 . \tag{1}$$

He replaced Eq. (1) by a "quantentheoretische Verwandlung" (quantum theoretical translation), namely

$$b_2(n, n-2) \cdot \exp[i\,\omega(n, n-2)\,t] = a_1(n, n-1) \cdot a_1(n-1, n-2) \cdot \exp\{i[\omega(n, n-1) + \omega(n-1, n-2)]\,t\} \tag{2}$$

or

$$b_2(n, n-2) = a_1(n, n-1) \cdot a_1(n-1, n-2) . \tag{3}$$

The multiplication rule (3) is necessary – in Heisenberg's words, "zwangsläufig" –, because Bohr's quantum condition

$$E_n - E_m = h\nu = h\frac{\omega}{2\pi} \tag{4}$$

implies

$$\omega(n, n-1) + \omega(n-1, n-2) = \omega(n, n-2) .$$

In this way Heisenberg justified his multiplication law for matrices, without knowing that it had been used by mathematicians long ago.

In the letter to Pauli, dated June 24, the main ideas of Heisenberg's later paper (No. 3, p. 382) were already present, even the "quantum condition":

$$h = 2\pi m \sum_\tau \{a^2(n+\tau, n)\,\omega(n+\tau, n) - a^2(n, n-\tau)\,\omega(n, n-\tau)\} . \tag{5}$$

(m denotes the mass of the oscillating system.)

He obtained Eq. (5) from the old, "classical" quantum condition,

$$\int p\,dq = n \cdot h = J , \tag{6}$$

by the same method of "systematic guessing" (or "sharpening the correspondence principle"), which he had already applied in the paper with Kramers (No. 2). That is, the condition (6) had to be rewritten, with $q = \sum_\tau a_\tau^2 \cdot \omega\tau)$, as

$$1 = 2\pi m \sum_\tau \tau \frac{\partial}{\partial J}(a_\tau^2 \cdot \tau\omega)\,, \tag{7}$$

and the differential quotient under the sum had to be replaced by a difference quotient in order to translate Eq. (7) into Eq. (5).

3. Heisenberg in Helgoland

In May 1925 Heisenberg tried to apply the method of "systematic guessing" to the hydrogen atom, hoping to obtain the intensities of the spectral lines. However, the hydrogen atom proved to be too difficult, and he turned instead to treat simple systems, such as the rotator and the anharmonic oscillator. With these he found the results mentioned above.

In June 1925 a heavy attack of hay fever forced Heisenberg to leave Göttingen, where he had returned for the summer semester, and to stay for two weeks on the rocky island of Helgoland [10]. Here, he tried to solve the anharmonic oscillator in one dimension, described by the equation of motion

$$\ddot{q} + \omega_0^2 q + \lambda q^2 = 0 \tag{8}$$

in a quantum theoretical translation. In classical theory, a solution of Eq. (8) would be

$$q = \lambda a_0 + a_1 \cdot \cos(\omega t) + \lambda a_2 \cdot \cos(2\omega t) + \lambda^2 a_3 \cdot \cos(3\omega t) + \ldots \tag{9}$$

i.e., a power series in the anharmonicity parameter λ (which is thought to be a small quantity). Heisenberg now replaced the terms of this series by a new set of terms, i.e.,

$$\lambda a(n,n);\; a(n,n-1)\cdot\cos[\omega(n,n-1)\cdot t];\; \lambda a(n,n-2)\cdot\cos[\omega(n,n-2)\cdot t];\;\ldots$$

and substituted these terms into the equation for q, Eq. (9). For calculating q^2, he made use of his multiplication rule (2). Thus he obtained equations between the coefficients $a(n,n)$, $a(n,n-1)$, etc., which did, however, not yet determine the quantum theoretical solution of Eq. (8) completely.

The final steps in the solution were as follows: First, Heisenberg inserted the expression for the coefficients into his quantum condition (5); thus, he fixed the coefficients $a(n,n-\tau)$ to a constant. Then he discovered that this constant was obtained from the assumption of a "lowest state" n_0. This lowest or ground state was defined by the condition that the transition amplitudes $a(n_0, n_0-\tau)$, $\tau = 1, 2, 3, \ldots$ are zero.

Heisenberg next calculated the energy of the anharmonic oscillator by inserting the quantum theoretical expansion for q into the energy expression [11]. To his great satisfaction, he found that in the energy expression terms corresponding to transitions $n \to n+\tau$ cancelled, i.e., that the quantum theoretical energy was a constant, at least as long as one neglects higher powers of λ (than λ^2). Many years later he told van der Waerden:

> In Helgoland, there was a moment in which it came to me like an enlightenment: I realized that the energy is constant in time. It was rather late in the night. I laboriously calculated the result, and it was correct. Then I climbed on a rock and watched the sunrise and I was happy. (*In Helgoland war ein Augenblick, in dem es mir wie eine Erleuchtung kam, als ich sah, daß die Energie zeitlich konstant war. Es war ziemlich spät in der Nacht. Ich rechnete es mühsam aus, und es stimmte. Da bin ich auf einen Felsen gestiegen und habe den Sonnenaufgang gesehen und war glücklich.*)

On his way back from Helgoland to Göttingen, Heisenberg visited Pauli in Hamburg. He had expected sharp criticism of his quantum theoretical attempts, but Pauli warmly encouraged him to continue along the same path. Soon afterwards, on June 21, Heisenberg wrote to Pauli: "In my attempt to fabricate quantum mechanics there is only slow progress, but I do not care anymore how far I move away from the theory of conditionally periodic systems [i.e., from the old quantum theory of Bohr and Sommerfeld]" [12]. Three days later, he wrote again to Pauli, explaining in detail his calculation of the anharmonic oscillator as well as the philosophical principle underlying the whole approach: that, in the quantum mechanical description of atomic phenomena, only observable quantities should enter.

4. Heisenberg's "Umdeutung" (Reinterpretation)

On July 9, Heisenberg sent the manuscript of a paper entitled "*Über die quantentheoretische Umdeutung kinematischer und mechanischer Beziehungen*" to Pauli. After the latter had expressed his full agreement with its contents, Heisenberg presented the manuscript to Max Born, who studied and submitted it to *Zeitschrift für Physik* (where it was received on July 29). This paper appeared in print in September 1925 and caused a revolution in physics (No. 3, p. 382).

In the introduction of his paper the author explained the fundamental idea of what he called quantum theoretical reinterpretation in the following words:

> It has become the practice to characterize this failure of the [old] quantum-theoretical rules as a deviation from classical mechanics, since the rules were essentially derived from classical mechanics. This characterization has, however, little meaning when one realizes that the Einstein-Bohr frequency condition (which is valid in all cases) already represents such a complete departure from classical mechanics, or rather (using the viewpoint of wave theory) from the kinematics underlying this mechanics, that even for the simplest quantum theoretical problems the validity of classical mechanics simply cannot be maintained. In this situation it seems sensible to discard all hope of observing hitherto unobservable quantities, such as the position and

period of the electron ... Instead it seems more reasonable to try to establish a theoretical quantum mechanics, analogous to classical mechanics, but in which only relations between observable quantities occur [13].

It is difficult to read Heisenberg's paper not only because of some unexplained notions and several tacitly made assumptions, but also because even the notation is inconsistent. In Sect. 1 for example, the author introduced certain amplitudes $\mathfrak{A}(n, n-\alpha)$, which are regarded as complex vectors determining the polarization and the phase of the emitted light (like in the previous paper No. 2) [14]. In Sect. 2, however, he changed the notation; i.e., he introduced, just as in his letters to Kronig and Pauli, complex scalar amplitudes, $a(n, n-\alpha)$ and $a(n, n+\alpha)$, forming an infinite array which should replace any classical quantity $x(t)$. The complex number $a(n, m)$ denotes, e.g., the amplitude of a virtual oscillator corresponding to the transition $n \rightarrow m$. In the classical theory, any variable $x(t)$, which is a periodic function of time, can be expanded into a Fourier series with coefficients $a_\alpha(n)$. Heisenberg replaced the latter by quantum theoretical amplitudes $a(n, n+\alpha)$. For the multiplication of two quantities he gave a multiplication rule for two arrays of quantities, $a(n, m)$ and $b(n, m)$, namely,

$$c(n, m) = \sum_s a(n, s) \cdot b(s, m)\,, \tag{11}$$

which represents nothing other than the rule of matrix multiplication (although the author did not recognize the fact). He had first formulated this rule for complex vector amplitudes in Sect. 1. However, the $\mathfrak{A}$ und $\mathfrak{B}$ in the final "quantum-theoretical" formula at the end of Sect. 1 were no longer considered to be complex vectors, but rather complex numbers. That procedure is indeed very confusing because, in all other cases, Heisenberg denoted vectors by gothic letters.

Right at the beginning of Sect. 2 the author explained a fundamental, highly original idea of his new mechanics. He assumed that the equation of motion,

$$\ddot{x} + f(x) = 0\,, \tag{12}$$

can be retained and that only the classical kinematic interpretation of x as a location depending on time had to be rejected. Instead of the function $x(t)$, he considered an array of terms,

$$a(n, m) \cdot \exp[\mathrm{i}\,\omega(n, m) \cdot t]\,,$$

in which $a(n, m)$ denote complex amplitudes, while $\omega(n, m)$ are the frequencies determined by Bohr's frequency condition Eq. (4) [15]. In the same Sect. 2, Heisenberg also "derived" his quantum condition Eq. (5) in the way we have outlined above, which must be considered the other decisive idea of the paper [16]. This quantum condition is, as Heisenberg stated in a footnote, equivalent to the sum rule of Werner Kuhn and Willy Thomas [17].

In Sect. 3, Heisenberg presented the solution of the anharmonic oscillator described by the equation of motion

$$\ddot{x} + \omega_0^2 x + \lambda x^3 = 0 \tag{13}$$

and, in addition, a preliminary treatment of the rotator, with application to the theory of band spectra and anomalous Zeeman effects [18].

5. Born's Conjecture on $pq - qp$

In his recollections Max Born wrote:

> After having sent Heisenberg's paper to the *Zeitschrift für Physik* for publication, I began to ponder about his symbolic multiplication, and was soon so involved in it that I thought the whole day and could hardly sleep at night. For I felt there was something fundamental behind it... And one morning ... I suddenly saw light: Heisenberg's symbolic multiplication was nothing but the matrix calculus, well-known to me since my student days from the lectures of Rosanes in Breslau.
>
> I found this by just simplifying the notation a little: instead of $q(n, n+\alpha)$... I wrote $q(n, m)$ and rewriting Heisenberg's form of Bohr's quantum conditions I recognized at once its formal significance. It meant that the two matrix products $\boldsymbol{pq}$ and $\boldsymbol{qp}$ are not identical... Closer inspection showed that Heisenberg's formula gave only the value of the diagonal elements ($m = n$) of the matrix $\boldsymbol{pq} - \boldsymbol{qp}$; it said that they were all equal and had the value $h/2\pi\mathrm{i}$. But what were the other elements $m \neq n$?
>
> Here my own constructive work began. Repeating Heisenberg's calculation in matrix notation, I soon convinced myself that the only reasonable value of the nondiagonal elements should be zero, and I wrote down the strange equation
>
> $$\boldsymbol{pq} - \boldsymbol{qp} = \frac{h}{2\pi\mathrm{i}}\mathbf{1},$$
>
> *where* $\mathbf{1}$ is the unit matrix. But this was only a guess, and my attempts to prove it failed [19].

Shortly afterwards, probably on June 19, Born asked his pupil to assist him in his work. He later reported:

> Jordan accepted and after only a few days brought me the solution of the problem: he showed that the canonical equations of motion, applied to the matrices $\boldsymbol{p}$ and $\boldsymbol{q}$, led to the result that the time derivative of $\boldsymbol{pq} - \boldsymbol{qp}$ must vanish, hence the matrix itself must be diagonal. Then we began to write a joint paper... [20].

In September 1925, Jordan sent Heisenberg (who as then in Copenhagen) a summary of the first two chapters of the joint paper with Born, containing the main principles of matrix mechanics, the matrix formulation of Heisenberg's quantum theoretical reformulation. On September 13, Heisenberg answered:

"Dear Jordan! Thank you very much for your letter and the proof [of energy conservation in matrix mechanics]. Yours and Born's work seems to me to be a very big step forward. (*Lieber Jordan! Vielen Dank für Ihren Brief und den Beweis. Ihre und Borns Arbeit scheint mir ein sehr großer Fortschritt.*)"

In the same letter to Jordan, Heisenberg sketched the first steps towards a perturbation theory, which he presented in a subsequent letter to Pauli, dated September 18, in a final form. It was based on the following definition of a "canonical transformation" in quantum mechanics: "Quite generally, every transformation which transforms $\boldsymbol{p}\cdot\boldsymbol{q}-\boldsymbol{q}\cdot\boldsymbol{p}$ into itself is canonical" [21].

It is possible to demonstrate that Heisenberg's canonical transformation can be written as

$$\boldsymbol{P}=\mathrm{e}^{S}\boldsymbol{p}\,\mathrm{e}^{-S}, \quad \boldsymbol{Q}=\mathrm{e}^{S}\boldsymbol{q}\,\mathrm{e}^{-S}. \tag{14}$$

Born, on the other hand, proposed a simpler and more general transformation, namely,

$$\boldsymbol{P}=\boldsymbol{S}\cdot\boldsymbol{p}\cdot\boldsymbol{S}^{-1}, \quad \boldsymbol{Q}=\boldsymbol{S}\cdot\boldsymbol{q}\cdot\boldsymbol{S}^{-1}, \tag{15}$$

which was used in Sect. 1.3 of the next paper on quantum mechanics, the socalled "Three-Men-Paper", written jointly by Born, Heisenberg, and Jordan, which we shall discuss now [22].

6. The "Three-Men-Paper"

A complete, consistent exposition of matrix mechanics was presented by Born, Heisenberg and Jordan in their exhaustive paper "Zur Quantenmechanik II" (No. 4, p. 397), which was received by *Zeitschrift für Physik* on November 16, 1925 [23]. The history of this paper is very complicated, especially because Heisenberg was in Copenhagen during most of the time it was composed (from September to the middle of October), and Born left Göttingen to visit the USA in late October 1925. Heisenberg was responsible for the final version of the Introduction and Chap. 1 and 2; Born wrote Chap. 3, Heisenberg elaborated the first part of Chap. 4; and the last Sect. 3 of Chap. 4 was produced by Jordan [24].

A rough idea of the context of the paper may be obtained by looking at the titles of the chapters and sections:

Chapter 1 Systems Having One Degree of Freedom
- § 1. Fundamental Principles
- § 2. The Canonical Equations, Energy Conservation and Frequency Condition
- § 3. Canonical Transformations
- § 4. Perturbation Theory
- § 5. Systems for which Time-Variables Enter Explicitly into the "Energy Function"

Chapter 2 Fundamentals of the Theory for Systems Having an Arbitrary Number of Degrees of Freedom

It is impossible to present here a short summary of the contents of the "Three-Men-Paper", which contains essential information going beyond the chapter and section titles listed above. However, we would like to draw attention to a review paper of Heisenberg, based on lectures presented to a mathematical audience which contains a summary of the material discussed in the "Three-Men-Paper" [25].

7. Pauli's Paper on the Hydrogen Spectrum

In late October 1925 Wolfgang Pauli succeeded in deriving the line spectrum of the hydrogen atom by matrixmechanical methods. His paper was received on January 17, 1926, by *Zeitschrift für Physik* and published later in March [26]. The effect of the paper was to help in convincing the majority of physicists that the "Göttingen quantum mechanics" was correct in its description of atomic phenomena.

8. Many-Body-Problem and Resonance

In the first half of 1926, the situation in quantum physics changed completely. Erwin Schrödinger developed his wave mechanics in a series of papers; he also proved that the Born-Heisenberg-Jordan matrices can be calculated with the help of wave mechanics and that the Göttingen theory and his own theory yield exactly the same energy values and spectral lines for atoms and atomic systems [27]. The latter result was also obtained by Pauli [28]. Heisenberg referred to the equivalence proofs in the introduction of a paper submitted in June 1926 on the many-body problem and resonance in quantum mechanics (paper No. 5, p. 456). Accordingly, in this paper he treated the problem of two coupled oscillators in two ways, first with the matrix method and next with Schrödinger's method, finding identical results in both cases: each term of the uncoupled system, with quantum numbers n and m (or m and n; where the quantum numbers denote the terms of oscillator 1 and 2) splits under the influence of the coupling into two terms, one having a symmetric and the other an antisymmetric eigenfunction in wave mechanics; for $m = n$ only the symmetric state occurs.

Heisenberg then applied the same consideration to the helium atom. Taking into account Pauli's exclusion principle for the electrons, he explained in a further paper (paper No. 2 in the next Group 4, p. 531), why orthohelium has only a triplet system and parhelium only a singlet system. In order to satisfy the exclusion principle, one has to restrict oneself to antisymmetric eigenfunctions, even in the case of several electrons as well [29]. In a later paper that also dealt with the many-body problem and resonance in quantum mechanics (paper No. 3 in Group 4, p. 551), Heisenberg applied the general principles and results of the preceding paper to special problems in the theory of multiplet spectra of atoms and bound spectra of molecules.

On September 26, 1926, at the 89th *Versammlung Deutscher Naturforscher und Ärzte*, Heisenberg presented a review on quantum mechanics, in which he reported on the results and successes thus far achieved in matrix and wave mechanics [30]. In his talk, he also addressed the problems of the physical interpretation of the theory, notably the question of whether the theory demanded quantum jumps, as the Göttingen proponents emphasized, or whether quantum jumps were not required, as Schrödinger suggested.

9. Uncertainty Relations and the Physical Interpretation of Quantum Mechanics

In April 1926, Heisenberg gave a colloquium talk on quantum mechanics in Berlin, at which he had the opportunity of discussing afterwards with Einstein the physical foundations of the new theory. Einstein had criticized the postulate that only observable quantities should be used in the theoretical description of atoms, arguing that it is the theory which decides what can be observed. A little later, Schrödinger proposed a continuum interpretation of wave mechanics which was not acceptable to the pioneers of quantum mechanics in Göttingen, Copenhagen, Cambridge and Hamburg [31]. Born, on the other hand, introduced the statistical interpretation of wave functions in two papers, dealing with the quantum-mechanical collision processes [32].

In the autumn of 1926, Heisenberg, then in Copenhagen, began to concern himself seriously with the problem of the physical interpretation of quantum mechanics. He first participated in the discussions with Schrödinger, who came to Copenhagen in September. After Bohr and Heisenberg failed to convince Schrödinger of their views, they sat down to develop systematically what was later called the "Copenhagen interpretation of quantum mechanics". Heisenberg first used his former matrix treatment of a system of two equal, interacting atoms to calculate the time averages of any given function of the energy; he found that the results agreed with those of a discontinuity theory in which only discrete amounts of energy are transferred between the atoms, and he demonstrated that this conclusion also remained valid in the case of several or even an infinite number of identical atoms (lattice) or in the case of two different atomic systems (paper No. 6, p. 472).

Heisenberg, in frequent and intense correspondence with Pauli in Hamburg, tried to decide the question of what can be observed in atomic systems by strictly referring to the matrix formulation. Even in describing the obviously continuous motion of electrons in a cloud chamber, he attempted to avoid the wave mechanical description and referred rather to the generalized scheme of quantum mechanics, which Dirac and Jordan developed in late 1926 [33]. Heisenberg, in particular, worked with Dirac's theory and derived with it the relation

$$p_1 \cdot q_1 \sim h\,, \tag{16}$$

expressing a limitation of the accuracy, q_1 and p_1, of simultaneous observations of the position variable and the corresponding momentum variable of a quantum mechanical system. This socalled uncertainty relation constitutes the main result of his next important paper on the physical (Heisenberg said "anschaulich") content of the quantum theoretical kinematics and mechanics, received on March 23, 1927, by *Zeitschrift für Physik* (paper No. 7, p. 478) [34].

Heisenberg described the last steps leading to and philosophical background of his paper in his autobiographical book *Physics and Beyond*:

> I now [i.e., in February 1927] concentrated all my efforts on the mathematical representation of the electron path in a cloud chamber, and when I realized fairly soon that the obstacles before me were quite insurmountable, I began to wonder whether we might not have been asking the wrong sort of question all along. But where had we gone wrong? The path of the electron through the cloud chamber obviously existed; one could easily observe it. The mathematical framework of quantum mechanics existed as well, and was much too convincing to allow for many changes. Hence it ought to be possible to establish a connection between the two, hard though it appeared to be.
>
> It must have been one evening after midnight when I suddenly remembered my conversation with Einstein and particularly his statement, "It is the theory which decides what we can observe". I was immediately convinced that the key to the gate that had been closed for so long must be sought right here. I decided to go on a nocturnal walk through Faelled Park and to think further about the matter. We had always said so glibly that the path of the electron in the cloud chamber could be observed. But perhaps what we really observed was something much less. Perhaps we merely saw a series of discrete and ill-defined spots through which the electron had passed. In fact, all we do see in the cloud chamber are individual water droplets which must certainly be much larger than the electron. The right question should therefore be: Can quantum mechanics represent the fact that an electron finds itself approximately in a given place and that it moves approximately with a given velocity, and can we make these approximations so close that they do not cause experimental difficulties?
>
> A brief calculation after my return to the Institute showed that one could indeed represent such situations mathematically, and that the approximations are governed by what would later be called the uncertainty principle of quantum mechanics: the product of the uncertainties in the measured values of the position and momentum (i.e., the product of mass and velocity) cannot be

smaller than Planck's constant. This formulation, I felt, established the much-needed bridge between the cloud chamber observations and the mathematics of quantum mechanics [35].

In the paper itself, Heisenberg discussed the meaning of the concepts of position, orbit of a microscopic particle, and energy in quantum mechanics (§ 1); he used the Dirac-Jordan transformation theory not only to obtain the uncertainty relation, but also to analyze a characteristic quantum-mechanical phenomenon, the Stern-Gerlach effect (§ 2), and to study the transition between micro- (quantum) and macro- (classical) mechanics (§ 3); finally, he treated the problem of observing the time moment at which a definite quantum jump occurs in a given system (§ 4). In a note added in proof, the author corrected a few statements in the paper, referring to "more recent investigations of Bohr" which had led to the "demand that we ascribe equal validity to the quite different experiences which show up in the corpuscular theory on the one hand, and in the wave theory on the other hand" (No. 7, pp. 503 – 504).

Bohr presented the results of his investigations in two lectures, given in the autumn of 1927 at the Congress in Como and at the Fifth Solvay Conference in Brussels [36], while Heisenberg's results were also presented at the Solvay Conference in a joint report with Born [37]. In any case, Heisenberg's paper on the uncertainty relation and Bohr's Como and Solvay lectures, which established what would later be called the principle of complementarity, provided the basis of the physical interpretation of quantum mechanics which subsequently came to dominate physics. In a course of lectures given at the University of Chicago in the spring of 1929, Heisenberg presented and analyzed in detail this view. It was published later as the book *The Physical Principles of the Quantum Theory* [38]. "The purpose of the book seems to me to be fulfilled", he wrote in the preface, "if it contributes somewhat to the diffusion of that 'Kopenhagener Geist der Quantentheorie'... which has directed the entire development of modern atomic physics".

The series of pioneering papers on the formulation and physical interpretation of quantum mechanics that Heisenberg published between the spring of 1925 and the spring of 1927 constitute an important, integral part of twentieth century physics. They initiated a transformation in the understanding of the nature of microscopic phenomena, which has not yet come to a conclusion.

References

1 R. Ladenburg: Die quantentheoretische Deutung der Zahl der Dispersionselektronen. Z. Phys. **4**, 451 – 468 (1921).
See the summary and English translation ("The quantum theoretical interpretation of the number of dispersion electrons") in *Sources of Quantum Mechanics, Edited with a Historical Introduction by B. L. van der Waerden* (North Holland, Amsterdam 1967) pp. 10 – 11, 139 – 157.

2 N. Bohr, H. A. Kramers, J. C. Slater: The quantum theory of radiation. Philos. Mag. (6) **47**, 785 – 802 (1924); reprinted in *Sources of Quantum Mechanics*, pp. 159 – 176. (The above quote appears on p. 163.)
German version: Über die Quantentheorie der Strahlung. Z. Phys. **24**, 69 – 87 (1924).

3 For an English translation of the Kramers-Heisenberg paper ("On the dispersion of radiation by atoms") and for a reprint of Kramers preceding notes on dispersion theory ("The law of dispersion and Bohr's theory of spectra"; "The quantum theory of dispersion") see *Sources of Quantum Mechanics*, pp. 223–251, and pp. 177–180, 199–201. The quote is from p. 226.

4 See paper No. 1, p. 345.

5 J. H. Van Vleck: The absorption of radiation by multiply periodic orbits, and its relation to the correspondence principle and the Rayleigh-Jeans law. Part I. Some extensions of the correspondence principle. Part II. Calculation of absorption by multiply periodic orbits. Phys. Rev. (2) **24**, 330–346, 347–365 (dated June 19, 1924, published in issue No. 4 of October 1924).

6 A. Einstein: Strahlungs-Emission und -Absorption nach der Quantentheorie. Verh. d. Deutsch. Phys. Ges. (2) **18**, 318–323 (1916); Zur Quantentheorie der Strahlung. Mitt. Phys. Ges. Zürich **16**, No. 18, 47–62 (1916) and Phys. Z. **18**, 121–128 (1917).
English translation of the latter paper: "On the Quantum Theory of Radiation" in *Sources of Quantum Mechanics*, pp. 63–77.

7 The importance of Van Vleck's paper was stressed by Jordan in a letter to Van der Waerden, dated December 1, 1961. (See *Sources of Quantum Mechanics*, pp. 17–18.)

8 See *Sources of Quantum Mechanics*, p. 25. The full text of Heisenberg's letter can be found in *Wolfgang Pauli: Wissenschaftlicher Briefwechsel/Scientific Correspondence, Band/Volume I: 1919–1929*, ed. by A. Hermann, K. von Meyenn, V. F. Weisskopf (Springer, New York, Heidelberg, Berlin 1979) pp. 219–221.

9 The full text of Heisenberg's letter to Kronig dealing with the reformulation of the (classical) Fourier series has been given by R. Kronig in his contribution (entitled "The turning point") to the Pauli memorial volume *Theoretical Physics in the Twentieth Century*, ed. by M. Fierz, V. F. Weisskopf (Interscience, New York, 1960) pp. 5–37, especially pp. 23–24.

10 For a detailed story of Heisenberg's discovery of quantum mechanics see also J. Mehra, M. Rechenberg: *The Historical Development of Quantum Theory, Volume 2. The Discovery of Quantum Mechanics 1925* (Springer, New York, Berlin, Heidelberg 1982) especially Chapter IV.

11 From the letters and other documents, it is not completely obvious whether Heisenberg evaluated in Helgoland the anharmonic oscillator with the equation of motion (8), or that having an anharmonic term proportional to q^3 in the equation of motion. In his published paper discussed below Heisenberg presented only the solution for the latter.

12 See *Sources of Quantum Mechanics*, p. 25. (The English translation is from Mehra and Rechenberg, *The Discovery of Quantum Mechanics 1925*, quoted in footnote 10, p. 263.)

13 See the English translation of paper 3: "Quantum Theoretical Re-Interpretation of Kinematic and Mechanical Relations", in *Sources of Quantum Mechanics*, pp. 261–276, especially pp. 261–262.

14 In *Sources of Quantum Mechanics*, pp. 30–32, van der Waerden has shown, following a suggestion by Th. Kuhn, that $\mathfrak{A}(n, n-\alpha)$ and the analogous $\mathfrak{A}(n, n+\alpha)$ are the same as the $\mathfrak{A}_e$ and $\mathfrak{A}_a$ which determine the emission and absorption of radiation by atoms in the paper of Kramers and Heisenberg. As we have seen, these $\mathfrak{A}_e$ and $\mathfrak{A}_a$ are vector amplitudes of virtual oscillators, multiplied by $-e$; but Heisenberg does not say this, avoiding the expression "virtual oscillator".

15 As we have mentioned earlier, the idea of using such arrays of terms $a(n, m) \cdot \exp[i\,\omega(n, m)\,t]$ to describe quantum mechanical variables occurred already in Heisenberg's letter to Kronig dated June 5, 1925.

16 Heisenberg used this quantum condition already in Helgoland to obtain energy conservation for the anharmonic oscillator.

17 For a proof of this equivalence see *Sources of Quantum Mechanics*, p. 31, and Mehra and Rechenberg: *The Discovery of Quantum Mechanics 1925*, quoted in footnote 10, pp. 242–247.

18 As has been mentioned above, Heisenberg presented the full solution of the anharmonic oscillator, Eq. (13), in his paper. Interestingly enough, in section 3, he began with an anharmonic oscillator, described by Eq. (8), such as he had considered in former letters, e.g., the one to Kronig dated June 5 or the one addressed to Pauli on June 24. The main reason for taking the oscillator with Eq. (13) was that its solution involved fewer terms. The classical Fourier series for the position x involved only terms with the period ωt, $3\,\omega t$, $5\,\omega t, \ldots$, which implied a similar simplification for the quantum theoretical reformulation; in the latter, only amplitudes $a(n, n-1)$, $a(n, n-3)$, $a(n, n-5)$, etc., occurred.

19 See Max Born: *My Life – Recollections of a Nobel Laureate* (Charles Scribner's Sons, New York 1978) pp. 217 – 218.
20 M. Born: *My Life*, p. 218.
21 Heisenberg to Pauli, September 18, 1925; see Pauli: *Wissenschaftlicher Briefwechsel I,* p. 239.
22 After having read a letter in which Heisenberg explained his perturbation theory, Born came to the conclusion that it was not correct and developed his own scheme, based on Eq. (15). The equivalence of Heisenberg's and Born's definition has been shown in *Sources of Quantum Mechanics*, p. 49.
23 An English translation of the Three-Men-Paper ("On Quantum Mechanics II") can be found in *Sources of Quantum Mechanics*, pp. 321 – 385.
24 See *Sources of Quantum Mechanics*, pp. 43 – 57, and J. Mehra, H. Rechenberg: *The Historical Development of Quantum Theory, Volume 3. The Formulation of Matrix Mechanics and Its Modifications 1925 – 1926* (Springer, New York, Berlin, Heidelberg 1982) pp. 321 – 385.
25 W. Heisenberg: Über quantentheoretische Kinematik und Mechanik. Math. Ann. **95**, 683 – 705 (received December 21, 1925, published 1926), reprinted as No. 3 in Volume B of *Collected Works*, pp. 29 – 51.
26 W. Pauli: Über das Wasserstoffspektrum vom Standpunkt der neuen Quantenmechanik. Z. Phys. **36**, 336 – 363 (received January 17, 1926, published in issue No. 5 of March 27, 1926); English translation: "On the Hydrogenspectrum from the Standpoint of the New Quantum Mechanics", in Sources *of Quantum Mechanics*, pp. 387 – 415.

Pauli applied a very clever trick to avoid the use of angle variables in the hydrogen problem. (Angle variables could not be treated in the matrix scheme of Born, Heisenberg and Jordan.) At about the same time Paul Dirac in Cambridge was able to extend the matrix scheme to include angle variables as well. His paper, "Quantum mechanics and a preliminary investigation of the hydrogen atom", Proc. Roy. Soc. (London) **A 110**, 561 – 579 (1926), was received on January 22, 1926 and published in the issue of March 1, 1926.
27 E. Schrödinger: Quantisierung als Eigenwertproblem (Erste Mitteilung). Ann. Phys. (4) **79**, 361 – 379 (1926), received January 27, 1926; "(Zweite Mitteilung)", ibid., 489 – 527, received February 23, 1926; "(Dritte Mitteilung"), Ann. Phys. (4) **80**, 437 – 490 (1926), received May 10, 1926; "(Vierte Mitteilung)", Ann. Phys. (4) **81**, 109 – 139 (1926), received June 21, 1926; Über das Verhältnis der Heisenberg-Born-Jordanschen Quantenmechanik zu der meinen. Ann. Phys. (4) **79**. 734 – 756 (1926), received March 18, 1926.
28 Letter from W. Pauli to P. Jordan, dated April 12, 1926. See B. L. van der Waerden's article, "From Matrix Mechanics and Wave Mechanics to Unified Quantum Mechanics", in: *The Physicist's Conception of Nature*, ed. by J. Mehra (Reidel, Dordrecht, Boston 1973), pp. 276 – 293.
29 The same conclusion was reached independently a little later by Paul Dirac in his paper "On the theory of quantum mechanics", Proc. Roy. Soc. London **A 112**, 661 – 677 (received August 26, 1926, published in issue No. A762 of October 1, 1926).
30 W. Heisenberg: Quantenmechanik. Die Naturwissenschaften **14**, 989 – 994 (published in the issue of November 5, 1926); reprinted as No. 4 in Volume B of *Collected Works*, pp. 52 – 57.
31 E. Schrödinger: "Quantisierung als Eigenwertproblem (Vierte Mitteilung)", quoted Ref. 27, in particular Sect. 7.
32 M. Born: Zur Quantenmechanik der Stoßvorgänge. Z. Phys. **37**, 863 – 867 (received June 25, 1926, published in issue No. 12 of July 10, 1926); Quantenmechanik der Stoßvorgänge. Z. Phys. **38**, 803 – 827 (received July 21, 1926, published in issue No. 11/12 of September 14, 1926).
33 The generalized scheme, the socalled "transformation theory", is contained in the papers of P.A.M. Dirac: The physical interpretation of the quantum dynamics. Proc. Roy. Soc. London **A 113**, 621 – 641 (received December 2, 1926, published in issue No. A765 of January 1, 1927), and of P. Jordan: Über eine neue Begründung der Quantenmechanik. Z. Phys. **40**, 809 – 838 (received December 18, 1926, published in issue No. 11/12 of January 18, 1927)

Dirac had been in Copenhagen since September 1926 and had had the opportunity of discussing physics with Bohr and Heisenberg. The aim of his paper was to extend Heisenberg's considerations on quantum fluctuations (paper No. 6, p. 472).
34 An English translation of this paper, "The Physical Content of Quantum Kinematics and Mechanics", can be found in: *Quantum Theory and Measurement*, ed. by J. A. Wheeler, W. H. Zurek (Princeton University Press, Princeton, N.J. 1983) pp. 62 – 84.

35 W. Heisenberg: *Physics and Beyond. Encounters and Conversations* (Harper & Row, New York, Evanston, London 1971) pp. 77 – 78. (Original German publication: *Der Teil und das Ganze. Gespräche im Umkreis der Atomphysik*, Piper, Munich 1969; reprinted in *Collected Works*, Volume C III, especially pp. 111 – 112).
36 N. Bohr: The quantum postulate and the recent development of atomic theory. Nature **121**, 580 – 590 (presented on September 16, 1927, at the International Conference of Physics at Como and as a report at the Fifth Solvay Conference in October 1927; published in the Supplements to the issue of April 14, 1928). German version: Das Quantenpostulat und die neuere Entwicklung der Atomphysik. Die Naturwissenschaften **16**, 245 – 257 (published in the issue of April 13, 1928).
37 M. Born, W. Heisenberg: "La mécanique des quanta", in *Électrons et Photons. Rapports et Discussions du Conseil de Physique Tenu à Bruxelles du 24 au 29 Octobre 1927 sous les Auspices de l'Institut International de Physique Solvay*, ed. by Institut International de Physique Solvay (Gauthier-Villars, Paris 1928) pp. 143 – 181; reprinted as paper No. 7 in Volume B of *Collected Works*, pp. 58 – 96.
38 W. Heisenberg: *The Physical Principles of the Quantum Theory*, translated into English by C. Eckardt, F. C. Hoyt (The University of Chicago Press, 1930); reprinted in Volume B of *Collected Works*, pp. 117 – 166.

Zeitschrift für Physik *31*, 617–626 (1925)

Über eine Anwendung des Korrespondenzprinzips auf die Frage nach der Polarisation des Fluoreszenzlichtes.

Von **W. Heisenberg,** zurzeit in Kopenhagen.

(Eingegangen am 30. November 1924.)

Das Ziel der vorliegenden Arbeit ist, zu zeigen, wie verschiedene empirische Regeln über Intensität und Polarisation von Spektrallinien aufgefaßt werden können als sinngemäße Verschärfungen des Korrespondenzprinzips, und wie dieser Gesichtspunkt dazu benutzt werden kann, die Resultate der Experimente über die Polarisation des Fluoreszenzlichtes quantitativ zu beschreiben.

In einer kürzlich erschienenen Note hat Bohr[1]) gezeigt, wie die Beobachtungen über die Polarisation des Fluoreszenzlichtes von Wood und Ellett[2]), Gaviola und Pringsheim[3]) und neuerdings von W. Hanle[4]), auf Grund der Quantentheorie gedeutet werden können durch eine genauere Betrachtung des Verhaltens der virtuellen Oszillatoren bei entarteten Systemen. Der hohe Polarisationsgrad bei Quecksilber und der Unterschied zwischen Quecksilber und Natrium konnten so im Prinzip quantentheoretisch beschrieben werden. Im folgenden soll als Ergänzung hierzu eine quantitative Beschreibung der Polarisation des Dispersions- und Fluoreszenzlichtes versucht werden auf Grund von Argumenten, die man durch eine Verschärfung der Anwendung des Korrespondenzprinzips auf den Übergang von entarteten zu nicht-entarteten Systemen erhalten kann. Dabei ergibt sich die Gelegenheit, auf die anderen bisher bekannten in ähnlicher Weise „scharfen" Anwendungen des Korrespondenzprinzips kurz einzugehen.

Abgesehen von der Grenze hoher Quantenzahlen gestattet das Korrespondenzprinzip im allgemeinen nur angenäherte Schlüsse über die Intensität und den Schwingungszustand der den möglichen Übergangsprozessen zugeordneten virtuellen Oszillatoren. Dies hat seinen Grund darin, daß die virtuellen Oszillatoren nur in einer sehr symbolischen Weise mit der Bewegung der Elektronen in den stationären Zuständen verknüpft sind. Besonders deutlich tritt dieser Umstand hervor, wenn wir es mit einem entarteten Problem zu tun haben, da dann, wie Bohr (l. c.) näher ausgeführt hat, den virtuellen Oszillatoren ein höherer Grad von Freiheit zukommt als der Bewegung in den stationären Zu-

1) N. Bohr, Naturw. **12**, 1115, 1924.
2) R. W. Wood und A. Ellett, Proc. Roy. Soc. **108**, 396, 1923.
3) E. Gaviola und P. Pringsheim, ZS. f. Phys. **25**, 367, 1924.
4) W. Hanle, ZS. f. Phys. **30**, 93, 1924.

ständen. Wenn sich in dieser Weise gezeigt hat, daß die virtuellen Oszillatoren die Strahlung des Atoms bestimmen, in gewissem Sinne unabhängig von der Bewegung der Elektronen des Atoms in dem betreffenden stationären Zustand, so kann man erwarten, daß sich die Analogie der virtuellen Oszillatoren mit den klassischen Strahlungsgrößen in manchen Fällen schärfer durchführen läßt, als dies bei einer Betrachtung der stationären Zustände allein möglich erschiene. So werden wir im folgenden zu zeigen versuchen, daß sich beim Übergang von einem entarteten zu einem nicht-entarteten System das Korrespondenzprinzip hinsichtlich der Beschreibung der virtuellen Oszillatoren so weit verschärfen läßt, daß eindeutige Schlüsse auf die quantentheoretische Strahlung, was Intensität und Polarisation betrifft, möglich sind.

Den einfachsten Fall einer solchen Verschärfungsmöglichkeit des Korrespondenzprinzips finden wir nach wichtigen Untersuchungen von Dorgelo, Ornstein und Burger[1]) bei den Intensitätsverhältnissen der Mehrfachlinien und ihrer Zeemankomponenten[2]). Die Entstehung der Multipletts kann durch ein modellmäßiges Analogon beschrieben werden, indem man annimmt, daß das Atom, welches in ein äußeres in einer Rosettenbahn laufendes Elektron (Hauptquantenzahl n, Impuls k) und einen Atomrest (Impuls r) zerfällt, eine Drehung um die Achse seines Gesamtimpulses j ausführt. Durch diese Drehung wird die vom Elektron ausgesandte Spektrallinie in drei Komponenten aufgespalten, von denen die eine dem Vektor des Gesamtimpulses parallel schwingt, während die beiden anderen aus entgegengesetzten zirkularen Schwingungen um diese Achse bestehen. In der klassischen Theorie ist es selbstverständlich, daß die Summe der Intensitäten dieser drei Linien nur um Größen von der Ordnung der Drehgeschwindigkeit ω_j um die Impulsachse abweicht von der Gesamtintensität der Linie, die bei der Drehgeschwindigkeit Null, also im Falle der Entartung, allein vorhanden wäre; daß ferner diese Summe der Intensitäten auch nur in Größen von derselben kleinen Ordnung abhängt von dem Winkel zwischen der Achse des Atomrestes und der des Elektronenimpulses. Denn die ausgestrahlte Gesamtenergie ist in der klassischen Theorie durch die Bahn des Elektrons bestimmt

[1]) H. Burger und H. Dorgelo, ZS. f. Phys. **23**, 258, 1924. L. Ornstein und H. Burger, ebenda **24**, 41, 1924.

[2]) L. Ornstein und H. Burger, ZS. f. Phys. **28**, 135, 1924; **29**, 241, 1924. Die erste Arbeit dieser Autoren über Zeemaneffektintensitäten enthält einen Versuch einer Verallgemeinerung der für Multipletts gültigen Regeln, welche kein Analogon in der klassischen Theorie hat und die auch bei allgemeiner Anwendung in Widerspruch zu stehen scheint zu der von Ornstein und Burger angegebenen, das Korrespondenzprinzip befriedigenden, zweiten Verallgemeinerung.

Zeitschrift für Physik *31*, 617–626 (1925)

und kann sich daher nicht mehr ändern als diese Bahn selbst. Nur die Verteilung der Gesamtenergie auf die drei Komponenten wird von dem Winkel zwischen der Achse des Atomrestes und der des Elektronenimpulses, also von j, wesentlich abhängen.

In der Quantentheorie konnte man von vornherein nichts darüber aussagen, ob und in welcher Weise die Übergangswahrscheinlichkeiten bei der Verwandlung eines entarteten in ein nicht-entartetes System unstetige Veränderungen erleiden. Wegen der grundsätzlichen Verschiedenheit der Übergangsmöglichkeiten im entarteten und im nichtentarteten Falle lassen sich die Übergangswahrscheinlichkeiten in beiden Fällen überhaupt nicht direkt vergleichen. Trotzdem wird durch das Korrespondenzprinzip die Annahme nahegelegt, daß auch im Falle der Quantentheorie die „gesamte" Übergangswahrscheinlichkeit, d. h. die Summe der zu den einzelnen Komponenten gehörigen Übergangswahrscheinlichkeiten wesentlich durch die Bewegung des Elektrons bestimmt ist und daher nur um Größen der Ordnung der Drehgeschwindigkeit von dem Werte bei Entartung und von dem Winkel zwischen der Achse r des Atomrestes und dem Impulsvektor k des Elektrons, also von j abhängt. Diese Annahme bedeutet offenbar nur eine naturgemäße Verschärfung des Korrespondenzprinzips, und, wie Ornstein und Burger gefunden haben, genügt sie in einfachen Fällen völlig zur Festlegung der Übergangswahrscheinlichkeiten. Betrachtet man nämlich viele Atome, so kann die oben gemachte Annahme offenbar dahin formuliert werden, daß die Summe der Intensitäten der Linien, die Sprüngen von einem bestimmten Multiplettniveau aus entsprechen, hinsichtlich der Abhängigkeit von j direkt proportional ist dem durch j charakterisierten statistischen Gewicht des betreffenden Niveaus. Da man diese Schlußweise wie in der klassischen Theorie sowohl auf Emission wie auf Absorption anwenden kann, erhält man im Falle der Dubletts eindeutige Resultate.

Zu gleicher Zeit sind die Ornstein-Burgerschen Regeln für die Theorie der Multiplettstruktur von größtem Interesse. Wir wissen ja aus den Zeemaneffekten[1]) und der Theorie der absoluten Intervalle der Dubletts und Tripletts[2]), daß eine einfache Theorie der Multiplettstruktur, wie wir sie oben zur Ableitung der Ornstein-Burgerschen Resultate benutzt haben, gar nicht möglich ist. Die Quantenzahl j scheint stets erst durch zwei Werte einen Zustand des Atoms zu definieren. und die Anwendbarkeit der Relativitätstheorie auf die absoluten Multi-

[1]) W. Heisenberg, ZS. f. Phys. **26**, 291, 1924.
[2]) A. Landé, ebenda **24**, 88, 1924; **25**, 46, 1924.

plettintervalle scheint in gewissem Sinne auf eine Vermischung von j und k hinzudeuten. Nun zeigen aber wieder die Ornstein-Burgerschen Regeln dadurch, daß gerade die Gestalt der Elektronenbahn im ganzen Multiplett die Intensität der Strahlung charakterisiert, daß die Quantenzahlen k und j doch, was ihren korrespondenzmäßigen Zusammenhang mit dem Charakter der Bewegung betrifft, in enger Analogie zur Quantentheorie der bedingt periodischen Systeme als Elektronenimpuls und Gesamtimpuls des Atoms gedeutet werden können, wie es auch früher schon aus der Gültigkeit der Auswahlprinzipien ($k \rightarrow k \pm 1$; $j \rightarrow j \pm 1, j$) geschlossen wurde.

Nach Ornstein und Burger (l. c.) läßt sich in ganz analoger Weise, wie im Falle der Multipletts, die oben besprochene Verschärfung des Korrespondenzprinzips auf die Intensitäten der Aufspaltungen im Zeemaneffekt anwenden. Auch hier scheint es nur naturgemäß, daß die „gesamte" Übergangswahrscheinlichkeit nur Änderungen von der Größenordnung des äußeren Magnetfeldes erleidet. Man kommt dann, wenn man die Aufspaltung einer bestimmten Multiplettlinie betrachtet, zu der Forderung, daß die Summe der Intensitäten der von einem bestimmten Niveau ausgehenden Linien von der Neigung der Atomachse gegen das Magnetfeld, d. h. also von der magnetischen Quantenzahl m unabhängig ist. Eindeutige Resultate kann man, wie Ornstein und Burger gezeigt haben, bei den ps-Linien der Dubletts und Tripletts erhalten, wenn man die Forderung hinzunimmt, daß die Summe der Intensitäten der senkrecht zum Feld polarisierten Komponenten gleich der Summe der Intensitäten der parallelen Komponenten sein soll.

So kommen wir zu einer zweiten oft besprochenen Möglichkeit, das Korrespondenzprinzip schärfer zu fassen, nämlich zum Problem der sogenannten spektroskopischen Stabilität. Spaltet eine Spektrallinie in einem schwachen elektrischen bzw. einem magnetischen Felde in eine Reihe parallel und senkrecht zum Felde polarisierter Komponenten auf — wir nehmen dabei an, daß die Bewegung des Elektrons in der Rosettenbahn keine durchgreifende Änderung, etwa wie das Wasserstoffatom in starkem elektrischen Feld, erleidet —, so ist es in der klassischen Theorie selbstverständlich, daß die Linie im ganzen unpolarisiert bleibt, d. h. daß die Summe der Intensitäten der senkrechten Komponenten gleich der Summe der Intensitäten der parallelen Komponenten ist (und zwar gilt dies bis auf Größen von der Ordnung des äußeren Feldes). In der Quantentheorie wäre es zunächst bei Betrachtung der stationären Zustände gar nicht zu erwarten, daß die Linie im ganzen unpolarisiert bleibt. Denn

Zeitschrift für Physik 31, 617–626 (1925)

während ohne äußeres Feld alle Richtungen der Atome im Raume gleich wahrscheinlich sind, werden beim Einschalten des Feldes die Atome in bestimmten ausgezeichneten Richtungen festgelegt und schon der Unterschied in der Statistik dieser beiden Fälle scheint zunächst eine Polarisation fast notwendig zu machen. Trotzdem haben wir allen Grund, anzunehmen, daß diese Polarisation nicht vorhanden ist. und daß vielmehr die für die Strahlung maßgebenden virtuellen Oszillatoren der Quantentheorie Gesetzen gehorchen, nach denen die engste Analogie zwischen der klassischen Theorie und der Quantentheorie gewahrt bleibt. Empirisch wird diese Meinung dadurch gestützt, daß eine Verschiedenheit der Intensität der senkrechten und parallelen Polarisation auch eine Verschiedenheit des Brechungsindex in der Richtung parallel und senkrecht zum Felde, also eine erhebliche von der Stärke des äußeren Feldes unabhängige Doppelbrechung[1]) der betreffenden Substanz mit sich bringen würde. Eine solche Doppelbrechung ist aber nicht beobachtet worden. Auch scheint es theoretisch befriedigender. wenn in diesem Problem, in welchem kein grundsätzlicher Widerspruch zwischen Quantentheorie und klassischer Theorie besteht, die Analogie zwischen beiden Theorien auch so eng wie möglich durchgeführt werden kann. Diese Verschärfung des Korrespondenzprinzips ist aber in etwas höherem Grade hypothetisch als die zuerst besprochene, weil es sich hier um ein statistisches Problem. um den Mittelwert über das Verhalten der virtuellen Oszillatoren vieler Atome handelt

Auf eine dritte Möglichkeit, aus der Analogie zwischen klassischer und Quantentheorie eindeutige Schlüsse zu ziehen. werden wir bei dem uns hier hauptsächlich interessierenden Problem der Polarisation des Resonanzlichtes und des gestreuten Lichtes geführt In der oben erwähnten Arbeit hat Bohr gezeigt, daß der hohe Polarisationsgrad des Fluoreszenzlichtes beim Woodschen Versuch in direktem Zusammenhang steht mit dem Problem der Entartung. Trotzdem kann man. wie im Falle der spektroskopischen Stabilität. auch hier das Verhalten der virtuellen Oszillatoren bei Entartung in einfacher Weise vergleichen mit ihrem Verhalten in einem äußeren Felde und kann bei einer besonderen Wahl dieses Feldes weiterhin diesen Vergleich dazu benutzen, den Polarisationsgrad des gestreuten Lichtes bzw. des Fluoreszenzlichtes quantitativ auszurechnen. Betrachten wir zuerst das klassische Analogon: Wenn auf Atome linear polarisiertes Licht fällt von einer Frequenz, die zunächst von allen Eigenfrequenzen der Atome weit verschieden ist (wir behandeln

[1]) O. Stern, ZS. f. Phys. 7, 249, 1922.

also den Fall der Streuung), so wird die Streustrahlung im allgemeinen aus Licht bestehen, das teilweise parallel, teilweise senkrecht zum elektrischen Vektor des einfallenden Lichtes polarisiert ist. Schaltet man nun ein äußeres Feld, z. B. ein Magnetfeld, parallel zur Richtung des elektrischen Vektors ein, so hat dies im allgemeinen zur Folge, daß die Eigenfrequenzen der Atome etwas geändert werden, und daß alle Atome eine gleichförmige Drehung um die Achse des äußeren Feldes erfahren. An der statistischen Verteilung und der Mitschwingung der Atome aber wird in erster Näherung nichts geändert, d. h. die Intensität des parallel bzw. senkrecht zum Vektor des einfallenden Lichtes polarisierten Streulichtes bleibt die gleiche wie ohne Feld. Dagegen ist der Charakter des senkrecht polarisierten Lichtes nicht mehr der eines gewöhnlichen Streulichtes. Sondern, wegen der Drehung des Atoms um die Achse des äußeren Feldes, wird das senkrecht polarisierte Streulicht ausgesandt in zwei Frequenzen, die sich von der Frequenz des anregenden Lichtes um die (positive oder negative) Drehgeschwindigkeit des Atoms um die Feldachse unterscheiden. Auf die Besonderheiten dieser Art Streulicht, die ihr quantentheoretisches Analogon in einem von Smekal[1]) in Anlehnung an die Lichtquantentheorie vorausgesagten Effekt besitzt, soll hier nicht näher eingegangen werden. Zur genaueren Diskussion verweisen wir auf eine (im Erscheinen begriffene) Arbeit von Kramers und dem Verfasser. Wesentlich für die obige Betrachtung ist natürlich, daß das äußere Feld parallel zum Lichtvektor angelegt wird, da sonst die erzwungene Schwingung der Atome im Lichte durch die Drehung der Atome im äußeren Felde einen völlig anderen Charakter bekäme. Besonders einfach ist bei normalem Zeemaneffekt der Fall eines Magnetfeldes parallel zum Lichtvektor, weil nach dem Larmorschen Theorem der Bewegungstypus bis auf die Drehung des Atoms um die Feldachse überhaupt nicht gestört wird. In der klassischen Theorie kann man also auf diese Weise den Fall der Entartung auf den der Nichtentartung zurückführen. Im Sinne des Korrespondenzprinzips liegt es daher wieder sehr nahe, dasselbe für den Fall der Quantentheorie zn postulieren, d. h. anzunehmen, daß der Polarisationsgrad des Streulichtes durch Anlegen eines magnetischen oder elektrischen Feldes parallel zum Lichtvektor nicht geändert wird. Obwohl in der Quantentheorie das Problem der Entartung völlig verschieden ist von dem des Atoms im äußeren Felde, so glauben wir doch, daß, ähnlich wie im Fall der spektroskopischen Stabilität, die virtuellen

[1]) A. Smekal, Naturw. **11**, 873, 1923.

Zeitschrift für Physik *31*, 617–626 (1925)

Oszillatoren Gesetzen unterworfen sind, die eine solche enge Analogie zwischen Quantentheorie und klassischer Theorie möglich machen. Damit haben wir aber auch die Möglichkeit, die gesuchte Polarisation der Streustrahlung quantitativ zu berechnen.

Untersuchen wir also jetzt quantentheoretisch die Polarisation des Streulichtes in einem Magnetfeld, das parallel zum elektrischen Vektor des (linear polarisierten) Lichtes liegt; hierdurch erhalten wir auch direkt den Wert der Polarisation ohne Feld. Die einfache Streustrahlung des Atoms von der Frequenz des einfallenden Lichtes ist, wie im klassischen Falle, vollständig parallel polarisiert. Der senkrecht schwingende Teil des Lichtes aber hat, wie im klassischen Analogon, Frequenzen, die um die (positive bzw. negative) Präzession des Atoms um die Feldachse von der Frequenz des einfallenden Lichtes verschieden sind. Dieses gestreute Licht muß also, wie oben erwähnt, gedeutet werden im Sinne der Quantenübergänge, auf deren Möglichkeit Smekal (l. c.) in Verbindung mit Überlegungen über den Comptoneffekt hingewiesen hat. In unserem Falle kommen hier Übergänge in Frage, bei denen die magnetische Quantenzahl des Atoms eine Änderung erleidet. Das Atom nimmt sozusagen das „Lichtquant" $h\nu$ auf, ändert seine Energie um $h\nu_M$ (ν_M = Frequenz der Atompräzession um die Achse des Magnetfeldes) und gibt daher (bis auf die Comptonkorrektion) das „Lichtquant" $h(\nu \pm \nu_M)$ ab. Die Intensitäten dieses Streulichtes können — für alle Einzelheiten verweisen wir auf die oben genannte Arbeit von Kramers und dem Verfasser — ähnlich wie in der Kramersschen Dispersionstheorie berechnet werden.

Mit einer solchen Betrachtung ist auch das Problem der Polarisation der Streustrahlung im entarteten Falle gelöst, wenn man die Annahme der völligen Korrespondenz zwischen klassischer und Quantentheorie hinsichtlich der Stabilität der Polarisation macht. Betrachten wir als speziellen einfachen Fall das Quecksilberatom im Normalzustand, so finden wir, daß wir 100 Proz. Polarisation bekommen. Denn da nur ein Niveau $m = 0$ vorhanden ist, können Sprünge von m gar nicht vorkommen. Ein senkrecht polarisiertes Licht ist dann offenbar nicht vorhanden, d. h. das gestreute Licht ist rein parallel polarisiert. Anders ist es schon beim Natriumatom, wo wir zwei magnetische Niveaus $m = +\frac{1}{2}$, $m = -\frac{1}{2}$ antreffen, also einen Teil senkrechter Polarisation des Streulichtes erwarten können.

Gehen wir jetzt von der Streuung außerhalb der Linie zur Resonanzfluoreszenz über. Während vom Standpunkte der klassischen Theorie aus die Fluoreszenz in gewissem Sinne nur als Spezialfall der Streuung erscheint, ist in der Quantentheorie bis jetzt wegen des Problems der Linienbreite eine so einfache Beschreibung der Fluoreszenz unmöglich; deshalb kann man auch die oben besprochenen, bei Dispersion gültigen einfachen Formeln nicht mehr verwenden. Trotzdem wird man erwarten können, daß sich eine enge Analogie zwischen Quantentheorie und klassischer Theorie auch im Falle der Fluoreszenz durchführen lassen wird, so daß die oben für die Streuung benutzte Stabilitätsforderung gültig bleibt. Wenn man in dieser Weise wieder durch Annahme eines Magnetfeldes parallel zum Lichtvektor das Problem der Entartung auf das der Nichtentartung zurückführt, so gestaltet sich die quantitative Rechnung sogar besonders einfach; denn jetzt können sowohl dem gewöhnlichen Streulichte als auch dem mit anderen Frequenzen auftretenden Fluoreszenzlichte Sprünge vom angeregten Zustand des Atoms aus zugeordnet werden. Der Polarisationsgrad der Linien mit Magnetfeld wird dann berechnet in der Weise, wie dies Breit[1]), Gaviola und Pringsheim (l. c.) im Falle der Natrium-D-Linien durchgeführt haben. Diese Verfasser nehmen nämlich an, daß die Resonanzstrahlung dieselben Intensitätsverhältnisse aufweist, wie eine spontane Ausstrahlung von dem durch Resonanz angeregten oberen Zustand. Betrachten wir also für die beiden D-Linien das Landésche Schema [g = magnetischer Aufspaltungsfaktor, m = magnetische Quantenzahl; die Zahlen neben den Übergangsstrichen bedeuten die Intensitäten der den Übergängen zugehörigen Linien (nach Ornstein-Burger)]:

Tabelle 1.

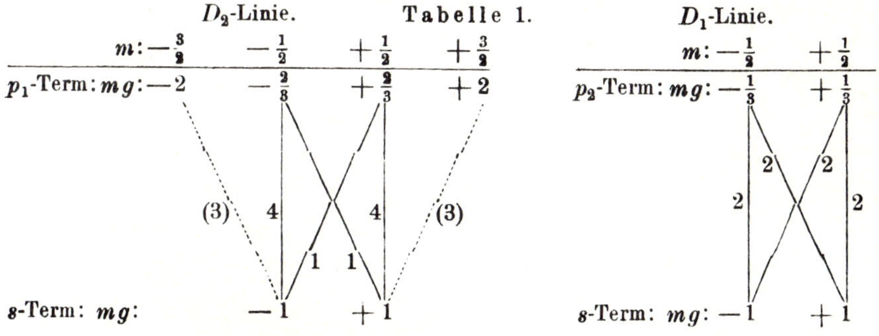

Im Falle der D_2-Linie werden, da das Licht parallel zum Magnetfeld schwingt, nur die Niveaus $m = -\frac{1}{2}, +\frac{1}{2}$ in p_1 angeregt, in der verursachten „spontanen" Strahlung haben wir nur die Linien $\pm\frac{1}{2} \rightarrow \pm\frac{1}{2}$. Das Intensitätsverhältnis zwischen parallelem und senkrechtem Licht ist nach Tabelle 1 4 : 1, der Polarisationsgrad also $\frac{4-1}{4+1} = 60$ Proz. Im

[1]) G. Breit, Phil. Mag. **47**, 832, 1924.

Zeitschrift für Physik *31*, 617–626 (1925)

Falle der D_1-Linie werden alle Niveaus angeregt, also haben wir völlige Depolarisation. Die Gesamtpolarisation von D_1 und D_2 zusammen hängt natürlich ab von dem Intensitätsverhältnis beider Linien, daher auch von der Stelle des Gefäßes, an der man mißt, d. h. von den Versuchsbedingungen. Für Quecksilber bekommt man offenbar, in naher Übereinstimmung mit Woods Versuchen, 100 Proz. Polarisation[1]). Das Landésche Schema ist:

m:	-1	0	$+1$
p_2-Term: mg:	$-\frac{3}{2}$	0	$+\frac{3}{2}$

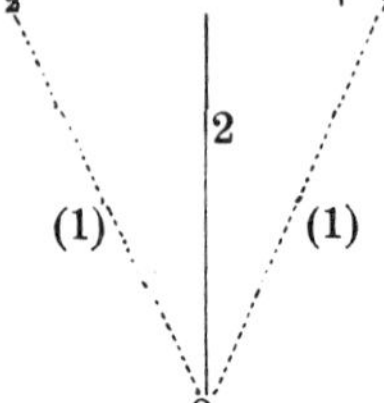

s-Term: mg: 0

Auch die obengenannte Stabilitätsforderung steht nach Wood und Ellett bei Hg in bester Übereinstimmung mit den Experimenten. Nicht erfüllt scheint sie zunächst bei NaD_2, doch kann man noch hoffen, daß sich diese Diskrepanz zwischen Theorie und Experiment durch die experimentell notwendig vorhandenen Störungen erklären läßt.

In ganz analoger Weise, wie wir hier durch eine Hypothese über die Stabilität der Polarisation eindeutige Resultate für den Fall linear polarisierten Lichtes erhalten haben, kann man auch für zirkulares Licht den Polarisationsgrad theoretisch ableiten, wenn man ein Magnetfeld parallel zur Schwingungsachse des einfallenden Lichtes eingeschaltet denkt. In diesem Falle bleibt klassisch der Polarisationsgrad bei Einschalten des Feldes wieder erhalten; es liegt nahe, quantentheoretisch dasselbe zu fordern, und wir haben dieselbe Berechnungsmöglichkeit wie im Falle des linearen Lichtes. Für den Quecksilbernormalzustand ergeben sich wieder 100 Proz. Polarisation. Dies war auch zu erwarten. Denn wenn im linearen Licht jedes Atom unabhängig von seiner Lage 100 Proz. Polarisation gibt, so muß dies wohl bedeuten, daß das Quecksilberatom wie ein isotroper Oszillator wirkt. Dieses Resultat befindet sich in

[1]) Anmerkung: Es ist interessant, mit diesen Werten für die Polarisation diejenigen zu vergleichen, die man erhielte, wenn man, wie dies von verschiedenen Autoren versucht wurde, das Verhalten der Atome ohne Magnetfeld gleichsetzen würde dem mittleren Verhalten von Atomen in Magnetfeldern in beliebigen, statistisch verteilten Richtungen. Dann ergäbe sich für Na D_2: 14 Proz., D_1: 0 Proz., $Hg_{Res.}$: 27 Proz. Man sieht aus diesem Vergleich, wie wesentlich die Unterscheidung von entarteten und nicht-entarteten Systemen für das vorliegende Problem ist.

Übereinstimmung damit, daß Hanle (l. c.) für jede Polarisationsart (linear, zirkular, elliptisch) praktisch völlige Polarisation des Fluoreszenzlichtes erhielt. In gleicher Weise ergeben sich alle Atome, die nur aus abgeschlossenen Elektronengruppen bestehen, als völlig optisch isotrop.

Ein interessantes Resultat scheint uns ferner, daß sich das Wasserstoffatom, wenn man die gewöhnliche Theorie des normalen Zeemaneffektes als gültig ansieht, im Normalzustand als völlig optisch isotrop erweist[1]). Das entsprechende Termschema lautet:

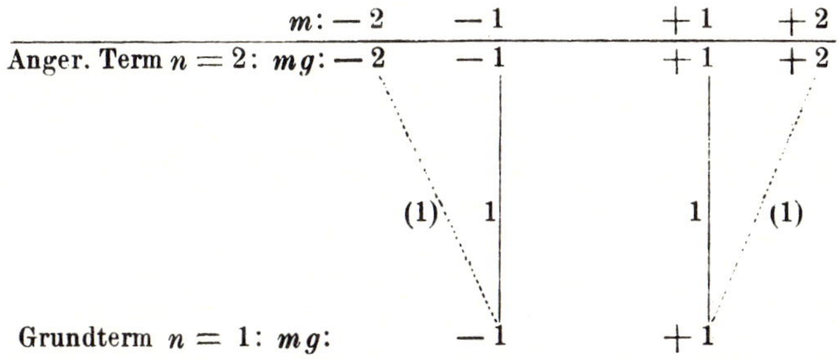

Der Ausschluß des Wertes $m = 0$, den ein Vergleich mit der Theorie des Starkeffektes des Wasserstoffspektrums zu verlangen scheint[2]), ist hier sehr wesentlich.

København, Universitetets Institut for teoretisk Fysik.

[1]) Es mag hier darauf hingewiesen werden, daß unter gewissen Voraussetzungen sich aus der „spektroskopischen Stabilität" die interessante Folgerung ergibt, daß vielleicht schon das Wasserstoffatom in konstanten äußeren Feldern nicht den Gesetzen der einfachen Mechanik und Quantentheorie gehorcht. In der Tat, denken wir uns Wasserstoffatome in einem Magnetfeld, und schalten wir ein sehr schwaches elektrisches Feld einmal parallel, einmal senkrecht zur Richtung des Magnetfeldes ein. Man kann leicht ausrechnen, daß nach der gewöhnlichen Mechanik die mittlere (wir meinen hier das Mittel über viele Atome) Wechselwirkungsenergie der Atome mit dem elektrischen Felde in beiden Fällen verschieden ausfällt. Dies steht im Widerspruch zur Forderung der spektroskopischen Stabilität, nach welcher, wie oben gesagt, die Dielektrizitätskonstante in beiden Richtungen, parallel und senkrecht zum Felde, dieselbe sein muß. Voraussetzung bei dieser Betrachtung ist allerdings, daß die im optischen Gebiet gültigen Dispersionsformeln bei der Frequenz $\nu = 0$ den richtigen Wert der Dielektrizitätskonstante ergeben, was wohl nicht sicher ist, aber doch um einer möglichst engen Analogie zwischen klassischer und Quantentheorie willen zunächst wahrscheinlich scheint.

[2]) N. Bohr, Quantentheorie der Linienspektren (Braunschweig 1922), S. 107.

Über die Streuung von Strahlung durch Atome.

Von **H. A. Kramers** und **W. Heisenberg** in Kopenhagen.

Mit sieben Abbildungen. (Eingegangen am 5. Januar 1924.)

Wenn ein Atom äußerer monochromatischer Strahlung von der Frequenz ν ausgesetzt ist, sendet es nicht nur sekundäre monochromatische Kugelwellen von der Frequenz ν aus, die mit der einfallenden Strahlung kohärent sind, sondern das Korrespondenzprinzip verlangt, daß im allgemeinen auch noch Kugelwellen von anderen Frequenzen ausgesandt werden. Diese Frequenzen sind alle von der Form $|\nu \pm \nu^*|$, wo $h\nu^*$ den Energieunterschied des Atoms im betrachteten und in irgend einem anderen Zustand bezeichnet. Die nicht kohärente Streustrahlung entspricht zum Teil gewissen Prozessen, die kürzlich von Smekal ins Auge gefaßt wurden anläßlich Betrachtungen, die an die Vorstellung von Lichtquanten anknüpfen. In der Abhandlung wird gezeigt, wie sich eine wellentheoretische Analyse der streuenden Wirkung des Atoms an der Hand des Korrespondenzprinzips in ungezwungener und scheinbar eindeutiger Weise durchführen läßt. Die Ausführungen bauen durchaus weiter auf der Auffassung der Verbindung der Wellenstrahlung des Atoms mit den stationären Zuständen, die in einer neuen Arbeit von Bohr, Kramers und Slater vertreten ist, und die Folgerungen, wenn sie sich bestätigen sollten, dürften eine interessante Stütze für diese Auffassung bilden.

§ 1. Einleitung. Die optischen Phänomene der Dispersion und Absorption, die bei Durchgang von monochromatischem Licht durch ein Gas auftreten, lassen sich auf Grund einer atomistischen Auffassung bekanntlich dahin deuten, daß jedes bestrahlte Atom sekundäre Kugelwellen aussendet, deren Frequenz dieselbe ist wie die des einfallenden Lichtes, und die mit diesem Lichte kohärent sind. Die Folgerung, daß nach dieser Auffassung eine schwache Streustrahlung in alle Richtungen ausgeschickt wird, hat sich in Rayleighs Theorie des Himmelblaus und auch in direkten Laboratoriumsversuchen glänzend bestätigt. Eine außerordentlich starke Stütze des theoretischen Bildes besteht vor allem in dem Umstand, daß die bezüglichen Beobachtungen eine Bestimmung der Avogadroschen Zahl gestatten.

Der auf der klassischen Elektrodynamik aufgebauten Elektronentheorie ist es gelungen, eine nähere theoretische Beschreibung der streuenden Wirkung der Atome zu geben. So hat die Vorstellung, daß im Atom quasielastisch gebundene Elektronen harmonische Schwingungen um einen Gleichgewichtszustand ausführen können, und daß diese Elektronen durch die elektrischen Kräfte im Strahlungsfelde in Mitschwingung versetzt werden, zu einer Dispersionstheorie geführt, die wesentliche Züge der beobachteten Dispersion wiedergibt, nicht nur im Gebiet der normalen Dispersion, sondern auch in den Gebieten anomaler Dispersion in der

Nähe der Absorptionslinien, deren Frequenzen den Eigenfrequenzen der Elektronen gleichgesetzt werden. Die Theorie verlangt, in Übereinstimmung mit den Beobachtungen, ausgeprägte Maxima in der Intensität der Streustrahlung eben bei diesen Frequenzen (Resonanzstrahlung).

Nichtsdestoweniger stößt der Versuch, eine genaue Deutung der Dispersionserscheinungen auf Grund der klassischen Theorie zu geben, bekanntlich auf Schwierigkeiten, die eng mit den Schwierigkeiten zusammenhängen, die einer Deutung der Spektren der Elemente nach dieser Theorie entgegenstehen, deren Lösung aber durch die Quantentheorie der Linienspektren angezeigt ist. Damit ist uns die Aufgabe gestellt, die Streuungs- und Dispersionswirkungen des Atoms im Anschluß an das quantentheoretische Bild des Atombaues zu beschreiben. Nach diesem Bilde ist nun das Auftreten einer Spektrallinie nicht an die Anwesenheit von elastisch schwingenden Elektronen geknüpft, sondern an Übergänge von einem stationären Zustand nach einem anderen. Das Bohrsche Korrespondenzprinzip gibt jedoch einen bedeutungsvollen Hinweis auf die Möglichkeit, die Reaktion des Atoms auf das Strahlungsfeld mittels klassischer Begriffe zu beschreiben. In einer vor kurzem erschienenen Arbeit von Bohr, Kramers und Slater[1]) wurde in großen Zügen angedeutet, wie eine solche Beschreibung sich in verhältnismäßig einfacher Weise durchführen ließe. Charakteristisch für diese Theorie ist vor allem die Annahme, daß die Reaktion des Atoms auf das Strahlungsfeld in erster Linie als eine Reaktion des Atoms in einem bestimmten stationären Zustand aufzufassen sei; die Übergänge zwischen zwei stationären Zuständen sollen sehr kurz dauern, und die nähere Beschaffenheit dieser Übergänge soll bei der Beschreibung der optischen Phänomene keine Rolle spielen. Der erste Schritt zur Beschreibung dessen, was stattfindet, wenn ein Atom in irgend einem stationären Zustand von monochromatischem Lichte bestrahlt wird, würde sich im Anschluß an eine solche Auffassung am einfachsten folgendermaßen gestalten. Es mögen ebene monochromatische polarisierte Wellen auf das Atom auffallen; der elektrische Vektor $\mathfrak{E}(t)$ dieser Wellen sei an der Stelle des Atoms durch den reellen Teil eines Vektors dargestellt:

$$\mathfrak{E}(t) = R(\mathfrak{E}\, e^{2\pi i \nu t}), \tag{1}$$

wo die Komponenten des zeitunabhängigen Vektors $\mathfrak{E}$ im allgemeinen komplexe Größen sind, und wo ν sowohl hier wie überall später in dieser Arbeit eine positive Größe ist. Unter dem Einfluß dieser Wellen sendet

[1]) ZS. f. Phys. **24**, 69, 1924; Phil. Mag. **47**, 785, 1924.

nun das Atom Kugelwellen in den umgebenden Raum aus. Das Moment $\mathfrak{P}(t)$ des schwingenden Dipols, durch den sich diese Kugelwellen darstellen lassen, sei ebenfalls gegeben durch den reellen Teil eines Ausdrucks:

$$\mathfrak{P}(t) = \boldsymbol{R}(\mathfrak{P} e^{2\pi i \nu t}), \tag{2}$$

wo $\mathfrak{P}$ ein im allgemeinen komplexer Vektor ist, der für einen gegebenen stationären Zustand von ν und $\mathfrak{E}$ abhängt; seine Richtung hängt von der Richtung von $\mathfrak{E}$ ab, während sein absoluter Betrag, jedenfalls im Grenzfall schwacher Bestrahlung, dem absoluten Betrag von $\mathfrak{E}$ proportional ist, d. h. $\mathfrak{P}$ ist eine lineare Vektorfunktion von $\mathfrak{E}$. Der Ausdruck (2) habe Gültigkeit, solange sich das Atom in dem betrachteten stationären Zustand aufhält.

Die hier gemachte Annahme erlaubt in weitgehendem Maße, den Erscheinungen von Dispersion, Absorption und Streuung von Licht durch Dämpfe gerecht zu werden, wie es aus den im Anfang dieser Arbeit erwähnten Umständen erhellt. Was speziell die Gesamtintensität des Streulichtes betrifft, so wäre man geneigt, die pro Zeiteinheit gestreute Energie S allgemein einfach gleich

$$S = \frac{(2\pi\nu)^4}{3c^3}(\mathfrak{P}\bar{\mathfrak{P}}) \tag{3}$$

zu setzen, wo $\bar{\mathfrak{P}}$ den zu $\mathfrak{P}$ konjugiert komplexen Vektor darstellt. In solchen Fällen, wo das Atom einen oder mehrere spontane Sprünge nach stationären Zuständen niedrigeren Energieinhaltes ausführen kann, muß man aber darauf vorbereitet sein, daß dieser Ausdruck nicht das Richtige trifft. Nach den Gesichtspunkten in der oben erwähnten Abhandlung von Bohr, Kramers und Slater wirkt nämlich auch bei Abwesenheit äußerer Strahlung das Atom in einem solchen Zustand als eine Quelle von Kugelwellen, deren Frequenzen ν_q jedem Sprunge mittels der Bohrschen Frequenzbedingung zugeordnet sind (spontane Ausstrahlung). Der einfachste Ansatz zur Beschreibung dieser Strahlung ist die Annahme, daß das Atom wie ein klassischer Dipol wirkt, dessen Moment durch den reellen Teil des Ausdrucks

$$\sum_q \mathfrak{A}_q e^{2\pi i \nu_q t} \tag{4}$$

dargestellt wird, wo die Amplitudenvektoren $\mathfrak{A}$ mit den Einsteinschen Wahrscheinlichkeitskoeffizienten a_q durch die Beziehung

$$a_q h \nu_q = \frac{(2\pi\nu_q)^4}{3c^3}(\mathfrak{A}_q \bar{\mathfrak{A}}_q) \tag{5}$$

verknüpft sind. Im Falle von Bestrahlung mit monochromatischem Lichte wird die Strahlung (4) im allgemeinen mit der Strahlung (2) zu Interferenzen Anlaß geben, die wegen der endlichen Lebensdauer des Atoms das Auftreten von weiteren Gliedern im Ausdruck für die Energie der Streustrahlung bedingen, die unter Umständen nicht gegen das Glied (3) vernachlässigbar sind (vgl. S. 705). Ohne Rücksicht auf die Frage nach den Gültigkeitsgrenzen des Ansatzes (2) hat einer von uns vor einiger Zeit das Problem in Angriff genommen, wie sich die Abhängigkeit der Größe $\mathfrak{P}$ von der Frequenz ν gestalten dürfte[1]). Die zwei Leitgedanken bei dieser Untersuchung waren einerseits die erfahrungsgemäße Anwendbarkeit der sich aus der klassischen Dispersionstheorie ergebenden Formeln für $\mathfrak{P}$, wenn man sich im Atom klassische Oszillatoren angebracht denkt, deren Eigenfrequenzen mit den Frequenzen der Absorptionslinien übereinstimmen, andererseits aber das Korrespondenzprinzip. Nach diesem Prinzip besteht ein enger Zusammenhang zwischen dem tatsächlichen Verhalten eines Atomsystems und der Wirkungsweise des Systems, so wie man es nach der klassischen Elektronentheorie auf Grund seiner Struktur erwarten würde. Im besonderen verlangt das Prinzip, daß im Gebiete der hohen Quantenzahlen die tatsächlichen Eigenschaften des Atoms sich mit Hilfe der klassischen elektrodynamischen Gesetze asymptotisch beschreiben lassen. Von dieser Forderung ausgehend war es, durch Vergleich der klassischen Dispersionsformeln mit dem klassischen Verhalten eines vielfach periodischen Systems einfallender Strahlung gegenüber, möglich, eine der Quantentheorie angepaßte Dispersionsformel aufzustellen. Im Falle, daß das Atom sich im Normalzustande befindet, deckt sich diese Formel mit einer schon früher von Ladenburg auf Grund andersartiger Betrachtungen vorgeschlagenen Formel[2]).

Der Zweck dieser Arbeit ist, zu zeigen, wie die nähere Verfolgung dieses Korrespondenzgedankens zum überraschenden Resultat führt, daß der Ansatz (2) für die Reaktion des Atoms auf die einfallende Strahlung zu eng gefaßt ist, und im allgemeinen folgendermaßen durch eine Reihe von Termen zu ergänzen ist:

$$\mathfrak{P}(t) = R\left\{\mathfrak{P} e^{2\pi i \nu t} + \sum_k \mathfrak{P}_k e^{2\pi i(\nu+\nu_k)t} + \sum_l \mathfrak{P}_l e^{2\pi i(\nu-\nu_l)t}\right\}, \quad (6)$$

wo $h\nu_k$ oder $h\nu_l$ den Energieunterschied des Atoms in zwei stationären Zuständen bedeutet, von denen der eine immer mit dem augenblicklichen Zustand des Atoms identisch ist, während die Vektoren $\mathfrak{P}_k$ und $\mathfrak{P}_l$ wiederum von $\mathfrak{E}$ und ν abhängen, von der ersteren Größe auch wieder in

[1]) H. A. Kramers, Nature **113**, 673, 1924; **114**, 310, 1924.

[2]) R. Ladenburg, ZS. f. Phys. **4**, 451, 1921. Siehe auch R. Ladenburg und F. Reiche, Naturw. **11**, 584, 1923.

der Form einer linearen Vektorfunktion. In Worten ausgedrückt besagt dieses Resultat: Unter Einfluß einer Bestrahlung mit monochromatischem Licht sendet ein Atom nicht nur kohärente Kugelwellen von derselben Frequenz aus, wie die des einfallenden Lichtes, sondern auch Systeme von nicht kohärenten Kugelwellen, deren Frequenzen sich als Kombinationen jener Frequenz mit anderen Frequenzen, die denkbaren Übergängen nach anderen stationären Zuständen entsprechen, darstellen lassen. Diese hinzukommenden Systeme von Kugelwellen müssen offenbar als gestreutes Licht auftreten; zur Dispersion und Absorption des einfallenden Lichtes können sie aber nicht beitragen.

Smekal[1]) ist schon vor einiger Zeit durch eine an die Vorstellung von Lichtquanten anknüpfende Überlegung genau zu demselben Resultat gelangt, nämlich daß beim Atom eine Streustrahlung von der Frequenz $\nu + \nu_k$ oder $\nu - \nu_l$ auftreten können müsse. Smekals Gedankengang können wir etwa folgendermaßen wiedergeben. Die Absorption bzw. Emission von Licht durch ein Atom läßt sich als ein Prozeß beschreiben, bei dem ein Lichtquantum von der Frequenz ν von einem Atom aufgenommen bzw. abgegeben wird; das Atom geht dadurch in einen höheren bzw. niedrigeren stationären Zustand über und ändert seine Energie um einen Betrag $h\nu$, und seinen Impuls um einen Betrag $h\nu/c$. Die gewöhnliche Streuung von Licht durch ein Atom läßt sich andererseits als die gleichzeitige Aufnahme eines Lichtquantums der Frequenz ν und Abgabe eines Lichtquantums der Frequenz ν' beschreiben. Das Atom geht dabei nicht in einen anderen stationären Zustand über, erleidet aber im allgemeinen eine Geschwindigkeitsänderung. In einem beliebig gewählten Bezugssystem sind die Frequenzen ν und ν' im allgemeinen verschieden. Im besonderen gilt das von einem Bezugssystem, in dem das Atom anfänglich ruht (Comptoneffekt). Verallgemeinernd spricht Smekal nun die Vermutung aus, daß es auch Prozesse im Atom geben müsse, bei denen gleichzeitig ein Lichtquantum aufgenommen und abgegeben wird, bei denen aber, im Gegensatz zu den eben erwähnten Streuungssprüngen, nicht nur der Bewegungszustand des Atoms sich ändert, sondern das Atom auch noch in einen anderen stationären Zustand übergeht. Vernachlässigen wir zunächst die kleine Geschwindigkeitsänderung des Atoms beim Sprung, und bezeichnen wir die Energieänderung des Atoms beim Sprung mit $h\nu_k$ oder $h\nu_l$, je nachdem diese Änderung in negativer oder

[1]) A. Smekal, Naturw. **11**, 873, 1923.

positiver Richtung vor sich geht, so wird die Frequenz des beim Prozeß ausgesandten Lichtquantums offenbar durch $\nu + \nu_k$ oder $\nu - \nu_l$ gegeben sein, wo ν die Frequenz des einfallenden Lichtquantums bedeutet. Dieses Resultat ist dahin zu deuten, daß, bei Bestrahlung mit Licht von der Frequenz ν, Licht von den Frequenzen $\nu + \nu_k$ oder $\nu - \nu_l$ vom Atom ausgesandt wird, während gleichzeitig die Atome eine Wahrscheinlichkeit dafür erhalten, ihren Energieinhalt um $h\nu_k$ zu vermindern oder um $h\nu_l$ zu vermehren.

Das Rechnen mit Lichtquanten hat vor allem seine Bedeutung darin, daß es uns instand setzt, die makroskopischen Erhaltungssätze von Energie und Impuls in einfacher und lehrreicher Weise mit den Vorstellungen der Quantentheorie in Verbindung zu bringen. Der Natur der Sache nach erlauben solche Betrachtungen aber keine Schlüsse über eine eventuelle korpuskulare Struktur des Lichtes zu ziehen, indem wir immer verlangen müssen, daß die so gewonnenen Resultate in widerspruchsfreier Weise mit der wellentheoretischen Beschreibung der optischen Phänomene in Übereinstimmung gebracht werden können. Man sieht ohne weiteres, daß diese Forderung in unserem Falle erfüllt ist, und daß der wellentheoretische Ansatz (6) eben dem Smekalschen Resultat entspricht. Es sei jedoch schon an dieser Stelle erwähnt, daß unsere späteren Betrachtungen zeigen werden, daß die von Smekal angegebenen Prozesse nicht die einzigen sind, die sich der streuenden Wirkung der Atome zuordnen lassen. Um zu Vollständigkeit zu gelangen, müssen noch solche Prozesse hinzugenommen werden, bei denen, um in der Sprache der Lichtquanten zu bleiben, das Atom unter Einfluß der Bestrahlung zur Aussendung von zwei Lichtquanten angeregt wird; das eine hat die Frequenz ν der einfallenden Strahlung, das andere eine solche Frequenz ν', daß der Verlust an Energie $h(\nu + \nu')$ einem Übergang des Atoms nach einem Zustand niedriger Energie entspricht.

Es ist in dieser Verbindung von Interesse, zu betonen, daß auch, wenn wir die Impulsänderungen des Atoms bei den Sprüngen nicht vernachlässigen, die fragliche Forderung der wellentheoretischen Beschreibbarkeit der Phänomene erfüllt ist. Führen wir nämlich ein Bezugssystem ein, in dem die Größe des Impulses des Atoms in seinem stationären Zustande gleich $h\nu/c$, während seine Richtung der des einfallenden Lichtes entgegengesetzt ist (bei den von Smekal nicht erörterten Prozessen soll er jedoch dieselbe Richtung haben), so ergibt sich nach der Lichtquantenrechnung die Frequenz des Streulichtes, das einem bestimmten Sprung entspricht, als dieselbe in allen Richtungen des Raumes, in Übereinstimmung mit der wellentheoretischen Auffassung eines Zuges monochromatischer Wellen, die ihre Quelle innerhalb eines sehr kleinen Raumabschnitts besitzen. Auf den eigentümlichen Umstand, daß das Zentrum dieser Kugelwellen gegen das erregte Atom bewegt ist, wollen wir hier nicht näher eingehen. Nur sei erwähnt, daß es ein starkes Argument dafür beibringt, daß Ansätze von der Art (2), (4) und (6) nicht strenge Gültigkeit besitzen können, und daß noch eine passende Modifikation an ihnen angebracht werden muß. Auf die folgenden Überlegungen werden solche Modifikationen jedoch nur unwesentlichen Einfluß haben.

Der Gedanke, die von Smekal angegebene Streuungswirkung der Atome bei äußerer Bestrahlung korrespondenzmäßig mit der nach der klassischen Theorie zu erwartenden streuenden Wirkung des Atomsystems in Verbindung zu bringen, war zuerst Kramers in Verbindung mit seiner Arbeit über die Dispersionstheorie eingefallen. Die Ausarbeitung dieses Gedankens, die in der vorliegenden Arbeit gegeben wird, ist durch eine gemeinsame Diskussion der Verfasser erfolgt.

§ 2. **Der Einfluß äußerer Bestrahlung auf ein Periodizitätssystem nach der klassischen Theorie.** Wir betrachten ein nichtentartetes Periodizitätssystem, dessen Bewegung sich durch die kanonischen Uniformisierungsvariablen $J_1 \ldots J_s$, $w_1 \ldots w_\tau$ beschreiben läßt. Das elektrische Moment des Systems als Funktion dieser Variablen sei durch folgende mehrfache Fourierreihe dargestellt:

$$\mathfrak{M}(t) = \sum_{\tau_1 \ldots \tau_s} \tfrac{1}{2} \mathfrak{C}_{\tau_1 \ldots \tau_s} e^{2\pi i(\tau_1 w_1 + \cdots \tau_s w_s)}. \tag{7}$$

Die Summation ist über alle positiven und negativen Werte der ganzen Zahlen $\tau_1 \ldots \tau_s$ zu erstrecken. Die Koeffizienten $\mathfrak{C}$ sind komplexe Vektoren, deren Komponenten nur von $J_1 \ldots J_s$ abhängen. Bezeichnen wir wieder konjugiert komplexe Größen durch einen oben angebrachten Strich, so hat man

$$\mathfrak{C}_{\tau_1 \ldots \tau_s} = \overline{\mathfrak{C}}_{-\tau_1 \ldots -\tau_s}. \tag{8}$$

Die Energie H des Systems hängt ebenfalls nur von den J ab. Die Fundamentalfrequenzen seien mit

$$\omega_k = \frac{\partial H}{\partial I_k} \quad (k = 1 \ldots s) \tag{9}$$

bezeichnet. Wir führen die folgende Abkürzung für einen oft auftretenden Differentialoperator ein:

$$\frac{\partial}{\partial J} = \tau_1 \frac{\partial}{\partial J_1} + \cdots \tau_s \frac{\partial}{\partial J_s}, \tag{10}$$

und weiter führen wir für die Frequenzen der in der Bewegung auftretenden harmonischen Komponenten die Abkürzung

$$\omega = \tau_1 \omega_1 + \cdots \tau_s \omega_s = \frac{\partial H}{\partial J} \qquad (11)$$

ein. Unsere Aufgabe besteht darin, das elektrische Moment des Systems als Funktion der Zeit zu finden, wenn das Atom einem ebenen monochromatischen Zuge von Lichtwellen ausgesetzt ist, deren Wellenlänge groß ist verglichen mit den Abmessungen des Systems.

Der Lichtvektor des einfallenden monochromatischen Lichtes sei wiederum durch den Ausdruck (1) gegeben. Es sei $J_1^* \ldots J_s^*$, $w_1^* \ldots w_s^*$ ein neues System von uniformisierenden Variablen, das durch eine infinitesimale Kontakttransformation

$$J_k^* - J_k = \frac{\partial K}{\partial w_k^*}, \quad w_k^* - w_k = -\frac{\partial K}{\partial J_k^*} \quad (k = 1 \ldots s) \qquad (12)$$

aus den alten Variablen hervorgeht. Es ist dann möglich, die Funktion $K(J_1^* \ldots J_k^*, w_1^* \ldots w_s^*, t)$ so zu wählen, daß in erster Annäherung die J_k^* unabhängig von der Zeit werden, während die w_k^* linear mit der Zeit anwachsen, und zwar so, daß $d w_k^*/dt = \omega_k$ ist. Man findet, daß die Funktion K folgendermaßen als der reelle Teil eines Komplexausdrucks geschrieben werden kann:

$$K = \boldsymbol{R}\left\{\sum_{\tau_1 \ldots \tau_s} -\frac{1}{2}\,\frac{(\mathfrak{E}\,\mathfrak{C}_{\tau_1 \ldots \tau_s})}{2\pi i(\omega + \nu)}\, e^{2\pi i(\tau_1 w_1^* + \cdots \tau_s w_s^* + \nu t)}\right\}, \qquad (13)$$

wo $\mathfrak{C}$ und ω hier dieselben Funktionen von den J^* darstellen, wie vorher von den J. Führen wir in (7) die durch (12) und (13) definierten neuen Uniformisierungsvariablen ein, und ersetzen wir weiter die w_k^* durch die Ausdrücke $\omega_k t$, so bekommen wir schließlich für das elektrische Moment des Atoms als Funktion von der Zeit den Ausdruck:

$$\mathfrak{M}(t) = \mathfrak{M}_0(t) + \mathfrak{M}_1(t), \quad \text{wo} \quad \mathfrak{M}_0(t) = \sum_{\tau_1 \ldots \tau_s} \tfrac{1}{2}\,\mathfrak{C}_{\tau_1 \ldots \tau_s}\, e^{2\pi i \omega t}. \qquad (14)$$

Hier und im folgenden denken wir uns die Sterne bei den neuen Uniformisierungsvariablen wieder weggelassen; $\mathfrak{M}_0$ entspricht der Bewegung des ungestörten Atoms. $\mathfrak{M}_1$ kann in folgender Weise als der reelle Teil einer $2s$-fachen Summe geschrieben werden:

$$\mathfrak{M}_1(t) = \boldsymbol{R}\left\{\sum_{\tau_1 \ldots \tau_s}\sum_{\tau_1' \ldots \tau_s'} \frac{1}{4}\left[\frac{\partial \mathfrak{C}}{\partial J'}\, e^{2\pi i \omega t} \cdot \frac{(\mathfrak{E}\,\mathfrak{C}')}{\omega' + \nu}\, e^{2\pi i(\omega' + \nu)t} \right.\right.$$
$$\left.\left. - \mathfrak{C}\, e^{2\pi i \omega t}\, \frac{\partial}{\partial J}\left(\frac{(\mathfrak{E}\,\mathfrak{C}')}{\omega' + \nu}\right) e^{2\pi i(\omega' + \nu)t}\right]\right\}. \qquad (15)$$

Die Summation ist über alle Paare von Kombinationen von ganzzahligen Werten von $\tau_1 \ldots \tau_s$, $\tau_1' \ldots \tau_s'$ zu erstrecken; $\mathfrak{C}$ und $\mathfrak{C}'$ sind als Abkürzung für $\mathfrak{C}_{\tau_1 \ldots \tau_s}$ und $\mathfrak{C}_{\tau_1' \ldots \tau_s'}$ geschrieben; $\partial/\partial J'$ in Analogie zu der Bezeichnung (10) als Abkürzung für $\tau_1' \partial/\partial J_1 + \cdots \tau_s' \partial/\partial J_s$, und ω' als Abkürzung für $\tau_1' \omega_1 + \cdots \tau_s' \omega_s$. Den Ausdruck (15) wollen wir nun dadurch umformen, daß wir erstens jedesmal diejenigen Terme zusammenfassen, für welche die Summen

$$\tau_1 + \tau_1' = \tau_1^0 \ldots \tau_s + \tau_s' = \tau_s^0 \tag{16}$$

denselben Wert haben. Setzen wir zur Abkürzung

$$\tau_1^0 \omega_1 + \cdots \tau_s^0 \omega_s = \omega_s^0, \tag{17}$$

so erhalten wir in dieser Weise:

$$\mathfrak{M}_1(t) = R \left\{ \sum_{\tau_1^0 \ldots \tau_s^0} \sum_{\tau_1 \ldots \tau_s} \frac{1}{4} \left[\frac{\partial \mathfrak{C}}{\partial J'} \cdot \frac{(\mathfrak{C}\mathfrak{C}')}{\omega' + \nu} - \mathfrak{C} \frac{\partial}{\partial J} \left(\frac{(\mathfrak{C}\mathfrak{C}')}{\omega' + \nu} \right) \right] e^{2\pi i (\omega^0 + \nu) t} \right\}. \tag{18}$$

Bei der Summation über $\tau_1 \ldots \tau_s$ ist für $\tau_1' \ldots \tau_s'$ jedesmal das durch (16) gegebene Wertesystem einzusetzen. Wir machen noch darauf aufmerksam, daß ω' und ω^0 sowohl positive wie negative Werte annehmen können, da die Summationen über alle positiven und negativen Werte für τ_k und τ_k^0 zu erstrecken sind. Für die Gültigkeit dieser Formel ist es offenbar eine Voraussetzung, daß die Frequenz ν des einfallenden Lichtes mit keiner der Frequenzen ω in der ungestörten Bewegung zusammenfällt.

Die Formel (18) besagt, daß das System unter dem Einfluß des einfallenden Lichtes eine Streustrahlung aussenden wird, deren Intensität der Intensität des einfallenden Lichtes proportional ist; in harmonische Komponenten zerlegt, enthält sie sowohl die Frequenz ν des einfallenden Lichtes wie Frequenzen, die sich als die Summe oder Differenz von ν und einer Frequenz ω^0 von der in (17) gegebenen Gestalt darstellen lassen. Die Frequenz ω^0 selbst braucht in der Bewegung des ungestörten Systems nicht vorzukommen. Vielmehr ersieht man aus (15), daß ω^0 immer von der Form $+|\omega| \pm |\omega'|$ ist, wo $|\omega|$ und $|\omega'|$ zwei in der ungestörten Bewegung tatsächlich vorkommende Frequenzen sind.

§ 3. Die Quantentheorie und die kohärente Streustrahlung. Stellen wir uns nun auf den Boden der Quantentheorie der Periodizitätssysteme, so haben wir es mit einer diskreten Mannigfaltigkeit von stationären Zuständen zu tun, die durch die Quantenbedingungen

$$J_k = n_k h \tag{19}$$

gegeben sind. Die Strahlung, welche das ungestörte System in einem gewissen stationären Zustand ausschickt, entspricht möglichen Übergängen nach stationären Zuständen niedriger Energie. Nach dem Korrespondenzprinzip ist sie trotzdem als sinngemäßes Analogon von der nach der klassischen Theorie zu erwartenden Strahlung aufzufassen. Während diese aus dem Ausdruck (14) für das schwingende elektrische Moment $\mathfrak{M}_0$ des ungestörten Atoms abgeleitet werden kann, ist die quantentheoretische Strahlung als von dem etwa durch einen Ausdruck der Form (4) dargestellten schwingenden Moment herrührend darzustellen. Jede Frequenz ν_q in diesem Ausdruck korrespondiert dabei mit einer klassischen Frequenz $\tau_1 \omega_1 + \cdots \tau_s \omega_s$, und zwar so, daß

$$\tau_k = n_k^{(1)} - n_k^{(2)}, \tag{20}$$

wo $n_k^{(1)}$ und $n_k^{(2)}$ die Werte der Quantenzahlen im Anfangszustand bzw. Endzustand darstellen. Die Größe der klassischen Frequenz

$$\omega = \left(\tau_1 \frac{\partial}{\partial J_1} + \cdots \tau_s \frac{\partial}{\partial J_s}\right) H = \frac{\partial H}{\partial J} \tag{21}$$

fällt mit der Größe der entsprechenden quantentheoretischen Frequenz ν_q nicht zusammen, weil diese ja durch

$$\nu_q = \frac{1}{h}(H^{(1)} - H^{(2)}) \tag{22}$$

gegeben ist. In der Grenze hoher Quantenzahlen läßt sich dieser Ausdruck aber annähernd in der Form:

$$\nu_q = \frac{\Delta H}{h} = \left(\frac{\Delta J_1}{h}\frac{\partial}{\partial J_1} + \cdots + \frac{\Delta J_s}{h}\frac{\partial}{\partial J_s}\right) H \tag{23}$$

schreiben, woraus wegen (19) und (20) sofort folgt, daß in dieser Grenze die quantentheoretischen Frequenzen asymptotisch mit den klassischen Frequenzen übereinstimmen. Im Gebiet niedriger Quantenzahlen stellt ν_q einen einfachen Mittelwert der korrespondierenden ω dar.

In der Grenze großer Quantenzahlen müssen weiter die Amplituden $\mathfrak{A}_q$ der harmonischen Komponenten der Strahlung mit den Amplituden $\mathfrak{C}$ der klassischen Schwingungen asymptotisch übereinstimmen, während man im Gebiet niedriger Quanten $\mathfrak{A}$ symbolisch als eine Art Mittelwert von $\mathfrak{C}$ ansehen kann. Den komplexen Vektor $\mathfrak{A}_q$ könnte man als die für den betreffenden Übergang charakteristische Amplitude bezeichnen. Sein Wert ist mit Rücksicht auf die Willkür in der Phase bis auf einen komplexen Faktor vom absoluten Betrag 1 festgelegt, und enthält also fünf Konstanten. Mit dem Einsteinschen Koeffizienten a_q für die

Wahrscheinlichkeit eines spontanen Übergangs hängt er offenbar durch die Relation (5):

$$a_q h \nu_q = \frac{(2\pi\nu_q)^4}{3c^3}(\mathfrak{A}_q \overline{\mathfrak{A}}_q) \tag{24}$$

zusammen.

Wie Bohr[1]) neuerdings betont hat, ist im Falle eines entarteten Systems der Polarisationszustand der ausgesandten Strahlung nicht eindeutig durch den Anfangszustand des betreffenden Übergangs festgelegt, so daß hier eine eindeutig bestimmte charakteristische Amplitude sich nicht definieren läßt. Wir haben uns aber in dieser Abhandlung ausschließlich auf die Betrachtung nicht-entarteter Systeme beschränkt, wo der Charakter der spontanen Strahlung voraussichtlich immer eindeutig durch den Zustand des Atoms bestimmt ist.

Unsere Aufgabe ist nun, einen der klassischen Formel (18) analogen quantentheoretischen Ausdruck für die streuende Wirkung des Systems bei äußerer Bestrahlung aufzustellen, und zwar in solcher Weise, daß diese Streuung in der Grenze großer Quantenzahlen asymptotisch mit der klassischen Streuung übereinstimmt. Wir werden zeigen, daß solches sich dadurch erreichen läßt, daß man, in analoger Weise wie es von Bohr im Falle der Frequenzen getan wurde, die in (18) auftretenden Differentialquotienten als Differenzen zweier Größen deutet, und zwar in solcher Weise, daß man zwanglos Formeln erhält, in die nur die für die Übergänge charakteristischen Frequenzen und charakteristischen Amplituden eingehen, während alle auf die mathematische Theorie der Periodizitätssysteme bezüglichen Symbole verschwunden sind.

Wir wollen anfangen mit jenem Teil des Streulichtes, das dieselbe Frequenz ν besitzt wie die einfallende Strahlung. Dieser entspricht im klassischen Fall denjenigen Termen in (18), für die $\tau_1^0 = \cdots \tau_s^0 = 0$, $\tau_k' = -\tau_k$, und die also zu einem streuenden Moment

$$\mathfrak{M}_{kl}(t) = R \sum_{\tau_k} \frac{1}{4}\left\{\frac{\partial \mathfrak{C}}{\partial J} \cdot \frac{(\mathfrak{E}\overline{\mathfrak{C}})}{\omega - \nu} + \mathfrak{C}\frac{\partial}{\partial J}\left(\frac{(\mathfrak{E}\overline{\mathfrak{C}})}{\omega - \nu}\right)\right\} e^{2\pi i \nu t} \tag{25}$$

gehören. Der Ausdruck in den geschweiften Klammern läßt sich offenbar in der Form eines einfachen Differentialquotienten schreiben. Nehmen wir weiter jedesmal zwei Terme zusammen, für welche die Größen τ_k numerisch gleiche aber entgegengesetzte Werte annehmen, so nimmt (25) folgende Form an:

$$\mathfrak{M}_{kl}(\nu) = R \sum_{\tau_k}{}' \frac{1}{4}\frac{\partial}{\partial J}\left\{\frac{\mathfrak{C}(\mathfrak{E}\overline{\mathfrak{C}})}{\omega - \nu} + \frac{\overline{\mathfrak{C}}(\mathfrak{E}\mathfrak{C})}{\omega + \nu}\right\} e^{2\pi i \nu t}. \tag{26}$$

---

[1]) N. Bohr, Naturw. **12**, 1115, 1924.

wo der Strich am Summationszeichen andeuten soll, daß nur über solche τ-Kombinationen zu summieren ist, für die ω positiv ausfällt. Einen quantentheoretischen Ausdruck für das in einem gegebenen stationären Zustande wirksame streuende Moment mit der Frequenz ν, der in der Grenze großer Quantenzahlen mit (26) asymptotisch zusammenfällt und der gleichzeitig der Erfahrung gerecht wird, erhält man nun, indem man schreibt

$$\left.\begin{aligned}\mathfrak{M}_{qu}(\nu) = R\Bigg[\sum_a \frac{1}{4h}\left(\frac{\mathfrak{A}_a(\mathfrak{E}\bar{\mathfrak{A}}_a)}{\nu_a - \nu} + \frac{\bar{\mathfrak{A}}_a(\mathfrak{E}\mathfrak{A}_a)}{\nu_a + \nu}\right) \\ - \sum_e \frac{1}{4h}\left(\frac{\mathfrak{A}_e(\mathfrak{E}\bar{\mathfrak{A}}_e)}{\nu_e - \nu} + \frac{\bar{\mathfrak{A}}_e(\mathfrak{E}\mathfrak{A}_e)}{\nu_e + \nu}\right)\Bigg] e^{2\pi i \nu t},\end{aligned}\right\} \quad (27)$$

wo die erste Summation über alle Frequenzen ν_a zu erstrecken ist, für die das System selektive Absorption aufweist, während die zweite Summation sich über alle in der spontanen Ausstrahlung vorhandenen Frequenzen ν_e erstreckt. Die Größen $\mathfrak{A}_a$ und $\mathfrak{A}_e$ sind die für die Absorptions- und Emissionsübergänge charakteristischen Amplituden. Die Formel (27) geht aus (26) hervor, indem man den in jedem Term in (26) auftretenden Differentialquotienten als den Unterschied zweier Größen zu deuten versucht, die sich auf zwei Bewegungszustände beziehen, für die die Werte der Größen $J_1 \cdots J_s$ sich um $\tau_1 h \cdots \tau_s h$ unterscheiden. Es hat dabei keinen Sinn, speziell zwei stationäre Zustände zu betrachten, einerseits weil man in dieser Weise keine Größe bekommen würde, die sich in natürlicher Weise auf die Reaktion des Atoms in einem bestimmten stationären Zustande bezieht, andererseits weil den Werten der Amplituden $\mathfrak{C}$ in den stationären Zuständen selbst keine besondere Bedeutung beizulegen ist. Vielmehr ist es ein symbolischer Mittelwert von $\mathfrak{C}$, erstreckt über den Bereich zwischen zwei stationären Zuständen, dem eine Bedeutung beizulegen ist, nämlich wie oben erwähnt, die Bedeutung als charakteristische Amplitude $\mathfrak{A}$ des entsprechenden Übergangs. So kommt man dazu, den in (26) auftretenden Differentialquotienten als den durch h dividierten Unterschied zwischen zwei Größen zu deuten, die sich auf zwei durch $\tau_1 \cdots \tau_s$ charakterisierte Übergänge beziehen, wobei der betrachtete stationäre Zustand den Endzustand des einen und den Anfangszustand des anderen Übergangs bildet.

Die Fig. 1 dient zur Illustration dieses Verfahrens. Das System habe zwei Freiheitsgrade, und die Bewegungszustände seien durch Punkte in einer J_1, J_2-Ebene, der Zeichenebene der Figur, dargestellt. Die stationären Zustände bilden ein Punktgitter, und P sei der stationäre

Zustand, dessen Reaktion wir untersuchen wollen, während Q und R zwei stationäre Zustände darstellen, deren $\mathcal{J}$-Werte $h\tau_1$ und $h\tau_2$ größer bzw. kleiner sind als für P. Der in (26) auftretende Differentialquotient ist dann gedeutet worden als der Unterschied zwischen einer sich auf Übergang a und einer sich auf Übergang e beziehenden Größe.

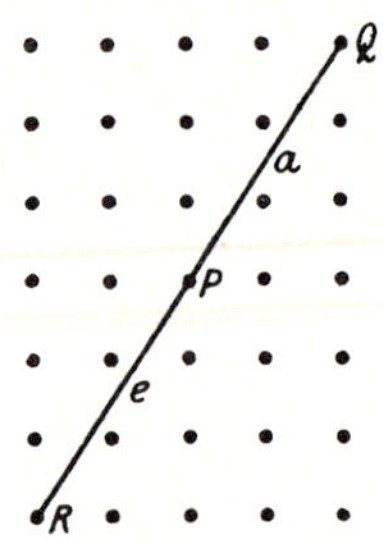

Fig. 1.

Für den einfachen Fall, wo der Vektor $\mathfrak{E}$ und alle Vektoren $\mathfrak{A}_a$ und $\mathfrak{A}_e$ reell und einander parallel sind, d. h. wo das einfallende Licht linear polarisiert, und wo der elektrische Vektor in allen den Übergängen a und e entsprechenden Strahlungen dem Lichtvektor parallel ist, wollen wir den Ausdruck (27) noch in etwas anderer Weise schreiben. Wir führen die Abklingungszeit τ_ν eines klassisch mit der Frequenz ν schwingenden Elektrons ein:

$$\tau_\nu = \frac{3\,c^3 m}{8\,\pi^2 e^2 \nu^2} \tag{28}$$

und definieren die „Stärke" eines Übergangs als die durch

$$f = a\,\tau_\nu$$

gegebene Zahl, wo a den in (24) auftretenden Einsteinschen Wahrscheinlichkeitskoeffizienten bezeichnet. Für (27) können wir nun schreiben:

$$\mathfrak{M}_{qu}(t) = \mathfrak{E}\,\frac{e^2}{4\,\pi^2 m}\left(\sum_a \frac{f_a}{\nu_a^2 - \nu^2} - \sum_e \frac{f_e}{\nu_e^2 - \nu^2}\right)\cos 2\,\pi\,\nu\,t. \tag{29}$$

Diese Formel, welche die Ähnlichkeit mit den klassischen Formeln deutlich hervortreten läßt, wurde von Kramers in seiner ersten Notiz [1]) über die Quantentheorie der Dispersion gegeben, während in einer zweiten Notiz [2]) die hier gegebene Ableitung kurz skizziert wurde. Die Glieder, welche sich auf die Absorptionslinien beziehen, entsprechen der früher von Ladenburg angegebenen Formel.

Im allgemeinen Falle, wo die Vektoren $\mathfrak{E}$ und $\mathfrak{A}$ nicht reell und nicht einander parallel sind, fällt die Richtung von $\mathfrak{M}$ nicht mehr mit der von $\mathfrak{E}$ zusammen, und die Möglichkeit einer so einfachen Schreibweise wie (29) besteht im allgemeinen nicht.

Wenn die Frequenz der einfallenden Strahlung sich der Frequenz ν_a einer Absorptionslinie oder der Frequenz ν_e einer Emissionslinie nähert, steigt das Moment stark an, aber in der nächsten Nähe dieser Frequenz

[1]) Nature **113**, 673, 1924.

[2]) Nature **114**, 310, 1924. Siehe auch J. H. van Vleck, Phys. Rev. **24**, 344, 1924.

muß die Formel (27), wenn sie auch sonst richtig sein mag, ihre Gültigkeit offenbar verlieren, ebenso wie die klassische Formel aufhört gültig zu sein, wenn ν sich ω nähert. Im Falle von naher Koinzidenz mit einer Absorptionslinie bekommen wir aber jedenfalls das Resultat, daß Licht von der Frequenz dieser Linie überaus stark gestreut wird. Ein Teil der von Wood und anderen beobachteten Resonanzstrahlung bei Metalldämpfen muß sicherlich diesen starken kohärenten Streuwellen zugeschrieben werden, von deren Anwesenheit uns übrigens schon die Tatsache der Absorption selber, wie auch die der metallischen Spiegelung bei hohem Druck Zeugnis ablegt. Teilweise wird die Resonanzstrahlung aber auch von den angeregten Zuständen herrühren, in die einige Atome durch die Bestrahlung gelangt sind. Auf die Resonanzstrahlung wollen wir hier nicht näher eingehen: sie wurde hier nur erwähnt, um zu betonen, wie die Form des Ausdrucks (27) geeignet ist, den Charakter der Absorptionslinien als singulärer Stellen bei der Streuung formelmäßig darzustellen.

Wenn andererseits ν nahe mit einer Emissionsfrequenz ν_e zusammenfällt, wird zwar der Ausdruck (27) sehr groß, aber wegen des Vorhandenseins einer spontanen Strahlung von der Frequenz ν_e und wegen unserer Unwissenheit, was die Phase des Streulichtes bei naher Resonanz betrifft, können wir nicht ohne weiteres schließen, daß durch die Bestrahlung die Kugelwellen der Frequenz ν_e verstärkt werden; gewisse Argumente, auf die wir hier nicht eingehen wollen, deuten gerade auf das Gegenteil hin.

Daß die durch (27) dargestellte Streustrahlung mit der einfallenden Strahlung kohärent ist, ersieht man sofort aus dem Umstand, daß in jedem Term sowohl die Amplitude $\mathfrak{A}$ wie die konjugierte Amplitude $\mathfrak{A}$ auftritt, wodurch die Unbestimmtheit in der Phase von $\mathfrak{A}$ selbst wegfällt. Diese Kohärenz ist die Ursache der Dispersion, und man findet leicht, daß der einfallende Strahl sich im allgemeinen in zwei polarisierte Bestandteile teilt, denen zwei verschiedene Brechungsexponenten entsprechen. Die Glieder unter dem zweiten Summationszeichen in (27) oder (29) entsprechen einer negativen Dispersion, welche der Einsteinschen „negativen Absorption" an den Stellen $\nu = \nu_e$ entspricht, genau so wie die gewöhnliche oder positive Dispersion den gewöhnlichen Absorptionslinien an den Stellen $\nu = \nu_a$ entspricht[1]).

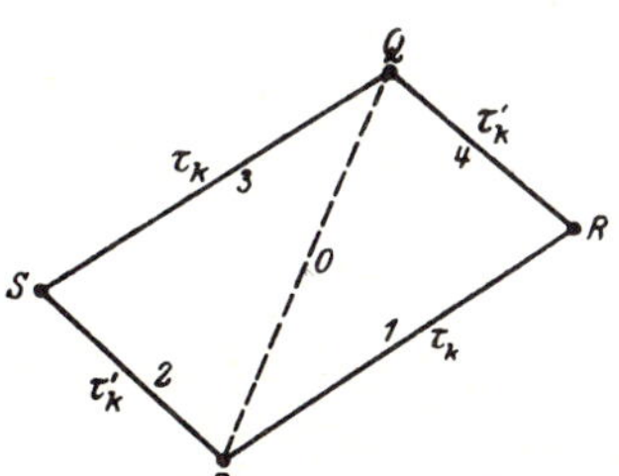

Fig. 2.

[1]) Wentzel hat neuerdings in einer interessanten Abhandlung (ZS. f. Phys. **29**, 306, 1924) eine quantentheoretische Behandlung der Dispersion zu geben ver-

§ 4. Die nicht-kohärente Streustrahlung. Wir schreiten nun dazu, den anderen Termen in der klassischen Formel (18), für die ω^0 von Null verschieden ist, eine korrespondenzmäßige quantentheoretische Deutung zu geben. Wir betrachten dazu die Fig. 2, die sich wieder auf ein System von zwei Freiheitsgraden bezieht. Es mögen die Punkte P, Q, R und S in der J_1, J_2-Ebene vier stationäre Zustände darstellen, und zwar so, daß, wenn dem Zustand P die Quantenzahlen $n_k = (n_k)_P$ zukommen, die Zustände Q, R, S durch die folgenden Quantenzahlen gekennzeichnet sind:

$$\left.\begin{aligned} (n_k)_Q = (n_k)_P + \tau_k + \tau_k' = (n_k)_P + \tau_k^0, \quad (n_k)_R = (n_k)_P + \tau_k, \\ (n_k)_S = (n_k)_P + \tau_k', \end{aligned}\right\} \quad (30)$$

wo die Systeme von ganzen Zahlen τ_k, τ_k' und τ_k^0 dieselbe Bedeutung haben wie in den Formeln (16) und (17). Die verschiedenen Übergänge sind durch die Zahlen 0, 1, 2, 3 und 4 bezeichnet. Die Übergange 1 und 3, sowie 2 und 4 sollen spontane Strahlungsübergänge darstellen, d. h. ihre charakteristischen Amplituden sollen von Null verschieden sein. Dagegen braucht PQ keinen möglichen Übergang darzustellen, d. h. $\mathfrak{A}_0$ kann eventuell gleich Null sein. Die den betrachteten Übergängen quanten-

sucht, die mit der hier behandelten nur wenig gemeinsam hat. Wentzel befürwortet eine Auffassung, nach der die quantentheoretische Dispersionsformel im allgemeinen nicht in der einfachen Helmholtz-Kettelerschen Form dargestellt werden kann, sondern als eine Art verzerrte klassische Dispersionsformel angesehen werden muß. Diese Möglichkeit verdient alle Aufmerksamkeit, denn bei unserer korrespondenzmäßigen Betrachtung konnte von einer strengen Ableitung der Formel für das induzierte Streumoment keine Rede sein. Trotzdem gibt es vorläufig u. E. keine experimentellen Gründe, die Gültigkeit einer einfachen Formel von der Art (27) oder (29) anzuzweifeln. Wentzel führt zwar als Beispiel die Dispersion beim Helium an, weil hier die „effektive" Absorptionsfrequenz auf der kurzwelligen Seite der Grenze der Absorptionsserie liegt. (Sie ist fast 5 Proz. größer als die Grenzfrequenz.) Eine solche Sachlage spricht aber keineswegs gegen die klassischen Formeln, denn zur Dispersion tragen nicht nur die Absorptionslinien des Heliums bei, sondern auch die kontinuierliche Absorption, die sich jenseits der Seriengrenze in die Richtung kurzer Wellen erstreckt. Eine einfache Rechnung, die sich auf theoretische Überlegungen über die Größe dieser kontinuierlichen Absorption stützt (vgl. H. A. Kramers, Phil. Mag. **46**, 836, 1923) oder auch auf eine Extrapolation auf den Fall des Heliums der empirischen Formeln für die Absorption von Röntgenstrahlen, zeigt nämlich, daß bei Anwendung der klassischen Formel der Einfluß dieser kontinuierlichen Absorption auf die Dispersion im optischen Gebiete derselbe ist wie der einer Absorptionslinie, deren Frequenz etwa 1,2 mal so groß ist wie die Frequenz der Seriengrenze, und deren „Intensität" mehrere Einheiten betragen kann, d. h. einer Anzahl von mehreren Dispersionselektronen entsprechen dürfte. Im Falle des Heliums gestattet das Experiment daher zunächst nicht, auf ein Versagen der klassischen Dispersionsformel zu schließen. (Vgl. auch K. F. Herzfeld und K. L. Wolf, Ann. d. Phys. **76**, 71, 1925.)

theoretisch entsprechenden Frequenzen seien mit ν_0, ν_1, ν_2, ν_3 und ν_4 bezeichnet.

Erstens ist nun die Exponentialfunktion in (18) folgendermaßen zu deuten:

$$e^{2\pi i(\omega^0 + \nu)t} \sim e^{2\pi i(\nu^0 + \nu)t}. \tag{31}$$

Eine quantentheoretische Deutung des in (18) innerhalb der geschweiften Klammern auftretenden Ausdrucks, bei der die Differentialquotienten in sinngemäßer Weise durch durch h dividierte Differenzen, die Frequenzen ω' in der Bewegung durch quantentheoretische Frequenzen, und die Amplituden $\mathfrak{C}$ durch die charakteristischen Amplituden $\mathfrak{A}$ der quantentheoretischen Übergänge ersetzt werden, ergibt sich weiter folgendermaßen in scheinbar eindeutiger Weise:

$$\left.\begin{aligned} \frac{\partial \mathfrak{C}}{\partial J'} \cdot \frac{(\mathfrak{E}\mathfrak{C}')}{\omega' + \nu} &\sim \frac{\mathfrak{A}_3 - \mathfrak{A}_1}{h} \cdot \frac{1}{2}\left(\frac{(\mathfrak{E}\mathfrak{A}_4)}{\nu_4 + \nu} + \frac{(\mathfrak{E}\mathfrak{A}_2)}{\nu_2 + \nu}\right), \\ \mathfrak{C} \cdot \frac{\partial}{\partial J}\left(\frac{(\mathfrak{E}\mathfrak{C}')}{\omega' + \nu}\right) &\sim \frac{\mathfrak{A}_3 + \mathfrak{A}_1}{2} \cdot \frac{1}{h}\left(\frac{(\mathfrak{E}\mathfrak{A}_4)}{\nu_4 + \nu} - \frac{(\mathfrak{E}\mathfrak{A}_2)}{\nu_2 + \nu}\right). \end{aligned}\right\} \tag{32}$$

Bei der Subtraktion heben sich eine Anzahl von Termen weg, und man bekommt als Deutung für den in (18) eingeklammerten Ausdruck:

$$\{\tau_k, \tau_k'\} \sim \frac{1}{h}\left\{-\frac{\mathfrak{A}_1(\mathfrak{E}\mathfrak{A}_4)}{\nu_4 + \nu} + \frac{\mathfrak{A}_3(\mathfrak{E}\mathfrak{A}_2)}{\nu_2 + \nu}\right\}. \tag{33}$$

Bei dieser Umformung war es stillschweigend angenommen, daß die Frequenzen ω, ω' und ω^0 alle positiv waren, und, in quantentheoretischer Analogie dazu, daß die Frequenzen ν_0, ν_1, ν_2, ν_3, ν_4, die durch die Formeln:

$$h\nu_0 = H(Q) - H(P), \quad \left.\begin{aligned} h\nu_1 &= H(R) - H(P), & h\nu_3 &= H(Q) - H(S), \\ h\nu_2 &= H(S) - H(P), & h\nu_4 &= H(Q) - H(R) \end{aligned}\right\} \tag{34}$$

definiert sind, alle positiv ausfallen. Jedesmal wo eine dieser Frequenzen negativ wird, ist in Formel (33) für die zugehörige charakteristische Amplitude $\mathfrak{A}$ der konjugierte Vektor $\mathfrak{A}$ einzusetzen.

Da nun gemäß (18) summiert werden soll, ist es wichtig zu bemerken, daß derjenige Term in (18), der sich durch Vertauschung der Werte von τ_k und τ_k' ergibt, sich wieder auf genau dasselbe Quadrupel von stationären Zuständen P, Q, R, S beziehen läßt. Bei dieser Vertauschung bekommen wir für die Größe innerhalb der geschweiften Klammern einen Ausdruck, der sich aus (33) ergibt, wenn man 1 mit 2, und 3 mit 4 vertauscht, so daß wir schreiben können:

$$\{\tau_k', \tau_k\} \sim \frac{1}{h}\left\{-\frac{\mathfrak{A}_2(\mathfrak{E}\mathfrak{A}_3)}{\nu_3 + \nu} + \frac{\mathfrak{A}_4(\mathfrak{A}\mathfrak{E}_1)}{\nu_1 + \nu}\right\}. \tag{35}$$

Für die Summe der zwei Terme in (18), die sich ihrer quantentheoretischen Deutung nach auf das Viereck P, Q, R, S beziehen lassen, bekommen wir also:

$$\{\tau_k, \tau_k'\} + \{\tau_k', \tau_k\} \sim \frac{1}{h}\left\{-\frac{\mathfrak{A}_1(\mathfrak{E}\mathfrak{A}_4)}{\nu_4+\nu} + \frac{\mathfrak{A}_4(\mathfrak{E}\mathfrak{A}_1)}{\nu_1+\nu} - \frac{\mathfrak{A}_2(\mathfrak{E}\mathfrak{A}_3)}{\nu_3+\nu} + \frac{\mathfrak{A}_3(\mathfrak{E}\mathfrak{A}_2)}{\nu_2+\nu}\right\}. \tag{36}$$

Wir definieren nun folgendermaßen einen komplexen Vektor $\mathfrak{M}(P, Q:R)$, der sich jedem Tripel von stationären Zuständen P, Q, R zuordnen läßt:

$$\mathfrak{M}(P, Q:R) = \frac{1}{4h}\left\{\frac{\mathfrak{A}_q(\mathfrak{E}\mathfrak{A}_p)}{\nu_p+\nu} - \frac{\mathfrak{A}_p(\mathfrak{E}\mathfrak{A}_q)}{\nu_q+\nu}\right\} e^{2\pi i(\nu_0+\nu)t}. \tag{37}$$

Die kleinen Buchstaben p, q, o beziehen sich auf die Übergänge RP, QR: QP (vgl. Fig. 3). Die Frequenzen ν_p, ν_q und ν_o in (37) sind durch die Relationen:

$$h\nu_p = H(R) - H(P), \quad h\nu_q = H(Q) - H(R), \quad h\nu_o = H(Q) - H(P) \tag{38}$$

definiert und können also gegebenenfalls negativ auftreten. Sobald ν_p (bzw. ν_q) negativ auftritt, ist die charakteristische Amplitude $\mathfrak{A}_p$ (bzw. $\mathfrak{A}_q$) des entsprechenden Übergangs durch den konjugierten Vektor $\mathfrak{A}_p$ (bzw. $\mathfrak{A}_q$) zu ersetzen.

Der Ausdruck auf der rechten Seite von (36), mulipliziert mit $\frac{1}{4}e^{2\pi i\nu t}$, läßt sich offenbar als die Summe von $\mathfrak{M}(P, Q:R)$ und $\mathfrak{M}(P, Q:S)$ darstellen, und wir werden daher nach (18) erwarten, daß sich das streuende Moment der Atome als eine Summe von Termen darstellen läßt, von denen jeder den Ausdruck $R\,\mathfrak{M}(P, Q:R)$ besitzt. Eine Schwierigkeit besteht nun zunächst darin, daß man nicht weiß, ob dieser Ausdruck sich auf die Reaktion des Atoms im Zustande P oder im Zustande Q bezieht ($\Re$ kommt nicht in Betracht). Eine Lösung dieser Frage erreicht man nur, indem man Betrachtungen heranzieht, die den Inhalt des Korrespondenzprinzips nicht unmittelbar berühren. Etwas Ähnliches trifft man bei der Frage nach der spontanen Ausstrahlung bei einem ungestörten Periodizitätssystem. Diese besteht ja aus harmonischen Komponenten, die je an eine Kombination zweier stationären Zustände geknüpft sind, und deren Frequenzen (22) und Amplituden $\mathfrak{A}$ sich korrespondenzmäßig deuten lassen; das Korrespondenzprinzip gestattet uns aber zunächst nicht, zu schließen, in welchem der zwei Zustände das Atom diese spontane Strahlung ausschickt; nur eine an das Energieprinzip und an das Wesen

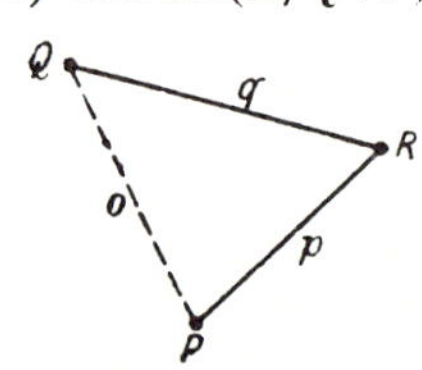

Fig. 3.

der Ausstrahlung anknüpfende Überlegung bringt hier die Entscheidung[1]), welche bekanntlich so lautet, daß die Ausstrahlung immer vom Zustande, wo der Energieinhalt der größere ist, stattfindet, in Übereinstimmung mit Bohrs Strahlungspostulat. Eine ähnliche Entscheidung müssen wir nun auch in unserem Falle treffen, und zwar werden wir annehmen, daß das Streumoment (37) sich auf den Zustand Q bezieht, wenn $\nu_0 + \nu$ positiv ist, während es sich auf den Zustand P bezieht, wenn $\nu_0 + \nu$ negativ ausfällt. Diese Behauptung läßt sich folgendermaßen begründen. Streustrahlung der Frequenz $\nu_0 + \nu$ bedeutet nach den fundamentalen Gesetzen, welche den Energieaustausch zwischen Strahlungsfeld und Atomen beherrschen, daß das Atom in dem betrachteten Zustand eine Wahrscheinlichkeit dafür besitzt, bei einer Zustandsänderung einen Energiebetrag $h(\nu_0 + \nu|)$ zu verlieren. Bestrahlung mit der Frequenz ν andererseits gibt immer auch Anlaß zu einer Streustrahlung der Frequenz ν, und das Zusammenwirken dieser zwei gibt zu einer solchen Reaktion auf das Strahlungsfeld Anlaß, daß dem Atom eine Wahrscheinlichkeit innewohnt, bei einer Zustandsänderung den Energiebetrag $h\nu$ entweder zu gewinnen oder zu verlieren. Eine tatsächliche Zustandsänderung des Atoms wird nun immer in einem Übergang nach einem anderen stationären Zustand bestehen; damit die Streustrahlung der Frequenz $|\nu + \nu_0|$ wirksam sein soll. um einen solchen Übergang herbeizuführen, muß sie also von einer Wirkung der Frequenz ν begleitet sein. Ist $(\nu + \nu_0)$ positiv, so ist dies nur dadurch möglich, daß das Atom gleichzeitig eine Energie $h\nu$ gewinnt, und eine Energie $h(\nu + \nu_0)$ verliert; beim Übergang verliert das Atom also im ganzen die Energiemenge $h\nu_0$, d. h. es muß vor dem Übergang im Zustand Q sein, und nachher im Zustand P. (Dies gilt sowohl für positive wie negative Werte von ν_0.) Ist dagegen $\nu + \nu_0$ negativ (hier ist also ν_0 immer negativ), so kann der Übergang nur dadurch zustande kommen, daß das Atom gleichzeitig sowohl eine Energiemenge $h\nu$ wie auch eine Energiemenge $h(-[\nu + \nu_0])$ verliert, also im ganzen eine Energie $-h\nu_0$. Vor dem Übergang muß das Atom also im Zustand P sein, nachher im Zustand Q. In dem zitierten Artikel von Smekal wurden Übergänge von der ersteren Art in Betracht gezogen. Übergänge der letzteren Art, die sich vom Standpunkte der Lichtquanten kaum so natürlich darbieten, wurden aber nicht berücksichtigt.

Mit Hilfe der Formeln (36) und (37) und der soeben getroffenen Entscheidung bezüglich des Atomzustandes, auf den sich (37) bezieht, sind

[1]) Siehe N. Bohr, ZS. f. Phys. **13**, 164, 1923.

wir jetzt imstande, eine quantentheoretische Deutung der Formel (18) in ihrer vollen Allgemeinheit anzugeben, und zwar bekommen wir in einem stationären Zustand P des Atoms für das streuende Moment. welches durch die äußere Strahlung induziert worden ist, den Ausdruck

$$\mathfrak{M}(t) = R\left\{\sum_Q \sum_R \mathfrak{M}(Q, P:R) + \sum_Q \sum_R \mathfrak{M}(P, Q:R)\right\}, \quad (39)$$

wo in beiden Summen über alle von P und Q verschiedenen stationären Zustände R des Atoms summiert werden soll. In der ersten Summe ist dann noch über alle stationären Zustände Q des Atoms zu summieren für die $H(Q) < H(P) + h\nu$; in der zweiten Summe über alle diejenigen Q-Zustände, für die $H(Q) < H(P) - h\nu$. Ebenso wie früher bedeutet $H(P)$ bzw. $H(Q)$ den Wert der Energie des Atoms vom Zustand P bzw. Q.

Wir wollen der Übersichtlichkeit wegen die allgemeinen Formeln (37) und (39) noch in etwas speziellerer Form schreiben, und zwar so, daß wir, unter ν^* immer eine positive Größe verstehend, denjenigen Beitrag $\mathfrak{M}(|\nu \pm \nu^*|)$ zum totalen Streumoment des Atoms im Zustand P an-

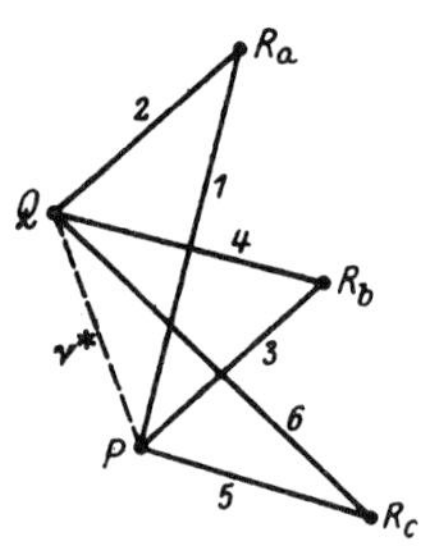

Fig. 4.

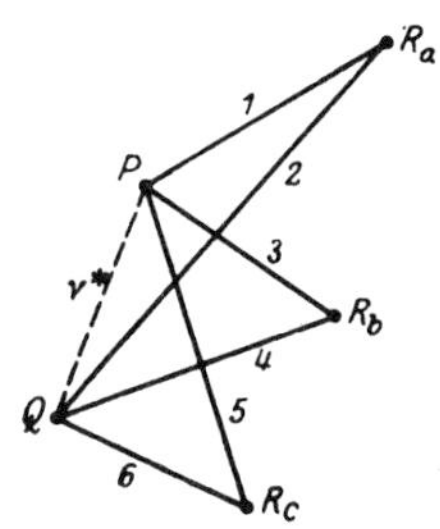

Fig. 5.

geben wollen, der, auf Grund der Anwesenheit eines Zustandes Q, einer bestimmten Frequenz $|\nu \pm \nu^*|$ im gestreuten Lichte entspricht. Es muß zwischen verschiedenen Fällen unterschieden werden.

Fall I. $H(Q) > H(P)$, $\qquad H(Q) - H(P) = h\nu^*$.

Der Zustand Q wirkt zur Streustrahlung mit, wenn $\nu > \nu^*$ und die Frequenz dieser Streustrahlung ist gleich $\nu - \nu^*$. Die von P und Q verschiedenen Zustände R verteilen sich in drei Gruppen R_b, R_a und R_c, je nachdem $H(R)$ größer als $H(Q)$ ist, kleiner als $H(Q)$ aber größer als $H(P)$, oder kleiner als $H(P)$. Die absoluten Werte der Frequenzen, die den Übergängen zwischen P oder Q einerseits und R andererseits entsprechen, seien mit ν_1, ν_2 usw., den Zahlen in der Fig. 4 entsprechend, bezeichnet. Es treten nur Terme auf, die in der ersten Summe in (39) auftreten, und für das streuende Moment $\mathfrak{M}(\nu - \nu^*)$ bekommen wir den Ausdruck

$$\mathfrak{M}(\nu - \nu^*) = R\,\frac{1}{4h}\left\{ \sum_{R_a} \left(\frac{\mathfrak{A}_2(\mathfrak{E}\,\overline{\mathfrak{A}}_1)}{\nu_1 - \nu} + \frac{\overline{\mathfrak{A}}_1(\mathfrak{E}\,\mathfrak{A}_2)}{\nu_2 + \nu} \right) \right.$$

$$\left. + \sum_{R_b} \left(\frac{\overline{\mathfrak{A}}_4(\mathfrak{E}\,\overline{\mathfrak{A}}_3)}{\nu_3 - \nu} - \frac{\overline{\mathfrak{A}}_3(\mathfrak{E}\,\overline{\mathfrak{A}}_4)}{\nu_4 - \nu} \right) + \sum_{R_c} \left(- \frac{\overline{\mathfrak{A}}_6(\mathfrak{E}\,\mathfrak{A}_5)}{\nu_5 + \nu} - \frac{\mathfrak{A}_5(\mathfrak{E}\,\overline{\mathfrak{A}}_6)}{\nu_6 - \nu} \right) \right\} e^{2\pi i(\nu - \nu^*)t} \quad (40)$$

Im Normalzustande des Atoms tritt nur Streustrahlung dieser Art auf.

Fall II. $H(Q) < H(P) \qquad H(P) - H(Q) = h\nu^*$.

a) Streustrahlung der Frequenz $\nu + \nu^*$ tritt für alle Werte der Frequenz ν auf. Die bezüglichen Terme in (39) gehören alle zur ersten Summe. Die von P und Q verschiedenen stationären Zustände zerfallen in Zustände R_a für die $H(R_a) > H(P)$, Zustände R_b für die $H(P) > H(R_b) > H(Q)$ und Zustände R_c für die $H(Q) > H(R_c)$. Die entsprechenden Frequenzen werden in der in der Fig. 5 angegebenen Weise bezeichnet und seien immer positiv gerechnet. Für das streuende Moment $\mathfrak{M}(\nu + \nu^*)$ bekommen wir den Ausdruck:

$$\mathfrak{M}(\nu + \nu^*) = R\,\frac{1}{4h}\left\{ \sum_{R_a} \left(\frac{\mathfrak{A}_2(\mathfrak{E}\,\overline{\mathfrak{A}}_1)}{\nu_1 - \nu} + \frac{\overline{\mathfrak{A}}_1(\mathfrak{E}\,\mathfrak{A}_2)}{\nu_2 + \nu} \right) \right.$$

$$\left. + \sum_{R_b} \left(- \frac{\mathfrak{A}_4(\mathfrak{E}\,\mathfrak{A}_3)}{\nu_3 + \nu} + \frac{\mathfrak{A}_3(\mathfrak{E}\,\mathfrak{A}_4)}{\nu_4 + \nu} \right) + \sum_{R_c} \left(- \frac{\overline{\mathfrak{A}}_6(\mathfrak{E}\,\mathfrak{A}_5)}{\nu_5 + \nu} - \frac{\mathfrak{A}_5(\mathfrak{E}\,\overline{\mathfrak{A}}_6)}{\nu_6 - \nu} \right) \right\} e^{2\pi i(\nu + \nu^*)t} \quad (41)$$

b) Streustrahlung der Frequenz $\nu^* - \nu$ tritt auf für alle Werte der bestrahlenden Frequenz ν, die kleiner sind als ν^*. Die bezüglichen Terme in (39) gehören alle zur zweiten Summe. Mit denselben Bezeichnungen wie im Falle IIa bekommen wir für das streuende Moment $\mathfrak{M}(\nu - \nu^*)$ den Ausdruck:

$$\mathfrak{M}(\nu - \nu^*) = R\,\frac{1}{4h}\left\{ \sum_{R_a} \left(\frac{\overline{\mathfrak{A}}_2(\mathfrak{E}\,\mathfrak{A}_1)}{\nu_1 + \nu} + \frac{\mathfrak{A}_1(\mathfrak{E}\,\overline{\mathfrak{A}}_2)}{\nu_2 - \nu} \right) \right.$$

$$\left. + \sum_{R_b} \left(- \frac{\mathfrak{A}_4(\mathfrak{E}\,\overline{\mathfrak{A}}_3)}{\nu_3 - \nu} + \frac{\mathfrak{A}_3(\mathfrak{E}\,\overline{\mathfrak{A}}_4)}{\nu_4 - \nu} \right) + \sum_{R_c} \left(- \frac{\mathfrak{A}_6(\mathfrak{E}\,\overline{\mathfrak{A}}_5)}{\nu_5 - \nu} - \frac{\overline{\mathfrak{A}}_5(\mathfrak{E}\,\mathfrak{A}_6)}{\nu_6 + \nu} \right) \right\} e^{2\pi i(\nu - \nu^*)t} \quad (42)$$

III. $H(Q) = H(P)$. Die Streustrahlung besitzt die Frequenz ν des einfallenden Lichtes.

a) Der Zustand Q ist nicht identisch mit dem Zustand P (vgl. Fig. 6). Dieser Fall geht sowohl aus I wie aus IIa hervor, wenn ν^* gleich Null wird. Die Summation über die Zustände R_b fällt natürlich weg.

Die Frequenzen ν_1 und ν_2 und auch die Frequenzen ν_5 und ν_6 fallen zusammen; wir bezeichnen sie mit ν_{12} bzw. ν_{56}.

$$\mathfrak{M}(\nu) = R\frac{1}{4h}\left\{\sum_{R_a}\left(\frac{\mathfrak{A}_2(\mathfrak{E}\overline{\mathfrak{A}}_1)}{\nu_{12}-\nu} + \frac{\overline{\mathfrak{A}}_1(\mathfrak{E}\mathfrak{A}_2)}{\nu_{12}+\nu}\right) + \sum_{R_b}\left(-\frac{\mathfrak{A}_5(\mathfrak{E}\overline{\mathfrak{A}}_6)}{\nu_{56}-\nu} - \frac{\overline{\mathfrak{A}}_6(\mathfrak{E}\mathfrak{A}_5)}{\nu_{56}+\nu}\right)\right\}e^{2\pi i\nu t} \tag{43}$$

Erinnern wir uns an die Bedeutung der Vektoren $\mathfrak{A}$, so sehen wir, daß wegen der Unbestimmtheit in der Phase der Amplituden $\mathfrak{A}$, das Streulicht nicht in Phasenbeziehung zum einfallenden Lichte steht.

b) Der Zustand Q ist identisch mit P (vgl. Fig. 7). Bezeichnen wir die Frequenzen der Absorptions- und Emissionslinien im Zustande P mit ν_a und ν_e, so bekommen wir:

$$\mathfrak{M}(\nu) = R\frac{1}{4h}\left\{\sum_{R_a}\left(\frac{\mathfrak{A}_a(\mathfrak{E}\overline{\mathfrak{A}}_a)}{\nu_a-\nu} + \frac{\overline{\mathfrak{A}}_a(\mathfrak{E}\mathfrak{A}_a)}{\nu_a+\nu}\right) - \sum_{R_e}\left(\frac{\mathfrak{A}_e(\mathfrak{E}\mathfrak{A}_e)}{\nu_e-\nu} + \frac{\mathfrak{A}_e(\mathfrak{E}\mathfrak{A}_e)}{\nu_e+\nu}\right)\right\}e^{2\pi i\nu t} \tag{44}$$

Die Unbestimmtheit in der Phase bei den Größen $\mathfrak{A}$ fällt in diesem Ausdruck offenbar nicht in Betracht, und die Streustrahlung ist deshalb

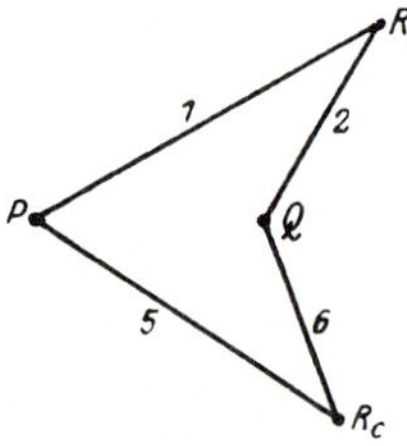

Fig. 6.

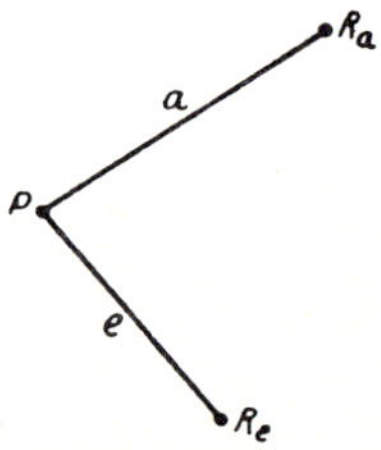

Fig. 7.

kohärent mit der einfallenden Strahlung. Die Formel (44) drückt genau dasselbe aus wie die Formel (27).

Es ist von besonderem Interesse, das singuläre Verhalten der Ausdrücke (40), (41) und (42) in der Nähe gewisser kritischer Werte der Frequenz zu studieren. So ersehen wir im Fall I aus Formel (40), daß das streuende Moment sehr großen Werten zustrebt, wenn die bestrahlende Frequenz sich dem Werte der Absorptionsfrequenz ν_1 nähert und die gestreute Frequenz sich dem Werte ν_2 nähert. Natürlich kann das streuende Moment nicht unendlich groß werden, so wie die Formel es verlangen würde; nach Analogie mit der klassischen Streuungsformel (18), die nur für Werte ν gilt, die von den Eigenfrequenzen ω des Systems verschieden sind, werden wir erwarten, daß die Formel (40) nur gilt, solange $\nu - \nu_1$ groß ist, verglichen mit der Linienbreite der Absorptionslinie. Trotzdem dürfen wir annehmen, daß die Formel (40) doch darin das richtige trifft, daß eben die Frequenz ν_1 eine kritische Stelle für die Streuung darstellt, und zwar so, daß bei Bestrahlung mit Licht, das die Frequenz ν_1 enthält, die Streustrahlung ein ausgeprägtes Maximum bei der Frequenz ν_2 besitzen wird. Ganz parallel damit geht nun die Tatsache, daß bei der Bestrahlung mit der Frequenz ν_1 das Atom eine Wahrscheinlichkeit dafür erhält, in den Zustand R_a überzugehen. In diesem Zustand angelangt, kann es unter Umständen spontan nach Q herunterfallen, und der Möglichkeit dieser Übergänge entspricht die Aussendung von Kugelwellen von der Frequenz ν_2 im Zustand R_a. Die Beobachtung der Streufrequenz ν_2 bei Bestrahlung mit ν_1 würde also von zwei verschiedenen Arten von Prozessen herrühren; teilweise wird die Streustrahlung von allen Atomen im Zustand P ausgesandt, den wir z. B. als dem Normalzustand entsprechend denken können, teilweise stammt sie von den wenigen Atomen, die durch die Bestrahlung in den Zustand R_a übergeführt worden sind, und nun unter anderen Frequenzen, auch die Frequenz ν_2 spontan ausstrahlen. Der Fall ist offenbar dem früher erwähnten Umstand analog, daß diejenige Streustrahlung, deren Frequenz mit der des einfallenden Lichtes übereinstimmt, teilweise von den Atomen im Normalzustand P, teilweise von den angeregten Atomen Q herrührt. Die zwei erwähnten Beiträge zur beobachtbaren Streustrahlung von der Frequenz ν_2 können voraussichtlich von derselben Größenordnung sein. Auf die Frage, wie groß ihr Verhältnis unter verschiedenen Umständen sein wird, eine Frage, die wieder eng mit der Struktur der Absorptionslinien zusammenhängt, wollen wir an dieser Stelle nicht näher eingehen.

Auf zwei wichtige Punkte sei indessen noch aufmerksam gemacht. Der erste betrifft das Verhältnis zwischen den Intensitäten, mit welchen bei Bestrahlung mit Licht, das ν_1 enthält, die Frequenzen ν_1 und ν_2 im Zustand P gestreut werden. Dieses Verhältnis kann natürlich nicht mittels der Formeln (40) und (44) berechnet werden, eben weil diese Formeln für sehr kleine Werte von $\nu - \nu_1$, die von derselben Größenordnung sind wie die Linienbreite, versagen. Wir wollen aber den Fall betrachten, wo die Frequenz ν des einfallenden Lichtes einen solchen Unterschied gegen ν_1 aufweist, daß $\nu - \nu_1$ zwar groß ist, verglichen mit der Linienbreite, daß aber gleichzeitig $\nu - \nu_1 = -\delta$ so klein ist, daß der bezügliche Term in den Ausdrücken (40) und (44) groß ist verglichen mit der Summe aller anderen Terme. Das Streumoment der Frequenz $\nu_1 - \delta$ wird dann annähernd gleich $R\,\frac{\mathfrak{A}_1(\mathfrak{E}\,\overline{\mathfrak{A}}_1)}{4h\,.\,\delta}\,e^{2\pi i(\nu_1-\delta)t}$ und das Streumoment der Frequenz $\nu_2 - \delta$ gleich $R\,\frac{\mathfrak{A}_2(\mathfrak{E}\,\overline{\mathfrak{A}}_1)}{4h\delta}\,e^{2\pi i(\nu_2-\delta)t}$. Das Intensitätsverhältnis dieser zwei Streustrahlungen wird also unabhängig von δ, und zwar gleich $\frac{\nu_1^4(\mathfrak{A}_1\overline{\mathfrak{A}}_1)}{\nu_2^4(\mathfrak{A}_2\overline{\mathfrak{A}}_2)}$ sein. Das ist aber genau das Verhältnis zwischen den Intensitäten, mit welchen im Zustand R_a das Atom die Frequenzen ν_1 und ν_2 ausstrahlt. Wir dürfen daher ganz allgemein erwarten, daß das Intensitätsverhältnis der Streustrahlungen, die bei der Bestrahlung eines Atoms mit Licht, das eine Absorptionsfrequenz enthält, auftreten werden, immer genau dasselbe sein wird wie das Intensitätsverhältnis der spontanen Strahlungen der entsprechenden Frequenzen, die in dem angeregten Zustand auftreten. Diese Erwartung oder Annahme bildet einen wesentlichen Punkt bei der Diskussion der Frage nach der Polarisation der Resonanzstrahlung bei Metalldämpfen[1]). Man wird einsehen, daß die obenstehenden Betrachtungen eine starke korrespondenzmäßige Stütze für diese Annahme herbeibringen, unabhängig von der Ausarbeitung näherer Vorstellungen über das Zustandekommen der Streustrahlung, so wie sie teilweise in der vorliegenden Arbeit versucht wurde.

Der zweite Punkt betrifft die Möglichkeit von gewissen Übergängen zwischen stationären Zuständen, mit denen das Vorkommen der Streustrahlung in unlöslicher Verbindung steht. Aus den Betrachtungen auf S. 698 geht es sofort hervor, daß in allen auf S. 699 bis 701 betrachteten Sonderfällen das Auftreten der Streustrahlung an die Möglichkeit eines Sprunges des Atoms vom Zustand P zum Zustand Q geknüpft ist, und zwar so, daß die Wahrscheinlichkeit zum Sprunge pro Zeiteinheit genau so groß ist wie die durch $h(|\nu \pm \nu^*|)$ dividierte Energie, welche in der Form von Streustrahlung der Frequenz $|\nu \pm \nu^*|$ pro Zeiteinheit ausgestrahlt wird. Enthält im Falle I die einfallende Strahlung die Absorptionsfrequenz ν_1, so wird die Wahrscheinlichkeit zu einem direkten Sprung nach Zustand Q also sehr groß. Nach dem Vorhergehenden dürfen wir erwarten, daß diese Wahrscheinlichkeit sich' zur Wahrscheinlichkeit

[1]) Vgl. N. Bohr, Naturw. **12**, 1115, 1924 und besonders W. Heisenberg, ZS. f. Phys. **31**, 617, 1925, wo eine quantitative Theorie der Beobachtungen auf Grundlage des Korrespondenzprinzips versucht ist.

zum Sprung nach einem anderen Zustand Q', der auch von R_a aus erreichbar ist, genau so verhalten werden wie die Wahrscheinlichkeiten. womit das Atom aus R_a teils nach Q und teils nach Q' springen kann.

Bisher sprachen wir nur von der kritischen Stelle, welche im Fall 1 die Frequenz ν_1 einer Absorptionslinie einnimmt in bezug auf die Streustrahlung. Aus Formel (40) entnehmen wir aber, daß auch, wenn ν sich der Frequenz ν_6 nähert, der Ausdruck für das streuende Moment stark anwächst. Dieser Umstand mutet zunächst sehr sonderbar an, weil man meinen könnte, daß bei Bestrahlung mit Licht, das die Frequenz ν_6 enthält, das Streulicht sehr stark werden muß, und daß also, wegen der Energiebilanz, das Atom eine Absorptionslinie von der Frequenz ν_6 aufweisen müsse. Eine solche Folgerung stände aber in offenbarem Gegensatz zur Formel (44), die die Tatsache zum Ausdruck bringt, daß für das Atom bei seiner Reaktion auf monochromatische Bestrahlung nur solche Frequenzen kritisch sind, die einem Übergang nach einem anderen stationären Zustand entsprechen. Es läßt sich aber einfach zeigen, daß Formel (40) sehr wohl die Verhältnisse richtig wiedergeben kann, ohne daß die Koinzidenz $\nu = \nu_6$ sich in irgend einer besonderen Weise bei der Absorption oder Dispersion des einfallenden Lichtes zu äußern braucht, wie es von (44) verlangt wird. Wenn ν sich ν_6 nähert, kommt die Frequenz $\nu - \nu^*$ der Streustrahlung mehr und mehr der Frequenz ν_5 nahe. Diese Frequenz ist aber von vornherein in der spontanen Ausstrahlung des Atoms enthalten. und zwar ist derjenige Teil im elektrischen Moment des Atoms, der dieser Frequenz entspricht, durch den Ausdruck $R\mathfrak{A}_5 e^{2\pi i \nu_5 t}$ gegeben. Für kleine Werte des Unterschieds $\nu - \nu_6 = -\delta$ bekommt das nach (40) induzierte Moment $R\left\{-\frac{\mathfrak{A}_5(\mathfrak{E}\mathfrak{A}_6)}{4h\delta} e^{2\pi i(\nu_5-\delta)t}\right\}$ also mehr und mehr die Fähigkeit, mit der vorhandenen spontanen Strahlung zu interferieren, und zwar so. daß im Ausdruck für die Intensität der Streuung noch Terme auftreten werden, die den Lichtvektor $\mathfrak{E}$ des einfallenden Lichtes in der ersten Potenz enthalten. Daß diese Terme trotz des Unterschiedes δ in den Frequenzen nicht verschwindend klein werden, hat seine natürliche Ursache darin, daß das Atom, eben auf Grund seiner spontanen Strahlung, nur eine endliche Lebensdauer besitzt. Die wirkliche Intensität der von der einfallenden Strahlung im Atom induzierten Streustrahlung hängt von der Phasenverknüpfung zwischen dem induzierten Streumoment und dem schon vorhandenem Moment der spontanen Ausstrahlung ab, und die Rechnung zeigt, daß bei geeigneter Phasenverbindung die Intensität der induzierten Streustrahlung bei einer bestrahlenden

Frequenz in der Nähe von $\nu = \nu_q$ keine Besonderheiten aufweisen wird, gleichzeitig damit, daß die Gleichung (40) zu Recht besteht. Eine kleine Rechnung möge dies verdeutlichen.

Die charakteristische Amplitude der spontanen Streustrahlung einer Frequenz ν_0 in einem stationären Zustand P sei einfachkeitshalber als reell angenommen, und ihr absoluter Wert sei p. Die mittlere Lebensdauer von P sei durch $1/a$ gegeben. Die Frequenz des induzierten Streumomentes sei $\nu_0 - \delta$, wo δ klein ist verglichen mit ν_0, aber groß verglichen mit a. Der absolute Wert ihrer Amplitude sei q. Der Augenblick, wo das Atom in dem Zustand P angelangt ist, sei mit $t = 0$ bezeichnet, und der Phasenunterschied zwischen den zwei Streumomenten zu diesem Zeitpunkte sei gleich φ. Die während der Lebensdauer T ausgestrahlte Energie ist dann dem Ausdruck

$$s(T, \varphi) = \int_0^T \{p \cos 2\pi \nu_0 t + q \cos [2\pi(\nu_0 - \delta) t + \varphi]\}^2 \, dt \tag{45}$$

proportional. Die im Mittel während der Lebensdauer des Zustandes P ausgestrahlte Energie ist dann dem folgenden Ausdrucke proportional:

$$S(\varphi) = a \int_0^\infty s(T, \varphi) e^{-aT} \, dT. \tag{46}$$

Die Rechnung ergibt, mit Vernachlässigung kleiner Größen von der Größenordnung δ/ν_0,

$$S(\varphi) = \frac{p^2}{2a} + \frac{pq}{a^2 + (2\pi\delta)^2}(a \cos \varphi + 2\pi\delta \sin \varphi) + \frac{q^2}{2a} \tag{47}$$

oder

$$S(\varphi) = \frac{p^2}{2a} + \frac{pq}{\sqrt{a^2 + (2\pi\delta)^2}} \sin(\varphi + \alpha) + \frac{q^2}{2a}, \quad \text{wo } \operatorname{tg} \alpha = \frac{a}{2\pi\delta}. \tag{48}$$

Es sei nun, in Übereinstimmung mit (40), das Moment q durch den Ausdruck

$$q = -\frac{Ep'p}{4h\delta} \tag{49}$$

gegeben, wo E die Amplitude des elektrischen Vektors in den einfallenden Wellen bezeichnet, während p' eine Größe ist von derselben Größenordnung wie p. Vernachlässigen wir dann noch Terme von der relativen Größenordnung δ^2/a^2, so nimmt (48) folgende Form an:

$$S(\varphi) = \frac{p^2}{2a} - \frac{p^2 p' E}{8\pi h} \cdot \frac{\sin(\varphi + \alpha)}{\delta^2} + \frac{p^2 p'^2 E^2}{32 h^2 a} \frac{1}{\delta^2}. \tag{50}$$

Aus diesem Ausdruck geht hervor, daß bei genügend schwacher Bestrahlung (E sehr klein) der zweite Term stets von derselben Größenordnung sein kann wie der dritte, weil für einen festgehaltenen Winkel $\varphi + \alpha$ beide Terme in derselben Weise mit δ variieren. Der dritte Term wäre genau durch den zweiten kompensiert, wenn $\sin(\varphi + \alpha)$ im Mittel gleich $\pi p' E / 4 h a$ wäre. Damit unsere Betrachtungen Gültigkeit haben, muß diese Größe klein gegen 1 sein. Vergleich mit (49), und die Überlegung, daß für sehr kleine Werte von $2\pi\delta$ diese Größe in (49) etwa durch einen Ausdruck von der Form $\sqrt{a^2 + (2\pi\delta)^2}$ zu ersetzen ist, lehren uns also, daß unsere Betrachtungen nur gültig sein können, wenn die induzierten Streumomente kleiner sind als die Momente der spontanen Streustrahlungen im Atom, auch in solchen Fällen, wo für gewisse kritische Frequenzen das induzierte Moment ein Maximum erreicht. Eine solche Einschränkung dürfte von vornherein als naturgemäß erscheinen.

Unsere korrespondenzmäßigen Formeln (40) und (44) haben also dazu geführt, die Existenz einer Phasenverbindung von der betrachteten Art anzunehmen. Der Umstand, daß die Phase des induzierten Streumomentes außer von $\mathfrak{A}_5$ noch von $\overline{\mathfrak{A}}_6$ abhängt, und daß dieser Vektor in dem stationären Zustand P der Natur der Sache nach vollständig unbestimmt ist was seine Phase anbelangt, zeigt, daß diese Annahme kaum als unnatürlich bezeichnet werden kann.

Bisher haben wir uns nur mit den kritischen Werten der Frequenz ν der einfallenden Strahlung in Fall I (S. 699) beschäftigt, wo die Streustrahlung die Frequenz $\nu - \nu^*$ besitzt, und mit einem Übergang nach einem Zustande höherer Energie verknüpft ist. Die Diskussion gestaltet sich aber wesentlich ähnlich wie in den Fällen II a und II b. In Fall II a gibt es zweierlei Arten von kritischen Frequenzen. Die eine wird wiederum von den Absorptionsfrequenzen ν_1 gebildet. Bestrahlung mit Licht, das die Frequenz ν_1 enthält, gibt Anlaß zu einer Streustrahlung, die bei der Frequenz ν_2 ein Intensitätsmaximum erreicht (vgl. Fig. 5). Parallel damit geht der Umstand, daß die Atome durch die Bestrahlung in den Zustand R_a gehoben werden können, und von hieraus wird ν_2 spontan ausgestrahlt. Weiter ist ν_6 eine kritische Frequenz, die aber nicht zu irgendwelchem Maximum in der Intensität der Streustrahlung der Frequenz ν_5, die schon von vornherein als spontane Strahlung vorhanden ist, Anlaß geben darf. In Fall II b kommen nur die kritischen Frequenzen ν_3 und ν_4 in Betracht (ν_2 und ν_5 sind immer größer als ν^*, während ν immer kleiner als ν^* sein soll). Bestrahlung mit Licht, das ν_3 enthält, gibt Anlaß zu einem Streulichte, das ein Maximum bei ν_4 besitzt. Dies entspricht dem Umstand, daß das Atom durch die Bestrahlung eine erhöhte Wahrscheinlichkeit erlangt, in den Zustand R überzugehen, und in diesem Zustande wird eben ν_4 spontan ausgestrahlt. Andererseits wird bei Bestrahlung mit ν_4 ein starkes Streumoment mit Frequenz ν_3 im Atom auftreten. Da wir aber annehmen müssen, daß das Streumoment mit der Frequenz der einfallenden Strahlung bei ν_4 kein Maximum aufweist, darf die Intensität der Streuausstrahlung auch kein Intensitätsmaximum erreichen bei der Frequenz ν_3, die schon von vornherein im Atom anwesend ist.

Zusammenfassend kann man sagen, daß ein Atom bei Bestrahlung mit Licht, das eine Absorptions- bzw. Emissionsfrequenz des Atoms enthält, die Strahlungseigenschaften desjenigen Zustandes nachahmt, der zusammen mit dem gegebenen Zustand die betreffende Absorptions- bzw. Emissionsfrequenz charakterisiert.

§ 5. Schlußbemerkungen. Man muß verlangen, daß die in dieser Arbeit betrachteten Arten von Streustrahlung und die Sprünge zwischen stationären Zuständen, zu denen sie Anlaß geben, sowohl die Energieverteilung in der Hohlraumstrahlung sowie die statistische Gleichgewichtsverteilung der Atome unverändert lassen. Die Ansätze, womit sich dies erreichen läßt, werden offenbar eine große Ähnlichkeit aufweisen müssen mit den Ansätzen, mit deren Hilfe es Pauli gelungen ist, das Wärmegleichgewicht zwischen freien Elektronen und Hohlraumstrahlung zu beschreiben[1]). Im Fall I müßte z. B. die Wahrscheinlichkeit für einen direkten Sprung von P nach Q durch eine Reihe von Termen dargestellt werden, die unter anderen einen Term proportional mit der Strahlungsdichte $\varrho(\nu_1)$ bei der Frequenz ν_1, und einen Term proportional mit dem Produkte $\varrho(\nu_1)\,\varrho(\nu_2)$ enthält. Dagegen darf, wie aus dem Obigen hervorgeht, kein Term auftreten, der $\varrho(\nu_6)$ proportional ist.

Die Betrachtungen in dieser Arbeit zeigen, daß es kaum möglich ist, der korrespondenzmäßigen Forderung, daß die in einem Atomsystem durch äußere Bestrahlung induzierte Streustrahlung in der Grenze großer Quantenzahlen mit der von der klassischen Theorie geforderten Streustrahlung übereinstimmt, in anderer Weise Genüge zu leisten als eben durch eine Formel von der Art (39), wo $\mathfrak{M}$ den durch (37) gekennzeichneten Charakter besitzt. Selbst wenn die Beschreibung des tatsächlich vorhandenen Streumomentes einen weniger einfachen Ausdruck für $\mathfrak{M}$ verlangen sollte, so entgeht man doch schwerlich der Konsequenz, daß das Moment der Streustrahlung einen besonders großen Wert annehmen wird, jedesmal wo eine Bedingung von der Form $\nu + \nu_p = 0$ oder $\nu + \nu_q = 0$ erfüllt ist.

[1]) W. Pauli jr., Zs. f. Phys. 18, 272, 1923.

Weiter scheint es auch nicht möglich, die Formeln so einzurichten, daß die Korrespondenzforderung erfüllt ist, ohne daß die kritischen Werte für die Frequenz ν in allgemeinen Übergängen entsprechen können, für die weder Anfangszustand noch Endzustand mit dem betrachteten Zustande des Atoms zusammenfallen. Die Schwierigkeit, die dadurch scheinbar entsteht für die Frage des Energieaustausches zwischen Atomen und Strahlungsfeld, läßt sich, wie gezeigt, in natürlicher Weise lösen, wenn man annimmt, daß es eine spontane Ausstrahlung des Atoms gibt, die während der ganzen Lebensdauer des betrachteten Zustandes wirksam ist. Man kann daraus schließen, daß eine Auffassung, nach der die streuende Wirkung eines Atoms bei äußerer Bestrahlung als die Wirkung des Atoms in einem bestimmten stationären Zustande anzusehen ist, notwendig die Auffassung mit sich führt, daß auch die spontane Strahlung eines Atoms als eine Strahlung in einem bestimmten stationären Zustande beschrieben werden muß, und nicht etwa als eine Wirksamkeit, die das Atom nur bei einem Übergange zwischen stationären Zuständen entfaltet. Das ist aber gerade die von Slater eingeführte Hypothese, die den Anlaß zu den von Bohr, Kramers und Slater näher entwickelten Ansichten gab. Wie auf S. 682 erwähnt, kann der allgemeine Grundgedanke dieser Ansichten eben dahin formuliert werden, daß die Rolle, welche das Atom bei den optischen Phänomenen spielt, immer auf Wechselwirkungen zwischen dem Strahlungsfeld und dem Atom in irgend einem stationären Zustande zurückgeführt werden kann: die eigentümlichen Prozesse, die wir als „Übergänge von einem stationären Zustand nach einem anderen“ beschreiben, darf man dabei als sehr kurz dauernd ansehen, und über ihre Beschaffenheit geben die optischen Phänomene, soweit sie heute bekannt und analysiert sind, keinen Aufschluß.

Kopenhagen, Institut for teoretisk Fysik. Dezember 1924.

Zeitschrift für Physik *33*, 879–893 (1925)

Über quantentheoretische Umdeutung kinematischer und mechanischer Beziehungen.

Von **W. Heisenberg** in Göttingen.

(Eingegangen am 29. Juli 1925.)

In der Arbeit soll versucht werden, Grundlagen zu gewinnen für eine quantentheoretische Mechanik, die ausschließlich auf Beziehungen zwischen prinzipiell beobachtbaren Größen basiert ist.

Bekanntlich läßt sich gegen die formalen Regeln, die allgemein in der Quantentheorie zur Berechnung beobachtbarer Größen (z. B. der Energie im Wasserstoffatom) benutzt werden, der schwerwiegende Einwand erheben, daß jene Rechenregeln als wesentlichen Bestandteil Beziehungen enthalten zwischen Größen, die scheinbar prinzipiell nicht beobachtet werden können (wie z. B. Ort, Umlaufszeit des Elektrons), daß also jenen Regeln offenbar jedes anschauliche physikalische Fundament mangelt, wenn man nicht immer noch an der Hoffnung festhalten will, daß jene bis jetzt unbeobachtbaren Größen später vielleicht experimentell zugänglich gemacht werden könnten. Diese Hoffnung könnte als berechtigt angesehen werden, wenn die genannten Regeln in sich konsequent und auf einen bestimmt umgrenzten Bereich quantentheoretischer Probleme anwendbar wären. Die Erfahrung zeigt aber, daß sich nur das Wasserstoffatom und der Starkeffekt dieses Atoms jenen formalen Regeln der Quantentheorie fügen, daß aber schon beim Problem der „gekreuzten Felder“ (Wasserstoffatom in elektrischem und magnetischem Feld verschiedener Richtung) fundamentale Schwierigkeiten auftreten, daß die Reaktion der Atome auf periodisch wechselnde Felder sicherlich nicht durch die genannten Regeln beschrieben werden kann, und daß schließlich eine Ausdehnung der Quantenregeln auf die Behandlung der Atome mit mehreren Elektronen sich als unmöglich erwiesen hat. Es ist üblich geworden, dieses Versagen der quantentheoretischen Regeln, die ja wesentlich durch die Anwendung der klassischen Mechanik charakterisiert waren, als Abweichung von der klassischen Mechanik zu bezeichnen. Diese Bezeichnung kann aber wohl kaum als sinngemäß angesehen werden, wenn man bedenkt, daß schon die (ja ganz allgemein gültige) Einstein-Bohrsche Frequenzbedingung eine so völlige Absage an die klassische Mechanik oder besser, vom Standpunkt der Wellentheorie aus, an die dieser Mechanik zugrunde liegende Kinematik darstellt, daß auch bei den einfachsten quantentheoretischen Problemen an eine Gültigkeit der klassischen Mechanik schlechterdings nicht gedacht werden kann. Bei dieser Sachlage scheint es geratener, jene Hoffnung auf eine Beobachtung der bisher unbeobachtbaren Größen (wie Lage, Umlaufszeit des Elektrons) ganz aufzugeben, gleichzeitig also einzuräumen, daß die teilweise Übereinstimmung der genannten Quantenregeln mit der Erfahrung mehr oder weniger zufällig sei, und zu versuchen, eine der klassischen Mechanik analoge quantentheoretische Mechanik auszubilden, in welcher nur Beziehungen zwischen beobachtbaren Größen vorkommen. Als die wichtigsten ersten Ansätze zu einer solchen quantentheoretischen Mechanik kann man neben der Frequenzbedingung die Kramerssche Dispersionstheorie[1]) und die auf dieser Theorie weiterbauenden Arbeiten[2]) ansehen. Im folgenden wollen wir einige neue quantenmechanische Beziehungen herauszustellen suchen und zur vollständigen Behandlung einiger spezieller Probleme benutzen. Wir werden uns dabei auf Probleme von einem Freiheitsgrade beschränken.

§ 1. In der klassischen Theorie ist die Strahlung eines bewegten Elektrons $\left(\text{in der Wellenzone, d. h. } \mathfrak{E} \backsim \mathfrak{H} \backsim \frac{1}{r}\right)$ nicht allein durch die Ausdrücke:

$$\mathfrak{E} = \frac{e}{r^3 c^2}[\mathfrak{r}[\mathfrak{r}\dot{\mathfrak{v}}]],$$

$$\mathfrak{H} = \frac{e}{r^2 c^2}[\dot{\mathfrak{v}}\mathfrak{r}]$$

gegeben, sondern es kommen in nächster Näherung noch Glieder hinzu, z. B. von der Form

$$\frac{e}{r c^3}\dot{\mathfrak{v}}\mathfrak{v},$$

die man als „Quadrupolstrahlung" bezeichnen kann, in noch höherer Näherung Glieder z. B. der Form

$$\frac{e}{r c^4}\dot{\mathfrak{v}}\mathfrak{v}^2;$$

in dieser Weise läßt sich die Näherung beliebig weit treiben. (Im Vorhergehenden bedeuteten: $\mathfrak{E}$, $\mathfrak{H}$ die Feldstärken im Aufpunkt, e die Ladung des Elektrons, $\mathfrak{r}$ den Abstand des Elektrons vom Aufpunkt, $\mathfrak{v}$ die Geschwindigkeit des Elektrons.)

Man kann sich fragen, wie jene höheren Glieder in der Quantentheorie aussehen müßten. Da in der klassischen Theorie die höheren

[1]) H. v. Kramers, Nature **113**, 673, 1924.

[2]) M. Born, ZS. f. Phys. **26**, 379, 1924. H. A. Kramers und W. Heisenberg, ZS. f. Phys. **31**, 681, 1925. M. Born und P. Jordan, ZS. f. Phys. (Im Erscheinen.)

Zeitschrift für Physik *33*, 879–893 (1925)

Näherungen einfach berechnet werden können, wenn die Bewegung des Elektrons bzw. ihre Fourierdarstellung gegeben ist, so wird man in der Quantentheorie Ähnliches erwarten. Diese Frage hat nichts mit Elektrodynamik zu tun, sondern sie ist, dies scheint uns besonders wichtig, rein kinematischer Natur; wir können sie in einfachster Form folgendermaßen stellen: Gegeben sei eine an Stelle der klassischen Größe $x(t)$ tretende quantentheoretische Größe; welche quantentheoretische Größe tritt dann an Stelle von $x(t)^2$?

Bevor wir diese Frage beantworten können, müssen wir uns daran erinnern, daß es in der Quantentheorie nicht möglich war, dem Elektron einen Punkt im Raum als Funktion der Zeit mittels beobachtbarer Größen zuzuordnen. Wohl aber kann dem Elektron auch in der Quantentheorie eine Ausstrahlung zugeordnet werden; diese Strahlung wird beschrieben erstens durch die Frequenzen, die als Funktionen zweier Variablen auftreten, quantentheoretisch in der Gestalt:

$$\nu(n, n-\alpha) = \frac{1}{h}\{W(n) - W(n-\alpha)\},$$

in der klassischen Theorie in der Form:

$$\nu(n, \alpha) = \alpha \cdot \nu(n) = \alpha \frac{1}{h} \frac{dW}{dn}.$$

(Hierin ist $n \cdot h = J$, einer der kanonischen Konstanten, gesetzt.)

Als charakteristisch für den Vergleich der klassischen mit der Quantentheorie hinsichtlich der Frequenzen kann man die Kombinationsrelationen anschreiben:

Klassisch:

$$\nu(n, \alpha) + \nu(n, \beta) = \nu(n, \alpha + \beta).$$

Quantentheoretisch:

$$\nu(n, n-\alpha) + \nu(n-\alpha, n-\alpha-\beta) = \nu(n, n-\alpha-\beta)$$

bzw. $$\nu(n-\beta, n-\alpha-\beta) + \nu(n, n-\beta) = \nu(n, n-\alpha-\beta).$$

Neben den Frequenzen sind zweitens zur Beschreibung der Strahlung notwendig die Amplituden; die Amplituden können als komplexe Vektoren (mit je sechs unabhängigen Bestimmungsstücken) aufgefaßt werden und bestimmen Polarisation und Phase. Auch sie sind Funktionen der zwei Variablen n und α, so daß der betreffende Teil der Strahlung durch den folgenden Ausdruck dargestellt wird:

Quantentheoretisch:

$$Re\{\mathfrak{A}(n, n-\alpha)\, e^{i\,\omega(n, n-\alpha)t}\}. \tag{1}$$

Klassisch:

$$Re\{\mathfrak{A}_\alpha(n)\, e^{i\omega(n) \cdot \alpha t}\}. \tag{2}$$

Der (in $\mathfrak{A}$ enthaltenen) Phase scheint zunächst eine physikalische Bedeutung in der Quantentheorie nicht zuzukommen, da die Frequenzen der Quantentheorie mit ihren Oberschwingungen im allgemeinen nicht kommensurabel sind. Wir werden aber sofort sehen, daß die Phase auch in der Quantentheorie eine bestimmte, der in der klassischen Theorie analoge Bedeutung hat. Betrachten wir jetzt eine bestimmte Größe $x(t)$ in der klassischen Theorie, so kann man sie repräsentiert denken durch eine Gesamtheit von Größen der Form

$$\mathfrak{A}_\alpha(n)\, e^{i\omega(n).\alpha t},$$

die, je nachdem die Bewegung periodisch ist oder nicht, zu einer Summe oder zu einem Integral vereinigt $x(t)$ darstellen:

$$\left.\begin{aligned} x(n, t) &= \sum_{-\infty}^{+\infty}{}^{\alpha}\, \mathfrak{A}_\alpha(n)\, e^{i\omega(n).\alpha t} \\ \text{bzw.}\quad x(n, t) &= \int_{-\infty}^{+\infty} \mathfrak{A}_\alpha(n)\, e^{i\omega(n)\alpha t}\, d\alpha. \end{aligned}\right\} \tag{2a}$$

Eine solche Vereinigung der entsprechenden quantentheoretischen Größen scheint wegen der Gleichberechtigung der Größen n, $n-\alpha$ nicht ohne Willkür möglich und deshalb nicht sinnvoll; wohl aber kann man die Gesamtheit der Größen

$$\mathfrak{A}(n, n-\alpha)\, e^{i\omega(n, n-\alpha)t}$$

als Repräsentant der Größe $x(t)$ auffassen und dann die oben gestellte Frage zu beantworten suchen: Wodurch wird die Größe $x(t)^2$ repräsentiert?

Die Antwort lautet klassisch offenbar so:

$$\mathfrak{B}_\beta(n)\, e^{i\omega(n)\beta t} = \sum_{-\infty}^{+\infty}{}^{\alpha}\, \mathfrak{A}_\alpha \mathfrak{A}_{\beta-\alpha}\, e^{i\omega(n)(\alpha+\beta-\alpha)t} \tag{3}$$

bzw.

$$= \int_{-\infty}^{+\infty} \mathfrak{A}_\alpha \mathfrak{A}_{\beta-\alpha}\, e^{i\omega(n)(\alpha+\beta-\alpha)t}\, d\alpha, \tag{4}$$

wobei dann

$$x(t)^2 = \sum_{-\infty}^{+\infty}{}^{\beta}\, \mathfrak{B}_\beta(n)\, e^{i\omega(n)\beta t} \tag{5}$$

bzw.

$$= \int_{-\infty}^{+\infty} \mathfrak{B}_\beta(n)\, e^{i\omega(n)\beta t}\, d\beta. \tag{6}$$

Zeitschrift für Physik 33, 879–893 (1925)

Quantentheoretisch scheint es die einfachste und natürlichste Annahme, die Beziehungen (3, 4) durch die folgenden zu ersetzen:

$$\mathfrak{B}(n, n-\beta) e^{i\omega(n, n-\beta)t} = \sum_{-\infty}^{+\infty}{}_{\alpha} \mathfrak{A}(n, n-\alpha)\, \mathfrak{A}(n-\alpha, n-\beta)\, e^{i\omega(n, n-\beta)t} \quad (7)$$

bzw.

$$= \int_{-\infty}^{+\infty} d\alpha\, \mathfrak{A}(n, n-\alpha)\, \mathfrak{A}(n-\alpha, n-\beta)\, e^{i\omega(n, n-\beta)t}; \quad (8)$$

und zwar ergibt sich diese Art der Zusammensetzung nahezu zwangläufig aus der Kombinationsrelation der Frequenzen. Macht man diese Annahme (7) und (8), so erkennt man auch, daß die Phasen der quantentheoretischen $\mathfrak{A}$ eine ebenso große physikalische Bedeutung haben wie die in der klassischen Theorie: nur der Anfangspunkt der Zeit und daher eine allen $\mathfrak{A}$ gemeinsame Phasenkonstante ist willkürlich und ohne physikalische Bedeutung; doch die Phase der einzelnen $\mathfrak{A}$ geht wesentlich in die Größe $\mathfrak{B}$ ein[1]). Eine geometrische Interpretation solcher quantentheoretischer Phasenbeziehungen in Analogie zur klassischen Theorie scheint zunächst kaum möglich.

Fragen wir weiter nach dem Repräsentant der Größe $x(t)^3$, so finden wir ohne Schwierigkeit:

Klassisch:

$$\mathfrak{C}(n, \gamma) = \sum_{-\infty}^{+\infty} \sum_{-\infty}^{+\infty}{}_{\alpha, \beta}\, \mathfrak{A}_{\alpha}(n)\, \mathfrak{A}_{\beta}(n)\, \mathfrak{A}_{\gamma-\alpha-\beta}(n). \quad (9)$$

Quantentheoretisch:

$$\mathfrak{C}(n, n-\gamma) = \sum_{-\infty}^{+\infty} \sum_{-\infty}^{+\infty}{}_{\alpha, \beta}\, \mathfrak{A}(n, n-\alpha)\, \mathfrak{A}(n-\alpha, n-\alpha-\beta)\, \mathfrak{A}(n-\alpha-\beta, n-\gamma) \quad (10)$$

bzw. die entsprechenden Integrale.

In ähnlicher Weise lassen sich alle Größen der Form $x(t)^n$ quantentheoretisch darstellen, und wenn irgend eine Funktion $f[x(t)]$ gegeben ist, so kann man offenbar immer dann, wenn diese Funktion nach Potenzreihen in x entwickelbar ist, das quantentheoretische Analogon finden. Eine wesentliche Schwierigkeit entsteht jedoch, wenn wir zwei Größen $x(t)$, $y(t)$ betrachten und nach dem Produkt $x(t)\, y(t)$ fragen.

[1]) Vgl. auch H. A. Kramers und W. Heisenberg, l. c. In die dort benutzten Ausdrücke für das induzierte Streumoment gehen die Phasen wesentlich ein.

Sei $x(t)$ durch $\mathfrak{A}$, $y(t)$ durch $\mathfrak{B}$ charakterisiert, so ergibt sich als Darstellung von $x(t) \cdot y(t)$:

Klassisch:

$$\mathfrak{C}_\beta(n) = \sum_{-\infty}^{+\infty}{}^\alpha \mathfrak{A}_\alpha(n) \mathfrak{B}_{\beta-\alpha}(n).$$

Quantentheoretisch:

$$\mathfrak{C}(n, n-\beta) = \sum_{-\infty}^{+\infty}{}^\alpha \mathfrak{A}(n, n-\alpha) \mathfrak{B}(n-\alpha, n-\beta).$$

Während klassisch $x(t) \cdot y(t)$ stets gleich $y(t)\, x(t)$ wird, braucht dies in der Quantentheorie im allgemeinen nicht der Fall zu sein. — In speziellen Fällen, z. B. bei der Bildung von $x(t) \cdot x(t)^2$, tritt diese Schwierigkeit nicht auf.

Wenn es sich, wie in der zu Beginn dieses Paragraphen gestellten Frage, um Bildungen der Form

$$v(t)\, \dot{v}(t)$$

handelt, so wird man quantentheoretisch $v\dot{v}$ ersetzen sollen durch $\frac{v\dot{v} + \dot{v}v}{2}$, um zu erreichen, daß $v\dot{v}$ als Differentialquotient von $\frac{v^2}{2}$ auftritt. In ähnlicher Weise lassen sich wohl stets naturgemäße quantentheoretische Mittelwerte angeben, die allerdings in noch höherem Grade hypothetisch sind als die Formeln (7) und (8).

Abgesehen von der eben geschilderten Schwierigkeit dürften Formeln vom Typus (7), (8) allgemein genügen, um auch die Wechselwirkung der Elektronen in einem Atom durch die charakteristischen Amplituden der Elektronen auszudrücken.

§ 2. Nach diesen Überlegungen, welche die Kinematik der Quantentheorie zum Gegenstand hatten, werden wir zum mechanischen Problem übergehen, das auf die Bestimmung der $\mathfrak{A}$, ν, W aus den gegebenen Kräften des Systems abzielt. In der bisherigen Theorie wird dieses Problem gelöst in zwei Schritten:

1. Integration der Bewegungsgleichung

$$\ddot{x} + f(x) = 0. \tag{11}$$

2. Bestimmung der Konstante bei periodischen Bewegungen durch

$$\oint p\, dq = \oint m\dot{x}\, dx = J (= nh). \tag{12}$$

Wenn man sich vornimmt, eine quantentheoretische Mechanik aufzubauen, welche der klassischen möglichst analog ist, so liegt es wohl sehr nahe, die Bewegungsgleichung (11) direkt in die Quantentheorie zu übernehmen, wobei es nur notwendig ist — um nicht vom sicheren Fun-

Zeitschrift für Physik 33, 879–893 (1925)

dament der prinzipiell beobachtbaren Größen abzugehen —, an Stelle der Größen $\ddot{x}, f(x)$ ihre aus § 1 bekannten quantentheoretischen Repräsentanten zu setzen. In der klassischen Theorie ist es möglich, die Lösung von (11) durch Ansatz von x in Fourierreihen bzw. Fourierintegralen mit unbestimmten Koeffizienten (und Frequenzen) zu suchen; allerdings erhalten wir dann im allgemeinen unendlich viele Gleichungen mit unendlich vielen Unbekannten bzw. Integralgleichungen, die sich nur in speziellen Fällen zu einfachen Rekursionsformeln für die $\mathfrak{A}$ umgestalten lassen. In der Quantentheorie sind wir jedoch vorläufig auf diese Art der Lösung von (11) angewiesen, da sich, wie oben besprochen, keine der Funktion $x(n, t)$ direkt analoge quantentheoretische Funktion definieren ließ.

Dies hat zur Folge, daß die quantentheoretische Lösung von (11) zunächst nur in den einfachsten Fällen durchführbar ist. Bevor wir auf solche einfache Beispiele eingehen, sei noch die quantentheoretische Bestimmung der Konstante nach (12) hergeleitet. Wir nehmen also an, daß die Bewegung (klassisch) periodisch sei:

$$x = \sum_{-\infty}^{+\infty}{}^{\alpha}\, a_\alpha(n)\, e^{i\alpha\omega_n t}; \tag{13}$$

dann ist

$$m\dot{x} = m \sum_{-\infty}^{+\infty} a_\alpha(n) \,.\, i\alpha\omega_n e^{i\alpha\omega_n t}$$

und

$$\oint m\dot{x}\,dx = \oint m\dot{x}^2\,dt = 2\pi m \sum_{-\infty}^{+\infty}{}^{\alpha}\, a_\alpha(n)\, a_{-\alpha}(n)\, \alpha^2 \omega_n.$$

Da ferner $a_{-\alpha}(n) = \overline{a_\alpha(n)}$ ist (x soll reell sein), so folgt

$$\oint m\dot{x}^2\,dt = 2\pi m \sum_{-\infty}^{+\infty}{}^{\alpha}\, |a_\alpha(n)|^2\, \alpha^2\omega_n. \tag{14}$$

Dieses Phasenintegral hat man bisher meist gleich einem ganzen Vielfachen von h, also gleich $n.h$ gesetzt; eine solche Bedingung fügt sich aber nicht nur sehr gezwungen der mechanischen Rechnung ein, sie erscheint auch selbst vom bisherigen Standpunkt aus im Sinne des Korrespondenzprinzips willkürlich; denn korrespondenzmäßig sind die J nur bis auf eine additive Konstante als ganzzahlige Vielfache von h festgelegt, und an Stelle von (14) hätte naturgemäß zu treten:

$$\frac{d}{dn}(nh) = \frac{d}{dn} \cdot \oint m\dot{x}^2\,dt,$$

das heißt

$$h = 2\pi m \cdot \sum_{-\infty}^{+\infty}{}^{\alpha}\, \alpha \frac{d}{dn}(\alpha\omega_n \cdot |a_\alpha|^2). \tag{15}$$

Eine solche Bedingung legt allerdings die a_α dann auch nur bis auf eine Konstante fest, und diese Unbestimmtheit hat empirisch in dem Auftreten von halben Quantenzahlen zu Schwierigkeiten Anlaß gegeben.

Fragen wir nach einer (14) und (15) entsprechenden quantentheoretischen Beziehung zwischen beobachtbaren Größen, so stellt sich die vermißte Eindeutigkeit von selbst wieder her.

Zwar besitzt eben nur Gleichung (15) eine an die Kramerssche Dispersionstheorie anknüpfende einfache quantentheoretische Verwandlung[1]):

$$h = 4\pi m \sum_0^\infty \alpha \{|a(n, n+\alpha)|^2 \omega(n, n+\alpha) - |a(n, n-\alpha)|^2 \omega(n, n-\alpha)\}, \quad (16)$$

doch diese Beziehung genügt hier zur eindeutigen Bestimmung der a; denn die in den Größen a zunächst unbestimmte Konstante wird von selbst durch die Bedingung festgelegt, daß es einen Normalzustand geben solle, von dem aus keine Strahlung mehr stattfindet; sei der Normalzustand mit n_0 bezeichnet, so sollen also alle

$$a(n_0, n_0 - \alpha) = 0 \text{ (für } \alpha > 0)$$

sein. Die Frage nach halbzahliger oder ganzzahliger Quantelung dürfte daher in einer quantentheoretischen Mechanik, die nur Beziehungen zwischen beobachtbaren Größen benutzt, nicht auftreten können.

Die Gleichungen (11) und (16) zusammen enthalten, wenn sie sich lösen lassen, eine vollständige Bestimmung nicht nur der Frequenzen und Energien, sondern auch der quantentheoretischen Übergangswahrscheinlichkeiten. Die wirkliche mathematische Durchführung gelingt jedoch zunächst nur in den einfachsten Fällen; eine besondere Komplikation entsteht auch bei vielen Systemen, wie z. B. beim Wasserstoffatom, dadurch, daß die Lösungen teils periodischen, teils aperiodischen Bewegungen entsprechen, was zur Folge hat, daß die quantentheoretischen Reihen (7), (8) und die Gleichung (16) stets in eine Summe und ein Integral zerfallen. Quantenmechanisch läßt sich eine Trennung in „periodische und aperiodische Bewegungen" im allgemeinen nicht durchführen.

Trotzdem könnte man vielleicht die Gleichungen (11) und (16) wenigstens prinzipiell als befriedigende Lösung des mechanischen Problems ansehen, wenn sich zeigen ließe, daß diese Lösung übereinstimmt bzw. nicht in Widerspruch steht mit den bisher bekannten quantenmechanischen Beziehungen; daß also eine kleine Störung eines mechanischen Problems zu Zusatzgliedern in der Energie bzw. in den Frequenzen Anlaß gibt, die

[1]) Diese Beziehung wurde schon auf Grund von Betrachtungen von Dispersion gegeben von W. Kuhn, ZS. f. Phys. 33, 408, 1925, und Thomas, Naturw. 13, 1925.

eben den von Kramers und Born gefundenen Ausdrücken entsprechen — im Gegensatz zu denen, welche die klassische Theorie liefern würde. Ferner müßte untersucht werden, ob im allgemeinen der Gleichung (11) auch in der hier vorgeschlagenen quantentheoretischen Auffassung ein Energieintegral $m\frac{\dot{x}^2}{2} + U(x) = \text{const}$ entspricht und ob die so gewonnene Energie — ähnlich, wie klassisch gilt: $\nu = \frac{\partial W}{\partial J}$ — der Bedingung genügt: $\Delta W = h.\nu$. Eine allgemeine Beantwortung dieser Fragen erst könnte den inneren Zusammenhang der bisherigen quantenmechanischen Versuche dartun und zu einer konsequent nur mit beobachtbaren Größen operierenden Quantenmechanik führen. Abgesehen von einer allgemeinen Beziehung zwischen der Kramersschen Dispersionsformel und den Gleichungen (11) und (16) können wir die oben gestellten Fragen nur in den ganz speziellen, durch einfache Rekursion lösbaren Fällen beantworten.

Jene allgemeine Beziehung zwischen der Kramersschen Dispersionstheorie und unseren Gleichungen (11), (16) besteht darin, daß aus Gleichung (11) (d. h. ihrem quantentheoretischen Analogon) ebenso wie in der klassischen Theorie folgt, daß sich das schwingende Elektron gegenüber Licht, das viel kurzwelliger ist als alle Eigenschwingungen des Systems, wie ein freies Elektron verhält. Dieses Resultat folgt auch aus der Kramersschen Theorie, wenn man noch Gleichung (16) berücksichtigt. In der Tat findet Kramers für das durch die Welle $E\cos 2\pi\nu t$ induzierte Moment:

$$M = e^2 E\cos 2\pi\nu t \cdot \frac{2}{h}\sum_0^\infty{}_\alpha \left\{\frac{|a(n,n+\alpha)|^2\nu(n,n+\alpha)}{\nu^2(n,n+\alpha)-\nu^2} - \frac{|a(n,n-\alpha)|^2\nu(n,n-\alpha)}{\nu^2(n,n-\alpha)-\nu^2}\right\},$$

also für $\nu \gg \nu(n,n+\alpha)$

$$M = -\frac{2Ee^2\cos 2\pi\nu t}{\nu^2 . h}\sum_0^\infty{}_\alpha \left\{|a(n,n+\alpha)|^2\nu(n,n+\alpha) - |a(n,n-\alpha)|^2\nu(n,n-\alpha)\right\},$$

was wegen (16) übergeht in

$$M = -\frac{e^2 E\cos 2\pi\nu t}{\nu^2 . 4\pi^2 m}.$$

§ 3. Als einfachstes Beispiel soll im folgenden der anharmonische Oszillator behandelt werden:

$$\ddot{x} + \omega_0^2 x + \lambda x^2 = 0. \tag{17}$$

Klassisch läßt sich diese Gleichung befriedigen durch einen Ansatz der Form

$$x = \lambda a_0 + a_1 \cos \omega t + \lambda a_2 \cos 2\omega t + \lambda^2 a_3 \cos 3\omega t + \cdots \lambda^{\tau-1} a_\tau \cos \tau\omega t,$$

wobei die a Potenzreihen in λ sind, die mit einem von λ freien Gliede beginnen. Wir versuchen quantentheoretisch einen analogen Ansatz und repräsentieren x durch Glieder der Form

$$\lambda a(n, n); \quad a(n, n-1) \cos \omega(n, n-1) t; \quad \lambda a(n, n-2) \cos \omega(n, n-2) t;$$
$$\ldots \lambda^{\tau-1} a(n, n-\tau) \cos \omega(n, n-\tau) t \ldots$$

Die Rekursionsformeln zur Bestimmung der a und ω lauten (bis auf Glieder der Ordnung λ) nach Gleichung (3), (4) bzw. (7), (8):

Klassisch:

$$\left.\begin{aligned} \omega_0^2 a_0(n) + \frac{a_1^2(n)}{2} &= 0; \\ -\omega^2 + \omega_0^2 &= 0; \\ (-4\omega^2 + \omega_0^2) a_2(n) + \frac{a_1^2}{2} &= 0; \\ (-9\omega^2 + \omega_0^2) a_3(n) + a_1 a_2 &= 0; \\ \cdots\cdots\cdots\cdots & \end{aligned}\right\} \tag{18}$$

Quantentheoretisch:

$$\left.\begin{aligned} \omega_0^2 a_0(n) + \frac{a^2(n+1, n) + a^2(n, n-1)}{4} &= 0; \\ -\omega^2(n, n-1) + \omega_0^2 &= 0; \\ (-\omega^2(n, n-2) + \omega_0^2) a(n, n-2) + \frac{a(n, n-1)\, a(n-1, n-2)}{2} &= 0; \\ (-\omega^2(n, n-3) + \omega_0^2)\, a(n, n-3) & \\ + \frac{a(n, n-1)\, a(n-1, n-3)}{2} + \frac{a(n, n-2)\, a(n-2, n-3)}{2} &= 0; \\ \cdots\cdots\cdots\cdots & \end{aligned}\right\} \tag{19}$$

Hierzu kommt die Quantenbedingung:

Klassisch ($J = nh$):

$$1 = 2\pi m \frac{d}{dJ} \sum_{-\infty}^{+\infty} \tau^2 \frac{|a_\tau|^2 \omega}{4}.$$

Quantentheoretisch:

$$h = \pi m \sum_{0}^{\infty} \left[|a(n+\tau, n)|^2 \omega(n+\tau, n) - |a(n, n-\tau)|^2 \omega(n, n-\tau) \right].$$

Dies ergibt in erster Näherung, sowohl klassisch wie quantentheoretisch:

$$a_1^2(n) \quad \text{bzw.} \quad a^2(n, n-1) = \frac{(n + \text{const})\, h}{\pi m \omega_0}. \tag{20}$$

Zeitschrift für Physik *33*, 879–893 (1925)

Quantentheoretisch läßt sich die Konstante in (20) bestimmen durch die Bedingung, daß $a(n_0, n_0 - 1)$ im Normalzustand Null sein solle. Numerieren wir die n so, daß n im Normalzustand gleich Null wird, also $n_0 = 0$, so folgt

$$a^2(n, n-1) = \frac{n h}{\pi m \omega_0}.$$

Aus den Rekursionsgleichungen (18) folgt dann, daß in der klassischen Theorie a_τ (in erster Näherung in λ) von der Form wird $\varkappa(\tau) n^{\frac{\tau}{2}}$, wo $\varkappa(\tau)$ einen von n unabhängigen Faktor darstellt. In der Quantentheorie ergibt sich aus (19)

$$a(n, n-\tau) = \varkappa(\tau) \sqrt{\frac{n!}{(n-\tau)!}}, \tag{21}$$

wobei $\varkappa(\tau)$ denselben, von n unabhängigen Proportionalitätsfaktor darstellt. Für große Werte von n geht natürlich der quantentheoretische Wert von a_τ asymptotisch in den klassischen über.

Für die Energie liegt es nahe, den klassischen Ansatz

$$\frac{m \dot{x}^2}{2} + m \omega_0^2 \frac{x^2}{2} + \frac{m \lambda}{3} x^3 = W$$

zu versuchen, der in der hier durchgerechneten Näherung auch quantentheoretisch wirklich konstant ist und nach (19), (20) und (21) den Wert hat:

Klassisch:

$$W = \frac{n h \omega_0}{2 \pi}. \tag{22}$$

Quantentheoretisch [nach (7), (8)]:

$$W = \frac{(n + \frac{1}{2}) h \omega_0}{2 \pi} \tag{23}$$

(bis auf Größen der Ordnung λ^2).

Nach dieser Auffassung ist also schon beim harmonischen Oszillator die Energie nicht durch die „klassische Mechanik", d. h. (22) darstellbar, sondern sie hat die Form (23).

Die genauere Durchrechnung auch der höheren Näherungen in W, a, ω soll ausgeführt werden am einfacheren Beispiel des anharmonischen Oszillators vom Typus:

$$\ddot{x} + \omega_0^2 x + \lambda x^3 = 0.$$

Klassisch kann man hier setzen:

$$x = a_1 \cos \omega t + \lambda a_3 \cos 3 \omega t + \lambda^2 a_5 \cos 5 \omega t + \cdots,$$

analog versuchen wir quantentheoretisch den Ansatz

$$a(n, n-1) \cos \omega(n, n-1) t; \quad \lambda a(n, n-3) \cos \omega(n, n-3) t; \quad \ldots$$

Die Größen a sind wieder Potenzreihen in λ, deren erstes Glied, wie in (21), die Form hat:

$$a(n, n-\tau) = \varkappa(\tau) \sqrt{\frac{n!}{(n-\tau)!}},$$

wie man durch Ausrechnen der den Gleichungen (18), (19) entsprechenden Gleichungen erhält.

Führt man die Berechnung von ω, a nach (18), (19) bis zur Näherung λ^2 bzw. λ durch, so erhält man:

$$\omega(n, n-1) = \omega_0 + \lambda \cdot \frac{3nh}{8\pi\omega_0^2 m} - \lambda^2 \cdot \frac{3h^2}{256\omega_0^5 m^2 \pi^2}(17n^2 + 7) + \cdots \quad (24)$$

$$a(n, n-1) = \sqrt{\frac{nh}{\pi\omega_0 m}}\left(1 - \lambda \frac{3nh}{16\pi\omega_0^3 m} + \cdots\right). \quad (25)$$

$$a(n, n-3) = \frac{1}{32}\sqrt{\frac{h^3}{\pi^3\omega_0^7 m^3} n(n-1)(n-2)}\left(1 - \lambda \frac{39(n-1)h}{32\pi\omega_0^3 m}\right). \quad (26)$$

Die Energie, die als das konstante Glied von

$$m\frac{\dot{x}^2}{2} + m\omega_0^2 \frac{x^2}{2} + \frac{m\lambda}{4} x^4$$

definiert ist (daß die periodischen Glieder wirklich alle Null sind, konnte ich nicht allgemein beweisen, in den durchgerechneten Gliedern war es der Fall), ergibt sich zu

$$W = \frac{(n+\frac{1}{2})h\omega_0}{2\pi} + \lambda \cdot \frac{3(n^2+n+\frac{1}{2})h^2}{8 \cdot 4\pi^2\omega_0^2 \cdot m}$$
$$- \lambda^2 \cdot \frac{h^3}{512\pi^3\omega_0^5 m^2}\left(17n^3 + \frac{51}{2}n^2 + \frac{59}{2}n + \frac{21}{2}\right). \quad (27)$$

Diese Energie kann man auch noch nach dem Kramers-Bornschen Verfahren berechnen, indem man das Glied $\frac{m\lambda}{4} x^4$ als Störungsglied zum harmonischen Oszillator auffaßt. Man kommt dann wirklich wieder genau zum Resultat (27), was mir eine bemerkenswerte Stütze für die zugrundegelegten quantenmechanischen Gleichungen zu sein scheint. Ferner erfüllt die nach (27) berechnete Energie die Formel [vgl. (24)]:

$$\frac{\omega(n, n-1)}{2\pi} = \frac{1}{h} \cdot [W(n) - W(n-1)],$$

welche ebenfalls als notwendige Bedingung für die Möglichkeit einer den Gleichungen (11) und (16) entsprechenden Bestimmung der Übergangswahrscheinlichkeiten zu betrachten ist.

Zum Schluß sei der Rotator als Beispiel angeführt und auf die Beziehung der Gleichungen (7), (8) zu den Intensitätsformeln beim Zeemaneffekt[1]) und bei den Multipletts[2]) hingewiesen.

Sei der Rotator repräsentiert durch ein Elektron, das im konstanten Abstand a um einen Kern kreist. Die „Bewegungsgleichungen" besagen dann klassisch wie quantentheoretisch nur, daß das Elektron im konstanten Abstand a eine ebene, gleichförmige Rotation um den Kern beschreibt mit der Winkelgeschwindigkeit ω. Die „Quantenbedingung" (16) ergibt nach (12):

$$h = \frac{d}{dn}(2\pi m a^2 \omega),$$

nach (16):

$$h = 2\pi m\{a^2\omega(n+1, n) - a^2\omega(n, n-1)\},$$

woraus in beiden Fällen folgt:

$$\omega(n, n-1) = \frac{h\,.\,(n + \text{const})}{2\pi m a^2}.$$

Die Bedingung, daß im Normalzustand ($n_0 = 0$) die Strahlung verschwinden solle, führt zu der Formel:

$$\omega(n, n-1) = \frac{h\,.\,n}{2\pi m a^2}. \tag{28}$$

Die Energie wird

$$W = \frac{m}{2}v^2$$

oder nach (7), (8)

$$W = \frac{m}{2}a^2 \cdot \frac{\omega^2(n, n-1) + \omega^2(n+1, n)}{2} = \frac{h^2}{8\pi^2 m a^2}(n^2 + n + \tfrac{1}{2}), \tag{29}$$

was wieder der Beziehung $\omega(n, n-1) = \frac{2\pi}{h}[W(n) - W(n-1)]$ genügt. Als Stütze für die von der bisher üblichen Theorie abweichenden Formeln (28) und (29) kann es angesehen werden, daß viele Bandenspektren (auch solche, bei denen die Existenz eines Elektronenimpulses unwahrscheinlich ist) nach Kratzer[3]) Formeln vom Typus (28), (29) (die man bisher der klassisch-mechanischen Theorie zuliebe durch halbzahlige Quantelung zu erklären suchte) zu fordern scheinen.

---

[1]) Goudsmit und R. de L. Kronig, Naturw. **13**, 90, 1925; H. Hönl, ZS. f. Phys. **31**, 340, 1925.

[2]) R. de L. Kronig, ZS. f. Phys. **31**, 885, 1925; A. Sommerfeld und H. Hönl, Sitzungsber. d. Preuß. Akad. d. Wiss. 1925, S. 141; H. N. Russell, Nature **115**, 835, 1925.

[3]) Vgl. z. B. A. Kratzer, Sitzungsber. d. Bayr. Akad. 1922, S. 107.

Um beim Rotator zu den Goudsmit-Kronig-Hönlschen Formeln zu gelangen, müssen wir das Gebiet der Probleme mit einem Freiheitsgrad verlassen und annehmen, daß der Rotator, in irgendwelcher Richtung im Raume, um die Achse z eines äußeren Feldes eine sehr langsame Präzession $\mathfrak{o}$ ausführe. Die dieser Präzession entsprechende Quantenzahl heiße m. Dann wird die Bewegung repräsentiert durch die Größen

$$\begin{aligned} z:&\quad a(n, n-1; m, m)\cos\omega(n, n-1)t; \\ x+iy:&\quad b(n, n-1; m, m-1)\, e^{i[\omega(n,n-1)+\mathfrak{o}]t}; \\ &\quad b(n, n-1; m-1, m)\, e^{i[-\omega(n,n-1)+\mathfrak{o}]t}. \end{aligned}$$

Die Bewegungsgleichungen lauten einfach:

$$x^2 + y^2 + z^2 = a^2,$$

was nach (7) zu den Gleichungen[1]) Anlaß gibt:

$$\begin{aligned} &\tfrac{1}{2}\{\tfrac{1}{2}a^2(n,n-1;m,m) + b^2(n,n-1;m,m-1) + b^2(n,n-1;m,m+1) \\ &+\tfrac{1}{2}a^2(n+1,n;m,m) + b^2(n+1,n;m-1,m) + b^2(n+1,n;m+1,m)\} = a^2. \quad (30) \end{aligned}$$

$$\begin{aligned} &\tfrac{1}{2}a(n, n-1; m, m)\, a(n-1, n-2; m, m) \\ &= b(n, n-1; m, m+1)\, b(n-1, n-2; m+1, m) \\ &+ b(n, n-1; m, m-1)\, b(n-1, n-2; m-1, m). \quad (31) \end{aligned}$$

Hierzu kommt nach (16) die Quantenbedingung:

$$\begin{aligned} &2\pi m\{b^2(n, n-1; m, m-1)\,\omega(n, n-1) \\ &- b^2(n, n-1; m-1, m)\,\omega(n, n-1)\} = (m + \text{const})\, h. \quad (32) \end{aligned}$$

Die diesen Gleichungen entsprechenden klassischen Beziehungen:

$$\left.\begin{aligned} \tfrac{1}{2}a_0^2 + b_1^2 + b_{-1}^2 &= a^2; \\ \tfrac{1}{4}a_0^2 &= b_1 b_{-1}; \\ 2\pi m(b_{+1}^2 - b_{-1}^2)\omega &= (m + \text{const})\, h \end{aligned}\right\} \quad (33)$$

genügen (bis auf die unbestimmte Konstante bei m) zur eindeutigen Festlegung der a_0, b_1, b_{-1}.

Die am einfachsten sich darbietende Lösung der quantentheoretischen Gleichungen (30), (31), (32) lautet:

$$b(n, n-1; m, m-1) = a\sqrt{\frac{(n+m+1)(n+m)}{4(n+\frac{1}{2})n}};$$

$$b(n, n-1; m-1, m) = a\sqrt{\frac{(n-m)(n-m+1)}{4(n+\frac{1}{2})n}};$$

$$a(n, n-1; m, m) = a\sqrt{\frac{(n+m+1)(n-m)}{(n+\frac{1}{2})n}}.$$

[1]) Die Gleichung (30) ist im wesentlichen identisch mit den Ornstein-Burgerschen Summenregeln.

Zeitschrift für Physik *33*, 879–893 (1925)

Diese Ausdrücke stimmen mit den Formeln von Goudsmit, Kronig und Hönl überein; man kann jedoch nicht einfach einsehen, daß diese Ausdrücke die einzige Lösung von (30), (31), (32) darstellen — was mir jedoch bei Beachtung der Randbedingungen (Verschwinden der a, b am „Rande", vgl. die oben zitierten Arbeiten von Kronig, Sommerfeld und Hönl, Russell) wahrscheinlich scheint.

Eine der hier angestellten ähnliche Überlegung führt auch bei den Intensitätsformeln der Multipletts zu dem Ergebnis, daß die genannten Intensitätsregeln mit Gleichung (7) und (16) im Einklang stehen. Dieses Resultat dürfte wiederum als Stütze insbesondere für die Richtigkeit der kinematischen Gleichung (7) anzusprechen sein.

Ob eine Methode zur Bestimmung quantentheoretischer Daten durch Beziehungen zwischen beobachtbaren Größen, wie die hier vorgeschlagene, schon in prinzipieller Hinsicht als befriedigend angesehen werden könnte, oder ob diese Methode doch noch einen viel zu groben Angriff auf das physikalische, zunächst offenbar sehr verwickelte Problem einer quantentheoretischen Mechanik darstellt, wird sich erst durch eine tiefergehende mathematische Untersuchung der hier sehr oberflächlich benutzten Methode erkennen lassen.

Göttingen, Institut für theoretische Physik.

Zur Quantenmechanik. II.

Von **M. Born, W. Heisenberg** und **P. Jordan** in Göttingen.

(Eingegangen am 16. November 1925.)

Die aus Heisenbergs Ansätzen in Teil I dieser Arbeit entwickelte Quantenmechanik wird auf Systeme von beliebig vielen Freiheitsgraden ausgedehnt. Die Störungstheorie wird für nicht entartete und eine große Klasse entarteter Systeme durchgeführt und ihr Zusammenhang mit der Eigenwerttheorie Hermitescher Formen nachgewiesen. Die gewonnenen Resultate werden zur Ableitung der Sätze über Impuls und Drehimpuls und zur Ableitung von Auswahlregeln und Intensitätsformeln benutzt. Schließlich werden die Ansätze der Theorie auf die Statistik der Eigenschwingungen eines Hohlraumes angewendet.

Einleitung. Die vorliegende Arbeit versucht den weiteren Ausbau der Theorie einer allgemeinen quantentheoretischen Mechanik, deren physikalische und mathematische Grundlagen in zwei vorausgegangenen Arbeiten der Verfasser[1]) dargestellt sind. Es erwies sich als möglich, die genannte Theorie auf Systeme von mehreren Freiheitsgraden zu erweitern[2]) (Kap. 2) und durch Einführung der „kanonischen Transformationen" das Problem der Integration der Bewegungsgleichungen auf bekannte mathematische Fragestellungen zurückzuführen; dabei ergab sich mittels dieser Theorie der kanonischen Transformationen einerseits eine Störungstheorie (Kap. 1, § 4), die eine weitgehende Ähnlichkeit mit der klassischen Störungstheorie aufweist, andererseits ein Zusammenhang der Quantenmechanik mit der mathematisch so hochentwickelten Theorie der quadratischen Formen unendlich vieler Variablen (Kap. 3). — Bevor wir aber auf die Darstellung dieser weiteren Entwicklung der Theorie eingehen, werden wir ihren physikalischen Inhalt genauer zu umgrenzen suchen.

Der Ausgangspunkt der versuchten Theorie war die Überzeugung, daß es nicht möglich sein werde, der Schwierigkeiten, die uns in der Quantentheorie gerade in den letzten Jahren auf Schritt und Tritt begegneten, Herr zu werden, ehe für die Mechanik der Atom- und Elektronenbewegungen ein mathematisches System von Beziehungen zwischen prinzipiell beobachtbaren Größen zur Verfügung stände von ähnlicher

---

[1]) W. Heisenberg, ZS. f. Phys. **33**, 879, 1925. M. Born und P. Jordan, ZS. f. Phys. **34**, 858, 1925. Im folgenden als (Teil) I zitiert.

[2]) Anm. bei der Korr. In einer inzwischen erschienenen Arbeit von P. Dirac (Proc. Roy. Soc. London **109**, 642, 1925) sind unabhängig einige der in Teil I und in dieser Arbeit enthaltenen Gesetzmäßigkeiten und weitere neue Folgerungen aus der Theorie angegeben worden.

Einfachheit und Einheit, wie das System der klassischen Mechanik. Ein solches System von quantentheoretischen Beziehungen zwischen beobachtbaren Größen wird allerdings gegenüber der bisherigen Quantentheorie den Mangel aufweisen müssen, daß es nicht unmittelbar geometrisch anschaulich interpretiert werden kann, da ja die Elektronenbewegungen nicht in den uns geläufigen Begriffen von Raum und Zeit beschrieben werden können; es wird ein charakteristischer Zug der neuen Theorie sein, daß sie ebensosehr eine Abänderung der bisherigen Kinematik wie der bisherigen Mechanik darstellt; es wird aber ein wichtiger Vorzug dieser Quantenmechanik darin bestehen, daß die Grundpostulate der Quantentheorie einen vollkommen organischen Bestandteil dieser Mechanik ausmachen, daß also z. B. die Existenz diskreter stationärer Zustände für die neue Theorie ebenso natürlich ist, wie etwa die Existenz diskreter Eigenschwingungsfrequenzen für die klassische Theorie (vgl. Kap. 3). Wenn man eben die fundamentalen, durch die quantentheoretischen Grundpostulate gegebenen Unterschiede zwischen Quantentheorie und klassischer Theorie im Auge behält, so scheint uns ein Formalismus, wie der in den beiden oben zitierten Arbeiten und im folgenden versuchte, wenn er sich als richtig erweisen sollte, eine Quantenmechanik darzustellen, die der klassischen so ähnlich ist, wie man nur irgendwie hoffen konnte. Wir erinnern hier nur an die Gültigkeit von Energie- und Impulssatz und an die Form der Bewegungsgleichungen (Kap. 1, § 2). Dieser Ähnlichkeit der neuen Teorie mit der klassischen entspricht es auch, daß von einem selbständigen Korrespondenzprinzip neben dieser Theorie wohl nicht die Rede sein kann; vielmehr kann die Theorie selbst als exakte Formulierung des Bohrschen Korrespondenzgedankens aufgefaßt werden. Es wird eine wichtige Aufgabe für die weitere Entwicklung der Theorie sein, die Art dieser Korrespondenz genauer zu untersuchen und den Übergang von der symbolischen Quantengeometrie in die anschauliche klassische Geometrie zu beschreiben. Im Hinblick auf diese Frage scheint es uns ein besonders wesentlicher Zug der neuen Theorie, daß in ihr die kontinuierlichen Spektra und die Linienspektra der Atome gleichberechtigt nebeneinander auftreten, d. h. als Lösung ein und derselben Bewegungsgleichung erscheinen und mathematisch eng miteinander verknüpft sind (vgl. Kap. 3, § 3); eine Unterscheidung zwischen „gequantelten“ und „ungequantelten“ Bewegungen verliert in dieser Theorie offenbar jeden Sinn, da in ihr nicht von einer Quantenbedingung die Rede ist, die bestimmte Bewegungen aus einer großen Anzahl von möglichen aussondert; an Stelle dieser Bedingung tritt vielmehr eine quantenmechanische Grundgleichung (Kap. 1, § 1), die für alle möglichen Bewegungen Geltung hat und die notwendig ist, um dem Bewegungsproblem überhaupt einen bestimmten Sinn zu geben.

Obwohl wir nun gerne aus der mathematischen Einheitlichkeit und Einfachheit der hier versuchten Theorie schließen möchten, daß sie schon wesentliche Züge der wirklichen Verhältnisse beim Problem des Atombaues wiedergibt, so muß man sich doch darüber klar sein, daß die Theorie eine Lösung der prinzipiellen Schwierigkeiten der Quantentheorie noch nicht geben kann. Die Kräfte, die dem Strahlungswiderstand der klassischen Theorie entsprechen, sind noch nicht in die Theorie eingearbeitet und für den Zusammenhang des Problems der Kopplung mit der hier versuchten Quantenmechanik sind nur einige undeutliche Anzeichen vorhanden (vgl. Kap. 1, § 5). Trotzdem scheint es, als ob diese prinzipiellen quantentheoretischen Schwierigkeiten vom Standpunkt der neuen Theorie aus ein anderes Aussehen zeigten, als bisher und als ob doch eine mehr als bisher begründete Hoffnung zur späteren Lösung dieser Probleme bestünde. Denken wir z. B. an die Frage der Stoßprozesse. Auf die grundsätzlichen Schwierigkeiten, die in der bisherigen Theorie einer Vereinigung der Grundpostulate der Quantentheorie mit der Gültigkeit des Energiesatzes bei schnellen Stößen im Wege stehen, hat in letzter Zeit besonders Bohr[1]) hingewiesen. In der hier versuchten Theorie ergeben sich nun sowohl die Grundpostulate, wie der Energiesatz als mathematische Folgen der quantenmechanischen Gleichungen und die Ergebnisse der Franck-Hertzschen Stoßversuche scheinen so eine naturgemäße mathematische Konsequenz der Theorie; daher darf man hoffen, daß bei einer künftigen Behandlung der Stoßprobleme auf Grund der Quantenmechanik eben wegen des organischen Zusammenhanges der Grundpostulate mit dieser Mechanik Schwierigkeiten der erwähnten Art nicht auftreten werden.

Der Fragenkomplex der anomalen Zeemaneffekte zeigt zunächst vom Standpunkt der hier versuchten Theorie aus kaum ein anderes Aussehen als bisher. Der in den Grundvoraussetzungen dieser Theorie enthaltene innige Zusammenhang der „aperiodischen“ und der „periodischen Bahnen“ bringt zwar mit sich, daß wir nicht sicher sein können, daß das Larmorsche Theorem allgemein gilt (Kap. 4, § 2); die Voraussetzungen für die Gültigkeit dieses Theorems sind beim Oszillator, nicht aber ohne weiteres beim Kernatom erfüllt. Doch ist es nicht wahrscheinlich, daß dieser Gesichtspunkt zu einer Deutung der anomalen Zeemaneffekte führen

[1]) N. Bohr, ZS. f. Phys. **34**, 142, 1925.

kann; vielmehr dürfte die Quantenmechanik bei den Zeemaneffekten mit denselben Schwierigkeiten zu kämpfen haben wie die bisherige Theorie; das Problem der anomalen Zeemaneffekte ist aber neuerdings durch eine Note von Uhlenbeck und Goudsmit[1]) in ein neues Stadium getreten. Diese Verfasser machen die Annahme, daß das Elektron selbst ein mechanisches und magnetisches Moment besitze (deren Verhältnis doppelt so groß sein solle, wie bei Atomen), daß es also eigentlich gar keine anomalen Zeemaneffekte gebe; durch diese Annahme fallen die Schwierigkeiten bei den statistischen Gewichten fort und es ergibt sich eine qualitative Deutung sämtlicher mit dem Problem der Multiplettstruktur und der Zeemaneffekte zusammenhängenden Phänomene; die Frage, ob man dadurch schon eine quantitative Deutung dieser Erscheinungen erhält, kann allerdings erst durch genauere Untersuchungen mit den Methoden der Quantenmechanik beantwortet werden; einige Resultate des Kap. 4 scheinen bezüglich der Zeemaneffekte diese Hoffnung auf die Möglichkeit einer späteren quantitativen Deutung zu bestärken.

Schließlich haben wir noch versucht, ein bekanntes statistisches Problem mit den durch die Theorie gegebenen neuen Methoden zu behandeln: Bekanntlich kann man durch Quantelung der Eigenschwingungen eines (in spiegelnde Wände eingeschlossenen) Hohlraums nach den bisherigen Methoden zu Ergebnissen kommen, die eine gewisse Ähnlichkeit mit den Ansätzen der Lichtquantentheorie aufweisen und die eine Ableitung der Planckschen Formel gestatten. Man erhält aber, wie stets von Einstein[2]) hervorgehoben wurde, bei einer solchen halbklassischen Behandlung der Hohlraumstrahlung einen falschen Wert für das mittlere Schwankungsquadrat der Energie in einem Teilvolumen. Dieses Ergebnis muß als besonders schwerwiegender Einwand gegen die bisherigen Methoden der Quantentheorie angesehen werden, weil es sich einerseits hier um ein Versagen der Theorie schon beim einfachen Problem des harmonischen Oszillators handelt, und weil andererseits die genannte Schwierigkeit auch bei jeder Statistik der Eigenschwingungen irgend eines mechanischen Systems, z. B. eines Kristallgitters, auftreten würde. Wir haben nun gefunden, daß die auf Grund der Kinematik und Mechanik der hier versuchten Theorie durchgeführte entsprechende Rechnung zum richtigen Werte des Schwankungsquadrates, wie zur Planckschen Formel führt, was wohl als wichtige Stütze für die hier versuchte Quantenmechanik anzusehen ist.

---

[1]) G. Uhlenbeck und S. Goudsmit, Naturwiss. **13**, 953, 1925.

[2]) A. Einstein, Phys. ZS. **10**, 185, 817, 1909.

Kapitel 1. Systeme von einem Freiheitsgrad.

§ 1. Grundprinzipien. I. Eine quantentheoretische Größe a — sei es Koordinate oder Impuls oder irgend eine Funktion beider — wird repräsentiert durch die Gesamtheit der Größen

$$a(nm)\, e^{2\pi i \nu(nm) t} \tag{1}$$

oder auch unter Weglassung des für alle zum System gehörigen Größen gleichen (nur von den Indizes n und m abhängigen) Faktors $e^{2\pi i \nu(nm)t}$ durch die Gesamtheit der Zahlen

$$a(nm). \tag{2}$$

Wir können also von einer (übrigens unendlichen) „Matrix“ a sprechen.

II. Die Rechenoperationen, wie Addition, Multiplikation der quantentheoretischen Größen sind entsprechend den für Matrizen gültigen Rechenregeln definiert.

III. Gegeben sei eine durch Additionen und Multiplikationen von Matrizen definierte Funktion $\boldsymbol{f}(\boldsymbol{x}_1, \boldsymbol{x}_2, \ldots \boldsymbol{x}_s)$, wo die $\boldsymbol{x}_1, \boldsymbol{x}_2, \ldots \boldsymbol{x}_s$ quantentheoretische Größen bedeuten. Dann führen wir zwei Arten von Differentialquotienten der Funktion $\boldsymbol{f}$ nach einer der Größen $\boldsymbol{x}$ (etwa $\boldsymbol{x}_1$) ein:

a) Differentialquotient erster Art:

$$\frac{\partial \boldsymbol{f}}{\partial \boldsymbol{x}_1} = \lim_{\alpha \to 0} \frac{\boldsymbol{f}(\boldsymbol{x}_1 + \alpha \mathbf{1}, \boldsymbol{x}_2, \ldots, \boldsymbol{x}_s) - \boldsymbol{f}(\boldsymbol{x}_1, \boldsymbol{x}_2, \ldots, \boldsymbol{x}_s)}{\alpha}, \tag{3}$$

wo α eine Zahl und $\mathbf{1}$ die durch

$$\mathbf{1} = (\delta_{nm}), \quad \delta_{nm} = \begin{cases} 1 \text{ für } n = m \\ 0 \quad \text{„} \quad n \neq m \end{cases}$$

definierte Einheitsmatrix ist.

b) Differentialquotient zweiter Art, definiert durch [1])

$$\frac{\partial \boldsymbol{f}}{\partial \boldsymbol{x}_1}(nm) = \frac{\partial D(\boldsymbol{f})}{\partial x_1(mn)}, \tag{4}$$

wo $D(\boldsymbol{f})$ die Diagonalsumme der Matrix $\boldsymbol{f}$ bedeutet.

Äußerlich werden wir die beiden Arten von Differentiationen durch den Bruchstrich [dicker Bruchstrich für a), dünner für b)] unterscheiden.

Die Differentiation zweiter Art wurde in Teil I ausschließlich benutzt, da sie eine einfache Formulierung des Variationsprinzips der Quantenmechanik ermöglicht und daher als naturgemäß erscheint. Für manche Rechnungen ist jedoch der Differentialquotient erster Art bequemer zu handhaben. Allgemein mag bemerkt werden, daß die Einführung eines Differentialquotienten in der Quantenmechanik etwas

[1]) Vgl. Teil I.

künstlich ist und daß die Operationen der linken Seite von (6) das naturgemäße Analogon zum Differentialquotienten der klassischen Theorie darstellen. Es ist für die Formulierung der kanonischen Gleichungen wichtig, festzustellen, daß für die Energiefunktion $H(pq)$ die beiden Arten (3) und (4) der Differentiation identisch werden[1]).

IV. Das Rechnen mit den quantentheoretischen Größen würde wegen der Nichtgültigkeit des kommutativen Gesetzes der Multiplikation in gewissem Sinne unbestimmt bleiben, wenn nicht der Wert von $pq - qp$ vorgeschrieben würde[2]). Wir führen daher als fundamentale quantenmechanische Relation ein:

$$pq - qp = \frac{h}{2\pi i} \mathbf{1}. \tag{5}$$

Auf die korrespondenzmäßig-physikalische Bedeutung dieser Relation werden wir später zu sprechen kommen. An dieser Stelle scheint es uns wichtig, hervorzuheben, daß Gleichung (5), Kap. 1, die einzige unter den Grundgleichungen der hier versuchten Quantenmechanik ist, in welchen die Planck sche Konstante h vorkommt. Es ist befriedigend, daß die Konstante h an dieser Stelle schon in so einfacher Form in die Grundlagen der Theorie eingeht; außerdem erkennt man aus (5), Kap. 1, daß die neue Theorie im Limes $h = 0$ in die klassische übergehen dürfte, wie es physikalisch gefordert werden muß.

---

[1]) In der Tat wurden ja für die Energiefunktion H in Teil I nicht irgendwelche Funktionen etwa der Form

$$H^* = \sum a_{sr} p^s q^r$$

zugelassen, sondern durch symmetrisierte Funktionen ersetzt, die zu denselben Hamiltonschen Gleichungen Anlaß gaben:

$$H = \sum a_{sr} \frac{1}{s+1} \sum_{l=0}^{s} p^{s-l} q^r p^l.$$

Für diese symmetrisierten Funktionen H aber gilt nach den in Teil I abgeleiteten Formeln

$$\frac{\partial H}{\partial p} = \sum a_{sr} \frac{1}{s+1} \left\{ \sum_{l=0}^{s-1} (s-l) p^{s-1-l} q^r p^l + \sum_{l=1}^{s} l p^{s-l} q^r p^{l-1} \right\}$$

$$= \sum a_{sr} \sum_{l=0}^{s-1} p^{s-1-l} q^r p^l = \frac{\partial H}{\partial p}.$$

$$\frac{\partial H}{\partial q} = \sum a_{sr} \frac{r}{s+1} \sum_{l=0}^{s} p^{s-l} q^{r-1} p^l = \sum a_{sr} \sum_{j=0}^{r-1} q^{r-1-j} p^s q^j = \frac{\partial H}{\partial q}.$$

[2]) Die Bewegungsgleichungen lassen lediglich erkennen, daß diese Differenz eine Diagonalmatrix sein muß.

Aus der Gleichung (5), Kap. 1 kann noch eine später wichtige Beziehung abgeleitet werden:

Sei $f(pq)$ irgend eine Funktion von p und q, so gilt:

$$\left.\begin{aligned} fq - qf &= \frac{\partial f}{\partial p}\frac{h}{2\pi i}, \\ pf - fp &= \frac{\partial f}{\partial q}\frac{h}{2\pi i}. \end{aligned}\right\} \tag{6}$$

Denn nehmen wir einmal an, diese Gleichungen seien richtig für irgend zwei Funktionen φ und ψ, dann sind sie auch richtig für $\varphi + \psi$ und $\varphi \cdot \psi$. Für $\varphi + \psi$ ist dies trivial, für $\varphi \cdot \psi$ ergibt eine leichte Rechnung:

$$\begin{aligned} \varphi \cdot \psi q - q \varphi \psi &= \varphi(\psi q - q \psi) + (\varphi q - q \varphi)\psi \\ &= \left(\varphi \frac{\partial \psi}{\partial p} + \frac{\partial \varphi}{\partial p}\psi\right)\frac{h}{2\pi i} = \frac{\partial(\varphi\psi)}{\partial p}\frac{h}{2\pi i}; \end{aligned}$$

analog für $p\varphi\psi - \varphi\psi p$.

Nun gilt die Relation (6) für p und q, also auch für jede Funktion f, die formal nach Potenzen von p und q entwickelbar ist.

§ 2. Die kanonischen Gleichungen, Energiesatz und Frequenzbedingung Seien jetzt eine Energiefunktion $H(pq)$ und die zugehörigen kanonischen Gleichungen

$$p = -\frac{\partial H}{\partial q}; \quad q = \frac{\partial H}{\partial p} \tag{7}$$

gegeben. Aus dem Kombinationsprinzip für

$$\nu(nm) + \nu(mk) = \nu(nk) \tag{8}$$

folgt, daß ν dargestellt werden kann in der Form

$$\nu(nm) = \frac{W_n - W_m}{h}. \tag{9}$$

Wir führen jetzt eine quantentheoretische Größe W ein als „Termgröße", definiert durch

$$W(nm) = \begin{cases} W_n \text{ für } n = m \\ 0 \quad „ \quad n \neq m. \end{cases}$$

W ist also eine Diagonalmatrix.

Dann gilt für irgend eine quantentheoretische Größe:

$$\dot{a} = \frac{2\pi i}{h}(Wa - aW). \tag{10}$$

In der Tat war ja (vgl. Teil 1) $\dot{a}$ definiert durch

$$a(nm) = 2\pi i \nu(nm) a(nm).$$

Zu den Grundpfeilern der Theorie, die wir hier aufzubauen suchen, gehört der Energiesatz ($H =$ const) und die Frequenzbedingung $\left(\nu(nm) = \frac{H_n - H_m}{h};\ H_n = W_n + \text{const}\right)$.

Den Beweis für diese beiden Sätze führen wir, indem wir die Beziehungen (6) und (10) in Gl. (7) (Kap. I) einsetzen.

Dann ergibt sich:

$$\left.\begin{aligned} Wq - qW = Hq - qH \\ Wp - pW = Hp - pH \end{aligned}\right\} \tag{11}$$

oder auch

$$(W - H)q - q(W - H) = 0,$$
$$(W - H)p - p(W - H) = 0.$$

Die Größe $W - H$ ist also mit p und q, daher auch mit jeder Funktion von p, q, insbesondere auch mit H vertauschbar:

$$(W - H)H - H(W - H) = 0.$$

Daraus folgt nach (10), Kap. 1:

$$\dot{H} = 0. \tag{12}$$

Damit ist der Energiesatz bewiesen, H ist als Diagonalmatrix $H(nm) = \delta_{nm} H_n$ erkannt. Aus (11), Kap. 1, folgt nun unmittelbar auch die Frequenzbedingung:

$$q(nm)(H_n - H_m) = q(nm)(W_n - W_m), \tag{13}$$

d. h.

$$\frac{H_n - H_m}{h} = \nu(nm). \tag{14}$$

Wir haben bisher aus den kanonischen Gleichungen mit Hilfe der Grundgleichung (5), Kap. 1, Energiesatz und Frequenzbedingung bewiesen. Nachträglich aber können wir den Beweis auch umkehren. Wir wissen, daß Energiesatz und Frequenzbedingung richtig sind. Wenn also die Energiefunktion H als analytische Funktion irgendwelcher Variablen P, Q gegeben ist, so gelten immer dann, wenn

$$PQ - QP = \frac{h}{2\pi i}\mathbf{1}$$

ist, die kanonischen Gleichungen:

$$\dot{Q} = \frac{\partial H}{\partial P}, \quad \dot{P} = -\frac{\partial H}{\partial Q}. \tag{15}$$

Dies folgt unmittelbar daraus, daß die Größen $PH - HP$ bzw. $HQ - QH$ in doppelter Weise, nämlich entsprechend (6), Kap. 1, und entsprechend Gleichung (10), Kap. 1, interpretiert werden können.

§ 3. Kanonische Transformationen. Unter einer „kanonischen Transformation“ der Variablen p, q in neue Variable P, Q wird man nach dem Vorhergehenden eine Transformation verstehen, bei welcher

$$pq - qp = PQ - QP = \frac{h}{2\pi i} \tag{16}$$

ist; denn dann gelten für P, Q wie für p, q die kanonischen Gleichungen (7), Kap. 1, bzw. (15), Kap. 1.

Eine allgemeine Transformation, die dieser Bedingung genügt, heißt

$$\left.\begin{aligned} P &= SpS^{-1} \\ Q &= SqS^{-1} \end{aligned}\right\}, \tag{17}$$

wo S eine beliebige quantentheoretische Größe bedeutet; wir möchten vermuten, daß (17), Kap. 1, sogar die allgemeinste kanonische Transformation darstellt. Die Transformation (17), Kap. 1, hat noch die einfache Eigenschaft, daß für irgend eine Funktion $f(P, Q)$ gilt:

$$f(P, Q) = Sf(p, q)S^{-1}, \tag{18}$$

wobei $f(p, q)$ aus $f(P, Q)$ dadurch hervorgeht, daß P durch p, Q durch q unter Beibehaltung der Funktionsform ersetzt wird. Der Beweis dieser Behauptung für Funktionen im Sinne unserer Definition folgt unmittelbar aus der Bemerkung, daß der Satz für Summe und Produkt mit Summanden bzw. Faktoren p, q gilt.

Die Wichtigkeit der kanonischen Transformation beruht auf folgendem Satze: Wenn irgend ein Wertepaar p_0, q_0 gegeben ist, das der Gleichung (15), Kap. 1, genügt, so kann man das Problem der Integration der kanonischen Gleichungen für eine Energiefunktion $H(pq)$ reduzieren auf das folgende Problem: Es ist eine Funktion S so zu bestimmen, daß mit

$$p = Sp_0S^{-1}, \quad q = Sq_0S^{-1} \tag{19}$$

die Funktion

$$H(pq) = SH(p_0q_0)S^{-1} = W \tag{20}$$

eine Diagonalmatrix wird. Gleichung (20), Kap. 1, ist das Analogon zur Hamiltonschen partiellen Differentialgleichung; S entspricht in gewisser Weise der Wirkungsfunktion.

§ 4. Störungstheorie. Es sei vorgelegt das durch die Energiefunktion definierte mechanische Problem:

$$H = H_0(pq) + \lambda H_1(pq) + \lambda^2 H_2(pq) + \cdots \tag{21}$$

Wir nehmen das durch die Energiefunktion $H_0(pq)$ definierte mechanische Problem als gelöst an; es seien also Lösungen p_0, q_0 bekannt, die der Bedingung $p_0q_0 - q_0p_0 = \frac{h}{2\pi i}\mathbf{1}$ genügen und

$H_0(p_0 q_0) = W_0$ zu einer Diagonalmatrix machen. Dann suchen wir eine Transformationsfunktion S so zu bestimmen, daß

$$p = S p_0 S^{-1}, \quad q = S q_0 S^{-1}, \tag{22}$$

und daß

$$H(pq) = S H(p_0 q_0) S^{-1} = W,$$

d. h. gleich einer Diagonalmatrix wird. Wir versuchen zur Lösung den Ansatz:

$$S = 1 + \lambda S_1 + \lambda^2 S_2 + \cdots \tag{23}$$

Dann ist

$$S^{-1} = 1 - \lambda S_1 + \lambda^2 (S_1^2 - S_2) + \lambda^3 \ldots \tag{24}$$

Nehmen wir für H den Ausdruck (21), Kap. 1, so können wir nach Potenzen von λ ordnen und erhalten die Näherungsgleichungen:

$$\left.\begin{aligned} H_0(p_0 q_0) &= W_0 \\ S_1 H_0 - H_0 S_1 + H_1 &= W_1 \\ S_2 H_0 - H_0 S_2 + H_0 S_1^2 - S_1 H_0 S_1 + S_1 H_1 - H_1 S_1 + H_2 &= W_2 \\ \cdots\cdots\cdots\cdots\cdots\cdots\cdots\cdots \\ S_r H_0 - H_0 S_r + F_r(H_0, \ldots, H_r, S_0, \ldots, S_{r-1}) &= W_r \end{aligned}\right\} \tag{25}$$

wobei $H_0, H_1, \ldots$ stets mit den Argumenten p_0, q_0 zu nehmen sind.

Die erste Gleichung (25), Kap. 1, ist erfüllt. Die übrigen lassen sich der Reihe nach auflösen, und zwar in ganz analoger Weise wie in der klassischen Theorie: Man bildet erst den Mittelwert zur Festlegung der Energiekonstante und kann dann ohne weiteres die Lösung hinschreiben:

$$\left.\begin{aligned} W_r &= \overline{F_r}, \\ S_r(mn) &= \frac{F_r(mn)}{h\nu_0(mn)}(1 - \delta_{nm}), \end{aligned}\right\} \tag{26}$$

wobei $\nu_0(nm)$ die Frequenzen der ungestörten Bewegung sind.
Diese Lösung genügt der Bedingung

$$S \cdot \tilde{S}^* = 1, \tag{27}$$

wo der Cirkumflex die Vertauschung von Zeilen und Spalten (Transposition) bedeutet und der Stern Übergang zur konjugierten komplexen Größe. Da wir auf diese Relation später von einem allgemeineren Standpunkt aus zurückkommen werden, wollen wir sie hier nur für die erste Näherung bestätigen, die wir sogleich ausrechnen werden; für diese lautet sie

$$S_1 + \tilde{S}_1^* = 0. \tag{28}$$

Die Bedeutung der Gleichung (8), Kap. 1, beruht darauf, daß aus ihr der Hermitesche Charakter der Matrizen $\boldsymbol{p}$, $\boldsymbol{q}$ folgt; denn nach (22), Kap. 1, wird [1])

$$\boldsymbol{q}^* = \boldsymbol{S}^* \boldsymbol{q}_0^* \boldsymbol{S}^{*-1} = \tilde{\boldsymbol{S}}^{-1} \tilde{\boldsymbol{q}}_0 \tilde{\boldsymbol{S}} = \tilde{\boldsymbol{q}},$$

und analog für $\boldsymbol{p}$.

In erster Näherung folgt aus (26), Kap. 1, wie in der klassischen Theorie:

$$\boldsymbol{W}_1 = \overline{\boldsymbol{H}}_1; \tag{29}$$

sodann

$$S_1(mn) = \frac{H_1(mn)}{h\nu_0(mn)}(1 - \delta_{mn}). \tag{30}$$

Dieser Ausdruck erfüllt in der Tat die Bedingung (28), Kap. 1, wegen der Voraussetzung, daß H_1 eine Hermitesche Form ist. Nunmehr kann man die Energie in zweiter Näherung berechnen und findet:

$$W_2 = \overline{H}_2 + \frac{1}{h} \sum_l{}' \frac{H_1(nl)\, H_1(ln)}{\nu_0(nl)}, \tag{31}$$

wo der Akzent am Summenzeichen bedeutet, daß die Glieder mit verschwindendem Nenner ($l = n$) fortzulassen sind.

In dieser Weise kann man fortfahren und sukzessive alle Glieder der Reihen $\boldsymbol{W}$ und $\boldsymbol{S}$ bestimmen. Setzt man die Reihe für $\boldsymbol{S}$ in (22), Kap. 1, ein, so erhält man die Entwicklungen

$$\boldsymbol{q} = \boldsymbol{q}_0 + \lambda \boldsymbol{q}_1 + \lambda^2 \boldsymbol{q}_2 + \cdots,$$
$$\boldsymbol{p} = \boldsymbol{p}_0 + \lambda \boldsymbol{p}_1 + \lambda^2 \boldsymbol{p}_1 + \cdots$$

mit bekannten Koeffizienten. So lautet z. B. die erste Näherung

$$\boldsymbol{q}_1 = \boldsymbol{S}_1 \boldsymbol{q}_0 - \boldsymbol{q}_0 \boldsymbol{S}_1,$$
$$\boldsymbol{p}_1 = \boldsymbol{S}_1 \boldsymbol{p}_0 - \boldsymbol{p}_0 \boldsymbol{S}_1;$$

oder ausführlich:

$$\left.\begin{aligned} q_1(mn) &= \frac{1}{h} \sum_k{}' \left(\frac{H_1(mk)\, q_0(kn)}{\nu_0(mk)} - \frac{q_0(mk)\, H_1(kn)}{\nu_0(kn)} \right) \\ p_1(mn) &= \frac{1}{h} \sum_k{}' \left(\frac{H_1(mk)\, q_0(kn)}{\nu_0(mk)} - \frac{q_0(mk)\, H_1(kn)}{\nu_0(kn)} \right) \end{aligned}\right\}. \tag{32}$$

Die Formeln (32), Kap. 1, bedeuten die Ergebnisse der Kramersschen Dispersionstheorie [2]) im Grenzfall unendlich kleiner Frequenz des äußeren Feldes; diese Möglichkeit einer einfachen Ableitung der sonst auf Grund von Korrespondenzbetrachtungen gewonnenen Formeln scheint uns sehr

---

[1]) Man beachte die Rechenregel $\widetilde{(\boldsymbol{a}\boldsymbol{b})} = \tilde{\boldsymbol{b}}\tilde{\boldsymbol{a}}$.

[2]) H. A. Kramers, Nature **113**, 673, 1924; **114**, 310, 1924. Vgl. auch R. Ladenburg, ZS. f. Phys. **4**, 451, 1921; R. Ladenburg und F. Reiche, Naturwiss. **11**, 584, 1923.

zugunsten der hier versuchten Theorie zu sprechen. Gleichung (31), Kap. 1, wurde von Born[1]) durch Umdeutung der entsprechenden klassischen Formeln erhalten. Die Glieder $m = n$ der Gleichung (32), Kap. 1, entsprechen der Kramersschen Formel für das gewöhnliche Dispersionslicht, die anderen Glieder $m \neq n$ entsprechen den von Kramers und Heisenberg[2]) angegebenen Formeln für das „Streulicht der Kombinationsfrequenzen". Die letzteren Ausdrücke wurden von Pauli[3]) zur Berechnung der Intensitäten der in äußeren elektrischen Feldern auftretenden (sonst „verbotenen") Übergänge bei Hg benutzt. Für die Ableitung der allgemeinen Dispersionsformeln (wenn die Frequenz des äußeren Feldes nicht verschwindet) sind noch allgemeinere Betrachtungen über die Wirkung zeitlich veränderlicher äußerer Kräfte notwendig, zu denen wir jetzt übergehen werden.

§ 5. Systeme, bei denen die Zeit explizite in der „Energiefunktion" vorkommt. Die Behandlung der quantenmechanischen Wirkung äußerer Kräfte, die explizite von der Zeit abhängen, scheint uns deshalb von besonderem Interesse, weil bei diesem Problem einige Unterschiede zwischen der quantentheoretischen und der klassischen Mechanik charakteristisch zu Tage kommen. Das Problem der Wirkung zeitlich veränderlicher äußerer Kräfte ist aufzufassen als Grenzfall des Problems der Wechselwirkung zweier Systeme, wobei der Einfluß der Wechselwirkung auf das eine System (es heiße A) so gering ist, daß die Wirkungen auf das andere System (B) durch diesen Einfluß nicht verändert werden. Betrachten wir also jetzt vom Standpunkt der Quantenmechanik die Kopplung zweier Systeme A, B; die Hamiltonsche Funktion zerfalle in drei Teile H_A, λH_B und $\varepsilon \lambda H_{AB}$ (λ sei ein zunächst willkürlicher Parameter, ε eine kleine Größe). Das System A sei bekannt. Zur Berechnung der Bewegungen von B genügt es in der klassischen Theorie, wenn man für die Koordinaten von B die Bewegungsgleichungen [aus der Hamiltonschen Funktion $\lambda\,(H_B + \varepsilon H_{AB})$] aufstellt, wobei man für die Koordinaten von A ihre Lösungen als Funktion der Zeit (für die bestimmten vorliegenden Werte der Konstanten in A) einsetzt. Hierdurch tritt eben bei Vernachlässigung der Rückwirkung neben der Konstanten von A nur die Zeit als neue Variable im Störungsproblem für B auf. In der Quantenmechanik liegen die Verhältnisse ebenso, wenn wir uns auf die Störungen erster Ordnung (d. h. die mit ε proportionalen Glieder

[1]) M. Born, ZS. f. Phys. **26**, 379, 1924.
[2]) H. A. Kramers und W. Heisenberg, ZS. f. Phys. **31**, 681, 1925,
[3]) W. Pauli, Verhandl. d. dän. Akad. d. Wiss. (Im Erscheinen.)

in den Koordinaten, Impulsen usw. von B) beschränken. Anders aber ist es bei den Störungen höherer Ordnung. Denn bei der Berechnung der Störungen höherer Ordnung kommen Produktbildungen vor aus Größen, von denen mehr als eine die Koordinaten von A implizite enthält. Dies bedeutet aber nach den Regeln der quantenmechanischen Produktbildung, daß es keineswegs genügt, die „äußeren Kräfte als Funktionen der Zeit" für die bestimmten Werte der Konstanten in A zu kennen, sondern diese äußeren Kräfte müssen uns für alle Werte der Konstanten bekannt sein. Damit scheint aber der Begriff der äußeren Kräfte eigentlich den Sinn zu verlieren. Die Auflösung dieser Schwierigkeit scheint uns in der Bemerkung zu liegen, daß die Rückwirkung selbst zu Gliedern der Ordnung $\lambda\varepsilon^2$ in den Koordinaten von B Anlaß gibt, daß also eine gleichzeitige Vernachlässigung der Rückwirkung und Berechnung der Glieder mit ε^2 in B nur dann einen Sinn hat, wenn auch λ als sehr klein betrachtet werden kann, d. h. physikalisch, wenn eine Abänderung der Größen in A um Beträge von der Ordnung der entsprechenden Größen in B keine merkliche Änderung der Wirkung von A auf B herbeiführen würde. In dieser Annäherung aber läßt sich auch die quantenmechanische Produktbildung und daher die Berechnung der Störungen höherer Ordnung in ε wieder durchführen, und zwar reduzieren sich die Regeln dieser Produktbildung einfach auf die der klassischen Multiplikation, da ja die in H_{AB} eingehenden Koordinaten, Amplituden und Frequenzen in dieser Näherung nicht von den Konstanten in A abhängen. In diesem Sinne könnte z. B. die Wirkung eines starken elektromagnetischen Wechselfeldes auf ein Atom durchaus als Wirkung einer „äußeren Kraft" unter Vernachlässigung der Rückwirkung berechnet werden, da die Energie des Feldes im Vergleich zu der des Atoms als unendlich angesehen werden kann. Auch die Wirkung einer α-Partikel auf die Elektronen eines Atoms könnte wegen der relativ großen Energie der α-Partikel als „äußere Kraft", wie in der klassischen Theorie aufgefaßt werden und auch die Fourierentwicklung der dabei auf die Elektronen wirkenden Kraft nach der Zeit wäre in dieser Näherung die klassische. Aber die Wirkung der Kräfte eines einzelnen Atoms auf ein anderes kann niemals als Wirkung äußerer Kräfte gedeutet werden — d. h. nur in den Gliedern erster Ordnung, in denen eine solche Deutung stets möglich ist —, da für die Glieder höherer Ordnung die Vernachlässigung der Rückwirkung zu falschen Resultaten führen würde.

Das Ergebnis unserer Überlegungen fassen wir dahin zusammen: Unter gewissen Voraussetzungen hat es, wie in der klassischen Theorie, einen Sinn, von der Wirkung bestimmter zeitlich veränderlicher Kräfte auf das Atom zu sprechen. In diesem Falle gelten für das Rechnen mit der explizite auftretenden Zeit die Rechenregeln der klassischen Theorie: Sei z. B. das äußere Kraftfeld periodisch mit der Periode ν_0, so lautet das allgemeine Glied einer Koordinate q:

$$q(mn, \tau) \, . \, e^{2\pi i [\nu(mn) + \tau\nu_0] t}, \tag{33}$$

das allgemeine Glied von q^2:

$$\sum_{k, \tau'} q(mk, \tau - \tau') q(kn, \tau') e^{2\pi i [\nu(mn) + \tau\nu_0] t}. \tag{34}$$

Der Fall der zeitlich veränderlichen äußeren Kräfte scheint uns deshalb ein bemerkenswertes Beispiel für den korrespondenzmäßigen Übergang der quantentheoretischen Kinematik in die klassische.

Wenn es sich nur um die Berechnung der Wirkungen erster Ordnung der äußeren Kräfte handelt, so bleiben die durch die folgenden Rechnungen zu gewinnenden Resultate auch richtig, wenn die am Anfang genannten Voraussetzungen nicht erfüllt sind — in genauer Analogie zur klassischen Theorie.

Nach den vorausgehenden Überlegungen läßt sich die mathematische Behandlung der Systeme, in denen (bei Gültigkeit der erwähnten Voraussetzungen) die Zeit explizite auftritt, einfach in Analogie zu den entsprechenden klassischen Methoden durchführen. Nehmen wir wieder an, die äußere Kraft sei periodisch in der Zeit mit der Periode ν_0, die Hamiltonsche Funktion sei[1])

$$H = H(p_k q_k, \cos 2\pi\nu_0 t). \tag{35}$$

Dann führen wir einen neuen Freiheitsgrad mit den Variabeln q', p' ein und nehmen als Hamiltonsche Funktion des neuen Problems, in dem die Zeit nicht mehr explizite vorkommt:

$$H' = H(p_k, q_k; q') + 2\pi\nu_0 \sqrt{1 - q'^2}\, p'. \tag{36}$$

Die kanonischen Gleichungen für p_k, q_k bleiben dabei die bisherigen, nur steht für $\cos 2\pi\nu_0 t$ stets q'. Die neu dazu kommenden Gleichungen heißen:

$$\left.\begin{aligned} \dot{q}' &= \frac{\partial H'}{\partial p'} = 2\pi\nu_0 \sqrt{1 - q'^2}, \\ \dot{p}' &= -\frac{\partial H'}{\partial q'} = -\frac{\partial H}{\partial q'} + 2\pi\nu_0 \frac{q'}{\sqrt{1 - q'^2}} \cdot p'. \end{aligned}\right\} \tag{37}$$

1) Wir nehmen hier einen Augenblick die Ergebnisse des nächsten Kapitels über Systeme mit mehreren Freiheitsgraden vorweg.

Die erste Gleichung sagt aus, daß q' wirklich (bis auf willkürliche Wahl des Anfangspunktes der Zeit) gleich $\cos 2\pi\nu_0 t$ wird, so daß die kanonischen Gleichungen für p_k, q_k dieselbe Form haben, wie beim früheren Problem; die zweite Gleichung (37), Kap. 1, gibt eine Bestimmung von p'. Durch (36), Kap. 1, ist also das Problem (35), Kap. 1, wirklich auf die sonst behandelten Fälle zurückgeführt.

Uns interessiert vor allem die Frage, welche Veränderungen wir an den Störungsformeln (25), Kap. 1, vorzunehmen haben, wenn die Zeit explizite in H_1, H_2, ..., nicht aber in H_0 auftritt. Eine einfache Überlegung zeigt, daß die Störungsformeln für unseren Fall aus den früheren dadurch hervorgehen, daß überall, wo früher ein Glied der Form $H_0 S_r - S_r H_0$ auftrat, jetzt $H_0 S_r - S_r H_0 + \frac{h}{2\pi i}\frac{\partial S_r}{\partial t}$ gesetzt wird. (H_0 kommt nur in solchen Verbindungen vor). Also heißen die niedrigsten Ordnungen der neuen Störungsformeln:

$$\left.\begin{aligned} H_0(p_0 q_0) &= W_0,\\ S_1 H_0 - H_0 S_1 - \frac{h}{2\pi i}\frac{\partial S_1}{\partial t} + H_1 &= W_1,\\ S_2 H_0 - H_0 S_2 - \frac{h}{2\pi i}\frac{\partial S_2}{\partial t} + \left(H_0 S_1 - S_1 H_0 + \frac{h}{2\pi i}\frac{\partial S_1}{\partial t}\right) S_1 &\\ + S_1 H_1 - H_1 S_1 + H_2 &= W_2,\\ \cdots\cdots\cdots\cdots\cdots\cdots& \end{aligned}\right\} \quad (38)$$

Wir möchten vermuten, daß diese Formeln (38), Kap. 1, auch dann gelten, wenn die Annahme, daß die äußeren Kräfte periodisch in der Zeit seien, nicht zutrifft — obwohl wir diese Annahme bei der Ableitung der Formeln benutzt haben.

Die Gleichungen erster Ordnung der Formeln (38), Kap. 1, die ja auch noch richtig bleiben, wenn die Voraussetzungen der „äußeren Kräfte" nicht mehr erfüllt sind, beantworten zusammen mit (22), Kap. 1:

$$q = q_0 + \lambda (S_1 q_0 - q_0 S_1),$$
$$p = p_0 + \lambda (S_1 p_0 - p_0 S_1),$$

die Fragen der Dispersionstheorie in einem allgemeinen Sinne. In der Tat, setzen wir:

$$H_1 = E \, . \, e q_0 \cos 2\pi\nu_0 t,$$

so folgt

$$\left.\begin{aligned} H_1(mn,\ 1) &= \frac{Ee}{2} q_0(mn), & H_1(mn,\ -1) &= \frac{Ee}{2} q_0(mn),\\ S_1(mn,\ 1) &= \frac{Ee}{2h}\frac{q_0(mn)}{\nu_0(mn)+\nu_0}, & S_1(mn,\ -1) &= \frac{Ee}{2h}\frac{q_0(mn)}{\nu_0(mn)-\nu_0}. \end{aligned}\right\} \quad (39)$$

Daraus folgt [vgl. (22), Kap. 1]

$$q_1(mn, +1) = \frac{Ee}{2h} \sum_k \left(\frac{q_0(mk)\, q_0(kn)}{\nu_0(mk) + \nu_0} - \frac{q_0(mk)\, q_0(kn)}{\nu_0(kn) + \nu_0} \right). \tag{40}$$

Wenn wir noch annehmen, daß wir kartesische Koordinaten, d. h. $p = m\dot{q}$ haben:

$$q_1(mn, 1) = \frac{Ee}{2h \cdot 2\pi i m} \sum_k \frac{q_0(mk)\, p_0(kn) - p_0(mk)\, q_0(kn)}{(\nu_0(mk) + \nu_0)(\nu_0(kn) + \nu_0)}; \tag{41}$$

ebenso

$$q_1(mn, -1) = \frac{Ee}{2h \cdot 2\pi i m} \sum_k \frac{q_0(mk)\, p_0(kn) - p_0(mk)\, q_0(kn)}{(\nu_0(mk) - \nu_0)(\nu_0(kn) - \nu_0)}. \tag{42}$$

Die Gleichungen (40), (41), (42), Kap. 1, stimmen überein mit den Formeln der Kramersschen Dispersionstheorie [1]). Besonders interessant scheint noch der Fall sehr hoher Frequenz des einfallenden Lichtes $|\nu_0| \gg |\nu_0(mk)|$ bzw. $|\nu_0(kn)|$. Dann erhält man in erster Näherung:

$$\mathbf{q}_1 = -\frac{Ee}{h\, 2\pi i \nu_0^2 m} (\mathbf{p}_0 \mathbf{q}_0 - \mathbf{q}_0 \mathbf{p}_0) \cos 2\pi \nu_0 t$$

oder wegen (5), Kap. 1:

$$\mathbf{q}_1 = +\frac{Ee}{4\pi^2 m \nu_0^2} \cos 2\pi \nu_0 t. \tag{43}$$

Dieses Ergebnis bedeutet, daß eben die quantenmechanische Vertauschungsrelation (5), Kap. 1, schließlich verursacht, daß für hinreichend hohe Frequenzen das Elektron sich bei der Streuung wie ein freies Elektron verhält. Das Streulicht der Frequenzen $\nu_0(mn) + \nu_0\ (m \neq n)$ verschwindet, das Streulicht der Frequenz ν_0 hat die bei Streuung durch ein freies Elektron zu erwartende Intensität [2]).

Kapitel 2. Grundlagen der Theorie der Systeme von beliebig vielen Freiheitsgraden.

§ 1. **Die kanonischen Bewegungsgleichungen; Störungstheorie bei nichtentarteten Systemen.** Bei $f > 1$ Freiheitsgraden liegt es nahe, statt einer Darstellung der quantentheoretischen Größen durch zweidimensionale Matrizen eine solche durch $2f$-dimensionale Matrizen zu wählen, entsprechend der $2f$-dimensionalen Mannigfaltigkeit der stationären Zustände im klassischen J-Raum:

$$\left.\begin{aligned} \mathbf{q}_k &= (q_k(n_1 \ldots n_f,\ m_1 \ldots m_f)), \\ \mathbf{p}_k &= (p_k(n_1 \ldots n_f,\ m_1 \ldots m_f)). \end{aligned}\right\} \tag{1}$$

[1]) Vgl. die Diskussion der für $\nu_0 = 0$ gewonnenen Resultate auf S. 567.

[2]) Vgl. die Arbeiten von W. Kuhn, ZS. f. Phys. **33**, 408, 1925; W. Thomas, Naturwiss. **13**, 627, 1925; F. Reiche und W. Thomas, ZS. f. Phys. **34**, 510, 1925.

Doch ist diese Darstellung, wenngleich unter Umständen sehr bequem und übersichtlich, durchaus nicht notwendig. Auch bei mehreren Freiheitsgraden werden die dynamischen Grundgleichungen die Form von Matrizengleichungen haben; aber die Matrizen können auch zweidimensional geschrieben werden, wie bisher. Denn schon bei einem Freiheitsgrad zeigte sich, daß die in der Zeilenanordnung der Matrizen zum Ausdruck kommende Reihenfolge der stationären Zustände (im Gegensatz zur bisherigen Theorie) ganz beliebig und durch keine innere Eigenschaft des Systems bestimmt ist. Diese Bemerkung kann nun ohne weiteres auf die mehrdimensionalen Matrizen übertragen werden; man kann beliebige Umordnungen vornehmen, speziell auch die $2f$-dimensionalen Matrizen in zweidimensionale verwandeln. Denn die grundlegenden Definitionen der Addition und Multiplikation sowie der zeitlichen Differentiation sind offenbar ganz unabhängig von irgendwelchen Ordnungsbeziehungen zwischen den Zeigersystemen $n_1, n_2, \ldots, n_f$, welche einzeln genommen die Zustände, paarweise die Übergänge bezeichnen.

Es ist danach auch klar, daß die allgemeinen Regeln der Matrizenanalysis, wie sie in I., Kap. I, und in Kap. 1 dieser Arbeit dargelegt wurden, ohne weiteres auch in der Theorie der Systeme von mehreren Freiheitsgraden anzuwenden sind. Ebenso überträgt sich unmittelbar die Ableitung der Bewegungsgleichungen aus dem Variationsprinzip von I., Kap. II, so daß wir sogleich anschreiben können:

$$\dot{q}_k = \frac{\partial H}{\partial p_k}; \quad \dot{p}_k = -\frac{\partial H}{\partial q_k}. \tag{2}$$

Als wesentlich neue, über die Theorie der Systeme mit einem Freiheitsgrad hinausgehende Annahme kommen bei mehreren Freiheitsgraden die allgemeinen Vertauschungsrelationen der p_k und q_k hinzu. Ebenso wie bei einem Freiheitsgrad würde ja das Rechnen mit den quantentheoretischen Größen in gewisser Weise unbestimmt, wenn nicht die „Vertauschungsrelationen" angegeben würden.

Als naheliegende Verallgemeinerung von (5), Kap. 1, bieten sich folgende Gleichungen dar:

$$\left.\begin{aligned} p_k q_l - q_l p_k &= \frac{h}{2\pi i}\,\delta_{kl}, \\ p_k p_l - p_l p_k &= 0, \\ q_k q_l - q_l q_k &= 0. \end{aligned}\right\} \tag{3}$$

Wenn H die (symmetrisierte) Energiefunktion bedeutet, kann auf Grund dieser Beziehungen (2), Kap. 2, ersetzt werden durch

$$\dot{q}_k = \frac{\partial H}{\partial p_k}, \quad \dot{p}_k = -\frac{\partial H}{\partial q_k}. \tag{2'}$$

Ferner folgt aus diesen Relationen [1]), wie oben in Kap. 1 dieser Arbeit:

$$\left.\begin{aligned} p_k f(q_1 \cdots q_f, p_1 \cdots p_f) - f p_k &= \frac{h}{2\pi i}\frac{\partial f}{\partial q_k}, \\ f q_k - q_k f &= \frac{h}{2\pi i}\frac{\partial f}{\partial p_k}. \end{aligned}\right\} \tag{4}$$

Der Beweis von Energiesatz und Frequenzbedingung wird dann aus (2′) und (4), Kap. 2, wie in Kap. 1 geführt. Ebenso läßt sich auf Grund von (3) und (4) zeigen, daß immer dann, wenn für ein System P_k, Q_k die Relationen (3), Kap. 2, erfüllt sind und die Energiefunktion als analytische unktion der P_k und Q_k gegeben ist, die kanonischen Bewegungsgleichungen (2′), Kap. 2, gelten.

Man wird also eine Transformation der Variablen p_k, q_k in neue Variablen P_k, Q_k dann als „kanonisch" bezeichnen, wenn sie die Relation (3), Kap. 2, ungeändert läßt.

Eine sehr allgemeine Klasse solcher Transformationen ist wieder durch die Formeln

$$\left.\begin{aligned} P_k &= S p_k S^{-1}, \\ Q_k &= S q_k S^{-1} \end{aligned}\right\} \tag{5}$$

gegeben. Diese Transformation hat auch wieder die Eigenschaft, jede Funktion $f(PQ)$ überzuführen in

$$f(P_1, \ldots, P_f, Q_1, \ldots, Q_f) = S f(p_1, \ldots, p_f, q_1, \ldots, q_f) S^{-1}. \tag{6}$$

Wenn ein System $p_1^0, \ldots, p_f^0, q_1^0, \ldots, q_f^0$ bekannt ist, das den Gleichungen (3), Kap. 2, genügt, so reduziert sich das Problem der Integration der Gleichungen (2), Kap. 2, wieder auf das andere Problem: Es ist eine Funktion S zu suchen, die den Gleichungen

$$\left.\begin{aligned} p_k &= S p_k^0 S^{-1}, \\ q_k &= S q_k^0 S^{-1} \end{aligned}\right\} \tag{5a}$$

genügt, und die

$$H(pq) = S H(p^0 q^0) S^{-1} = W \tag{7}$$

in eine Diagonalmatrix überführt.

Gleichung (7) stellt wieder das Analogon zur Hamiltonschen partiellen Differentialgleichung dar.

---

[1]) Die physikalische Bedeutung dieser Relationen bei der Dispersionstheorie wird diskutiert von H. A. Kramers, Physika, Dez. 1925.

Die Gleichungen (3), Kap. 2, zusammen mit (2), Kap. 2, würden offenbar viel zu viel Forderungen an die p_k, q_k stellen, wenn alle diese Gleichungen voneinander unabhängig wären. Es muß als eine interessante mathematische Aufgabe angesehen werden, die Gleichungen (3) aus einem Minimum unabhängiger und widerspruchsfreier Voraussetzungen abzuleiten; doch soll auf diese Frage hier nicht eingegangen werden. Wir begnügen uns mit der Bemerkung, daß

$$\frac{d}{dt}\sum_k (p_k q_k - q_k p_k) = 0$$

aus den Bewegungsgleichungen (1), Kap. 2, allgemein zu folgern ist. Dagegen soll allgemein gezeigt werden, daß die Gleichungen (3), Kap. 2, zusammen mit den Bewegungsgleichungen (2), Kap. 2, bzw. der gleichwertigen Forderung (7), Kap. 2, erfüllbar sind (natürlich von singulären Ausnahmefällen abgesehen).

Dieser Beweis ist zu liefern im Zusammenhang mit der Verallgemeinerung der Störungstheorie von Kap. 1, § 4, für beliebig viele Freiheitsgrade. Wir denken uns die Energiefunktion $H(pq)$ derart als

$$H = H_0(pq) + \lambda H_1(pq) + \lambda^2 H_2(pq) + \cdots \tag{8}$$

geschrieben, daß

$$H_0(pq) = \sum_{k=1}^{f} H^{(k)}(p_k q_k)$$

wird. Für $\lambda = 0$ haben wir also f ungekoppelte Systeme von je einem Freiheitsgrad; die f Probleme

$$H = H^{(k)}(p_k q_k)$$

seien gelöst durch

$$q_k = q_k^0, \quad p_k = p_k^0$$

mit q_k^0, p_k^0 als zweidimensionalen Matrizen

$$q_k^0 : (q_k^0(nm)); \quad p_k^0 : (p_k^0(nm)). \tag{10}$$

Wenn wir diese f ungekoppelten Systeme formal als ein einziges System von f Freiheitsgraden betrachten, so sind die q_k^0 p_k^0 als $2f$-dimensionale Matrizen

$$\left.\begin{aligned} q_k^0 &= (q_k^0(n_1 \dots n_f;\quad m_1 \dots m_f)),\\ p_k^0 &= (p_k^0(n_1 \dots n_f;\quad m_1 \dots m_f)) \end{aligned}\right\} \tag{11}$$

darzustellen, für welche gilt:

$$q_k^0(n_1 \dots n_f;\quad m_1 \dots m_f) = \delta_k q_k^0(n_k m_k),$$
$$p_k^0(n_1 \dots n_f;\quad m_1 \dots m_f) = \delta_k p_k^0(n_k m_k),$$

wobei $\delta_k = 1$, wenn $n_j = m_j$ für alle j außer $j = k$, $\delta_k = 0$, wenn für irgend ein j ($\neq k$) nicht $n_j = m_j$ ist. Daraus aber erkennt man: Erstlich bleiben die Gleichungen

$$p_k^0 q_k^0 - q_k^0 p_k^0 = \frac{h}{2\pi i} \mathbf{1}, \tag{12}$$

ursprünglich für die zweidimensionalen Matrizen (10), Kap. 2, geltend, auch für die $2f$-dimensionalen Matrizen (11), Kap. 2, in Kraft. Zweitens ergeben sich die Beziehungen

$$\left.\begin{aligned} p_k^0 q_l^0 - q_l^0 p_k^0 &= 0 \text{ für } l \neq k, \\ p_k^0 p_l^0 - p_l^0 p_k^0 &= q_k^0 q_l^0 - q_l^0 q_k^0 = 0. \end{aligned}\right\} \tag{13}$$

Für $\lambda = 0$ gelten daher wirklich alle Gleichungen (13), Kap. 2. Es soll gezeigt werden, daß p, q derart bestimmt werden können, daß (3), Kap. 2, auch für die höheren Näherungen gleichzeitig mit $H = W$ erfüllt wird. Vorauszusetzen ist dabei wiederum, daß das System H_0 als nichtentartet gewählt wurde, d. h. daß unter den Diagonalgliedern von H_0 bei Einsetzen von $q = q^0$, $p = p^0$ keine zwei gleich sind. In diesem Falle haben wir nur wieder den Ansatz (5a), Kap. 2,

$$q_k = S q_k^0 S^{-1}; \quad p_k = S p_k^0 S^{-1} \tag{14}$$

zu machen und

$$S = 1 + \lambda S_1 + \lambda^2 S_2 + \cdots$$

derart zu bestimmen, daß die Gleichung $H = W$ erfüllt wird. Die Gleichungen (3), Kap. 2, sind dann sämtlich auch befriedigt, da sie durch (14) in (12), (13) übergehen. Also ist der verlangte Beweis geführt.

Die Gleichungen (3) sind invariant gegenüber einer linearen orthogonalen Transformation der q_k und p_k. Denn setzt man

$$q'_k = \sum_l a_{kl} q_l, \qquad \sum_l a_{kl} a_{jl} = \delta_{kj},$$
$$p'_k = \sum_l a_{kl} p_l,$$

so wird

$$p'_k q'_l - q'_l p'_k = \sum_{hj} a_{kh} a_{lj} (p_h q_j - q_j p_h) = \delta_{kl} \frac{h}{2\pi i}$$

und entsprechend für die anderen Relationen. Fordern wir daher die Bedingungen (3), Kap. 2, für ein kartesisches Koordinatensystem, so gelten sie auch in jedem andern kartesischen Koordinatensystem.

Anhangsweise sei hier noch darauf hingewiesen, daß ein wohlbekannter Satz der klassischen Mechanik sich auf Grund der Festsetzungen (3), Kap. 2, ohne weiteres in die neue Theorie überträgt. Es sei

$$H = E_{\text{kin}} + E_{\text{pot}} = \tfrac{1}{2} \sum_k \frac{p_k^2}{m_k} + E_{\text{pot}}, \tag{15}$$

und E_{pot} eine homogene Funktion der Koordinaten vom Grade n. Es wird dann nach (3), Kap. 2:

$$E_{\text{pot}} = \frac{1}{n} \sum_k \frac{\partial E_{\text{pot}}}{\partial q_k} q_k \tag{16}$$

und

$$\frac{d}{dt} \sum_k p_k q_k = \sum_k (\dot{p}_k q_k + p_k \dot{q}_k) = 2 E_{\text{kin}} - n E_{\text{pot}},$$

also für die Mittelwerte

$$\overline{E}_{\text{kin}} = \frac{n}{2} \overline{E}_{\text{pot}}. \tag{17}$$

Z. B. wird für $n = 2$ (harmonische Schwingungen) $\overline{E}_{\text{kin}} = \overline{E}_{\text{pot}}$ und für $n = -1$ (Coulombsche Kräfte) $\overline{E}_{\text{kin}} = -\frac{1}{2}\overline{E}_{\text{pot}}$.

§ 2. Entartete Systeme. Wir wollen nun die entarteten Systeme ins Auge fassen. Wird das Verschwinden einiger der Frequenzen $\nu(nm)$ zugelassen (der Einfachheit halber denken wir uns die Matrizen zweidimensional dargestellt), so kann die Energiekonstanz $\dot{H} = 0$ noch immer durch die in I und hier ausgeführten Überlegungen aus den Bewegungsgleichungen und den Vertauschungsregeln (3), Kap. 2, abgeleitet werden. Aber $\dot{H} = 0$ hat nicht mehr notwendig die Folge, daß H eine Diagonalmatrix ist, und der Beweis des Frequenzsatzes wird damit undurchführbar. Es sind also bei entarteten Systemen die Bewegungsgleichungen zusammen mit (3), Kap. 2, allein nicht ausreichend zur eindeutigen Festlegung der Eigenschaften des Systems, und es ist eine Verschärfung dieser Grundgleichungen nötig. Es liegt nahe anzunehmen, daß diese Verschärfung so zu fassen ist: Als Grundgleichungen sollen allgemein gewählt werden können die Vertauschungsrelationen und

$$H = W = \text{Diagonalmatrix}. \tag{18}$$

Durch diese Forderung ist offenbar die Gültigkeit der Frequenzbedingung auch für die entarteten Systeme gesichert. Sehr wahrscheinlich wird dabei auch (von singulären Fällen abgesehen) die Energie W stets eindeutig bestimmt sein. Dagegen werden die Koordinaten q_k nicht eindeutig festgelegt. Man kann sich, wenn man eine Lösung p_k, q_k von $H(pq) = W$ besitzt, neue Lösungen verschaffen durch den Ansatz

$$\left.\begin{aligned} p' &= S p S^{-1}, \\ q' &= S q S^{-1}. \end{aligned}\right\} \tag{19}$$

Damit wird

$$H(p'q') = W' = S W S^{-1},$$

und die Forderung $\mathbf{W}' = \mathbf{W}$ ergibt $\mathbf{W}\mathbf{S} - \mathbf{S}\mathbf{W} = \dot{\mathbf{S}} \frac{h}{2\pi i} = 0$, also

$$\mathbf{S} = \text{const.} \tag{20}$$

Betrachten wir zunächst dieses Ergebnis in seiner Bedeutung für die nichtentarteten Systeme. Hier muß nach (20), Kap. 2, die Matrix $\mathbf{S}$ eine Diagonalmatrix werden, und die Gleichungen (19), Kap. 2, bedeuten

$$\left.\begin{aligned} p'(nm) &= p(nm)\, S_n S_m^{-1}, \\ q'(nm) &= q(nm)\, S_n S_m^{-1}, \end{aligned}\right\} \tag{19'}$$

wenn wir kurz S_n für $S(nn)$ schreiben.

Die Unbestimmtheit der Lösung, die hierdurch angezeigt ist, wird wesentlich herabgemindert durch die Forderung, daß auch die neue Lösung $\mathbf{p}'\ \mathbf{q}'$ eine „reelle", durch Hermitesche Matrizen dargestellte Bewegung werden soll; denn das ergibt

$$|S_n S_m^{-1}| = |S_m S_m^{-1}|,$$

oder

$$|S_n| = |S_m|. \tag{21}$$

Die hier zutage tretende Unbestimmtheit bedeutet also eine Willkür der Phasenkonstanten; und zwar erhalten wir hier einen allgemeinen Beweis der schon in I aufgestellten Vermutung, daß in jedem Problem für jeden Zustand n je eine Phase φ_n unbestimmt bleibt. Man übersieht auch an Hand von (19'), in welcher Weise diese Phasen in die Elemente der Matrizen $\mathbf{p}$, $\mathbf{q}$ eingehen. In Teil I wurde weiter vermutet, daß außer dieser beschriebenen Phasenwillkür bei nichtentarteten Systemen keine Nichteindeutigkeit mehr zu erwarten sei. Nun könnten wir offenbar bei der Störungsrechnung von Kap. 1, § 4, zu jeder der „periodischen" Matrizen S_n noch eine konstante Matrix hinzu addieren. Doch bedeutet das natürlich nicht, daß in jeder Näherung neue unbestimmt bleibende Phasen hinzukommen. Man übersieht leicht, daß durch Ausnutzung dieser Möglichkeit keine allgemeinere Lösung $\mathbf{p}$, $\mathbf{q}$ gefunden werden könnte, sofern $\mathbf{p}^0$, $\mathbf{q}^0$ von vornherein mit unbestimmten Phasen angenommen war.

Gehen wir nun zu entarteten Systemen über, so können wir aus (20) nicht mehr folgern, daß $\mathbf{S}$ eine Diagonalmatrix sei, und es ergibt sich durch (19) wirklich die Möglichkeit, von $\mathbf{p}$, $\mathbf{q}$ wesentlich verschiedene Lösungen $\mathbf{p}'$, $\mathbf{q}'$ abzuleiten. Es scheint, daß diese Nichteindeutigkeit in der Natur der Sache liegt. Die entarteten Systeme besitzen offenbar eine Labilität, dank deren durch beliebig kleine Störungen endliche Änderungen der Koordinaten herbeigeführt werden können, und die darin ihren mathematischen Ausdruck findet, daß bei völliger Abwesenheit von Störungen die Lösung der dynamischen

Gleichungen teilweise unbestimmt bleibt. Natürlich werden bei jedem einzelnen wirklichen Atom die für die physikalischen Eigenschaften des Systems, insbesondere die Übergangswahrscheinlichkeiten maßgebenden Koordinaten stets durch äußere Störungen oder durch die Vorgeschichte des Systems eindeutig festgelegt sein.

Wir wollen nun den Einfluß beliebiger Störungen auf das entartete System untersuchen. Sei also

$$H(p\,q) = H_0 + \lambda H_1 + \lambda^2 H_2 + \cdots, \tag{22}$$

und sei p^0, q^0 eine beliebige, aber fest gewählte Lösung des ungestörten Problems:

$$H_0(p_0\,q_0) = W_0. \tag{23}$$

Der Ansatz

$$p = S\,p^0\,S^{-1},$$
$$q = S\,q^0\,S^{-1},$$

mit

$$S = S_0(1 + \lambda S_1 + \lambda^2 S_2 + \cdots), \tag{24}$$

$$S^{-1} = (1 - \lambda(S_1 + \lambda S_2 \ldots) + \lambda^2 \ldots)S_0^{-1}, \tag{25}$$

ergibt dann, wenn wir in H_0, $H_1 \ldots$ die Argumente p^0, q^0 auslassen:

$$S_0 H_0 S_0^{-1} = W_0, \tag{26}$$

$$S_0 S_1 H_0 S_0^{-1} - S_0 H_0 S_1 S_0^{-1} + S_0 H_1 S_0^{-1} = W_1, \tag{27}$$

$$S_0 S_2 H_0 S_0^{-1} - S_0 H_0 S_2 S_0^{-1} + S_0 F_2(H_0 H_1 H_2; S_1) S_0^{-1} = W_2, \tag{28}$$

. . . .

$$S_0 S_r H_0 S_0^{-1} - S_0 H_0 S_r S_0^{-1}$$
$$+ S_0 F_r(H_0 H_1 \ldots H_r, S_1 \ldots S_{r-1}) S_0^{-1} = W_r. \tag{29}$$

Das sind fast wieder die Gleichungen (26), Kap. 1, nur mit dem Unterschied, daß auf der linken Seite überall vorn mit S_0 und hinten mit S_0^{-1} multipliziert ist.

Die Gleichung (26), Kap. 2, ist bereits oben erörtert worden; es wird $S_0(n\,m)$ gleich Null außer für verschwindendes $\nu_0(n\,m)$. Die noch verbleibende Willkür in S_0 muß nun nach Möglichkeit ausgenutzt werden, um die nächste Gleichung lösbar zu machen. Es kann natürlich nicht von jeder Lösung von $H = H_0$, also insbesondere auch nicht von der ausgewählten p^0, q^0 erwartet werden, daß sie den Grenzfall $\lambda = 0$ der Lösung p, q des Problems (22), Kap. 2, liefert. Die Funktion S_0 soll dazu dienen, aus p^0, q^0 diejenige Lösung des entarteten Problems herzustellen, welche diese verlangte Eigenschaft besitzt.

Die Gleichung (27) kann umgeschrieben werden in

$$S_1 H_0 - H_0 S_1 + H_1 = S_0^{-1} W_1 S_0. \tag{30}$$

Um sie lösbar zu machen, muß S_0 so bestimmt werden, daß

$$\overline{H_1} = S_0^{-1} W_1 S_0 \tag{31}$$

mit einer Diagonalmatrix W_1 wird. Für die gleichzeitige Befriedigung dieser Gleichung und der durch (26), Kap. 2, gestellten Forderungen kann hier natürlich ebensowenig eine allgemeine Anweisung gegeben werden, wie für die Bestimmung der säkulären Störungen in der klassischen Theorie. Später werden wir aber mit einer neuen, algebraischen Methode zu einer einfachen Behandlung einer großen Klasse von Entartungen gelangen (Kap. 3).

Ist (31), Kap. 2, erfüllt, so kann (30), Kap. 2, wie in Kap. 1 gelöst werden. Willkürlich bleiben dabei diejenigen Glieder $S_1(nm)$ von S_1, für die $\nu_0(nm)$ verschwindet, und diese Unbestimmtheit muß benutzt werden, um in der nächsten Näherungsgleichung, die zu

$$S_2 H_0 - H_0 S_2 + F_2 = S_0^{-1} W_2 S_0 \tag{32}$$

umgeschrieben werden kann, die zur Lösbarkeit notwendige Beziehung

$$\overline{F_2(H_0, H_1, H_2; S_1)} = S_0^{-1} W_2 S_0 \tag{31'}$$

mit einer Diagonalmatrix W_2 zu erfüllen. Die Fortsetzung des Verfahrens ist klar.

Die Schwierigkeit liegt darin, daß man in jeder Näherung Gleichungen zu befriedigen hat durch Matrizen, die schon größtenteils festgelegt sind, so daß nicht zu übersehen ist, ob diese Gleichungen wirklich lösbar sein werden. Es besteht jedoch bekanntlich eine ganz entsprechende Schwierigkeit in der klassischen Theorie. Beseitigt werden diese Schwierigkeiten wenigstens für die höheren Näherungen, wenn in irgend einer Näherung das System nichtentartet wird.

Seien z. B. in

$$q = q^0 + \lambda q^{(1)} + \cdots,$$
$$p = p^0 + \lambda p^{(1)} + \cdots$$

$p^{(1)}$, $q^{(1)}$ wirklich bestimmt worden, so daß also mit

$$Q = q_0 + \lambda q^{(1)}$$
$$P = p^0 + \lambda p^{(1)}$$

gilt:

$$H(PQ) = W_0 + \lambda W_1 + \lambda^2 H_2' + \lambda^3 H_3' + \cdots,$$

und sei

$$\nu_0(nm) + \lambda \nu_1(nm) \neq 0 \text{ für } n \neq m.$$

Wenn wir $W_0 + \lambda W_1$ kurz mit H_0' bezeichnen und den Ansatz

$$p = S P S^{-1}$$
$$q = S Q S^{-1}$$

machen, so ist

$$S (H_0' + \lambda^2 H_2' + \lambda^3 H_3' + \cdots) S^{-1} = W$$

zu machen, was nach dem Verfahren von Kap. 1 durch

$$S = 1 + \lambda^2 S_2 + \lambda^3 S_3 + \cdots$$

erzielt werden kann. Die Verallgemeinerung dieser Betrachtungen für den Fall, daß erst mit der r-ten Näherung ein nicht entartetes System $W = W_0 + \lambda W_1 + \cdots + \lambda^r W_r$ erreicht wird, ergibt sich von selbst [1]).

Am Schluß scheint es uns noch wichtig, darauf hinzuweisen, daß die bekannten Konvergenzschwierigkeiten bei den Reihen der klassischen Störungstheorie, die in der Diskussion des Dreikörperproblems eine so große Rolle spielen, hier in der quantenmechanischen Störungstheorie nicht auftreten; vielmehr dürften hier die im Endlichen verlaufenden Bahnen im allgemeinen auch periodisch sein.

Kapitel 3. Zusammenhang mit der Theorie der Eigenwerte Hermitescher Formen.

§ 1. Allgemeine Methode. Im voranstehenden ist die Lösung der quantentheoretischen Grundgleichungen in möglichst engem Anschluß an die klassische Theorie entwickelt worden. Hinter diesem Formalismus der Störungstheorie verbirgt sich aber ein sehr einfacher, rein algebraischer Zusammenhang, und es lohnt sich wohl, diesen ans Licht zu ziehen. Denn abgesehen von der tieferen Einsicht in die mathematische Struktur der Theorie gewinnt man dabei auch den Vorteil, die von der Mathematik vorbereiteten Methoden und Ergebnisse verwerten zu können. So werden wir zu einer neuen Definition der Energiekonstanten („Terme") gelangen, die auch im Falle aperiodischer Bewegungen, also kontinuierlich veränderlicher Indizes gültig bleibt. Hierdurch gewinnt man die Aussicht, Methoden zur direkten Energieberechnung ohne explizite Lösung des Bewegungsproblems zu finden, die der Sommerfeldschen Methode der komplexen Integration in der bisherigen Theorie entsprechen. Sodann werden wir die Störungen einer großen Klasse entarteter Systeme vollständig behandeln können, was mit den vorher erörterten Störungsmethoden noch nicht möglich war.

---

[1]) Die analogen Fälle der klassischen Mechanik sind von M. Born und W. Heisenberg, Ann. d. Phys. **74**, 1, 1924, diskutiert worden.

Ist ein Problem von f Freiheitsgraden durch die Energiefunktion $H(p, q)$ vorgelegt, so können wir zunächst irgend ein System von Matrizen $p_k^0\, q_k^0$ derart wählen, daß jedenfalls die Vertauschungsregeln (3), Kap. 2, erfüllt sind; z. B. können die p_k, q_k eines Systems ungekoppelter harmonischer Oszillatoren genommen werden.

Dann kann man, wie oben Kap. 2, § 1, erwähnt, die dynamische Aufgabe, d. h. die Bestimmung der p_k, q_k, auch so formulieren:

Es soll eine Transformation $(p_k^0\, q_k^0) \rightarrow (p_k\, q_k)$ gefunden werden, welche die Gleichungen (3), Kap. 2, invariant läßt und zugleich die Energie in eine Diagonalmatrix verwandelt.

Man übersieht die Transformation von Matrizen am besten, wenn man sie als Koeffizientensysteme von linearen Transformationen bzw. bilinearen Formen auffaßt. Wir schicken daher einige bekannte Sätze der Algebra solcher Formen voraus.

Zu jeder Matrix $\mathbf{a} = (a(nm))$ gehört eine bilineare Form

$$A(xy) = \sum_{nm} a(nm)\, x_n\, y_m \tag{1}$$

zweier Reihen von Variablen $x_1, x_2, \ldots$ und $y_1, y_2, \ldots$ Ist die Matrix vom Hermiteschen Typus, d. h. ist die transponierte Matrix $\tilde{\mathbf{a}} = (a(mn))$ gleich der zur ursprünglichen konjugiert komplexen Matrix:

$$\tilde{\mathbf{a}} = \mathbf{a}^*, \quad a(mn) = a^*(nm), \tag{2}$$

so nimmt die Form A reelle Werte an, wenn man für die Variablen y_n die konjugiert komplexen Werte x_n setzt:

$$A(x\,x^*) = \sum_{nm} a(nm)\, x_n\, x_m^*. \tag{1a}$$

Es sei an die leicht zu beweisende Rechenregel erinnert:

$$\widetilde{(\mathbf{a}\mathbf{b})} = \tilde{\mathbf{b}}\,\tilde{\mathbf{a}}. \tag{3}$$

Wir unterwerfen nun die x_n einer linearen Transformation

$$x_n = \sum_{l} v(ln)\, y_l \tag{4}$$

mit der (komplexen) Matrix $\mathbf{v} = (v(ln))$.

Dann geht die Form A über in:

$$A(x\,x^*) = B(y\,y^*) = \sum_{nm} b(nm)\, y_n\, y_m^* \tag{5}$$

mit

$$b(nm) = \sum_{kl} v(nk)\, a(kl)\, v^*(ml),$$

oder in Matrizenschreibweise:

$$b = v\, a\, \tilde{v}^*. \tag{6}$$

Man sagt, die Matrix b gehe durch die Transformation v aus a hervor.

Die Matrix b ist wieder vom Hermiteschen Typus, denn es gilt nach (3), Kap. 3,

$$\tilde{b} = v^*\, \tilde{a}\, \tilde{v} = v^*\, a^*\, \tilde{v} = b^*. \tag{7}$$

Wir nennen die Matrix v orthogonal, wenn die zugehörige Transformation die Hermitesche Einheitsform

$$E(x\,x^*) = \sum_n x_n\, x_n^*$$

invariant läßt; nach dem soeben gewonnenen Resultat ist das dann und nur dann der Fall, wenn

$$v\,\tilde{v}^* = 1, \quad \text{oder} \quad \tilde{v}^* = v^{-1}. \tag{8}$$

So sind z. B. die in I, § 2, erwähnten Permutationsmatrizen reelle orthogonale Matrizen.

Bei endlicher Variablenzahl ist es bekanntlich immer möglich, eine Form orthogonal auf eine Summe von Quadraten zu transformieren [Hauptachsen-Transformation][1]):

$$A(x\,x^*) = \sum_n W_n\, y_n\, y_n^*. \tag{9}$$

Für die Matrizen bedeutet das: Es gibt eine Matrix, für die

$$v\,\tilde{v}^* = 1 \quad \text{und} \quad v\, a\, \tilde{v}^* = v\, a\, v^{-1} = W, \tag{10}$$

wo $W = (W_n\, \delta_{nm})$ eine Diagonalmatrix ist.

Bei unendlichen Matrizen gilt in allen bisher untersuchten Fällen ein analoger Satz; nur kann es vorkommen, daß rechter Hand der Index n außer einer Reihe diskreter Zahlen auch einen kontinuierlichen Wertebereich durchläuft, dem in (9) und in der Transformation (4) ein Integralbestandteil entspricht[2]).

Die Größen W_n heißen „Eigenwerte", ihre Gesamtheit ist das „mathematische" Spektrum der Form, bestehend aus „Punkt"- und

[1]) Wir nennen die Koeffizienten der transformierten Form W_n, weil sie in der Quantendynamik die Bedeutung der „Energie" haben.

[2]) Die Theorie der quadratischen (bzw. Hermiteschen) Formen von unendlich vielen Variablen ist bisher hauptsächlich nur für eine besondere Klasse („beschränkte" Formen) durchgeführt worden (D. Hilbert, Grundzüge einer allgemeinen Theorie der linearen Integralgleichungen; E. Hellinger, Crelles Journ. **136**, 1, 1910). Hier handelt es sich aber gerade um nicht beschränkte Formen. Trotzdem dürfen wir wohl annehmen, daß die Sätze in der Hauptsache ebenso lauten.

„Strecken“-Spektrum. Dieses ist, wie wir sehen werden, mit dem „Termspektrum“ der Physik identisch, während das „Frequenzspektrum“ durch Differenzbildung daraus entsteht.

Diese Hauptachsen-Transformation liefert uns nun unmittelbar die Lösung unseres dynamischen Problems: es sollte eine solche Transformation $(p^0 q^0) \rightarrow (p q)$ gefunden werden, welche die Gleichungen (3), Kap. 2, invariant läßt und zugleich die Energie in eine Diagonalmatrix überführt.

Denn nach dem obigen Satze der Algebra gibt es eine orthogonale Matrix S, für die also

$$S \tilde{S}^* = 1, \quad \tilde{S}^* S = 1 \tag{11}$$

ist, von der Art, daß durch die Transformation

$$\left.\begin{aligned} p_k &= S p_k^0 \tilde{S}^* = S p_k^0 S^{-1}, \\ q_k &= S q_k^0 \tilde{S}^* = S q_k^0 S^{-1} \end{aligned}\right\} \tag{12}$$

1. der Hermitesche Charakter von p_k^0, q_k^0 auch für die p_k, q_k erhalten bleibt, 2. die Gleichungen (3), Kap. 2, invariant sind, 3. die Energie in eine Diagonalmatrix

$$H(p q) = S H(p^0 q^0) S^{-1} = W \tag{13}$$

übergeführt wird.

Wir wollen die Frage der Eindeutigkeit dieser Lösung diskutieren, vor allem, ob nicht durch eine andere orthogonale Transformation T andere Energiewerte erzeugt werden können. Angenommen, es wäre

$$T H(p^0 q^0) T^{-1} = W'$$

eine von W verschiedene Diagonalmatrix. Dann hätte man

$$T S^{-1} S H S^{-1} S T^{-1} = T S^{-1} W (T S^{-1})^{-1},$$

und unsere Frage bedeutet, ob es möglich ist, aus einer Diagonalmatrix W eine andere W' durch

$$W' = M W M^{-1}, \quad M \tilde{M}^* = 1 \tag{14}$$

zu bilden, derart, daß W' nicht durch Permutation der Diagonalglieder aus W zu erhalten ist.

Die Gleichung (14), Kap. 3, aber läßt sich schreiben

$$W' M - M W = 0,$$

und bedeutet daher

$$M(nm)(W'_n - W_m) = 0. \tag{14a}$$

Aus der Orthogonalität von M folgt für $m = n$ insbesondere

$$\sum_k |M(nk)|^2 = 1, \quad \sum_k |M(kn)|^2 = 1;$$

folglich können bei festem n weder alle $M(nk)$ noch alle $M(kn)$ verschwinden. Dann ergibt aber (14a), Kap. 3, daß es zu jedem n sicherlich ein m gibt, für das $W'_n = W_m$ ist, d. h. alle W'_n kommen unter den W_m vor. Und ebenso folgt auch das Umgekehrte.

Daher führen alle aus dem Ansatz (12), Kap. 3, (bei bestimmten p_k^0, q_k^0) ableitbaren Lösungen auf dieselben Werte für die Energien der stationären Zustände im Einklang mit der in Kap. 2 ausgesprochenen Vermutung, daß die vorkommenden Energien durch die dynamischen Grundgleichungen stets eindeutig bestimmt sind.

Entartete Systeme werden dadurch charakterisiert sein, daß mehrfache Eigenwerte vorkommen. Die Mehrfachheit des Eigenwertes W_n, d. h. die Anzahl der linear unabhängigen Lösungen $v(ln)$ der Gleichung (4), Kap. 3, wird das statistische Gewicht des betreffenden Zustandes angeben.

Die Wichtigkeit der Gleichung (9), Kap. 3, für unsere physikalische Theorie beruht darauf, daß es in der Algebra endlicher oder beschränkter unendlicher Formen verschiedene Methoden [1]) gibt, die Eigenwerte einer Form zu bestimmen, ohne die Transformation wirklich durchzuführen. Man kann hoffen, daß solche Methoden bei der späteren Behandlung bestimmter physikalischer Systeme gute Dienste leisten werden.

§ 2. Anwendung auf die Störungstheorie. Im folgenden wollen wir zeigen, daß die hier dargelegte algebraische Auffassung des dynamischen Problems nicht nur zu genau den Formeln führt, die früher Kap. 1, § 4, im Anschluß an die Störungstheorie der klassischen Mechanik abgeleitet worden sind, sondern daß sie bei der Behandlung entarteter Systeme der bisherigen Theorie noch erheblich überlegen ist.

Wir nehmen also wieder an, $\boldsymbol{H}$ habe die Form

$$\boldsymbol{H} = \boldsymbol{H}_0 + \lambda \boldsymbol{H}_1 + \lambda^2 \boldsymbol{H}_2 + \cdots,$$

wo das durch $\boldsymbol{H}_0$ bestimmte dynamische Problem die Lösung $\boldsymbol{p}_k^0, \boldsymbol{q}_k^0$ hat. Diese Größen nehmen wir als die Ausgangskoordinaten, aus denen durch eine orthogonale Transformation $\boldsymbol{S}$ die $\boldsymbol{p}_k, \boldsymbol{q}_k$ zu finden sind. Die vorausgesetzte Form von $\boldsymbol{H}$ bedeutet natürlich grundsätzlich keine Beschränkung der Allgemeinheit, insofern man ja von $\boldsymbol{H}$ stets einen Anteil $\boldsymbol{H}_0$ von beliebiger, gewünschter Form abtrennen kann; nur wird die

[1]) Bei endlichen Formen sind die Eigenwerte die Wurzeln einer algebraischen Gleichung. Hier und auch bei beschränkten unendlichen Matrizen kann man sie z. B. nach dem Verfahren von Graeffe und Bernoulli bestimmen; siehe etwa C. Courant und D. Hilbert, Methoden der mathematischen Physik I, § 3, S. 14, 15. Berlin, Springer, 1924.

Konvergenz der Potenzreihe nach λ wesentlich von der geschickten Wahl von H_0 abhängen.

Um die Hermitesche Form

$$\sum_{mn} H_{mn} x_m x_n^*$$

auf ihre Hauptachsen zu transformieren, kann man bekanntlich so verfahren:

Man suche die linearen Gleichungen

$$W x_k - \sum_l H(kl)\, x_l = 0 \tag{15}$$

zu lösen; das ist nur möglich für gewisse Werte des Parameters W, nämlich $W = W_n$, wo W_n wieder die Eigenwerte (Energiewerte) bedeuten. Wir nehmen zunächst an, daß keine Entartung vorliegt, also alle W_n verschieden sind. Dann gehört zu jedem W_n eine bis auf einen Faktor bestimmte Lösung $x_k = x_{kn}$; es gelten also die Identitäten

$$W_n x_{kn} - \sum_l H(kl)\, x_{ln} = 0,$$

$$W_m x_{km}^* - \sum_l H^*(kl)\, x_{lm}^* = 0.$$

Multipliziert man die erste mit x_{km}^*, die zweite mit x_{kn} und summiert über k, so folgt durch Subtraktion wegen des Hermiteschen Charakters von H:

$$(W_n - W_m) \sum_k x_{kn} x_{km}^* = 0.$$

Durch geeignete Wahl des Proportionalitätsfaktors kann man es ferner erreichen, daß

$$\sum_k x_{kn} x_{kn}^* = 1$$

ist. Folglich bilden die x_{kn} eine orthogonale Matrix

$$\boldsymbol{S} = (x_{kn}).$$

Diese ist es gerade, welche die gegebene Form auf eine Quadratsumme transformiert; denn setzt man

$$x_k = \sum_n x_{kn} y_n$$

in die Form ein, so erhält man

$$\begin{aligned}\sum_{kl} H(kl)\, x_k x_l^* &= \sum_{kl} \sum_{mn} H(kl)\, x_{km} x_{ln}^* y_m y_n^* \\ &= \sum_{mn} \sum_l W_m x_{lm} x_{ln}^* y_m y_n^* \\ &= \sum_m W_m y_m y_m^*.\end{aligned}$$

Nach unserer Voraussetzung haben nun die Koeffizienten der Gleichungen (15), Kap. 3, die Form:

$$H(kl) = \delta_{kl} W_l^0 + \lambda H_1(kl) + \lambda^2 H_2(kl) + \cdots$$

Daher suchen wir die Lösung von (15), Kap. 3, durch Entwicklungen der Form:

$$\left.\begin{aligned} W &= W^0 + \lambda W^{(1)} + \lambda^2 W^{(2)} + \cdots \\ x_k &= x_k^0 + \lambda x_k^{(1)} + \lambda^2 x_k^{(2)} + \cdots \end{aligned}\right\} \tag{16}$$

zu gewinnen. Setzen wir das in (15), Kap. 3, ein, so bekommen wir die Näherungsgleichungen:

$$\left.\begin{aligned} &\text{a)}\ x_k^0 (W^0 - W_k^0) = 0, \\ &\text{b)}\ x_k^{(1)} (W^0 - W_k^0) = -x_k^0 W^{(1)} + \sum_l H^{(1)}(kl)\, x_l^0, \\ &\text{c)}\ x_k^{(2)} (W^0 - W_k^0) = -(x_k^{(1)} W^{(1)} + x_k^{(0)} W^{(2)}) \\ &\qquad + \sum_l \left(H^{(1)}(kl) x_l^{(1)} + H^{(2)}(kl) x_l^0\right). \end{aligned}\right\} \tag{17}$$

Aus (17a), Kap. 3, folgt, daß W gleich einem der W_k werden muß; denn sonst würden alle x_k^0 verschwinden, und dann würde man aus den folgenden Näherungsgleichungen auch das Verschwinden von $x_k^{(1)}, x_k^{(2)} \ldots$ der Reihe nach erschließen können.

Nehmen wir nun das Ausgangssystem als nicht entartet, also alle W_k^0 als verschieden an, so lautet die Lösung von (17 a), Kap. 3:

$$W = W_n^0; \quad x_{nn}^0 = y_n^0; \quad x_{kn}^0 = 0 \text{ für } k \neq n. \tag{18}$$

Dabei ist y_n^0 eine beliebige Zahl.

Setzen wir das in (17b), Kap. 3, ein, so hat man, je nachdem $k = n$ oder $k \neq n$ ist:

$$0 = y_n^0 (-W^{(1)} + H^{(1)}(nn)),$$
$$x_k^{(1)} (W_n^0 - W_k^0) = H^{(1)}(kn)\, y_n^0, \quad k \neq n.$$

Die Lösung lautet also:

$$W^{(1)} = H^{(1)}(nn); \quad x_{nn}^{(1)} = y_n^{(1)}; \quad x_{kn}^{(1)} = -\frac{H^{(1)}(kn)}{h\nu_0(kn)} y_n^0 \text{ für } k \neq n, \tag{19}$$

wo $y_n^{(1)}$ wiederum eine beliebige Zahl ist.

Nunmehr folgt ebenso aus (17c), Kap. 3:

$$\left.\begin{aligned} W^{(2)} &= H^{(2)}(nn) - \frac{1}{h} \sum_l{}' \frac{H^{(1)}(nl) H^{(1)}(ln)}{\nu_0(ln)}, \\ x_{nn}^{(2)} &= y_n^{(2)} \\ x_{kn}^{(2)} &= \left(\frac{1}{h^2} \sum_l{}' \frac{H^{(1)}(kl) H^{(1)}(ln)}{\nu_0(kn)\nu_0(ln)} - \frac{H^{(1)}(nn) H^{(1)}(kn)}{h^2 \nu_0(kn)^2} \right. \\ &\qquad \left. - \frac{H^{(2)}(kn)}{h\nu_0(kn)}\right) y_n^0 - \frac{H^{(1)}(kn)}{h\nu_0(kn)} y_n^{(1)}. \end{aligned}\right\} \tag{20}$$

Ebenso leicht gewinnt man die Lösung der dritten Näherung; wir geben nur den Energiewert an:

$$W^{(3)} = H^{(3)}(nn) - \frac{1}{h}\sum_l{}' \frac{H^{(1)}(nl)\,H^{(2)}(ln) + H^{(2)}(nl)\,H^{(1)}(ln)}{\nu_0(ln)}$$

$$+ \frac{1}{h^2}\left(\sum_{kl}{}' \frac{H^{(1)}(nl)\,H^{(1)}(lk)\,H^{(1)}(kn)}{\nu_0(ln)\,\nu_0(kn)} - H^{(1)}(nn)\sum_l{}' \frac{H^{(1)}(nl)\,H^{(1)}(ln)}{\nu_0(ln)^2}\right).$$

Die vorläufig willkürlichen Größen $y_n^{(0)}, y_n^{(1)}, \ldots$ dienen dazu, die Lösung zu normieren (orthogonal ist sie von selbst); die Bedingung

$$\sum_k x_{kn} x_{kn}^* = 1$$

liefert für

$$x_{kn} = \overset{0}{x}_{kn} + \lambda x_{kn}^{(1)} + \lambda^2 x_{kn}^{(2)} + \cdots$$

die Gleichungen:

$$\sum_k \overset{0}{x}_{kn} \overset{*0}{x}_{kn} = 1$$

$$\sum_k \left(\overset{0}{x}_{kn} x_{kn}^{*(1)} + x_{kn}^{(1)} \overset{*0}{x}_{kn}\right) = 0$$

. . . .

Setzt man hier die eben gefundene Lösung ein, so folgt der Reihe nach

$$|\overset{0}{y}_n|^2 = 1$$

$$\overset{0}{y}_n y_n^{*(1)} + \overset{*0}{y}_n y_n^{(1)} = 0$$

. . . .

Setzen wir nun

$$y_n^{(p)} = a_n^{(p)} e^{i\varphi_n(p)}, \tag{21}$$

so erhalten wir:

$$\left.\begin{aligned} \overset{0}{a}_n &= 1 \\ 2a_n^{(1)} \cos\left(\overset{0}{\varphi}_n - \varphi_n^{(1)}\right) &= 0 \\ &\cdots\cdots\cdots \\ 2a_n^{(r)} \cos\left(\overset{0}{\varphi}_n - \varphi_n^{(r)}\right) &= F^{(r)}\left(a^{(r-1)}, \varphi^{(r-1)}, \ldots\right). \end{aligned}\right\} \tag{22}$$

Man kann also die Phasenkonstanten $\overset{0}{\varphi}_n, \varphi_n^{(1)}, \ldots$ beliebig wählen; dann sind die $\overset{0}{a}_n, a_n^{(1)}, \ldots$ sukzessive berechenbar und eindeutig bestimmt. Das steht in Übereinstimmung mit unserem früher (§ 3) gefundenen Resultat, daß die Phasen der Diagonalglieder von $\boldsymbol{S}$ unbestimmt bleiben.

Setzt man die gefundenen Werte $\overset{0}{a}_n = 1, \ldots$ in (21) Kap. 3 und dies in (18), (19), (20), Kap. 3, ein, so sieht man, daß das früher durchgeführte „Störungsverfahren" gerade die Lösung geliefert hat, bei der die Phasen $\varphi_n^{(p)}$ verschwinden, also die Diagonalglieder von $\boldsymbol{S}$ reell sind.

Wir betrachten nun den Fall, daß das Ausgangssystem entartet ist, und zwar sei W_n^0 ein r-facher Eigenwert; das bedeutet, die Gleichung (17a), Kap. 3, hat die Lösung

$$\left.\begin{array}{c} W = W_n;\ \overset{0}{x}_{nn} = \overset{0}{y}_{1,n},\ \overset{0}{x}_{n,n+1} = y_{2,n} \cdots \\ \overset{0}{x}_{n,n+r-1} = \overset{0}{y}_{r,n}, \\ \overset{0}{x}_{kn} = 0 \text{ für } k \neq n, n+1, \ldots, n+r-1. \end{array}\right\} \quad (23)$$

Dann verschwindet die linke Seite von (17b), Kap. 3, für

$$k = n, n+1, \ldots, n+r-1;$$

das liefert (r) Gleichungen:

$$W^{(1)} \overset{0}{y}_{kn} - \sum_{l=r}^{r} H^{(1)}(n+k, n+l)\, \overset{0}{y}_{ln} = 0;\ k = 1, 2, \ldots, r, \quad (24)$$

deren Koeffizientenschema wieder vom Hermiteschen Typus ist.

Setzt man die Determinante gleich Null, so erhält man eine Säkulargleichung vom r-ten Grade für $W^{(1)}$:

$$\text{Det.}\ \big(W^{(1)}\delta_{kl} - H^{(1)}(n+k, n+l)\big) = 0, \quad (25)$$

deren Wurzeln sicherlich reell sind. Zu jeder Wurzel gehören eine oder mehrere unabhängige Lösungen der Gleichungen (24), Kap. 3.

Wählt man eine solche Lösung aus, so kann man für diese das Näherungsverfahren fortsetzen; doch soll das hier nicht weiter ausgeführt werden.

Es genügt uns, eingesehen zu haben, daß man mit unserer algebraischen Methode alle Entartungen von endlicher Vielfachheit beherrscht, d. h. sie auf die Lösung von algebraischen Gleichungen zurückführen kann. Wenn z. B. jeder Eigenwert zweifach ist, also zu jedem eine verschwindende Frequenz $\nu_0(nm)$ gehört, so führt das Störungsproblem auf eine quadratische Gleichung

$$\begin{vmatrix} W^{(1)} - H^{(1)}(n,n) & -H^{(1)}(n,n+1) \\ -H^{(1)}(n+1,n) & W^{(1)} - H^{(1)}(n+1,n+1) \end{vmatrix} = 0.$$

Dieser Fall liegt vor, wenn zwei ursprünglich gleiche, nicht entartete Systeme (wobei alle Frequenzen eines einzelnen Systems verschieden sein sollen) durch irgendwelche Kräfte gekoppelt werden.

Eine interessante Bedeutung hat ferner die Orthogonalitätsrelation

$$\sum_k \overset{0}{x}_{kn} \cdot \overset{*0}{x}_{kn} = 1$$

bei entarteten Systemen. Wegen (23) geht diese Relation über in

$$\sum_{l=1}^{r} \overset{0}{y}_{ln} \overset{*0}{y}_{ln} = 1.$$

Hieraus folgt, wenn m irgend eine Zahl der Reihe $n, n+1, \ldots n+r-1$, und k irgend eine Zahl außerhalb dieser Reihe bedeutet, daß die Summen

$$\sum_{m=n}^{n+r-1} p^0(mk)p^{*0}(mk),$$

$$\sum_{m=n}^{n+r-1} q^0(mk)q^{*0}(mk)$$

auch bei Entartung eindeutig bestimmt sind, d. h. daß diese Summen gegen die Transformationen, die nach (19), Kap. 2, bei Entartung aus gewissen Lösungen p, q neue, davon wesentlich verschiedene Lösungen $p'q'$ hervorgehen lassen, invariant sind. Dieses Ergebnis gibt eine mathematische Darstellung der sogenannten spektroskopischen Stabilität, die in den neueren Theorien über die Feinstrukturintensitäten (vgl. Kap. 4) eine wichtige Rolle gespielt hat.

§ 3. Kontinuierliche Spektra. Das gleichzeitige Auftreten von kontinuierlichen Spektra und Linienspektra als Lösungen derselben Bewegungsgleichungen und derselben Vertauschungsrelationen schien uns ein besonders wesentlicher Zug der neuen Theorie. Trotz dieses engen Zusammenhangs beider Arten von Spektra bestehen aber zwischen den kontinuierlichen und den diskreten Spektra mathematisch wie physikalisch charakteristische Unterschiede, entsprechend dem Unterschied zwischen Fourierreihe und Fourierintegral in der klassischen Theorie; es erscheint uns deshalb notwendig, auch die Behandlung der kontinuierlichen Spektra hier in groben Umrissen darzustellen. Die mathematische Theorie der bei unendlichen quadratischen Formen auftretenden Streckenspektra ist im Anschluß an die grundlegenden Untersuchungen von Hilbert ausführlich entwickelt worden von Hellinger (l. c.) für den Fall beschränkter quadratischer Formen. Wenn wir uns hier erlauben, die Ergebnisse Hellingers auf die bei uns auftretenden unbeschränkten Formen zu übertragen, so scheint uns dies deshalb berechtigt, weil die Methoden Hellingers offenbar vollständig dem physikalischen Sinne des gestellten Problems entsprechen.

Betrachten wir zunächst kurz das klassische Analogon unseres Problems, die aperiodische Bewegung und ihr Fourierintegral. Während in einer Fourierreihe zu einer Schwingung $e^{2\pi i \nu t}$ stets eine gewisse Amplitude $a(\nu)$ gehört, tritt beim Fourierintegral an Stelle von $a(\nu)$ eine Größe der Form $\varphi(\nu)\,d\nu$, wo man $\varphi(\nu)$ gewissermaßen als Amplitudendichte pro Frequenzintervall $d\nu$ bezeichnen kann. In ähnlicher Weise kann man, was physikalisch unmittelbar einleuchtet, alle Größen, wie Intensität, Polarisation usw., stets nur auf einen Frequenzbereich $d\nu$ zwischen ν und $\nu + d\nu$ beziehen, nicht aber auf eine bestimmte Frequenz selbst. Ganz ähnliche Verhältnisse werden wir auch in der Quantenmechanik zu erwarten haben. An Stelle der Größen $q(kl)$ werden Größen der Form $q(k, W)\,dW$ bzw. $q(W, W')\,dW dW'$ treten, je nachdem der eine der beiden Indizes oder beide im kontinuierlichen Gebiet liegen. Ja, an Stelle der Energie W selbst wird eine „Gesamtenergie" pro Intervall dW treten müssen, da ja im kontinuierlichen Gebiet die Wahrscheinlichkeit, daß das Atom eine ganz bestimmte Energie W hat, Null ist. Um über diese Fragen Klarheit zu schaffen, werden wir im folgenden die mathematische Theorie nach Hellinger kurz skizzieren.

Bei unendlichen quadratischen Formen kann der Fall eintreten, daß die Form

$$\sum_{mn} H(mn)\, x_m x_n^*$$

nicht durch eine orthogonale Substitution in die Gestalt $\sum_n W_n y_n y_n^*$ übergeführt werden kann. Dann dürfen wir in Analogie zu den Ergebnissen bei beschränkten Formen annehmen, daß es eine Darstellung mit kontinuierlichem Spektrum

$$\sum_{mn} H(mn)\, x_m x_n^* = \sum_n W_n y_n y_n^* + \int W(\varphi)\, y(\varphi) y^*(\varphi)\, d\varphi \qquad (26)$$

gibt, bei der die ursprünglichen Variablen durch eine „orthogonale Transformation" mit neuen Variablen $y(n)$, $y(\varphi)$ zusammenhängen; nur muß man genauer auseinandersetzen, was man hier unter einer orthogonalen Transformation versteht.

Betrachten wir wieder die linearen Gleichungen (15), Kap. 3,

$$W x_k - \sum_l H(kl)\, x_l = 0, \qquad (27)$$

so wird der betrachtete Fall eines Integralbestandteils in (26), Kap. 3, dann eintreten, wenn es nicht nur diskrete Werte W_n gibt, für die diese Gleichungen lösbar sind, sondern auch ein Kontinuum solcher Werte, eine oder mehrere „Strecken" der W-Achse (Streckenspektrum). Für irgend einen Punkt W dieses Kontinuums existiert eine Lösung $x_l(W)$ (oder mehrere, was wir der Einfachheit halber ausschließen wollen); für zwei solche W-Werte, W' und W'', gelten also die Gleichungen:

$$\left.\begin{aligned} W' x_k(W') - \sum_l H(kl)\, x_l(W') &= 0, \\ W'' x_k^*(W'') - \sum_l H^*(kl)\, x_l^*(W'') &= 0, \end{aligned}\right\} \qquad (28)$$

aus denen man, wie oben, schließt

$$(W' - W'') \sum_k x_k(W') x_k(W'') = 0. \tag{29}$$

Versucht man zu diesen Orthogonalitätsrelationen die Normierungsbedingung

$$\sum_k |x_k(W)|^2 = 1$$

hinzuzufügen, so sieht man, daß die Funktion der zwei Veränderlichen

$$\sum_k x_k(W') x_k(W'')$$

unstetig in einem schlimmen Sinne wäre, wenn sie überhaupt existierte. Tatsächlich konvergiert die betrachtete Summe nicht, stellt also auch keine Funktion dar.

Daher ist eine andere Art der Normierung notwendig. Nach Hellinger setzt man

$$\sum_k \left|\int x_k(W)\, d\, W\right|^2 = \varphi(W). \tag{30}$$

Die Reihe auf der linken Seite ist im allgemeinen konvergent und stellt eine monotone Funktion $\varphi(W)$ dar, die mit gewissen Einschränkungen willkürlich gewählt werden kann, da ja die $x_k(W)$ nur bis auf einen von k unabhängigen Faktor bestimmt sind. Auf die physikalische Bedeutung dieser Funktion $\varphi(W)$, durch welche die Lösungen $x_k(W)$ festgelegt werden, werden wir später eingehen. Hellinger nennt $\varphi(W)$ die „Basisfunktion" und zeigt, daß sich die Orthogonalitätsbedingungen in folgende Form bringen lassen: Es seien Δ_1 und Δ_2 irgend zwei Intervalle des Streckenspektrums und Δ_{12} das ihnen gemeinsame Teilstück (das auch fehlen kann); dann gilt:

$$\left.\begin{aligned}\sum_k \int_{\Delta_1} x_k(W')\, d\, W' \int_{\Delta_2} x_k(W'')\, d\, W'' &= \int_{\Delta_{12}} d\,\varphi(W)\\ &= \varphi(W^{(2)}) - \varphi(W^{(1)}),\end{aligned}\right\} \tag{31}$$

wo $W^{(1)}, W^{(2)}$ die Endpunkte von Δ_{12} sind. Rechts steht also Null, wenn die Strecken Δ_1, Δ_2 sich nicht überdecken.

Denkt man sich die Intervalle $\Delta_1, \Delta_2, \Delta_{12}$ sehr klein, so kann man symbolisch schreiben

$$\sum_k x_k(W')\, d\, W' . x_k(W'')\, d\, W'' = d\,\varphi(W). \tag{32}$$

Diese Beziehung legt den Gedanken nahe, allgemein mit den Größen $x_k(W)\, d\, W$ als „Differentiallösungen" von (27), Kap. 3, zu operieren, wobei man nur zu beachten hat, daß die betreffenden Gleichungen stets im Sinne von (31), Kap. 3, zu interpretieren sind; diese Differentiallösungen sind dann in gewöhnlicher Weise orthogonal, aber nicht auf 1, sondern auf das Differential der Basisfunktion $\varphi(W)$ normiert.

Die Gesamtheit der diskreten Werte x_{kn} und der in einem Index diskret, im anderen kontinuierlich verteilten $x_k(W)$ bildet die Elemente der „orthogonalen" Matrix:

$$S = (x_{kn}, x_k(W)\, d\, W),$$

die man schematisch so darstellen kann:

$$S =$$

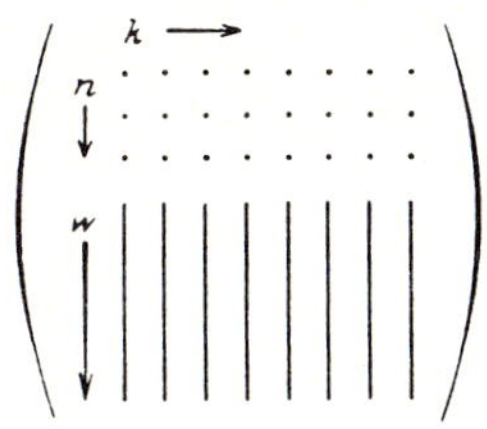

(33)

Die Orthogonalitäts- und Normalisierungsgleichungen für die ganze Matrix zerfallen in vier verschiedene Typen:

$$\left.\begin{array}{ll} \sum_k x_{km} x_{kn}^* = \delta_{mn}; & \\ \sum_k x_{kn} x_k^*(W)\, d\, W = 0; & \sum_k x_k(W)\, d\, W . x_{kn}^* = 0; \\ \sum_k x_k(W')\, d\, W'\, x_k^*(W'')\, d\, W'' = d\,\varphi. & \end{array}\right\} \quad (34)$$

Man kann aber auch die Orthogonalitätsrelationen für die Kolonnen anschreiben; diese lauten:

$$\sum_n x_{kn} x_{ln}^* + \int \frac{x_k(W)\, d\, W . x_l^*(W)\, d\, W}{d\,\varphi}$$

$$= \sum_n x_{kn} x_{l-}^* + \int \frac{d\, W}{\varphi'} x_k(W)\, x_l^*(W) = \delta_{kl}, \quad (35)$$

wo $\varphi' = \dfrac{d\,\varphi}{d\, W}$ gesetzt ist.

Mit Hilfe dieser Matrix hat man die Variablen x_n in neue y_n, $y(\varphi)\, d\,\varphi$ zu transformieren; man setze:

$$\left.\begin{array}{rl} y_n = & \sum_k x_{kn} . x_k \\ y(\varphi)\, d\,\varphi = & \sum_k x_k(W)\, d\, W . x_k. \end{array}\right\} \quad (36)$$

Dann ergibt eine einfache Rechnung:

$$\sum_n W_n y_n y_n^* + \int W(\varphi)\, y(\varphi) y^*(\varphi)\, d\varphi = \sum_{kl} H(kl)\, x_k x_l^*. \tag{37}$$

Damit ist die Hauptachsentransformation ausgeführt.

Untersuchen wir nun weiter, welche Darstellung der Koordinaten- und Impulsmatrizen man mit Hilfe dieser orthogonalen Transformation erhält, d. h. was die Gleichungen

$$\left.\begin{aligned} p &= S p_0 S^{-1} \\ q &= S q_0 S^{-1} \end{aligned}\right\} \tag{38}$$

oder allgemein

$$f(pq) = S f(p_0 q_0) S^{-1} \tag{39}$$

hier bedeuten. Wir finden z. B. für p vier Typen von Elementen:

$$\left.\begin{aligned} p(mn) &= \sum_{kl} x_{km}^* p^0(kl)\, x_{ln} \\ p(m, W)\, dW &= \sum_{kl} x_{km}^* p^0(kl)\, x_l(W)\, dW, \\ p(W, n)\, dW &= \sum_{kl} x_k^*(W)\, dW . p^0(kl)\, x_{ln}, \\ p(W', W'')\, dW' dW'' &= \sum_{kl} x_k^*(W')\, dW'\, p^0(kl)\, x_l(W'')\, dW''. \end{aligned}\right\} \tag{40}$$

In ähnlicher Weise werden allgemein entsprechend unserer früher ausgesprochenen Erwartung an Stelle der Amplituden $p(mn)$ im Falle eines kontinuierlich veränderlichen Index „Amplitudendichten" $p(mW)\,dW$ treten, die sich auf ein Intervall dW beziehen. Dabei ist es aber nicht notwendig, als den kontinuierlich veränderlichen Index eben die Energie zu nehmen. Man könnte statt der Energie z. B. die Größe $\varphi(W)$ einführen. An Stelle von $p(mW)\,dW$ würde dann $p(m\varphi)\dfrac{dW}{d\varphi}\,d\varphi$ treten. Schließlich wird die Energie W_n im kontinuierlichen Falle ersetzt durch die Größe $W(\varphi)\,d\varphi$. An Stelle der Energie des einzelnen Atoms tritt also eine Art Gesamtenergie pro Intervall dW. Daher bedeutet $d\varphi$ im wesentlichen die Anzahl der Atome, deren Energie zwischen W und $W + dW$ liegt oder die a priori-Wahrscheinlichkeit dafür, daß die Energie des Atoms zwischen W und $W + dW$ liegt. Hier erkennen wir den Unterschied zwischen den Fällen der diskreten stationären Zustände einerseits und der kontinuierlichen Zustandsmannigfaltigkeiten andererseits am deutlichsten und wir sehen einen einfachen Zusammenhang des Problems der statistischen Gewichte mit der Frage nach der Normierung der Lösung von (27), Kap. 3. Im Falle diskreter Zustände machen wir bei nicht mehrfachen Eigenwerten den einfachen physikalischen Ansatz, daß jeder Zustand das statistische Gewicht 1 haben soll. Dem entspricht es, daß wir eine Normierung der x_{kn} auf Grund der Forderung

$$\sum_k x_{kn} x^*_{kn} = 1$$

durchführten. Im Falle kontinuierlicher Zustandsmannigfaltigkeiten war eine so einfache Festlegung der a priori-Wahrscheinlichkeiten nicht möglich, zu ihrer Bestimmung und damit auch zur Bestimmung der Funktion φ sind eingehendere Untersuchungen des betreffenden Problems notwendig. Daher dürfte sich auch der Zusammenhang der Übergangswahrscheinlichkeiten mit den Amplituden im Falle kontinuierlicher Spektra etwas verwickelter gestalten, als bei den Linienspektra.

Die durch (40), Kap. 3, und entsprechende Formen dargestellten Matrizen von p, q oder $f(p, q)$ lassen sich allgemein durch nebenstehendes Schema anschaulich machen: Die physikalische Bedeutung dieses Schemas leuchtet ein.

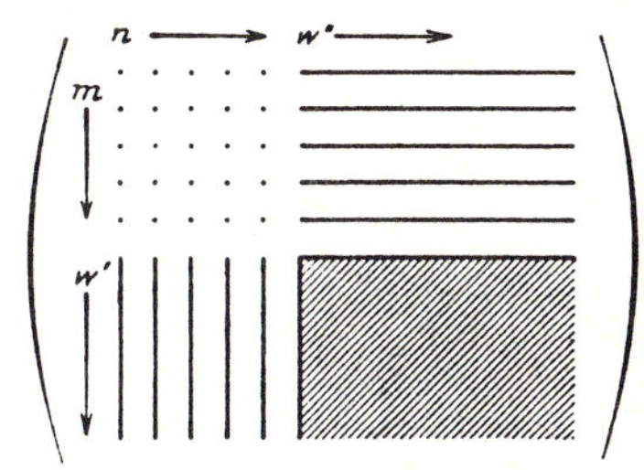

Es gibt vier Arten von „Übergängen", die etwa ein einfaches Analogon bilden zu den in der bisherigen Theorie des Wasserstoffatoms postulierten „Übergängen": 1. von Ellipse zu Ellipse, 2. von Ellipse zu Hyperbel, 3. von Hyperbel zu Ellipse, 4. von Hyperbel zu Hyperbel.

Gegen die Formeln (38) und (40), Kap. 3, kann noch eingewendet werden, daß die unendlichen Summen der rechten Seite sicher in manchen Fällen nicht konvergieren, also auch keine Funktion darstellen, da ja auch in der klassischen Theorie eine Darstellung einer Funktion $f(p, q)$ durch Fourierintegrale manchmal nicht möglich ist, z. B. dann nicht, wenn die betreffenden Funktionen f für große Zeiten linear mit der Zeit anwachsen (was im allgemeinen für die Koordinate der Fall sein wird). Auf diesen Einwand kann man aber erwidern, daß die beobachtbaren Wirkungen des Atoms (wie die Strahlung, die Kraft auf ein anderes Atom usw.) im allgemeinen nicht zu dieser Art von Funktionen gehören, daß also die ihnen entsprechenden Summen vom Typus der Formeln (40), Kap. 3, konvergieren dürften.

Kapitel 4. Physikalische Anwendungen der Theorie.

§ 1. Sätze über Impuls und Drehimpuls; Intensitätsformeln und Auswahlregeln. Als Anwendung der allgemeinen Theorie, wie sie im vorangehenden begründet wurde, sollen nun die bekannten Tatsachen bezüglich der „Quantelung" des Drehimpulses und einige damit zusammenhängende Gesetzmäßigkeiten abgeleitet werden.

Wir werden dabei zugleich einige charakteristische Beispiele für die Integration der quantenmechanischen Bewegungsgleichungen kennenlernen. Die früher besprochenen Störungsmethoden können natürlich erst dann erfolgreich angewandt werden, wenn eine Reihe besonders einfacher Beispiele, welche als ungestörte Systeme H_0 gewählt werden können, auf andere Weise integriert worden sind. Die quantenmechanischen Bewegungsgleichungen, wie sie aus der Komponentenzerlegung der Matrizengleichungen entspringen, bieten nun die besondere Schwierigkeit, daß — abgesehen vom Beispiel des harmonischen Oszillators — in jeder einzelnen Gleichung bereits unendlich viele Unbekannte auftreten. Ein im folgenden mehrfach gebrauchtes und wie es scheint, sehr häufig anwendbares Verfahren, diese Schwierigkeit zu überwinden, besteht im folgenden. Man sucht zunächst in Analogie zur klassischen Theorie Integrale der Bewegungsgleichungen, also Funktionen $A(p, q)$, welche auf Grund der Bewegungsgleichungen und der Vertauschungsregeln zeitlich konstant sind und daher bei nichtentarteten periodischen Systemen Diagonalmatrizen werden. Ist nun $\varphi(p, q)$ irgend eine Funktion, so kann die Differenz

$$\varphi A - A\varphi = \psi$$

vermittelst der Vertauschungsregeln berechnet werden; wenn A eine Diagonalmatrix ist, ergibt sich ein System von Gleichungen, die nur je endlich viele Unbekannte, nämlich je eine einzige Komponente der Matrizen φ und ψ (und je zwei Diagonalglieder von A) enthalten.

Ist in kartesischen Koordinaten $H = H'(p) + H''(q)$ — worin also auch die relativistische Mechanik enthalten ist —, so ist sofort zu sehen, daß die Komponenten des Drehimpulses $\mathfrak{M}$:

$$\left.\begin{aligned} M_x &= \sum_{k=1}^{f/3} (p_{ky} q_{kz} - q_{ky} p_{kz}), \\ M_y &= \sum_{k=1}^{f/3} (p_{kz} q_{kx} - q_{kz} p_{kx}), \\ M_z &= \sum_{k=1}^{f/3} (p_{kx} q_{ky} - q_{kx} p_{ky}) \end{aligned}\right\} \tag{1}$$

unter den gleichen allgemeinen Bedingungen konstant werden, wie in der klassischen Theorie. Denn es ergibt sich für die zeitliche Ableitung etwa von M_z eine Summe

$$\dot{M}_z = \varphi(q) + \psi(p),$$

und wegen der Vertauschbarkeit aller p untereinander und aller q untereinander verschwinden φ, ψ, unter denselben Bedingungen wie klassisch.

Die gleiche Bemerkung ist anzuwenden auf den Translationsimpuls

$$\mathfrak{p} = \sum_{k=1}^{f/3} \mathfrak{p}_k; \quad \text{d. h.} \quad p_x = \sum_{k=1}^{f/3} p_{kx}, \ldots, \tag{2}$$

der ebenfalls konstant wird. Also gilt auch der Schwerpunktsatz, wie in der klassischen Theorie.

Wir merken hier sogleich eine später zu benutzende Formel an, die aus den Vertauschungsrelationen (3), Kap. 2, abzuleiten ist. Es wird:

$$\begin{aligned} M_x M_y - M_y M_x &= \sum_{kl} \{(p_{ky} q_{kz} - q_{ky} p_{kz})(p_{lz} q_{lx} - q_{lz} p_{lx}) \\ &\qquad - (p_{kz} q_{kx} - q_{kz} p_{kx})(p_{ly} q_{lz} - q_{ly} p_{lz})\}, \\ &= \sum_{kl} \{p_{ky} q_{lx} (q_{kz} p_{lz} - p_{lz} q_{kz}) \\ &\qquad + q_{ky} p_{lx} (p_{kz} q_{lz} - q_{lz} p_{kz}), \\ &= \frac{h}{2\pi i} \sum_k (p_{kx} q_{ky} - q_{kx} p_{ky}), \end{aligned}$$

also

$$M_x M_y - M_y M_x = \varepsilon M_z, \quad \left(\varepsilon = \frac{h}{2\pi i}\right). \tag{3}$$

Man sieht übrigens aus dieser Formel unmittelbar, daß der Flächensatz wie in der klassischen Theorie stets für höchstens eine oder für alle drei Achsen gilt.

Für das Folgende wollen wir annehmen, daß das uns vorliegende Problem bei Behandlung nach den im vorigen Kapitel entwickelten Methoden zu diskreten Energiewerten (Punktspektrum) führt. Ist dann $\dot{M}_z = 0$ bei einem nichtentarteten System — dies wird z. B. der Fall sein, wenn Kräfte mit Symmetrie um die z-Achse auf das Atom wirken —, so muß M_z eine Diagonalmatrix werden; die einzelnen Diagonalglieder sind als Drehmomente des Atoms um die z-Achse für die einzelnen Zustände des Atoms anzusehen. Zur Untersuchung der Elektronenbewegungen in diesem Falle beachten wir zunächst, daß aus (1), Kap. 4,

$$q_{lz} M_z - M_z q_{lz} = 0 \tag{4}$$

folgt, was wegen $M_z(nm) = \delta_{nm} M_{zn}$ bedeutet:

$$q_{lz}(nm)(M_{zn} - M_{zm}) = 0. \tag{5}$$

Man sieht: Bei einem Quantensprung, bei dem sich das Drehmoment M_z ändert, liegt die „Schwingungsebene" der erzeugten „Kugelwelle" senkrecht zur z-Achse. Weiter wird

$$\left.\begin{aligned} q_{lx} M_z - M_z q_{lx} &= -\varepsilon q_{ly}, \\ q_{ly} M_z - M_z q_{ly} &= \varepsilon q_{lx}, \end{aligned}\right\} \tag{6}$$

also

$$\left.\begin{aligned} q_{lx}(nm)(M_{zn} - M_{zm}) &= -\varepsilon q_{ly}(nm), \\ q_{ly}(nm)(M_{zn} - M_{zm}) &= \varepsilon q_{lz}(nm). \end{aligned}\right\} \tag{7}$$

Bei Sprüngen ohne Änderung von M_z ist das ausgestrahlte Licht parallel der z-Achse linear polarisiert. Aus (7), Kap. 4, folgt weiter

$$\left\{(M_{zn} - M_{zm})^2 - \frac{h^2}{4\pi^2}\right\} q_{l\eta}(nm) = 0; \quad \eta = x, y. \tag{8}$$

Man schließt endlich: Bei jedem Quantensprung ändert sich M_{zn} um 0 oder um $\pm \frac{h}{2\pi}$. Das im letzteren Falle ausgestrahlte Licht ist nach (7), Kap. 4, zirkular polarisiert. Nach dem obigen Ergebnis betreffs der möglichen Änderungen von M_z kann M_{zn} dargestellt werden in der Form

$$M_{zn} = \frac{h}{2\pi}(n_1 + C), \quad n_1 = \cdots -2, -1, 0, 1, 2 \cdots \tag{9}$$

Gäbe es Zustände, deren Drehmoment nicht in diese Reihe paßt, so könnten zwischen diesen und den in (9), Kap. 4, gegebenen keine Übergänge und keinerlei Wechselwirkung eintreten. Man kann (9), Kap. 4, zum Anlaß nehmen, um eine Spaltung von n in zwei Komponenten durchzuführen, von denen die eine die in (9), Kap. 4, eingeführte Zahl n_1 ist, während die andere n_2 die verschiedenen n mit gleichem n_1 abzählt. Unsere Matrizen werden dann vierdimensional, und die gewonnenen Ergebnisse bezüglich der Elektronenbewegungen lassen sich so zusammenfassen:

$$q_{lz}(nm) = \delta_{n_1, m_1} q_{lz}(nm); \tag{10}$$

$$\left.\begin{aligned} q_{lx}(nm) &= \delta_{1, |n_1 - m_1|}\, q_{lx}(nm), \\ q_{ly}(nm) &= \delta_{1, |n_1 - m_1|}\, q_{ly}(nm); \end{aligned}\right\} \tag{10'}$$

$$q_{lx}(n_1, n_2;\ n_1 \pm 1, m_2) \mp i\, q_{ly}(n_1, n_2;\ n_1 \pm 1, m_2) = 0. \tag{10''}$$

Aus (4), Kap. 4, und (6), Kap. 4, folgt ferner, wenn $\mathfrak{q}_l^2 = \mathfrak{q}_l^2 = \mathfrak{q}_{lx}^2 + \mathfrak{q}_{ly}^2 + \mathfrak{q}_{lz}^2$ gesetzt wird:

$$\mathfrak{q}_l^2 \mathfrak{M}_z - \mathfrak{M}_z \mathfrak{q}_l^2 = 0. \tag{11}$$

Diese Beziehung bedeutet, daß $\mathfrak{q}_l^2$ in bezug auf die „Quantenzahl" n_1 eine Diagonalmatrix ist.

Die Relationen (4) bis (7), Kap. 4, und (10), (11), Kap. 4, sind auch dann richtig, wenn wir für $\mathfrak{q}_{lx}, \mathfrak{q}_{ly}, \mathfrak{q}_{lz}$ einsetzen:

$$\mathfrak{p}_{lx}, \mathfrak{p}_{ly}, \mathfrak{p}_{lz} \quad \text{oder auch} \quad \mathfrak{M}_x, \mathfrak{M}_y, \mathfrak{M}_z.$$

Also gilt insbesondere

$$\left.\begin{aligned} &M_x(nm) = \delta_{1, |n_1 - m_1|} M_x(nm); \quad M_y(nm) = \delta_{1, |n_1 - m_1|} M_y(nm), \\ &M_x(n_1, n_2;\ n_1 \pm 1, m_2) \mp i M_y(n_1, n_2;\ n_1 \pm 1, m_2) = 0. \end{aligned}\right\} \tag{12}$$

Ferner ist [vgl. Gleichung (1), Kap. 4] $\mathfrak{M}^2 = M^2 = M_x^2 + M_y^2 + M_z^2$ in bezug auf n_1 eine Diagonalmatrix; denn es gilt

$$M^2 M_z - M_z M^2 = 0. \tag{13}$$

Für ein System, in welchem alle drei Flächensätze gelten, können die konstanten Komponenten von $\mathfrak{M}$ gewiß nicht sämtlich Diagonalmatrizen sein. Denn sonst wären auf jede dieser Komponenten die oben für $M_z =$ Diagonalmatrix ausgeführten Betrachtungen anzuwenden, was zu Widersprüchen führen würde. Ein solches System ist also notwendig entartet.

Wir wollen nun ein System $H = H_0 + \lambda H_1 + \cdots$ von folgender Art betrachten: Für $\lambda = 0$ sollen alle drei Flächensätze gelten. Für $\lambda \neq 0$ soll das System nichtentartet sein; dabei soll die Konstanz von M_z bestehen bleiben. Die Energie H_0 hängt nicht von n_1 ab. Die Ergebnisse, die bei dieser Untersuchung für $\lambda \neq 0$ gewonnen werden, können zum Teil auch auf das entartete System H_0 übertragen werden, nämlich soweit, als sie unabhängig sind erstens von λ und zweitens von der ausgezeichneten Richtung z.

Die vorausgesetzte Entartung des Systems für $\lambda = 0$ wird dadurch ausgedrückt, daß $\dot{M}_x$, $\dot{M}_y$, $\frac{d}{dt}(M^2)$ keine Glieder mit der nullten Potenz von λ enthalten. Es wird also

$$\left.\begin{aligned} \nu_0(nm)\, M_\eta(nm) &= 0, \quad \eta = x, y; \\ \nu_0(nm)\, M^2(nm) &= 0. \end{aligned}\right\} \tag{14}$$

Da W_0 von der früher eingeführten Quantenzahl n_1 unabhängig ist, also $\nu_0(n_1, n_2; m_1, n_2) = 0$, während stets $\nu_0(n_1, n_2; m_1, m_2) \neq 0$ für $n_2 \neq m_2$ ist, so folgt aus (14), Kap. 4:

$$\left.\begin{aligned} M_\eta^0(nm) &= \delta_{n_2 m_2} M_\eta^0(nm), \\ M^{0^2}(nm) &= \delta_{n_2 m_2} M^{0^2}(nm). \end{aligned}\right\} \tag{15}$$

Das Quadrat des Gesamtimpulses M^{0^2} ist wegen (13), (15), Kap. 4, eine Diagonalmatrix. Die Doppelsumme, welche ein Element der Matrix M_x^0, M_y^0 darstellt, zieht sich in eine einfache Summe zusammen:

$$\left.\begin{aligned} &\sum_{k_1 k_2} M_x^0(n_1 n_2;\, k_1 k_2)\, M_y^0(k_1 k_2;\, m_1 m_2) \\ &= \delta_{n_2 m_2} \sum_{k_1} M_x^0(n_1 n_2;\, k_1 n_2)\, M_x^0(k_1 n_2;\, m_1 n_2), \end{aligned}\right\} \tag{16}$$

die wegen der endlichen Anzahl der bei festem n_2 möglichen n_1 (die Glieder von $M^{0^2} = M_x^{0^2} + M_y^{0^2} + M_z^{0^2} \geqq M_z^2$ hängen nicht von n_1 ab) nur eine endliche Anzahl von Summanden enthält. Wir können in (3), Kap. 4, angewandt für M_x^0, M_y^0, M_z^0, jeweils die zu einem bestimmten n_2 gehörigen Gleichungen nach n_1 summieren und erhalten bei festem n_2 [1]):

$$\sum_{n_1} M_z(n_1 n_2;\, m_1 n_2) = \sum_{n_1} (n_1 + C)\frac{h}{2\pi} = 0. \tag{17}$$

Wenn wir noch beachten, daß nach (12), Kap. 4, und (16), Kap. 4, die Summe (17), Kap. 4, für jede einzelne lückenlose Reihe der n_1 verschwindet, so folgt, daß die bei festem n_2 möglichen Werte von $n_1 + C$ eine lückenlose Reihe bilden und symmetrisch zu Null liegen, also notwendig entweder ganze Zahlen oder „halbe Zahlen", d. h. Zahlen der Reihe $\cdots -\frac{3}{2}, -\frac{1}{2}, \frac{1}{2}, \frac{3}{2}, \cdots$ sein müssen. Führen wir nachträglich für das Moment M_z um die z-Achse statt $(n_1 + C)\frac{h}{2\pi}$ die bisher in der Literatur übliche Bezeichnung $m\frac{h}{2\pi}$ ein, so haben wir also gezeigt, daß für m die Auswahlregel $m \rightarrow \begin{cases} m+1 \\ m \\ m-1 \end{cases}$ gilt, und daß m entweder „ganz"- oder „halbzahlig" ist.

Unser Ergebnis zeigt ferner, daß Verbote einzelner Zustände, wie sie z. B. in der bisherigen Theorie des Wasserstoffs notwendig waren, um Zusammenstöße des Elektrons mit dem Kerne zu verhüten, in der hier versuchten Theorie keinen Platz haben.

Wir werden nun über (5), Kap. 4, und (8), Kap. 4, hinausgehend auch das Auswahlprinzip für die „Quantenzahl des Gesamtimpulses", ferner die Intensitäten beim Zeemaneffekt aus den Grundgleichungen unserer Theorie herzuleiten suchen.

Erinnern wir uns an die klassische Theorie dieser Auswahlregeln: Es ist dort nur nötig, ein Koordinatensystem einzuführen, dessen Z-Achse mit der Richtung des Gesamtimpulses zusammenfällt, dann werden für die neuen Koordinaten dieselben Resultate hinsichtlich $\mathfrak{M}$ abgeleitet werden können, wie vorher für M_z. Konstruieren wir uns also klassisch ein solches Koordinatensystem x', y', z': Es wird jedenfalls gelten müssen:

$$z' = x\frac{M_x}{M} + y\frac{M_y}{M} + z\frac{M_z}{M},$$

damit die z'-Achse die Richtung des Gesamtimpulses habe. (Den Index ^{0} bei den Impulsen und Koordinaten werden wir im folgenden der Einfach-

[1]) Auf die Tatsache, daß bei endlicher Diagonalsumme $D(\boldsymbol{a}\boldsymbol{b})$ stets $D(\boldsymbol{a}\boldsymbol{b}) = D(\boldsymbol{b}\boldsymbol{a})$ ist, wurde schon in I. aufmerksam gemacht.

heit halber wieder weglassen; es beziehen sich die Rechnungen stets auf den Grenzfall $\lambda = 0$.) Ferner können wir es so einrichten, daß die x'-Achse in der x, y-Ebene liegt. Dadurch ist dann alles festgelegt und es gilt:

$$x' = y \frac{M_y}{\sqrt{M_x^2 + M_y^2}} - x \frac{M_y}{\sqrt{M_x^2 + M_y^2}},$$
$$y' = \frac{z(M_x^2 + M_y^2) - x M_z M_x - y M_z M_y}{M\sqrt{M_x^2 + M_y^2}}.$$

Versuchen wir nun ein analoges Verfahren in der Quantenmechanik. Wir führen die drei Größen ein:

$$\left.\begin{aligned} Z_l &= q_{lx} M_x + q_{ly} M_y + q_{lz} M_z, \\ X_l &= q_{ly} M_x - M_y q_{lx}, \\ Y_l &= M_x q_{lz} M_x + M_y q_{lz} M_y - q_{lx} M_z M_x - M_y M_z q_{ly}. \end{aligned}\right\} \quad (18)$$

Zur Ableitung der gesuchten Auswahlregeln brauchen wir noch einige Vertauschungsrelationen, die sich aus (4), Kap. (4), und (6), Kap. 4, ergeben $\left(\varepsilon = \frac{h}{2\pi i}\right)$:

$$q_{lx} M^2 - M^2 q_{lx} = 2\varepsilon (q_{lz} M_y - M_z q_{ly}) \quad (19)$$

und die durch zyklische Vertauschung hieraus hervorgehenden Gleichungen für q_{ly}, q_{lz}. Dann folgt[1]) aus (3), (4), (6) und (19), Kap. 4:

$$\left.\begin{aligned} X_l M^2 - M^2 X_l &= 2\varepsilon Y_l, \\ Y_l M^2 - M^2 Y_l &= \varepsilon (X_l M^2 + M^2 X_l), \\ Z_l M^2 - M^2 Z_l &= 0. \end{aligned}\right\} \quad (20)$$

[1]) Die erste und dritte Formel von (20), Kap. 4, ergibt sich aus ganz einfacher Rechnung. Die zweite Gleichung (20), Kap. 4, kann etwa folgendermaßen abgeleitet werden. Nach (18), Kap. 4, gilt:

$$Y_l = M_x q_{lz} M_x + M_y q_{ly} M_y - q_{lx} M_z M_x - M_y M_z q_{ly},$$

und wegen (6), Kap. 4:

$$\begin{aligned} Y_l &= q_{lz}(M_x^2 + M_y^2) - \varepsilon q_{ly} M_x + \varepsilon M_y q_{lx} + \varepsilon^2 q_{lz} \\ &\quad - q_{lx} M_z M_x - M_y M_z q_{ly} \\ &= q_{lz}(M^2 - M_z^2) - \varepsilon . X_l + \varepsilon^2 q_{lz} - q_{lx} M_z M_x - M_y M_z q_{ly}. \end{aligned}$$

Bei der Berechnung von $Y_l M^2 - M^2 Y_l$ ist jetzt zu beachten, daß M^2 mit M_x, M_y, M_z vertauschbar ist. Es folgt daher für den zweiten Teil der oben angeschriebenen Formel für Y_l:

$$(q_{lx} M_z M_x + M_y M_z q_{ly}) M^2 - M^2 (q_{lx} M_z M_x + M_y M_z q_{ly}) = \text{(vgl. 19, Kap. 4)}$$
$$= 2\varepsilon (q_{lz} M_y M_z M_x - M_z q_{ly} M_z M_x + M_y M_z q_{lx} M_z - M_y M_z M_x q_{lz}).$$

Ferner folgt aus den Vertauschungsrelationen, wenn man beachtet, daß nach (19, Kap. 4) $q_{lz} M^2 - M^2 q_{lz} = 2\varepsilon X_l$ ist:

$$q_{lz} M_y M_z M_x - M_y M_z M_x q_{lz} = \varepsilon (M_y M_z q_{ly} - q_{lx} M_z M_x),$$
$$M_y M_z q_{lx} M_z - M_z q_{ly} M_z M_x = - X_l . M_z^2 - \varepsilon (M_z q_{ly} M_y - q_{lx} M_x M_z),$$

Diese Gleichungen sind ganz analog zu den für die Auswahlregeln von M_z maßgebenden Relationen (4) und (6), Kap. 4; da wir weiter unten zeigen werden, daß sich die q_{lx}, q_{ly}, q_{lz} wirklich als lineare Funktionen mit für $\lambda = 0$ zeitlich konstanten Koeffizienten der X_l, Y_l, Z_l ausdrücken lassen, so können wir aus (20), Kap. 4, direkt die Auswahlregeln für M bestimmen. Da M^2 eine Diagonalmatrix ist, so folgt aus (20), Kap. 4:

$$\left.\begin{aligned} X_l(nm)(M_m^2 - M_n^2) &= -2\,\varepsilon\, Y_l(nm), \\ Y_l(nm)(M_m^2 - M_n^2) &= \varepsilon\, X_l(nm)(M_m^2 + M_n^2), \\ Z_l(nm)(M_m^2 - M_n^2) &= 0. \end{aligned}\right\} \tag{21}$$

Die letzte der Gleichungen (21), Kap. 4, sagt aus, daß in Z keine Schwingungen vorkommen, die einer Änderung von M^2 entsprechen. Aus den beiden ersten Gleichungen folgt

$$X_l(nm)\left\{(M_m^2 - M_n^2)^2 - \frac{h^2}{2\pi^2}(M_m^2 + M_n^2)\right\} = 0. \tag{22}$$

Setzen wir nun $M_m^2 = \left(\frac{h}{2\pi}\right)^2 (a_m^2 - \frac{1}{4})$ (wo a_m irgend eine Funktion der Quantenzahlen bedeutet), so ergibt (22), Kap. 4,

$$X_l(nm)\left((a_n - a_m)^2 - 1\right)\left((a_n + a_m)^2 - 1\right) = 0,$$

oder, wenn $X_l(nm)$ nicht verschwindet,

$$a_n = \pm a_m \pm 1. \tag{23}$$

Es bedeutet keine Beschränkung der Allgemeinheit, wenn wir a_m stets als positiv $\geqq \frac{1}{2}$ annehmen. Die a_m bilden also eine Reihe der Form $C, 1 + C, 2 + C, \ldots$, wo C eine Konstante $\geqq \frac{1}{2}$ vorstellt. Setzen wir $a_m = j + \frac{1}{2}$, so wird

$$M^2 = j(j+1)\cdot\left(\frac{h}{2\pi}\right)^2 \tag{24}$$

und für j gilt das Auswahlprinzip $j \to \begin{cases} j+1 \\ j \\ j-1 \end{cases}$.

Dieses Ergebnis erinnert formal an die in die Landésche g-Formel eingehenden Werte von M^2.

---

und schließlich ergibt sich die gesuchte Formel (20), Kap. 4:

$$\begin{aligned} Y_l M^2 - M^2 Y_l &= 2\varepsilon X_l(M^2 - M_z^2 + \varepsilon^2) - \varepsilon(X_l M^2 - M^2 X_l) + 2\varepsilon X_l M_z^2 \\ &\quad - 2\varepsilon^2(q_{lx} M_x M_z - q_{lx} M_z M_x + M_y M_z q_{ly} - M_z M_y q_{ly}) \\ &= 2\varepsilon X_l(M^2 - M_z^2 + \varepsilon^2) - \varepsilon(X_l M^2 - M^2 X_l) + 2\varepsilon X_l M_z^2 - 2\varepsilon^3 X_l \\ &= \varepsilon(X_l M^2 + M^2 X_l). \end{aligned}$$

Führen wir nun wieder für M_z die Bezeichnung $m\frac{h}{2\pi}$ ein, so entnehmen wir aus (12), Kap. 4, und den Relationen:

$$\mathsf{M}^2 = \mathsf{M}_x^2 + \mathsf{M}_y^2 + \mathsf{M}_z^2 \text{ und}$$
$$(\mathsf{M}_x + i\mathsf{M}_y)(\mathsf{M}_x - i\mathsf{M}_y) = \mathsf{M}_x^2 + \mathsf{M}_y^2 - i\varepsilon\mathsf{M}_z = \mathsf{M}^2 - \mathsf{M}_z^2 - i\varepsilon\mathsf{M}_z:$$

$$\left.\begin{aligned} M_x(j, m-1; j, m) + iM_y(j, m-1; j, m) &= \frac{h}{2\pi}\sqrt{j(j+1) - m(m-1)}, \\ M_x(j, m; j, m-1) - iM_y(j, m; j, m-1) &= \frac{h}{2\pi}\sqrt{j(j+1) - m(m-1)}. \end{aligned}\right\} \quad (25)$$

Der Maximalwert m_{max} von m bei einem gegebenen Werte von j ist dadurch charakterisiert, daß die Sprünge $m_{max} \rightarrow m_{max} + 1$ nicht vorkommen, d. h. daß für diese Sprünge z. B. die rechte Seite von (24), Kap. 4, verschwindet. Dies ergibt

$$j = m_{max}.$$

Also kann auch j nur „halb-" oder „ganzzahlig" sein.

Die Berechnung der Intensitätsformeln beim Zeemaneffekt, d. h. der Abhängigkeit q_{lx}, q_{ly}, q_{lz} von m erscheint jetzt sehr einfach. Wir entnehmen aus (18), Kap. 4, durch Auflösen nach q_{lx}, q_{ly}, q_{lz} die Relationen

$$\left.\begin{aligned} q_{lz} &= (Z_l\mathsf{M}_z + \varepsilon X_l + Y_l)\mathsf{M}^{-2}, \\ q_{lx} + iq_{ly} &= [Z_l - q_{lz}(\mathsf{M}_z + i\varepsilon) + iX_l](\mathsf{M}_x - i\mathsf{M}_y)^{-1}, \\ q_{lx} - iq_{ly} &= [Z_l - q_{lz}(\mathsf{M}_z - i\varepsilon) - iX_l](\mathsf{M}_x + i\mathsf{M}_y)^{-1}. \end{aligned}\right\} \quad (26)$$

Diese Gleichungen erbringen auch den früher versäumten Beweis, daß die q_{lx}, q_{ly}, q_{lz} dargestellt werden können als lineare Funktionen der X_l, Y_l, Z_l mit für $\lambda = 0$ zeitlich konstanten Koeffizienten. Zugleich enthalten die Gleichungen (26), Kap. 4, die gesuchten Intensitätsformeln. Um dies einzusehen, bemerken wir zunächst, daß die X_l, Y_l, Z_l in bezug auf m Diagonalmatrizen sind. Denn es gilt:

$$\left.\begin{aligned} X_l\mathsf{M}_z - \mathsf{M}_zX_l &= 0, \\ Y_l\mathsf{M}_z - \mathsf{M}_zY_l &= 0, \\ Z_l\mathsf{M}_z - \mathsf{M}_zZ_l &= 0. \end{aligned}\right\} \quad (27)$$

Jetzt zerfällt unser Problem in zwei Teile, in die Diskussion der Intensitäten bei den Sprüngen $j \rightarrow j$ und $j \rightarrow j - 1$ (die Sprünge $j \rightarrow j + 1$ geben dann nichts Neues). Wir behandeln zunächst die Übergänge $j \rightarrow j$. Für diese sind nach (20), Kap. 4, nur Glieder in Z_l

vorhanden. Wir nennen diese Glieder $Z_l(j; m)$. Dann ergibt (26), Kap. 4, unter Berücksichtigung von $M_z = m \frac{h}{2\pi}$ und (24), Kap. 4:

$$\left.\begin{aligned} q_{lz}(j, m) &= \frac{2\pi}{h} Z_l(j, m) \frac{m}{j(j+1)}, \\ (q_{lx}+i q_{ly})(j, m-1; j, m) &= \frac{2\pi}{h} Z_l(j, m-1) \sqrt{\frac{j(j+1)-m(m-1)}{j(j+1)}}, \\ (q_{lx}-i q_{ly})(j, m; j, m-1) &= \frac{2\pi}{h} Z_l(j, m) \sqrt{\frac{j(j+1)-m(m-1)}{j(j+1)}}. \end{aligned}\right\} \tag{28}$$

Um schließlich noch die Abhängigkeit der Größe $Z_l(j, m)$ von m zu erhalten, benutzen wir etwa die Relation

$$M_x q_{ly} - q_{ly} M_x = \varepsilon q_{lz}; \tag{29}$$

sie ergibt in unserem Falle, daß $Z_l(j, m)$ nicht von m abhängt. Wir erhalten so für die Übergänge $j \to j$:

$$\left.\begin{aligned} & q_{lz}(j, m) : (q_{lx}+i q_{ly})(j, m-1; j, m) : (q_{lx}-i q_{ly})(j, m; j, m-1) \\ & \quad = m : \sqrt{j(j+1)-m(m-1)} : \sqrt{j(j+1)-m(m-1)}. \end{aligned}\right\} \tag{30}$$

Analog behandeln wir die Sprünge $j \to j-1$. Für diese ist nach (21), Kap. 4, $X_l(j, m; j-1, m) = \frac{\varepsilon}{j} Y_l(j, m; j-1, m)$. Drücken wir aus (26), Kap. 4, die Intensitäten durch $X_l(j, m; j-1, m)$ aus, so ergibt sich:

$$\left.\begin{aligned} & q_{lz}(j, m; j-1, m) = i \frac{2\pi}{h} X_l(j, m; j-1, m) \frac{1}{j}, \\ & (q_{lx}+i q_{ly})(j, m-1; j-1, m) \\ & \qquad = i \frac{2\pi}{h} X_l(j, m-1; j-1, m-1) \frac{\sqrt{j-m}}{j\sqrt{j+m-1}}, \\ & (q_{lx}-i q_{ly})(j, m; j-1, m-1) \\ & \qquad = -i \frac{2\pi}{h} X_l(j, m; j-1, m) \frac{\sqrt{j+m-1}}{j\sqrt{j-m}}. \end{aligned}\right\} \tag{31}$$

Um schließlich noch die Abhängigkeit der Größe $X_l(j, m; j-1, m)$ von m festzulegen, benutzen wir wieder die Relation (29), Kap. 4, die uns hier nach einfacher Rechnung ergibt:

$$X_l(j, m; j-1, m) = A(j, j-1)\sqrt{j^2-m^2}. \tag{32}$$

Wir erhalten so

$$\left.\begin{aligned} & q_{lz}(j, m; j-1, m) : (q_{lx}+i q_{ly})(j, m-1; j-1, m) \\ & \quad : (q_{lx}-i q_{ly})(j, m; j-1, m-1) = \sqrt{j^2-m^2} : \sqrt{(j-m)(j-m+1)} \\ & \qquad : -\sqrt{(j+m)(j+m-1)}. \end{aligned}\right\} \tag{33}$$

Die Sprünge $j \rightarrow j+1$ geben im wesentlichen dieselben Intensitäten; es gilt dann:

$$\left.\begin{aligned} & q_{lz}(j, m; j+1, m) : (q_{lx} + i q_{ly})(j, m; j+1, m+1) \\ & : (q_{lx} - i q_{ly})(j, m+1; j+1, m) = \sqrt{(j+1)^2 - m^2} \\ & : \sqrt{(j+m+2)(j+m+1)} : -\sqrt{(j-m+1)(j-m)}. \end{aligned}\right\} \quad (34)$$

Die Formeln (30), (33), (34), Kap. 4, stimmen mit den auf korrespondenzmäßigem Wege gefundenen Intensitätsformeln [1]) überein.

Auf eine einfache Folgerung aus (21), Kap. 4, müssen wir noch hinweisen: Die Sprünge $\Delta j = 0$ kommen nur in der „Z_I-Richtung" vor. Wenn wir die Bewegung eines einzigen Elektrons um einen Kern, also den Fall des Wasserstoffatoms betrachten, so folgt direkt aus (1), Kap. 4, daß Z verschwindet. Also kommen dann Sprünge $\Delta j = 0$ überhaupt nicht vor.

§ 2. Der Zeemaneffekt. Wenn man die Lorentzsche Kraft eines Magnetfeldes $\mathfrak{H}$ auf das Elektron $e/c\,[\mathfrak{v}\,\mathfrak{H}]$ in die Quantenmechanik übernimmt, so scheint es zunächst selbstverständlich, daß sich für die Atome der normale Zeemaneffekt ergibt. Denn genau unter denselben Voraussetzungen, unter denen klassisch das Larmorsche Theorem für das Kernatom abgeleitet werden kann — nämlich Vernachlässigung der Glieder mit $\mathfrak{H}^2$ —, ergibt sich auch hier das Larmorsche Theorem. Trotzdem besteht ein gewisser Unterschied zwischen der Quantenmechanik und der klassischen Theorie in der Berechtigung der Vernachlässigung von $\mathfrak{H}^2$. In der klassischen Theorie ist die Vernachlässigung von $\mathfrak{H}^2$ sicher für die Bahnen kleiner Dimensionen erlaubt, sicher nicht erlaubt für sehr große Bahnen oder gar Hyperbelbahnen. In der Quantenmechanik sind alle diese Bahnen — die weit außen liegenden, wie die innersten — wegen der der Quantenmechanik eigentümlichen Kinematik so eng miteinander verknüpft, daß die Berechtigung zur Vernachlässigung der Größe $\mathfrak{H}^2$ nicht ohne weiteres einleuchtet. Sind doch selbst vom Normalzustand aus die Wahrscheinlichkeiten von Übergängen zu freien Elektronen beträchtlich.

Für den Oszillator sind wir also des normalen Zeemaneffekts sicher; für das Kernatom dagegen scheint es nicht völlig ausgeschlossen, daß der enge Zusammenhang der weit außen und weit innen liegenden Bahnen zu Ergebnissen führt, die etwas vom normalen Zeemaneffekt abweichen. Wir

[1]) S. Goudsmit und R. de L. Kronig, Naturwiss. **13**. 90, 1925; H. Hönl, ZS. f. Phys. **32**, 340, 1925.

müssen aber hervorheben, daß eine Reihe gewichtiger Gründe gegen die Möglichkeit einer Deutung der anomalen Zeemaneffekte auf dieser Grundlage sprechen. Vielmehr wird man vielleicht hoffen dürfen, daß die Uhlenbeck-Goudsmitsche Hypothese (vgl. S. 560) später eine quantitative Beschreibung der genannten Phänomene gestattet.

§ 3. Gekoppelte harmonische Resonatoren. Statistik der Wellenfelder. Ein System gekoppelter harmonischer Oszillatoren, gegeben durch

$$H = \frac{1}{2} \sum_{k=1}^{f} \frac{p_k^2}{m_k} + Q(q), \tag{35}$$

mit einer quadratischen Form $Q(q)$ der Koordinaten (mit Zahlen als Koeffizienten) stellt das denkbar einfachste System von mehreren Freiheitsgraden dar. Wie in Kap. 2, § 1, festgestellt wurde, bleiben die Vertauschungsregeln invariant bei gleichzeitiger orthogonaler Transformation der Koordinaten und Impulse. Es kann deshalb das System (35), Kap. 4, wie in der klassischen Theorie in ein System ungekoppelter Oszillatoren übergeführt werden. Insbesondere sind die Schwingungen eines Kristallgitters wie in der klassischen Theorie nach Eigenschwingungen zu zerlegen. Jede einzelne Eigenschwingung ist in der früher ausführlich erörterten Weise als einfacher linearer Oszillator zu behandeln, und die Zusammenfassung sämtlicher ungekoppelter Oszillatoren zu einem einzigen System hat in der in Kap. 2, § 1, erläuterten Weise zu geschehen. Dasselbe wird auch dann gelten, wenn wir zum Grenzfall eines Systems von unendlich vielen Freiheitsgraden übergehen und etwa die Schwingungen eines zum Kontinuum idealisierten elastischen Körpers oder endlich eines elektromagnetischen Hohlraums betrachten.

Auch in der bisherigen Quantentheorie sind die Schwingungen eines elektromagnetischen Hohlraums oft Gegenstand eingehender Untersuchungen gewesen. Denn einerseits handelt es sich hier eben um das denkbar einfachste, nach den bisherigen Methoden zu behandelnde Problem des harmonischen Oszillators, andererseits weist das bekannte Ergebnis, daß die Energie einer Eigenschwingung ein ganzzahliges Vielfaches von $h\nu$ sein sollte, eine formale Ähnlichkeit mit den Ansätzen der Lichtquantentheorie auf, und man hoffte deshalb durch die Behandlung der Hohlraumstrahlung Einblick in das Wesen der Lichtquanten zu bekommen. Allerdings ist es von vornherein klar, daß der eben geschilderte Angriff auf das Problem der Lichtquanten von der wesentlichsten Seite dieses Problems, nämlich von dem Phänomen der Kopplung entfernter Atome keineswegs Rechenschaft geben kann. Denn dieses Problem geht überhaupt nicht in unsere Fragestellung nach den Schwingungen eines Hohlraums ein. Trotzdem kann zwischen den Eigenschwingungen des Hohlraums und den einmal postulierten Lichtquanten eine so enge Zuordnung durchgeführt werden, daß jeder Statistik der Eigenschwingungen des Hohlraums auch eine bestimmte Statistik der Lichtquanten entspricht und umgekehrt.

Debye[1]) hat versucht, durch eine Verteilung individueller Lichtquanten auf die Eigenschwingungen des Hohlraums eine solche Statistik zu geben, und es gelang ihm, auf diese Weise die Plancksche Formel abzuleiten. Uns scheint jedoch eine solche Mischung wellentheoretischer und lichtquantenmäßiger Begriffe kaum dem Wesen des Problems zu entsprechen. Vielmehr glauben wir, daß es konsequent sei, die wellentheoretische Seite des Problems ganz von der Lichtquantentheorie zu trennen, also die wellentheoretische Statistik der Hohlraumstrahlung durchaus nach den allgemeinen, z. B. für quantentheoretische Atomsysteme geltenden statistischen Gesetzen zu behandeln. Die zugeordnete Lichtquantenstatistik ist dann, wie wir zeigen werden, die Bosesche Statistik[2]); dieses Ergebnis scheint nicht unnatürlich, da diese Statistik hier nichts mit der Annahme unabhängiger Lichtkorpuskeln zu tun hat, sondern als Übertragung der Statistik der Eigenschwingungen aufzufassen ist — was nur zeigt, daß eben die Annahme statistisch unabhängiger Lichtkorpuskeln nicht das Richtige treffen würde.

Für jede derartige Behandlung der Hohlraumstrahlung in der bisherigen Quantentheorie ergab sich aber die grundsätzliche Schwierigkeit, daß sie zwar zum Planckschen Strahlungsgesetz, nicht aber zum richtigen Mittelwert des Schwankungsquadrates der Energie in einem Teilvolumen führte. Es zeigt sich also, daß eine konsequente Behandlung der Eigenschwingungen eines mechanischen Systems oder eines elektromagnetischen Hohlraums nach der bisherigen Theorie zu den schwersten Widersprüchen führt. Wir hatten deshalb die Hoffnung, daß die veränderte Kinematik, die der hier versuchten Theorie zugrunde liegt, den richtigen Wert für die Interferenzschwankungen liefert, so daß die genannten Widersprüche fortfallen und sich eine konsequente Statistik der Hohlraumstrahlung als möglich erweist.

Die Zustände des Oszillatorensystems können gekennzeichnet werden durch „Quantenzahlen" $n_1, n_2, n_3, \ldots$ der einzelnen Oszillatoren, so daß

[1]) P. Debye, Ann. d. Phys. **33**, 1427, 1910. Vgl. auch P. Ehrenfest, Phys. ZS. **7**, 528, 1906.

[2]) S. N. Bose, ZS. f. Phys. **26**, 178, 1924.

die Energien der einzelnen Zustände bis auf eine additive Konstante gegeben sind durch

$$E_n = h \cdot \sum_k \nu_k n_k. \tag{36}$$

Die additive Konstante, die „Nullpunktsenergie", ist gleich

$$C = \tfrac{1}{2} h \sum_k \nu_k \tag{36'}$$

(sie wäre insbesondere im Grenzfall unendlich vieler Freiheitsgrade unendlich groß). Wir wollen die Größe E_n in (2), Kap. 4, weiterhin kurz als thermische Energie bezeichnen. Nach dem in Teil I Gesagten ist jedem der durch ein bestimmtes Wertesystem $n_1, n_2, n_3, \ldots$ gekennzeichneten Zustände des Systems das gleiche statistische Gewicht zuzuschreiben. Die Folgerungen, die sich hieraus ergeben, sind unmittelbar zu übersehen auf Grund der folgenden Bemerkung:

Pflanzen sich in einem s-dimensionalen isotropen Raumstück der Größe $V = l^s$ Wellen fort mit der Phasengeschwindigkeit v, so ist die Anzahl der Eigenschwingungen für den Frequenzbereich $d\nu$ gleich der Anzahl der „Zellen" im Bose-Einsteinschen Sinne für $d\nu$; und zwar gilt das für beliebige s, also auch etwa für schwingende Membranen oder Saiten. Denn wenn von Polarisationseigenschaften usw. abgesehen wird, so bestimmt sich die Anzahl der Eigenschwingungen für $d\nu$ durch Beantwortung der Frage, auf wieviele Weisen ganze positive Zahlen $m_1, \ldots m_s$ so gewählt werden können, daß das aus

$$\frac{2l}{v} \cdot \nu = \sqrt{m_1^2 + \cdots + m_s^2}$$

bestimmte ν in $d\nu$ fällt. Ist $K_s(a)$ das Volumen einer s-dimensionalen Kugel vom Radius a, so gibt es $\frac{V}{v^s} K_s(\nu)$ Eigenschwingungen einer Frequenz kleiner als ν. Andererseits ist die Anzahl der Zellen für $d\nu$ so zu bestimmen: Die Impulskomponenten $p_1, \ldots, p_s$ des Quants genügen der Gleichung

$$\frac{h\nu}{v} = \sqrt{p_1^2 + \cdots + p_s^2},$$

und die Größe der Zellen im $2s$-dimensionalen Phasenraum ist h^s. Daraus ersieht man, daß auch die Anzahl der zu Frequenzen kleiner als ν gehörigen Zellen gleich $\frac{V}{v^s} K_s(\nu)$ ist.

Es kann also, wie oben erwähnt, eine umkehrbar eindeutige Zuordnung der Zellen zu den Eigenschwingungen derart durchgeführt werden, daß die einzelnen Paare immer zum gleichen $d\nu$ gehören. Dabei kann übrigens diese Zuordnung noch derart durchgeführt werden, daß auch die Richtungen einer Eigenschwingung und der Lichtquanten der zugeordneten Zelle in den gleichen infinitesimalen Winkelbereich fallen. Nach (36), Kap. 4, ist dann die Quantenzahl eines Oszillators der Anzahl der Quanten in der zugehörigen Zelle gleichzusetzen. Jede Statistik der Lichtquanten ergibt eine zugeordnete Statistik der Eigenschwingungen und umgekehrt. Man sieht, daß die oben gemachte Feststellung über die Gewichte der Zustände des Oszillatorensystems durch diese Zuordnung unmittelbar in die Grundannahme der Bose-Einsteinschen Statistik übergeht. Die gleichwahrscheinlichen Komplexionen sind dadurch definiert, daß angegeben wird, wieviele Quanten in jeder Zelle sitzen[1]).

Nach der Debyeschen Statistik ist die Anzahl der mit r Quanten behafteten Oszillatoren (bis auf einen nur von ν abhängigen Faktor) gleich

$$\frac{1}{r} \cdot e^{-r\frac{h\nu}{kT}}, \tag{37}$$

und das Plancksche Gesetz kommt durch

$$\sum_{r=1}^{\infty} e^{-r\frac{h\nu}{kT}} = \frac{1}{e^{\frac{h\nu}{kT}} - 1}$$

zustande; unbefriedigenderweise gilt übrigens (37), Kap. 4, nur für $r > 0$ und gibt nicht auch die Anzahl der mit Null Quanten behafteten Oszillatoren an. Nach der neuen Auffassung tritt an Stelle von (37), Kap. 4, nach Bose der Ausdruck[2])

$$\left(1 - e^{-\frac{h\nu}{kT}}\right) e^{-r\frac{h\nu}{kT}}, \tag{38}$$

der in der Sprache der Lichtquantentheorie die Anzahl der „r-fach besetzten Zellen" gibt, und die Plancksche Formel folgt aus

$$\sum_{r=0}^{\infty} r\left(1 - e^{\frac{h\nu}{kT}}\right) e^{-r\frac{h\nu}{kT}} = \frac{1}{e^{\frac{h\nu}{kT}} - 1}.$$

[1]) A. Einstein, Sitzungsber. d. Preuß. Akad. d. Wiss. 1925, S. 3. Für die Beurteilung der Einsteinschen Hypothese, daß auch auf das ideale Gas diese Form der Statistik anzuwenden sei, können unsere Betrachtungen natürlich keinen neuen Gesichtspunkt ergeben.

[2]) Natürlich muß dieser Ausdruck auch beispielsweise für die elastischen Wellen in einem Kontinuum angenommen werden, wodurch eine gewisse Abänderung an einer von Schrödinger (Phys. ZS. **25**, 89, 1924) gegebenen Betrachtung über das thermische Gleichgewicht zwischen Licht- und Schallstrahlen nötig wird. Diese Abänderung ist leicht auszuführen in Analogie zum Wahrscheinlichkeitsansatz für den Comptoneffekt unter Annahme der Einsteinschen Gastheorie, wie er früher (P. Jordan, ZS. f. Phys. **33**, 649, 1925) mitgeteilt wurde.

Die zugeordnete Lichtquantenstatistik zur Debyeschen Schwingungsstatistik wird durch die von Wolfke[1]) und Bothe[2]) entwickelte Theorie dargestellt. Allerdings sprechen diese Verfasser nicht von r-fach besetzten Zellen, sondern bezeichnen (37), Kap. 4, als Anzahl der „r-quantigen Lichtquantenmoleküle".

Die erwähnte Unzulänglichkeit der klassischen Wellentheorie tritt bekanntlich bei der Untersuchung der Energieschwankungen im Strahlungsfeld folgendermaßen zutage. Kommuniziert ein Volumen V mit einem sehr großen Volumen derart, daß die Wellen eines schmalen Bereichs $\nu, \nu + d\nu$ ungehindert von einem ins andere laufen können, während für alle anderen Wellen die Volumina getrennt bleiben, und ist E die Energie der Wellen mit der Frequenz ν in V, so kann das Schwankungsquadrat $\overline{\Delta^2} = \overline{(E - \overline{E})^2}$ nach Einstein durch eine Umkehrung des Boltzmannschen Prinzips berechnet werden. Ist $z_\nu \, d\nu$ die auf die Volumeneinheit bezogene Anzahl der Eigenschwingungen (Zellen) für $d\nu$, so daß

$$\overline{E} = \frac{z_\nu h\nu}{e^{\frac{h\nu}{kT}} - 1} \cdot v \tag{39}$$

gilt, so ergibt sich

$$\overline{\Delta^2} = h\nu\overline{E} + \frac{\overline{E}^2}{z_\nu V}. \tag{40}$$

Berechnet man jedoch die Energieschwankungen aus den Interferenzen im Wellenfeld, so ergibt die klassische Theorie, wie Lorentz[3]) ausführlich nachgerechnet hat, nur den zweiten Summanden in (40), Kap. 4. Dieser Widerspruch besteht natürlich ganz allgemein auch für die Wellen etwa in einem Kristallgitter oder einem elastischen Kontinuum. Sein Ursprung ist nach Ehrenfest[4]) darin zu suchen, daß in der Einsteinschen Überlegung Additivität der Entropien von V und dem großen Volumen vorausgesetzt wurde. Diese Additivität der Entropien besteht aber nach der klassischen Theorie der Eigenschwingungen nur im Gültigkeitsbereich des Rayleigh-Jeansschen Gesetzes. Eben das Nichtbestehen statistischer Unabhängigkeit der Teilvolumina im allgemeinen Falle ist ein so unnatürliches Ergebnis der bisherigen Theorie der Hohl-

[1]) M. Wolfke, Phys. ZS. **22**, 375, 1921.

[2]) W. Bothe, ZS. f. Phys. **20**, 145, 1923; **23**, 214, 1924.

[3]) H. A. Lorentz, Les Théories Statistiques en Thermodynamique (Leipzig, 1916), S. 59.

[4]) P. Ehrenfest, Vortrag im Göttinger Seminar über Struktur der Materie, Sommer 1925. Der Inhalt dieses Vortrages ist uns bei unseren Überlegungen eine wertvolle Hilfe gewesen. Inzwischen veröffentlicht, ZS. f. Phys. **34**, 362, 1925.

raumstrahlung, daß man auf ein Versagen dieser Theorie schon beim einfachen Problem des harmonischen Oszillators schließen muß.

Wir wollen nun das Schwankungsquadrat $\overline{\Delta^2}$ aus den Interferenzen gemäß der Quantenmechanik berechnen. Zur Vermeidung rechnerischer Komplikationen, die das Wesen der Sache nicht berühren, beziehen wir uns auf den denkbar einfachsten Fall, nämlich eine eingespannte schwingende Saite. Es können übrigens alle wesentlichen Punkte der Rechnung ohne weiteres auf allgemeinere Fälle übertragen werden. Zunächst werde die klassische Behandlungsweise erläutert.

Die Länge der Saite sei l und $u(x, t)$ die seitliche Auslenkung. Bei Einführung der durch

$$u(x,t) = \sum_{k=1}^{\infty} q_k(t) \sin k \frac{\pi}{l} x, \tag{41}$$

oder

$$q_k(t) = \frac{2}{l} \int_0^l u(x,t) \sin k \frac{\pi}{l} x \, . \, dx \tag{41'}$$

gegebenen Fourierkoeffizienten $q_k(t)$ als Koordinaten geht die Energie der Saite in eine Quadratsumme über. Es wird nämlich bei geeigneter Wahl der Einheiten

$$H = \frac{1}{2} \int_0^l \left\{ u^2 + \left(\frac{\partial u}{\partial x}\right)^2 \right\} dx = \frac{l}{4} \sum_{k=1}^{\infty} \left\{ \dot{q}_k(t)^2 + \left(k \frac{\pi}{l}\right)^2 q_k(t)^2 \right\}. \tag{42}$$

Für die Energie E auf einem Abschnitt $(0, a)$ der Saite erhalten wir allgemeiner

$$E = \frac{1}{2} \int_0^a \sum_{j,k=1}^{\infty} \left\{ \dot{q}_j \dot{q}_k \sin j \frac{\pi}{l} x \sin k \frac{\pi}{l} x \right.$$

$$\left. + q_j q_k j k \left(\frac{\pi}{l}\right)^2 \cos j \frac{\pi}{l} \cos k \frac{\pi}{l} x \right\} dx. \tag{43}$$

Nehmen wir in (43), Kap. 4, nur die Glieder mit $j = k$, so erhalten wir unter der ausdrücklichen Voraussetzung, daß alle in Betracht kommenden Wellenlängen klein gegen a seien, gerade den Wert $\frac{a}{l} H$. Man sieht daraus: Die Differenz

$$\Delta = E - \overline{E},$$

worin der Querstrich die Mitteilung über die Phasen φ_k in

$$q_k = a_k \cos(\omega_k t + \varphi_k); \quad \omega_k = k \frac{\pi}{l} \tag{44}$$

bedeutet, geht aus (43), Kap. 4, hervor, indem die Summanden mit $j = k$ ausgelassen werden. Dieses Phasenmittel ist mit dem Zeitmittel identisch. Man erhält dann durch Ausführung der Integration

$$\Delta = \frac{1}{4} \sum_{\substack{j,k=1 \\ j \neq k}}^{\infty} \left\{ \dot{q}_j \dot{q}_k K_{jk} + jk\, q_j q_k \left(\frac{\pi}{l}\right)^2 K'_{jk} \right\} \tag{45}$$

mit

$$\left.\begin{aligned} K_{jk} &= \frac{\sin (j-k)\frac{\pi}{l} a}{(j-k)\frac{\pi}{l}} - \frac{\sin (j+k)\frac{\pi}{l} a}{(j+k)\frac{\pi}{l}} \\ &= \frac{\sin (\omega_j - \omega_k) a}{\omega_j - \omega_k} - \frac{\sin (\omega_j + \omega_k) a}{\omega_j + \omega_k}, \\ K'_{jk} &= \frac{\sin (j-k)\frac{\pi}{l} a}{(j-k)\frac{\pi}{l}} + \frac{\sin (j+k)\frac{\pi}{l} a}{(j+k)\frac{\pi}{l}} \\ &= \frac{\sin (\omega_j - \omega_k) a}{\omega_j - \omega_k} + \frac{\sin (\omega_j + \omega_k) a}{\omega_j + \omega_k}. \end{aligned}\right\} \tag{45'}$$

Das Quadrat $\overline{\Delta^2}$ soll in Rücksicht auf die spätere quantenmechanische Rechnung ausführlich angeschrieben werden. Es ist

$$\Delta^2 = (\Delta_1 + \Delta_2)^2 = \Delta_1^{\,2} + \Delta_2^{\,2} + \Delta_1 \Delta_2 + \Delta_2 \Delta_1 \tag{46}$$

mit

$$\Delta_1^{\,2} + \Delta_2^{\,2} = \frac{1}{16} \sum_{\substack{j,k=1 \\ j \neq k}}^{\infty} \sum_{\substack{\iota,\varkappa=1 \\ \iota \neq \varkappa}}^{\infty} \left\{ \dot{q}_j \dot{q}_k \dot{q}_\iota \dot{q}_\varkappa K_{jk} K_{\iota\varkappa} \right. \\ \left. + jk\iota\varkappa \left(\frac{\pi}{l}\right)^4 q_j q_k q_\iota q_k K'_{jk} K'_{\iota k}; \right. \tag{46'}$$

$$\Delta_1 \Delta_2 + \Delta_2 \Delta_1 = \frac{1}{16} \sum_{\substack{j,k=1 \\ j \neq k}}^{\infty} \sum_{\substack{\iota,\varkappa=1 \\ \iota \neq \varkappa}}^{\infty} \left(\frac{\pi}{l}\right)^2 \{ jk\, q_j q_k \dot{q}_\iota \dot{q}_\varkappa K'_{jk} K_{\iota\varkappa} \\ + \iota\varkappa\, \dot{q}_j \dot{q}_k q_\iota q_\varkappa K_{jk} K'_{\iota\varkappa} \}. \tag{46''}$$

Aus (44), Kap. 4, folgt $\overline{\Delta_1 \Delta_2 + \Delta_2 \Delta_1} = 0$ und

$$\overline{\Delta^2} = \overline{\Delta_1^{\,2}} + \overline{\Delta_2^{\,2}} = \frac{1}{8} \sum_{j,k=1}^{\infty} \left\{ \overline{\dot{q}_j^2}\, \overline{\dot{q}_k^2}\, K_{jk}^2 + j^2 k^2 \left(\frac{\pi}{l}\right)^4 \overline{q_j^2}\, \overline{q_k^2}\, K'^2_{jk} \right\}. \tag{47}$$

Lassen wir nun die Seitenlänge l sehr groß werden, so rücken die ω_k nach (44), Kap. 4, immer enger zusammen, so daß die Summe (47) in ein Integral übergeht:

$$\overline{\Delta^2} = \overline{\Delta_1^{\,2}} + \overline{\Delta_2^{\,2}} = \frac{1}{8} \int_0^\infty \int_0^\infty d\omega_j\, d\omega_k \frac{l^2}{\pi^2} \left\{ \overline{\dot{q}_j^2}\, \overline{\dot{q}_k^2}\, K_{jk}^2 + j^2 k^2 \left(\frac{\pi}{l}\right)^4 \overline{q_j^2}\, \overline{q_k^2}\, K'^2_{jk} \right\}. \tag{47'}$$

Endlich nehmen wir auch das „Volum" a als sehr groß an und machen Gebrauch von der Beziehung

$$\lim_{a \to \infty} \frac{1}{a} \int_{-\Omega}^{\Omega'} \frac{\sin^2 \omega \alpha}{\omega^2} f(\omega)\, d\omega = \pi f(0) \quad \text{für} \quad \Omega, \Omega' > 0. \tag{48}$$

Man sieht dann, daß nur die ersten Summanden $\frac{\sin(\omega_j - \omega_k)\, a}{\omega_j - \omega_k}$ in (45) einen in Betracht kommenden Beitrag liefern; und zwar ergibt sich in (47'):

$$\overline{\Delta^2} = \frac{a\, l}{8\pi} \int_0^\infty d\omega \left\{ (\overline{\dot{q}_\omega^2})^2 + (\omega^2 \overline{q_\omega^2})^2 \right\}. \tag{49}$$

Andererseits wird die mittlere Energie im Volumen a nach (42), Kap. 4, gleich

$$\overline{E} = \frac{a}{l} \cdot \frac{l}{4} \cdot \int_0^\infty d\omega \frac{l}{\pi} \cdot \left\{ \overline{\dot{q}_\omega^2} + \omega^2 \overline{q_\omega^2} \right\} = \frac{a\, l}{4\pi} \int_0^\infty d\omega \left\{ \overline{\dot{q}_\omega^2} + \omega^2 \overline{q_\omega^2} \right\}. \tag{50}$$

Dabei gilt

$$\overline{\dot{q}_\omega^2} = \omega^2 \overline{q_\omega^2}, \tag{51}$$

eine Beziehung, die — woran gleich hier erinnert sei — nach Kap. 1 auch in der Quantenmechanik gültig bleibt. Um zu den in (39), (40), Kap. 4, gebrauchten Größen $\overline{\Delta^2}$, $\overline{E}$ überzugehen, haben wir in (49), (50), Kap. 4 nur die auf $d\nu = \frac{d\omega}{2\pi}$ bezüglichen Anteile zu entnehmen und diese durch $d\nu$ zu dividieren. Dann ergibt sich mit $v = a$:

$$\overline{\Delta^2} = \frac{\overline{E}^2}{2v} \tag{52}$$

Aus (44), Kap. 4, entnimmt man, daß in unserem Falle $z_\nu = 2$ ist; denn es wird

$$d\omega_k = 2\pi\, d\nu_k = \frac{\pi}{l}\, dk.$$

Daher gibt also (52), Kap. 4, in der Tat gerade das zweite Glied in (40), Kap. 4.

Beim Übergang zur Quantenmechanik sind (41), (41'), (42), (43), Kap. 4, als Matrizengleichungen für u, H, q, E aufzufassen. Dabei bleibt jedoch x eine Zahl; denn betrachten wir statt der kontinuierlichen Saite eine elastische Punktreihe, so bedeutet x die (mit der Gitterkonstanten multiplizierte) Nummer jeweils eines bestimmten Punktes.

Die Matrix q_k hat $2f$ Dimensionen, wenn f die Anzahl der Eigenschwingungen ist; bei der elastischen Saite also unendlich viele. Die Komponenten $q_k(nm)$ von q_k verschwinden sämtlich, außer denen mit

$$\left.\begin{array}{l} n_j - m_j = 0 \text{ für } j \neq k, \\ n_k - m_k = \pm 1. \end{array}\right\} \tag{53}$$

Das Phasenmittel einer Matrix ist diejenige Diagonalmatrix, die mit der Diagonale der betreffenden Matrix übereinstimmt. Aus (53), Kap. 4, können zum Teil ähnliche Folgerungen gezogen werden, wie aus (44), Kap. 4. Die Überlegungen, die früher zu (46), 46'), (46'') führten, bleiben für die Quantenmechanik erhalten. Auch gelten für die Diagonalmatrix $\overline{\boldsymbol{\Delta}_1^2 + \boldsymbol{\Delta}_2^2}$ die Formeln (47), (47'), Kap. 4, mit Matrizen q_k und endlich wird entsprechend (52), Kap. 4, wenn wir die zu einem bestimmten ν gehörigen Teile von $\overline{\boldsymbol{\Delta}^2}$ als $\overline{\Delta^2}$ bezeichnen:

$$\overline{\Delta_1^2 + \Delta_2^2} = \frac{\overline{E^{*2}}}{2v}. \tag{52'}$$

Darin ist gemäß (49), (50), (51), Kap. 4, E^* nicht mehr die mittlere thermische Energie, sondern die Summe von dieser und der Nullpunktsenergie; nach den elementaren Oszillatorformeln ist

$$\overline{E^*} = h\nu \cdot v + \overline{E},$$

$$\overline{\Delta_1^2 + \Delta_2^2} = \frac{1}{2}(h\nu)^2 v + h\nu \overline{E} + \frac{\overline{E^2}}{2v}; \tag{54}$$

denn die Nullpunktsenergie für $d\nu$ wird gleich

$$\frac{v}{l} \cdot \frac{h\nu}{2} \cdot l z_\nu d\nu = h\nu \,.\, V d\nu.$$

Wir haben nun noch $\boldsymbol{\Delta}_1 \boldsymbol{\Delta}_2 + \boldsymbol{\Delta}_2 \boldsymbol{\Delta}_1$ zu betrachten. Indem wir diese Größe ganz entsprechend behandeln wie $\overline{\boldsymbol{\Delta}_1^2 + \boldsymbol{\Delta}_2^2}$, erhalten wir entsprechend zu (49), Kap. 4, den Ausdruck

$$\overline{\boldsymbol{\Delta}_1 \boldsymbol{\Delta}_2 + \boldsymbol{\Delta}_2 \boldsymbol{\Delta}_1} = \frac{a l^2}{8\pi} \int_0^\infty d\omega \cdot \omega^2 \left\{ (q_\omega \dot{q}_\omega)^2 + (\dot{q}_\omega q_\omega)^2 \right\}.$$

Nach den Vertauschungsregeln wird nun aber, da zufolge (42), Kap. 4, die Größe $l/2$ als „Masse" der Resonatoren anzusehen ist:

$$-q_j \dot{q}_j(nn) = \dot{q}_j q_j(nn) = \frac{1}{2} \cdot \frac{2}{l} \cdot \frac{h}{2\pi i} = \frac{h}{2 l \pi i}.$$

Folglich wird der zu $d\nu$ gehörige Anteil $\overline{\Delta_1 \Delta_2 + \Delta_2 \Delta_1}$ von $\overline{\boldsymbol{\Delta}_1 \boldsymbol{\Delta}_2 + \boldsymbol{\Delta}_2 \boldsymbol{\Delta}_1}$ nach Division durch $d\nu$ gleich

$$\overline{\Delta_1 \Delta_2 + \Delta_2 \Delta_1} = -\tfrac{1}{2}(h\nu)^2 V,$$

und mit (54) folgt in der Tat

$$\overline{\Delta^2} = h\nu \bar{E} + \frac{\bar{E}^2}{z_\nu V} \tag{55}$$

in Übereinstimmung mit (40), Kap. 4.

Wenn man bedenkt, daß die hier behandelte Frage doch ziemlich weit entfernt liegt von den Problemen, aus deren Untersuchung die Quantenmechanik erwachsen ist, so wird man das mit (55), Kap. 4, erzielte Ergebnis als besonders ermutigend für den weiteren Ausbau der Theorie betrachten.

Man würde nach dem oben erwähnten Ergebnis von Ehrenfest die interferenzmäßige Berechnung der Schwankungen ersparen können — und zugleich die Gewißheit gewinnen, daß auch bei anderen, ähnlichen Fragestellungen keine Widersprüche möglich sind —, wenn man unmittelbar die Additivität der Entropien der Teilvolumina in der Quantenmechanik der Wellenfelder nachweisen könnte. Daß die Additivität wirklich allgemein besteht, möchten wir nach unserem obigen Ergebnis vermuten.

Die Gründe für das Auftreten des von der klassischen Theorie nicht gelieferten Gliedes in (55), Kap. 4, sind offenbar mit den Gründen für das Auftreten der Nullpunktsenergie eng verwandt. In beiden Fällen liegt der wesentliche Unterschied der hier versuchten Theorie von der bisherigen nicht in einer Verschiedenheit der mechanischen Gesetze, sondern in der für diese Theorie charakteristischen Kinematik. Man könnte sogar in der Formel (55), Kap. 4, in die ja gar keine mechanischen Prinzipien eingehen, eines der anschaulichsten Beispiele für die Verschiedenheit der quantentheoretischen Kinematik von der bisherigen erblicken.

Wenn sich die hier versuchte Quantenmechanik als schon in wesentlichen Zügen richtig erweisen sollte, so wäre wohl ganz allgemein als der wichtigste Fortschritt gegenüber der bisherigen Theorie eben dies zu bezeichnen: daß in dieser Theorie Kinematik und Mechanik wieder in eine so enge Verbindung gebracht sind, wie etwa Kinematik und Mechanik in der klassischen Theorie, und daß die fundamentalen neuen Gesichtspunkte, welche aus den Grundpostulaten der Quantentheorie für die mechanischen Begriffe und die Begriffe von Raum und Zeit folgen, in der Kinematik ebenso wie in der Mechanik und in der Verbindung von Kinematik und Mechanik einen adäquaten Ausdruck finden.

Zeitschrift für Physik *38*, 411–426 (1926)

Mehrkörperproblem und Resonanz in der Quantenmechanik.

Von **W. Heisenberg** in Kopenhagen.

(Eingegangen am 11. Juni 1926.)

Die Arbeit versucht, eine Grundlage für die quantenmechanische Behandlung des Mehrkörperproblems zu geben. Zu diesem Zwecke wird ein für die Quantenmechanik des Mehrkörperproblems charakteristisches Resonanzphänomen ausführlich untersucht und ein Zusammenhang der auf Grund dieser Untersuchung gewonnenen Resultate mit der Einstein-Boseschen Abzählung und dem Paulischen Verbot äquivalenter Bahnen hergestellt.

Die Quantenmechanik ist bisher nur auf Systeme, die aus einem beweglichen Massenpunkt bestehen, angewandt worden. An dieser Beschränkung waren in erster Linie die mathematischen Schwierigkeiten schuld, die einer Berechnung der einzelnen Amplitudenelemente bisher im Wege standen. In der letzten Zeit ist in dieser Frage ein außerordentlicher Fortschritt erzielt worden durch die bedeutungsvollen Untersuchungen, in denen Schrödinger[1]), ausgehend von der de Broglieschen[2]) Wellentheorie der Materie, einen neuen mathematisch wesentlich bequemeren Zugang zum Gebiet der Quantenmechanik entdeckt hat. Ebenso, wie seinerzeit eine große formale Ähnlichkeit der klassischen Mechanik mit einer geometrischen Optik in mehrdimensionalen Räumen von Hamilton aufgedeckt und zur Grundlage für die wirksamste mathematische Behandlungsweise klassischer Probleme ausgebaut wurde, so besteht nach Schrödinger eine große formale Ähnlichkeit der Quantenmechanik mit einer Wellenoptik in mehrdimensionalen Räumen, die auch hier zur wirksamsten mathematischen Behandlungsweise quantenmechanischer Probleme führt. Bei einem System von f Freiheitsgraden ersetzt Schrödinger das quantenmechanische Problem durch eine Eigenwertaufgabe in einem Raum von f Dimensionen; die Energien der stationären Zustände sind die Eigenwerte des Problems; die Matrizenelemente erscheinen als Koeffizienten einer Entwicklung nach Eigenfunktionen; wenn die Eigenfunktionen gefunden sind, können die Amplitudenelemente durch reine Quadraturen ermittelt werden. In der physikalischen Interpretation des mathematischen Formalismus besteht jedoch ein Unterschied

[1]) E. Schrödinger, Ann. d. Phys. **79**, 361, 489, 734, 1926.
[2]) L. de Broglie, Ann. de phys. (10) **3**, 22, 1925.

zwischen Schrödingers und unserer Auffassung. Wegen der mathematischen Äquivalenz des Schrödingerschen Verfahrens mit der Quantenmechanik könnte allerdings die Frage nach dem den Gleichungen zugrunde liegenden physikalischen Geschehen einstweilen als eine unsere Anschauung betreffende Zweckmäßigkeitsfrage betrachtet werden; aber nur solange wir nicht versuchen, auf Grund der einmal gewählten anschaulichen Bilder die Grundlage dieser Quantentheorie zu erweitern. Um ein spezielles Beispiel zu nennen: Für die hier zu behandelnden Mehrkörperprobleme möchten wir einen direkten Anschluß an die der Quantenmechanik zugrunde liegenden Gesichtspunkte verlangen, während Schrödingers Darstellungsweise [1]) wesentliche Änderungen der bisherigen Gleichungen als möglich erscheinen läßt. Aber selbst, wenn man sich nur auf die Diskussion der Anschauungsfrage beschränkt, so glaube ich, daß es Fälle gibt, in denen die Wellenvorstellung gezwungener ist, als die Veranschaulichung durch die Bewegung von Korpuskeln in Räumen, in denen eine nichtklassische Kinematik gilt. Man denke etwa an die Vorstellung des rotierenden Elektrons. Soviel ich sehen kann, stellt auch das Schrödingersche Verfahren nicht eine konsequente Wellentheorie der Materie im Sinne de Broglies dar. Der Übergang zum Raum von f Dimensionen und die Berechnung der Wellengeschwindigkeit aus der gegenseitigen potentiellen Energie von Partikeln bedeutet eine Anleihe bei den Begriffen der Korpuskulartheorie. Selbst wenn sich eine konsequente Wellentheorie der Materie im gewöhnlichen dreidimensionalen Raum entwickeln ließe, dem Programm de Broglies und Einsteins entsprechend, so wäre dadurch kaum eine erschöpfende Beschreibung der atomistischen Vorgänge in unseren gewöhnlichen Raum-Zeitbegriffen gewonnen. Eben in Anbetracht der immer mehr zutage kommenden engen Analogie von Licht und Materie möchte man glauben, daß eine solche Wellentheorie der Materie ebensowenig eine vollständige Beschreibung unserer atomistischen Erfahrungen ermöglichte wie die Wellentheorie des Lichtes eine Deutung der optischen Erfahrungen. In Anbetracht dieser Analogie scheint es mir eine der wichtigsten Seiten der Quantenmechanik, daß sie auf der Korpuskularvorstellung der Materie basiert ist; freilich handelt es sich dabei nicht um eine Beschreibung der Bewegungen von Korpuskeln in unseren gewöhnlichen Raum-Zeitbegriffen. Dies konnte man auch kaum erwarten; denn selbst wenn sich die Korpuskeln als Singularitäten der metrischen Struktur des

[1]) Vgl. E. Schrödinger, l. c., Fußnote auf S. 750.

Zeitschrift für Physik 38, 411–426 (1926)

Raumes herausstellen sollten, wie es der Wunsch der Kontinuumstheorien ist, so wäre dies wohl keine Beschreibung in unseren gewöhnlichen Raum-Zeitbegriffen — es sei denn, man rechnet einen Raum, dessen Maßbestimmung von der Euklidischen wesentlich abweicht, zu den „gewöhnlichen" Räumen.

Doch sollen hier nicht länger die physikalischen Schwierigkeiten diskutiert werden, die beim jetzigen Stande der Theorie doch nicht gelöst werden können. Wir sind damit zufrieden, festzustellen, daß das Schrödingersche Verfahren und die Quantenmechanik sich gegenseitig ergänzen, sowohl in der mathematischen Durchführung wie in der Schöpfung anschaulicher Bilder und Analogien, die es uns ermöglichen, tiefer als bisher in das physikalische Wesen der Vorgänge in sehr kleinen Räumen einzudringen.

Das Ziel unserer Untersuchung ist die quantenmechanische Behandlung der Systeme, die aus mehreren Massenpunkten bestehen. Eine solche Behandlung scheint zunächst auf erhebliche Schwierigkeiten zu stoßen: Die Züge der de Broglieschen Wellentheorie, die zur Einstein-Boseschen[1]) Statistik führen, scheinen kein Analogon in der Quantenmechanik zu besitzen; Zusatzregeln, wie etwa das Paulische Verbot äquivalenter Bahnen[2]), haben in dieser Form im mathematischen Schema der Quantenmechanik keinen Platz. Man könnte also an ein Versagen der Quantenmechanik beim Problem der äquivalenten Bahnen denken. Schließlich mag noch an eine bekannte Schwierigkeit bei der quantitativen Deutung der Spektra erinnert werden: Der Abstand zwischen Singlett- und Triplettsystem in den Spektra der Erdalkalien und im Heliumspektrum ist um Größenordnungen zu weit, als daß er einfach als Unterschied in der magnetischen Wechselwirkungsenergie zweier rotierenden Elektronen aufgefaßt werden könnte.

Das Ziel der folgenden Untersuchungen ist eine genauere Analyse der Aussagen, die man bei konsequenter Anwendung der Quantenmechanik über das Mehrelektronenproblem machen kann. Um das Resultat vorweg zu nehmen: Die oben erwähnten Schwierigkeiten lösen sich bei dieser Analyse ganz von selbst, und es läßt sich ein Zusammenhang zwischen der Bose-Einsteinschen Abzählung und der Quantenmechanik herstellen.

---

[1]) N. S. Bose, ZS. f. Phys. **26**, 178, 1924. Die für uns in Betracht kommende Anwendung dieser Statistik auf die Materieteilchen ist gegeben von A. Einstein, Sitzungsber. d. preuß. Akad. d. Wiss. 1924, S. 261; 1925, S. 3, 8.

[2]) W. Pauli, ZS. f. Phys. **31**, 765, 1925.

§ 1. Das denkbar einfachste Mehrkörperproblem ist ein System zweier gekoppelter Oszillatoren. Bekanntlich läßt sich ein solches System, solange die Wechselwirkungsenergie eine quadratische Funktion der Koordinaten ist, stets zerlegen in zwei ungekoppelte Oszillatoren. Man kann nicht zweifeln, daß dieses letztere Problem nach der Quantenmechanik ohne neue Annahmen behandelt werden kann; also auch das erstere. Zugleich besitzt dieses Beispiel alle charakteristischen Eigenschaften der quantentheoretischen Mehrkörperprobleme, und es lassen sich durch die genauere Analyse dieses einfachen Modells schon alle Resultate ableiten, die später zur Erklärung der Spektren benötigt werden. Das Oszillatorbeispiel hat noch den weiteren Vorteil, daß kaum Unterschiede zwischen der Behandlung nach der klassischen Theorie, der bisherigen Quantentheorie und der Quantenmechanik bestehen; zu jedem quantentheoretischen Resultat gibt es hier ein einfaches klassisch-mechanisches Analogon. Auf dieses klassisch-mechanische Analogon werden wir später bei der Behandlung allgemeinerer quantenmechanischer Systeme mehr oder weniger verzichten müssen.

Es ist ein charakteristischer Zug der Atomsysteme, daß die Teilsysteme, aus denen sie zusammengesetzt werden können, nämlich die Elektronen, gleich und gleichen Kräften unterworfen sind. Um diesen Zug in unserem Beispiel wiederzufinden, setzen wir als Hamiltonsche Funktion an:

$$H = \frac{1}{2m} p_1^2 + \frac{m}{2} \omega^2 q_1^2 + \frac{1}{2m} p_2^2 + \frac{m}{2} \omega^2 q_2^2 + m \lambda q_1 q_2; \tag{1}$$

die Frequenzen und Massen der beiden zu koppelnden Oszillatoren werden also als gleich angenommen. In (1) bedeuten $q_1\, q_2$ die Koordinaten, p_1, p_2 die Impulse, m und ω Masse bzw. Frequenz der Oszillatoren, λ die Konstante der Wechselwirkung. Durch die bekannte Transformation:

$$q_1' = \frac{1}{\sqrt{2}} (q_1 + q_2); \quad q_2' = \frac{1}{\sqrt{2}} (q_1 - q_2) \tag{2}$$

geht (1) über in

$$H = \frac{1}{2m} p_1'^2 + \frac{m}{2} \omega_1'^2 q_1'^2 + \frac{1}{2m} p_2'^2 + \frac{m}{2} \omega_2'^2 q_2'^2, \tag{3}$$

wobei

$$\omega_1'^2 = \omega^2 + \lambda, \quad \omega_2'^2 = \omega^2 - \lambda. \tag{4}$$

H zerfällt jetzt additiv in zwei Oszillatorenenergien, die den beiden „Hauptschwingungen" entsprechen. Ist nur die erste q_1' angeregt, so schwingen die beiden Partikel in gleicher Phase nach der gleichen Seite,

Zeitschrift für Physik *38*, 411–426 (1926)

ist nur q_2' angeregt, so schwingen sie gleichzeitig nach entgegengesetzten Seiten oder, in anderen Worten, mit einer Phasendifferenz π (Fig. 1).

Die Energien der stationären Zustände des ganzen Systems werden dargestellt durch die Gleichung

$$H_{n'_1, n'_2} = \frac{\omega_1' h}{2\pi}\left(n_1' + \frac{1}{2}\right) + \frac{\omega_2' h}{2\pi}\left(n_2' + \frac{1}{2}\right), \tag{5}$$

wo n_1' und n_2' ganze Zahlen sind. Bezeichnet man einen Term durch das Symbol $n_1' n_2'$, so kommt man zum Termschema von Fig. 2. Auf die Unterscheidung von $+$ und $\cdot$ in Fig. 2 soll einstweilen nicht geachtet werden.

Wir wollen jetzt die möglichen Übergangsprozesse untersuchen. Um eine möglichst enge Analogie zu den Atomsystemen herzustellen, nehmen wir an, daß die beiden Oszillatoren aus geladenen Teilchen bestehen, die auf der gleichen Linie um den gleichen, entgegengesetzt zu den Teilchen geladenen Punkt schwingen können. Das elektrische Dipolmoment ist dann im wesentlichen gegeben durch $q_1 + q_2$. Daraus folgt, was aus Fig. 1 unmittelbar einleuchtet, daß nur die erste Hauptschwingung ein elektrisches Moment hat. Es kommen in dieser Näherung also nur Übergänge von n_1' um 1, also in Fig. 2 nur Übergänge vertikaler Richtung vor. Durch das Dipolmoment ist aber die Strahlung nur in erster Näherung gegeben. Auch die Quadrupole und die höheren Pole geben zu Strahlung geringerer Größenordnung Anlaß. Diese höheren Glieder in der Ausstrahlung sind durch homogene symmetrische Funktionen zweiten, dritten usw. Grades in q_1 und q_2 und deren zeitlichen Differentialquotienten gegeben. Ersetzt man q_1 und q_2 nach (2) durch q_1' und q_2', so entstehen homogene Funktionen zweiten, dritten usw. Grades in q_1' und q_2', die wegen der Symmetrie in q_1 und q_2 die Koordinate q_2' samt deren zeitlichen Differentialquotienten nur in einer geraden Anzahl von Faktoren enthalten können. Dies bedeutet, daß sich auch bei Einrechnung aller höheren Näherungen in der Strahlung n_2' stets nur um eine gerade Zahl ändern kann. Das Termschema der Fig. 2 kann also in zwei Teilsysteme ($+$ und .) zerlegt werden, derart, daß nur Kombinationen innerhalb der Systeme $+$ bzw. ., nie aber Interkombinationen

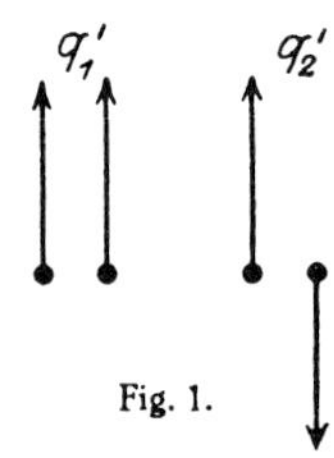

Fig. 1.

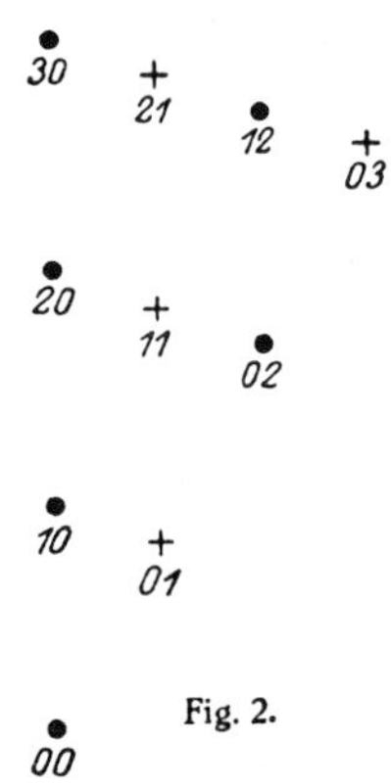

Fig. 2.

möglich sind. Nicht einmal durch Stoßprozesse ist ein Übergang vom einen zum anderen System möglich, da ja die Wahrscheinlichkeit eines Überganges durch Stoß als Kombination von Anregungen aller (Dipol-, Quadrupol- usw.) Momente betrachtet werden kann. Das Nichtvorhandensein der Interkombinationen ist aber fest geknüpft an die ursprüngliche Gleichheit der zu koppelnden Oszillatoren. Sobald bezüglich der Masse oder der Frequenz ein kleiner Unterschied zwischen beiden Oszillatoren vorhanden war, treten Interkombinationen eben von der relativen Stärke jenes (relativen) Unterschiedes auf. Nehmen wir aber einmal die ursprüngliche Gleichheit der Oszillatoren an. Dann tritt wegen jenes Mangels an Interkombinationen in der quantenmechanischen Behandlung des Problems eine charakteristische Unbestimmtheit auf. Sind in der Natur beide Termsysteme + und ., oder nur das eine System ., oder + realisiert? Von einer quantenmechanischen Lösung eines Problems wird ja nur verlangt, daß das ihr entsprechende Termschema „geschlossen“ sei, d. h. daß es alle und nur die Terme enthalte, die mit irgendwelchen von ihnen kombinieren können, und daß alle Übergangswahrscheinlichkeiten zu nicht vorhandenen Termen verschwinden. Es ist also sowohl das System der ., wie das der + für sich eine quantenmechanische Lösung des Problems (1), ebenso die Kombination beider Systeme. Diese Unbestimmtheit der quantenmechanischen Lösung scheint mir das wesentlichste Ergebnis dieser Untersuchung. Sie gibt ebensoviel Freiheit, daß die Forderungen der Bose-Einsteinschen Abzählung und das Paulische Verbot äquivalenter Bahnen ungezwungen dem System der Quantenmechanik eingefügt werden können. Vor der genaueren Untersuchung dieses Sachverhalts möchte ich zeigen, daß gekoppelte Systeme sich in der Quantenmechanik stets ähnlich wie das eben erläuterte Beispiel verhalten.

§ 2. In der klassischen Theorie können zwei periodisch schwingende Systeme nur dann in eigentliche Resonanz treten, wenn die Frequenz des einzelnen Systems von der Energie des Systems unabhängig und für beide Systeme ungefähr gleich ist. Man kann von Resonanz in diesem Sinne also nur bei harmonischen Oszillatoren sprechen. In der Quantenmechanik treten — im Einklang mit den allgemeinsten Erfahrungen — zwei Atomsysteme immer dann in Resonanz, wenn die Absorptionsfrequenz des einen Systems mit der Emissionsfrequenz des anderen oder umgekehrt übereinstimmt; wegen des den quantenmechanischen Gleichungen innewohnenden Charakters von Linearität ist also die Resonanz in der Quantenmechanik ein viel allgemeineres Phänomen als in der klassischen Theorie.

Zeitschrift für Physik 38, 411–426 (1926)

Zur näheren Untersuchung nehmen wir zwei völlig gleiche Systeme a und b von je f Freiheitsgraden an, die durch eine in beiden Systemen symmetrische Wechselwirkungsenergie λH^1 gekoppelt werden. Die einzelnen Systeme a und b sollen nicht entartet sein. Die Energien ihrer stationären Zustände sind durch H_n^a bzw. H_m^b gegeben.

Faßt man zunächst ohne Berücksichtigung der Wechselwirkung die beiden Systeme zu einem einzigen zusammen, so ist die Gesamtenergie des stationären Zustandes „n, m“ gegeben durch

$$H_{nm} = H_n^a + H_m^b. \tag{6}$$

Das Gesamtsystem besitzt nun die für die Resonanz charakteristische Entartung: wegen

$$H_{nm} = H_{mn} \tag{7}$$

ist jeder Eigenwert doppelt, mit Ausnahme der Eigenwerte, für die $n = m$. In anderen Worten: Es tritt immer dann Resonanz ein, wenn die beiden Systeme ursprünglich nicht im gleichen Zustand waren; dann ergibt die Vertauschung der beiden Systeme die gleiche Energie; nur bei äquivalenten Zuständen der Teilsysteme fällt die Resonanz (oder die Entartung) fort. Ein Beispiel für ein in dieser Weise entstehendes Termspektrum ist schematisch in Fig. 3 gegeben. In dem durch die Wechselwirkung gestörten System wird die Entartung aufgehoben. Es entsteht eine säkulare Schwebung, bei der die Energie zwischen den beiden Teilsystemen hin und her pulsiert. Formal ist der Zusammenhang folgender: Die Zusatzenergie W^1 des gestörten Systems ist in erster Näherung gegeben durch den Zeitmittelwert von H^1 über die ungestörte Bewegung. Dieser Mittelwert wird im allgemeinen noch Glieder enthalten, die Übergängen entsprechen, bei denen die Systeme a und b den Platz tauschen.

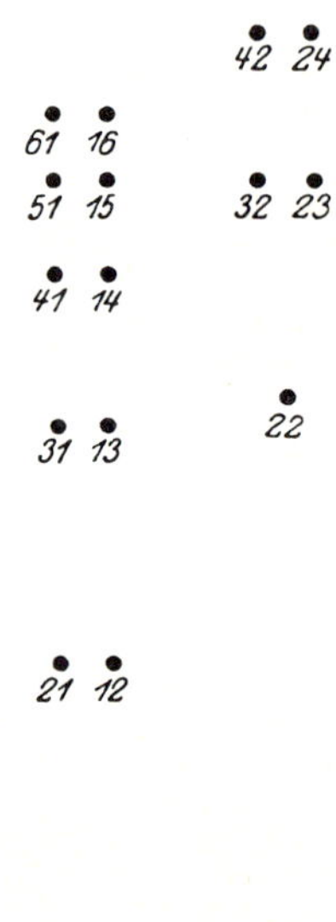

Fig. 3.

Man muß daher eine kanonische Transformation ausführen so, daß W^1 eine Diagonalmatrix wird. Die Rechnung ist in der Arbeit[1]) Quantenmechanik II, S. 589, angegeben. Die kanonische Transformation soll sein

$$W^1 = S^{-1} \overline{H^1} S, \tag{8}$$

$$q' = S^{-1} q S, \tag{9}$$

. . . .

[1]) M. Born, W. Heisenberg und P. Jordan, ZS. f. Phys. 35, 557, 1926.

S ist eine Matrix, die wie $\overline{H^1}$ nur Diagonalglieder enthält und Glieder, die Übergängen zwischen Zuständen gleicher Energie entsprechen. Für die nicht entarteten Zustände sind die Diagonalglieder 1. Es sind nun die zwei linearen Gleichungen mit den beiden Unbekannten S_{nm}, S_{mn} zu lösen:

$$\left.\begin{aligned} W^1 S_{nm} - H^1(nm, nm) S_{nm} - H^1(nm, mn) S_{mn} &= 0, \\ -H^1(mn, nm) S_{nm} + W^1 S_{mn} - H^1(mn, mn) S_{mn} &= 0. \end{aligned}\right\} \quad (10)$$

Hierin ist wegen der Symmetrie von H^1 in den Systemen a und b

$$H^1(nm, nm) = H^1(mn, mn), \quad H^1(nm, mn) = H^1(mn, nm). \quad (11)$$

Es ist zweckmäßig, die beiden durch Nullsetzen der Determinante aus (10) folgenden Lösungen für W^1 wieder mit nm und mn zu numerieren. Es muß aber beachtet werden, daß diese Zahlen dann den Quantenzahlen n' der Hauptschwingungen in § 1 entsprechen und nicht die Zustände der einzelnen Teilsysteme charakterisieren. Vielmehr führen in jedem Zustand die beiden Teilsysteme die gleichen Bewegungen (in verschiedener Phase) aus. Die Lösungen von (10) lauten unter Berücksichtigung von (11):

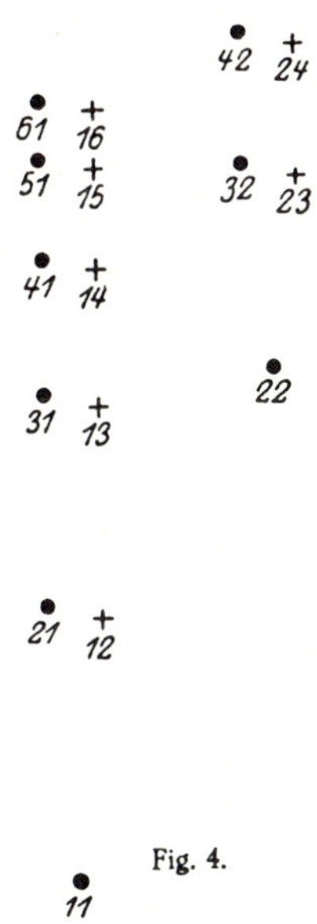

Fig. 4.

$$\left.\begin{aligned} W^1_{nm} &= H^1(nm, nm) + H^1(nm, mn), \\ S_{nm, nm} &= \frac{1}{\sqrt{2}}, \quad S_{mn, nm} = \frac{1}{\sqrt{2}}. \\ W^1_{mn} &= H^1(nm, nm) - H^1(nm, mn). \\ S_{nm, mn} &= \frac{1}{\sqrt{2}}, \quad S_{mn, mn} = -\frac{1}{\sqrt{2}}. \end{aligned}\right\} \quad (12)$$

Die Wirkung der Kopplung besteht zunächst darin, daß alle Energiewerte verschieden werden. Das Termspektrum ist jetzt durch Fig. 4 schematisch dargestellt.

Dieses Termspektrum läßt sich wieder — und dies ist das entscheidende Resultat — einteilen in zwei Termreihen, die in keiner Weise miteinander kombinieren können (+ und . in Fig. 4). Die Ausstrahlung wird nämlich wieder durch eine Funktion der p und q dargestellt sein, die bei der Vertauschung der beiden Teilsysteme den Wert

Zeitschrift für Physik 38, 411–426 (1926)

nicht ändert. Sei diese Funktion f, so gilt für deren Matrizenelemente nach Ausführung der kanonischen Transformation (8):

$$f' = S^{-1} f S, \tag{13}$$

also wegen der Symmetrie von f:

1. Wenn $n_1 \neq m$, $n_2 \neq m$, $n \neq m_1$, $n \neq m_2$,

$$\left.\begin{aligned}
f'_{n_1 m, n_2 m} &= \tfrac{1}{2}(f_{n_1 m, n_2 m} + f_{m n_1, m n_2} + f_{n_1 m, m n_2} + f_{m n_1, n_2 m}) \\
&= f_{n_1 m, n_2 m} + f_{n_1 m, m n_2}, \\
f'_{n m_1, n m_2} &= \tfrac{1}{2}(f_{n m_1, n m_2} + f_{m_1 n, m_2 n} - f_{n m_1, m_2 n} - f_{m_1 n, n m_2}) \\
&= f_{n m_1, n m_2} - f_{n m_1, m_2 n}. \\
f'_{n_1 m, m n_2} &= \tfrac{1}{2}(f_{n_1 m, n_2 m} - f_{m n_1, m n_2} + f_{m n_1, n_2 m} - f_{n_1 m, m n_2}) = 0. \\
f'_{n m_1, m_2 n} &= \tfrac{1}{2}(f_{m_1 n, m_2 n} - f_{n m_1, n m_2} + f_{m_1 n, n m_2} - f_{n m_1, m_2 n}) = 0.
\end{aligned}\right\} \tag{14}$$

2. Für Kombinationen mit Zuständen, in denen die Teilsysteme in „äquivalenten Zuständen“ [1]) sind:

$$\left.\begin{aligned}
f'_{n_1 m, mm} &= \frac{1}{\sqrt{2}}(f_{n_1 m, mm} + f_{m n_1, mm}) = \sqrt{2} f_{n_1 m, mm} \\
f'_{m m_1, mm} &= \frac{1}{\sqrt{2}}(f_{m_1 m, mm} - f_{m m_1, mm}) = 0.
\end{aligned}\right\} \tag{15}$$

Die Interkombinationen zwischen den Termsystemen + und · verschwinden also. Die Linienintensitäten innerhalb der beiden Termreihen sind in erster Näherung dieselben, wie die Intensitäten zwischen den entsprechenden ursprünglichen Termen des einzelnen Teilsystems (a und b); denn die Amplituden vom Typus $f_{n_1 m; m n_2}$, die gleichzeitigen Übergängen beider Systeme entsprechen, verschwinden in erster Näherung. Nur bei Kombinationen mit denjenigen Zuständen, die äquivalenten Zuständen der Teilsysteme a und b entsprechen, ist nach (15) die Intensität in der einen Termreihe doppelt so groß, wie im ursprünglichen System, die andere Termreihe enthält keine solchen Zustände. Die „Gesamtintensität“ wird also in erster Näherung durch die Resonanz nicht geändert.

Die Einteilung in die beiden nicht kombinierenden Termreihen kann durch äußere Einwirkungen nicht verändert werden. Offenbar herrschen also hier genau dieselben Verhältnisse wie im speziellen Beispiel in § 1. Das Nichtauftreten der Interkombinationen ist wieder gebunden an die

[1]) Hier und in der ganzen Arbeit bezieht sich der Ausdruck „in äquivalenten Zuständen“ auf das ungestörte System. Im gestörten System führen ja die Teilsysteme stets gleiche Bewegungen mit verschiedener Phase aus.

Gleichheit der zu koppelnden Systeme; sobald das eine System vor dem anderen ausgezeichnet wird, treten Interkombinationen auf. Ferner hat die quantenmechanische Lösung wieder den oben betonten Grad von Unbestimmtheit; sowohl das System + bzw. · für sich allein, als auch die Kombination beider hat als vollständige Lösung des Problems zu gelten.

Es mag auch von Interesse sein, das eben geschilderte Resonanzproblem vom Standpunkt des Schrödingerschen Verfahrens aus zu betrachten. In den ursprünglichen Teilsystemen wird die normierte Schrödingersche Eigenfunktion φ_n^a bzw. φ_m^b, zu den Zuständen mit den Energien H_n^a bzw. H_m^b gehörig, als bekannt angenommen. Es gilt z. B. für das Matrizenelement $n_1 n_2$ der Koordinate q_k^a des Systems a:

$$q_k^a(n_1, n_2) = \int \dots \int q_k^a \varphi_{n_1}^a \overline{\varphi_{n_2}^a} \, d q_1^a \dots d q_f^a.$$

(Der Strich über φ bedeutet: konjugiert komplexer Wert.)

Faßt man beide Systeme zu einem zusammen, so gehört zur Energie $H_{n,m} = H_n^a + H_m^b$ die Eigenfunktion $\varphi_n^a \varphi_m^b$. Die zu den Lösungen (12) gehörigen Eigenfunktionen des gestörten Systems gehen aus denen des ungestörten durch eine lineare Transformation mit der Matrix S hervor. Es gehört also zu

$$W_{nm}^1 \quad \text{die Eigenfunktion} \quad \frac{1}{\sqrt{2}}(\varphi_n^a \varphi_m^b + \varphi_m^a \varphi_n^b) \tag{16}$$

und

$$W_{mn}^1 \quad \text{die Eigenfunktion} \quad \frac{1}{\sqrt{2}}(\varphi_n^a \varphi_m^b - \varphi_m^a \varphi_n^b).$$

Das Nichtvorhandensein von Interkombinationen zwischen den beiden Teilsystemen folgt nun einfach daraus, daß ein Integral über eine in a und b symmetrische Funktion oder einen symmetrischen Operator f vom Typus

$$\int f \tfrac{1}{2} (\varphi_n^a \varphi_m^b + \varphi_m^a \varphi_n^b) (\overline{\varphi_n^a} \overline{\varphi_m^b} - \overline{\varphi_m^a} \overline{\varphi_n^b}) \, d q_1^a \dots d q_f^b \tag{17}$$

das Vorzeichen wechselt, wenn die Systeme a und b vertauscht werden, aber dabei doch den gleichen Wert behält: es muß also Null sein.

§ 3. Im folgenden soll als Beispiel die Anwendung dieser allgemeinen Theorie auf das Heliumatom kurz skizziert werden; auf die quantitative Durchführung möchte ich später zurückkommen.

1. Wir nehmen an, die Elektronen seien Punktladungen ohne Magnet und Drehimpuls. Dann entsteht ohne Wechselwirkung im wesentlichen das Schema der Fig. 3, mit Wechselwirkung das der Fig. 4. Die beiden

Zeitschrift für Physik *38*, 411–426 (1926)

Systeme, in die das Termspektrum zerfällt, sind Par- und Orthohelium. Interkombinationen zwischen ihnen sind einstweilen nicht möglich. Da der Energieunterschied zwischen zwei entsprechenden Termen von Par- und Orthohelium zurückgeführt wird auf eine durch die Coulombschen Abstoßungskräfte der Elektronen hervorgerufene Resonanzschwebung, ist es verständlich, daß er im allgemeinen von der gleichen Größenordnung ist wie die durch die Abschirmung verursachte Abweichung des betreffenden Terms vom Wasserstoffterm. Ferner folgt aus Fig. 4, daß der 1 *S*-Term nur in einem der beiden Systeme vorhanden ist. Daß dieses System dasjenige ist, dessen Energiewerte im allgemeinen höher liegen, folgt aus der Rechnung. Die Übergangswahrscheinlichkeiten innerhalb des Ortho- bzw. Parheliumsystems sind in erster Näherung die gleichen wie die entsprechenden in Wasserstoff. Nur die Parhe-Übergänge zum 1 *S*-Term sollten in gröbster Annäherung doppelt so häufig sein.

Wollte man sich von der Bewegung der Elektronen im Atom ein der quantenmechanischen Lösung einigermaßen entsprechendes anschauliches Bild machen, so müßte man sich hier etwa vorstellen, daß die beiden Elektronen periodisch in kontinuierlicher Weise die Plätze tauschen, in Analogie zu den Energieschwebungen beim oben erwähnten Oszillatorbeispiel, wobei die Periode dieser Schwebung eben durch den Abstand des Orthoheliumterms vom entsprechenden Parheliumterm gegeben ist.

2. Wir denken uns entsprechend der Compton-Uhlenbeck-Goudsmitschen Hypothese[1]) die Elektronen als kleine magnetische Kreisel, denen wir zunächst eine bestimmte für die beiden Elektronen verschiedene Achsenrichtung vorschreiben wollen. Dann ändert sich am Termschema von Fig. 4 qualitativ nichts, aber es treten schwache Interkombinationen zwischen Ortho- und Parhelium auf von der Stärke der Wechselwirkung zwischen Magnet und Bahn. Denn die beiden Elektronen sind jetzt nicht mehr gleichberechtigt.

3. Wir lassen die Richtungen der Elektronenmagnete willkürlich. Dann ergibt die Rechnung, daß jeder Term des bisherigen Systems aufspaltet in vier Terme, dem statistischen Gewicht der Elektronenmagnete entsprechend und daß das Termschema wie früher in zwei vollständig getrennte Teile zerfällt; denn die Elektronen sind jetzt wieder gleichberechtigt. Aber die Teilung ist anders als früher (vgl. Fig. 5 + und ·).

---

[1]) A. H. Compton, Journ. Frankl. Inst. **192**, 145, 1924; E. Uhlenbeck und S. Goudsmit, Naturwiss. **13**, 953, 1925.

Das eine System stellt ein Termspektrum dar, bei dem Orthohelium ein Triplett-, Parhelium ein Singlettsystem ist. das andere ein Termspektrum, bei dem Orthohelium ein Singlett- und Parhelium ein Triplettsystem ist. Interkombinationen zwischen Par- und Orthohelium sind vorhanden, wie in 2. von der Stärke der Wechselwirkung zwischen Bahn und Elektronenmagnet. Dagegen gibt es keine Interkombinationen zwischen den Systemen + und ·. Man kann den physikalischen Sachverhalt etwa kurz so kennzeichen: Zwischen beiden Elektronen besteht zunächst eine grobe elektrische Resonanz; diese wird durch die Magnete gestört, und es treten die Übergänge von Ortho- nach Parhelium auf. Aber auch bei Berücksichtigung der Magnete besteht eine feinere exakte Resonanz der Elektronen, die zur Teilung in die beiden oben charakterisierten Termreihen Anlaß gibt. Wie die Rechnung zu den hier beschriebenen Resultaten führt, soll in einer demnächst erscheinenden Arbeit erläutert werden.

§ 4. Es ist beim Heliumspektrum ein empirisches Faktum, daß nur das eine System (in Fig. 5) vorhanden ist, und, soviel wir hier sehen können, wenigstens qualitativ, mit dem He-Spektrum übereinstimmt; das andere System ist in der Natur nicht realisiert. Diese Tatsache scheint mir — wenn wir annehmen, daß sich unsere für zwei Systeme abgeleiteten Resultate auch auf den Fall beliebig vieler Systeme verallgemeinern lassen — den eigentlichen Zusammenhang zwischen der hervorgehobenen quantenmechanischen Unbestimmtheit einerseits und der Paulischen Regel und der Einstein-Boseschen Abzählung andererseits zu bedeuten. Wenn nur das eine (·) der beiden Systeme in der Natur vorkommt, so gibt dies einerseits zu einer Reduktion der statistischen Gewichte eben in dem von Bose vorgeschlagenen Sinne Anlaß; andererseits ist bei der richtigen Wahl des Systems Paulis Verbot äquivalenter Bahnen von selbst erfüllt. Die Verallgemeinerung dieser Überlegungen auf Systeme, die aus n gleichen Teilsystemen bestehen, wird daher in folgender Weise zu denken sein: Den $n!$ Vertauschungen der n Systeme entsprechend gibt es ohne Wechselwirkung im allgemeinen je $n!$ gleiche Eigenwerte. Durch die Wechselwirkung wird die Entartung aufgehoben, das Termsystem zerfällt wegen Resonanz in $n!$ Teilsysteme. Unter ihnen gibt es ein System, das keine äquivalenten Bahnen enthält und nicht mit den anderen Systemen kombinieren kann. Dieses System kommt in der

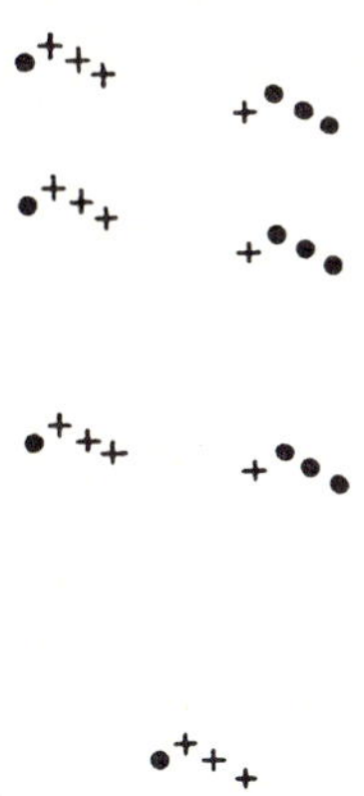

Fig. 5.

Zeitschrift für Physik 38, 411–426 (1926)

Natur allein vor und stellt die wirkliche Lösung dar. Zugleich entspricht die Reduktion der statistischen Gewichte von $n!$ auf 1 der Bose-Einsteinschen Abzählung. Die hier gegebene Formulierung dieser Abzählung geht aber über die Bose-Einsteinsche insofern hinaus, als sie die Wahl eines ganz bestimmten Systens aus den $n!$ Lösungen vorschreibt, nämlich desjenigen Systems, das keine äquivalenten Teilsysteme enthält und daher mit Paulis Verbot im Einklang ist. Eine Begründung dafür, daß gerade nur dieses eine System unter allen möglichen quantenmechanischen Lösungen vorkommt, kann von der einfachen quantenmechanischen Rechnung aus wohl nicht gegeben werden. Es scheint mir aber ein wichtiges Resultat dieser Untersuchung, daß Paulis Verbot und die Einsteinsche Statistik den gleichen Ursprung haben, und daß sie der Quantenmechanik nicht widersprechen. Auch ein von Einstein mehrfach betontes Paradoxon hat ein Analogon in unseren Überlegungen: Wenn die zu koppelnden Teilsysteme voneinander verschieden sind, so muß für sie die klassische Statistik gelten; im Prinzip bis herab zu unendlich kleinen Unterschieden. Trotzdem wird die Abzählung für gleiche Systeme ganz anders. Für verschiedene Systeme muß auch nach den hier durchgeführten Rechnungen stets die klassische Abzählung bestehen bleiben, da Übergänge zwischen den $n!$ Teilsystemen vorkommen; also kann kein Teilsystem ausgeschlossen werden. Die Übergänge werden aber bei abnehmenden Unterschieden der Partikel immer seltener. Werden die den Übergängen entsprechenden Amplituden kleiner als eine endliche, durch die Schärfe des betreffenden Zustandes definierte Größe, so besteht eine logische Möglichkeit, die Übergänge ganz auszuschließen und die Abzählung zu ändern. Es sei noch betont, daß nach den hier ausgeführten Überlegungen eine endliche Wechselwirkung der Systeme für die Abänderung der Abzählung eine notwendige Voraussetzung ist. Wenn die Perioden der dem Resonanzeffekt entsprechenden Energiepulsationen länger sind als die Lebensdauer, so verlieren die oben ausgeführten Rechnungen den Sinn.

Wir haben oben hervorgehoben, daß der Ausschluß aller Termsysteme bis auf ein bestimmtes ohne Verletzung der Gesetze der Quantenmechanik möglich ist. Es muß aber darauf hingewiesen werden, daß dieser Ausschluß doch gewisse sehr charakteristische Beschränkungen mit sich bringt. Er bedeutet nämlich, daß es keinen physikalischen Sinn hat, über die Bewegung oder die diese Bewegung darstellende Matrix eines einzelnen Elektrons oder über die Matrix irgend einer nichtsymmetrischen Funktion der Elektronen in einem Atomsystem zu sprechen. Eine solche Matrix würde nämlich im allgemeinen Glieder enthalten, die

Übergängen von einem Termsystem zu anderen, also auch zu nicht vorkommenden Termsystemen entsprechen. Deshalb kann z. B. auch den Vertauschungsrelationen in ihrer gewöhnlichen Form im allgemeinen kein physikalischer Sinn beigelegt werden; wohl aber allen symmetrisierten Formen dieser Bedingungen. Es handelt sich indessen hier nur um eine formelle Beschränkung, denn alle prinzipiell beobachtbaren Größen sind ihrem Wesen nach symmetrische Funktionen der Elektronen.

Der hier geschilderte Zusammenhang müßte nur als Programm angesehen werden, solange die mathematische Behandlung der Systeme, die aus n gleichen Teilsystemen bestehen, nicht durchgeführt ist. Ich möchte als Ergänzung daher kurz angeben, wie dasjenige Termsystem, das keine äquivalenten Zustände der Teilsysteme mit sich führt und das als einzige Lösung übrig bleiben soll, allgemein eindeutig konstruiert werden kann.

Die n völlig gleichen Teilsysteme mögen sich in den durch die „Quantenzahlen" $m_1, m_2 \ldots m_n$ festgelegten stationären Zuständen befinden. Dann haben alle die Zustände des Gesamtsystems, die aus dem eben betrachteten durch irgend eine Umstellung der $m_1 \ldots m_n$ hervorgehen, die gleiche Energie. Die Wechselwirkungsenergie H^1 wird in erster Näherung nur Glieder enthalten, die Übergängen von höchstens zweien der Teilsysteme entsprechen, wenn H^1 sich additiv aus Wechselwirkungen zwischen je zwei Teilsystemen zusammensetzt. Um den Zeitmittelwert der Störungsenergie W^1 zu einer Diagonalmatrix zu machen, muß wie in § 2 eine kanonische Transformation ausgeführt werden. Um sie zu finden, ist in bekannter Weise ein System von $n!$ Gleichungen mit $n!$ Unbekannten S_k zu lösen. Wir bezeichnen dasjenige Glied in H^1, das keinem Übergang eines der Teilsysteme entspricht, mit H^1_α; diejenigen Glieder, die Vertauschungen zweier der Zahlen $m_1 \ldots m_n$ entsprechen [dies sind im ganzen $\frac{n(n-1)}{2}$ Glieder], mit $H^1_\beta, H^1_\gamma, \ldots H^1_\nu$; dann ordnen wir die $n!$ Zustände gleicher Energie so, daß an erster Stelle der Zustand $m_1 \ldots m_n$ kommt, dann alle Zustände, die aus ihm durch eine „Transposition" hervorgehen; dann die, zu deren Herstellung aus dem ersten Zustand zwei Transpositionen notwendig sind usw. Die das System $n!$ linearer Gleichungen charakterisierende Determinante hat dann etwa die Form

$$\begin{vmatrix} W - H^1_\alpha & -H^1_\beta & -H^1_\gamma \ldots -H^1_\nu & 0 \ldots 0 \\ -H^1_\delta & W - H^1_\alpha & 0 \ldots\ldots 0 & -H^1_\varepsilon \ldots \\ -H^1_\eta & 0 & W - H^1_\alpha \, 0 \ldots 0 & -H^1_\zeta \ldots \\ \cdots & \cdots & \cdots & \cdots \\ \cdots & \cdots & \cdots & \cdots \end{vmatrix} \tag{18}$$

Zeitschrift für Physik *38*, 411–426 (1926)

Eine Lösung dieses Gleichungssystems lautet:

$$W^1 = H^1_\alpha - H^1_\beta - H^1_\gamma - \dots - H^1_\nu; \quad S_k = \frac{1}{\sqrt{n!}} \cdot (-1)^{\delta_k}, \tag{19}$$

wo δ_k diejenige Anzahl von Transpositionen bedeutet, die nötig sind, um den Zustand $m_1 \dots m_n$ in den mit der Nummer k bezeichneten Zustand überzuführen. Die zu dieser Lösung gehörige Schrödingersche Eigenfunktion ist also gegeben durch

$$\varphi = \frac{1}{\sqrt{n!}} \sum (-1)^{\delta_k} \cdot \varphi_1(m^k_\alpha)\, \varphi_2(m^k_\beta) \dots \varphi_n(m^k_\nu), \tag{20}$$

wo wieder δ_k die Anzahl von Transpositionen bedeutet, die nötig sind, um die Reihe $m_1 m_2 \dots m_n$ in $m^k_\alpha m^k_\beta \dots m^k_\nu$ zu verwandeln. Diese Funktion φ hat die besondere Eigenschaft, daß sie bei Vertauschung der Quantenzahlen zweier Teilsysteme das Vorzeichen wechselt. Hieraus folgt sofort, daß das durch (20) charakterisierte Termsystem nicht mit Zuständen, bei denen zwei oder mehr Teilsysteme in äquivalenten Zuständen sind, kombinieren kann. Die Eigenfunktion ψ eines solchen Zustandes müßte nämlich notwendig gegen Vertauschungen der äquivalenten Teilsysteme invariant sein. Ist f eine die Strahlung darstellende und daher notwendig symmetrische Funktion (oder ein symmetrischer Operator) der Koordinaten der Teilsysteme, so wird das Integral

$$\int f \psi \varphi \, d\Omega = 0, \tag{21}$$

da es bei Vertauschung zweier äquivalenter Teilsysteme das Vorzeichen wechseln müßte, ohne seinen Wert zu ändern.

Man kann durch Induktion (Schluß von n auf $n + 1$) einsehen, daß die Terme vom Typus (20) wirklich ein geschlossenes System bilden, d. h. mit keinem anderen Term kombinieren, und daß das durch (20) gegebene Termsystem das einzige ist, das keine äquivalenten Zustände der Teilsysteme enthält. Allerdings ist mir eine strenge Durchführung dieses Beweises noch nicht gelungen. Von dieser Ungenauigkeit abgesehen, kann (20) als die definitive Konstruktion des die Lösung darstellenden Termsystems betrachtet werden.

Nach dieser mehr mathematischen Ergänzung komme ich zum physikalischen Inhalt dieser Untersuchung zurück. Der Zusammenhang zwischen der Einstein-Boseschen Statistik gleicher Partikel und der Quantenmechanik besteht also in der Auswahl einer bestimmten quantenmechanischen Lösung unter vielen möglichen Lösungen. Eine solche Auswahl bedeutet für hinreichend kleine Wechselwirkung im wesentlichen eine Phasenbeziehung zwischen den Teilsystemen oder Partikeln. Vielleicht könnte man durch eine nähere Untersuchung dieser Phasenbeziehung zu Resultaten kommen, die direkt als Analogon zu den Interferenzen der de Broglieschen Wellen aufgefaßt werden können. Möglicherweise führt also eine nähere Verfolgung der Gründe, um derentwillen die eine Lösung ausgewählt wird, schon in die großen physikalischen Schwierigkeiten, die bei den Kopplungsproblemen auftreten. Mir scheint aber aus der vorliegenden Untersuchung zu folgen, daß wir wahrscheinlich diese Schwierigkeiten nicht zu lösen brauchen, um die Spektra der Atome mit mehreren Elektronen auszurechnen. Diese Spektra dürften vielmehr schon durch die Quantenmechanik bestimmt sein.

Zeitschrift für Physik *40*, 501–506 (1926)

Schwankungserscheinungen und Quantenmechanik.

Von **W. Heisenberg** in Kopenhagen.

Mit 1 Abbildung. (Eingegangen am 6. November 1926.)

Die vorliegende Note versucht zu zeigen, daß die Quantenmechanik mit den von der Diskontinuumstheorie geforderten Schwankungsformeln stets in Übereinstimmung ist.

Die große physikalische Bedeutung der Schwankungserscheinungen beruht darauf, daß sie zu den anschaulichsten und unmittelbarsten Auswirkungen der bei Vorgängen in kleinen Räumen und Zeiten auftretenden Diskontinuitäten gehören. Nach Einstein kann z. B. die Brownsche Bewegung als direkte Folge der atomistischen Struktur der Materie angesehen werden, die Energie- und Impulsschwankungen im Strahlungshohlraum führen unmittelbar zur Einsteinschen Lichtquantenvorstellung, die Energieschwankungen etwa in einem Kristallgitter stehen in engster Verbindung mit der Existenz diskreter stationärer Zustände eines mechanischen Systems. Da die Theorie der mechanischen Systeme durch die Quantenmechanik einer quantitativen Behandlung zugänglich gemacht worden ist, wird zwischen der Quantenmechanik und jener letztgenannten Art von Schwankungserscheinungen ein sehr enger Zusammenhang bestehen müssen. Dieser Zusammenhang soll im folgenden klargestellt werden. Die folgende Note bedeutet also eine Erweiterung und Präzisierung der Überlegungen, die Born, Jordan und der Verfasser[1]) früher gelegentlich der allgemeinen Entwicklung der Quantenmechanik dargestellt haben[2]). Eine tiefere Untersuchung der damals gefundenen Zusammenhänge schien mir nützlich, da neuerdings von manchen Seiten Zweifel an der Existenz der Diskontinuitäten laut geworden sind.

§ 1. Der mathematische Teil der folgenden Überlegung schließt sich an eine Arbeit des Verfassers[3]) über quantenmechanische Resonanz an; ich übernehme hier die Resultate jener Arbeit. Es sei zunächst dieses Problem gestellt: Zwei gleichartige Atome a und b mögen sich

---

[1]) M. Born, W. Heisenberg und P. Jordan, Quantenmechanik II. ZS. f. Phys. **35**, 557, 1926, Kap. 4, § 3.

[2]) Die Kritik von A. Smekal, ZS. f. Phys. **37**, 319, 1926, bezieht sich nicht auf den hier behandelten Fall mechanischer Systeme; ob eine solche Kritik im Falle der Hohlraumstrahlung zu fruchtbaren Resultaten führen kann, läßt sich zurzeit nicht entscheiden; ich möchte annehmen, daß der Fall des Hohlraums dem des Kristallgitters vollkommen analog ist.

[3]) W. Heisenberg, Mehrkörperproblem und Resonanz in der Quantenmechanik, ZS. f. Phys. **38**, 411, 1926; im folgenden als (l. c.) zitiert.

(bei Vernachlässigung aller Strahlungskräfte) zunächst in den Zuständen n und m, d. h. $W^a = E_n$, $W^b = E_m$ befinden; sie werden durch eine sehr kleine Wechselwirkung gekoppelt. Dann besteht Resonanz zwischen beiden Atomen; diese Resonanz läßt sich anschaulich auf zwei verschiedene Weisen beschreiben:

1. Es werden im Laufe der Zeit mit einer gewissen Häufigkeit Energiesprünge vorkommen, derart, daß sich gleichzeitig W^a diskontinuierlich von E_n nach E_m und W^b von E_m nach E_n oder umgekehrt ändert. In anderen Worten, im Laufe der Zeit wird das „Lichtquantum“ [1]) $(E_n - E_m)$ immer wieder einmal vom Atom a zum Atom b übergehen, nach irgendeiner Zeit wieder zu a zurückkehren. Im Mittel wird das Lichtquantum aus Symmetriegründen die Hälfte der Zeit im Atom a, die andere Hälfte im Atom b sein.

2. Die Resonanz ist als korrespondenzmäßiges Analogon der klassischen Wechselwirkung zwischen zwei Oszillatoren anzusehen. Deshalb wird die Energie mit einer langsamen Schwebungsfrequenz zwischen den Atomen a und b hin- und herpulsieren. Die Energie eines Atoms ist eine harmonisch-periodische Funktion der Zeit. In dieser Form widerspricht die Beschreibung II der Beschreibung I vollständig. Durch die Quantenmechanik wird jedoch diese Korrespondenzbeschreibung wesentlich modifiziert und einer mathematischen Behandlung zugänglich gemacht: Es hat, bei einem bestimmten Zustand des Gesamtsystems, keinen Sinn, von der Energie eines Atoms als Funktion der Zeit zu sprechen. Nur die Zeitmittelwerte irgend einer von der Zeit abhängigen Größe haben einen physikalischen Sinn bei einem bestimmten Zustand des Gesamtsystems. Die Energie eines Atoms ist im hier betrachteten Falle formal darstellbar durch eine Matrix, die einer harmonischen Funktion der Zeit entspricht. Die harmonischen Glieder dieser Matrix sind aber je zwei Zuständen des Gesamtsystems zugeordnet. Solange keine Übergänge des Gesamtsystems in Betracht gezogen werden, sind nur die obigen Zeitmittelwerte prinzipiell beobachtbar.

Es soll gezeigt werden, daß für alle prinzipiell beobachtbaren Effekte die Beschreibung II der Beschreibung I äquivalent ist. Die erste Frage sei zunächst, ob es möglich sein wird, etwa durch Stoßprozesse mit einem der beiden Atome Energiewerte experimentell festzustellen, die irgendwo zwischen E_n und E_m liegen. Die Antwort lautet von beiden Beschreibungen aus: nein. Im Falle II braucht man nur die Grund-

[1]) Man sollte vielleicht besser „Schallquantum“ sagen, da es sich um mechanische Schwingungszustände handelt.

Zeitschrift für Physik *40*, 501–506 (1926)

definitionen der Quantenmechanik auf das Gesamtsystem (der beiden Atome a und b) anzuwenden, um einzusehen, daß (bis auf Größen von der Ordnung der Wechselwirkung) nur dieselben Energiedifferenzen $E_n - E_m$ wie ohne Wechselwirkung bei Stoßprozessen übertragen werden können. Um weiterzukommen, sei die quantentheoretische Matrix für die Energie eines einzelnen Atoms (etwa a) ausgerechnet: Alle Matrizen des Systems mit Wechselwirkung gehen aus denen des ungestörten Systems durch eine kanonische Transformation mit der Matrix S hervor [l. c. Gleichung (8) und (7)]:

$$\left.\begin{aligned} \boldsymbol{W}' &= \boldsymbol{S}^{-1}\boldsymbol{W}\boldsymbol{S}, \\ \boldsymbol{q}' &= \boldsymbol{S}^{-1}\boldsymbol{q}\boldsymbol{S}, \end{aligned}\right\} \tag{1}$$

hierin ist $\boldsymbol{S}$ [l. c. Gleichung (12)]:

$$S_{nm,\,nm} = \frac{1}{\sqrt{2}};\; S_{nm,\,mn} = \frac{1}{\sqrt{2}};\; S_{mn,\,nm} = \frac{1}{\sqrt{2}};\; S_{mn,\,mn} = -\frac{1}{\sqrt{2}}. \tag{2}$$

Bezeichnet man also mit $\boldsymbol{W}^a$ die Energie des Atoms a im ungestörten System, mit $\boldsymbol{W}'^a$ die im gestörten, so wird

$$\boldsymbol{W}'^a = \boldsymbol{S}^{-1}\boldsymbol{W}^a\boldsymbol{S}. \tag{3}$$

Dies ist nach (2) eine Matrix mit den Gliedern:

$$\left.\begin{aligned} W'^a_{nm,\,nm} &= \tfrac{1}{2}(E_n + E_m); \quad & W'^a_{nm,\,mn} &= \tfrac{1}{2}(E_n - E_m); \\ W'^a_{mn,\,nm} &= \tfrac{1}{2}(E_n - E_m); \quad & W'^a_{mn,\,mn} &= \tfrac{1}{2}(E_n + E_m). \end{aligned}\right\} \tag{4}$$

Als prinzipiell beobachtbare Größen in einem bestimmten Zustand, z. B. nm des Gesamtsystems, werden anzusehen sein: der Zeitmittelwert von $\boldsymbol{W}^a$ selbst, das mittlere Schwankungsquadrat der Energie, irgendwelche Schwankungsmittelwerte. Ganz allgemein können alle diese Schwankungen zurückgeführt werden auf die Zeitmittelwerte irgend einer Funktion von $\boldsymbol{W}: \boldsymbol{f}(\boldsymbol{W})$ (z. B. etwa den Mittelwert von $\boldsymbol{W}^2$, $\boldsymbol{W}^4$ usw.). Wir rechnen daher den Zeitmittelwert solch einer Funktion $\boldsymbol{f}$ aus: Es wird

$$\boldsymbol{f}' = \boldsymbol{S}^{-1}\boldsymbol{f}\boldsymbol{S}, \tag{5}$$

also

$$f'_{nm,\,nm} = \tfrac{1}{2}\big(f(E_n) + f(E_m)\big);$$

ebenso wird

$$f'_{mn,\,mn} = \tfrac{1}{2}\big(f(E_n) + f(E_m)\big).$$

Man sieht ohne weiteres, daß diese Werte mit denen, die aus der Auffassung I gewonnen werden können, völlig identisch sind. Man könnte die Frage sogar umgekehrt stellen: Gibt es eine Funktion $E(t)$ derart, daß der Mittelwert irgend einer beliebigen Funktion $f(E)$ der Gleichung (15) genügt:

$$\overline{f(E(t))} = \tfrac{1}{2}\big(f(E_n) + f(E_m)\big).$$

Man wird dann wohl finden, daß nur Funktionen von dem in Fig. 1 dargestellten Typus diese Eigenschaft haben, und zwar dann, wenn die Gesamtlänge der oberen Stücke der Kurve der Gesamtlänge der unteren Stücke im Mittel gleich ist, was genau den Annahmen I entspricht.

Das Resultat ist also, daß die Quantenmechanik in dem hier behandelten Falle hinsichtlich aller Schwankungsgrößen zu den gleichen Resultaten kommt wie die Diskontinuumsvorstellung, in anderen Worten, es zeigt sich, daß die Tatsache der Diskontinuität in ungezwungener Weise dem System der Quantenmechanik eingefügt ist. Es scheint, als ob die Quantenmechanik ebensoviel und ebensowenig über die diskontinuierlichen Vorgänge auszusagen gestattete, als wirklich prüfbar ist. Die Zeitpunkte der Übergänge, überhaupt die Übergänge selbst kommen in diesem Schema nicht vor. Es ist also unsere Meinung, daß in der Fig. 1 nur die Gesamtlänge der oberen und unteren Stücke der Kurve eine physikalische Bedeutung hat.

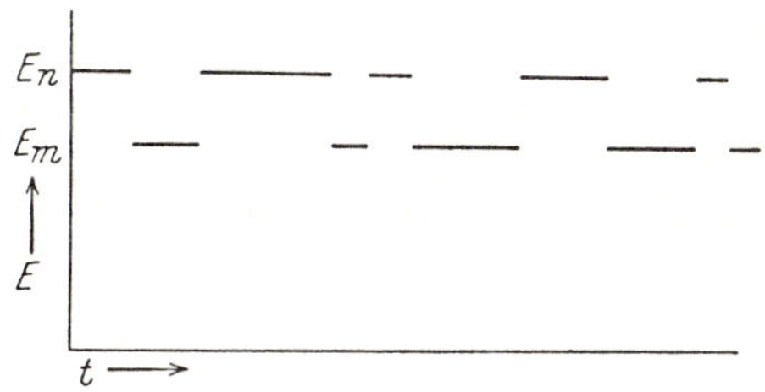

Fig. 1.

Es mag noch hinzugefügt werden, daß alle Rechnungen nur in der Näherung durchgeführt wurden, in der die Wechselwirkung der beiden Atome als verschwindend klein betrachtet werden kann. Dies ist aber eben die Näherung, in der das Problem überhaupt definiert ist. Der Begriff „Energie eines Atoms“ hat ja nur so lange einen bestimmten Sinn, als die Wechselwirkungsenergie vernachlässigt werden kann.

§ 2. Der mathematische Grund für die Übereinstimmung der quantenmechanischen Schwankungsmittelwerte mit denen der Diskontinuumstheorie ist die Form der kanonischen Transformation (1), (5). Diese Form hat andererseits in der Quantenmechanik so allgemein Geltung, daß sich die oben angestellten Schwankungsüberlegungen auf die allgemeinsten vorkommenden Fälle erweitern lassen. Zunächst ist die Annahme von § 1, daß a und b zwei Atome seien, unwesentlich; es können irgendwelche gleichartige mechanische Systeme sein. Ferner bleiben alle Betrachtungen auch dann richtig, wenn die Wechselwirkung mehrerer gleichartiger mechanischer Systeme betrachtet wird. Es weist dann das ungestörte System eine höhere Entartung auf. Die Form (1), (5) der kanonischen Transformation bleibt jedoch stets erhalten, nur die Matrix S selbst wird anders. Auf die Werte von S kam es für die Übereinstimmung der Diskontinuumsschwankungen mit den quantenmechanischen

Schwankungen gar nicht an, nur die Form der kanonischen Transformation war wesentlich. Bezeichnet man etwa die Zustände irgendeines gestörten mechanischen Systems mit „α", die Zustände des ungestörten Systems mit „β". Dann bedeutet eine kanonische Transformation (1), (5) im Sinne der Diskontinuumsvorstellung folgendes: Befindet sich das gestörte System im Zustand α, so ist $|S_{\alpha\beta}|^2$ die Wahrscheinlichkeit dafür, daß (bei Stoßprozessen, bei plötzlichem Aufhören der Störung usw.) das System im Zustand β betroffen wird. Für jede Funktion etwa von W^a gilt z. B. nach (5):

$$f_\alpha(W^a) = \sum_\beta |S_{\alpha\beta}|^2 f(W_\beta^a), \tag{6}$$

wo W_β^a denjenigen Wert von W^a bezeichnet, den das System a im Zustand β annimmt. Nach allgemeinen Prinzipien gilt natürlich

$$\sum_\beta |S_{\alpha\beta}|^2 = 1. \tag{7}$$

Als zweite Erweiterung kann die Annahme verlassen werden, daß es sich um gleichartige mechanische Systeme handelt. Es soll nur vorausgesetzt werden, daß die gleiche Energiedifferenz $E_n - E_m$ in allen Systemen vorkommt, da es sich sonst (eben für die Systeme, in denen sie nicht vorkommt) gar nicht um einen Fall von Resonanz handeln würde. In den Worten der Diskontinuumsvorstellung: es soll das Lichtquant $E_n - E_m$ in allen Systemen Platz finden können. Die Rechnungen des § 1 können auf diesen Fall ungeändert übertragen werden.

Die dritte Erweiterung, die vorgenommen werden soll, bezieht sich auf die Art der Größe, deren Schwankungen untersucht werden. Die bisherigen Überlegungen werden nicht geändert, wenn statt der Energie des Teilsystems irgend eine andere Größe betrachtet wird, die im ungestörten Falle durch eine Diagonalmatrix dargestellt werden kann. Zu diesen Größen gehört z. B. der Gesamtdrehimpuls, der Drehimpuls um eine feste Achse usw. Die Schwankungsmittelwerte aller dieser Größen stimmen nach der quantenmechanischen Rechnung von § 1 mit den aus der Diskontinuumsvorstellung folgenden Schwankungsmittelwerten überein. Sofern man überhaupt über die Energie, den Drehimpuls usw. eines mechanischen Systems als Funktion der Zeit sprechen will, kommen nach der Quantenmechanik nur Funktionen von dem in Fig. 1 gezeichneten Typus in Betracht. Als Wahrscheinlichkeitskoeffizienten gehen übrigens immer die gleichen Größen $|S_{\alpha\beta}|^2$ ein.

Endlich werden die Überlegungen des § 1 auch für den Fall aperiodischer Bewegungen ihre Gültigkeit beibehalten, da die Form der kanonischen Transformation (1), (5) ja allgemeine Gültigkeit hat. Als Beispiel mag zum Schlusse noch angegeben werden, in welcher Weise die von Born, Jordan und dem Verfasser (l. c.) angegebenen Rechnungen über Schwankungen in einem Kristallgitter in den hier ausgeführten Überlegungen enthalten sind. Es handelt sich um Energieschwankungen in einem kleinen Teilvolumen eines Kristalls. Im ungestörten System soll das Teilvolumen aus dem Kristall herausgeschnitten sein. Dann besteht im gestörten System Resonanz hinsichtlich aller derjenigen Eigenschwingungen, deren Frequenz für Kristall und Teilvolumen im ungestörten System die gleiche ist. Dies ist angenähert der Fall für alle Eigenschwingungen, bei denen die Wellenlängen klein gegen die Lineardimensionen des kleinen Teilvolumens sind. Nur für solche Eigenschwingungen hat also das Schwankungsproblem einen bestimmten Sinn. Aus den Überlegungen des § 1 kann ohne Rechnung vorausgesehen werden, daß die Berechnung des mittleren Schwankungsquadrats und aller höheren Schwankungsmittelwerte nach der Quantenmechanik das gleiche Resultat geben muß wie die Bose-Einsteinsche Lichtquantenstatistik. (Eben die letztgenannte Statistik deswegen, weil auch in der Quantenmechanik stets nur nach den Energiewerten der stationären Zustände gefragt wird, vgl. l. c. Quantenmechanik II.) Für den Fall des mittleren Schwankungsquadrats ist die Rechnung explizit in der oben genannten Arbeit „Quantenmechanik II“ durchgeführt worden.

Die hier durchgeführten Rechnungen scheinen mir ein Argument dafür, daß eine kontinuierliche Interpretation des quantenmechanischen Formalismus, also auch der de Broglie-Schrödingerschen Wellen, nicht dem Wesen der bekannten formalen Zusammenhänge entsprechen würde. Vielmehr ist nach diesen Rechnungen die Tatsache der Diskontinuitäten harmonisch im mathematischen Schema der Quantenmechanik enthalten.

Zeitschrift für Physik *43*, 172–198 (1927)

Über den anschaulichen Inhalt der quantentheoretischen Kinematik und Mechanik.

Von **W. Heisenberg** in Kopenhagen.

Mit 2 Abbildungen. (Eingegangen am 23. März 1927.)

In der vorliegenden Arbeit werden zunächst exakte Definitionen der Worte: Ort, Geschwindigkeit, Energie usw. (z. B. des Elektrons) aufgestellt, die auch in der Quantenmechanik Gültigkeit behalten, und es wird gezeigt, daß kanonisch konjugierte Größen simultan nur mit einer charakteristischen Ungenauigkeit bestimmt werden können (§ 1). Diese Ungenauigkeit ist der eigentliche Grund für das Auftreten statistischer Zusammenhänge in der Quantenmechanik. Ihre mathematische Formulierung gelingt mittels der Dirac-Jordanschen Theorie (§ 2). Von den so gewonnenen Grundsätzen ausgehend wird gezeigt, wie die makroskopischen Vorgänge aus der Quantenmechanik heraus verstanden werden können (§ 3). Zur Erläuterung der Theorie werden einige besondere Gedankenexperimente diskutiert (§ 4).

Eine physikalische Theorie glauben wir dann anschaulich zu verstehen, wenn wir uns in allen einfachen Fällen die experimentellen Konsequenzen dieser Theorie qualitativ denken können, und wenn wir gleichzeitig erkannt haben, daß die Anwendung der Theorie niemals innere Widersprüche enthält. Zum Beispiel glauben wir die Einsteinsche Vorstellung vom geschlossenen dreidimensionalen Raum anschaulich zu verstehen, weil für uns die experimentellen Konsequenzen dieser Vorstellung widerspruchsfrei denkbar sind. Freilich widersprechen diese Konsequenzen unseren gewohnten anschaulichen Raum–Zeitbegriffen. Wir können uns aber davon überzeugen, daß die Möglichkeit der Anwendung dieser gewohnten Raum—Zeitbegriffe auf sehr große Räume weder aus unseren Denkgesetzen noch aus der Erfahrung gefolgert werden kann. Die anschauliche Deutung der Quantenmechanik ist bisher noch voll innerer Widersprüche, die sich im Kampf der Meinungen um Diskontinuums- und Kontinuumstheorie, Korpuskeln und Wellen auswirken. Schon daraus möchte man schließen, daß eine Deutung der Quantenmechanik mit den gewohnten kinematischen und mechanischen Begriffen jedenfalls nicht möglich ist. Die Quantenmechanik war ja gerade aus dem Versuch entstanden, mit jenen gewohnten kinematischen Begriffen zu brechen und an ihre Stelle Beziehungen zwischen konkreten experimentell gegebenen Zahlen zu setzen. Da dies gelungen scheint, wird andererseits das mathematische Schema der Quantenmechanik auch keiner Revision bedürfen. Ebensowenig wird eine Revision der Raum—Zeitgeometrie für kleine Räume und Zeiten notwendig sein, da wir durch Wahl hinreichend schwerer Massen die quantenmechanischen Gesetze den klassischen beliebig annähern können, auch wenn es sich um noch so kleine Räume und Zeiten handelt. Aber daß eine Revision der kinematischen und mechanischen Begriffe notwendig ist, scheint aus den Grundgleichungen der Quantenmechanik unmittelbar zu folgen. Wenn eine bestimmte Masse m gegeben ist, hat es in unserer gewohnten Anschauung einen einfach verständlichen Sinn, vom Ort und der Geschwindigkeit des Schwerpunkts dieser Masse m zu sprechen. In der Quantenmechanik aber soll eine Relation $pq - qp = \frac{h}{2\pi i}$ zwischen Masse, Ort und Geschwindigkeit bestehen. Wir haben also guten Grund, gegen die kritiklose Anwendung jener Worte „Ort" und „Geschwindigkeit" Verdacht zu schöpfen. Wenn man zugibt, daß für Vorgänge in sehr kleinen Räumen und Zeiten Diskontinuitäten irgendwie typisch sind, so ist ein Versagen eben der Begriffe „Ort" und „Geschwindigkeit" sogar unmittelbar

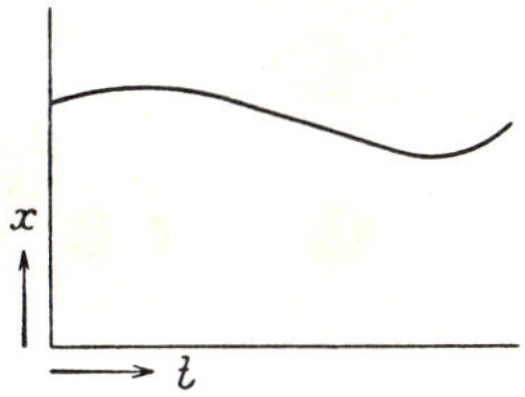

Fig. 1.

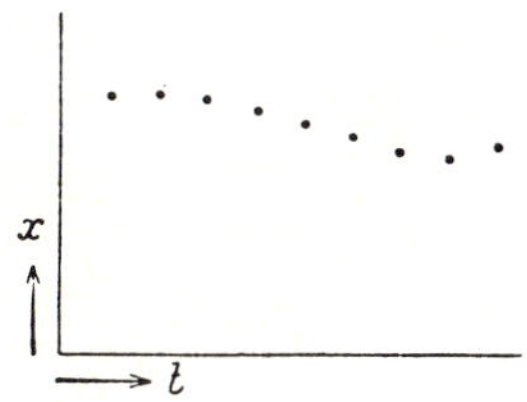

Fig. 2.

plausibel: Denkt man z. B. an die eindimensionale Bewegung eines Massenpunktes, so wird man in einer Kontinuumstheorie eine Bahnkurve $x(t)$ für die Bahn des Teilchens (genauer: dessen Schwerpunktes) zeichnen können (Fig. 1), die Tangente gibt jeweils die Geschwindigkeit. In einer Diskontinuumstheorie dagegen wird etwa an Stelle dieser Kurve eine Reihe von Punkten endlichen Abstandes treten (Fig. 2). In diesem Falle ist es offenbar sinnlos, von der Geschwindigkeit an einem bestimmten Orte zu sprechen, weil ja die Geschwindigkeit erst durch zwei Orte definiert werden kann und weil folglich umgekehrt zu jedem Punkt je zwei verschiedene Geschwindigkeiten gehören.

Es entsteht daher die Frage, ob es nicht durch eine genauere Analyse jener kinematischen und mechanischen Begriffe möglich sei, die bis jetzt in der anschaulichen Deutung der Quantenmechanik bestehenden Widersprüche aufzuklären und zu einem anschaulichen Verständnis der quantenmechanischen Relationen zu kommen [1]).

---

[1]) Die vorliegende Arbeit ist aus Bestrebungen und Wünschen entstanden, denen schon viel früher, vor dem Entstehen der Quantenmechanik, andere Forscher deutlichen Ausdruck gegeben haben. Ich erinnere hier besonders an Bohrs Ar-

§ 1. Die Begriffe: Ort, Bahn, Geschwindigkeit, Energie. Um das quantenmechanische Verhalten irgend eines Gegenstandes verfolgen zu können, muß man die Masse dieses Gegenstandes und die Wechselwirkungskräfte mit irgendwelchen Feldern und anderen Gegenständen kennen. Nur dann kann die Hamiltonsche Funktion des quantenmechanischen Systems aufgestellt werden. [Die folgenden Überlegungen sollen sich im allgemeinen auf die nichtrelativistische Quantenmechanik beziehen, da die Gesetze der quantentheoretischen Elektrodynamik noch sehr unvollständig bekannt sind] [1]). Über die „Gestalt" des Gegenstandes ist irgendwelche weitere Aussage unnötig, am zweckmäßigsten bezeichnet man die Gesamtheit jener Wechselwirkungskräfte mit dem Worte Gestalt.

Wenn man sich darüber klar werden will, was unter dem Worte „Ort des Gegenstandes", z. B. des Elektrons (relativ zu einem gegebenen Bezugssystem), zu verstehen sei, so muß man bestimmte Experimente angeben, mit deren Hilfe man den „Ort des Elektrons" zu messen gedenkt; anders hat dieses Wort keinen Sinn. An solchen Experimenten, die im Prinzip den „Ort des Elektrons" sogar beliebig genau zu bestimmen gestatten, ist kein Mangel, z. B.: Man beleuchte das Elektron und betrachte es unter einem Mikroskop. Die höchste erreichbare Genauigkeit der Ortsbestimmung ist hier im wesentlichen durch die Wellenlänge des benutzten Lichtes gegeben. Man wird aber im Prinzip etwa ein Γ-Strahl-Mikroskop bauen und mit diesem die Ortsbestimmung so genau durchführen können, wie man will. Es ist indessen bei dieser Bestimmung ein Nebenumstand wesentlich: der Comptoneffekt. Jede Beobachtung des vom Elektron kommenden Streulichtes setzt einen lichtelektrischen Effekt (im Auge, auf der photographischen Platte, in der Photozelle) voraus, kann also auch so gedeutet werden, daß ein Lichtquant das Elektron trifft, an diesem reflektiert oder abgebeugt wird und dann durch die Linsen des Mikro-

---

beiten über die Grundpostulate der Quantentheorie (z. B. ZS. f. Phys. **13**, 117, 1923) und Einsteins Diskussionen über den Zusammenhang zwischen Wellenfeld und Lichtquanten. Am klarsten sind in neuester Zeit die hier besprochenen Probleme diskutiert und die auftretenden Fragen teilweise beantwortet worden von W. Pauli (Quantentheorie, Handb. d. Phys., Bd. XXIII, weiterhin als l. c. zitiert); durch die Quantenmechanik hat sich an der Formulierung dieser Probleme durch Pauli nur wenig geändert. Es ist mir auch eine besondere Freude, an dieser Stelle Herrn W. Pauli für die vielfache Anregung zu danken, die ich aus gemeinsamen mündlichen und schriftlichen Diskussionen empfangen habe, und die zu der vorliegenden Arbeit wesentlich beigetragen hat.

[1]) In jüngster Zeit sind jedoch auf diesem Gebiet große Fortschritte erzielt worden durch Arbeiten von P. Dirac [Proc. Roy. Soc. (A) **114**, 243, 1927 und später erscheinende Untersuchungen].

skops nochmal abgelenkt den Photoeffekt auslöst. Im Augenblick der Ortsbestimmung, also dem Augenblick, in dem das Lichtquant vom Elektron abgebeugt wird, verändert das Elektron seinen Impuls unstetig. Diese Änderung ist um so größer, je kleiner die Wellenlänge des benutzten Lichtes, d. h. je genauer die Ortsbestimmung ist. In dem Moment, in dem der Ort des Elektrons bekannt ist, kann daher sein Impuls nur bis auf Größen, die jener unstetigen Änderung entsprechen, bekannt sein; also je genauer der Ort bestimmt ist, desto ungenauer ist der Impuls bekannt und umgekehrt; hierin erblicken wir eine direkte anschauliche Erläuterung der Relation $\mathbf{pq} - \mathbf{qp} = \frac{h}{2\pi i}$. Sei q_1 die Genauigkeit, mit der der Wert q bekannt ist (q_1 ist etwa der mittlere Fehler von q), also hier die Wellenlänge des Lichtes, p_1 die Genauigkeit, mit der der Wert p bestimmbar ist, also hier die unstetige Änderung von p beim Comptoneffekt, so stehen nach elementaren Formeln des Comptoneffekts p_1 und q_1 in der Beziehung

$$p_1 q_1 \sim h. \tag{1}$$

Daß diese Beziehung (1) in direkter mathematischer Verbindung mit der Vertauschungsrelation $\mathbf{pq} - \mathbf{qp} = \frac{h}{2\pi i}$ steht, wird später gezeigt werden. Hier sei darauf hingewiesen, daß Gleichung (1) der präzise Ausdruck für die Tatsachen ist, die man früher durch Einteilung des Phasenraumes in Zellen der Größe h zu beschreiben suchte.

Zur Bestimmung des Elektronenortes kann man auch andere Experimente, z. B. Stoßversuche vornehmen. Eine genaue Messung des Ortes erfordert Stöße mit sehr schnellen Partikeln, da bei langsamen Elektronen die Beugungserscheinungen, die nach Einstein eine Folge der de Broglie-Wellen sind (siehe z. B. Ramsauereffekt) eine genaue Bestimmung des Ortes verhindern. Bei einer genauen Ortsmessung ändert sich der Impuls des Elektrons also wieder unstetig und eine einfache Abschätzung der Genauigkeiten mit den Formeln der de Broglieschen Wellen gibt wieder die Relation (1).

Durch diese Diskussion scheint der Begriff „Ort des Elektrons" klar genug definiert und es sei nur noch ein Wort über die „Größe" des Elektrons hinzugefügt. Wenn zwei sehr schnelle Teilchen im sehr kurzen Zeitintervall Δt hintereinander das Elektron treffen, so liegen die durch die beiden Teilchen definierten Orte des Elektrons einander sehr nahe in einem Abstand Δl. Aus den Gesetzen, die bei α-Strahlen beobachtet sind, schließen wir, daß sich Δl bis auf Größen der Ordnung 10^{-12} cm herabdrücken läßt, wenn nur Δt hinreichend klein und die Teilchen hinreichend schnell gewählt werden. Diesen Sinn hat es, wenn wir sagen, das Elektron sei eine Korpuskel, deren Radius nicht größer als 10^{-12} cm ist.

Gehen wir nun über zum Begriff „Bahn des Elektrons". Unter Bahn verstehen wir eine Reihe von Raumpunkten (in einem gegebenen Bezugssystem), die das Elektron als „Orte" nacheinander annimmt. Da wir schon wissen, was unter „Ort zu einer bestimmten Zeit" zu verstehen sei, treten hier keine neuen Schwierigkeiten auf. Trotzdem ist leicht einzusehen, daß z. B. der oft gebrauchte Ausdruck: die „1 S-Bahn des Elektrons im Wasserstoffatom" von unserem Gesichtspunkt aus keinen Sinn hat. Um diese 1 S-„Bahn" zu messen, müßte nämlich das Atom mit Licht beleuchtet werden, dessen Wellenlänge jedenfalls erheblich kürzer als 10^{-8} cm ist. Von solchem Licht aber genügt ein einziges Lichtquant, um das Elektron völlig aus seiner „Bahn" zu werfen (weshalb von einer solchen Bahn immer nur ein einziger Raumpunkt definiert werden kann), das Wort „Bahn" hat hier also keinen vernünftigen Sinn. Dies kann ohne Kenntnis der neueren Theorien schon einfach aus den experimentellen Möglichkeiten gefolgert werden.

Dagegen kann die gedachte Ortsmessung an vielen Atomen im 1 S-Zustand ausgeführt werden. (Atome in einem gegebenen „stationären" Zustand lassen sich z. B. durch den Stern-Gerlachversuch im Prinzip isolieren.) Es muß also für einen bestimmten Zustand z. B. 1 S des Atoms eine Wahrscheinlichkeitsfunktion für die Orte des Elektrons geben, die dem Mittelwert der klassischen Bahn über alle Phasen entspricht und die durch Messungen beliebig genau feststellbar ist. Nach Born[1]) ist diese Funktion durch $\psi_{1S}(q)\,\overline{\psi}_{1S}(q)$ gegeben, wenn $\psi_{1S}(q)$ die zum Zustand 1 S gehörige Schrödingersche Wellenfunktion bedeutet. Mit Dirac[1]) und Jordan[1]) möchte ich im Hinblick auf spätere Ver-

[1]) Die statistische Bedeutung der de Broglie-Wellen wurde zuerst formuliert von A. Einstein (Sitzungsber. d. preuß. Akad. d. Wiss. 1925, S. 3). Dieses statistische Element in der Quantenmechanik spielt dann eine wesentliche Rolle bei M. Born, W. Heisenberg und P. Jordan, Quantenmechanik II (ZS. f. Phys. **35**, 557, 1926), bes. Kap. 4, § 3, und P. Jordan (ZS. f. Phys. **37**, 376, 1926); es wird in einer grundlegenden Arbeit von M. Born (ZS. f. Phys. **38**, 803, 1926) mathematisch analysiert und zur Deutung der Stoßphänomene benutzt. Die Begründung des Wahrscheinlichkeitsansatzes aus der Transformationstheorie der Matrizen findet sich in den Arbeiten: W. Heisenberg (ZS. f. Phys. **40**, 501, 1926), P. Jordan (ebenda **40**, 661, 1926), W. Pauli (Anm. in ZS. f. Phys. **41**, 81, 1927), P. Dirac (Proc. Roy. Soc. (A) **113**, 621, 1926), P. Jordan (ZS. f. Phys. **40**, 809, 1926). Allgemein ist die statistische Seite der Quantenmechanik diskutiert bei P. Jordan (Naturwiss. **15**, 105, 1927) und M. Born (Naturwiss. **15**, 238, 1927).

allgemeinerungen sagen: Die Wahrscheinlichkeit ist gegeben durch $S(1\,S, q)\,\bar{S}(1\,S, q)$, wo $S(1\,S, q)$ diejenige Kolonne der Transformationsmatrix $S(E, q)$ von E nach q bedeutet, die zu $E = E_{1S}$ gehört ($E =$ Energie).

Darin, daß in der Quantentheorie zu einem bestimmten Zustand, z. B. 1 S, nur die Wahrscheinlichkeitsfunktion des Elektronenortes angegeben werden kann, mag man mit Born und Jordan einen charakteristisch statistischen Zug der Quantentheorie im Gegensatz zur klassischen Theorie erblicken. Man kann aber, wenn man will, mit Dirac auch sagen, daß die Statistik durch unsere Experimente hereingebracht sei. Denn offenbar wäre auch in der klassischen Theorie nur die Wahrscheinlichkeit eines bestimmten Elektronenortes angebbar, solange wir die Phasen des Atoms nicht kennen. Der Unterschied zwischen klassischer und Quantenmechanik besteht vielmehr darin: Klassisch können wir uns durch vorausgehende Experimente immer die Phase bestimmt denken. In Wirklichkeit ist dies aber unmöglich, weil jedes Experiment zur Bestimmung der Phase das Atom zerstört bzw. verändert. In einem bestimmten stationären „Zustand" des Atoms sind die Phasen prinzipiell unbestimmt, was man als direkte Erläuterung der bekannten Gleichungen

$$\mathbf{E}t - t\mathbf{E} = \frac{h}{2\pi i} \quad \text{oder} \quad \mathbf{J}w - w\mathbf{J} = \frac{h}{2\pi i}$$

ansehen kann. ($\mathbf{J} =$ Wirkungsvariable, $w =$ Winkelvariable.)

Das Wort „Geschwindigkeit" eines Gegenstandes läßt sich durch Messungen leicht definieren, wenn es sich um kräftefreie Bewegungen handelt. Man kann z. B. den Gegenstand mit rotem Licht beleuchten und durch den Dopplereffekt des gestreuten Lichtes die Geschwindigkeit des Teilchens ermitteln. Die Bestimmung der Geschwindigkeit wird um so genauer, je langwelliger das benutzte Licht ist, da dann die Geschwindigkeitsänderung des Teilchens pro Lichtquant durch Comptoneffekt um so geringer wird. Die Ortsbestimmung wird entsprechend ungenau, wie es der Gleichung (1) entspricht. Wenn die Geschwindigkeit des Elektrons im Atom in einem bestimmten Augenblick gemessen werden soll, so wird man etwa in diesem Augenblick die Kernladung und die Kräfte von den übrigen Elektronen plötzlich verschwinden lassen, so daß die Bewegung von da ab kräftefrei erfolgt, und wird dann die oben angegebene Bestimmung durchführen. Wieder kann man sich, wie oben, leicht überzeugen, daß eine Funktion $p(t)$ für einen gegebenen Zustand eines Atoms, z. B. 1 S, nicht definiert werden kann. Dagegen gibt es

Zeitschrift für Physik 43, 172–198 (1927)

wieder eine Wahrscheinlichkeitsfunktion von p in diesem Zustand, die nach Dirac und Jordan den Wert $S(1\,S, p)\,\overline{S}(1\,S, p)$ hat. $S(1\,S, p)$ bedeutet wieder diejenige Kolonne der Transformationsmatrix $S(E, p)$ von $\mathbf{E}$ nach $\mathbf{p}$, die zu $E = E_{1S}$ gehört.

Schließlich sei noch auf die Experimente hingewiesen, welche gestatten, die Energie oder die Werte der Wirkungsvariablen J zu messen; solche Experimente sind besonders wichtig, da wir nur mit ihrer Hilfe definieren können, was wir meinen, wenn wir von der diskontinuierlichen Änderung der Energie und der J sprechen. Die Franck-Hertzschen Stoßversuche gestatten, die Energiemessung der Atome wegen der Gültigkeit des Energiesatzes in der Quantentheorie zurückzuführen auf die Energiemessung geradlinig sich bewegender Elektronen. Diese Messung läßt sich im Prinzip beliebig genau durchführen, wenn man nur auf die gleichzeitige Bestimmung des Elektronenortes, d. h. der Phase verzichtet (vgl. oben die Bestimmung von p), der Relation $\mathbf{E}\mathbf{t} - \mathbf{t}\mathbf{E} = \frac{h}{2\pi i}$ entsprechend. Der Stern-Gerlachversuch gestattet die Bestimmung des magnetischen oder eines mittleren elektrischen Moments des Atoms, also die Messung von Größen, die allein von den Wirkungsvariablen J abhängen. Die Phasen bleiben prinzipiell unbestimmt. Ebensowenig wie es sinnvoll ist, von der Frequenz einer Lichtwelle in einem bestimmten Augenblick zu sprechen, kann von der Energie eines Atoms in einem bestimmten Moment gesprochen werden. Dem entspricht im Stern-Gerlachversuch der Umstand, daß die Genauigkeit der Energiemessung um so geringer wird, je kürzer die Zeitspanne ist, in der die Atome unter dem Einfluß der ablenkenden Kraft stehen [1]). Eine obere Grenze für die ablenkende Kraft ist nämlich dadurch gegeben, daß die potentielle Energie jener ablenkenden Kraft innerhalb des Strahlenbündels nur um Beträge variieren darf, die erheblich kleiner sind als die Energiedifferenzen der stationären Zustände, wenn eine Bestimmung der Energie der stationären Zustände möglich sein soll. Sei E_1 ein Energiebetrag, der dieser Bedingung genügt (E_1 gibt zugleich die Genauigkeit jener Energiemessung an), so ist also E_1/d der Höchstwert der ablenkenden Kraft, wenn d die Breite des Strahlenbündels (meßbar durch die Weite der benutzten Blende) bedeutet. Die Winkelablenkung des Atomstrahls ist dann $\frac{E_1 t_1}{d p}$, wo t_1 die Zeitspanne bezeichnet, in der die Atome unter Einfluß der ablenkenden

[1]) Vgl. hierzu W. Pauli, l. c. S. 61.

Kraft stehen, p den Impuls der Atome in der Strahlrichtung. Diese Ablenkung muß mindestens gleicher Größenordnung sein wie die natürliche durch Beugung an der Blende hervorgebrachte Verbreiterung des Strahls, damit eine Messung möglich sei. Die Winkelablenkung durch Beugung ist etwa λ/d, wenn λ die de Brogliesche Wellenlänge bezeichnet, also

$$\frac{\lambda}{d} \sim \frac{E_1 t_1}{dp} \quad \text{oder da} \quad \lambda = \frac{h}{p},$$

$$E_1 t_1 \sim h. \tag{2}$$

Diese Gleichung entspricht der Gleichung (1) und zeigt, wie eine genaue Energiebestimmung nur durch eine entsprechende Ungenauigkeit in der Zeit erreicht werden kann.

§ 2. Die Dirac-Jordansche Theorie. Die Resultate des vorhergehenden Abschnitts möchte man zusammenfassen und verallgemeinern in dieser Behauptung: Alle Begriffe, die in der klassischen Theorie zur Beschreibung eines mechanischen Systems verwendet werden, lassen sich auch für atomare Vorgänge analog den klassischen Begriffen exakt definieren. Die Experimente, die solcher Definition dienen, tragen aber rein erfahrungsgemäß eine Unbestimmtheit in sich, wenn wir von ihnen die simultane Bestimmung zweier kanonisch konjugierten Größen verlangen. Der Grad dieser Unbestimmtheit ist durch die (auf irgendwelche kanonisch konjugierten Größen erweiterte) Relation (1) gegeben. Es liegt nahe, hier die Quantentheorie mit der speziellen Relativitätstheorie zu vergleichen. Nach der Relativitätstheorie läßt sich das Wort „gleichzeitig" nicht anders definieren, als durch Experimente, in welche die Ausbreitungsgeschwindigkeit des Lichts wesentlich eingeht. Gäbe es eine „schärfere" Definition der Gleichzeitigkeit, also z. B. Signale, die sich unendlich schnell fortpflanzen, so wäre die Relativitätstheorie unmöglich. Weil es solche Signale aber nicht gibt, weil vielmehr in der Definition der Gleichzeitigkeit schon die Lichtgeschwindigkeit vorkommt, ist Raum geschaffen für das Postulat der konstanten Lichtgeschwindigkeit, deshalb steht dieses Postulat mit dem sinngemäßen Gebrauch der Wörter „Ort, Geschwindigkeit, Zeit" nicht in Widerspruch. Ähnlich steht es mit der Definition der Begriffe: „Elektronenort, Geschwindigkeit" in der Quantentheorie. Alle Experimente, die wir zur Definition dieser Worte verwenden können, enthalten notwendig die durch Gleichung (1) angegebene Ungenauigkeit, wenn sie auch den einzelnen Begriff p, q exakt zu definieren gestatten. Gäbe es Experimente, die gleichzeitig eine „schärfere" Bestimmung von p und q

ermöglichen, als es der Gleichung (1) entspricht, so wäre die Quantenmechanik unmöglich. Diese Ungenauigkeit, die durch Gleichung (1) festgelegt ist, schafft also erst Raum für die Gültigkeit der Beziehungen, die in den quantenmechanischen Vertauschungsrelationen

$$pq - qp = \frac{h}{2\pi i}$$

ihren prägnanten Ausdruck finden; sie ermöglicht diese Gleichung, ohne daß der physikalische Sinn der Größen p und q geändert werden müßte.

Für diejenigen physikalischen Phänomene, deren quantentheoretische Formulierung noch unbekannt ist (z. B. die Elektrodynamik), bedeutet Gleichung (1) eine Forderung, die zum Auffinden der neuen Gesetze nützlich sein mag. Für die Quantenmechanik läßt sich Gleichung (1) durch eine geringfügige Verallgemeinerung aus der Dirac-Jordanschen Formulierung herleiten. Wenn wir für den bestimmten Wert η irgend eines Parameters den Ort q des Elektrons zu q' bestimmen mit einer Genauigkeit q_1, so können wir dieses Faktum durch eine Wahrscheinlichkeitsamplitude $S(\eta, q)$ zum Ausdruck bringen, die nur in einem Gebiet der ungefähren Größe q_1 um q' von Null merklich verschieden ist. Insbesondere kann man z. B. setzen

$$S(\eta, q) \operatorname{prop} e^{-\frac{(q-q')^2}{2 q_1^2} - \frac{2\pi i}{h} p'(q-q')}, \text{ also } S\overline{S} \operatorname{prop} e^{-\frac{(q-q')^2}{q_1^2}}. \qquad (3)$$

Dann gilt für die zu p gehörige Wahrscheinlichkeitsamplitude

$$S(\eta, p) = \int S(\eta, q)\, S(q, p)\, dq. \qquad (4)$$

Für $S(q, p)$ kann nach Jordan gesetzt werden

$$S(q, p) = e^{\frac{2\pi i p q}{h}}. \qquad (5)$$

Dann wird nach (4) $S(\eta, p)$ nur für Werte von p, für welche $\frac{2\pi(p-p')q_1}{h}$ nicht wesentlich größer als 1 ist, merklich von Null verschieden sein. Insbesondere gilt im Falle (3):

$$S(\eta, p) \operatorname{prop} \int e^{\frac{2\pi i(p-p')q}{h} - \frac{(q'-q)^2}{2 q_1^2}}\, dq,$$

d. h.

$$S(\eta, p) \operatorname{prop} e^{-\frac{(p-p')^2}{2 p_1^2} + \frac{2\pi i}{h} q'(p-p')}, \text{ also } S\overline{S} \operatorname{prop} e^{-\frac{(p-p')^2}{p_1^2}},$$

wo

$$p_1 q_1 = \frac{h}{2\pi}. \qquad (6)$$

Die Annahme (3) für $S(\eta, q)$ entspricht also dem experimentellen Faktum, daß der Wert p' für p, der Wert q' für q [mit der Genauigkeitsbeschränkung (6)] gemessen wurde.

Rein mathematisch ist für die Dirac-Jordansche Formulierung der Quantenmechanik charakteristisch, daß die Relationen zwischen $\mathfrak{p}$, $\mathfrak{q}$, E usw. als Gleichungen zwischen sehr allgemeinen Matrizen geschrieben werden können, derart, daß irgend eine vorgegebene quantentheoretische Größe als Diagonalmatrix erscheint. Die Möglichkeit einer solchen Schreibweise leuchtet ein, wenn man sich die Matrizen anschaulich als Tensoren (z. B. Trägheitsmomente) in mehrdimensionalen Räumen deutet, zwischen denen mathematische Beziehungen bestehen. Man kann die Achsen des Koordinatensystems, in dem man diese mathematischen Beziehungen ausdrückt, immer in die Hauptachsen eines dieser Tensoren legen. Schließlich kann man die mathematische Beziehung zwischen zwei Tensoren A und B auch immer durch die Transformationsformeln charakterisieren, die ein nach den Hauptachsen von A orientiertes Koordinatensystem in ein anderes überführen, das nach den Hauptachsen von B orientiert ist. Die letztere Formulierung entspricht der Schrödingerschen Theorie. Als die eigentlich „invariante“, von allen Koordinatensystemen unabhängige Formulierung der Quantenmechanik wird man dagegen die Diracsche Schreibweise der q-Zahlen ansehen. Wenn wir aus jenem mathematischen Schema physikalische Resultate ableiten wollen, so müssen wir den quantentheoretischen Größen, also den Matrizen (oder „Tensoren“ im mehrdimensionalen Raum) Zahlen zuordnen. Dies ist so zu verstehen, daß in jenem mehrdimensionalen Raum eine bestimmte Richtung willkürlich vorgegeben wird (nämlich durch die Art des angestellten Experiments festgesetzt wird) und gefragt wird, welches der „Wert“ der Matrix (z. B. in jenem Bilde der Wert des Trägheitsmoments) in dieser vorgegebenen Richtung sei. Diese Frage hat nur dann einen eindeutigen Sinn, wenn die vorgegebene Richtung mit der Richtung einer der Hauptachsen jener Matrix zusammenfällt; in diesem Falle gibt es eine exakte Antwort auf die gestellte Frage. Aber auch, wenn die vorgegebene Richtung nur wenig abweicht von der einer der Hauptachsen der Matrix, so kann man noch mit einer gewissen durch die relative Neigung gegebenen Ungenauigkeit, mit einem gewissen wahrscheinlichen Fehler von dem „Wert“ der Matrix in der vorgegebenen Richtung sprechen. Man kann also sagen: Jeder quantentheoretischen Größe oder Matrix läßt sich eine Zahl, die ihren „Wert“ angibt, mit einem bestimmten wahrscheinlichen Fehler zuordnen; der wahrscheinliche Fehler hängt vom

Zeitschrift für Physik 43, 172–198 (1927)

Koordinatensystem ab; für jede quantentheoretische Größe gibt es je ein Koordinatensystem, in dem der wahrscheinliche Fehler für diese Größe verschwindet. Ein bestimmtes Experiment kann also niemals über alle quantentheoretischen Größen genaue Auskunft geben, vielmehr teilt es in einer für das Experiment charakteristischen Weise die physikalischen Größen in „bekannte" und „unbekannte" (oder: mehr und weniger genau bekannte Größen) ein. Die Resultate zweier Experimente lassen sich nur dann exakt auseinander herleiten, wenn die beiden Experimente die physikalischen Größen in gleicher Weise in „bekannte" und „unbekannte" einteilen (d. h. wenn die Tensoren in jenem mehrfach zur Veranschaulichung gebrauchten mehrdimensionalen Raum in beiden Experimenten von der gleichen Richtung aus „angesehen" werden). Bewirken zwei Experimente verschiedene Einteilungen in „Bekanntes" und „Unbekanntes", so läßt sich der Zusammenhang der Resultate jener Experimente füglich nur statistisch angeben.

Zur genaueren Diskussion dieses statistischen Zusammenhangs sei ein Gedankenexperiment vorgenommen. Ein Stern-Gerlachscher Atomstrahl werde zunächst durch ein Feld F_1 geschickt, das so stark inhomogen in der Strahlrichtung ist, daß es merklich viele Übergänge durch „Schüttelwirkung" hervorruft. Dann laufe der Atomstrahl eine Weile frei, in einem bestimmten Abstand von F_1 aber beginne ein zweites Feld F_2, ähnlich inhomogen wie F_1. Zwischen F_1 und F_2 und hinter F_2 sei es möglich, die Anzahl der Atome in den verschiedenen stationären Zuständen durch ein eventuell angelegtes Magnetfeld zu messen. Die Strahlungskräfte der Atome seien Null gesetzt. Wenn wir wissen, daß ein Atom im Zustand der Energie E_n war, bevor es F_1 passierte, so können wir dieses experimentelle Faktum dadurch zum Ausdruck bringen, daß wir dem Atom eine Wellenfunktion — z. B. im p-Raum — mit der bestimmten Energie E_n und der unbestimmten Phase β_n

$$S(E_n, p) = \psi(E_n, p)\, e^{-\frac{2\pi i E_n (\alpha + \beta_n)}{h}}$$

zuordnen. Nach dem Durchqueren des Feldes F_1 wird sich diese Funktion verwandelt haben in[1])

$$S(E_n, p) \xrightarrow{F_1} \sum_m c_{nm}\, \psi(E_m, p)\, e^{-\frac{2\pi i E_m (\alpha + \beta_m)}{h}} \tag{7}$$

[1]) Vgl. P. Dirac, Proc. Roy. Soc. (A) 112, 661, 1926 und M. Born, ZS. f. Phys. 40, 167, 1926.

Hierin seien die β_m irgendwie willkürlich festgesetzt, so daß die c_{nm} durch F_1 eindeutig bestimmt sind. Die Matrix c_{nm} transformiert die Energiewerte vor dem Durchgang durch F_1 auf die nach dem Durchgang durch F_1. Führen wir hinter F_1 eine Bestimmung der stationären Zustände z. B. durch ein inhomogenes Magnetfeld aus, so werden wir mit einer Wahrscheinlichkeit $c_{nm}\bar{c}_{nm}$ finden, daß das Atom vom Zustand n in den Zustand m übergegangen ist. Wenn wir experimentell feststellen, daß das Atom eben in den Zustand m wirklich übergegangen sei, so werden wir ihm zur Berechnung alles Folgenden nicht die Funktion $\sum\limits_m c_{nm} S_m$, sondern eben die Funktion S_m mit unbestimmter Phase zuzuordnen haben; durch die experimentelle Feststellung: „Zustand m" wählen wir aus der Fülle der verschiedenen Möglichkeiten (c_{nm}) eine bestimmte: m aus, zerstören aber gleichzeitig, wie nachher erläutert wird, alles, was an Phasenbeziehungen noch in den Größen c_{nm} enthalten war. Beim Durchgang des Atomstrahls durch F_2 wiederholt sich das gleiche wie bei F_1. Es seien d_{nm} die Koeffizienten der Transformationsmatrix, die die Energien vor F_2 auf die nach F_2 transformieren. Wenn zwischen F_1 und F_2 keine Bestimmung des Zustandes vorgenommen wird, so verwandelt sich die Eigenfunktion nach folgendem Schema:

$$S(E_n, p) \xrightarrow{F_1} \sum_m c_{nm} S(E_m, p) \xrightarrow{F_2} \sum_m \sum_l c_{nm} d_{ml} S(E_l, p). \tag{8}$$

Es sei $\sum\limits_m c_{nm} d_{ml} = e_{nl}$ gesetzt. Wird der stationäre Zustand des Atoms hinter F_2 festgestellt, so wird man mit einer Wahrscheinlichkeit $e_{nl}\bar{e}_{nl}$ den Zustand l finden. Wenn dagegen zwischen F_1 und F_2 die Feststellung: „Zustand m" gemacht wurde, so wird die Wahrscheinlichkeit für „l" hinter F_2 durch $d_{ml}\bar{d}_{ml}$ gegeben sein. Bei mehrfacher Wiederholung des ganzen Experiments (wobei jedesmal zwischen F_1 und F_2 der Zustand bestimmt werde) wird man also hinter F_2 den Zustand l mit der relativen Häufigkeit $Z_{nl} = \sum\limits_m c_{nm}\bar{c}_{nm} d_{ml}\bar{d}_{ml}$ beobachten. Dieser Ausdruck stimmt nicht überein mit $e_{nl}\bar{e}_{nl}$. Jordan (l. c.) hat deswegen von einer „Interferenz der Wahrscheinlichkeiten" gesprochen. Dem möchte ich mich aber nicht anschließen. Denn die beiden Experimente, die zu $e_{nl}\bar{e}_{nl}$ bzw. Z_{nl} führen, sind ja physikalisch wirklich verschieden. In einem Falle erleidet das Atom zwischen F_1 und F_2 keine Störung, im anderen wird es durch die Apparate, die eine Feststellung des stationären Zustandes ermöglichen, gestört. Diese Apparate haben zur Folge, daß sich die „Phase" des Atoms um prinzipiell unkontrollierbare Beträge ändert, ebenso, wie sich bei einer Bestimmung des Elektronenortes der Impuls ändert (vgl. § 1). Das Magnetfeld zur Bestimmung des Zustandes zwischen F_1 und F_2 wird die Eigenwerte E verstimmen, bei der Beobachtung der Bahn des Atomstrahls werden (ich denke etwa an Wilsonaufnahmen) die Atome statistisch verschieden und unkontrollierbar gebremst usf. Dies hat zur Folge, daß die endgültige Transformationsmatrix e_{nl} (von den Energiewerten vor dem Eintreten in F_1 auf die nach dem Austreten aus F_2) nicht mehr durch $\sum_m c_{nm} d_{ml}$ gegeben ist, sondern jedes Glied der Summe hat noch einen unbekannten Phasenfaktor. Wir können also nur erwarten, daß der Mittelwert von $e_{nl}\bar{e}_{nl}$ über alle diese eventuellen Phasenänderungen gleich Z_{nl} ist. Eine einfache Rechnung ergibt, daß dies der Fall ist. — Wir können also nach gewissen statistischen Regeln von einem Experiment auf die möglichen Resultate eines anderen schließen. Das andere Experiment selbst wählt aus der Fülle der Möglichkeiten eine ganz bestimmte aus und beschränkt dadurch für alle späteren Experimente die Möglichkeiten. Eine solche Deutung der Gleichung für die Transformationsmatrix S oder der Schrödingerschen Wellengleichung ist nur deshalb möglich, weil die Summe von Lösungen wieder eine Lösung darstellt. Darin erblicken wir den tiefen Sinn der Linearität der Schrödingerschen Gleichungen; deswegen können sie nur als Gleichungen für Wellen im Phasenraum verstanden werden und deswegen möchten wir jeden Versuch, diese Gleichungen z. B. im relativistischen Falle (bei mehreren Elektronen) durch nichtlineare zu ersetzen, für aussichtslos halten.

§ 3. Der Übergang von der Mikro- zur Makromechanik. Durch die in den vorausgehenden Abschnitten durchgeführte Analyse der Worte „Elektronenort", „Geschwindigkeit", „Energie" usw. scheinen mir die Begriffe der quantentheoretischen Kinematik und Mechanik hinreichend geklärt, so daß ein anschauliches Verständnis auch der makroskopischen Vorgänge vom Standpunkt der Quantenmechanik aus möglich sein muß. Der Übergang von der Mikro- zur Makromechanik ist schon von Schrödinger[1]) behandelt worden, aber ich glaube nicht, daß die Schrödingersche Überlegung das Wesen des Problems trifft, und zwar aus folgenden Gründen: Nach Schrödinger soll in hohen Anregungszuständen eine Summe von Eigenschwingungen ein nicht allzu großes Wellenpaket ergeben können, das seinerseits unter periodischen Änderungen seiner Größe die periodischen Bewegungen des klassischen „Elektrons"

[1]) E. Schrödinger, Naturwiss. 14, 664, 1926.

ausführt. Hiergegen ist folgendes einzuwenden: Wenn das Wellenpaket solche Eigenschaften hätte, wie sie hier beschrieben wurden, so wäre die vom Atom ausgesandte Strahlung in eine Fourierreihe entwickelbar, bei der die Frequenzen der Oberschwingungen ganzzahlige Vielfache einer Grundfrequenz sind. Die Frequenzen der vom Atom ausgesandten Spektrallinien sind aber nach der Quantenmechanik nie ganzzahlige Vielfache einer Grundfrequenz — ausgenommen den Spezialfall des harmonischen Oszillators. Schrödingers Überlegung ist also nur für den von ihm behandelten harmonischen Oszillator durchführbar, in allen anderen Fällen breitet sich im Laufe der Zeit ein Wellenpaket über den ganzen Raum in der Umgebung des Atoms aus. Je höher der Anregungszustand des Atoms ist, desto langsamer erfolgt jene Zerstreuung des Wellenpakets. Aber wenn man lange genug wartet, wird sie eintreten. Das oben angeführte Argument über die vom Atom ausgesandte Strahlung läßt sich zunächst gegen alle Versuche anwenden, die einen direkten Übergang der Quantenmechanik in die klassische für hohe Quantenzahlen erstreben. Man hat deshalb früher versucht, jenem Argument durch Hinweis auf die natürliche Strahlungsbreite der stationären Zustände zu entgehen; sicherlich zu Unrecht, denn erstens ist dieser Ausweg schon beim Wasserstoffatom wegen der geringen Strahlung in hohen Zuständen versperrt, zweitens muß der Übergang der Quantenmechanik in die klassische auch ohne Anleihe bei der Elektrodynamik verständlich sein. Auf diese bekannten Schwierigkeiten, die einer direkten Verbindung der Quantentheorie mit der klassischen im Wege stehen, hat schon früher Bohr[1]) mehrfach hingewiesen. Wir haben sie nur deswegen wieder so ausführlich erläutert, weil sie neuerdings in Vergessenheit zu geraten scheinen.

Ich glaube, daß man die Entstehung der klassischen „Bahn" prägnant so formulieren kann: Die „Bahn" entsteht erst dadurch, daß wir sie beobachten: Sei z. B. ein Atom im 1000. Anregungszustand gegeben. Die Bahndimensionen sind hier schon relativ groß, so daß es im Sinne von § 1 genügt, die Bestimmung des Elektronenortes mit verhältnismäßig langwelligem Licht vorzunehmen. Wenn die Bestimmung des Ortes nicht allzu ungenau sein soll, so wird der Comptonrückstoß zur Folge haben, daß das Atom sich nach dem Stoß in irgend einem Zustand zwischen, sagen wir, dem 950. und 1050. befindet; gleichzeitig kann der Impuls des Elektrons mit einer aus (1) bestimmbaren Genauigkeit aus dem Dopplereffekt geschlossen werden. Das so gegebene ex-

[1]) N. Bohr, Grundpostulate der Quantentheorie, l. c.

Zeitschrift für Physik *43*, 172–198 (1927)

perimentelle Faktum kann man durch ein Wellenpaket — besser Wahrscheinlichkeitspaket — im q-Raum von einer durch die Wellenlänge des benutzten Lichtes gegebenen Größe, zusammengesetzt im wesentlichen aus Eigenfunktionen zwischen der 950. und der 1050. Eigenfunktion, und durch ein entsprechendes Paket im p-Raum charakterisieren. Nach einiger Zeit werde eine neue Ortsbestimmung mit der gleichen Genauigkeit ausgeführt. Ihr Resultat läßt sich nach § 2 nur statistisch angeben, als wahrscheinliche Orte kommen alle innerhalb des nun schon verbreiterten Wellenpakets mit berechenbarer Wahrscheinlichkeit in Betracht. Dies wäre in der klassischen Theorie keineswegs anders, denn auch in der klassischen Theorie wäre das Resultat der zweiten Ortsbestimmung wegen der Unsicherheit der ersten Bestimmung nur statistisch angebbar: auch die Systembahnen der klassischen Theorie würden sich ähnlich ausbreiten wie das Wellenpaket. Allerdings sind die statistischen Gesetze selbst in der Quantenmechanik und in der klassischen Theorie verschieden. Die zweite Ortsbestimmung wählt aus der Fülle der Möglichkeiten eine bestimmte „q“ aus und beschränkt für alle folgenden Bestimmungen die Möglichkeiten. Nach der zweiten Ortsbestimmung können die Resultate späterer Messungen nur berechnet werden, indem man dem Elektron wieder ein „kleineres“ Wellenpaket der Größe λ (Wellenlänge des zur Beobachtung benutzten Lichtes) zuordnet. Jede Ortsbestimmung reduziert also das Wellenpaket wieder auf seine ursprüngliche Größe λ. Die „Werte“ der Variablen p und q sind während aller Versuche mit einer gewissen Genauigkeit bekannt. Daß die Werte von p und q innerhalb dieser Genauigkeitsgrenzen den klassischen Bewegungsgleichungen Folge leisten, kann direkt aus den quantenmechanischen Gesetzen

$$\dot{p} = -\frac{\partial H}{\partial q}; \qquad \dot{q} = \frac{\partial H}{\partial p} \tag{9}$$

geschlossen werden. Die Bahn kann aber, wie gesagt, nur statistisch aus den Anfangsbedingungen berechnet werden, was man als Folge der prinzipiellen Ungenauigkeit der Anfangsbedingungen betrachten kann. Die statistischen Gesetze sind für die Quantenmechanik und die klassische Theorie verschieden; dies kann unter gewissen Bedingungen zu groben makroskopischen Unterschieden zwischen klassischer und Quantentheorie führen. Bevor ich ein Beispiel hierfür diskutiere, möchte ich an einem einfachen mechanischen System: der kräftefreien Bewegung eines Massenpunktes, zeigen, wie der oben diskutierte Übergang zur klassischen Theorie mathematisch zu formulieren ist. Die Bewegungsgleichungen lauten (bei eindimensionaler Bewegung)

$$H = \frac{1}{2m} p^2; \quad \dot{q} = \frac{1}{m} p; \quad \dot{p} = 0. \tag{10}$$

Da die Zeit als Parameter (als „c-Zahl") behandelt werden kann, wenn keine von der Zeit abhängigen äußeren Kräfte vorkommen, so lautet die Lösung dieser Gleichungen:

$$q = \frac{1}{m} p_0 t + q_0; \qquad p = p_0, \tag{11}$$

wo p_0 und q_0 Impuls und Ort zur Zeit $t = 0$ bedeuten. Zur Zeit $t = 0$ werde [siehe Gleichung (3) bis (6)] der Wert $q_0 = q'$ mit der Genauigkeit q_1, $p_0 = p'$ mit der Genauigkeit p_1 gemessen. Um aus den „Werten" von p_0 und q_0 auf die „Werte" von q zur Zeit t zu schließen, muß nach Dirac und Jordan diejenige Transformationsfunktion gefunden werden, die alle Matrizen, bei denen q_0 als Diagonalmatrix erscheint, in solche transformiert, bei denen q als Diagonalmatrix erscheint. p_0 kann in dem Matrixschema, in dem q_0 als Diagonalmatix erscheint, durch den Operator $\frac{h}{2\pi i} \frac{\partial}{\partial q_0}$ ersetzt werden. Nach Dirac [l. c. Gleichung (11)] gilt dann für die gesuchte Transformationsamplitude $S(q_0, q)$ die Differentialgleichung:

$$\left\{\frac{t}{m} \frac{h}{2\pi i} \frac{\partial}{\partial q_0} + q_0\right\} S(q_0, q) = q\, S(q_0, q) \tag{12}$$

$$\frac{t}{m} \frac{h}{2\pi i} \frac{\partial S}{\partial q_0} = \left(q_0 - q\right) S(q_0, q)$$

$$S(q_0, q) = \text{const} . e^{\frac{2\pi i m \int \left(q - q_0\right) d q_0}{h . t}} \tag{13}$$

$S\bar{S}$ ist also von q_0 unabhängig, d. h. wenn zur Zeit $t = 0$ q_0 exakt bekannt ist, so sind zu irgendwelcher Zeit $t > 0$ alle Werte von q gleich wahrscheinlich, d. h. die Wahrscheinlichkeit, daß q in einem endlichen Bereich liegt, ist überhaupt Null. Dies ist ja anschaulich auch ohne weiteres klar. Denn die exakte Bestimmung von q_0 führt zu unendlich großem Comptonrückstoß. Das gleiche würde natürlich für jedes beliebige mechanische System gelten. Wenn aber q_0 zur Zeit $t = 0$ nur mit einer Genauigkeit q_1 und p_0 mit der Genauigkeit p_1 bekannt war [vgl. Gleichung (3)]

$$S(\eta, q_0) = \text{const} . e^{-\frac{(q_0 - q')^2}{2 q_1^2} - \frac{2\pi i}{h} p'(q_0 - q')},$$

Zeitschrift für Physik *43*, 172–198 (1927)

so wird die Wahrscheinlichkeitsfunktion für q nach der Formel

$$S(\eta, q) = \int S(\eta, q_0)\, S(q_0, q)\, d q_0$$

zu berechnen sein. Es ergibt sich

$$S(\eta, q) = \text{const.} \int e^{\frac{2\pi i m}{t h}\left[q_0\left(q - \frac{t}{m} p'\right) - \frac{q_0^2}{2}\right] - \frac{(q' - q_0)^2}{2 q_1^2}} d q_0. \tag{14}$$

Führt man die Abkürzung

$$\beta = \frac{t h}{2 \pi m q_1^2} \tag{15}$$

ein, so wird der Exponent in (14)

$$-\frac{1}{2 q_1^2}\left\{q_0^2\left(1 + \frac{i}{\beta}\right) - 2 q_0\left(q' + \frac{i}{\beta}\left(q - \frac{t}{m} p'\right)\right) + q'^2\right\}.$$

Das Glied mit q'^2 kann in den konstanten (von q unabhängigen Faktor) einbezogen werden und die Integration ergibt

$$S(\eta, q) = \text{const.}\, e^{\frac{1}{2 q_1^2} \frac{\left[q' + \frac{i}{\beta}\left(q - \frac{t}{m} p'\right)\right]^2}{1 + \frac{i}{\beta}}}, \tag{16}$$

$$= \text{const.}\, e^{-\frac{\left(q - \frac{t}{m} p' - i \beta q'\right)^2 \left(1 - \frac{i}{\beta}\right)}{2 q_1^2 (1 + \beta^2)}}$$

Daraus folgt

$$S(\eta, q)\, \overline{S(\eta, q)} = \text{const.}\, e^{-\frac{\left(q - \frac{t}{m} p' - q'\right)^2}{q_1^2 (1 + \beta^2)}}. \tag{17}$$

Das Elektron befindet sich also zur Zeit t an der Stelle $\frac{t}{m} p' + q'$ mit einer Genauigkeit $q_1 \sqrt{1 + \beta^2}$. Das „Wellenpaket“ oder besser „Wahrscheinlichkeitspaket“ hat sich um den Faktor $\sqrt{1 + \beta^2}$ vergrößert. β ist nach (15) proportional der Zeit t, umgekehrt proportional der Masse — dies ist unmittelbar plausibel — und umgekehrt proportional q_1^2. Eine allzu große Genauigkeit in q_0 hat eine große Ungenauigkeit in p_0 zur Folge und führt deshalb auch zu einer großen Ungenauigkeit in q. Der Parameter η, den wir oben aus formalen Gründen eingeführt hatten, könnte hier in allen Formeln weggelassen werden, da er in die Rechnung nicht eingeht.

Als Beispiel dafür, daß der Unterschied der klassischen statistischen Gesetze von den quantentheoretischen unter Umständen zu groben makroskopischen Unterschieden zwischen den Resultaten beider Theorien führt, sei die Reflexion eines Elektronenstromes an einem Gitter kurz diskutiert. Wenn die Gitterkonstante von der Größenordnung der de Broglieschen Wellenlänge der Elektronen ist, so erfolgt die Reflexion in bestimmten diskreten Raumrichtungen, wie die Reflexion von Licht an einem Gitter. Die klassische Theorie gibt hier grob makroskopisch etwas anderes. Trotzdem können wir keineswegs an der Bahn eines einzelnen Elektrons einen Widerspruch gegen die klassische Theorie feststellen. Wir könnten es, wenn wir das Elektron etwa auf eine bestimmte Stelle eines Gitterstrichs lenken könnten und dann feststellen, daß die Reflexion dort unklassisch erfolgt. Wenn wir den Ort des Elektrons aber so genau bestimmen wollen, daß wir sagen können, auf welche Stelle eines Gitterstrichs es trifft, so bekommt das Elektron durch diese Ortsbestimmung eine große Geschwindigkeit, die de Brogliesche Wellenlänge des Elektrons wird um so viel kleiner, daß nun die Reflexion wirklich in dieser Näherung in der klassisch vorgeschriebenen Richtung erfolgen kann und wird, ohne den quantentheoretischen Gesetzen zu widersprechen.

§ 4. Diskussion einiger besonderen Gedankenexperimente. Nach der hier versuchten anschaulichen Deutung der Quantentheorie müssen die Zeitpunkte der Übergänge, der „Quantensprünge" ebenso konkret, durch Messungen feststellbar sein, wie etwa die Energien in stationären Zuständen. Die Genauigkeit, mit der ein solcher Zeitpunkt ermittelbar ist, wird nach Gleichung (2) durch $\frac{h}{\Delta E}$ gegeben sein[1]), wenn ΔE die Änderung der Energie beim Quantensprung bedeutet. Wir denken etwa an folgendes Experiment: Ein Atom, zur Zeit $t = 0$ im Zustand 2, möge durch Strahlung in den Normalzustand 1 übergehen. Dem Atom kann dann etwa analog zu Gleichung (7) die Eigenfunktion

$$S(t,p) = e^{-\alpha t}\psi(E_2,p)\,e^{-\frac{2\pi i E_2 t}{h}} + \sqrt{1-e^{-2\alpha t}}\,\psi(E_1,p)\,e^{-\frac{2\pi i E_1 t}{h}} \quad (18)$$

zugeordnet werden, wenn wir annehmen, daß die Strahlungsdämpfung sich in einem Faktor der Form $e^{-\alpha t}$ in den Eigenfunktionen äußert (die wirkliche Abhängigkeit ist vielleicht nicht so einfach). Dieses Atom werde zur Messung seiner Energie durch ein inhomogenes Magnetfeld geschickt, wie dies beim Stern-Gerlachversuch üblich ist, doch soll das unhomogene Feld dem Atomstrahl ein langes Stück Weges folgen. Die jeweilige Beschleunigung wird man etwa dadurch messen, daß man die ganze Strecke, die der Atomstrahl im Magnetfeld durchmißt, in kleine

[1]) Vgl. W. Pauli, l. c. S. 12.

Zeitschrift für Physik *43*, 172–198 (1927)

Teilstrecken einteilt, an deren Ende man jeweils die Ablenkung des Strahles feststellt. Je nach der Geschwindigkeit des Atomstrahles entspricht der Einteilung in Teilstrecken am Atom eine Einteilung in kleine Zeitintervalle Δt. Nach § 1, Gleichung (2) entspricht dem Intervall Δt eine Genauigkeit in der Energie von $\frac{h}{\Delta t}$. Die Wahrscheinlichkeit, eine bestimmte Energie E zu messen, läßt sich direkt schließen aus $S(p, E)$ und wird daher im Intervall von $n\Delta t$ bis $(n+1)\Delta t$ berechnet durch:

$$\underset{n\Delta t \rightarrow (n+1)\Delta t}{S(p, E)} = \int_{n\Delta t}^{(n+1)\Delta t} S(p, t)\, e^{\frac{2\pi i E t}{h}}\, dt.$$

Wenn zur Zeit $(n+1)\Delta t$ die Feststellung: „Zustand 2" gemacht wird, so ist dem Atom für alles spätere nicht mehr die Eigenfunktion (18) zuzuordnen, sondern eine, die aus (18) hervorgeht, wenn man t durch $t-(n+1)\Delta t$ ersetzt. Stellt man dagegen fest: „Zustand 1", so ist dem Atom von da ab die Eigenfunktion

$$\psi(E_1, p)\, e^{-\frac{2\pi i E_1 t}{h}}$$

zuzuordnen. Man wird also zunächst in einer Reihe von Intervallen Δt beobachten: „Zustand 2", dann dauernd „Zustand 1". Damit eine Unterscheidung der beiden Zustände noch möglich sei, darf Δt nicht unter $\frac{h}{\Delta E}$ herabgedrückt werden. Mit dieser Genauigkeit ist also der Zeitpunkt des Übergangs bestimmbar. Ein Experiment von der eben geschilderten Art meinen wir ganz im Sinne der alten von Planck, Einstein und Bohr begründeten Auffassung der Quantentheorie, wenn wir von der diskontinuierlichen Änderung der Energie sprechen. Da ein solches Experiment prinzipiell durchführbar ist, muß eine Einigung über seinen Ausgang möglich sein.

In Bohrs Grundpostulaten der Quantentheorie hat die Energie eines Atoms ebenso, wie die Werte der Wirkungsvariabeln J vor anderen Bestimmungsstücken (Ort des Elektrons usw.) den Vorzug, daß sich ihr Zahlwert stets angeben läßt. Diese Vorzugsstellung, die die Energie den anderen quantenmechanischen Größen gegenüber einnimmt, verdankt sie indessen nur dem Umstand, daß sie bei abgeschlossenen Systemen ein Integral der Bewegungsgleichungen darstellt (für die Energiematrix gilt $\mathbf{E} =$ const); bei nicht abgeschlossenen Systemen wird dagegen die Energie sich vor keiner anderen quantenmechanischen

Größe auszeichnen. Insbesondere wird man Experimente angeben können, bei denen die Phasen w des Atoms exakt meßbar sind, bei denen dann aber die Energie prinzipiell unbestimmt bleibt, einer Relation $Jw - wJ = \frac{h}{2\pi i}$ oder $J_1 w_1 \sim h$ entsprechend. Ein solches Experiment stellt z. B. die Resonanzfluoreszenz dar. Bestrahlt man ein Atom mit einer Eigenfrequenz, sagen wir $\nu_{12} = \frac{E_2 - E_1}{h}$, so schwingt das Atom in Phase mit der äußeren Strahlung, wobei es prinzipiell keinen Sinn hat, zu fragen, in welchem Zustand E_1 oder E_2 das Atom so schwingt. Die Phasenbeziehung zwischen Atom und äußerer Strahlung läßt sich z. B. durch die Phasenbeziehung vieler Atome untereinander (Woods Versuche) feststellen. Will man von Experimenten mit Strahlung lieber absehen, so kann man die Phasenbeziehung auch so messen, daß man genaue Ortsbestimmungen im Sinne des § 1 des Elektrons zu verschiedenen Zeiten relativ zur Phase des eingestrahlten Lichtes (an vielen Atomen) vornimmt. Dem einzelnen Atom wird etwa die „Wellenfunktion"

$$S(q, t) = c_2 \psi_2 (E_2, q)\, e^{-\frac{2\pi i (E_2 t + \beta)}{h}} + \sqrt{1 - c_2^2}\, \psi_1 (E_1, q)\, e^{-\frac{2\pi i E_1 t}{h}} \qquad (19)$$

zugeordnet werden können; hierin hängt c_2 von der Stärke und β von der Phase des eingestrahlten Lichtes ab. Die Wahrscheinlichkeit eines bestimmten Ortes q ist also

$$S(q, t)\, \overline{S(q, t)} = c_2^2 \psi_2 \overline{\psi_2} + (1 - c_2^2)\, \psi_1 \overline{\psi_1}$$
$$+ c_2 \sqrt{1 - c_2^2} \left(\psi_2 \overline{\psi_1}\, e^{-\frac{2\pi i}{h}[(E_2 - E_1) t + \beta]} + \overline{\psi_2} \psi_1\, e^{+\frac{2\pi i}{h}[(E_2 - E_1) t + \beta]} \right) \qquad (20)$$

Das periodische Glied in (20) ist vom unperiodischen experimentell trennbar, da die Ortsbestimmungen bei verschiedenen Phasen des eingestrahlten Lichtes ausgeführt werden können.

In einem bekannten von Bohr angegebenen Gedankenexperiment werden die Atome eines Stern-Gerlachschen Atomstrahls zunächst an einer bestimmten Stelle durch eingestrahltes Licht zur Resonanzfluoreszenz erregt. Nach einem Stück Weges durchlaufen sie ein inhomogenes Magnetfeld; die von den Atomen ausgehende Strahlung kann während des ganzen Weges, vor und hinter dem Magnetfeld, beobachtet werden. Bevor die Atome in das Magnetfeld kommen, besteht gewöhnliche Resonanzfluoreszenz, d. h. analog zur Dispersionstheorie muß angenommen werden, daß alle Atome in Phase mit dem einfallenden Licht Kugelwellen aussenden. Diese letzte Auffassung steht zunächst im Gegensatz zu dem,

Zeitschrift für Physik 43, 172–198 (1927)

was eine grobe Anwendung der Lichtquantentheorie oder der quantentheoretischen Grundregeln ergibt: denn aus ihr würde man schließen, daß nur wenige Atome in den „oberen Zustand" durch Aufnahme eines Lichtquants gehoben werden, die gesamte Resonanzstrahlung käme also von wenigen intensivstrahlenden erregten Zentren. Es lag daher früher nahe, zu sagen: die Lichtquantenauffassung darf hier nur für die Energie-Impulsbilanz herangezogen werden, „in Wirklichkeit" strahlen alle Atome im unteren Zustand schwach und kohärent Kugelwellen aus. Nachdem die Atome das Magnetfeld passiert haben, kann aber kaum ein Zweifel sein, daß der Atomstrahl sich in zwei Strahlen geteilt hat, von denen der eine den Atomen im oberen, der andere den Atomen im unteren Zustand entspricht. Wenn nun die Atome im unteren Zustand strahlen, so läge hier eine grobe Verletzung des Energiesatzes vor, denn die gesamte Anregungsenergie steckt in dem Atomstrahl mit den Atomen im oberen Zustand. Vielmehr kann kein Zweifel darüber sein, daß hinter dem Magnetfeld nur der eine Atomstrahl mit den oberen Zuständen Licht — und zwar unkohärentes Licht — der wenigen intensiv strahlenden Atome im oberen Zustand aussendet. Wie Bohr gezeigt hat, macht dieses Gedankenexperiment besonders deutlich, welche Vorsicht manchmal bei der Anwendung des Begriffs: „stationärer Zustand" nötig ist. Von der hier entwickelten Auffassung der Quantentheorie aus läßt sich eine Diskussion des Bohrschen Experiments ohne Schwierigkeiten durchführen. In dem äußeren Strahlungsfelde sind die Phasen der Atome bestimmt, also hat es keinen Sinn, von der Energie des Atoms zu sprechen. Auch nachdem das Atom das Strahlungsfeld verlassen hat, kann man nicht sagen, daß es sich in einem bestimmten stationären Zustand befände, sofern man nach den Kohärenzeigenschaften der Strahlung fragt. Man kann aber Experimente anstellen, zu prüfen, in welchem Zustand das Atom sei; das Resultat dieses Experiments läßt sich nur statistisch angeben. Ein solches Experiment wird durch das inhomogene Magnetfeld wirklich durchgeführt. Hinter dem Magnetfeld sind die Energien der Atome bestimmt, also die Phasen unbestimmt. Die Strahlung erfolgt hier inkohärent und nur von den Atomen im oberen Zustand. Das Magnetfeld bestimmt die Energien und zerstört daher die Phasenbeziehung. Das Bohrsche Gedankenexperiment ist eine sehr schöne Erläuterung der Tatsache, daß auch die Energie des Atoms „in Wirklichkeit" keine Zahl, sondern eine Matrix ist. Der Erhaltungssatz gilt für die Energiematrix und deswegen auch für den Wert der Energie so genau, als dieser jeweils gemessen wird. Rechnerisch läßt sich die Aufhebung der Phasenbeziehung etwa so verfolgen: Seien Q die Koordinaten des Atomschwerpunktes, so wird man dem Atom statt (19) die Eigenfunktion

$$S(Q,t)\,S(q,t) = S(Q,q,t) \tag{21}$$

zuordnen, wo $S(Q,t)$ eine Funktion ist, die [wie $S(\eta, q)$ in (16)] nur in einer kleinen Umgebung eines Punktes im Q-Raum von Null verschieden ist und sich mit der Geschwindigkeit der Atome in der Strahlrichtung fortpflanzt. Die Wahrscheinlichkeit einer relativen Amplitude q für irgendwelche Werte Q ist gegeben durch das Integral von

$$S(Q,q,t)\,S(Q,q,t) \text{ über } Q, \text{ d. h. durch (20).}$$

Die Eigenfunktion (21) wird sich aber im Magnetfeld berechenbar verändern und sich wegen der verschiedenen Ablenkung der Atome im oberen und unteren Zustand hinter dem Magnetfeld verwandelt haben in

$$S(Q,q,t) = c_2\,S_2(Q,t)\,\psi_2(E_2,q)\,e^{\frac{2\pi i(E_2 t+\beta)}{h}} + \sqrt{1-c_2^2}\,S_1(Q,t)\,\psi_1(E_1,q)\,e^{\frac{2\pi i E_1 t}{h}} \tag{22}$$

$S_1(Q,t)$ und $S_2(Q,t)$ werden Funktionen des Q-Raumes sein, die nur in einer kleinen Umgebung eines Punktes von Null verschieden sind; aber dieser Punkt ist für S_1 ein anderer, als für S_2. $S_1 S_2$ ist also überall Null. Die Wahrscheinlichkeit einer relativen Amplitude q und eines bestimmten Wertes Q ist daher

$$S(Q,q,t)\,\overline{S}(Q,q,t) = c_2^2\,S_2\,\overline{S}_2\,\psi_2\,\overline{\psi_2} + (1-c_2^2)\,S_1\,\overline{S}_1\,\psi_1\,\overline{\psi}_1. \tag{23}$$

Das periodische Glied aus (20) ist verschwunden, und damit die Möglichkeit, eine Phasenbeziehung zu messen. Das Resultat der statistischen Ortsbestimmung wird immer dasselbe sein, gleichgültig, bei welcher Phase des einfallenden Lichtes sie vorgenommen werde. Wir dürfen annehmen, daß Experimente mit Strahlung, deren Theorie ja noch nicht durchgeführt ist, die gleichen Resultate über die Phasenbeziehungen der Atome zum einfallenden Licht ergeben werden.

Zum Schluß sei noch der Zusammenhang der Gleichung (2) $E_1 t_1 \sim h$ mit einem Problemkomplex studiert, den Ehrenfest und andere Forscher[1]) an Hand des Bohrschen Korrespondenzprinzips in zwei wichtigen Arbeiten diskutiert haben[2]). Ehrenfest und Tolman sprechen von „schwacher Quantisierung", wenn eine gequantelte periodische Bewegung durch Quantensprünge oder andere Störungen unter-

[1]) P. Ehrenfest und G. Breit, ZS. f. Phys. 9, 207, 1922; und P. Ehrenfest und R. C. Tolman, Phys. Rev. 24, 287, 1924; siehe auch die Diskussion bei N. Bohr, Grundpostulate der Quantentheorie l. c.

[2]) Auf diesen Zusammenhang hat mich Herr W. Pauli hingewiesen.

Zeitschrift für Physik 43, 172–198 (1927)

brochen wird in Zeitintervallen, die nicht als sehr lange im Verhältnis zur Periode des Systems angesehen werden können. Es sollen in diesem Falle nicht nur die exakten quantenmäßigen Energiewerte vorkommen, sondern mit einer geringeren qualitativ angebbaren a priori-Wahrscheinlichkeit auch Energiewerte, die nicht allzu weit von den quantenmäßigen Werten abweichen. In der Quantenmechanik ist dieses Verhalten so zu deuten: Da die Energie durch die äußeren Störungen oder die Quantensprünge wirklich verändert wird, so muß jede Energiemessung, sofern sie eindeutig sein soll, sich in einer Zeit zwischen zwei Störungen abspielen. Dadurch ist eine obere Grenze für t_1 im Sinne von § 1 gegeben. Den Energiewert E_0 eines gequantelten Zustandes messen wir also auch nur mit einer Genauigkeit $E_1 \sim \frac{h}{t_1}$. Dabei hat die Frage, ob das System solche Energiewerte E, die von E_0 abweichen, „wirklich" mit dem entsprechend kleineren statistischen Gewicht annehme, oder ob ihre experimentelle Feststellung nur an der Ungenauigkeit der Messung liege, prinzipiell keinen Sinn. Ist t_1 kleiner als die Periode des Systems, so hat es keinen Sinn mehr, von diskreten stationären Zuständen oder diskreten Energiewerten zu sprechen.

Ehrenfest und Breit (l. c.) machen in ähnlichem Zusammenhang auf das folgende Paradoxon aufmerksam: Ein Rotator, den wir uns etwa als Zahnrad denken wollen, sei mit einer Vorrichtung versehen, die nach f Umdrehungen des Rades die Drehrichtung gerade umkehrt. Das Zahnrad greife etwa in eine Zahnstange ein, die ihrerseits zwischen zwei Klötzen linear verschiebbar ist; die Klötze zwingen nach einer bestimmten Anzahl Drehungen die Stange und damit das Rad zur Umkehr. Die wahre Periode T des Systems ist lang im Verhältnis zur Umlaufszeit t des Rades; die diskreten Energiestufen liegen entsprechend dicht, und zwar um so dichter, je größer T ist. Da vom Standpunkt der konsequenten Quantentheorie aus alle stationären Zustände gleiches statistisches Gewicht haben, werden für hinreichend großes T praktisch alle Energiewerte mit gleicher Häufigkeit vorkommen — im Gegensatz zu dem, was für den Rotator zu erwarten wäre. Dieses Paradoxon wird durch Betrachtung von unseren Gesichtspunkten aus zunächst noch verschärft. Um nämlich festzustellen, ob das System die zum reinen Rotator gehörigen diskreten Energiewerte allein oder besonders häufig annehmen wird, oder ob es mit gleicher Wahrscheinlichkeit alle möglichen Werte (d. h. Werte, die den kleinen Energiestufen $\frac{h}{T}$ entsprechen) annimmt, genügt eine Zeit t_1, die klein im Verhältnis zu T (aber $\gg t$) ist; d. h. obwohl die große Periode für solche Messungen gar nicht in Wirksamkeit tritt, äußert sie sich scheinbar darin, daß alle möglichen Energiewerte auftreten können. Wir sind der Ansicht, daß solche Experimente zur Bestimmung der Gesamtenergie des Systems auch wirklich alle möglichen Energiewerte gleichwahrscheinlich liefern würden; und zwar ist an diesem Ergebnis nicht die große Periode T, sondern die linear verschiebbare Stange schuld. Selbst wenn sich das System einmal in einem Zustand befindet, dessen Energie der Rotatorquantelung entspricht, so kann es durch äußere Kräfte, die an der Stange angreifen, leicht in solche übergeführt werden, die der Rotatorquantelung nicht entsprechen[1]). Das gekoppelte System: Rotator und Stange, zeigt eben ganz andere Periodizitätseigenschaften als der Rotator. Die Lösung des Paradoxons liegt vielmehr im folgenden: Wenn wir die Energie des Rotators allein messen wollen, müssen wir erst die Kopplung zwischen Rotator und Stange lösen. In der klassischen Theorie könnte bei hinreichend kleiner Masse der Stange die Lösung der Kopplung ohne Energieänderung geschehen, deshalb könnte dort die Energie des Gesamtsystems der des Rotators (bei kleiner Masse der Stange) gleichgesetzt werden. In der Quantenmechanik ist die Wechselwirkungsenergie zwischen Stange und Rad mindestens von der gleichen Größenordnung, wie eine Energiestufe des Rotators (auch bei kleiner Masse der Stange bleibt für die elastische Wechselwirkung zwischen Rad und Stange eine hohe Nullpunktsenergie!); bei Lösung der Kopplung stellen sich für Stange und Rad einzeln ihre quantenmäßigen Energiewerte her. Sofern wir also die Energiewerte des Rotators allein messen können, finden wir stets mit der durch das Experiment gegebenen Genauigkeit die quantenmäßigen Energiewerte. Auch bei verschwindender Masse der Stange ist aber die Energie des gekoppelten Systems von der Energie des Rotators verschieden; die Energie des gekoppelten Systems kann alle möglichen (durch die T-Quantelung zugelassenen) Werte gleichwahrscheinlich annehmen.

Die quantentheoretische Kinematik und Mechanik ist von der gewöhnlichen weitgehend verschieden. Die Anwendbarkeit der klassischen kinematischen und mechanischen Begriffe kann aber weder aus unseren Denkgesetzen noch aus der Erfahrung gefolgert werden; zu diesem Schluß

[1]) Dies kann nach Ehrenfest und Breit nicht oder nur sehr selten geschehen durch Kräfte, die am Rad angreifen.

gibt uns die Relation (1) $p_1 q_1 \sim h$ das Recht. Da Impuls, Ort, Energie usw. eines Elektrons exakt definierte Begriffe sind, braucht man sich nicht daran zu stoßen, daß die fundamentale Gleichung (1) nur eine qualitative Aussage enthält. Da wir uns ferner die experimentellen Konsequenzen der Theorie in allen einfachen Fällen qualitativ denken können, wird man die Quantenmechanik nicht mehr als unanschaulich und abstrakt[1]) ansehen müssen. Freilich möchte man, wenn man dies zugibt, auch die quantitativen Gesetze der Quantenmechanik direkt aus den anschaulichen Grundlagen, d. h. im wesentlichen der Relation (1) ableiten können. Jordan hat deswegen versucht, die Gleichung

$$S(q\,q'') = \int S(q\,q')\,S(q'\,q'')\,d\,q'$$

als Wahrscheinlichkeitsrelation zu deuten. Dieser Deutung können wir uns aber nicht anschließen (§ 2). Vielmehr glauben wir, daß die quantitativen Gesetze aus den anschaulichen Grundlagen heraus einstweilen nur nach dem Prinzip der größtmöglichen Einfachheit verstanden werden können. Wenn z. B. die X-Koordinate des Elektrons keine „Zahl" mehr ist, wie nach Gleichung (1) experimentell geschlossen werden kann, dann ist es die denkbar einfachste Annahme [die nicht mit (1) im Widerspruch steht], daß diese X-Koordinate ein Diagonalglied einer Matrix sei, deren Nichtdiagonalglieder sich in einer Ungenauigkeit bzw. bei Transformationen in anderen Weisen (vgl. z. B. § 4) äußern. Die Aussage, daß etwa die Geschwindigkeit in der X-Richtung „in Wirklichkeit" keine Zahl, sondern Diagonalglied einer Matrix sei, ist vielleicht nicht abstrakter und unanschaulicher, als die Feststellung, daß die elektrische Feldstärke „in Wirklichkeit" der Zeitanteil eines antisymmetrischen Tensors der Raumzeitwelt sei. Das Wort „in Wirklichkeit" wird hier ebenso sehr und ebenso wenig berechtigt sein, wie bei irgend einer mathematischen Beschreibung natürlicher Vorgänge. Sobald man zugibt, daß alle quantentheoretischen Größen „in Wirklichkeit" Matrizen seien, folgen die quantitativen Gesetze ohne Schwierigkeiten.

1) Schrödinger bezeichnet die Quantenmechanik als formale Theorie von abschreckender, ja abstoßender Unanschaulichkeit und Abstraktheit. Sicher wird man den Wert der mathematischen (und insofern anschaulichen) Durchdringung der quantenmechanischen Gesetze, die Schrödingers Theorie geleistet hat, nicht hoch genug einschätzen können. In den prinzipiellen, physikalischen Fragen hat aber meines Erachtens die populäre Anschaulichkeit der Wellenmechanik vom geraden Wege abgeführt, der durch die Arbeiten Einsteins und de Broglies einerseits, durch die Arbeiten Bohrs und die Quantenmechanik andererseits vorgezeichnet war.

Wenn man annimmt, daß die hier versuchte Deutung der Quantenmechanik schon in wesentlichen Punkten richtig ist, so mag es erlaubt sein, in wenigen Worten auf ihre prinzipiellen Konsequenzen einzugehen. Daß die Quantentheorie im Gegensatz zur klassischen eine wesentlich statistische Theorie sei in dem Sinne, daß aus exakt gegebenen Daten nur statistische Schlüsse gezogen werden könnten, haben wir nicht angenommen. Gegen solche Annahmen sprechen ja z. B. auch die bekannten Experimente von Geiger und Bothe. Vielmehr gelten in allen Fällen, in denen in der klassischen Theorie Relationen bestehen zwischen Größen, die wirklich alle exakt meßbar sind, die entsprechenden exakten Relationen auch in der Quantentheorie (Impuls- und Energiesatz). Aber an der scharfen Formulierung des Kausalgesetzes: „Wenn wir die Gegenwart genau kennen, können wir die Zukunft berechnen", ist nicht der Nachsatz, sondern die Voraussetzung falsch. Wir können die Gegenwart in allen Bestimmungsstücken prinzipiell nicht kennenlernen. Deshalb ist alles Wahrnehmen eine Auswahl aus einer Fülle von Möglichkeiten und eine Beschränkung des zukünftig Möglichen. Da nun der statistische Charakter der Quantentheorie so eng an die Ungenauigkeit aller Wahrnehmung geknüpft ist, könnte man zu der Vermutung verleitet werden, daß sich hinter der wahrgenommenen statistischen Welt noch eine „wirkliche" Welt verberge, in der das Kausalgesetz gilt. Aber solche Spekulationen scheinen uns, das betonen wir ausdrücklich, unfruchtbar und sinnlos. Die Physik soll nur den Zusammenhang der Wahrnehmungen formal beschreiben. Vielmehr kann man den wahren Sachverhalt viel besser so charakterisieren: Weil alle Experimente den Gesetzen der Quantenmechanik und damit der Gleichung (1) unterworfen sind, so wird durch die Quantenmechanik die Ungültigkeit des Kausalgesetzes definitiv festgestellt.

Nachtrag bei der Korrektur. Nach Abschluß der vorliegenden Arbeit haben neuere Untersuchungen von Bohr zu Gesichtspunkten geführt, die eine wesentliche Vertiefung und Verfeinerung der in dieser Arbeit versuchten Analyse der quantenmechanischen Zusammenhänge zulassen. In diesem Zusammenhang hat mich Bohr darauf aufmerksam gemacht, daß ich in einigen Diskussionen dieser Arbeit wesentliche Punkte übersehen hatte. Vor allem beruht die Unsicherheit in der Beobachtung nicht ausschließlich auf dem Vorkommen von Diskontinuitäten, sondern hängt direkt zusammen mit der Forderung, den verschiedenen Erfahrungen gleichzeitig gerecht zu werden, die in der Korpuskulartheorie einerseits,

Zeitschrift für Physik *43*, 172–198 (1927)

der Wellentheorie andererseits zum Ausdruck kommen. Z. B. ist bei Benutzung eines gedachten *Γ*-Strahlmikroskops die notwendige Divergenz des Strahlenbündels in Betracht zu ziehen; diese erst hat zur Folge, daß bei der Beobachtung des Elektronenortes die Richtung des Comptonrückstoßes nur mit einer Ungenauigkeit bekannt ist, die dann zur Relation (1) führt. Ferner ist nicht genügend betont, daß die einfache Theorie des Comptoneffekts in Strenge nur auf freie Elektronen anwendbar ist. Die daraus folgende Vorsicht bei Anwendung der Unsicherheitsrelation ist, wie Prof. Bohr klargestellt hat, unter anderem wesentlich für eine allseitige Diskussion des Übergangs von Mikro- zu Makromechanik. Schließlich sind die Betrachtungen über die Resonanzfluoreszenz nicht ganz korrekt, weil der Zusammenhang zwischen der Phase des Lichtes und der der Elektronenbewegung nicht so einfach ist, wie angenommen. Dafür, daß ich die genannten neueren Untersuchungen Bohrs, die in einer Arbeit über den begrifflichen Aufbau der Quantentheorie bald erscheinen werden, im Entstehen kennenlernen und diskutieren durfte, bin ich Herrn Prof. Bohr zu herzlichem Danke verpflichtet.

Kopenhagen, Institut für theoret. Physik der Universität.

Group 4

Applications of Quantum Mechanics (1926–1933)

Applications of Quantum Mechanics (1926–1933)

An Annotation by Ch. P. Enz, Geneva

1. Introduction

The discovery of quantum mechanics between 1925 and 1926 was almost simultaneously followed by a movement to "set off for the new land" ("*Der Aufbruch in das neue Land*", title of Chap. 6 of Ref. [1]) to explore the new theory with respect to its physical interpretation and its formal structure. Heisenberg played an important part in both these stages (see the preceding introduction by B. L. van der Waerden).

In his exploratory role, Heisenberg mainly contributed as coauthor of the "three-men-paper" ("*3-Männerarbeit*", see Ref. [2], letter [102], and Ref. [3], p. 95) with Born and Jordan [4] but also as Pauli's partner in an exchange of letters on the development of a formalism of action and angle variables (see Ref. [2], letters [117] and [118], and Ref. [3], Sect. V.4). This endeavor was never published because the problem had just been solved in a most elegant way by Dirac [5].

In a third stage devoted to the "settling of the new land", quantum mechanics was applied to the physical problems left unexplained by the old quantum theory. Here again, Heisenberg made important contributions which are described in this introduction.

Prominent among the unsolved physical problems was the anomalous Zeeman effect with which Heisenberg had struggled since the beginning of his scientific career (see the annotation to Group 2 by F. Hund), which he at last succeeded in solving in 1926 (Sect. 2). A more difficult problem, because it involves two electrons, was the singlet-triplet splitting in the spectrum of the helium atom, which Heisenberg solved in the same year (Sect. 3). The third application concerned an old many-electron problem, namely, the explanation of ferromagnetism; this was settled by Heisenberg in 1928 (Sect. 4).

After these major breakthroughs, Heisenberg's interest turned to quantum field theory, cosmic radiation, and nuclear structure. Only occasionally did he come back to particular problems suggested by experiments such as electrons and holes in atoms and in solids (Sect. 5) and incoherent x-ray scattering (Sect. 6).

2. Action-Angle Variables and the Anomalous Zeeman Effect

The action-angle variables proved their usefulness in Heisenberg's application of the new theory to the anomalous Zeeman effect (see Ref. [3], Sect. V.5). In this work (paper No. 1, p. 516), which he published with Jordan, the spectroscopic effect of the Zeeman energy H_1, of the spin-orbit coupling term H_2 and of Sommerfeld's relativistic correction term H_3 [6] was calculated in first-order perturbation theory. H_2 and H_3 contained the orbital averages $\langle r^{-3}\rangle$ and $\langle (W_0 + Ze^2/r)^2\rangle$ respectively, and the only way at that time to calculate $\langle r^{-2}\rangle$ was by making use of the action-angle variables. The calculation of $\langle r^{-3}\rangle$, on the other hand, could have been based on Pauli's brilliant matrix mechanical solution of the hydrogen atom [7]. However, the authors preferred to rely on the simpler, although quite unphysical, two-dimensional hydrogen model investigated by Dirac at the end of Ref. [5]. Of course, Schrödinger's theory would have offered a more systematic method of calculating these averages (see Ref. [8], p. 103). However, the techniques described in Ref. [8] were not yet available when Heisenberg and Jordan submitted their paper on March 16, 1926, just 2 days before Schrödinger submitted his third note [9].

Already in a letter to Pauli on December 24, 1925 (Ref. [2], letter [112]), Heisenberg had noted that the problem of the anomalous Zeeman effect could casily be solved if the averages $\langle r^{-2}\rangle$ and $\langle r^{-3}\rangle$ were known. The crucial condition at that time, however, was the acceptance of Goudsmit and Uhlenbeck's model of the electron spin [10], which had actually been proposed by A. H. Compton as early as 1921 [11]. The work on this problem was hampered by Pauli's well-known opposition to the spin idea. When Pauli finally agreed in a letter to Bohr (Ref. [2], letter [127]), the road was open for the exploration of all the other problems in which electron spin plays a crucial role (see the following sections).

3. Exchange and the Helium Atom

Early in May 1926, Heisenberg informed Pauli that he had found "a rather decisive argument that your prohibition of equivalent orbitals is related to singlet-triplet spacing" (see the postcard [132] in Ref. [2]). This recognition of the role of Pauli's exclusion principle was the key to Heisenberg's success with atoms having more than one electron and, most importantly, for his two famous papers on the helium spectrum (see Ref. [3], Sect. V.6) and ferromagnetism (see Sect. 4). Both these problems arose from the realization that magnetic (spin-spin) interaction energies were far too small to serve as an explanation; in both cases, the subtle combination of Coulomb repulsion and Pauli principle leading to exchange was the answer discovered by Heisenberg.

In a first paper (No. 5 of Group 3, p. 456) submitted on June 11, 1926, Heisenberg "tries to give the foundation for the quantum mechanical treatment of the many-body problem". This paper already gives the qualitative solution of the problem of singlet-triplet splitting by introducing the notion of quantum mechanical resonance due to exchange degeneracy of nonequivalent states. However, its historical significance lies in the fact that it stresses for the first time the implication of the exclusion principle for identical particles (electrons), namely, the determinantal form of n-particle states and the reduction by a factor $n!$ of the statistical weight of these states [12]. Although he does not quite succeed [13], Heisenberg also tries to show by induction from n to $n+1$ that a system which is described by a determinantal wave function "is the only one which contains no equivalent states of the partial systems" (l.c., p. 470).

The main paper on two-electron systems (No. 2, p. 531) was submitted on July 24, 1926, only $1\frac{1}{2}$ months after the first. Heisenberg took this investigation on the helium atom as an opportunity to acquaint himself with Schrödinger's new methods (see Ref. [1], p. 104; English translation, p. 72); indeed, he does a detailed, even numerical calculation. "Why should one not once in a while ride a steam roller" he writes to Pauli in the letter announcing this paper (Ref. [2], p. 328). The main result is that the energy of the para- (antisymmetric) state is higher than that of the ortho- (symmetric) state by twice the exchange integral $H^1(vw, wv)$, while the perturbation of the mean of the ortho- and para-levels is given by the Coulomb integral $H^1(vw, vw)$. In a first part, numerical values of these integrals are given for $Z = 2$(He) and $Z = 3(Li^+)$, v being the $n = 1$ S-state and w the P- and D-states, with $n = 2$ and 3. In spite of having taken into account shielding and the Rydberg and polarisation corrections, Heisenberg was not satisfied with the attained accuracy (see his concluding remarks). In fact, as can be seen from Ref. [8], this problem has seen a long sequel of approximation methods.

In a second part of the paper No. 2, the effect of electron spin ("*Elektronenmagnet*") through spin-orbit coupling (H_1) and spin-spin dipolar coupling (H_2) is analysed in a discussion which on p. 544 is somewhat heavy: Parallel spins (states a and d) have "equivalent partial systems" ($m_1 = m_2$) and hence have no resonance degeneracy, while antiparallel spins (states b and c) do, since they are nonequivalent ($m_1 \neq m_2$). Yet, a and d combine with the symmetric part $b+c$ of the "resonance problem" b, c to form the triplet $a, b+c, d$ (not a, b, c as stated), while the singlet is $b-c$ (not c). This discussion is interesting because it clearly shows the correspondence between the ortho-para separation of the orbital states and the triplet-singlet separation of the spin states.

While for two electrons the consequences of the above correspondence are still trivial, they cease to be so in the case of three or more particles, which is the subject of the third paper of this trilogy (paper No. 3, p. 551) [14]. The reason is that for three or more particles, parallel spins are unavoidable and totally prevent antisymmetric spin states. Most significantly, Heisenberg notes that "for [spin] values $s > 1/2$, such a consideration would not be realizable" (l.c., p. 565). With this in mind, the correspondence noted by Heisenberg may be considered a precursor of Dirac's fundamental theorem, which was the basis for his vector model [15] (see Sect. 4). Nonetheless, this third paper is noteworthy mainly because it is Heisenberg's first experience with group theory and, together with the simultaneous first paper by Wigner on the same subject [16], inaugurates a year of fascination with group theory in *Zeitschrift für Physik*, to which we will come back in the next section.

4. Exchange and Ferromagnetism

By introducing his idea of a molecular field [17] into Langevin's theory of the magnetism of classical dipoles, Pierre Weiss was able to account for most of the phenomenology of ferromagnets [18]. However, he himself realised that this field could not arise from magnetic interactions. On the other hand, the absence of electric dipole moments in the ground state of atoms [19] suggests, as Heisenberg observes in the introduction to his fundamental paper "that the electric interactions between two atoms should be proportional to the square of the cosine of their mutual angle rather than proportional to the cosine, contrary to the assumptions of Weiss' theory" (No. 4, p. 580).

Heisenberg's paper, submitted on May 20, 1928, provided the breakthrough sought for 20 years. He had the lucky inspiration to see that, as he says in § 1, "the empirical results with ferromagnetism place us into a very similar situation as the one we had once encountered earlier with the spectrum of the helium atom" (p. 581). Indeed, the exchange interaction was once again the key to the puzzle, an idea Heisenberg had already expressed in a letter to Pauli in November 1926 (see Ref. [2], p. 353). The problem stated in § 2 is to find the distribution of energy levels ("*Termwerte*") of $2n$ localized electrons (one valence electron per atom) to first-order in the coupling by the exchange integral J_0 between each electron and its z nearest neighbors. Couplings among more distant electrons are negligible because of the exponential decrease of the exchange integrals.

In a second step, the partition function is calculated. Here, Heisenberg makes, as he admits, the "somewhat arbitrary" assumption of a Gaussian distribution around the mean energy E_σ with variance $\overline{\Delta E_\sigma^2}$. To calculate E_σ and $\overline{\Delta E_\sigma^2}$, Heisenberg follows the group theory fashion that blossomed in *Zeitschrift für Physik* all through 1927 and which, as we saw in Sect. 3, had been inaugurated by himself and Wigner. Today, both the terminology and the technique developed in this literature are unfamiliar to most physicists. As to terminology, the reason for the difficulty is that, during this hectic activity of 1927, older mathematics literature was apparently overlooked (see footnote 16 to Chap. V, p. 408 of Ref. [20]), while the introduction of the spin-Hamiltonian concept by Dirac [15] made the calculation of group characters unnecessary (see below).

The "Partitio" in Eqs. (4) and (3) of Heisenberg's paper No. 4 are, respectively, a Young pattern and its dual (see Chap. V, § 13 of Ref. [20]), restricted to two-dimensional representations (spin 1/2). The rows of the Young pattern characterise the cycles of a permutation of the $2n$ electrons, and the dual has rows and columns interchanged. Permutations of the electron's spin independent of its orbit means, of course, that spin-orbit interactions are neglected in this approach. As to the calculation of the characters in Eq. (13) they follow easily, e.g., from Eq. (13.15), p. 138 of Ref. [21].

This last quoted formula, however, as with all calculations in Heisenberg's paper, is valid only for spin 1/2 (one valence electron per atom). In his contribution to Sommerfeld's 60th birthday [22], Heisenberg has generalized his theory to y valence electrons per atom, which are assumed to obey Hund's rule ("*ein Elektron in einer Quantenzelle*"), i.e., to give rise to an atomic spin of $y/2$. However, as van Vleck has remarked in Footnote 16 on p. 329 of his famous book [23], this elaborate calculation of $\overline{\Delta E_\sigma^2}$ is wrong, since Heisenberg has omitted the contribution of two independent transpositions T_{12} and T'_{12} in Eqs. (8) and (9) of Ref. [22].

The motivation for this generalization to $y > 1$ was that, in his paper No. 4, Heisenberg did not get the Langevin function [18] $\coth x - 1/x$, but $\tanh x$. The latter followed from the logarithmic derivative of the partition function S in

Eq. (22a) on p. 591. In his second paper [22], the number of terms $f(m)$ in the expression (16) for S is defined through the identity

$$S \propto \sum_{m=-ny}^{+ny} f(m) z^m = \left(\sum_{k=-y/2}^{+y/2} z^k \right)^{2n} . \tag{1}$$

For $y = 1$, this identity reduces to Eq. (22a) of paper No. 4, becoming nothing other than the coefficient of x^m for $\lambda_1 = \lambda_2 = \ldots = 1\,(\Sigma\lambda_i = 2n)$ in Eq. (13.14) of Ref. [21]. This is a remarkably clever way of transforming the partition function into a sum over the magnetic quantum number m. But in spite of the above-mentioned deviation from the Langevin function, Heisenberg was able to reproduce the main features of Weiss' theory and thus give an explanation of ferromagnetism in terms of the exchange integral J_0.

From the point of view of perturbation theory, the Gaussian distribution introduced by Heisenberg in paper No. 4 is not convincing. Indeed, after integration over it, the partition function S, Eq. (20), has the familiar canonical form with weight f_σ, except for the term coming from the variance $\overline{\Delta E_\sigma^2}$, which is second order in J_0, while perturbation theory had been carried only to first order in J_0. In fact, in his exhaustive report to the Solvay Council on magnetism [24] of October 1930, Pauli starts precisely at this point, namely, with Eq. (20) of paper 4, leaving out the questionable term. Using a different method, Pauli reproduces Heisenberg's result for spin 1/2, Eq. (24) except for the terms of second order in $\beta = zJ_0/kT$ (see Eqs. (64) – (66) of Ref. [24]). This critique also shows that Heisenberg's condition for ferromagnetism, $z \geqslant 8$ (see paper No. 4, p. 592), is not to be taken seriously, although, as he points out, it is satisfied for Fe, Co, and Ni. Heisenberg reported his work on ferromagnetism in a series of letters to Pauli predating the submission of the manuscript (Ref. [2], letters [192] – [196] and [198]).

Van Vleck follows in his book [23] Heisenberg's statistical evaluation more closely, except for using the method of steepest descent. However, his presentation of Heisenberg's theory is of considerable historical importance because, stimulated by Dirac [15], he was the first to use consistently the spin Hamiltonian (Eq. (16) of Ref. [23])

$$\mathscr{H}_{\text{spin}} = \sum_{\langle i,j \rangle} J_{ij} \boldsymbol{S}_i \cdot \boldsymbol{S}_j , \tag{2}$$

where $\langle i,j \rangle$ are pairs of nearest neighbors (in the above notation $J_{ij} = J_0$). Most importantly, in § 78, van Vleck gives a proof of Eqs. (14) and (17) of paper No. 4 "derived by Heisenberg with the rather involved machinery of group theory" showing "that Dirac's kinematical interpretation of the exchange effect... frees us from the need of using this" (Ref. [23], p. 340). The frustration of the "group pest" (see Ref. [20], preface to the second edition), combined with the popularity of van Vleck's book precipitated the new look of Eq. (2) above. Nonetheless, it is still surprising that this expression could become known as "Heisenberg Hamiltonian" or "Heisenberg model" [18], since it is surely vain to look for it in paper No. 4 and in Ref. [22]. As far as I can trace it (see also Ref. [25], p. 16), this name was also promoted by van Vleck, who uses "Heisenberg model" in the title of the third section of his review of 1945 [26].

Although the spin Hamiltonian became quite popular, its justification remained controversial. In his thorough review, Herring assures us that "doubts that have been expressed in the literature about this form are unfounded" [25]. Even more controversial is the history of the sign of the exchange integral J_0: Heisenberg, who was fully aware of the difficulty of this question, came to the conclusion that "for sufficiently high principal quantum numbers J_0 is certainly positive" ($n \gtrsim 3$) (see paper No. 4, p. 596). Herring, however, observes that, for interatomic spacings large enough for the spin Hamiltonian (2) to be valid, J_{ij} is almost certainly negative (antiferromagnetic) whenever i and j are nearest neighbor atoms of the same kind (see Ref. [25], pp. 6, 59). In fact, magnetism has turned out to be an even more complex phenomenon than Heisenberg had anticipated on p. 116 of Ref. [22]: new exchange mechanism have turned out to be important (see Ref. [18], p. 681).

In Ref. [22], Heisenberg made the interesting prediction that the elastic constants of a ferromagnet should change when traversing the Curie temperature. There is indeed experimental evidence in Ni [27] and in some Fe – Ni alloys [28], but the theoretical understanding is still limited [29]. In the later paper No. 5, Heisenberg made a similar prediction concerning the stress-strain relation of iron: the curve should show a discontinuous change of slope at the strain corresponding to the (spontaneous) saturation magnetostriction [30]. Here, no experimental results are known to me.

This third paper on magnetism by Heisenberg (No. 5, p. 598) is purely phenomenological and is based on the much weaker magnetic (as opposed to the electric exchange) interaction between the atoms and on the ensuing formation of domains [17, 18]. The walls separating these domains are the subject of the work in progress by Bloch [31] mentioned on p. 599. In a short review which slightly predates this third paper, Heisenberg describes the actual knowledge of ferromagnetism to an industrial audience [32].

5. Electrons and Holes

One of the applications of quantum mechanics and, in particular, of the exclusion principle concerned the symmetry between holes and electrons in a filled atomic shell or in a filled energy band of a crystal: paper No. 7 (p. 610). In this problem, Heisenberg made use of the technique of second quantization, which was familiar to him through his work with Pauli on quantum field theory [33]. The second-quantized Hamiltonian expressed in terms of holes, Eq. (15) of paper No. 7, contains additional one-particle operators arising from commutations of two-particle operators. For an N-electron shell and n electrons, this leads to the wave equation (17) or (37) for the $N - n$ holes, which is nothing else than the Hartree-Fock approximation (see e.g. Eq. (3.11) of Ref. [34]), containing the energy A, Eq. (16), of the closed shell in the same approximation (see Eq. (3.8) of Ref. [34]). Heisenberg seems to have been unaware of this connection with the work of Hartree [35] and Fock [36]. The Hartree term $C(r_1)$ is just the Coulomb potential, while the Fock term $D(r_1)$ is the nonlocal exchange potential. For a given N, and neglecting spin interactions, A merely leads to a shift of the spectrum of n holes relative to that of n electrons. On the other hand, if one approximates the effect of the potentials $C(r_1)$ and $D(r_1)$ by an effective nuclear charge, this leads to a factor of order one, by which the n-hole spectrum is multiplied relative to the n-electron spectrum. A numerical comparison of the spectra of Ti and Ni confirms this conclusion.

In a second part of paper No. 7, the electron-hole symmetry is applied to a metal with the "anomalous" Hall effect (see Ref. [18], p. 236). In the simplest case, such a metal or a semiconductor has an almost filled conduction band in which electrical current is carried by holes. By introducing a periodic potential eV_0, the Coulomb interaction between conduction electrons is accounted for, meaning that there is no need for two-particle operators in this case. Using a simple form of energy band, Eq. (44), and proceeding from Bloch functions, Eq. (43), to Wannier functions, Eq. (45) (see Ref. [18], p. 187), Heisenberg shows that, for holes, the Hall coefficient has a sign opposite that of electrons.

6. Incoherent X-Ray Scattering

It was Heisenberg's colleague in Leipzig, Peter Debye, who suggested that the Thomas-Fermi approximation used for the coherent, or phase-dependent, part of x-ray scattering should be generalizable to the incoherent part. In paper No. 8 (p. 627) Heisenberg devised an elegant and simple solution entirely in the spirit of Thomas and Fermi (see e.g. Ref. [19], p. 271). The latter amounts to treating the electrons in the atom as free, except for the condition of binding, namely, that the total energy in each point r be negative,

$$\frac{p^2}{2m} - e\varphi(r) < 0 \tag{3}$$

$\varphi(r)$ being the Thomas-Fermi potential. Free electrons means, in particular, that all orbital states are doubly occupied, corresponding to both spin directions, except perhaps for the highest. Obviously, this is not a realistic approximation, since it is the exact opposite of Hund's rule, which is the consequence of the Coulomb repulsion. Yet here, it is consistent. With double occupancy, use can be made of the closure of the spin states which leads to the $\delta_{\sigma\sigma'}$, in Eq. (15) of paper No. 8. At this point, the generalized Thomas-Fermi hypothesis (17) appears natural, and the result is straightforward.

A subsequent note by Bewilogua [37] gave a numerical table of Heisenberg's formula; later, in a short review, Heisenberg returned to the definition of coherent and incoherent radiation [38]. This publication, unfortunately, is hampered by ambiguous notation, since in Eqs. (13) and (14) X_n stands both for a coordinate of the nth nucleus and for a wave function in state n. Although incoherent scattering has remained a secondary issue of x-ray research, it is interesting to note that Heisenberg's formula has survived in the literature [39].

References

1 W. Heisenberg: *Der Teil und das Ganze* (Piper, München 1969); English translation by A. J. Pomerans, *Physics and Beyond* (Allen and Unwin, London 1971).

2 *Wolfgang Pauli, Scientific Correspondence with Bohr, Einstein, Heisenberg, a.o., Volume I: 1919–1929*, ed. by A. Hermann, K. v. Meyenn, V. F. Weisskopf (Springer, New York 1979).

3 J. Mehra, H. Rechenberg: *The Historical Development of Quantum Theory* (Springer, New York 1982) Vol. 3.

4 See paper No. 4 of Group 3 (p. 397).

5 P. A. M. Dirac: Proc. Roy. Soc. (London) **A 110**, 561 (1926).

6 A. Sommerfeld: Sitzungsbericht Bayer. Akad. Wiss. (1915) p. 459.

7 W. Pauli: Z. Phys. **36**, 336 (1926).

8 H. A. Bethe, E. E. Salpeter: In *Encyclopedia of Physics*, ed. by S. Flügge (Springer, Berlin 1957) Vol. XXXV, p. 88.

9 E. Schrödinger: Ann. Phys. (Leipzig) **79**, 734 (1926).

10 G. Uhlenbeck, S. Goudsmit: Die Naturwissenschaften **13**, 953 (1925).

11 A. H. Compton: J. Franklin Inst. **192**, 145 (1921).

12 Dirac's paper in Proc. Roy. Soc. **A 112**, 661 (1926) where he introduces determinantal wave functions and derives Fermi-Dirac statistics was submitted on August 26, 1926.

13 For a general proof, see W. Pauli: *General Principles of Quantum Mechanics*, transl. by P. Achuthan, K. Venkatesan (Springer, Berlin, Heidelberg 1980) p. 119.

14 The correct W-values in Table (23) seem to be $2A+2C$, $2A-C$, $2A-C$, corresponding to $A=B$, $C=D$.

15 P. A. M. Dirac: Proc. Roy. Soc. **A 123**, 714 (1929); *The Principles of Quantum Mechanics*, 4th ed. (Oxford University Press, London 1958) § 58.

16 E. Wigner: Z. Phys. **40**, 492 (1926).

17 P. Weiss: J. Phys. (Paris) (4) **6**, 661 (1907); Phys. Z. **9**, 358 (1908).

18 See e.g., N. W. Ashcroft, N. D. Mermin: *Solid State Physics* (Holt, Rinehart and Winston, New York 1976) Chaps. 31–33.

19 See e.g., L. I. Schiff: *Quantum Mechanics*, Second ed. (McGraw-Hill, New York 1955) p. 160.

20 H. Weyl: *The Theory of Groups and Quantum Mechanics*, translated by H. P. Robertson (Dover, New York 1950).

21 E. P. Wigner: *Group Theory and Its Application to the Quantum Mechanics of Atomic Spectra*, translated by J. J. Griffin (Academic, New York 1959).

22 W. Heisenberg: In *Probleme der modernen Physik – Arnold Sommerfeld zum 60. Geburtstag gewidmet von seinen Schülern*, ed. by P. Debye (Hirzel, Stuttgart 1928) p. 114; reprinted as No. 6 of Volume B of *Collected Works*.

23 J. H. van Vleck: *The Theory of Electric and Magnetic Susceptibilities* (Oxford University Press, London 1932).

24 W. Pauli: In *Le Magnétisme*, 6ème Conseil de Physique Solvay (Gauthier-Villars, Paris 1932) p. 175.

25 C. Herring: In *Magnetism*, ed. by G. T. Rado, H. Suhl (Academic, New York 1966) Vol. IIB, p. 1.

26 J. H. van Vleck: Rev. Mod. Phys. **17**, 27 (1945). In Rev. Mod. Phys. **7**, 167 (1935), van Vleck still uses the name "Dirac vector model".

27 B. Golding, M. Barmatz: Phys. Rev. Lett. **23**, 223 (1969).

28 M. Shimizu: J. Phys. Soc. Jpn. **45**, 1520 (1978); G. Hausch, H. Warlimont: Phys. Lett. **41 A**, 437 (1972); Acta Met. **21**, 401 (1973).

29 D. J. Kim: J. Phys. Soc. Jpn. **40**, 1250 (1976); Solid State Commun. **30**, 249 (1979).

30 See e.g. W. J. Carr, Jr.: In *Encyclopedia of Physics*, ed. by S. Flügge (Springer, Berlin 1966) Vol. XVIII/2, p. 274, Chap. VI;
J. Kanamori: In *Magnetism*, ed. by G. T. Rado, H. Suhl (Academic, New York 1966) Vol. I, p. 127.

31 F. Bloch: Z. Phys. **74**, 295 (1932).

32 W. Heisenberg: Metallwirtschaft **9**, 843 (1930); reprinted as No. 9 of Volume B of *Collected Works*.

33 W. Heisenberg, W. Pauli: Z. Phys. **56**, 1 (1929); **59**, 168 (1930), to be reprinted in Volume A II of *Collected Works*.
34 D. J. Thouless: *The Quantum Mechanics of Many-Body Systems* (Academic, New York 1961).
35 D. R. Hartree: Proc. Cambridge Philos. Soc. **24**, 89 (1928).
36 V. I. Fock: Z. Phys. **61**, 126 (1930)
37 L. Bewilogua: Phys. Z. **32**, 740 (1931).
38 W. Heisenberg: In *Ergebnisse der technischen Röntgenkunde* (Akad. Verlagsgesellschaft Leipzig 1933) Vol. III, p. 26; reprinted as No. 11 of Volume B of *Collected Works*.
39 R. W. James: *The Optical Principles of the Diffraction of X-Rays* (G. Bell, London 1967) pp. 125 and 462 (note that in Eq. (9.4) the first curly bracket should be squared).

Anwendung der Quantenmechanik auf das Problem der anomalen Zeemaneffekte.

Von **W. Heisenberg** und **P. Jordan** in Göttingen.

(Eingegangen am 16. März 1926.)

Von Uhlenbeck und Goudsmit ist zur Erklärung der anomalen Zeemaneffekte die Comptonsche Hypothese des rotierenden Elektrons herangezogen worden. Die vorliegende Arbeit untersucht das quantenmechanische Verhalten des durch diese Hypothese charakterisierten Atommodells. Das Ergebnis ist, daß die Zeemaneffekte und die Feinstrukturen der Dublettspektren durch die genannte Hypothese vollständig erklärt werden können.

Eine Untersuchung des magnetischen Verhaltens von Atomsystemen lehrt, daß auch nach den Gesetzen der Quantenmechanik Atomsysteme, die aus Punktladungen aufgebaut sind, stets normalen Zeemaneffekt zeigen müssen.

Zur Erklärung der anomalen Zeemaneffekte haben Uhlenbeck und Goudsmit die Hypothese herangezogen [1]), daß jedes einzelne Elektron Träger eines magnetischen Moments $\mathfrak{m}$ und eines entsprechenden mechanischen Drehimpulses $\mathfrak{s}$ sein solle. Dabei sollen $\mathfrak{m}$ und $\mathfrak{s}$ durch die Beziehung

$$\mathfrak{m} = \frac{e}{mc}\mathfrak{s} \tag{1}$$

verknüpft sein. Der Quotient von magnetischem und mechanischem Moment soll sich also von dem für Atomsysteme mit Punktladungen gültigen Werte $\frac{e}{2mc}$ um den Faktor 2 unterscheiden. Auf die Frage, welche Argumente sich vom Standpunkt der Elektrodynamik aus für und wider diese Hypothese anführen lassen, soll hier nicht eingegangen werden. Vielmehr soll im folgenden das quantenmechanische Verhalten des Uhlenbeck-Goudsmitschen Modells untersucht und das Ergebnis mit der Erfahrung verglichen werden. Bekanntlich führte die Anwendung der früher üblichen Quantenregeln auf dieses Modell zu Widersprüchen mit der Erfahrung.

---

[1]) Die Hypothese des rotierenden Elektrons stammt schon von A. Compton, Journ. Frankl. Inst. **192**, 145, 1921. Die Anwendung dieser Hypothese auf das uns hier interessierende Problem der Zeemaneffekte wurde jedoch erst von E. Uhlenbeck und S. Goudsmit, Naturwiss. **13**, Heft 47, 1925, angegeben.

§ 1. Die Hamiltonsche Funktion des Modells. Wir nehmen im folgenden an, daß ein Elektron von der Ladung $-e$, dem magnetischen Moment $\mathfrak{m}$ und dem Drehimpuls $\mathfrak{s}$[1] $\left(\mathfrak{m} = \frac{e}{mc}\mathfrak{s}\right)$ um einen Z-fach positiv geladenen schweren Kern kreist; der Drehimpuls dieser Bewegung heiße $\mathfrak{k}$. Von außen möge ein magnetisches Feld $\mathfrak{H}$ die Bewegung des Elektrons stören. Magnetisch verhält sich dieses Modell offenbar genau wie das von Pauli und Landé vorgeschlagene, das bei der formalen Ordnung der komplizierten Spektren so große Dienste geleistet hat. Die beim Zusammenwirken mehrerer Valenzelektronen auftretenden Feinstrukturen und Zeemaneffekte lassen sich in den meisten Fällen auf die Feinstruktur und die Zeemaneffekte des oben beschriebenen einfachen Modells zurückführen.

Die Bewegung des Elektrons wird, wenn man vom Einfluß der Relativität, der Wirkung des äußeren Feldes und der Wirkung von $\mathfrak{m}$ absieht, durch die Pauli-Diracsche[2]) Theorie des Wasserstoffatoms gegeben.

Die hinzukommende Störungsenergie zerfällt in drei Teile,

$$H = H_1 + H_2 + H_3:$$

1. Der vom äußeren Feld $\mathfrak{H}$ herrührende Teil ist nach bekannten Regeln gegeben durch

$$H_1 = \mathfrak{H} \cdot \left(\frac{e}{2mc}\mathfrak{k}\right) + \mathfrak{H}\left(\frac{e}{mc}\mathfrak{s}\right) = \frac{e}{2mc}\mathfrak{H}(\mathfrak{k} + 2\mathfrak{s}). \qquad (2)$$

2. Betrachtet man den Schwerpunkt des Elektrons als ruhend, den Kern als um das Elektron kreisend, so erzeugt der Kern am Orte des Elektrons das Magnetfeld

$$\mathfrak{H}_i = \frac{eZ}{c}\frac{[\mathfrak{r}\mathfrak{v}]}{r^3} = \frac{eZ}{mc}\frac{\mathfrak{k}}{r^3}.$$

Diesem Feld entspricht eine Larmorpräzession des Impulses $\mathfrak{s}$ vom Betrag $\frac{e}{mc}\overline{\mathfrak{H}_i}$. Nach Thomas[3]) haben wir jedoch zu beachten, daß dies

---

[1]) Nach der Compton-Uhlenbeck-Goudsmitschen Hypothese ist für das einzelne Elektron ein ganz bestimmter $\mathfrak{s}$-Impuls, nämlich quantenmechanisch $\mathfrak{s}^2 = \left(\frac{h}{2\pi}\right)^2 s(s+1)$; $s = \frac{1}{2}$ einzusetzen. Wir lassen hier aber $\mathfrak{s}$ zunächst unbestimmt, um auch die durch Kopplung mehrerer Elektronenmagnete entstehenden Multipletts (Tripletts, Quartetts usw.) mitbehandeln zu können.

[2]) W. Pauli jr., ZS. f. Phys. **36**, 336, 1926; P. Dirac, Proc. Roy. Soc. London **110**, März 1926.

[3]) L. H. Thomas, Nature **117**, 514, 1926.

die Larmorpräzession ist nur in dem eben betrachteten System, in dem der Schwerpunkt des Elektrons ruht. Um die Präzession zu bekommen in dem System, in dem der Kern oder besser der Schwerpunkt des ganzen Atoms ruht, muß noch eine Lorentztransformation ausgeführt werden. Für die Lamorpräzession in diesem letzteren, uns eigentlich interessierenden System ergibt sich so nach Thomas der Wert $\frac{e^2 Z}{2 m^2 c^2} \overline{\frac{1}{r^3}} \mathfrak{k}$. Diesem Wert entspricht in der Hamiltonschen Funktion offenbar ein Glied

$$H_2 = \frac{e^2 Z}{2 m^2 c^2} \overline{\frac{1}{r^3}} \mathfrak{k} \mathfrak{s}. \tag{3}$$

3. Die relativistische Massenveränderlichkeit gibt nach Sommerfelds Theorie zu einer Zusatzenergie vom Betrag

$$H_3 = -\frac{1}{2 m c^2} \left[W_0^2 + 2 e^2 Z W_0 \overline{\frac{1}{r}} + e^4 Z^2 \overline{\frac{1}{r^2}} \right] \tag{4}$$

Anlaß. Die Striche über den von r abhängigen Gliedern bedeuten Mittelung über die ungestörte Bewegung. Wir werden im folgenden annehmen, daß die Störungsfunktion H in der Quantenmechanik dieselbe Form hat wie in der klassischen Mechanik und Elektrodynamik. Als Begründung für diese Annahme läßt sich anführen, daß alle in H vorkommenden Größen vertauschbar sind und daß daher nach dem Korrespondenzprinzip kaum Formen für H in Betracht kommen dürften, die wesentlich von der hier abgeleiteten abweichen. Eine zwangläufige Begründung der hier angegebenen Störungsfunktionen läßt sich nicht geben, solange eine konsequent durchgeführte quantentheoretische Elektrodynamik fehlt.

§ 2. Gedankengang der Störungsrechnung. Bei der nun folgenden quantenmechanischen Rechnung dürfen wir annehmen, daß im ungestörten System die Absolutbeträge von $\mathfrak{k}$ und $\mathfrak{s}$ gequantelt, d. h. Diagonalmatrizen sind. Diese Annahme könnte für

$$\mathfrak{k} \left[\mathfrak{k}^2 = \left(\frac{h}{2\pi}\right) k (k+1) \right]$$

als unberechtigt angesehen werden, da im ungestörten System die bekannte Entartung von k besteht. Da aber in der Störungsenergie H selbst nur $\mathfrak{k}$, nicht aber die zu $\mathfrak{k}$ konjugierte Perihellänge auftritt, so wird bei Mitberücksichtigung von H doch die vorausgenommene Quantelung von $|\mathfrak{k}|$ eintreten; dies bedeutet physikalisch, daß nach dem hier zu untersuchenden Modell der Zeemaneffekt des Wasserstoffs dem der Alkaliatome vollkommen analog ist; bei den Alkaliatomen ist $|\mathfrak{k}|$ schon durch die Wechselwirkung mit den anderen Elektronen festgelegt. Eine analoge Betrachtung läßt sich für die Komponente M_z des Gesamtimpulses $\mathfrak{M}$ des Atoms in der Feldrichtung anstellen. Das Ausgangssystem ist zwar hinsichtlich $M_z = \frac{h}{2\pi} m$ entartet; da aber in der Störungsenergie H nur m, nicht aber die zu m konjugierte Winkelvariable auftritt, so wird durch H die Quantelung von M_z doch eintreten. Wir werden also unsere Rechnungen vereinfachen können durch die Annahme, daß im Ausgangssystem $|\mathfrak{k}|$, $|\mathfrak{s}|$, M_z nicht entartet und daher als Diagonalmatrizen quantenmäßig festgelegt seien.

Dann ist das Ausgangssystem nur noch hinsichtlich einer Koordinate entartet (vgl. die vollkommen analoge Behandlung des Modells nach der klassischen Mechanik[1]). Wir können diese Koordinate charakterisieren durch die Komponente $\mathfrak{s}_z = \frac{h}{2\pi} m_s$ des Eigenimpulses $\mathfrak{s}$ des Elektrons und die konjugierte Winkelvariable. Wir können sie aber auch charakterisieren durch den Gesamtimpuls $\mathfrak{M}$ und die zu ihm kanonisch konjugierte Variable.

Das Störungsverfahren der Quantenmechanik für entartete Systeme läßt sich folgendermaßen skizzieren[2]):

Gegeben sei irgend eine Lösung p^0, q^0 des ungestörten Problems, ferner die Störungsfunktion in Abhängigkeit von den Koordinaten des ungestörten Problems. Wäre das Ausgangssystem nicht entartet, so würde die der Störung entsprechende Zusatzenergie W durch den Zeitmittelwert H der Störungsfunktion über die ungestörte Bewegung gegeben sein. Dieser Mittelwert H ist dann von selbst Diagonalmatrix. Ist jedoch das Ausgangssystem entartet, d. h. fallen etwa die Energiewerte der Zustände $n + 1 \ldots n + r$ zusammen, so enthält der Mittelwert H der Störungsenergie noch Glieder, die Übergängen zwischen den Zuständen $n + 1 \ldots n + r$ entsprechen, d. h. H ist keine Diagonalmatrix.

In diesem Falle soll mit den $p^0 q^0$ eine kanonische Transformation

$$\left.\begin{aligned} p' &= S^{-1} p^0 S, \\ q' &= S^{-1} q^0 S \end{aligned}\right\} \tag{5}$$

vorgenommen werden, derart, daß

$$W = S^{-1} H S \tag{6}$$

[1]) Z. B. bei W. Pauli, ZS. f. Phys. **16**, 155. 1923; **20**, 371, 1924.

[2]) M. Born, W. Heisenberg und P. Jordan, ZS. f. Phys. **35**, 557, 1926. Siehe bes. Kap. 3, § 2.

eine Diagonalmatrix wird. Die Transformationsmatrix S enthält dabei, wie H, nur Glieder, die Übergängen zwischen Zuständen der Reihe $n+1, n+2 \ldots n+r$ entsprechen, und Diagonalglieder. Die Transformationsfunktion S kann aufgefunden werden, indem man die r Gleichungen mit r Unbekannten

$$W S_k - \sum_l H_{kl} S_l = 0 \quad (k, l = n+1 \ldots n+r) \tag{7}$$

zu lösen sucht. Diese Lösungen existieren für r verschiedene Werte von W, die „Eigenwerte" des Problems, zugleich die Zusatzenergien des gestörten Systems. Bedeute * Übergang zur konjugiert-komplexen Größe, $\sim$ Vertauschung der Indizes, so gilt für irgend zwei Eigenwerte W_n, W_m:

$$\left.\begin{aligned} W_n S_{kn} - \sum_l H_{kl} S_{ln} &= 0, \\ W_m S^*_{km} - \sum_l H^*_{kl} S^*_{lm} &= 0; \end{aligned}\right\} \tag{8}$$

also

$$(W_n - W_m) \sum_{k=n+1}^{n+r} S_{kn} S^*_{km} = 0.$$

Normiert man noch durch

$$\sum_{k=n+1}^{n+r} S_{kn} S^*_{kn} = 1, \tag{9}$$

so gilt $S \cdot \tilde{S}^* = 1$, und S ist die gesuchte Transformationsmatrix. Durch Einsetzen in (5) erhält man in dieser Näherung die Koordinaten des gestörten Systems.

§ 3. Durchführung der Rechnung. Die Anwendung dieses Verfahrens auf das hier zu behandelnde Problem führt auf folgende allgemeine Rechnung:

1. Der Teil H_3 der Störungsenergie kann, da er die entarteten Koordinaten nicht enthält, zunächst weggelassen und nachträglich als additive Konstante hinzugefügt werden.

2. Für die Impulse $\mathfrak{k}$ und $\mathfrak{s}$ gelten nach den allgemeinen Regeln der Quantenmechanik (l. c.) die Relationen [1]):

$$\left.\begin{aligned} \mathfrak{k}^2 = \left(\frac{h}{2\pi}\right)^2 k(k+1), \quad \mathfrak{s}^2 = \left(\frac{h}{2\pi}\right)^2 s(s+1), \\ k_x k_y - k_y k_x = -\varepsilon k_z{}^{1)} \left(\varepsilon = \frac{h}{2\pi i}\right), \\ [\mathfrak{k}\mathfrak{k}] = -\varepsilon \mathfrak{k}; \quad [\mathfrak{s}\mathfrak{s}] = -\varepsilon \mathfrak{s} \end{aligned}\right\} \tag{10}$$

oder einfacher

[1]) Da in der oben zitierten Arbeit die Drehimpulse mit dem entgegengesetzten Vorzeichen definiert waren, galt dort

$$M_x M_y - M_y M_x = \varepsilon M_z.$$

(die eckigen Klammern bedeuten das vektorielle Produkt). Jede Komponente von $\mathfrak{k}$ ist mit jeder Komponente von $\mathfrak{s}$ vertauschbar.

Setzt man

$$k_z = m_k \frac{h}{2\pi}, \quad s_z = m_s \frac{h}{2\pi},$$

so wird

$$\left.\begin{aligned}
(k_x + i k_y)(k, m_k - 1;\ k, m_k) &= \frac{h}{2\pi}\sqrt{k(k+1) - m_k(m_k - 1)},\\
(k_x - i k_y)(k, m_k;\ k\, m_k - 1) &= \frac{h}{2\pi}\sqrt{k(k+1) - m_k(m_k - 1)};\\
(s_x + i s_y)(s, m_s - 1;\ s, m_s) &= \frac{h}{2\pi}\sqrt{s(s+1) - m_s(m_s - 1)},\\
(s_x - i s_y)(s, m_s;\ s, m_s - 1) &= \frac{h}{2\pi}\sqrt{s(s+1) - m_s(m_s - 1)}.
\end{aligned}\right\} \tag{11}$$

Führt man jetzt statt m_k die Variable m durch die Gleichung $m = m_k + m_s$ ein, so ist zu m_s jetzt kanonisch konjugiert die Differenz der bisher zu m_k und m_s kanonisch konjugierten „Knotenlängen" (vgl. die oben zitierte Rechnung nach der klassischen Mechanik). Also gilt

$$H_1 + H_2 = \frac{e}{2mc}\mathfrak{H}(\mathfrak{k} + 2\mathfrak{s}) + \frac{e^2 Z}{2m^2c^2}\overline{\frac{1}{r^3}}\cdot\mathfrak{k}\mathfrak{s}. \tag{12}$$

Mit den Abkürzungen

$$\frac{e}{2mc}|\mathfrak{H}|\frac{h}{2\pi} = \mu$$

und

$$\frac{1}{2}\frac{e^2 Z}{m^2c^2}\overline{\frac{1}{r^3}}\left(\frac{h}{2\pi}\right)^2 = \lambda$$

folgt:

$$H_1 + H_2 = \mu(k_z + 2s_z) + \lambda\left(k_z s_z + \tfrac{1}{2}(k_x + i k_y)(s_x - i s_y) + \tfrac{1}{2}(k_x - i k_y)(s_x + i s_y)\right)$$

und

$$\left.\begin{aligned}
&(H_1 + H_2)(m_s, m_s) = \mu(m + m_s) + \lambda m_s(m - m_s),\\
&(H_1 + H_2)(m_s, m_s - 1)\\
&= \tfrac{1}{2}\lambda\sqrt{[s(s+1) - m_s(m_s - 1)][k(k+1) - (m - m_s)(m - m_s + 1)]},\\
&(H_1 + H_2)(m_s - 1.\ m_s)\\
&= \tfrac{1}{2}\lambda\sqrt{[s(s+1) - m_s(m_s - 1)][k(k+1) - (m - m_s)(m - m_s + 1)]}.
\end{aligned}\right\} \tag{13}$$

Die Indizes m, k, s sind als konstant auf der linken Seite dieser Gleichungen weggelassen worden.

3. Die Zahl der zu einem gegebenen Wertesystem k, s, m gehörigen Werte von m_s bestimmt sich durch die Bedingungen:

$$\left.\begin{array}{ll} & -s \leqq m_s \leqq +s \\ \text{und} & \\ & -k \leqq m - m_s \leqq +k \\ \text{oder} & \\ & k + m \geqq m_s \geqq -k + m. \end{array}\right\} \qquad (14)$$

Die Transformationsfunktion S erhält man nach (7) durch Lösung der linearen Gleichungen

$$W S_r - \Sigma H_{rl} S_l = 0, \qquad (15)$$

wobei die Indizes $r\,l$ alle Werte von m_s durchlaufen, die bei einem gegebenen Wertesystem k, s, m möglich sind. Die Eigenwerte von W ergeben sich durch Nullsetzen der Determinante mit den Gliedern $\delta_{rl} W - H_{rl}$. Bezeichnet man mit m_1 den kleinsten, mit m_2 den größten Wert von m_s, der bei gegebenem k, s, m möglich ist, so ergibt sich Gleichung (16).

$$0 = \begin{vmatrix} W-\mu(m+m_1)-\lambda m_1(m-m_1) & -\frac{1}{2}\lambda\sqrt{[s(s+1)-m_1(m_1+1)][k(k+1)-(m-m_1)(m-m_1-1)]} & 0 & 0\ldots 0 \\ -\frac{1}{2}\lambda\sqrt{[s(s+1)-m_1(m_1+1)][k(k+1)-(m-m_1)(m-m_1-1)]} & W-\mu(m+m_1+1)-\lambda(m_1+1)(m-m_1-1) & -\frac{1}{2}\lambda\sqrt{} & 0\ldots 0 \\ 0 & -\frac{1}{2}\lambda\sqrt{[s(s+1)-(m_1+1)(m_1+2)][k(k+1)-(m-m_1-1)(m-m_1-2)]} & W-\frac{1}{2}\lambda\sqrt{} & \ldots 0 \\ \cdots & \cdots & \cdots & \cdots \\ \cdots & \cdots & \cdots & \cdots \\ \cdots & \cdots & \cdots & \cdots \end{vmatrix} \qquad (16)$$

Wir haben also eine algebraische Gleichung $(m_2 - m_1 + 1)$-ten Grades für W mit in k, s, m rationalen Koeffizienten. Die Summe der Wurzeln ist gleich dem negativen Koeffizienten des zweiten Gliedes und daher gegeben durch

$$\sum_{n=m_1}^{m_2} W_n = \sum_{m=m_1}^{m_2} \mu(m + m_1) + \lambda m_1 (m - m_1). \qquad (17)$$

Daß die Summe $\sum\limits_{n=m}^{m_2} W_n$ linear in λ und μ sei, ist die Aussage des sogenannten „Summationsprinzips" der Zeemaneffekte.

4. Um die Resultate der quantenmechanischen Rechnung in allen Einzelheiten verfolgen zu können, wird es zweckmäßig sein, ein spezielles Beispiel zu untersuchen. Wir wählen dazu das Dublettmodell, d. h. $s = \frac{1}{2}$.

Die möglichen Werte von m_s sind hier im allgemeinen $\pm\frac{1}{2}$, doch für $m = k + \frac{1}{2}$

ist m_s nur des Wertes $+\frac{1}{2}$, für $m = k - \frac{1}{2}$ ist m_s nur des Wertes $-\frac{1}{2}$ fähig. An Stelle von (16) tritt also im allgemeinen die Gleichung:

$$\begin{vmatrix} W - \mu(m-\frac{1}{2}) + \lambda(m+\frac{1}{2})\frac{1}{2} & -\frac{1}{2}\lambda\sqrt{k(k+1)-(m+\frac{1}{2})(m-\frac{1}{2})} \\ -\frac{1}{2}\lambda\sqrt{k(k+1)-(m+\frac{1}{2})(m-\frac{1}{2})} & W - \mu(m+\frac{1}{2}) - \lambda\frac{1}{2}(m-\frac{1}{2}) \end{vmatrix} = 0 \quad (18)$$

oder

$$W^2 - \left(2\mu m - \frac{\lambda}{2}\right)W + \mu^2(m^2 - \tfrac{1}{4}) - \mu\lambda\cdot m - \frac{\lambda^2}{4}k(k+1) = 0. \quad (19)$$

$$W = \mu m - \frac{\lambda}{4} \pm \frac{1}{2}\sqrt{\mu^2 + 2\mu\lambda . m + \lambda^2(k+\tfrac{1}{2})^2}. \quad (20)$$

Für $m = k + \frac{1}{2}$ dagegen ergibt sich $m_s = \frac{1}{2}$ und

$$W = \mu(m + \tfrac{1}{2}) + \frac{\lambda}{2}(m - \tfrac{1}{2}), \quad (21)$$

für $m = -k - \frac{1}{2}$ folgt $m_s = -\frac{1}{2}$ und

$$W = \mu(m - \tfrac{1}{2}) - \frac{\lambda}{2}(m + \tfrac{1}{2}). \quad (22)$$

Führt man noch die Abkürzung

$$v = \frac{\lambda}{\mu}(k + \tfrac{1}{2})$$

ein, so wird

$$\left.\begin{aligned} W &= \mu\left(m - \frac{v}{k+\frac{1}{2}} \pm \sqrt{1 + 2\frac{m}{k+\frac{1}{2}}v + v^2}\right) \\ &\text{bzw.} \\ W_{m=k+\frac{1}{2}} &= \mu\left[m\left(1 + \frac{v}{2k+1}\right) - \frac{v}{k+\frac{1}{2}} + \frac{1}{2}\right] \\ &= \mu\left[k + 1 + v\left(\frac{1}{2} - \frac{1}{k+\frac{1}{2}}\right)\right], \\ W_{m=-k-\frac{1}{2}} &= \mu\left[m\left(1 - \frac{v}{2k+\frac{1}{2}}\right) - \frac{v}{k+\frac{1}{2}} - \frac{1}{2}\right] \\ &= \mu\left[-k - 1 + v\left(\frac{1}{2} - \frac{1}{k+\frac{1}{2}}\right)\right]. \end{aligned}\right\} \quad (23)$$

Die Gleichungen (23) stimmen mit den aus der Voigtschen Kopplungstheorie [1]) bekannten Dublettformeln überein.

5. Wir gehen zur Berechnung der Intensitäten über. Zur Bestimmung der Transformationsfunktion S lösen wir die Gleichung:

$$\left.\begin{aligned} W S_{-\frac{1}{2}} - (H_1 + H_2)(-\tfrac{1}{2}, -\tfrac{1}{2})S_{-\frac{1}{2}} \\ -(H_1 + H_2)(-\tfrac{1}{2}, +\tfrac{1}{2})S_{+\frac{1}{2}} = 0. \end{aligned}\right\} \quad (24)$$

[1]) Vgl. z. B. A. Sommerfeld, ZS. f. Phys. 8, 257, 1922.

Es ergibt sich:

$$\left.\begin{aligned} S_{+\frac{1}{2}} &= C\cdot\left(W-\mu\left(m-\tfrac{1}{2}\right)+\frac{\lambda}{2}\left(m+\tfrac{1}{2}\right)\right), \\ S_{-\frac{1}{2}} &= C\cdot\frac{\lambda}{2}\sqrt{k(k+1)-\left(m^2-\tfrac{1}{4}\right)}, \end{aligned}\right\} \tag{25}$$

wo C eine zunächst willkürliche Konstante darstellt. Unterscheidet man wieder die beiden Werte von W:

$$„W_{+\frac{1}{2}}\text{“ und } „W_{-\frac{1}{2}}\text{“},$$

so folgt:

$$\left.\begin{aligned} S_{+\frac{1}{2},\,+\frac{1}{2}} &= C_{+\frac{1}{2}}\cdot\left(W_{+\frac{1}{2}}-\mu\left(m-\tfrac{1}{2}\right)+\frac{\lambda}{2}\left(m+\tfrac{1}{2}\right)\right) \\ &= \tfrac{1}{2}\,C_{+\frac{1}{2}}\left(\mu+\lambda m+\sqrt{\mu^2+2\lambda\mu m+\lambda^2\left(k+\tfrac{1}{2}\right)^2}\right), \\ S_{-\frac{1}{2},\,+\frac{1}{2}} &= C_{+\frac{1}{2}}\frac{\lambda}{2}\sqrt{k(k+1)-\left(m^2-\tfrac{1}{4}\right)}, \\ S_{+\frac{1}{2},\,-\frac{1}{2}} &= \tfrac{1}{2}\,C_{-\frac{1}{2}}\cdot\left(\mu+\lambda m-\sqrt{\mu^2+2\lambda\mu m+\lambda^2\left(k+\tfrac{1}{2}\right)^2}\right), \\ S_{-\frac{1}{2},\,-\frac{1}{2}} &= C_{-\frac{1}{2}}\frac{\lambda}{2}\sqrt{k(k+1)-\left(m^2-\tfrac{1}{4}\right)}. \end{aligned}\right\} \tag{26}$$

Aus der Normierungsbedingung (9) folgt schließlich:

$$\left.\begin{aligned} |C_{+\frac{1}{2}}| &= \frac{1}{\sqrt{\tfrac{1}{2}\left(\mu+\lambda m+\sqrt{\mu^2+2\lambda\mu m+\lambda^2\left(k+\tfrac{1}{2}\right)^2}\right)\cdot\sqrt{\mu^2+2\lambda\mu m+\lambda^2\left(k+\tfrac{1}{2}\right)^2}}}, \\ |C_{-\frac{1}{2}}| &= \frac{1}{\sqrt{\tfrac{1}{2}\left(-\mu-\lambda m+\sqrt{\mu^2+2\lambda\mu m+\lambda^2\left(k+\tfrac{1}{2}\right)^2}\right)\cdot\sqrt{\mu^2+2\lambda\mu+\lambda^2\left(k+\tfrac{1}{2}\right)^2}}}. \end{aligned}\right\} \tag{27}$$

Für den Spezialfall $m = \pm\left(k+\frac{1}{2}\right)$ ergibt sich naturgemäß [vgl. (21) und (22)], weil hier keine Entartung vorliegt:

$$\left.\begin{aligned} S_{+\frac{1}{2},\,+\frac{1}{2}} &= 1, \quad S_{+\frac{1}{2},\,-\frac{1}{2}} = 0, \\ S_{-\frac{1}{2},\,-\frac{1}{2}} &= 1, \quad S_{-\frac{1}{2},\,+\frac{1}{2}} = 0. \end{aligned}\right\} \tag{28}$$

Die eigentliche Berechnung der Intensitäten geschieht jetzt durch Einsetzen von (26), (27), (28) in die Transformation (5). Für p^0, q^0 sind die Lösungen des ungestörten Systems einzufügen. Wir haben dabei zu beachten, daß die Koordinaten q^0 des Elektrons Diagonalmatrizen bezüglich m_s sind.

Ferner genügt es, die Übergänge $k \to k-1$ zu betrachten, da die Übergänge $k \to k+1$ nichts Neues geben.

Aus der Arbeit von Born, Heisenberg und Jordan [l. c. Kap. 4, Gl. (33)] entnehmen wir:

$$\left.\begin{aligned} q_z^0\,(k,\ m,\ m_s;\ k-1,\ m,\ m_s) &= A\,(k)\sqrt{k^2-(m-m_s)^2}\,,\\ (q_x^0+i\,q_y^0)\,(k,\ m-1,\ m_s;\ k-1,\ m,\ m_s)&\\ &= A\,(k)\sqrt{(k-m+m_s)\,(k-m+m_s+1)}\,,\\ (q_x^0-i\,q_y^0)\,(k,\ m,\ m_s;\ k-1,\ m-1,\ m_s)&\\ &= A\,(k)\sqrt{(k+m-m_s)\,(k+m-m_s-1)}\,. \end{aligned}\right\}\qquad (29)$$

$A\,(k)$ bedeutet eine nur von k abhängige Größe.

Durch Einsetzen von (26) bis (29) in (5) erhält man durch elementares Ausrechnen die gesuchten Intensitäten. Die allgemeinen Formeln sind aber ziemlich kompliziert. Wir geben im folgenden das Resultat für den Spezialfall des D-Linientypus an, betrachten also Übergänge $k=1\rightarrow k=0$. Es ergibt sich aus (5):

$$\left.\begin{aligned}
|q_z'|^2\,(1,\ \tfrac{1}{2},\ \tfrac{1}{2};\ 0,\ \tfrac{1}{2},\ \tfrac{1}{2}) &= \tfrac{1}{2}\,|A^2(1)|\left(1+\frac{\mu+\frac{1}{2}\lambda}{\sqrt{\mu^2+\lambda\mu+\frac{9}{4}\lambda^2}}\right),\\
|q_z'|^2\,(1,\ \tfrac{1}{2},\ -\tfrac{1}{2};\ 0,\ \tfrac{1}{2},\ \tfrac{1}{2}) &= \tfrac{1}{2}\,|A^2(1)|\left(1-\frac{\mu+\frac{1}{2}\lambda}{\sqrt{\mu^2+\lambda\mu+\frac{9}{4}\lambda^2}}\right),\\
|q_z'|^2\,(1,\ -\tfrac{1}{2},\ +\tfrac{1}{2};\ 0,\ -\tfrac{1}{2},\ -\tfrac{1}{2}) &= \tfrac{1}{2}\,|A^2(1)|\left(1+\frac{-\mu+\frac{1}{2}\lambda}{\sqrt{\mu^2-\lambda\mu+\frac{9}{4}\lambda^2}}\right),\\
|q_z'|^2\,(1,\ -\tfrac{1}{2},\ -\tfrac{1}{2};\ 0,\ -\tfrac{1}{2},\ -\tfrac{1}{2}) &= \tfrac{1}{2}\,|A^2(1)|\left(1-\frac{-\mu+\frac{1}{2}\lambda}{\sqrt{\mu^2-\lambda\mu+\frac{9}{4}\lambda^2}}\right).\\
|q_x'+iq_y'|^2\,(1,\ -\tfrac{1}{2},\ \tfrac{1}{2};\ 0,\ \tfrac{1}{2},\ \tfrac{1}{2}) &= |A^2(1)|\left(1+\frac{\mu-\frac{1}{2}\lambda}{\sqrt{\mu^2-\lambda\mu+\frac{9}{4}\lambda^2}}\right),\\
|q_x'+iq_y'|^2\,(1,\ -\tfrac{1}{2},\ -\tfrac{1}{2};\ 0,\ \tfrac{1}{2},\ \tfrac{1}{2}) &= |A^2(1)|\left(1-\frac{\mu-\frac{1}{2}\lambda}{\sqrt{\mu^2-\lambda\mu+\frac{9}{4}\lambda^2}}\right),\\
|q_x'+iq_y'|^2\,(1,\ -\tfrac{3}{2},\ -\tfrac{1}{2};\ 0,\ -\tfrac{1}{2},\ -\tfrac{1}{2}) &= |A^2(1)|\,.\,2,\\
|q_x'-iq_y'|^2\,(1,\ \tfrac{3}{2},\ \tfrac{1}{2};\ 0,\ \tfrac{1}{2},\ \tfrac{1}{2}) &= |A^2(1)|\,.\,2,\\
|q_x'-iq_y'|^2\,(1,\ \tfrac{1}{2},\ \tfrac{1}{2};\ 0,\ -\tfrac{1}{2},\ -\tfrac{1}{2}) &= |A^2(1)|\,.\left(1-\frac{\mu+\frac{1}{2}\lambda}{\sqrt{\mu^2+\lambda\mu+\frac{9}{4}\lambda^2}}\right),\\
|q_x'-iq_y'|^2\,(1,\ +\tfrac{1}{2},\ -\tfrac{1}{2};\ 0,\ -\tfrac{1}{2},\ -\tfrac{1}{2}) &= |A^2(1)|\,.\left(1+\frac{\mu+\frac{1}{2}\lambda}{\sqrt{\mu^2+\lambda\mu+\frac{9}{4}\lambda^2}}\right).
\end{aligned}\right\}\qquad (30)$$

Auch diese Intensitätsformeln stimmen überein mit den aus der Voigtschen Theorie hergeleiteten (vgl. A. Sommerfeld, l. c. S. 266).

§ 4. Spezielle Behandlung der Grenzfälle $\lambda \ll \mu$ und $\mu \ll \lambda$. Um den Vergleich der empirischen Ergebnisse mit der Theorie zu erleichtern, wird es zweckmäßig sein, die Resultate der Theorie für die Grenzfälle $\lambda \ll \mu$, $\lambda \gg \mu$ besonders herzuleiten. Der Grenzfall $\lambda \ll \mu$ ist ohne weiteres aus den Rechnungen des vorigen Abschnitts zu erhalten. In erster Näherung (bis auf Größen der Ordnung λ^2) z. B. zerfällt die Determinante (16) in das Produkt der Diagonalglieder, und es wird

$$W = (H_1 + H_2)(m_s, m_s).$$

Zur Berechnung des Grenzfalles $\mu \ll \lambda$ aber sind neue Überlegungen notwendig. Wir setzen zunächst $\mu = 0$; dann wird $H_1 = 0$, und es bleibt von H_2 ein Glied proportional $\mathfrak{k}\mathfrak{s}$. Nun wird es zweckmäßig sein, den Gesamtimpuls $\mathfrak{M}$ des Atoms

$$\mathfrak{M} = \mathfrak{k} + \mathfrak{s}$$

einzuführen. Dann gilt wegen der Vertauschbarkeit von $\mathfrak{k}$ und $\mathfrak{s}$

$$\mathfrak{M}^2 = \mathfrak{k}^2 + \mathfrak{s}^2 + 2\,\mathfrak{k}\mathfrak{s}. \tag{31}$$

Da andererseits

$$\mathfrak{M}^2 = \left(\frac{h}{2\pi}\right)^2 j(j+1),$$

so wird

$$\left.\begin{aligned} \left(\frac{2\pi}{h}\right)^2 \mathfrak{k}\mathfrak{s} &= \tfrac{1}{2}\bigl(j(j+1) - k(k+1) - s(s+1)\bigr) \\ \text{und}\qquad H_2 &= \tfrac{1}{2}\lambda\bigl(j(j+1) - k(k+1) - s(s+1)\bigr). \end{aligned}\right\} \tag{32}$$

H ist als Funktion von j Diagonalmatrix.

Für kleine Werte von μ können wir jetzt das durch (32) charakterisierte System als das „ungestörte" betrachten. Im ungestörten System führt also jetzt das Atom eine Präzession um die Achse des Gesamtimpulses aus. Die Energiewerte des gestörten Systems sind gegeben durch die zeitlichen Mittelwerte von H_1 über die ungestörte Bewegung. Denkt man sich $\mathfrak{k}$ und $\mathfrak{s}$ zerlegt in je eine Komponente parallel und eine senkrecht zu $\mathfrak{M}$, so wird die letztere eben wegen jener Präzession bei der Mittelung wegfallen und nur die erstere einen Beitrag zu H_1 liefern.

Wir können diese der klassischen Mechanik entlehnte Betrachtungsweise in die Quantenmechanik übernehmen, da alle in Betracht kommenden Größen vertauschbar sind.

Es wird die in Richtung von $\mathfrak{M}$ genommene

$$(\text{Komponente von } \mathfrak{k}) = \frac{(\mathfrak{M}\,.\,\mathfrak{k})}{\mathfrak{M}^2}\cdot\mathfrak{M}$$

und die

$$(\text{Komponente von } \mathfrak{s}) = \frac{(\mathfrak{M}\,\mathfrak{s})}{\mathfrak{M}^2}\cdot\mathfrak{M},$$

also

$$\left.\begin{aligned}\overline{H_1} &= \frac{e}{2mc}\dot{\mathfrak{H}}\,.\,\mathfrak{M}\left(\frac{(\mathfrak{M}\,\mathfrak{k})}{\mathfrak{M}^2} + 2\,\frac{(\mathfrak{M}\,\mathfrak{s})}{\mathfrak{M}^2}\right)\\ &= \frac{e}{2mc}\mathfrak{H}\,.\,\mathfrak{M}\left(1 + \frac{(\mathfrak{M}\,\mathfrak{s})}{\mathfrak{M}^2}\right)\\ &\doteq \mu\,.\,m\left(1 + \frac{j(j+1) - k(k+1) + s(s+1)}{2j(j+1)}\right),\end{aligned}\right\} \tag{33}$$

schließlich allgemein für $\mu \ll \lambda$:

$$\left.\begin{aligned}\overline{H_1 + H_2} = \mu\,.\,m\left(1 + \frac{j(j+1) - k(k+1) + s(s+1)}{2j(j+1)}\right)\\ + \tfrac{1}{2}\lambda\big(j(j+1) - k(k+1) - s(s+1)\big).\end{aligned}\right\} \tag{34}$$

Gleichung (34) stimmt überein mit den Landéschen Formeln (g- und γ-Werte, „Intervallproportionen").

§ 5. Berechnung der Feinstruktur ohne Feld. Die bisherigen Rechnungen haben allgemein den Beweis erbracht, daß die Uhlenbeck-Goudsmitsche Hypothese die Zeemaneffekte sowie die Intervallproportionen in Übereinstimmung mit der Erfahrung wiedergibt.

Zur Entscheidung der Frage, ob die zugrunde gelegte Hypothese auch die absoluten Werte der Intervalle richtig wiedergibt, müssen noch die Werte von λ und H_3 ausgerechnet werden.

Es handelt sich also um die Berechnung der Mittelwerte

$$\overline{\frac{1}{r}},\quad \overline{\frac{1}{r^2}},\quad \overline{\frac{1}{r^3}}.$$

Wir werden dieser Rechnung das zweidimensionale Wasserstoffatom[1]) zugrunde legen; es gilt dann für die Energie des ungestörten Atoms:

$$H_0 = \frac{1}{2m}(p_x^2 + p_y^2) - \frac{e^2 Z}{r}.$$

$$\left.\begin{aligned} p_x x - x p_x = \frac{h}{2\pi i};\quad p_y y - y p_y = \frac{h}{2\pi i};\\ x y - y x = 0;\qquad p_x p_y - p_y p_x = 0.\end{aligned}\right\} \tag{35}$$

[1]) Die exakten Berechnungen der Mittelwerte für den dreidimensionalen Fall sind von W. Pauli ausgeführt worden und geben dasselbe Resultat wie die obigen Rechnungen.

Führt man Polarkoordinaten ein nach den Formeln:

$$\left.\begin{aligned} r^2 = x^2 + y^2; \quad p_r = m\dot{r}; \quad \varphi = \operatorname{arctg}\frac{y}{x}; \\ p_\varphi = m(x\dot{y} - y\dot{x}) = m r^2 \dot{\varphi}, \end{aligned}\right\} \tag{36}$$

so gilt

$$\left.\begin{aligned} H_0 = \frac{1}{2m}\left[p_r^2 + \frac{1}{r^2}\left(p_\varphi^2 - \frac{1}{4}\left(\frac{h}{2\pi}\right)^2\right)\right] - \frac{e^2 Z}{r}; \\ p_r r - r p_r = \frac{h}{2\pi i}; \quad p_\varphi \varphi - \varphi p_\varphi = \frac{h}{2\pi i}; \\ r\varphi - \varphi r = 0; \quad p_r p_\varphi - p_\varphi p_r = 0. \end{aligned}\right\} \tag{36 a}$$

Nach der mehrfach zitierten Arbeit Quantenmechanik II [S. 600, Gleichung (17)] ist p_φ gequantelt

$$p_\varphi = m_0 \frac{h}{2\pi},$$

wobei wir, um mit Paulis Ergebnissen (l. c.) in Einklang zu kommen, m_0 als halbzahlig voraussetzen; und zwar wird $m_0 - \frac{1}{2}$ mit dem oben eingeführten k identisch. Die Hamiltonsche Funktion hat nämlich beim dreidimensionalen Problem nach Pauli die Form

$$H_0 = \frac{1}{2m}\left(p_r^2 + \frac{1}{r}\mathfrak{k}^2\right) - \frac{e^2 Z}{r}. \tag{37}$$

Will man (36 a) und (37) zur Übereinstimmung bringen, so folgt:

$$\mathfrak{k}^2 = \left(\frac{h}{2\pi}\right)^2 k(k+1) = p_\varphi^2 - \frac{1}{4}\left(\frac{h}{2\pi}\right)^2 = \left(\frac{h}{2\pi}\right)^2\left(m_0^2 - \frac{1}{4}\right).$$

Der Mittelwert $\overline{\frac{1}{r}}$ ergibt sich nun zunächst aus der Gleichung [Quantenmechanik II, S. 577, Gleichung (17)]:

$$-\overline{\frac{Z e^2}{r}} = \overline{E_{\text{pot}}} = -2\,\overline{E_{\text{kin}}} = 2\,W_0. \tag{38}$$

Hierin ist

$$W_0 = H_0 = -\frac{R h Z^2}{n^2},$$

wo n eine ganze Zahl bedeutet. Ferner schließt man nach Pauli aus den Bewegungsgleichungen:

$$\frac{d}{dt}p_r = -\frac{\partial H_0}{\partial r} = -\frac{1}{m r^3}\left[p_\varphi^2 - \frac{1}{4}\left(\frac{h}{2\pi}\right)^2\right] + \frac{e^2 Z}{r^2}; \tag{39}$$

also durch zeitliche Mittelung:

$$\overline{\frac{e^2 Z}{r^2}} = \overline{\frac{1}{m r^3}} \cdot \left[p_\varphi^2 - \frac{1}{4}\left(\frac{h}{2\pi}\right)^2\right]. \tag{40}$$

Schließlich gilt nach (36)

und

$$\left.\begin{aligned} m r^2 \dot{\varphi} &= p_\varphi \\ \overline{\frac{p_\varphi}{m r^2}} &= \overline{\dot{\varphi}}. \end{aligned}\right\} \tag{41}$$

Denkt man sich nun die zur Hauptquantenzahl n bzw. zu der ihr entsprechenden Wirkungsvariablen $J = n h$ konjugierte Winkelvariable[1]) eingeführt, so wird, analog der klassischen Theorie, gelten

$$\overline{\dot{\varphi}} = 2\pi \dot{w} = -2\pi \frac{\partial H_0}{\partial J} = +\frac{4\pi R Z^2}{n^3}. \tag{42}$$

Aus den Gleichungen (38), (40), (41) und (42) ergibt sich schließlich, unter Benutzung der Beziehung $p_\varphi = \frac{h}{2\pi}(k + \frac{1}{2})$:

$$\left.\begin{aligned} \overline{\frac{1}{r}} &= \frac{1}{e^2} \cdot \frac{2 R h Z}{n^2}; \\ \overline{\frac{1}{r^2}} &= \frac{m}{p_\varphi} \cdot \overline{\dot{\varphi}} = \frac{m}{\frac{h}{2\pi}(k+\frac{1}{2})} \cdot \frac{4\pi R Z^2}{n^3} = \frac{8\pi^2 m R Z^2}{h(k+\frac{1}{2}) n^3}; \\ \overline{\frac{1}{r^3}} &= \overline{\frac{1}{r^2}} \cdot \frac{e^2 Z m}{p_\varphi^2 - \frac{1}{4}\left(\frac{h}{2\pi}\right)^2} = \frac{m^2 e^2 R Z^3 . 32\pi^4}{k(k+\frac{1}{2})(k+1) n^3 h^3}. \end{aligned}\right\} \tag{43}$$

Bei Abwesenheit eines äußeren Magnetfeldes wird also nach (2) bis (4), (34) und (43) die gesamte Störungsenergie gegeben durch:

$$\left.\begin{aligned} H_2 + H_3 &= \frac{1}{2}\frac{e^2 Z}{m^2 c^2} \cdot \frac{4\pi^2 m^2 e^2 R Z^3 \big(j(j+1) - k(k+1) - s(s+1)\big)}{h k (k+\frac{1}{2})(k+1) n^3} \\ &\qquad - \frac{1}{2 m c^2}\left(-\frac{3 R^2 h^2 Z^4}{n^4} + \frac{8\pi^2 m e^4 R Z^4}{h(k+\frac{1}{2}) n^3}\right) \\ &= \frac{2 R^2 h^2 Z^4}{n^3 m c^2}\left(\frac{j(j+1) - k(k+1) - s(s+1)}{2k(k+\frac{1}{2})(k+1)} \right. \\ &\qquad \left. - \frac{1}{k+\frac{1}{2}} + \frac{3}{4n}\right). \end{aligned}\right\} \tag{44}$$

Für Dublettatome ($s = \frac{1}{2}$), d. h. für die Spektren des Wasserstoffs, der Alkalien und für die Röntgenspektren, ist eine genaue empirische Prüfung dieser Formel möglich: Erfahrungsgemäß fallen hier (bei Ver-

[1]) Die Berechtigung zur Einführung solcher Variablen ist durch die Arbeit von Born und Wiener (ZS. f. Phys. **36**, 174, 1926) und die oben angeführte Arbeit von Dirac erwiesen.

nachlässigung der Wechselwirkung der Elektronen aufeinander) zwei Energieniveaus von verschiedenem k, aber gleichem j zusammen. Der Abstand zweier Niveaus von gleichem k und verschiedenem j wird durch die Sommerfeldsche Feinstrukturformel gegeben.

In Gleichung (44) sind die Werte $s = \frac{1}{2}$, $j = k \pm \frac{1}{2}$ einzusetzen; es ergibt sich

$$\left.\begin{aligned}
&\text{für } k = j - \tfrac{1}{2}:\\
&H_2 + H_3 = \frac{2R^2h^2Z^4}{n^3mc^2}\left(\frac{1}{2j(j+\frac{1}{2})} - \frac{1}{j} + \frac{3}{4n}\right)\\
&\qquad\quad = \frac{2R^2h^2Z^4}{n^3mc^2}\left(-\frac{1}{j+\frac{1}{2}} + \frac{3}{4n}\right);\\
&\text{für } k = j + \tfrac{1}{2}:\\
&H_2 + H_3 = \frac{2R^2h^2Z^4}{n^3mc^2}\left(-\frac{1}{2(j+\frac{1}{2})(j+1)} - \frac{1}{j+1} + \frac{3}{4n}\right)\\
&\qquad\quad = \frac{2R^2h^2Z^4}{n^3mc^2}\left(-\frac{1}{j+\frac{1}{2}} + \frac{3}{4n}\right).
\end{aligned}\right\} \tag{45}$$

Also allgemein für $s = \frac{1}{2}$:

$$H_2 + H_3 = \frac{2R^2h^2Z^4}{n^3mc^2}\left(-\frac{1}{j+\frac{1}{2}} + \frac{3}{4n}\right). \tag{46}$$

Die Formel (46) gibt die Erfahrungstatsachen vollständig wieder. Insbesondere folgt aus dem Nichtauftreten von k in Gleichung (46), daß die „Abschirmungsdubletts“ durch die Uhlenbeck-Goudsmitsche Theorie erklärt werden. Ferner stimmen die Aufspaltungen der magnetischen Dubletts mit den aus der Sommerfeldschen Feinstrukturformel gewonnenen überein.

Obwohl die Frage, wie weit die Grundannahmen (2) bis (4) der hier dargestellten Theorie frei von Willkür sind, noch nicht entschieden werden kann, so wird man die Ergebnisse unserer Rechnungen doch als wichtige Stütze für die Compton-Uhlenbeck-Goudsmitsche Hypothese einerseits, für die Quantenmechanik andererseits ansehen können.

Zeitschrift für Physik 39, 499–518 (1926)

Über die Spektra von Atomsystemen mit zwei Elektronen.

Von W. Heisenberg in Kopenhagen.

(Eingegangen am 24. Juli 1926.)

Als spezielles Beispiel für die Quantenmechanik der Mehrkörperprobleme, zu deren Behandlung durch das Studium eines charakteristischen Resonanzphänomens die Grundlage gegeben ist, sollen in dieser Arbeit die Atomsysteme mit zwei Elektronen und deren Spektra untersucht werden. Es ergibt sich eine befriedigende Darstellung der wesentlichsten Eigenschaften dieser Spektra.

Das Heliumatom und die ihm verwandten Ionen Li^+, Be^{++} usw. sind mehrfach Gegenstand eingehender Untersuchungen vom Standpunkt der früheren Quantentheorie aus gewesen[1]). Die Behandlung dieses Problems nach der neueren Quantenmechanik ist bisher nicht durchgeführt worden. Als Grundlage für diese Behandlung sehen wir die Kenntnis eines für die Mehrkörperprobleme der Quantenmechanik charakteristischen Resonanzphänomens an, das vor kurzem vom Verfasser in dieser Zeitschrift behandelt worden ist[2]). Die folgenden Rechnungen sollen eine Übersicht über die Bewegungen der Elektronen im genannten Atom geben, soweit man quantenmechanisch von Bewegungen sprechen kann, und sie sollen das Spektrum von He und Li^+ qualitativ und in grober Näherung quantitativ ableiten aus den Gesetzen der Quantenmechanik. Das Verfahren, das bei diesen Rechnungen benutzt wurde, kann als konsequente Anwendung der einfachsten Störungstheorie bezeichnet werden. Die Reihen der Störungstheorie konvergieren jedoch gut nur für die hochangeregten Zustände $(D, d, F, f, \ldots)$ des Atoms; für die (P, p)-Terme schon wird die hier durchgerechnete erste Näherung dieser Reihen recht ungenau, für die (S, s)-Terme oder den Normalzustand versagt sie ganz. Um zu einer scharfen quantitativen Berechnung des Spektrums zu kommen, wird man verfeinerte Methoden ausbilden müssen, die schnell konvergente Reihen liefern, etwa ähnlich der Kramersschen Behandlung des He-Normalzustandes auf Grund der früheren Quantentheorie. Das Ziel dieser Arbeit ist jedoch nur eine Ableitung derjenigen Resultate, die sich bei systematischer Anwendung der gewöhnlichen Störungstheorie gewinnen lassen.

---

1) A. Landé, Phys. ZS. **20**, 228, 1919; N. Bohr, Göttinger Vorträge (unpubliziert); H. A. Kramers, ZS. f. Phys. **13**, 312, 1923; J. H. van Vleck, Phys. Rev. **21**, 372, 1923; M. Born und W. Heisenberg, ZS. f. Phys. **26**, 216, 1924.

2) W. Heisenberg, Mehrkörperproblem und Resonanz in der Quantenmechanik, ZS. f. Phys. **38**, 411, 1926.

Zeitschrift für Physik 39, 499–518 (1926)

I. Berechnung des Spektrums ohne Berücksichtigung der Elektronenmagnete.

§ 1. Entsprechend den Rechnungen in der oben angeführten Arbeit des Verfassers muß zuerst das Verhalten des „ungestörten" Atoms bekannt sein, bei dem die Wechselwirkung zwischen den Elektronen nicht berücksichtigt ist. Man könnte also, um völlig konsequent zu sein, als ungestörtes System ein Atom mit zwei Elektronen betrachten, bei dem beide Elektronen unter dem Einfluß der Kernladung Z stehen, ohne aufeinander eine Wechselwirkung auszuüben. Ein solches Vorgehen wäre aber sehr unzweckmäßig, denn eben bei den angeregten Zuständen des Atoms wirkt näherungsweise immer auf das eine Elektron die Kernladung Z, auf das andere wegen der Abschirmung die Kernladung $Z - 1$. In der klassisch-mechanischen Rechnung war daher das naturgemäße Ausgangssystem ein Atom, bei dem auf ein bestimmtes „inneres" Elektron die Kernladung Z, auf das andere die Kernladung $Z - 1$ wirkte. In der Quantenmechanik würde ein solches Ausgangssystem keine vernünftige Annäherung darstellen, da ja die beiden Elektronen periodisch den Platz tauschen, weshalb einmal das eine, einmal das andere Elektron unter dem Einfluß der Ladung Z oder $Z - 1$ steht. Vielmehr scheint für die Quantenmechanik am meisten geeignet ein Ausgangssystem, bei dem für jedes Elektron in der Nähe des Kerns die Ladung Z, in einiger Entfernung die Ladung $Z - 1$ für die Kraft maßgebend ist. Man denke sich also um den Kern eine gleichmäßig geladene Kugelschale (Gesamtladung e, Radius r_0) gelegt. Dann ist im ungestörten Atom die potentielle Energie eines Elektrons:

$$-\frac{e^2 Z}{r} + f(r), \text{ wo } f(r) = \begin{cases} \dfrac{e^2}{r_0} \text{ von } r = 0 \text{ bis } r = r_0 \\ \dfrac{e^2}{r} \text{ von } r = r_0 \text{ bis } r = \infty \end{cases}. \tag{1}$$

Der hier eingehende Radius r_0 muß nachträglich so bestimmt werden, daß die Reihen der Störungstheorie möglichst gut konvergieren.

Für eine genaue quantitative Rechnung der Termwerte wäre es also jetzt nötig, zu dem durch die potentielle Energie (1) definierten mechanischen System die Energiewerte und die Schrödingerschen Eigenfunktionen genau zu bestimmen, was ohne mathematische Schwierigkeiten durchgeführt werden kann. Für unsere Näherungsrechnung können wir diese Bestimmung aber ganz entbehren, da die Lösung von (1) praktisch genau genug übereinstimmt mit der Lösung des durch die Kernladung Z bzw. $Z - 1$ definierten mechanischen Systems, je nach der Lage von r_0

relativ zum „Bahnradius". Denken wir zum Beispiel an eine p-Bahn. Nach der klassischen Theorie wäre die Lösung von (1) identisch mit der eines Systems der Ladung $Z - 1$, solange $r_0 < 2 \frac{a}{Z-1}$ ist $\left(a = \frac{h^2}{4\pi^2 m e^2}\right)$. In der Quantenmechanik besteht diese Identität nicht mehr, doch sind die Unterschiede zwischen beiden Lösungen in mehrfacher Hinsicht sehr gering. Bei den Lösungen von (1), bei denen — um die einfache aus der früheren Theorie entlehnte Bezeichnungsweise zu gebrauchen — der Bahnradius wesentlich kleiner als r_0 ist, muß beachtet werden, daß die Energie sich um das additive Glied e^2/r_0 von der Energie des durch die Kernladung Z charakterisierten Systems unterscheidet.

Zur Vereinfachung der folgenden Rechnung können wir ferner annehmen, daß im ungestörten System die Projektion des Gesamtimpulses auf eine vorgegebene äußere Richtung und der Gesamtimpuls selbst „gequantelt", d. h. Diagonalmatrizen seien. Es werden also Zusatzfelder fingiert, die schon im ungestörten System die Entartung von j und m aufheben; dies ist berechtigt, da die Quantelung von j durch die Störungsenergie, die von m durch ein beliebig kleines äußeres Feld doch eintreten wird. Im ungestörten System ist dann noch im allgemeinen eine dreifache Entartung vorhanden, neben der Resonanzentartung die der Elektronenimpulse k_1 und k_2 [1]). In dem speziellen Falle jedoch, in dem das eine Elektron sich in der Grundbahn $n = 1$ befindet, fallen die beiden Impulsentartungen fort und es bleibt nur noch die Resonanzentartung übrig; dies folgt aus einer einfachen Abzählung der ungestörten Zustände mit gleicher Energie. Empirisch sind nur Terme von diesem letztgenannten Typus bekannt. Wir dürfen uns daher auch mit der Berechnung der Terme dieser Art begnügen.

Im ungestörten System (1) befindet sich also das eine Elektron in einer 1 S-Bahn. Schließen wir diejenigen Zustände des Gesamtsystems, bei denen das andere Elektron in einer S-Bahn läuft, aus — für solche Zustände ist die hier versuchte Näherungsmethode doch nicht brauchbar —, so wird r_0 stets erheblich größer als der „Radius der 1 S-Bahn" gewählt werden können. Die Energie dieses einen Elektrons ist also in erster Näherung gegeben durch $-\frac{RhZ^2}{1^2} + \frac{e^2}{r_0}$, in nächster Näherung durch

$$-\frac{RhZ^2}{1^2} + \frac{e^2}{r_0} + \overline{\left(f(r) - \frac{e^2}{r_0}\right)} \qquad (2)$$

[1]) Tatsächlich führen Lösungen von (1), die gleichen n-, doch verschiedenen k-Werten entsprechen, zu etwas verschiedenen Energien. Diese Energieunterschiede sind aber im allgemeinen so klein, daß es konsequenter erscheint, sie zur Störungsenergie zu rechnen und das ungestörte System als entartet zu betrachten.

Zeitschrift für Physik *39*, 499–518 (1926)

Der Strich bedeutet zeitliche Mittelung, genommen über die ungestörte Bewegung in einer Bahn für $n = 1$ und Ladung Z. Analog wird die Energie des anderen Elektrons im ungestörten System bestimmt durch

$$-\frac{Rh(Z-1)^2}{n^2} + \overline{\left(f(r) - \frac{e^2}{r}\right)}. \tag{3}$$

r_0 ist hier wesentlich kleiner als der Bahnradius. Gemittelt wird über eine durch die Hauptquantenzahl n, einen bestimmten Impuls k und die Ladung $(Z - 1)$ charakterisierte Bahn. Vertauschung der beiden Elektronen gibt die gleiche Gesamtenergie.

Die Störungsenergie ist gegeben durch die Coulombsche Abstoßung der beiden Elektronen, vermindert um die Abweichungen des Potentials (1) vom Potential eines Kerns mit der Ladung Z. Dies ergibt

$$H^1 = \frac{e^2}{r_{12}} - f(r_1) - f(r_2). \tag{4}$$

r_{12} bedeutet den Abstand zwischen beiden Elektronen, r_1 bzw. r_2 den Abstand des Elektrons 1 bzw. 2 vom Kern.

§ 2. Betrachten wir nun einen ganz speziellen Zustand des ungestörten Atoms, charakterisiert durch den Gesamtimpuls j und dessen Projektion m auf eine feste Achse; das eine Elektron befinde sich in einer Bahn $n_1 = 1$, das andere in einer Bahn $n_2 = n \geqq 2$. Der Drehimpuls k seiner Bahn ist, da das andere Elektron in einer S-Bahn läuft, gleich dem Gesamtimpuls des Atoms:

$$k = j. \tag{5}$$

Das System ist entartet nur noch insofern, als die Vertauschung der beiden Elektronen die gleiche Gesamtenergie ergibt und kann ganz nach der l. c. angegebenen Methode behandelt werden. Bezeichnet man den durch die Quantenzahlen $n = 1$, $k = 0$, $m = 0$ charakterisierten Zustand mit v, den durch $n = n$, $k = j$, $m = m$ charakterisierten Zustand mit w und bezeichnet ferner durch den ersten bzw. zweiten Index den Zustand des ersten bzw. zweiten Elektrons, so beruht nach den Überlegungen l. c., Gleichungen (10) bis (12), die Berechnung der säkularen Bewegungen im wesentlichen auf der Ausrechnung der Glieder

$$H^1(v\,w;\, v\,w) \quad \text{und} \quad H^1(v\,w;\, w\,v). \tag{6}$$

Diese beiden Glieder werden von m nicht abhängen, da ja ohne äußeres Feld keine Richtung im Raum ausgezeichnet sein kann. Also spaltet jeder Term des ungestörten Systems auf in zwei Terme; nur diejenigen Terme des ungestörten Systems, bei denen die Elektronen in äquivalenten Bahnen waren, bleiben einfach. Wir nennen die beiden

Termsysteme, in die das Termspektrum zerfällt, O(rtho)-System und P(ara)-System und bezeichnen mit diesem letzten Namen dasjenige System, das Zustände mit äquivalenten Bahnen enthält. Durch Berechnung der Werte (6) müssen wir zeigen, daß die Energiewerte des P-Systems größer sind als die des O-Systems, und daß der berechnete Abstand zwischen beiden dem empirischen Abstand von Ortho- und Parheliumtermen bzw. Ortho- und Paralithiumtermen näherungsweise entspricht. Daß die beiden Termsysteme nicht kombinieren, wurde l. c. bewiesen.

Die Berechnung der Größen (6) geschieht am einfachsten nach der Schrödingerschen Methode der Eigenfunktionen[1]). Bei einem wasserstoffähnlichen Atom der Kernladung Z läßt sich die zu den Quantenzahlen n, k, m gehörige Schrödingersche Eigenfunktion ψ darstellen als Produkt dreier Funktionen, von denen jede nur von einer der drei Variabeln r, ϑ, φ abhängt:

$$\psi_{n,k,m} = X_{nk}(r) \cdot Y_{km}(\vartheta)\, Z_m(\varphi). \tag{7}$$

Bezeichnet man mit $d\,\Omega$ das Raumelement $d\,r\,r^2\,d\,\vartheta \sin\vartheta\,d\,\varphi$, mit s, t irgend zwei „stationäre Zustände" oder Eigenwerte, so gilt bei geeigneter Normierung von ψ die Orthogonalitätsrelation:

$$\int \psi_s \overline{\psi}_t\, d\,\Omega = \delta_{st} = \begin{cases} 1 \text{ für } s = t \\ 0 \text{ für } s \neq t. \end{cases} \tag{7a}$$

Der Strich über ψ_t bedeutet: konjugiert komplexer Wert. Die Funktionen X, Y, Z lauten nach Schrödinger[2]):

$$\left.\begin{aligned}
X_{n,k}(r) &= \sqrt{\left(\frac{2Z}{na}\right)^3 \frac{(n-k-1)!}{(n+k)!\,2n}} \cdot e^{-\frac{\xi}{2}}\, \xi^k \sum_0^\infty \frac{(-\xi)^s}{s!} \binom{n+k}{n-k-1-s}, \\
&\qquad \text{wo } a = \frac{h^2}{4\pi^2 m e^2}, \quad \xi = \frac{2Z}{na} \cdot r; \\
Y_{k,m}(\vartheta) &= \sqrt{\left(k+\frac{1}{2}\right)\frac{(k-m)!}{(k+m)!}}\, P_{km}(\cos\vartheta), \text{ wo in bekannter Weise} \\
&\qquad P_{km}(t) = \sqrt{1-t^2}^{\,m} \frac{d^m}{dt^m} P_k(t), \\
&\qquad P_k(t) = \frac{1}{2^k k!} \frac{d^k}{dt^k} (1-t^2)^k; \\
Z_m(\varphi) &= \frac{e^{im\varphi}}{\sqrt{2\pi}}.
\end{aligned}\right\} \tag{8}$$

---

[1]) E. Schrödinger, Ann. d. Phys. **79**, 361, 489, 734, 1926.

[2]) Derselbe, ebenda S. 361. Die Normierungskonstanten entnehme ich einer Arbeit von I. Waller, ZS. f. Phys. **38**, 635, 1926.

Zeitschrift für Physik 39, 499–518 (1926)

Die Berechnung der Größen (6) geschieht jetzt nach den Formeln:

$$\left.\begin{aligned} \text{a)}\quad H^1(v\,w,\,v\,w) &= \int\cdots\int H^1\,\psi_{v,Z}(r_1,\vartheta_1,\varphi_1)\,\psi_{w,Z-1}(r_2,\vartheta_2,\varphi_2) \\ &\quad \overline{\psi}_{v,Z}(r_1,\vartheta_1,\varphi_1)\,\overline{\psi}_{w,Z-1}(r_2,\vartheta_2,\varphi_2)\,d\Omega_1\,d\Omega_2, \\ \text{b)}\quad H^1(v\,w,\,w\,v) &= \int\cdots\int H^1\,\psi_{v,Z}(r_1,\vartheta_1,\varphi_1)\,\psi_{w,Z-1}(r_2,\vartheta_2,\varphi_2) \\ &\quad \overline{\psi}_{w,Z-1}(r_1,\vartheta_1,\varphi_1)\,\overline{\psi}_{v,Z}(r_2,\vartheta_2,\varphi_2)\,d\Omega_1\,d\Omega_2. \end{aligned}\right\} \quad (9)$$

Hierbei wird also angenommen, daß die Eigenfunktionen des Problems (1) von den wasserstoffähnlichen Eigenfunktionen nicht sehr verschieden seien. Von diesen beiden Werten gibt der zweite: $H^1(v\,w,\,w\,v)$ nach den l. c. ausgeführten Überlegungen den halben Termabstand zwischen O- und P-System im „gestörten" System an. Der erste Wert: $H^1(v\,w,\,v\,w)$ gibt den Abstand der Mitte zwischen Ortho- und Paraterm vom Termwert des ungestörten Systems an. Es scheint zweckmäßig, zu diesem Abstand die Abweichung der Energie des ungestörten Atoms (2) und (3) vom Wert $-\frac{R h Z^2}{1^2} - \frac{R h (Z-1)^2}{n^2}$ zu addieren. Dann erhält man den Abstand der genannten Termmitte von $-\frac{R h Z^2}{1^2} - \frac{R h (Z-1)^2}{n^2}$ als

$$H^{1\prime}(v\,w,v\,w) = \int\cdots\int\left(\frac{e^2}{r_{12}} - \frac{e^2}{r_2}\right) |\,\psi^2_{v,Z}(r_1,\vartheta_1,\varphi_1)\,| \cdot |\,\psi^2_{w,Z-1}(r_2,\vartheta_2,\varphi_2)\,|\,d\Omega_1\,d\Omega_2. \quad (10)$$

In dieser Formel kommt r_0 nicht mehr vor. Die Energieänderung des gestörten Systems erster Ordnung ist also von der Wahl von r_0 unabhängig. Allerdings kann die Formel (10) für den Abstand der genannten Termmitte von $-\left(\frac{R h Z^2}{1^2} + \frac{R h (Z-1)^2}{n^2}\right)$ keinen auch nur näherungsweise richtigen Wert liefern. Wegen der Symmetrie der 1 S-Bahn verschwindet nämlich dieser Wert (10) nahezu, und die quadratischen Glieder der nächsten Ordnung der Störungsrechnung, die der Wirkung der sogenannten „Polarisation" entsprechen, geben einen größeren Beitrag zu dem genannten Abstand als das Glied (10). Dagegen stellt für den Abstand zwischen O- und P-System die Formel (9) wirklich die erste Approximation dar.

Bei der numerischen Ausrechnung der Größen (6) wird es zweckmäßig sein, in (9) zuerst die Integration über die Winkel und dann die über die Radien auszuführen. Es ergibt sich zunächst nach (8):

$$\left.\begin{aligned}&\iint\left(\frac{e^2}{r_{12}}-\frac{e^2}{r_2}\right)Y_{0,0}^2(\vartheta_1)\,Z_{0,0}^2(\varphi_1)\,d\varphi_1\,d\vartheta_1\sin\vartheta_1\\ &=\iint\left(\frac{e^2}{\sqrt{r_1^2+r_2^2-2\,r_1 r_2\cos(1,2)}}-\frac{e^2}{r_2}\right)\frac{1}{4\pi}\sin\vartheta_1\,d\vartheta_1\,d\varphi_1\\ &=\frac{e^2}{r_{(1,2)}}-\frac{e^2}{r_2},\ \text{wo für } r_{(1,2)} \text{ gilt: } r_{(1,2)}=\begin{cases}r_1 \text{ für } r_1>r_2\\ r_2 \text{ für } r_2>r_1.\end{cases}\end{aligned}\right\}\quad(11)$$

Die Integration über die Winkel ϑ_2, φ_2 ergibt den Faktor 1, und es bleibt

$$\left.\begin{aligned}&H^{1'}(v\,w,\,v\,w)\\ &=\iint d\,r_1 r_1^2\,d\,r_2 r_2^2\,X_{1,0;Z}^2(r_1)\,X_{n,k;Z-1}^2(r_2)\left(\frac{e^2}{r_{(1,2)}}-\frac{e^2}{r_2}\right).\end{aligned}\right\}\quad(12)$$

Bei Einführung der Abkürzung $\xi=\frac{2Z}{a}r$ ergibt sich

$$\left.\begin{aligned}H^{1'}(v\,w,\,v\,w)&=\int_0^\infty d\,r_2 r_2^2\,X_{n,k;Z-1}^2(r_2)\\ &\quad\cdot\frac{1}{2}\int_0^\infty e^{-\xi_1}\xi_1^2\,d\,\xi_1\left(\frac{e^2}{\xi_{(1,2)}}-\frac{e^2}{\xi_2}\right)\frac{2Z}{a}\\ &=-\frac{Z e^2}{a}\int_0^\infty d\,r_2 r_2^2\,\frac{e^{-\xi_2}(2+\xi_2)}{\xi_2}\,X_{n,k;Z-1}^2(r_2).\end{aligned}\right\}\quad(13)$$

In den folgenden Rechnungen sollen die gesuchten Mittelwerte nicht für alle n und k exakt ausgerechnet werden; es läßt sich nämlich leicht zeigen, daß die gesuchten Mittelwerte stets zu Energieformeln vom Rydberg-Ritzschen Typus (näherungsweise) Anlaß geben. Es genügt daher, bei einem gegebenen Werte von k den Mittelwert für die beiden kleinsten Werte von n ($k+1$ und $k+2$) auszurechnen und die übrigen Mittelwerte nach der Rydberg-Ritzschen Formel zu extrapolieren. Es seien daher die aus (8) hervorgehenden Formeln angeführt:

$$\left.\begin{aligned}X_{k+1,k}&=\sqrt{\left(\frac{2Z}{(k+1)a}\right)^3\frac{1}{(2k+2)!}}\,e^{-\frac{\xi}{2}}\xi^k,\ \text{wo}\ \xi=\frac{2Z}{(k+1)a}r,\\ X_{k+2,k}&=\sqrt{\left(\frac{2Z}{(k+2)a}\right)^3\frac{1}{(2k+2)!\,(2k+4)}}\,e^{-\frac{\xi}{2}}\xi^k(2k+2-\xi),\\ &\qquad\xi=\frac{2Z}{(k+2)a}r.\end{aligned}\right\}(14)$$

Beachtet man, daß für den Zustand n, k die Ladung $Z - 1$ maßgebend ist, und setzt man $\varepsilon = \frac{Z-1}{Z \cdot n}$, so ergibt sich

$$\left.\begin{aligned} & H^{1'}_{n=k+1}(vw, vw) \\ & = -\frac{Ze^2}{a} \cdot \varepsilon^{3+2k} \frac{1}{(2k+2)!} \int_0^\infty d\xi_2 \, e^{-\xi_2(1+\varepsilon)} \xi_2^{2k+1} (2+\xi_2) \\ & = -2Rh \cdot Z \frac{(Z-1)^{2k+3}}{(Z(k+2)-1)^{2k+3}} \left(1 + \frac{Z(k+2)-1}{Z(k+1)^2}\right) \end{aligned}\right\} \quad (15)$$

und

$$\left.\begin{aligned} & H^{1'}_{n=k+2}(vw, vw) = -\frac{Ze^2}{a} \varepsilon^{3+2k} \\ & \cdot \frac{1}{(2k+2)!(2k+4)} \int_0^\infty d\xi_2 \, e^{-\xi_2(1+\varepsilon)} \xi_2^{2k+1} (2+\xi_2)(2k+2-\xi_2\varepsilon)^2 \\ & = -2ZRh \frac{(Z-1)^{2k+3}}{(Z(k+3)-1)^{2k+3}} \frac{4(k+1)(k+2)+(2k+6)\varepsilon^2+2\varepsilon^3}{(1+\varepsilon)^2 \cdot (2k+4)}. \end{aligned}\right\} \quad (16)$$

Die numerische Rechnung ergibt für $\frac{H^{1'}(vw, vw)}{Rh}$:

Tabelle 1.

		$k = 1$	$k = 2$
$Z = 2$	$n = k + 1$ . .	— 0,0020	— $6{,}7 \cdot 10^{-6}$
	$n = k + 2$ . .	— 0,00070	—
$Z = 3$	$n = k + 1$ . .	— 0,0098	— $5{,}5 \cdot 10^{-5}$
	$n = k + 2$ . .	— 0,0032	—

Von den Mittelwerten für $k = 2$ ist nur je einer angegeben, da sich herausstellt, daß die zum Wert $k = 2$ gehörigen Änderungen von $H^{1'}$ durch die Ritzkorrektion in der hier angestrebten Approximation vernachlässigt werden können. Für Werte von $k \geqslant 3$ gibt $H^{1'}$ keinen merklichen Beitrag zur Rydbergkorrektion. Der von $H^{1'}$ herrührende Beitrag zur Termkorrektion δ, definiert durch $E = -\frac{RhZ^2}{1^2} - \frac{Rh(Z-1)^2}{(n+\delta)^2}$, wird:

Tabelle 2.

		$k = 1$	$k = 2$
$Z = 2$	$n = 2$	— 0,0080	— $9 \cdot 10^{-5}$
	$n = 3$	— 0,0095	—
$Z = 3$	$n = 2$	— 0,0098	— $1{,}9 \cdot 10^{-4}$
	$n = 3$	— 0,0108	—

Das Ritzsche Glied ist sehr klein im Verhältnis zur ganzen Korrektion.

Wir gehen nun zur Berechnung der Größe $H^1(vw, wv)$ über und nehmen zuerst die Integration über die Winkel vor. H^1 kann bei dieser Integration ersetzt werden durch das Glied $\frac{e^2}{r_{12}}$, da alle anderen Glieder nur von den Koordinaten je eines Elektrons abhängen, also zu einem Fourierglied, das Übergängen beider Elektronen entspricht, keinen Beitrag geben können.

Sei x eine Größe < 1, so handelt es sich also zunächst um das Integral

$$\left.\frac{1}{4\pi}\oint\oint\frac{Y_{km}(\vartheta_1)Y_{km}(\vartheta_2)Z_m(\varphi_1)Z_m(\varphi_2)\,d\vartheta_1\sin\vartheta_1\,d\varphi_1\,d\vartheta_2\sin\vartheta_2\,d\varphi_2}{\sqrt{1-2x(\cos\vartheta_1\cos\vartheta_2+\sin\vartheta_1\sin\vartheta_2\cos(\varphi_1-\varphi_2)+x^2}}\right\}\quad(17)$$

Dieses Integral ist wegen (8) zurückführbar auf das bei Kugelfunktionentwicklungen bekannte Poissonsche Integral[1]) und hat den Wert

$$\frac{x^k}{2k+1}. \quad (18)$$

Bezeichnet man wieder mit $r_{(1,2)}$ den größeren, mit $r_{(2,1)}$ den kleineren der beiden Werte r_1 und r_2, so wird also

$$\left.\begin{aligned}&H^1(vw, wv)\\&=\int_0^\infty\int_0^\infty dr_1\,r_1^2\,dr_2\,r_2^2\,\frac{e^2\,r_{(2,1)}^k}{r_{(1,2)}^{k+1}}\cdot X_{nk}(r_1)X_{10}(r_1)X_{nk}(r_2)X_{10}(r_2).\end{aligned}\right\}\quad(19)$$

Für den Spezialfall $n = k+1$ ergibt sich, wenn wieder $\xi = \frac{2Z}{a}r$ und $\varepsilon = \frac{Z-1}{Z(k+1)}$ gesetzt wird:

$$\left.\begin{aligned}&H^1(vw, wv)\\&=\frac{2e^2Z}{a}\cdot\frac{\varepsilon^{2k+3}}{2.(2k+2)!}\,\frac{1}{(2k+1)}\int_0^\infty\int_0^\infty\frac{\xi_{(2,1)}^k}{\xi_{(1,2)}^{k+1}}\,\xi_1^k e^{-\xi_1\frac{1+\varepsilon}{2}}\,\xi_2^k e^{-\xi_2\frac{1+\varepsilon}{2}}\,d\xi_1\,\xi_1^2\,d\xi_2\,\xi_2^2\\&=\frac{e^2Z}{a}\cdot\frac{\varepsilon^{2k+3}}{(2k+2)!\,(2k+1)}\int_0^\infty d\xi_1\,\xi_1\,e^{-\xi_1\cdot\frac{1+\varepsilon}{2}}\left\{\int_0^{\xi_1}d\xi_2\,\xi_2^{2k+2}\,e^{-\xi_2\frac{1+\varepsilon}{2}}\right.\\&\qquad\left.+\,\xi_1^{2k+1}\int_{\xi_1}^\infty d\xi_2\,\xi_2\,e^{-\xi_2\frac{1+\varepsilon}{2}}\right\}.\end{aligned}\right\}\quad(20)$$

[1]) Vgl. z. B. Courant-Hilbert, Methoden der mathematischen Physik, Berlin 1924, 1, 423.

Zeitschrift für Physik *39*, 499–518 (1926)

Die Ausführung dieses Integrals ergibt $\left(\varepsilon = \frac{Z-1}{(k+1)Z}\right)$:

$$H^1_{n=k+1}(vw, wv) = Rh \,.\, Z \frac{8\,\varepsilon^{2k+3}}{(1+\varepsilon)^{2k+5}} \frac{2k+5}{2k+1}. \tag{21}$$

Analog folgt für $n = k+2 \left(\varepsilon = \frac{Z-1}{Z(k+2)}\right)$:

$$H^1_{n=k+2}(vw, wv) = \frac{e^2 Z}{a} \frac{\varepsilon^{2k+3}}{(2k+2)!\,(2k+4)} \int\limits_0^\infty d\xi_1\, \xi_1^{2+k}(2k+2-\xi_1\varepsilon)\, e^{-\frac{1+\varepsilon}{2}\xi_1}$$

$$\cdot \int\limits_0^\infty d\xi_2\, \xi_2^{2+k} \cdot \frac{\xi^k_{(2,1)}}{\xi^{k+1}_{(1,2)}} (2k+2-\xi_2\varepsilon)\, \varepsilon^{-\frac{1+\varepsilon}{2}\xi_2}$$

und

$$\left.\begin{aligned} &H^1_{n=k+2}(vw, wv) = ZRh \frac{8\,\varepsilon^{2k+3}}{(1+\varepsilon)^{2k+5}} \\ &\frac{\left\{(2k+2)^2(2k+5) - 2\left(\frac{\varepsilon}{1+\varepsilon}\right)(2k+2)(4k^2+20k+25) + \left(\frac{\varepsilon}{1+\varepsilon}\right)^2(2k+3)(4k^2+26k+44)\right\}}{(2k+4)(2k+1)}. \end{aligned}\right\} \tag{22}$$

Aus diesen Rechnungen folgt zunächst, daß $H^1(vw, wv)$ stets positiv ist. Nach den l. c. angestellten Überlegungen bedeutet dies, daß das *P*-System, also jenes Termsystem, das Zustände mit äquivalenten Elektronen enthält, stets die höher liegenden Termwerte besitzt.

Die numerischen Werte für $\frac{H^1(vw, wv)}{Rh}$ sind:

Tabelle 3.

	$k=1$	$k=2$	$k=3$
$Z=2$ $n=k+1$	0,007 65	0,000 025 7	$5{,}25 \,.\, 10^{-8}$
$Z=2$ $n=k+2$	0,002 46	0,000 015 0	$4{,}31 \,.\, 10^{-8}$
$Z=3$ $n=k+1$	0,030 7	0,000 189	$6{,}95 \,.\, 10^{-7}$
$Z=3$ $n=k+2$	0,009 35	0,000 108	$5{,}72 \,.\, 10^{-7}$

Der Unterschied zwischen Ortho- und Paraterm ist in dieser Näherung nach den l. c. angestellten Überlegungen gleich $2\,H^1(vw, wv)$.

Aus den oben angegebenen Zahlen ist in der folgenden Tabelle die Differenz der Rydberg-Ritzkorrektionen von Para- und Orthotermen nach der Formel $\frac{n^3}{(Z-1)^2} \cdot \frac{H^1(vw, wv)}{Rh} = \Delta\delta$ berechnet und mit den empirischen Werten verglichen. Die empirischen Werte für den Abstand zwischen Ortho- und Paralithium entnehme ich Messungen von S. Werner[1]):

Tabelle 4.

		$\Delta\delta$ für					
		$2p-2P$	$3p-3P$	$3d-3D$	$4d-4D$	$4f-4F$	$5f-5F$
He	ber.	0,061	0,067	0,000 69	0,000 96	$3{,}4 \cdot 10^{-6}$	$5{,}4 \cdot 10^{-6}$
	gem.	0,075	0,079	0,000 44	0,000 63	—	—
Li^+	ber.	0,061	0,063	0,001 3	0,001 7	$1{,}1 \cdot 10^{-5}$	$1{,}8 \cdot 10^{-5}$
	gem.	0,067	0,069	0,001 0	0,001 4	—	—

Bei den P-Termen kann man nicht erwarten, daß die Formeln (21) und (22) ein recht genaues Resultat geben. Bei den D-Termen mag ein Teil der schlechten Übereinstimmung daran liegen, daß es sich um Größen handelt, die nahe an der Grenze der Meßgenauigkeit liegen; es mag aber der theoretische Wert auch noch durch die quadratischen Glieder, d. h. in der zweiten Näherung des Störungsverfahrens, erhebliche Änderungen erfahren. Es handelt sich bei den hier durchgeführten Rechnungen eben um eine Abschätzung. Das qualitative Verhalten der Rydberg-Ritz-korrektion als Funktion der Quantenzahlen wird durch die Theorie wohl befriedigend wiedergegeben.

§ 3. Um neben dem relativen Abstand des O- und P-Systems auch die absolute Lage der Terme etwa relativ zu den Wasserstofftermen zu errechnen, muß man zur zweiten Näherung in der Störungsenergie übergehen. Eine exakte Durchführung dieser zweiten Näherung wäre aber sehr mühsam, daher habe ich mich auch hier mit einer rohen Abschätzung der Glieder zweiter Ordnung begnügt. Für große Werte von k ist es anschaulich ohne weiteres einleuchtend, daß der größte Beitrag zum Gliede zweiter Ordnung von der sogenannten „Polarisation“ des Atominneren herrührt. Neben der Polarisationswirkung sind noch andere Glieder vorhanden — höheren induzierten Polen (Quadrupolen usw.) entsprechend —, die sowohl die absolute Lage der Terme, wie die relative Lage von O- und P-Systemtermen merklich beeinflussen. Die Wirkung des induzierten Dipols jedoch beeinflußt die relative Lage der beiden Termsysteme nicht. Näherungsweise gibt also das Polarisationsglied (nach

[1]) Herr S. Werner hat mir die Resultate dieser Messungen freundlicherweise noch vor der Publikation mitgeteilt. Siehe auch die inzwischen erschienene Arbeit von H. Schüler, ZS. f. Phys. **37**, 568, 1926. Das Spektrum von Ortholithium$^+$ wurde angegeben von H. Schüler, Naturw. **12**, 579, 1924; Ann. d. Phys. **76**, 292, 1925 und S. Werner, Nature, Febr. 1924, Nature, Okt. 1925.

Hinzufügung des in den Tabellen 1 und 2 berechneten Gliedes erster Ordnung) direkt den Abstand der Mitte zwischen den Termen des *O*- und *P*-Systems und dem betreffenden Wasserstoffterm an. Für große Werte von k kann man das Polarisationsglied den Rechnungen von I. Waller (l. c.) entnehmen. Für kleine Werte von k jedoch stellen, wie Waller bemerkt, seine Rechnungen keine Näherung für das gesuchte Glied dar. Man kann sich dies wieder anschaulich klar machen. Waller berechnet die polarisierende Wirkung eines homogenen elektrischen Feldes auf das Atominnere. Solange das induzierende Elektron weit vom Kern entfernt ist, mag die von ihm auf das Atominnere wirkende Kraft als homogenes elektrisches Feld angesehen werden. Wird sein Abstand vom Kern jedoch von der gleichen Größenordnung, wie der des inneren Elektrons, so ist eine solche Approximation durch ein homogenes Feld unrichtig, der induzierte Dipol wird wesentlich kleiner als der durch Wirkung eines homogenen Kraftfeldes berechnete. Wenn man diese Abnahme der Polarisierbarkeit mit abnehmendem k näher studiert auf Grund der Glieder zweiter Ordnung, so kommt man zu dem Resultat, daß schon für *d*-, *D*-, *f*-, *F*- usw. Terme die Werte Wallers eine sehr gute Annäherung darstellen, daß aber für die *P*-Terme das Polarisationsglied in Wirklichkeit nur etwa den dritten Teil des aus der Wallerschen Rechnung folgenden Wertes beträgt. Die zu diesem Resultat führende Rechnung will ich hier nicht angeben. Als Beitrag der Polarisationswirkung zur Rydberg-Ritzkorrektion entnehmen wir also aus der Arbeit von Waller (l. c., S. 646) die Werte (die Werte für *P*-Terme haben wir durch 3 dividiert):

	P	*D*	*F*
He	∼ — 0,018	— 0,0025	— 0,000 43
Li^+	∼ — 0,014	— 0,0021	— 0,000 34

Wir addieren hierzu den von den Gliedern $H^{1'}(vw, vw)$ herrührenden Beitrag erster Ordnung und vergleichen dies mit der gemessenen Rydbergkorrektion der Mitte zwischen Ortho- und Paraterm.

Tabelle 5.

		P	*D*	*F*
He	ber. . . .	∼ — 0,028	— 0,0026	— 0,000 43
	gem. . . .	— 0,029	— 0,0025	—
Li^+	ber. . . .	∼ — 0,025	— 0,0023	— 0,000 34
	gem. . . .	— 0,020	— 0,0022	∼ — 0,000 2

Die theoretischen Werte für P-Terme sind auf Grund der Abschätzung, durch die sie gewonnen sind, erheblich ungenauer, als es bei der guten Übereinstimmung zwischen Theorie und Beobachtung den Anschein hat.

Durch die vorhergehenden Rechnungen sind wir zu einer allgemeinen Übersicht über die Lage der Terme des gesuchten Spektrums gekommen. Bei diesen Abschätzungen mußten die S-Terme sowie diejenigen Terme, die angeregten Zuständen beider Elektronen entsprechen, ausgeschlossen werden. Für die Übergangswahrscheinlichkeiten folgt aus den l. c. ausgeführten Rechnungen, daß bei dem in diesem 1. Teil behandelten Atommodell, das aus Punktladungen aufgebaut ist, Interkombinationen zwischen O- und P-System nicht vorkommen; die Übergangswahrscheinlichkeiten innerhalb des einzelnen Systems weichen (ausgenommen die Übergänge zum 1 S-Term!) von denen des Wasserstoffs nicht mehr ab, als die Energie der betreffenden stationären Zustände von der entsprechenden Wasserstoffenergie. Die Behandlung des aus Punktladungen aufgebauten Atoms ist deshalb für uns zu einem vorläufigen Abschluß gebracht, und wir gehen zur Behandlung des aus magnetischen Elektronen gebauten Modells und zur Diskussion der Feinstruktur über.

II. Das Spektrum bei Berücksichtigung der Elektronenmagnete.

§ 1. Zur ersten Orientierung sei das folgende Problem behandelt: Der Schwerpunkt der beiden Elektronen sei fest, die Drehimpulse $\mathfrak{s}_1$, $\mathfrak{s}_2$ der Elektronen („Quantenzahl" $s_1 = s_2 = \frac{1}{2}$) jedoch der Richtung nach unbestimmt. Zwischen jedem der beiden Elektronenmagnete und irgend einer festen äußeren Kraft $\mathfrak{K}$ bestehe eine Wechselwirkung, die durch das Glied $H_1(\mathfrak{K}, \mathfrak{s})$ in der Hamiltonschen Funktion beschrieben werde. Ferner soll die Wechselwirkung zwischen beiden Magneten durch $H_2(\mathfrak{s}_1, \mathfrak{s}_2)$ gegeben sein. Wesentlich ist, daß sich die beiden Elektronenmagnete in jeder Beziehung gleich verhalten. Die Lösung dieses Problems nach der früheren Quantentheorie würde offenbar so lauten: die beiden Elektronenmagnete stellen sich parallel oder antiparallel — unabhängig vom Verhältnis der Größenordnung von H_1 und H_2, da die Wirkung der äußeren Kraft auf beide Magnete die gleiche ist —, die Orientierung des Gesamtsystems im äußeren Felde $\mathfrak{K}$ gibt wegen $|\mathfrak{s}_1| = |\mathfrak{s}_2| = \frac{1}{2}\frac{h}{2\pi}$ zu einem „Triplett" und einem „Singlett" Anlaß.

Ein ganz analoges Resultat ergibt sich aus der quantenmechanischen Behandlung. Die Projektionen von $\mathfrak{s}_1$, $\mathfrak{s}_2$ auf die Richtung der äußeren

Zeitschrift für Physik *39*, 499–518 (1926)

Kraft $\mathfrak{K}$ sollen mit m_1, m_2 bezeichnet werden. Die Größe $m_1 + m_2 = m$ kann wieder von vornherein als Diagonalmatrix angesehen werden. Es gibt also im ungestörten System vier Zustände, die wir mit a, b, c, d numerieren wollen.

m_1	m_2	m	
$\frac{1}{2}$	$\frac{1}{2}$	1	a
$\frac{1}{2}$	$-\frac{1}{2}$	0	b
$-\frac{1}{2}$	$\frac{1}{2}$	0	c
$-\frac{1}{2}$	$-\frac{1}{2}$	-1	d

Die Zustände a und d sind wegen der Quantisierung von m offenbar schon im ungestörten System festgelegt und haben die Energie $H_1(aa) + H_2(aa)$ bzw. $H_1(dd) + H_2(dd)$, wobei $H_2(aa) = H_2(dd)$. Zwischen den Zuständen b und c jedoch besteht die Resonanzentartung, wie sie l. c. beschrieben worden ist. Von diesem Gesichtspunkt aus sind a und d Zustände mit äquivalenten Teilsystemen. Sie kombinieren daher auch nur mit einer der beiden Lösungen des Resonanzproblems bc, und die Teilung der vier Zustände in zwei Systeme von ein und drei Zuständen erscheint hier als Analogon zur Einteilung in O- und P-System in Teil I. Die Energie der Zustände „b, c" des gestörten Systems wird wieder nach l. c., Gleichung (12) [hierbei ist zu beachten, das $H_1(bc) = 0$]

$$„b“ : H_1(b,b) + H_2(bb) + H_2(bc) \text{ bzw. } „c“ : H_1(bb) + H_2(bb) - H_2(bc). \quad (23)$$

Wesentlich dafür, daß die Resonanzeinteilung mit der Einteilung in „Multipletts" (Singlett-Triplett) übereinstimmt, ist der Umstand, daß für die Elektronen $s = \frac{1}{2}$ ist.

Denken wir uns nun wieder nach l. c., Gleichung (16) die Schrödingerschen Eigenfunktionen dieses Problems aufgeschrieben, so werden die zu den „Triplett"-Zuständen a, b, d gehörigen Eigenfunktionen in den Koordinaten der beiden Elektronenmagnete symmetrisch sein, die zum „Singlett" c gehörige Eigenfunktion wird bei Vertauschung der Elektronen das Vorzeichen wechseln, sagen wir „antisymmetrisch" sein.

An diesen Betrachtungen ändert sich prinzipiell nichts, wenn man zum eigentlichen Problem, dem Heliumatom mit magnetischen Elektronen, übergeht. Solange die durch H_1 und H_2 charakterisierten Wechselwirkungsenergien erheblich kleiner sind als die Energien und Energiedifferenzen des im 1. Teile behandelten Modells, so kann die eben durchgeführte Rechnung direkt als Lösung der Gleichung der säkularen Bewegungen für das gestörte System angesehen werden, wobei als ungestörtes System eben jeder beliebige Zustand des Modells im 1. Teile in Betracht kommt. Nur die Orientierung der (parallelen oder antiparallelen) Vektoren $\mathfrak{s}_1$ und $\mathfrak{s}_2$ erfolgt nicht nach einer äußeren Kraft $\mathfrak{K}$, sondern durch Impulszusammensetzung mit dem Impulsmoment des Modells vom 1. Teile nach der gewöhnlichen Strukturregel.

Der Grund, weshalb sich die Vektoren $\mathfrak{s}_1$ und $\mathfrak{s}_2$ parallel bzw. antiparallel stellen, wie es die gewöhnliche Strukturregel fordert, ist also nicht etwa eine große Wechselwirkung zwischen $\mathfrak{s}_1$ und $\mathfrak{s}_2$, sondern es ist der Umstand, daß auf beide Magnete durchaus die gleiche Kraft wirkt; denn die beiden Elektronen des im ersten Teil behandelten Modells führen ja im Laufe der Zeit genau die gleichen Bewegungen (in verschiedener Phase) aus.

Auch die Einteilung des ganzen Termsystems in die beiden Teilsysteme, die nicht kombinieren, ergibt sich jetzt ohne Schwierigkeit. Die Schrödingersche Eigenfunktion eines Zustandes des ganzen Atoms ist gleich dem Produkt der Eigenfunktionen des Modells vom Teil I und der des eben besprochenen „Singlett-Triplettmodells". Da die Eigenfunktionen des O-Systems antisymmetrisch, die des P-Systems symmetrisch waren, werden die Eigenfunktionen des ganzen Atoms

symmetrisch für die Tripletts des P-Systems und die Singletts des O-Systems,

antisymmetrisch für die Singletts des P-Systems und die Tripletts des O-Systems.

Dadurch ist die Einteilung in die nichtkombinierenden Termsysteme gegeben; wählt man insbesondere dasjenige Teilsystem aus, das keine äquivalenten Bahnen enthält, so werden die Paraterme Singletts, die Orthoterme Tripletts. Diese Einteilung in nichtkombinierende Teilsysteme bleibt (vgl. l. c.) auch dann noch bestehen, wenn die Voraussetzung der oben angedeuteten Rechnungen nicht mehr erfüllt ist, d. h. wenn z. B. der O—P-Abstand nicht mehr groß gegenüber den Werten H_1 und H_2 ist. In diesem Falle hört aber die einfache Einteilung des Spektrums in Singletts und Tripletts auf.

Dieses Verhalten hat eine große Ähnlichkeit mit dem Verhalten eines Atoms ohne Resonanz und mit einer Wechselwirkung von der Größenordnung des Singlett-Triplettabstandes zwischen den Elektronenmagneten und es ist durchaus verständlich, daß die meisten charakteristischen Struktureigenschaften auf dieses letztere Modell zurückgeführt werden konnten.

Umgekehrt gibt auch das hier diskutierte Zweielektronenmodell alle feineren Züge in der Struktur z. B. der Erdalkalispektren wieder. Für

Zeitschrift für Physik 39, 499–518 (1926)

die Tripletts wird das bekannte „Kosinusgesetz“ immer dann gelten, wenn die magnetische Wechselwirkung zwischen Kern und Elektronen größer ist als die Wirkung der beiden Elektronenmagnete aufeinander. In einem äußeren Magnetfeld werden die Tripletts sich verhalten wie in den sonst zur Ableitung der Zeemaneffekte diskutierten Modellen.

§ 2. Man kann sich die Frage stellen, weshalb gerade He und Li^+ aus der Reihe der Erdalkalien insofern herausfallen, als für sie die Feinstruktur nicht dem gewöhnlichen Triplettschema entspricht. Der Grund dafür muß in der relativ großen Wechselwirkung der Elektronenmagnete aufeinander zu suchen sein [1]). Um die Feinstruktur von He und Li^+ näher zu untersuchen, schreiben wir denjenigen Teil der Hamiltonschen Funktion des ganzen Modells, der der Wirkung der Magnete entspricht, in derselben Näherung, wie es beim Wasserstoff üblich ist, an ($\mathfrak{r}_1 \mathfrak{v}_1$ bzw. $\mathfrak{r}_2 \mathfrak{v}_2$ seien Radius und Geschwindigkeit der Elektronen):

$$\left.\begin{aligned}
\Delta H = {} & \frac{e^2 Z}{m c^2} \frac{[\mathfrak{r}_1 \mathfrak{v}_1]}{r_1^3} \mathfrak{s}_1 - \frac{e^2}{m c^2} \frac{[\mathfrak{r}_1 - \mathfrak{r}_2, \mathfrak{v}_1 - \mathfrak{v}_2]}{r_{12}^3} \mathfrak{s}_1 \\
& + \frac{e^2 Z}{m c^2} \frac{[\mathfrak{r}_2 \mathfrak{v}_2]}{r_2^3} \mathfrak{s}_2 - \frac{e^2}{m c^2} \frac{[\mathfrak{r}_2 - \mathfrak{r}_1, \mathfrak{v}_2 - \mathfrak{v}_1]}{r_{12}^3} \mathfrak{s}_2 \\
& - \frac{e^2 Z}{2 m c^2} \frac{[\mathfrak{r}_1 \mathfrak{v}_1]}{r_1^3} \mathfrak{s}_1 + \frac{e^2}{2 m c^2} \frac{[\mathfrak{r}_1 - \mathfrak{r}_2, \mathfrak{v}_1]}{r_{12}^3} \mathfrak{s}_1 \\
& - \frac{e^2 Z}{2 m c^2} \frac{[\mathfrak{r}_2 \mathfrak{v}_2]}{r_2^3} \mathfrak{s}_2 + \frac{e^2}{2 m c^2} \frac{[\mathfrak{r}_2 - \mathfrak{r}_1, \mathfrak{v}_2]}{r_{12}^3} \mathfrak{s}_2 \\
& + \frac{e^2}{m^2 c^2} \frac{-3 (\mathfrak{s}_1, \mathfrak{r}_2 - \mathfrak{r}_1)(\mathfrak{s}_2, \mathfrak{r}_2 - \mathfrak{r}_1) + (\mathfrak{s}_1 \mathfrak{s}_2) r_{12}^2}{r_{12}^5}
\end{aligned}\right\} \qquad (24)$$

Die erste und zweite Zeile entspricht der magnetischen Wechselwirkung in demjenigen Bezugssystem, in dem das betreffende Elektron ruht. Zeile 3 und 4 gibt die nach Thomas [2]) von der Lorentztransformation herrührenden Glieder [die hier direkt verwendet werden können, da uns nur der zeitliche Mittelwert von (24) interessiert]. Zeile 5 gibt die Wechselwirkung zwischen beiden Magneten an. Es muß jetzt der Mittelwert von (24) über die ungestörte Bewegung gebildet werden.

$$\Delta \overline{H} = \int \cdots \int d\Omega_1 \, d\Omega_2 \, \tfrac{1}{2} [\varphi_1(v) \varphi_2(w) - \varphi_1(w) \varphi_2(v)]^2 \Delta H.$$

Dieser Mittelwert zerfällt wieder in zwei und zwei Glieder vom Typus (9a) und (9b). Die letzteren zwei Glieder entsprechen der Resonanzwirkung und dürften nur einen Bruchteil der ersten Glieder ausmachen. Zur

[1]) Den Hinweis hierauf, sowie viele wertvolle Anregungen in gemeinsamen Diskussionen verdanke ich Herrn N. Bohr.

[2]) L. H. Thomas, Nature **117**, 514, 1926.

groben Abschätzung begnügen wir uns mit der Berechnung der ersten Glieder, die der zweiten scheint auch rechnerisch recht kompliziert. Bei den ersten Gliedern ist zunächst der Mittelwert für das Elektron 1 in einer 1 S-Bahn vorzunehmen, für das andere in einer n, k-Bahn, dann sind die beiden Elektronen zu vertauschen und man muß zwischen beiden Werten das arithmetische Mittel nehmen. Es mag in dieser Näherung erlaubt sein, den kleineren Radius, z. B. $\mathfrak{r}_1$ gegen $\mathfrak{r}_2$ bzw. umgekehrt zu streichen. Mittelt man ferner in der fünften Zeile von (24) zunächst nach der klassischen Theorie über die „k-Präzession" des äußeren Elektrons, so kann man $\overline{\dfrac{-3(\mathfrak{s}_1\mathfrak{r}_1)(\mathfrak{s}_2\mathfrak{r}_1)+(\mathfrak{s}_1\mathfrak{s}_2)r_1^2}{r_1^5}}$ ersetzen durch $\overline{\dfrac{\frac{3}{2}(\mathfrak{s}_1\bar{\mathfrak{k}}_1)(\mathfrak{s}_2\bar{\mathfrak{k}}_1)-\frac{1}{2}(\mathfrak{s}_1\mathfrak{s}_2)}{r_1^3}}$ (bzw. mit $\mathfrak{k}_2$), wo $\bar{\mathfrak{k}}_1$ bzw. $\bar{\mathfrak{k}}_2$ den Einheitsvektor in der Richtung $\mathfrak{k}_1$ bzw. $\mathfrak{k}_2$ bezeichnet. Wir nehmen an, daß die quantenmechanische Mittelung zum gleichen Resultat führt. Dann entsteht aus (24) wegen $k_{1S} = 0$:

$$\Delta H = \frac{e^2}{2m^2c^2}\left(\frac{1}{r^3}\right)_{n,k}\left\{(Z-3).\,\mathfrak{k}\,\frac{\mathfrak{s}_1+\mathfrak{s}_2}{2}+3\,|\mathfrak{s}_1|\cos(\mathfrak{s}_1\mathfrak{k}).\,|\mathfrak{s}_2|\cos(\mathfrak{s}_2\mathfrak{k})-(\mathfrak{s}_1\mathfrak{s}_2)\right\}. \quad (25)$$

Das letzte Glied der Klammer gibt zur Feinstruktur innerhalb des Tripletts keinen Beitrag, da es nur von der relativen Lage von $\mathfrak{s}_1$ und $\mathfrak{s}_2$ abhängt.

Um das zweite Glied der Klammer auszurechnen, bemerken wir, daß klassisch:

$$|\mathfrak{s}_1|\cos(\mathfrak{s}_1\mathfrak{k}) = |\mathfrak{s}_1|\,.\,\{\cos(\mathfrak{k}\mathfrak{s})\cos(\mathfrak{s}_1\mathfrak{s}) - \sin(\mathfrak{k}\mathfrak{s})\sin(\mathfrak{s}_1\mathfrak{s})\cos\psi\}, \quad (26)$$

wo $\mathfrak{s} = \mathfrak{s}_1 + \mathfrak{s}_2$ und ψ der zu diesem $\mathfrak{s}$ kanonisch konjugierte Winkel[1]) ist. Analog

$$|\mathfrak{s}_2|\cos(\mathfrak{s}_2\mathfrak{k}) = |\mathfrak{s}_2|\,\{\cos(\mathfrak{k}\mathfrak{s})\cos(\mathfrak{s}_2\mathfrak{s}) + \sin(\mathfrak{k}\mathfrak{s})\sin(\mathfrak{s}_2\mathfrak{s})\cos\psi\}. \quad (27)$$

Also

$$\overline{|\mathfrak{s}_1|\cos(\mathfrak{s}_1\mathfrak{k})\,|\mathfrak{s}_2|\cos(\mathfrak{s}_2\mathfrak{k})} = |\mathfrak{s}_1|\,|\mathfrak{s}_2|\,\{\cos^2(\mathfrak{k}\mathfrak{s})\cos(\mathfrak{s}_1\mathfrak{s})\cos(\mathfrak{s}_2\mathfrak{s}) - \tfrac{1}{2}\sin^2(\mathfrak{k}\mathfrak{s})\sin(\mathfrak{s}_1\mathfrak{s})\sin(\mathfrak{s}_2\mathfrak{s})\}. \quad (28)$$

Diese Rechnung läßt sich auf Grund der Kronigschen[2]) Formeln in die Quantenmechanik übertragen.

Es ist zu diesem Zwecke am einfachsten, unter Beachtung der Normierung unsere Vektoren $\mathfrak{s}_1$, $\mathfrak{s}_2$ mit Kronigs Vektoren K_1 und K_0, unser $\mathfrak{s}$

---

[1]) Vgl. z. B. W. Heisenberg, ZS. f. Phys. **26**, 291, 1924.

[2]) R. de L. Kronig, ZS. f. Phys. **33**, 261, 1925. Der Beweis dieser Formeln auf Grund der Quantenmechanik wurde gegeben von P. Dirac, Proc. Roy. Soc. London **111**, 281, 1926.

Zeitschrift für Physik *39*, 499–518 (1926)

mit Kronigs L, unser $\mathfrak{k}$ mit Kronigs J(!), unser j mit Kronigs R zu identifizieren. Dann gibt eine naheliegende Erweiterung der Kronigschen Tabelle 2 (l. c., S. 265) die gesuchten Mittelwerte. Das Resultat ist für $s_1 = s_2 = \frac{1}{2}$, $s = 1$:

$$\overline{|\mathfrak{s}_1|\cos(\mathfrak{s}_1\mathfrak{k})|\mathfrak{s}_2|\cos(\mathfrak{s}_2\mathfrak{k})} = \left(\frac{h}{2\pi}\right)^2 \left\{\frac{[j(j+1)-k(k+1)-s(s+1)]^2}{16\,k(k+1)} - \frac{[-(k+\frac{1}{2})^2+(j+s+\frac{1}{2})^2][(k+\frac{1}{2})^2-(j-s+\frac{1}{2})^2]}{16\,k(k+1)}\right\}. \quad (29)$$

Daraus wieder für $s_1 = s_2 = \frac{1}{2}$, $s = 1$:

$$\Delta H = \left(\frac{h}{2\pi}\right)^2 \frac{e^2}{2\,m^2c^2}\overline{\left(\frac{1}{r^3}\right)}_{n,k}\left\{\frac{Z-3}{4}[j(j+1)-k(k+1)-2] + 3\,\frac{[j(j+1)-k(k+1)-2]^2-[(j+\frac{3}{2})^2-(k+\frac{1}{2})^2][(k+\frac{1}{2})^2-(j-\frac{1}{2})^2]}{16\,k(k+1)} - \frac{1}{4}\right\}. \quad (30)$$

Da die Feinstruktur nur für den Term $2p$ hinreichend bekannt ist, setzen wir ein $k = 1$, $j = 2, 1, 0$, und erhalten

$$\Delta H_{2p} = \left(\frac{h}{2\pi}\right)^2 \frac{e^2}{2\,m^2c^2}\overline{\left(\frac{1}{r^3}\right)}_{2p}\left\{(Z-3)\left(\frac{1}{2}, -\frac{1}{2}, -1\right) + \left(\frac{3}{8}, -\frac{3}{8}, \frac{3}{2}\right) - \frac{1}{4}\right\}. \quad (31)$$

Als Spezialfall ergibt sich für

$$Z = 2;\quad \Delta H_{2p} = \left(\frac{h}{2\pi}\right)^2 \frac{e^2}{2\,m^2c^2}\overline{\left(\frac{1}{r^3}\right)}_{2p}\left(-\frac{3}{8}, -\frac{1}{8}, \frac{9}{4}\right),$$

$$Z = 3;\quad \Delta H_{2p} = \left(\frac{h}{2\pi}\right)^2 \frac{e^2}{2\,m^2c^2}\overline{\left(\frac{1}{r^3}\right)}_{2p}\left(\frac{1}{8}, -\frac{5}{8}. \frac{5}{4}\right).$$

Aus (31) schließt man zunächst: Für große Werte von Z gilt das „Kosinusgesetz“, wie es erwartet werden muß. Für kleine Werte von Z jedoch überwiegen die anderen Glieder. Wenn wir auf die bekannten Intensitätsverhältnisse 5 : 3 : 1 beim $2p$-Triplett achten, erhalten wir die folgenden Aufspaltungsbilder, etwa für $2p - 2s$[1]):

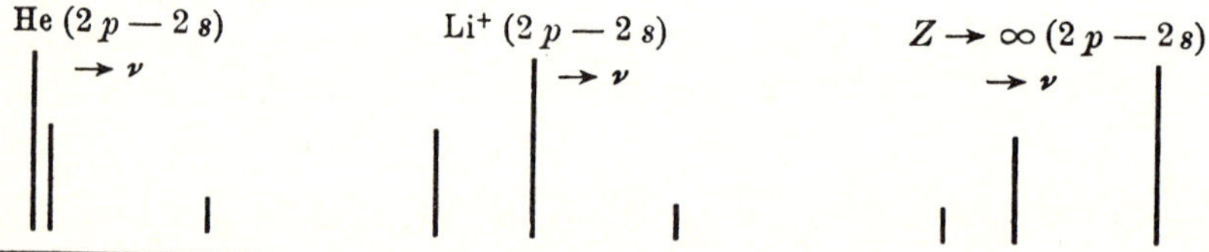

[1]) Auf die Möglichkeit, das Heliumdublett als verzerrtes Triplett zu deuten, hat schon T. C. Slater, Proc. Nat. Acad. Amer. **11**, 732, 1925, hingewiesen. Vgl. insbesondere auch die Intensitätsmessungen von L. S. Ornstein und H. C. Burger, ZS. f. Phys. **26**, 57, 1924 und D. Burger, ebenda **38**, 437, 1926.

Wenn man zunächst von den absoluten Werten absieht, gibt die Figur links wohl die empirische Feinstruktur beim He vollständig wieder. Die beiden starken Linien links schmelzen bei der Messung zu einer zusammen und geben mit der Komponente rechts das Orthoheliumdublett mit der bekannten Intensitätsanomalie. Auch die zweite Figur steht mit der empirischen Feinstruktur der Li^+-Linie 5485 nach Schüler (l. c.) in befriedigender Übereinstimmung, wenn man zum Vergleich das allgemeine Verhalten des empirischen Aufspaltungsbildes heranzieht. Von Schüler werden jedoch noch eine Reihe weiterer Komponenten dieser Linie angegeben. Es mag dies damit in Verbindung gebracht werden, daß MacLennan und Ainslie[1]) auch bei der roten Lithiumlinie des Bogenspektrums eine kompliziertere Feinstruktur als bei gewöhnlichen Dublettlinien beobachtet haben. Ob diese besonderen Feinstrukturen, wie MacLennan annimmt, als Isotopeneffekte gedeutet werden können, scheint wohl zweifelhaft. Man wird eher, wie Schüler, an eine wirkliche Kernstruktur denken. In der Tat würde ein magnetisches Moment des Kernes von derselben Größenordnung, wie das Moment des Elektrons einen Effekt der entsprechenden Größenordnung ergeben. Jedenfalls aber wird man annehmen dürfen, daß das unserer Theorie zugrunde gelegte Modell keine solche besondere Feinstruktur zeigen würde.

Die Berechnung der absoluten Größe der oben angegebenen Feinstruktur hängt von der Bestimmung des Wertes $\overline{\left(\frac{1}{r^3}\right)}_{2p}$ ab. Obwohl der $2p$-Term selbst wenig von dem entsprechenden Wasserstoffterm abweicht, kann sich diese Abweichung bei einem Mittelwert von $1/r$ in der dritten Potenz schon sehr stark bemerkbar machen, denn für $k = 0$ wird $\overline{\frac{1}{r^3}} = \infty$. Der für wasserstoffähnliche Atome gültige Wert von $\overline{\frac{1}{r^3}}$ ist also sicher eine sehr schlechte Annäherung. Andererseits ist eine genaue Berechnung von $\overline{\frac{1}{r^3}}$ recht kompliziert. Zur vorläufigen Abschätzung setzen wir daher doch die dem wasserstoffähnlichen Atom entsprechenden Werte ein — wir erwarten dann, daß die so gewonnenen theoretischen Werte erheblich zu klein werden — und vergleichen den Abstand der äußersten Komponenten mit dem gemessenen Werte:

[1]) T. C. MacLennan und D. S. Ainslie, Proc. Roy. Soc. London **101**, 342, 1922.

Zeitschrift für Physik 39, 499–518 (1926)

Für wasserstoffähnliche Spektra ist

$$\left(\frac{1}{r^3}\right)_{2p} = Rh \, . \, \alpha^2 \frac{Z^3}{24} \quad (\alpha = \text{Sommerfelds Feinstrukturkonstante}). \tag{32}$$

Der Abstand zwischen den äußersten Komponenten des Feinstrukturbildes wird also für

$$\text{He}: \; Rh \, . \, \alpha^2 \frac{(Z-1)^3}{24} \cdot \frac{21}{8} = 0{,}62\,\text{cm}^{-1}; \;\; \text{beob. } 1{,}07\,\text{cm}^{-1},$$

$$\text{Li}^+: \; Rh \, . \, \alpha^2 \frac{(Z-1)^3}{24} \cdot \frac{15}{8} = 3{,}65\,\text{cm}^{-1}; \;\; \text{beob. } 4{,}4\,\text{cm}^{-1}.$$

Ein Teil der Diskrepanz mag auch auf die Vernachlässigung des Resonanzgliedes bei der Mittelung von (24) zurückzuführen sein.

Durch die hier durchgeführten Rechnungen sollte gezeigt werden, daß die Quantenmechanik auch für Atome mit zwei Elektronen eine qualitative Beschreibung des Spektrums bis in die feineren Einzelheiten ermöglicht, und daß sie das Verhalten der Terme als Funktion der Quantenzahlen näherungsweise zu bestimmen gestattet. Diese Betrachtungen bedürfen nach zwei Richtungen hin noch sehr dringend einer Ergänzung. Für die Auswahl des einen Termsystems, das keine Zustände mit äquivalenten Bahnen der Elektronen enthält, fehlt bis jetzt eine klare Begründung. Man kann hoffen, daß man beim Versuch zur Beseitigung dieses Mangels auf tieferliegende Zusammenhänge stoßen wird. Die andere Lücke der hier durchgeführten Überlegungen besteht in dem Mangel an Genauigkeit. Man wird verlangen müssen, daß die theoretischen Termwerte so genau berechnet werden, daß die Übereinstimmung dieser Termwerte mit den experimentellen bis in die letzten Dezimalen geprüft werden kann. Die Methoden der quantenmechanischen Störungstheorie müssen also nach und nach verfeinert werden, bis man zu Verfahren kommt, die, wie die Verfahren der Astronomie, ohne allzu große Rechenarbeit eine wirklich quantitative Bestimmung der gesuchten Konstanten ermöglichen.

Kopenhagen, Institut for teoretisk Fysik, Juli 1926.

Zeitschrift für Physik *41*, 239–267 (1927)

Mehrkörperprobleme und Resonanz in der Quantenmechanik. II.

Von **Werner Heisenberg** in Kopenhagen.

Mit 1 Abbildung. (Eingegangen am 22. Dezember 1926.)

Die Gesetze der Quantenmechanik werden auf die Atome mit mehreren Elektronen angewendet. Insbesondere wird die Frage untersucht, in welcher Weise das quantenmechanische Resonanzphänomen mit der Multiplettstruktur der Termsysteme komplizierter Atome zusammenhängt. Das gleiche Resonanzphänomen wird in der Theorie der Moleküle und ihrer Bandenspektra studiert und führt dort zu einer Deutung der bekannten Intensitätsanomalie in gewissen Rotationsbanden.

Die Spektra der Atomsysteme mit zwei Elektronen können durch die Quantenmechanik — mit Sicherheit wenigstens näherungsweise — quantitativ beschrieben werden [1]. Der wesentliche neue Gesichtspunkt, der gegenüber den Atomsystemen mit einem Elektron beim Mehrkörperproblem auftrat, war ein Resonanzphänomen, das zu einer Einteilung des gesamten Termsystems führte in nicht kombinierende Teilsysteme, von denen jedes für sich eine „vollständige" quantenmechanische Lösung bildete. Von diesen Teilsystemen entsprach eines dem Paulischen Prinzip [2], nach dem keine äquivalenten Bahnen der Elektronen im Atom vorkommen können. Dieses Teilsystem wurde als physikalische Lösung willkürlich ausgewählt. Eine solche Auswahl einer einzelnen Lösung ermöglicht Abänderungen der Statistik von der Art, wie sie etwa von Bose und Einstein [3] vorgeschlagen wurde. Zu ähnlichen Resultaten ist unabhängig Dirac [4] gekommen. Dirac betonte besonders, daß man ganz allgemein die dem Paulischen Prinzip entsprechende Lösung auswählen sollte, und er gelangte dadurch zu einer Statistik des idealen Gases, die in wesentlichen Zügen mit der Fermischen Theorie der Gasentartung [5] übereinstimmt. Durch diese Überlegungen ist im Prinzip die quantenmechanische Lösung für Atomsysteme mit vielen Elektronen gegeben. Um aber zu zeigen, daß die so erhaltenen Lösungen wirklich den empirisch beobachteten Spektren dieser Atome qualitativ entsprechen, sind noch allgemeine Betrachtungen notwendig.

---

[1]) W. Heisenberg, ZS. f. Phys. **38**, 411, 1926, im folgenden als Teil I zitiert; **39**, 499, 1926.

[2]) W. Pauli, ebenda **31**, 765, 1925.

[3]) N. S. Bose, ebenda **26**, 178, 1924; A. Einstein, Sitzungsber. d. Preuß. Akad. 1924, S. 261 und 1925, S. 3.

[4]) P. Dirac, Proc. Roy. Soc. London (A) **112**, 661, 1926.

[5]) E. Fermi, Lincei Rend. **3**, 145, 1926.

I. Allgemeine Grundlagen.

Die folgenden Rechnungen können auf Grund der bisher üblichen Formen der Quantenmechanik — also mittels Matrizen, die den klassischen Fourierreihen entsprechen, einerseits, mittels der Schrödingerschen Wellen im Phasenraum andererseits — durchgeführt werden. Die formalen Zusammenhänge — auch hier bei den Mehrkörperproblemen — erhalten aber durch eine allgemeinere mathematische Fassung der Quantenmechanik, wie sie kürzlich von Dirac und Jordan[1]) ausgearbeitet wurde, eine mehr allgemeine und symmetrische Darstellung. Ich wiederhole hier die wichtigsten Grundbegriffe:

Die Lösung eines quantenmechanischen Problems ist gegeben durch die Lösung des Eigenwertproblems [vgl. Quantenmehanik II, Kap. 3, Gleichung (13)]

$$S^{-1} H S = W \quad \text{oder} \quad HS - SW = 0; \quad S\tilde{S}^* = 1, \tag{1}$$

hierin bedeutet H die Hamiltonsche Funktion des Systems als Funktion der p und q, W die die Energie darstellende Diagonalmatrix, deren Diagonalglieder die Eigenwerte des Problems darstellen, S ist die gesuchte Transformationsmatrix. In der ursprünglichen Fassung von (1) sollten p und q in H wieder aufgefaßt werden als Matrizen, die Fourierreihen entsprachen, gewonnen durch Lösung irgend eines geeignet gewählten Ausgangsproblems. Gleichung (1) war dann im allgemeinen eine komplizierte lineare Differenzengleichung unendlich hoher Ordnung für die S_{nl}, deren Eigenwerte W_n die Energien der stationären Zustände ergaben. Nach Dirac können allgemeinere Matrizen eingeführt werden. Diese Matrizen können z. B. so gewählt werden, daß alle $q\,(q_1, \ldots q_f)$ des Systems Diagonalmatrizen werden; wegen $p_k q_k - q_k p_k = \frac{h}{2\pi i}$ sind dann natürlich die $p\,(p_1, \ldots p_f)$ sicher keine Diagonalmatrizen. In gleicher Weise mag man irgend ein anderes System kanonischer Variabeln $\xi, \eta\; (\xi_1 \ldots \xi_f,\; \eta_1 \ldots \eta_f)$ einführen, das den Vertauschungsrelationen

---

1) P. Dirac, Proc. Roy. Soc. London (im Erscheinen). Herrn Dirac bin ich für die Möglichkeit, die Arbeit vor der Publikation kennenzulernen, zu Dank verpflichtet. Inzwischen habe ich durch freundliche briefliche Mitteilung erfahren, daß auch Herr Jordan [ZS. f. Phys. (im Erscheinen)] unabhängig Resultate erhalten hat, die denen der Diracschen Arbeit und denen einer vorausgehenden Note des Verfassers [ZS. f. Phys. (im Erscheinen)] äquivalent sind. Einige jener Resultate waren dem Verfassser auch früher und unabhängig schon durch Herrn W. Pauli mitgeteilt worden. Vgl. auch die wichtigen Untersuchungen von K. Lanczos, ZS. f. Phys. **35**, 812, 1926. Anmerkung bei der Korrektur: Siehe auch die Arbeit von F. London, ZS. f. Phys. **40**, 193, 1926.

genügt, und kann z. B. alle ξ als Diagonalmatrizen wählen. In $\boldsymbol{S}$ treten dann als Indizes auf die Nummer n des Zustandes mit der Energie W_n und die Zahlen ξ.

$$\boldsymbol{S}: (S(n, \xi)) \quad \text{oder} \quad (S_n(\xi)). \tag{2}$$

Dann ist Gleichung (1) äquivalent[1]) mit

$$\left[H\left(\xi, \frac{h}{2\pi i}\frac{\partial}{\partial \xi}\right) - W_n\right] S_n(\xi) = 0, \tag{3}$$

woraus $S_n(\xi)$ bestimmt werden kann. Zu (3) kommt noch die Orthogonalitäts- und Normalisierungsrelation:

$$\left.\begin{aligned} &\boldsymbol{S}\tilde{\boldsymbol{S}}^* = 1 \quad \text{oder} \quad \sum\nolimits^{\xi} S_n(\xi)\, S_m^*(\xi) = \delta_{nm} \\ \text{bzw.}\quad &\int S_n(\xi)\, S_m^*(\xi)\, d\xi = \delta_{nm} = \begin{matrix} 1 \text{ für } n = m, \\ 0 \text{ für } n \neq m, \end{matrix} \end{aligned}\right\} \tag{4}$$

je nachdem ξ diskreter oder kontinuierlicher Werte fähig ist. Ausführlich geschrieben lauten die letzten Gleichungen

$$\sum\nolimits^{\xi_1, \xi_2 \ldots \xi_f} S_n(\xi_1 \ldots \xi_f)\, S_m^*(\xi_1 \ldots \xi_f) = \delta_{nm}$$

bzw.

$$\int S_n(\xi)\, S_m^*(\xi)\, d\xi_1 \ldots d\xi_f = \delta_{nm}.$$

Im Spezialfall ξ gleich den Lagenkoordinaten q, η gleich den Impulskoordinaten p geht (3) über in die Schrödingersche Differentialgleichung. Im Spezialfall $\xi = J$, $\eta = w$, wo die J und w Wirkungs- und Winkelvariable irgend eines geeignet gewählten Systems bezeichnen, geht (3) über in die ursprüngliche Fassung der Gleichung (1). $\boldsymbol{S}$ ist im allgemeinen eine Matrix, deren einer Index n diskontinuierlich und deren anderer Index ξ kontinuierlich veränderlich ist. Im Spezialfall $\xi = q$, $\eta = p$ ist eine Kolonne dieser Matrix: $S_n(q)$ die Schrödingersche Eigenfunktion. Die Größen $S_n(\xi)$ und $S_n^*(\xi)$ fassen alle Aussagen zusammen, die sich auf den einen stationären Zustand n des Atoms beziehen. Die üblichen auf die Energie als Diagonalmatrix bezogenen Matrizenelemente einer Funktion $\boldsymbol{f}(\xi, \eta)$ werden gegeben durch $\boldsymbol{S}^{-1}\boldsymbol{f}\boldsymbol{S}$ oder

$$\left.\begin{aligned} &f_{nm} = \sum\nolimits^{\xi} S_n^*(\xi)\, f\left(\xi, \frac{h}{2\pi i}\frac{\partial}{\partial \xi}\right) S_m(\xi) \\ \text{bzw.}\quad &f_{nm} = \int S_n^*(\xi)\, f\left(\xi, \frac{h}{2\pi i}\frac{\partial}{\partial \xi}\right) S_m(\xi)\, d\xi. \end{aligned}\right\} \tag{5}$$

[1]) Die Äquivalenz der Matrizen mit Operatoren wurde schon von M. Born und N. Wiener, ZS. f. Phys. **36**, 174, 1926 entdeckt. Dieser Zusammenhang spielt dann eine wesentliche Rolle in der Untersuchung von E. Schrödinger, Ann. d. Phys. **79**, 734, 1926, und wird in voller Allgemeinheit erkannt und ausgenutzt durch P. Dirac, l. c.

Zeitschrift für Physik *41*, 239–267 (1927)

Physikalisch bedeutet das Produkt:

$$S_n^*(\xi^0)\, S_n(\xi^0)$$

bzw. im kontinuierlichen Fall

$$S_n^*(\xi^0)\, S_n(\xi^0)\, d\,\xi$$

die Wahrscheinlichkeit dafür, daß im stationären Zustand n die Variabeln ξ die Werte ξ^0 annehmen bzw. im kontinuierlichen Fall, daß die ξ zwischen ξ^0 und $\xi^0 + d\,\xi^0$ liegen [1]). Die Größen $S_n(\xi)$ seien allgemein die „Eigenfunktionen" des quantentheoretischen Problems genannt. Sie sind Funktionen so vieler Variabeln, als das Problem Freiheitsgrade hat: $S_n(\xi_1, \ldots \xi_f)$. Ist die Gleichung (1) bzw. (3) in den verschiedenen Freiheitsgraden separierbar, so kann S als Produkt von Funktionen dargestellt werden, die zu den einzelnen Freiheitsgraden gehören.

$$S = S_n^1(\xi_1)\, S_n^2(\xi_2) \ldots S^f(\xi_f). \tag{6}$$

Für die folgenden Rechnungen ist besonders wichtig der Fall eines Elektrons, bei dem jedoch das Magnetmoment, das nach der Hypothese von Uhlenbeck und Goudsmit [2]) von der Eigenrotation des Elektrons herrührt, nicht vernachlässigt wird. Ein solches Elektron hat vier Freiheitsgrade [der Gesamtimpuls des „Spins" sei, da er fest gegeben ist $s = {}^1/_2$ nicht mitgezählt [3])], z. B. die drei Lagenkoordinaten des Schwerpunktes, und die Lage der „Knotenlinie" der Eigenrotation, zu dieser konjugiert die Projektion m_s von s auf eine vorgegebene Achse. Wenn zunächst die Wechselwirkung zwischen dem Elektronenmagnet und der Bewegung des Elektronenschwerpunktes vernachlässigt wird, so läßt sich S in der Form schreiben

$$S(\xi_1\, \xi_2\, \xi_3)\, S^4(\xi_4).$$

Wenn für ξ_4 die eben genannte Lage der Knotenlinie gewählt wird, so würde die Bestimmung von S^4 die Behandlung des Kreiselektrons nach der Schrödingerschen Theorie erfordern. Dieses Problem hat aber noch gewisse formale Schwierigkeiten. Da aber die Wahl von ξ_4 ganz frei steht, und da andererseits die quantenmechanische Formulierung des Kreiselelektrons in der Sprache der Matrizen bekannt ist, so wird eine solche Bestimmung von $S^4(q_4)$ überflüssig; man kann z. B. ξ_4 als die Projektion m_s des Impulses s auf eine feste äußere Achse wählen, dann wird S^4 die Einheitsmatrix; m_s nimmt nur die beiden Werte $\pm {}^1/_2$ an.

[1]) W. Heisenberg, Schwankungserscheinungen und Quantenmechanik, l. c.; siehe auch P. Jordan, l. c.

[2]) E. Uhlenbeck und S. Goudsmit, Naturw. **13**, Heft 47, 1925.

[3]) Die Impulsgrößen seien im folgenden der Einfachheit halber immer durch ihre Quantenzahl gegeben.

II. Atomsysteme mit mehreren Elektronen.

§ 1. Durch die Forschungen der letzten Jahre sind die Spektra der komplizierteren Atome systematisch geordnet worden, derart, daß zu jedem „Term" korrespondenzmäßig im Sinne von Bohrs Theorie der Spektra und des periodischen Systems eine bestimmte zugehörige Bewegung der Elektronen im Atom angegeben werden kann. Als die wichtigsten Schritte zu dieser Systematik der Spektra mag man bezeichnen: Sommerfelds n, k, j-Ordnung der Terme, Landés Theorie der Zeemaneffekte, Cataláns Entdeckung der Multipletts, Paulis oben genanntes Prinzip über den Ausschluß äquivalenter Bahnen, das Russell-Saunderssche Modell, Hunds Systematik der Spektren der höheren Perioden des periodischen Systems und die Uhlenbeck-Goudsmitsche Hypothese vom Drehimpuls des Elektrons. Der durch diese Entwicklung gekennzeichnete Zweig der Quantentheorie hat kürzlich durch Hund eine zusammenfassende Darstellung gefunden; für alle Einzelheiten verweise ich daher auf dieses Buch [1]).

In den Modellvorstellungen, die solcher Termordnung zugrunde liegen, wurde immer eine ganz bestimmte Analogie der quantentheoretischen Gesetze zur klassischen Mechanik ausgearbeitet, die man als „adiabatische" Korrespondenz bezeichnen mag. Die Bewegung eines Elektrons wurde bestimmt durch die Kräfte vom Kern und von den anderen Elektronen, wobei jedoch über die Bewegung der anderen Elektronen Mittelwerte gebildet wurden; das feinere Wechselspiel der Elektronen im Atom wurde nicht berücksichtigt. Da zwischen den Elektronen im Atom ein kontinuierlicher Energieaustausch — wie er in den klassischen Modellen in Wirklichkeit stets eintreten würde — wegen der quantentheoretischen Grundpostulate (vgl. die Franck-Hertzschen Stoßversuche) nicht zu erwarten war [2]), so konnte man hoffen, daß die eben genannte grobe „adiabatische" Korrespondenz schon wesentliche Züge der wirklichen Gesetzmäßigkeiten ergeben würde; diese Hoffnung hat sich auch durch die Ordnung der Spektra bestätigt. Eine quantitative Durchführung jenes korrespondenzmäßigen Zusammenhanges ist aber natürlich erst durch die Quantenmechanik möglich.

Dieser grobe Korrespondenzzusammenhang konnte aber andererseits sicher keine vollständige Deutung des wirklichen quantentheoretischen Geschehens geben, weil ja in der Quantenmechanik auch ein Analogon

[1]) F. Hund, Linienspektra und periodisches System der Elemente. Berlin, Springer, 1927.

[2]) N. Bohr, Nature **113**, 223, 1924.

Zeitschrift für Physik *41*, 239–267 (1927)

zu dem feineren mechanischen Wechselspiel der Elektronen bestehen mußte. Diese Seite der Quantenmechanik läßt sich aber gar nicht mehr auf Grund mechanischer Bilder verfolgen. Denn im feineren Wechselspiel der Elektronen im Atom treten, wie bei der Wechselwirkung eines Elektrons mit äußerer Strahlung die für die Quantentheorie charakteristischen Züge von Diskontinuität direkt in Erscheinung [1]). Die Quantenmechanik war gerade aus dem Bestreben entstanden, die Wechselwirkung eines Elektrons mit äußerer Strahlung einerseits, die mit Elektronen andererseits den gleichen allgemeinen Gesetzmäßigkeiten zu unterwerfen. Die Quantenmechanik ermöglicht also eine einfache Behandlung jenes feineren Wechselspiels und führt so auf das hier zu behandelnde Resonanzphänomen.

Das Fehlen des Analogons zum feineren Wechselspiel der Elektronen in jenem groben korrespondenzmäßigen Bilde mußte zu Widersprüchen zwischen Modell und Erfahrung führen. Dieser Widerspruch lag in der Existenz der Multipletts. Die Multipletts kommen bekanntlich dadurch zustande, daß die Elektronenmagnete im Sinne der Uhlenbeck-Goudsmitschen Hypothese sich parallel bzw. antiparallel stellen und sich dann im ganzen mit dem Drehimpuls l der Elektronen zum Gesamtimpuls j zusammensetzen. Im mechanischen Modell würden sich die Elektronenmagnete nicht parallel stellen, da zwischen ihnen nur außerordentlich kleine Wechselwirkungen bestehen. Selbst wenn sehr große Wechselwirkungen vorhanden wären, so hingen diese Wirkungen nicht nur von der relativen Richtung der Magnete, sondern auch von der relativen Lage der Elektronenschwerpunkte und insbesondere von den durch diese Lagen ausgezeichneten Richtungen ab; und dann könnten sich die Magnete nicht parallel stellen. Wir müssen also erwarten, daß für die Deutung der Multiplettstruktur und der damit zusammenhängenden Gesetzmäßigkeiten die quantenmechanische Resonanz eine entscheidende Rolle spielt.

§ 2. Da die Kräfte, welche die Elektronenmagnete mit der Bewegung der Elektronenschwerpunkte koppeln, im allgemeinen klein sind, so mag es berechtigt sein, diese Kräfte in erster Annäherung zu vernachlässigen. Aus den empirischen Ergebnissen ist für diesen Grenzfall (d. h. wenn die Multiplettaufspaltung sehr klein gegen den Abstand verschiedener Multipletts ist) zu schließen, daß das ganze Termsystem eines Atoms eingeteilt werden kann in Teilsysteme, zwischen denen Interkombinationen nicht vorkommen. (Vgl. Ortho- und Parasystem im Heliumspektrum).

[1]) Vgl. z. B. W. Heisenberg, Schwankungserscheinungen und Quantenmechanik, l. c.

Wir führen zunächst diese Einteilung in nicht kombinierende Termsysteme für n punktsymmetrische Elektronen aus. Dabei gehen wir von einem „ungestörten" System aus, bei dem die Wechselwirkung zwischen den Elektronen Null ist und betrachten die Wechselwirkung als Störung. Im ungestörten System mögen sich die Elektronen in verschiedenen stationären Zuständen befinden, zu denen die Eigenfunktionen $l, m, \ldots p$ gehören. Die Nummer desjenigen Elektrons, von dessen Koordinaten eine Eigenfunktion abhängt, sei als Index der betreffenden Eigenfunktionen angehängt (z. B. $l_1, m_2 \ldots p_n$). Die Eigenfunktion des Gesamtsystems kann im ungestörten Problem als Produkt aller Eigenfunktionen der einzelnen Elektronen geschrieben werden, z. B. $l_1 m_2 \ldots p_n$. Das ungestörte Problem weist $n!$-fache Entartung auf, da Vertauschung der Elektronen zu gleichen Energiewerten des Gesamtsystems führt. Als Eigenfunktionen kommen also irgendwelche lineare Aggregate solcher Produkte in Betracht.

Wenn eine Einteilung des Termsystems des ungestörten Problems in nicht kombinierende Teilsysteme gefunden ist, so kann diese Einteilung durch irgendwelche Wechselwirkung der Elektronen — vorausgesetzt, daß alle Elektronen gleich sind und daher gleiche Kräfte aufeinander ausüben — nicht gestört werden. Die Einteilung in nicht kombinierende Termsysteme ist also ganz unabhängig von der Art der Wechselwirkung; zugleich ist sie der entscheidende Schritt zur Lösung des Problems mit Wechselwirkung. Die Einteilung in nicht kombinierende Termsysteme läßt sich mathematisch in zwei Weisen formulieren:

1. Eine Reihe nicht kombinierender Termsysteme ist dadurch definiert, daß eine Eigenfunktion irgend eines dieser Systeme, mit einer symmetrischen Funktion der Elektronen multipliziert, nach Eigenfunktionen nur des gleichen Systems entwickelbar ist. Oder:

2. Seien u und v Eigenfunktionen zweier verschiedener Systeme, so soll das Integral $\int u^* v f d\xi$ [s. Gleichung (5)] verschwinden, wenn f irgend eine symmetrische Funktion der Elektronen ist.

Wir versuchen nun, solche Systeme zu konstruieren und beginnen mit der Eigenfunktion

$$u = l_1 m_2 \ldots p_n. \tag{7}$$

Alle Eigenfunktionen $l, m, \ldots p$ seien verschieden. Es gibt im ungestörten System $n!$ Terme gleicher Energie; die zugehörigen Eigenfunktionen gehen aus (7) hervor, wenn die die Elektronennummer bezeichnenden Indizes in irgend einer Weise permutiert werden. Eine Substitution in

Zeitschrift für Physik *41*, 239–267 (1927)

den Nummern der Elektronen heiße S und sei, wie dies in der Gruppentheorie üblich ist, als

$$S = \begin{pmatrix} 1 & 2 & \dots & n \\ a_1 & a_2 & \dots & a_n \end{pmatrix} \tag{8}$$

geschrieben. Mit Su bezeichnen wir dann die Eigenfunktion

$$Su = l_{a_1} m_{a_2} \dots p_{a_n}. \tag{9}$$

Unter der Potenz einer Substitution versteht man die mehrfache Anwendung derselben Substitution; unter $S^0 = 1$ die „identische" Substitution. Es gibt stets eine kleinste Potenz S^ν von S für die ($\nu > 0$)

$$S^\nu = 1 \tag{10}$$

ergibt. ν heißt dann der Grad der Substitution S. Das uns hier gestellte Problem nicht kombinierender Termsysteme steht nun in engstem Zusammenhang mit der Theorie der Permutationsgruppen [1]. Es sei ω eine primitive Wurzel der Gleichung

$$\omega^\nu = 1. \tag{11}$$

Dann bilden wir die Summen von Eigenfunktionen:

$$\left.\begin{aligned} U_1 &= \frac{1}{\sqrt{\nu}}\{u + Su + S^2 u + \dots + S^{\nu-1} u\}, \\ U_\omega &= \frac{1}{\sqrt{\nu}}\{u + \omega Su + \omega^2 S^2 u + \dots + \omega^{\nu-1} S^{\nu-1} u\}, \\ &\dots\dots\dots\dots\dots\dots\dots\dots \\ U_{\omega^{\nu-1}} &= \frac{1}{\sqrt{\nu}}\{u + \omega^{\nu-1} Su + \omega^{2(\nu-1)} S^2 u + \dots + \omega^{(\nu-1)^2} S^{\nu-1} u\} \end{aligned}\right\} \tag{12}$$

(Der Faktor $\frac{1}{\sqrt{\nu}}$ ist eingefügt, damit die Eigenfunktionen U normiert sind, wenn die Eigenfunktionen u es waren.)

Diese ν Eigenfunktionen U werden im allgemeinen nur einen Teil der linear unabhängigen Eigenfunktionen umfassen. ν ist nach bekannten Sätzen stets ein Teiler von $n!$. Wir wählen dann irgend eine Eigenfunktion w, die aus (7) durch Permutation hervorgeht, die aber nicht unter den Funktionen u, Su, ... $S^{\nu-1}u$ enthalten ist und wenden auf sie den gleichen Prozeß mit der gleichen Substitution S an, wie vorher auf u. So erhält man neue Eigenfunktionen W_1, W_ω ... $W_{\omega^{\nu-1}}$. Dieses Verfahren wird so lange fortgesetzt, bis die $n!$ Permutationen von (7)

[1] Vgl. z. B. J. Serret, Handbuch der höheren Algebra, Bd. 2, S. 260. Leipzig 1879.

erschöpft sind. Faßt man alle Terme, die zu einer bestimmten Potenz von ω: ω^λ gehören, zu einem Termsystem Γ_{ω^λ} zusammen — Γ_{ω^λ} mag nun auch Terme mit irgendwelchen anderen Funktionen, als $l, m, \ldots p$ enthalten —, so gilt der Satz:

Die Systeme $\Gamma_1, \Gamma_\omega, \ldots \Gamma_{\omega^{\nu-1}}$ kombinieren nicht miteinander.

Um dies einzusehen, bildet man das Integral [oder die entsprechende Summe, vgl. (5)]

$$\int \gamma^*_{\omega^\lambda} \gamma'_{\omega^\mu} f\, d\xi, \tag{13}$$

wo γ_{ω^λ}, γ'_{ω^μ} irgendwelche Eigenfunktionen des Systems Γ_{ω^λ} bzw. Γ_{ω^μ}, f eine symmetrische Funktion der Elektronen bedeuten. Wenn unter dem Integral (13) die Elektronennummern irgendwie permutiert werden, kann sich der Wert des Integrals nicht ändern. Wir wenden auf den Ausdruck unter dem Integral speziell die Permutation S an. Aus den Gleichungen (12) folgt allgemein, daß $S \cdot \gamma_{\omega^\lambda} = \omega^{-\lambda} \gamma_{\omega^\lambda}$ ist. Also geht, bei Anwendung der Substitution S, (13) über in

$$\omega^{+\lambda-\mu} \int \gamma^*_{\omega^\lambda} \gamma'_{\omega^\mu} f\, d\xi$$

und es folgt

$$(1 - \omega^{+\lambda-\mu}) \int \gamma^*_{\omega^\lambda} \gamma'_{\omega^\mu} f\, d\xi = 0. \tag{14}$$

Also ist das Integral der linken Seite, das die Kombinationen bestimmt, Null für $\lambda \neq \mu$, d. h. die Systeme Γ_{ω^λ}, Γ_{ω^μ} kombinieren nicht miteinander, es sind nur Kombinationen innerhalb eines Systems Γ möglich.

Die Einteilung in nicht kombinierende Termsysteme kann unter Umständen noch weiter fortgesetzt werden so, daß einige oder alle Systeme Γ wieder in verschiedene Teilsysteme zerfallen. Wir denken an eine bestimmte Eigenfunktion des Systems Γ_{ω^λ}:

$$U_{\omega^\lambda} = \frac{1}{\sqrt{\nu}} \{u + \omega^\lambda S u + \omega^{2\lambda} S^2 u + \cdots \omega^{(\nu-1)\lambda} S^{\nu-1} u\}. \tag{15}$$

Wenn es eine neue Substitution T vom Grade μ gibt, derart, daß

$$\begin{aligned} &T(1 + \omega^\lambda S + \omega^{2\lambda} S^2 + \cdots \omega^{(\nu-1)\lambda} S^{\nu-1}) \\ &= (1 + \omega^\lambda S + \omega^{2\lambda} S^2 + \cdots \omega^{(\nu-1)\lambda} S^{\nu-1}) T, \end{aligned} \tag{16}$$

so bilden wir die Funktionen ($\varepsilon^\mu = 1$):

$$\begin{aligned} Z_{\omega^\lambda, \varepsilon^k} = \frac{1}{\sqrt{\nu}} \{ &U_{\omega^\lambda} + \varepsilon^k T U_{\omega^\lambda} \\ &+ \varepsilon^{2k} T^2 U_{\omega^\lambda} + \cdots + \varepsilon^{(\mu-1)k} T^{\mu-1} U_{\omega^\lambda} \} \end{aligned} \tag{17}$$

und teilen in dieser Weise das Termsystem Γ_{ω^λ} in μ neue Systeme, die nicht kombinieren; denn nur je Terme mit gleichem Exponenten von ε kombinieren dann nach Gleichung (13). Die Bedingung (16) muß aber für die Substitution T erfüllt sein, damit die Eigenfunktion $T\,U_{\omega^\lambda}$ wieder zum System Γ_{ω^λ} gehört.

Die Gleichung (16) ist für alle Systeme Γ_{ω^λ} erfüllt, wenn T mit S vertauschbar ist, d. h. wenn

$$ST = TS \tag{18}$$

ist. Die elementare Gruppentheorie gibt allgemeine Methoden zum Auffinden von Substitutionen T, die dieser Bedingung genügen, wenn S gegeben ist. Wenn ω^λ eine primitive ν-te Wurzel von 1 ist, muß T sogar der Bedingung (18) genügen, um (16) genügen zu können.

Wenn ω^λ keine primitive ν-te Wurzel von 1 ist, so wird es jedoch im allgemeinen Substitutionen T geben, die nicht mit S vertauschbar sind, aber doch der Gleichung (16) genügen. Im allgemeinen werden also die Termsysteme Γ_{ω^λ}, die zu nicht primitiven Wurzeln ω^λ gehören, einer weiteren Unterteilung fähig sein als die Systeme, die zu primitiven Einheitswurzeln gehören.

Es mag noch darauf hingewiesen werden, daß die identische Substitution 1 und irgendwelche Potenz S^r von S, für T eingesetzt, sicher der Gleichung (18) genügen, daß sie aber nicht zu einer neuen Einteilung der Systeme Γ_{ω^λ} Anlaß geben.

Die Frage, in wieviele nicht kombinierende Termsysteme das Termschema eines Atoms mit n Elektronen zerfällt, kann von den hier durchgeführten Überlegungen aus nicht in dieser Allgemeinheit beantwortet werden. Aber wir können wohl nach dem angegebenen Verfahren in jedem bestimmten Falle die Einteilung finden. Da es in praktischen Fällen von Wichtigkeit sein wird, diese Einteilung in möglichst wenigen Schritten durchzuführen, so beanspruchen am meisten die Substitutionen hohen Grades unser Interesse, da sie zu einer Einteilung in entsprechend viele Termsysteme Anlaß geben. Den Grad einer Substitution kann man am einfachsten beurteilen, wenn man die Substitution in ihre Zyklen auflöst. Eine zyklische Substitution kann in der Form $(a_1\, a_2 \ldots a_z)$ geschrieben werden und bedeutet dann in der oben gebrauchten Schreibweise (8)

$$\begin{pmatrix} a_1 & a_2 & \ldots & a_z \\ a_2 & a_3 & \ldots & a_1 \end{pmatrix}. \tag{19}$$

Der Grad dieser Substitution ist gleich der Anzahl ihrer Elemente z. Jede Substitution läßt sich als Produkt mehrerer auf verschiedene Elemente bezogenen zyklischen Substitutionen darstellen. Bezeichnet also $z_1, z_2, \ldots$ den Grad der verschiedenen Zyklen, in die sich S auflösen läßt, so ist

$$z_1 + z_2 + \cdots = n$$

und das kleinste gemeinsame Vielfache von $z_1, z_2, \ldots$ gibt den Grad von S an. Im Falle z. B. von 12 Elektronen in verschiedenen Quantenbahnen wären für S Substitutionen hohen Grades von folgendem Typus anwendbar:

$$S = (1\,2\,3\,4\,5)\,(6\,7\,8\,9\,10\,11\,12).$$

Der Grad von S ist hier $5 . 7 = 35$ und das oben beschriebene Verfahren gibt 35 nicht kombinierende Termsysteme. Ferner

$$S = (1\,2\,3)\,(4\,5\,6\,7)\,(8\,9\,10\,11\,12).$$

Der Grad von S ist $3 . 4 . 5$, es ergibt sich eine Einteilung in 60 nicht kombinierende Systeme. Dies ist offenbar die höchste Anzahl nicht kombinierender Systeme, die sich in einem Schritt mit der oben beschriebenen Methode für $n = 12$ erreichen läßt.

Wenn einige der Eigenfunktionen $l, m, \ldots p$ gleich sind — also im Falle äquivalenter Elektronen —, so ist die Anzahl der verschiedenen Terme gleicher Energie nicht mehr $n!$, sondern $\frac{n!}{n_\alpha!\, n_\beta! \ldots}$, wo $n_\alpha, n_\beta \ldots$ die Anzahl der Elektronen im Zustand α, β bezeichnet. In diesem Falle hat man, um das oben geschilderte Verfahren anwenden zu können, Substitutionen S zu wählen, deren Grad ein Teiler von $\frac{n!}{n_\alpha!\, n_\beta! \ldots}$ ist, und mit diesen die Einteilung in nicht kombinierende Terme durchzuführen.

Um ein Beispiel für diese Methoden zu geben, sei die Rechnung für das Lithiumatom, $n = 3$, durchgeführt[1]). Es ist die Eigenfunktion des Ausgangssystems:

$$u = l_1 m_2 p_3.$$

Wir nehmen zunächst an, daß l, m und p verschieden seien. Die Substitution höchsten Grades ist die zyklische, etwa $(1\,2\,3)$. Sei ε eine primitive dritte Wurzel von 1, so erhält man zunächst nach (12)

$$\left.\begin{aligned} U_1 &= \frac{1}{\sqrt{3}}\,\{l_1 m_2 p_3 + l_2 m_3 p_1 + l_3 m_1 p_2\} \\ U_\varepsilon &= \frac{1}{\sqrt{3}}\,\{l_1 m_2 p_3 + \varepsilon\, l_2 m_3 p_1 + \varepsilon^2\, l_3 m_1 p_2\} \\ U_{\varepsilon^2} &= \frac{1}{\sqrt{3}}\,\{l_1 m_2 p_3 + \varepsilon^2\, l_2 m_3 p_1 + \varepsilon\, l_3 m_1 p_2\}. \end{aligned}\right\} \tag{20}$$

[1]) Anm. bei der Korrektur. In einer inzwischen erschienenen Arbeit ist E. Wigner, ZS. f. Phys. **40**, 492, 1926 für das Problem der drei Elektronen unabhängig zu Resultaten gekommen, die den hier gegebenen äquivalent sind.

Zeitschrift für Physik 41, 239–267 (1927)

Unter den Eigenfunktionen (20) kam etwa $m_1 l_2 p_3$ noch nicht vor. Es ergibt sich also eine weitere Reihe W_1, W_ε, W_{ε^2}. Die drei nicht kombinierenden Termsysteme lassen sich am einfachsten unter der Form zusammenfassen:

$$\Gamma_\eta = \left\{ \begin{aligned} \gamma_\eta^{\mathrm{I}} &= \frac{1}{\sqrt{3}} \{l_1 m_2 p_3 + \eta\, l_2 m_3 p_1 + \eta^2 l_3 m_1 p_2\}, \\ \gamma_\eta^{\mathrm{II}} &= \frac{1}{\sqrt{3}} \{m_1 l_2 p_3 + \eta\, m_2 l_3 p_1 + \eta^2 m_3 l_1 p_2\}, \end{aligned} \right\} \tag{21}$$

wo η für die drei Systeme bzw. die Werte 1, ε, ε^2 annimmt. Die Termsysteme Γ_ε und Γ_{ε^2} können nicht mehr weiter eingeteilt werden, da mit der Substitution (1 2 3) nur die Potenzen dieser Substitution vertauschbar sind. Γ_1 dagegen gehört zu einer nicht primitiven Einheitswurzel. Wählt man als Substitution T irgend eine Transposition, so ist für T Gleichung (16) erfüllt. Der Grad der Substitution T ist 2, die beiden nicht kombinierenden Termsysteme, in die Γ_1 noch eingeteilt wird, gehören zu den Quadratwurzeln von 1: + 1 bzw. — 1. Die Eigenfunktionen sind

$$\Gamma_{1,1} : \frac{1}{\sqrt{2}} (\gamma_1^{\mathrm{I}} + \gamma_1^{\mathrm{II}}); \quad \Gamma_{1,-1} : \frac{1}{\sqrt{2}} (\gamma_1^{\mathrm{I}} - \gamma_1^{\mathrm{II}}). \tag{21a}$$

$\Gamma_{1,1}$ und $\Gamma_{1,-1}$ sind als das symmetrische bzw. das antisymmetrische Termsystem wohlbekannt.

In den Systemen $\Gamma_{1,1}$ und $\Gamma_{1,-1}$ ist das Problem des ungestörten Atoms nicht mehr entartet, da $\Gamma_{1,1}$ bzw. $\Gamma_{1,-1}$ nur je einen Term einer bestimmten Energie im ungestörten Atom enthalten. Innerhalb der Termsysteme Γ_ε und Γ_{ε^2} ist aber insofern eine Entartung vorhanden, als je zwei Terme gleicher Energie im ungestörten System vorkommen. Diese Entartung wird durch die Wechselwirkung aufgehoben in einer Weise, die von der Art der Wechselwirkung abhängt.

Die Matrix der über die ungestörte Bewegung gemittelten Wechselwirkungsenergie hat vier wesentlich verschiedene Glieder, ein „Mittelwertsglied": $W_{l,m,p;\,l,m,p} = A$ und drei „Resonanzglieder", die Vertauschungen je zweier Elektronen entsprechen:

$$W_{l,m,p;\,m,l,p} = B; \quad W_{l,m,p;\,p,m,l} = C; \quad W_{l,m,p,\,l,p,m} = D.$$

Die Wechselwirkung führt zu einem Störungsproblem, das nach bekannten Prinzipien durch die folgende Determinante gegeben ist:

$$\begin{vmatrix} W - A & -(B + \eta\, C + \eta^2 D) \\ -(B + \eta^2 C + \eta\, D) & W - A. \end{vmatrix}$$

Dies ergibt die sechs Lösungen (22):

Γ_1	Γ_ε	Γ_{ε^2}
$W = A + B + D + C$, E. F. $\frac{1}{\sqrt{2}}(\gamma_1^{\mathrm{I}} + \gamma_1^{\mathrm{II}})$ $\Big\}\ \Gamma_{1,1}$	$W = A + R$, E. F. $\frac{1}{\sqrt{2}}(e^{i\varphi}\gamma_\varepsilon^{\mathrm{I}} + e^{-i\varphi}\gamma_\varepsilon^{\mathrm{II}})$	$W = A + R$, E. F. $\frac{1}{\sqrt{2}}(e^{-i\varphi}\gamma_{\varepsilon^2}^{\mathrm{I}} + e^{i\varphi}\gamma_{\varepsilon^2}^{\mathrm{II}})$
$W = A - B - D - C$, E. F. $\frac{1}{\sqrt{2}}(\gamma_1 - \gamma_1^{\mathrm{II}})$ $\Big\}\ \Gamma_{1,-1}$	$W = A - R$, E. F. $\frac{1}{\sqrt{2}}(e^{i\varphi}\gamma_\varepsilon^{\mathrm{I}} - e^{-i\varphi}\gamma_\varepsilon^{\mathrm{II}})$	$W = A - R$, E. F. $\frac{1}{\sqrt{2}}(e^{-i\varphi}\gamma_{\varepsilon^2}^{\mathrm{I}} - e^{i\varphi}\gamma_{\varepsilon^2}^{\mathrm{II}})$

Hierbei ist $B + \eta C + \eta^2 D = R e^{2i\varphi}$ gesetzt.

Im Termsystem (22) besteht eine charakteristische Entartung noch insofern, als je ein Term des Systems Γ_ε die gleiche Energie hat, wie ein Term des Systems Γ_{ε^2}. Formal liegt diese Entartung daran, daß die Matrixelemente der Wechselwirkungsenergie B, C, D reell sind. Sie sind dies für jede beliebige Art von Wechselwirkung und die eben genannte Entartung ist eine wesentliche Seite des Resonanzphänomens. Gelegentlich der Diskussion der Multiplettstruktur der Termsysteme werden wir hierauf noch zurückkommen. Hier sei darauf hingewiesen, daß diese Entartung zur Folge hat, daß die Einteilung in nicht kombinierende Termsysteme bis zum gewissen Grade unbestimmt sein kann.

Wenn zwei Elektronen äquivalent sind, etwa $l = m$, so gibt die Γ-Einteilung schon ein vollständiges Termschema mit je einer Eigenfunktion $\gamma_\eta^{\mathrm{I}} = \gamma_\eta^{\mathrm{II}}$. Entartungen treten dann innerhalb eines Systems Γ_η nicht mehr auf, ferner ist $A = D$, $B = C$; man erhält die Lösungen

$\Gamma_{1,1}$	Γ_ε	Γ_{ε^2}
$W = A + 2B$, E. F. γ_1	$W = A - B$, E. F. γ_ε	$W = A - B$, E. F. γ_{ε^2}

(23)

§ 3. Wenn nach den oben geschilderten Methoden das Termspektrum des Atoms bei Annahme kugelsymmetrischer Elektronen bestimmt worden ist, so können nachträglich die Veränderungen berücksichtigt werden, die eintreten, wenn die Wechselwirkung der Elektronenmagnete mit der Schwerpunktsbewegung der Elektronen nicht mehr vernachlässigt werden kann. Wir wollen zunächst die Tatsache der Multiplettstruktur, d. h. die Tatsache, daß die Elektronenmagnete sich stets „parallel" oder „antiparallel" stellen, aus solchen Rechnungen herleiten.

Die Eigenfunktionen des Gesamtsystems, das aus Elektronen mit Eigenrotation besteht, werden, solange die oben genannte Magnetwechsel-

Zeitschrift für Physik *41*, 239–267 (1927)

wirkung klein ist, sich als Produkte der Eigenfunktionen der Schwerpunktsbewegung einerseits, der Magnetbewegung andererseits darstellen lassen. Wenn mehrere Eigenfunktionen der Schwerpunktsbewegung zu dem gleichem Energiewert gehören — wie dies nach den obigen Rechnungen auch bei Berücksichtigung der Wechselwirkung manchmal der Fall sein wird —, so kommen lineare Aggregate von Produkten der Schwerpunktseigenfunktionen und der Magneteigenfunktionen für die Gesamteigenfunktion in Betracht.

Die Eigenfunktionen des Gesamtsystems sollen antisymmetrisch in den Elektronen sein. Wir gehen von einem ungestörten System aus, bei dem die Wechselwirkung der Magnete mit der Schwerpunktsbewegung Null ist und zählen in diesem ungestörten System die Stellungen der Magnete relativ zu einer willkürlich gewählten Richtung. Die Wechselwirkungsenergie zwischen Magneten und Elektronen wird betrachtet als Störungsenergie, von der der zeitliche Mittelwert über die Bewegung der Elektronenschwerpunkte die säkularen Bewegungen der Magnete bestimmt. Die Störungsenergie ist symmetrisch in den Elektronen. Da ferner die Eigenfunktion des Gesamtsystems antisymmetrisch ist, so ist auch jener Zeitmittelwert der Störungsenergie symmetrisch in den Elektronen. Die Wechselwirkung zwischen dem Magneten eines Elektrons und den übrigen Elektronen und deren Magneten hängt ab von der Stellung dieses Magnets relativ zu den durch die Lagen aller Elektronen ausgezeichneten Richtungen und relativ zu den Stellungen der anderen Magnete. Nun führen wegen der Antisymmetrie der Gesamteigenfunktion alle Elektronen die gleiche „Bewegung" (in verschiedenen Phasen) aus; also kann jener Mittelwert der Störungsenergie nur abhängen: erstens von der Richtung der Magnete relativ zu einer durch die Schwerpunktsbewegung der Elektronen ausgezeichneten Gesamtrichtung — dies kann in Atomen nur die Richtung des Gesamtdrehimpulses der Elektronen sein, in zweiatomigen Molekülen etwa käme auch die Kernverbindungslinie in Betracht —, und zweitens von der relativen Richtung der Magnete. Es hat nun zunächst den Anschein, als ob die endgültige Einstellung der Magnete vom Verhältnis jener zwei Teile der gemittelten Störungsenergie abhänge. Dies ist aber nicht der Fall. Vielmehr setzen sich die Drehimpulse s_i der Magnete unabhängig von jenem Verhältnis stets zu einem gequantelten Impuls s nach der gewöhnlichen Strukturregel zusammen, d. h. sie stellen sich „parallel" oder „antiparallel". Um dies einzusehen, behandeln wir ein etwas einfacheres mechanisches Problem: An Stelle der Wechselwirkung, die von der Richtung des Gesamtimpulses $\mathfrak{l}$

der Elektronen abhängt, führen wir eine äußere Kraft ein; wir setzen die Hamiltonsche Funktion additiv zusammen aus zwei Gliedern, von denen das erste der Wirkung der äußeren Kraft entspricht, das zweite stellt eine Wechselwirkung zwischen den Magneten dar, die nur von der gegenseitigen Richtung der Magnete abhängt. Die Hamiltonsche Funktion des eigentlichen Störungsproblems setzt sich allerdings nicht additiv aus diesen beiden Teilen zusammen. Zwar entspricht die Wechselwirkung der Magnete mit der Schwerpunktsbewegung ganz dem, was hier äußere Kraft genannt wurde. Aber die (gemittelte) Wechselwirkung der Magnete untereinander hängt gleichzeitig von der gegenseitigen Richtung, wie von der Richtung relativ zur Achse l ab. Man kann aber leicht einsehen: Wenn im Falle der Additivität jener beiden Glieder die oben aufgestellte Behauptung über die Zusammensetzung von s_i richtig ist, gilt sie auch allgemein.

Gehen wir also zur quantenmechanischen Behandlung jenes vereinfachten Problems über:

Im „ungestörten" System, in dem die gegenseitige Wechselwirkung der Magnete vernachlässigt ist, hat jeder Magnet zwei mögliche Stellungen relativ zur äußeren, als axialsymmetrisch gedachten Kraft $m_s = \pm \frac{1}{2}$; diesen zwei Stellungen entsprechen zwei Eigenfunktionen $S^4(\xi_4)$, die wir der Einfachheit halber mit α und β bezeichnen. Die Nummer des Elektrons, zu dem diese Eigenfunktionen gehören, d. h. von dessen Variablen sie abhängen, sei als Index angefügt: $\alpha_1, \alpha_2, \ldots \alpha_n$; $\beta_1, \beta_2, \ldots \beta_n$. Es besteht im ungestörten System Entartung insofern, als alle Zustände gleiche Energie besitzen, die zu einem gegebenen Werte von $\sum m_s$, d. h. zu einem bestimmten Werte des Gesamtimpulses um die Achse des äußeren Feldes gehören; denn bei gegebener Anzahl der Magnete n ist durch $m = \sum m_s$ die Anzahl n_α bzw. n_β der Zustände mit den Eigenfunktionen α und β gegeben:

$$n_\alpha + n_\beta = n; \quad \tfrac{1}{2}(n_\alpha - n_\beta) = m. \tag{24}$$

Für Werte $s > \frac{1}{2}$ ließe sich eine solche Betrachtung nicht durchführen. Die Tatsache der Multiplettstruktur der komplizierten Spektra ist an den speziellen Wert $s = \frac{1}{2}$ geknüpft. Die äußere Kraft stellt an das ungestörte Problem keine andere Forderung als die, daß $\sum_n m_s = m$ gequantelt sein soll, und diese Forderung gerät auch mit einer noch so hohen Wechselwirkungsenergie der Elektronenmagnete nicht in Widerspruch. Umgekehrt: Die wirkliche Aufteilung der Systeme ursprünglich gleicher Energie — ihre Anzahl ist $\dfrac{n!}{n_\alpha!\, n_\beta!}$ — in Systeme verschiedener

Zeitschrift für Physik *41*, 239–267 (1927)

Energie auf Grund einer noch so kleinen gegenseitigen Wechselwirkung erfolgt ganz unabhängig von der äußeren Kraft, d. h. sie entspricht der Parallel- und Antiparallelstellung der Magnete.

Die eben genannte Entartung bei verschwindender Wechselwirkung ist völlig analog zur Entartung in den Rechnungen von § 2 und kann insofern als Resonanzentartung bezeichnet werden. Die Resonanzschwebung, die ihr entspricht, ist aber eben die einfache Präzession der Vektoren s_i um die Achse s des Gesamtdrehimpulses der Magnete. Mit dem Wort Resonanz darf man also hier nicht die Vorstellung von einem komplizierten Wechselspiel in den Bewegungen der Magnete verbinden.

Zur wirklichen Ausrechnung der Eigenfunktionen, die zu den so entstehenden Multipletts gehören, seien noch einige Formeln angegeben: Die Wechselwirkungsenergie zwischen irgend zweien der Elektronenmagnete, etwa 1 und 2, wird eine Matrix sein, die Diagonalglieder und je zwei Glieder, die Vertauschungen der Elektronen entsprechen, enthält. In anderen Worten: Es gibt für zwei Elektronen vier Zustände, charakterisiert durch:

	I	II	III	IV
m_1	$\frac{1}{2}$	$\frac{1}{2}$	$-\frac{1}{2}$	$-\frac{1}{2}$
m_2	$\frac{1}{2}$	$-\frac{1}{2}$	$\frac{1}{2}$	$-\frac{1}{2}$

Die Matrix der Wechselwirkungsenergie enthält nur Übergänge bei konstantem m, da sie nur von der relativen Richtung von 1 und 2 abhängt, d. h. nur Übergänge von II nach III. Ferner ist $H_{\mathrm{I,I}} = H_{\mathrm{IV,IV}} = a$; $H_{\mathrm{II,II}} = H_{\mathrm{III,III}} = b$; $H_{\mathrm{III,II}} = H_{\mathrm{II,III}} = c$. Zwischen diesen drei Konstanten besteht noch die Beziehung

$$a = b + c. \tag{25}$$

Dies ist leicht zu sehen aus einer Behandlung des Problems der zwei Elektronen. Die Gesamtenergie W des Systems (unter Weglassung der äußeren Kräfte) ist im Zustand $m = 1$ (also $m_1 = m_2 = \frac{1}{2}$) offenbar $H_{\mathrm{I,I}} = a$ (Eigenfunktion $\alpha_1 \alpha_2$), ebenso im Zustand $m = -1$ ($m_1 = m_2 = -\frac{1}{2}$) $H_{\mathrm{IV,IV}} = a$ (Eigenfunktion $\beta_1 \beta_2$). Im Zustand $m = 0$ besteht Entartung, die nach bekannten Methoden zur Gleichung (l. c.)

$$W = S^{-1} H S$$

und zur Determinante

$$\begin{vmatrix} W - H_{\mathrm{II,II}} & -H_{\mathrm{II,III}} \\ -H_{\mathrm{III,II}} & W - H_{\mathrm{III,III}} \end{vmatrix} = 0 \tag{26}$$

führt. Es ergibt sich

$$(W-b)^2 = c^2, \quad W = b + c \quad \text{und} \quad W = b - c.$$

Der durch $W = b + c$ charakterisierte Zustand hat die Eigenfunktion $\frac{1}{\sqrt{2}}(\alpha_1 \beta_2 + \alpha_2 \beta_1)$; sie ist symmetrisch in den Elektronen 1 und 2 und gehört deshalb zum Triplettsystem. $W = b - c$ gehört zur Eigenfunktion $\frac{1}{\sqrt{2}}(\alpha_1 \beta_2 - \alpha_2 \beta_1)$ und bildet das Singlettsystem. Solange keine äußeren Kräfte auf die Magnete wirken, haben die drei Zustände des Tripletts gleiche Energie, also $a = b + c$. Wir drücken in den folgenden Rechnungen stets b durch a und c aus.

Als Beispiel mag der Fall dreier Elektronenmagnete im äußeren Kraftfeld und mit Wechselwirkung hier angeschrieben werden. Es gibt einen Zustand $m = \frac{3}{2}$ ohne Entartung. Der Energieausdruck der äußeren Kraft in bezug auf einen Elektronenmagnet mit $m_s = \frac{1}{2}$ mag mit K_α, der in bezug auf $m_s = -\frac{1}{2}$ mit K_β bezeichnet werden. Zu $m = \frac{3}{2}$ gehört dann die Energie $3K_\alpha + 3a$, die Eigenfunktion $\alpha_1 \alpha_2 \alpha_3$. Im Zustand $m = \frac{1}{2}$ besteht dreifache Entartung, charakterisiert durch die drei Eigenfunktionen $\alpha_1 \alpha_2 \beta_3$, $\alpha_3 \alpha_1 \beta_2$, $\alpha_2 \alpha_3 \beta_1$, die zur gleichen Energie gehören. Die Energie, die von der äußeren Kraft herrührt, ist $2K_\alpha + K_\beta$. Bezeichnet man $W - 2K_\alpha - K_\beta - a - 2b = W - 2K_\alpha - K_\beta - 3a + 2c$ mit X, so ergibt sich die Determinante

$$\begin{matrix} (123) \\ (231) \\ (312) \end{matrix} \left| \begin{matrix} X & -c & -c \\ -c & X & -c \\ -c & -c & X \end{matrix} \right| \qquad (27)$$

Die Lösung ist:

$$\left.\begin{aligned} X_1 &= 2c; & &\text{Eigenfunktion } \delta_1 = \frac{1}{\sqrt{3}}(\alpha_1\alpha_2\beta_3 + \alpha_2\alpha_3\beta_1 + \alpha_3\alpha_1\beta_2) \\ X_2 &= X_3 = -c; & &\text{Eigenfunktionen } \delta_\varepsilon = \frac{1}{\sqrt{3}}(\alpha_1\alpha_2\beta_3 + \varepsilon\alpha_2\alpha_3\beta_1 + \varepsilon^2\alpha_3\alpha_1\beta_2) \end{aligned}\right\}, \quad (28)$$

wo ε eine der beiden primitiven Wurzeln der Gleichung $\varepsilon^3 = 1$ bedeutet. X_1 gehört zu einer in allen Elektronen symmetrischen Eigenfunktion und daher zum selben Termsystem wie die Lösung $m = \frac{3}{2}$, $\alpha_1 \alpha_2 \alpha_3$. Die Energie W ist hier

$$W = 2K_\alpha + K_\beta + 3a.$$

Bei verschwindender äußerer Kraft ist dies die Energie des Zustandes $m = \frac{3}{2}$. Die Rechnungen für $m = -\frac{1}{2}$ und $m = -\frac{3}{2}$ sind analog zu denen für $m = \frac{1}{2}$ und $m = \frac{3}{2}$, und wir erhalten für die Energie das in der folgenden Tabelle angegebene Resultat.

	Quartettsystem	Zwei Dublettsysteme
$m = \frac{3}{2}$	$W = 3a + 3K_\alpha$	
$m = \frac{1}{2}$	$W = 3a + 2K_\alpha + K_\beta$	$W = 3a - 3c + 2K_\alpha + K_\beta$
$m = -\frac{1}{2}$	$W = 3a + K_\alpha + 2K_\beta$	$W = 3a - 3c + K_\alpha + 2K_\beta$
$m = -\frac{3}{2}$	$W = 3a + 3K_\beta$	

(29)

Zwischen den beiden Dublettsystemen besteht Entartung, da sie zu gleichen Energiewerten gehören. Die Einteilung in δ_ε-Systeme, d. h. in Quartetts und Dubletts, ist der Einteilung in Γ-Systeme in Tab. 6 vollkommen analog. Quartetts und Dubletts kombinieren nicht miteinander.

Die Berechnung der Magneteigenfunktionen und der Multipletteinteilung für mehr als drei Elektronen geschieht wieder am einfachsten nach den Methoden von § 2. — Die Anzahl der verschiedenen vorhandenen Multipletts läßt sich nach der gewöhnlichen Strukturregel bei einer gegebenen Anzahl n von Elektronen leicht angeben. Die Anzahl der bei einem bestimmten Werte von m vorhandenen Terme ist nämlich nach (24)

$$\frac{n!}{n_\alpha!\, n_\beta!} = \frac{n!}{\left(\frac{n}{2} + m\right)!\left(\frac{n}{2} - m\right)!}. \tag{30}$$

Also gibt es stets:

$$\frac{n!}{\left(\frac{n}{2}+m\right)!\left(\frac{n}{2}-m\right)!} - \frac{n!}{\left(\frac{n}{2}+m+1\right)!\left(\frac{n}{2}-m-1\right)!} = \frac{n!\,(2m+1)}{\left(\frac{n}{2}+m+1\right)!\left(\frac{n}{2}-m\right)!}. \tag{31}$$

Multipletts der Multiplizität $2m$, wobei m ganz- oder halbzahlig ist, je nachdem $n/2$ ganz- oder halbzahlig ist. Verschiedene Multipletts gleicher Multiplizität haben stets gleiche Energie, solange die Elektronenschwerpunkte festgehalten werden.

Im Atom orientieren sich die Elektronenmagnete nicht nach einer äußeren Kraft, sondern nach der Achse des Gesamtdrehimpulses l der Elektronen. Gerät aber das Atom z. B. in ein starkes äußeres Magnetfeld — wir denken also an den Fall des Paschen-Backeffekts —, so erfolgt die Orientierung der Magnete relativ zu diesem äußeren Felde. In diesem Grenzfall lassen sich die Eigenfunktionen des Gesamtatoms darstellen als Produkte bzw. Summen von Produkten der Eigenfunktionen der Elektronenschwerpunkte und denen der Magnete.

Wir wollen für das Problem der drei Elektronen auch diese vollständige Termeinteilung bei Berücksichtigung der Elektronenmagnete angeben. Dann sollen die Lösungen von § 2, Gl. (22), mit den Lösungen (28) dieses Abschnitts zu Produkten so kombiniert werden, daß Lösungen entstehen, die in den Elektronennummern antisymmetrisch sind.

	Energie	Eigenfunktion
Keine äquivalenten Elektronen, d. h. m, l, p verschieden	Quartettsystem: $W = A - B - C - D$	$\delta_1 \cdot \frac{1}{\sqrt{2}}(\gamma_1^{\mathrm{I}} - \gamma_1^{\mathrm{II}})$
	1. Dublettsystem: $W = A - R$	$\frac{1}{2}\left\{\delta_\varepsilon(e^{-i\varphi}\gamma_{\varepsilon 2}^{\mathrm{I}} - e^{i\varphi}\gamma_{\varepsilon 2}^{\mathrm{II}}) + \delta_{\varepsilon 2}(e^{i\varphi}\gamma_\varepsilon^{\mathrm{I}} - e^{-i\varphi}\gamma_\varepsilon^{\mathrm{II}})\right\}$
	2. Dublettsystem: $W = A + R$	$\frac{1}{2}\left\{\delta_\varepsilon(e^{-i\varphi}\gamma_{\varepsilon 2}^{\mathrm{I}} + e^{i\varphi}\gamma_{\varepsilon 2}^{\mathrm{II}}) - \delta_{\varepsilon 2}(e^{i\varphi}\gamma_\varepsilon^{\mathrm{I}} + e^{-i\varphi}\gamma_\varepsilon^{\mathrm{II}})\right\}$
Zwei Elektronen „äquivalent“, d. h. $l = m$	Dublettsystem: $W = A - B$	$\frac{1}{\sqrt{2}}\left\{\delta_\varepsilon \gamma_{\varepsilon 2} - \delta_{\varepsilon 2}\gamma_\varepsilon\right\}$

(32)

Die Wechselwirkungen der Elektronenmagnete sind in den Energiewerten hier vernachlässigt. Die gleichen Betrachtungen sind auf alle Atome anwendbar, bei denen sich alle Elektronen, bis auf drei, in abgeschlossenen Gruppen befinden. Die Eigenfunktionen des Gesamtproblems sind nicht als Produkte, sondern als Summen von Produkten der Eigenfunktionen von § 2 und der von Gl. (28) darstellbar. Es kommt dies immer dann vor, wenn Eigenfunktionen verschiedener Systeme Γ zu gleichen Energiewerten führen.

Zum Vergleich von (32) mit dem empirischen Befund sei zunächst hervorgehoben, daß sich die verschiedenen Multipletts um Größen der Resonanzwechselwirkung in den Termwerten unterscheiden. Der Term höchster Multiplizität liegt am tiefsten, denn die Größen B, C, D sind, wie in der Theorie des Heliumspektrums (vgl. l. c., S. 508), im allgemeinen positiv. Empirisch liegt ganz allgemein der Term höchster Multiplizität bei gleichen Elektronenquantenzahlen am tiefsten, und dies kann ohne Schwierigkeiten hier auch allgemein gefolgert werden. Denn der Term höchster Multiplizität entspricht ganz allgemein der symmetrischen Eigenfunktion aller Elektronenmagnete, wenn die Eigenfunktionen $l, m \ldots p$ verschieden sind. (Für den Maximalwert $m = \frac{n}{2}$ ist nämlich die einzige zugehörige Eigenfunktion $\alpha_1 \alpha_2 \ldots \alpha_n$.) Also muß die mit dieser Eigenfunktion der Magnete zu multiplizierende Eigenfunktion **des** Problems der kugelsymmetrischen Elektronen notwendig die **anti-**

Zeitschrift für Physik *41*, 239–267 (1927)

symmetrische Eigenfunktion sein. Der ihr entsprechende Energiewert enthält stets alle Resonanzglieder mit negativem Zeichen [vgl. Teil I, Gl. (19)] und ist daher kleiner als alle anderen Energiewerte. Wenn einige Elektronen in abgeschlossenen Gruppen gebunden sind, so ist das eben angeführte Argument noch auf die übrigen Elektronen anwendbar, sofern die Resonanz mit den abgeschlossenen Schalen vernachlässigt werden kann.

Zum Schema (32) sei noch bemerkt, daß — analog zum Heliumproblem — im Atom schließlich Kombinationen zwischen Dublett- und Quartettsystemen nur auftreten mit einer relativen Stärke, die der relativen Weite der Multipletts im Verhältnis zu den Abständen verschiedener Multipletts entspricht. Dagegen kombinieren die verschiedenen Dublettsysteme miteinander mit normalen Intensitäten.

§ 4. In den vorhergehenden Abschnitten haben wir die Bewegungen der Elektronenschwerpunkte und der Magnete zuerst getrennt behandelt, und wir sind dann durch Zusammensetzen der erhaltenen Lösungen zu den Lösungen für das gegebene Atomproblem gelangt. Um das gestellte Problem von allen Seiten kennenzulernen, wird es wichtig sein, diesen Weg auch einmal in umgekehrter Richtung zu gehen und mit einem Atom, das aus magnetischen Elektronen besteht, zu beginnen.

Wir gehen aus vom Fall starker äußerer Felder, wobei die Wechselwirkung zwischen verschiedenen Elektronen ganz vernachlässigt sei. Dann befindet sich jedes Elektron für sich in einem stationären Zustand. Da die Wechselwirkung der Elektronenmagnete mit der Schwerpunktsbewegung im allgemeinen klein ist, läßt sich die Eigenfunktion eines Elektrons nach (6) im allgemeinen als Produkt darstellen. Wir verstehen unter $l, m, \ldots p$ wieder diejenigen Teile der Eigenfunktionen, die zur Schwerpunktsbewegung gehören, unter α und β diejenigen, die den Elektronenmagneten entsprechen. Die gesamte Eigenfunktion eines Elektrons hat dann also etwa die Form $l_1 \alpha_1$. Wenn die Wechselwirkung der Magnete mit der Schwerpunktsbewegung nicht klein ist, ist eine solche Produktdarstellung nicht möglich, wir schreiben dann einfach l_1^α.

In diesem letzten Falle ist das ungestörte System nicht entartet; die als Lösung auszuwählende antisymmetrische Eigenfunktion ist durch die zugehörige Determinante, etwa

$$\begin{vmatrix} l_1^\alpha \, m_1^\alpha \ldots p_1^\beta \\ l_2^\alpha \, m_2^\alpha \ldots p_2^\beta \\ \cdot \quad \cdot \quad \cdot \quad \cdot \\ \cdot \quad \cdot \quad \cdot \quad \cdot \\ l_n^\alpha \, m_n^\alpha \ldots p_n^\beta \end{vmatrix} \tag{33}$$

gegeben. Als Abkürzung für diese Determinante sei das Zeichen

$$\begin{array}{c|l} & l^{\alpha}\, m^{\alpha} \ldots p^{\beta} \\ \hline 1 & \\ 2 & \\ \vdots & \\ n & \end{array}$$

geschrieben. Wenn jedoch die Wechselwirkung der Magnete mit der Schwerpunktsbewegung der Elektronen vernachlässigt werden kann (d. h. Aufspaltung innerhalb der Multipletts klein gegen den Abstand verschiedener Multipletts), so ist das ungestörte System auch bei beliebig starken äußeren Kräften entartet. Man kann nämlich die „α" und „β" Stellungen der Magnete auf die Elektronenstellungen l, m ... in beliebiger Weise verteilen. Es tritt also hier die Entartung vom Magnetproblem des § 3 auf, die die Einteilung in Multipletts zur Folge hat. Wenn die Eigenfunktionen l, m, ... p alle verschieden sind, so ist die Anzahl der zum gleichen Energiewert des ungestörten Problems gehörigen Determinanten durch (24) und die Anzahl der vorkommenden Multipletts durch (31) gegeben. Wenn zwei von ihnen, etwa die beiden ersten, gleich sind, so werden die Determinanten

$$\begin{array}{c|l} & l^{\alpha}\, l^{\alpha} \ldots p^{\beta} \\ \hline 1 & \\ 2 & \\ \vdots & \\ n & \end{array} \quad \text{und} \quad \begin{array}{c|l} & l^{\beta}\, l^{\beta} \ldots p^{\beta} \\ \hline 1 & \\ 2 & \\ \vdots & \\ n & \end{array}$$

identisch Null, die anderen beiden Determinanten

$$\begin{array}{c|l} & l^{\alpha}\, l^{\beta} \ldots p^{\beta} \\ \hline 1 & \\ 2 & \\ \vdots & \\ n & \end{array} \quad \text{und} \quad \begin{array}{c|l} & l^{\beta}\, l^{\alpha} \ldots p^{\beta} \\ \hline 1 & \\ 2 & \\ \vdots & \\ n & \end{array}$$

werden bis aufs Vorzeichen gleich. Zu zwei gleichen Eigenfunktionen der Elektronen müssen also jedenfalls verschiedene Eigenfunktionen der Magnete α und β hinzutreten; „äquivalente Bahnen" kommen nicht vor. Die Anzahl der möglichen Multipletts ist gleich der Anzahl bei $n - 2a$ Elektronenmagneten im Sinne von § 3, wo a die Anzahl der vorhandenen Paare gleicher l, m ... bedeutet. Mehr als je zwei gleiche Eigenfunktionen l, m, ... kommen nicht vor.

Bei den Rechnungen in § 3 waren zunächst die Abstände zwischen den verschiedenen Multipletts durch die kleine Magnetwechselwirkung bedingt. Multipletts gleicher Vielfachheit hatten gleiche Energie. Jetzt sind von vornherein die Abstände der Multipletts durch die Resonanzwechselwirkung der Elektronen gegeben. — Der weitere Gang der Rechnung hängt von den speziellen Bedingungen des betrachteten Atomsystems ab.

Zuweilen ist es zweckmäßig, von einem bestimmten Term eines Atoms auszugehen und zu untersuchen, welche Zustände des Gesamtsystems entstehen, wenn an diesen einen Term ein Elektron in einer gegebenen Quantenbahn angelagert wird. Ein solches Verfahren führt etwa zu folgenden Rechnungen:

Das Ausgangssystem habe $n-1$ Elektronen und die antisymmetrische Eigenfunktion

$$U(1, 2, \ldots, n-1). \tag{34}$$

Entartungen seien durch äußere Felder aufgehoben. Diese Funktion gehört zu einem gewissen Wert $m = \sum_n m_s$ im Sinne des § 3. Wir denken etwa an einen positiven Wert von m, dann wird es stets einen anderen stationären Zustand mit der Eigenfunktion W geben, der z. B. zum Werte $\sum_n m_s = m - 1$ gehört und der, bei verschwindender Multiplettaufspaltung und bei verschwindenden äußeren Kräften die gleiche Energie hat, d. h. eben zum gleichen Multiplett gehört.

Es werde jetzt ein Elektron mit der Eigenfunktion k_n^β angelagert. Solange die Wechselwirkung dieses neuen Elektrons mit dem Atomsystem sehr klein ist, besteht keine Entartung. Die (antisymmetrische!) Eigenfunktion des ganzen Systems heißt

$$\frac{1}{\sqrt{n}}\{U(1, 2, \ldots n-1)\, k_n^\beta + (-1)^n U(2, 3, \ldots n)\, k_1^\beta \\ + (-1)^{2n} U(3, 4 \ldots n, 1)\, k_2^\beta + \cdots\} \tag{35}$$

Wird aber die Multiplettaufspaltung als klein gegen die genannte Wechselwirkungsenergie zu betrachten sein, so besteht Entartung, der Zustand mit der Eigenfunktion

$$\frac{1}{\sqrt{n}}\{W(1, 2, \ldots n-1)\, k_n^\alpha + (-1)^n W(2, 3 \ldots n)\, k_1^\alpha \\ + (-1)^{2n} W(3, 4, \ldots n, 1)\, k_2^\alpha + \cdots\} \tag{36}$$

gehört zur gleichen Energie des ungestörten Systems.

Das Störungsproblem führt auf eine einfache zweigliedrige Determinante, aus der die Energieaufspaltung leicht zu berechnen ist, wenn U

und W gegeben sind. Die Energieaufspaltung entspricht im wesentlichen einer Resonanzwechselwirkung des Elektrons mit dem ursprünglichen Atom. Gleichzeitig ist der eine der beiden entstehenden Terme Glied eines Multipletts der Vielfachheit $r - 1$, der andere gehört zu einem Multiplett der Vielfachheit $r + 1$, wenn der Term des Ausgangssystems zu einem Multiplett der Vielfachheit r gehörte.

Die Gleichungen (35) und (36) geben also in relativ einfacher Weise Aufschluß über die Lage der Multipletts in einem Atom, wenn die Struktur (und die zugehörigen Eigenfunktionen) des vorhergehenden Atoms im periodischen System oder des Ions bekannt ist.

Die hier durchgeführten Überlegungen sollten einen allgemeinen Überblick über die physikalischen Probleme, die bei Atomen mit mehreren Elektronen auftreten, und über die mathematischen Methoden zu ihrer Lösung geben. Vielleicht ist das hier versuchte Schema noch nicht genügend allgemein. Als gesicherte Ergebnisse möchte ich hervorheben: Einteilung aller Spektra in Multipletts, deren Aufspaltung durch die Wechselwirkung der Elektronenmagneten mit der Schwerpunktsbewegung der Elektronenmagneten gegeben ist (Kosinusgesetz und Intervallproportionen!); die Abstände verschiedener Multipletts bei gleichen Elektronenquantenzahlen beruhen auf Resonanzwechselwirkung; wie im Helium zerfallen die Termsysteme bei verschwindender Multiplettaufspaltung in nicht kombinierende Teilsysteme.

III. Der Resonanzeffekt in der Theorie der Bandenspektra.

In der Theorie der Bandenspektra hat die Quantenmechanik gegenüber der früheren Quantentheorie einen wesentlichen Unterschied nur insofern ergeben, als Oszillations- und Rotationsquantenzahl im Sinne der früheren Theorie in der Energieformel halbzahlig zu zählen sind[1]). Dieses Ergebnis genügt zwar zur Darstellung einer großen Reihe beobachteter Bandenspektra. Aber es gibt andere Bandenspektra, in denen es nicht ausreicht.

In vielen Banden ist ein Intensitätswechsel beobachtet, derart, daß die aufeinanderfolgenden Linien einer Rotationsbande abwechselnd stark und schwach sind, in anderen Banden fällt jede zweite Linie überhaupt

[1]) W. Heisenberg, ZS. f. Phys. **33**, 879, 1925. Die genauere Durchführung der Theorie der Bandenspektren auf Grund der Quantenmechanik verdankt man den Arbeiten von: L. Mensing, ZS. f. Phys. **36**, 814, 1926; O. Halpern, ebenda **38**, 8, 1926; D. Dennison, Phys. Rev. **28**, 318, 1926; E. Schrödinger, Ann. d. Phys. **79**, 489, 1926; E. Fues, Ann. d. Phys. **80**, 367, 1926; **81**, 281, 1926.

aus[1]). Dieses empirische Faktum hat bekanntlich zur Darstellung der Terme dieser Banden mit den berüchtigten Viertelsquantenzahlen Anlaß gegeben. So gut sich eine solche Darstellung zur Ordnung des empirischen Materials eignen mag[2]), sie widerspricht dem Korrespondenzprinzip, da sie Änderungen einer Rotationsquantenzahl $p \rightarrow p \pm \frac{1}{2}$ zuläßt. Diese anomalen Banden sind nur beobachtet worden bei zweiatomigen Molekülen mit gleichen Kernen[3]). Daher liegt es nahe, den Intensitätswechsel mit einer Resonanz der Kerne in Verbindung zu bringen[4]). Es sei daher hier das Modell einer zweiatomigen Molekel mit gleichen Kernen untersucht.

Die Eigenfunktion des gesamten Molekülproblems wird sich darstellen lassen als Produkt von vier Eigenfunktionen, von denen die erste zu den Elektronenbewegungen, die zweite zur Kernschwingung, die dritte zur Translation des Molekülschwerpunktes, endlich die vierte zur Rotation der Kerne gehört. Es handelt sich nun wieder um die Einteilung der Terme in zwei Systeme, von denen das eine Eigenfunktionen entspricht, die symmetrisch in den Koordinaten der Kerne sind, das andere gehört zu Eigenfunktionen, die bei Vertauschung der Kerne das Vorzeichen wechseln. Über den ersten Faktor der genannten Eigenfunktionen, der zu den Bewegungen der Elektronen im Molekül gehört, liegt eine ausführliche Untersuchung in einer Arbeit von Hund[5]) vor. Wir übernehmen aus ihr das Resultat, daß, von Entartungen abgesehen, die „Elektronenfunktion" stets entweder symmetrisch oder antisymmetrisch in den Koordinaten der Kerne ist. Die Eigenfunktion der Oszillationen ist, da man als einzige Variable, von der sie abhängt, den Abstand der Kerne wählen kann, sicher symmetrisch in den Koordinaten der Kerne. Ebenso ist die Eigenfunktion, die der Translation des Molekelschwerpunktes ent-

---

[1]) R. Mecke, Phys. ZS. **25**, 597, 1924; ZS. f. Phys. **28**, 261, 1924; **32**, 823, 1925; G. Dieke, ebenda **31**, 326, 1925; **33**, 161, 1925.

[2]) A. Kratzer, Ann. d. Phys. **71**, 72, 1923.

[3]) Schon R. Mecke, ZS. f. Phys. **31**, 709, 1925 und J. C. Slater, Nature **117**, 555, 1926 haben angenommen, daß der Intensitätswechsel mit der Symmetrie in den Kernen zusammenhinge. Sowohl die Überlegung Meckes, wie die Slaters ist aber von der hier versuchten in wesentlichen Punkten verschieden. Die Untersuchung Slaters stimmt mit der unseren hinsichtlich der korrespondenzmäßigen Zusammenhänge überein.

[4]) Vgl. auch die Untersuchung von J. H. van Vleck, Phys. Rev. **28**, 980, 1926.

[5]) F. Hund, ZS. f. Phys. (im Erscheinen). Für die Möglichkeit, diese Arbeit vor der Publikation zu studieren und für manche Diskussionen des oben gestellten Problems bin ich Herrn Hund zu Dank verpflichtet.

spricht, sicher symmetrisch in den Kernkoordinaten. Als die unabhängige Variable, von der die Eigenfunktion der Rotation des Moleküls abhängt, wählen wir den Winkel w um die Achse des Rotationsimpulses p der Kerne, d. h. den Winkel, den die Kernverbindungslinie mit der Knotenlinie — hier irgend einer festen Geraden in der Ebene, in der das Molekül rotiert — einschließt. Der Drehimpuls p ist ganzzahlig, die Energie wird aber proportional $(p+\frac{1}{2})^2$ bis auf eine additive Konstante. Die Eigenfunktion, die zu einem bestimmten Wert p der Rotationsquantenzahl gehört, heißt dann (nicht normiert)

$$e^{ipw}. \tag{37}$$

Vertauschung der Kerne bedeutet Änderung von w um π. Also sind alle Rotationsfunktionen (37) mit geradzahligem p symmetrisch, alle mit ungeradzahligem p antisymmetrisch in den Koordinaten der Kerne. Diese Betrachtung ließe sich natürlich auch mit Schrödingerschen Eigenfunktionen, d. h. z. B. mit den gewöhnlichen Polarwinkeln ϑ, φ durchführen; die Rechnungen werden dann ein wenig komplizierter. Schließlich erhalten wir folgende Einteilung in nicht kombinierende Termsysteme:

A. Symmetrisches System	B. Antisymmetrisches System
Symmetrische Elektronenfunktion und $p = 0, 2, 4 \ldots$ Antisymmetrische Elektronenfunktion und $p = 1, 3, 5 \ldots$	Symmetrische Elektronenfunktion und $p = 1, 3, 5 \ldots$ Antisymmetrische Elektronenfunktion und $p = 0, 2, 4 \ldots$

Diese Einleitung entspricht der Einteilung der Heliumterme in ein Para- und ein Orthosystem.

Die durch Kombination dieser Terme entstehenden Banden unterscheiden sich zunächst durch folgenden Punkt wesentlich von den üblichen Bandenspektra. Es gibt bei einem bestimmten Elektronensprung entweder nur einen Nullzweig $(p \to p)$ oder nur P- und R-Zweig $(p \to p \pm 1)$. Das erste ist der Fall, wenn der Elektronensprung zwischen zwei Elektronentermen gleicher Art (symmetrisch $\longleftrightarrow$ symmetrisch bzw. antisymmetrisch $\longleftrightarrow$ antisymmetrisch) stattfindet, das zweite, wenn er zwischen zwei Elektronentermen ungleicher Art (symmetrisch $\longleftrightarrow$ antisymmetrisch) stattfindet.

Für die weitere Diskussion sind verschiedene Fälle zu unterscheiden. Zunächst sei angenommen, daß die Kerne kugelsymmetrisch seien, d. h. außer den hier betrachteten Freiheitsgraden keine weiteren Freiheitsgrade, wie etwa eine Eigenrotation, besitzen. Dann kombinieren die oben

Zeitschrift für Physik 41, 239–267 (1927)

aufgeschriebenen beiden Systeme in keiner Weise, und man wird in Analogie zur Statistik der Elektronen und der Lichtquanten annehmen müssen, daß nur das eine der beiden Termsysteme in der Natur vorhanden ist. Man wird in erster Linie an das antisymmetrische System B denken, um die Analogie der Statistik der Protonen zur Elektronenstatistik vollständig zu machen. Man kann dessen aber nicht sicher sein, und es kommt für die folgenden Überlegungen nicht darauf an, welches System ausgewählt wird. Wenn nur eines der beiden Systeme vorhanden ist, so entstehen bei Kombination der Terme Rotationsbanden, bei denen immer jede zweite Linie ausfällt. Wenn z. B. das antisymmetrische System die endgültige Lösung gibt, so gehört zu einem Übergang von symmetrischer zu symmetrischer Elektronenfunktion nur ein Nullzweig mit den Übergängen ... $5 \rightarrow 5$, $3 \rightarrow 3$, $1 \rightarrow 1$. Ähnliches ist beobachtet in den Banden von Helium und Sauerstoff, und es scheint befriedigend, daß die Kerne von Helium (Atomgewicht 4) und Sauerstoff (Atomgewicht 16), als besonders stabile, „abgeschlossene“ Gebilde keine Eigenrotation besitzen. Der Ausschluß der einen der beiden Lösungen, d. h. die Reduktion der statistischen Gewichte um den Faktor 2 spielt bekanntlich auch in der statistischen Ableitung der chemischen Konstanten zweiatomiger Moleküle eine Rolle.

Nehmen wir als zweiten Fall an, daß die beiden Kerne des betrachteten Moleküls eine Eigenrotation, etwa wie die Elektronen, von der Größe $s = \frac{1}{2}$ besitzen, und nehmen wir an, daß etwa das antisymmetrische System als endgültige Lösung in Betracht kommt. Dann ist das oben angegebene Termschema dem Termschema des Heliumatoms vollkommen analog[1]). Beide Systeme A und B (Ortho- und Parasystem) werden vorkommen; sie zeigen ganz schwache Interkombinationen. Im einen (A) stehen die Kernmagnete antiparallel, im andern (B) stehen sie parallel. Alle Terme des Systems (B) haben deshalb das statistische Gewicht 3, die des Systems (A) haben das Gewicht 1. In einer bestimmten Bande — etwa dem Nullzweig, der zu einem Übergang: symm. $\leftrightarrow$ symm. Elektronenfunktion gehört — werden also alle Linien auftreten; die Linien ... $5 \rightarrow 5$, $3 \rightarrow 3$, $1 \rightarrow 1$ werden aber dreimal so intensiv sein (von der übrigen Abhängigkeit der Intensität von der Temperatur usw. abgesehen), wie die Linien ... $4 \rightarrow 4$, $2 \rightarrow 2$, $0 \rightarrow 0$. Wenn der Drehimpuls s größer als $\frac{1}{2}$ ist, so nähert sich das Verhältnis der statistischen Gewichte der beiden Termsysteme mehr der Einheit.

---

[1]) Den Hinweis hierauf verdanke ich Herrn F. Hund.

Solche Fälle von Intensitätswechsel scheinen empirisch bei N_2^+, vielleicht auch bei H_2 [1]) verwirklicht. Es scheint befriedigend, daß der Stickstoffkern (3 Heliumkerne + 2 Wasserstoffkerne + 1 Elektron) und der Wasserstoffkern selbst Eigenrotationen besitzen.

Eine etwas andere Art von Intensitätswechsel wird eintreten, wenn zwar den Kernen keine weiteren Freiheitsgrade mehr zukommen, wenn aber verschiedene Isotope, d. h. verschiedene Arten Kerne gleicher Ladung vorhanden sind. Dann werden alle die Moleküle, bei denen die beiden Kerne verschieden sind, gewöhnliche Bandenspektra aussenden, alle anderen Moleküle werden Banden der oben geschilderten Art aussenden, bei denen jede zweite Linie ausfällt. Der im ganzen beobachtete Intensitätswechsel hängt dann vom Mischungsverhältnis der Isotopen ab. Ein solcher Fall scheint experimentell nicht gefunden zu sein.

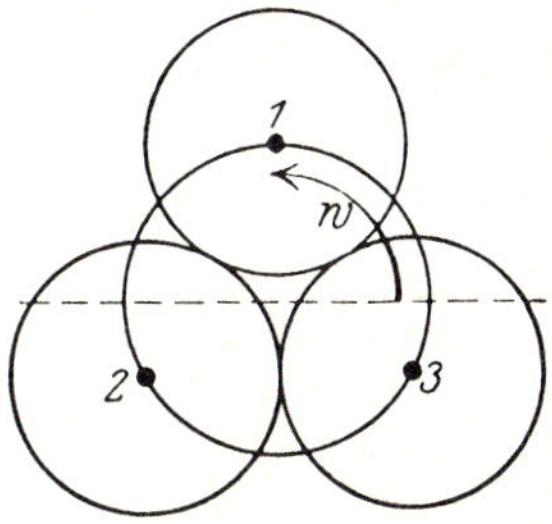

Fig. 1.

Die hier durchgeführten Überlegungen können nach den Methoden von II § 2 leicht auf den Fall mehratomiger Molekeln mit gleichen Kernen erweitert werden. Eine solche Untersuchung hat zwar wegen der Unübersichtlichkeit des experimentellen Materials einstweilen nur theoretisches Interesse. Aber da bei den Molekülen den in II § 2 angewandten Substitutionsgruppen eine mehr anschauliche und konkrete Bedeutung zukommt, sei ein spezielles Beispiel hier durchgeführt.

Wir denken an ein dreiatomiges Molekül, bestehend aus drei gleichen Atomen, die in einem gleichseitigen Dreieck angeordnet sind (Fig. 1). Ferner betrachten wir der Einfachheit halber nur Eigenfunktionen der Elektronen, die symmetrisch in den Kernen sind.

Die Eigenfunktionen der Translation und der Oszillationen sind stets symmetrisch in den Kernen; wir betrachten zunächst nur die Rotationen in der Molekülebene. Deren Impuls sei p und die zugehörige Eigenfunktion ist wieder e^{ipw}, wenn w, wie oben, den Winkel um die Achse von p bedeutet. Es muß nun nach II § 2 drei nicht kombinierende Termsysteme Γ_1, Γ_ε, Γ_{ε^2} $\left(\varepsilon = e^{\frac{2\pi i}{3}}\right)$ geben, die die Eigenschaft haben, daß ihre Eigenfunktionen sich mit 1 bzw. ε bzw. ε^2 multiplizieren, wenn auf diese Eigenfunktionen die Substitution S (1 2 3) angewendet wird.

[1]) G. Dieke, Versl. Amsterdam **33**, 390, 1924.

Zeitschrift für Physik *41*, 239–267 (1927)

Die zyklische Substitution (1 2 3) bedeutet nun eben eine Drehung des Moleküls um $\frac{2\pi}{3}$, also eine Zunahme von w um $\frac{2\pi}{3}$. Ferner gibt es Zustände des Moleküls, bei denen die Reihenfolge der Atome auf dem Kreis der Fig. 1 nicht 1 2 3, sondern 1 3 2 ist. Die Substitution (1 2 3) bedeutet hier eine Drehung um $\frac{4\pi}{3}$. Auch die Bandenterme dieser Anordnung lassen sich in die drei Systeme Γ_1, Γ_ε, Γ_{ε^2} einteilen und wir erhalten schließlich in vollkommener Analogie zu (22):

Γ_1	Γ_ε	Γ_{ε^2}
Anordnung 1 2 3, $p \equiv 0 \pmod 3$	Anordnung 1 2 3, $p \equiv 1 \pmod 3$	Anordnung 1 2 3, $p \equiv 2 \pmod 3$
Anordnung 1 3 2, $p \equiv 0 \pmod 3$	Anordnung 1 3 2, $p \equiv 2 \pmod 3$	Anordnung 1 3 2, $p \equiv 1 \pmod 3$

(38)

Wie in Tabelle 2 hat je ein Term von Γ_ε die gleiche Energie, wie ein anderer von Γ_{ε^2}; die Energie[1]) ist hier proportional p^2. Wenn die Kerne als punktsymmetrisch angenommen werden und die symmetrische bzw. antisymmetrische Lösung allein übrig bleibt, so bleibt nur je ein Term des Systems Γ_1 und wir erhalten Rotationsbanden, bei denen nur jede dritte Linie wirklich vorkommt, die zwei dazwischen liegenden Linien fallen immer aus. Bei Annahme weiterer Freiheitsgrade für die Kerne tritt ein entsprechender Intensitätswechsel ein.

An diesen Überlegungen ändert sich nichts Wesentliches, wenn die allgemeinste Rotationsbewegung der dreiatomigen Dreiecksmolekel betrachtet wird. Der Winkel w bleibt eine zyklische Koordinate und p bleibt konstant. Es kommt eine andere Rotationsquantenzahl hinzu, die dem gesamten Drehimpuls der Molekel entspricht; die Banden, zu denen sie Anlaß gibt, überlagern sich den oben betrachteten ohne Störung.

Wir haben in den vorhergehenden Rechnungen stets von einer Resonanz der Kerne gesprochen; dies ist berechtigt um der Analogie zu den Betrachtungen von II § 2 willen. Aber auch hier darf man beim Wort Resonanz nicht an eine komplizierte Schwebungsbewegung denken; allerdings beschreiben die Kerne gleiche Bahnen (in verschiedener Phase) und tauschen mit einer gewissen Schwebungsfrequenz den Platz. Aber sie tun dies eben ganz einfach auf Grund der Rotation und die Schwebungsfrequenz ist die Umlaufszahl. Diese Überlegung gibt auch die einfache

[1]) Vgl. D. Dennison, l. c.

Aufklärung eines Paradoxons, das beim Verständnis der Termeinteilung Schwierigkeiten machen könnte. Es scheint zunächst merkwürdig, daß auch (bei punktsymmetrischen Kernen) innerhalb des antisymmetrischen Systems, bei dem keine „äquivalenten Bahnen“ der Kerne vorkommen können, einfache Rotationen möglich sind, bei denen doch anschaulich die Kerne sicher in „äquivalenten“ Bahnen umlaufen. Die Lösung dieser Schwierigkeit liegt darin, daß ja der Ausdruck „äquivalente Bahnen“ oben nur für den Grenzfall verschwindender Wechselwirkung definiert war. Bindet man die Kerne durch eine Zusatzkraft an einen festen Punkt und löst adiabatisch die Wechselwirkung, so gehen die Rotationsbewegungen mit antisymmetrischer Eigenfunktion sicher nicht in „äquivalente Bewegungen“ der beiden Kerne über. — Die Auswahl des antisymmetrischen Systems kann eben hier durch die Forderung des Nichtvorkommens äquivalenter Bahnen nicht ungezwungen ersetzt werden.

Empirisches Material zur Prüfung der hier angestellten Überlegungen findet sich bis jetzt nur bei den Banden der zweiatomigen Moleküle. Man wird von einer Durchforschung dieses empirischen Materials vom Gesichtspunkt der Kernresonanz aus wichtige Aufschlüsse über die Eigenschaften der Kerne erhoffen dürfen.

Kopenhagen, Institut für theoretische Physik.

Zur Theorie des Ferromagnetismus.

Von **W. Heisenberg** in Leipzig.

Mit 1 Abbildung. (Eingegangen am 20. Mai 1928.)

Die Weissschen Molekularkräfte werden zurückgeführt auf ein quantenmechanisches Austauschphänomen; und zwar handelt es sich um diejenigen Austauschvorgänge, die in letzter Zeit von Heitler und London mit Erfolg zur Deutung der homöopolaren Valenzkräfte herangezogen worden sind.

Einleitung. Die ferromagnetischen Erscheinungen sind in formal befriedigender Weise durch die bekannte Weisssche Theorie* gedeutet worden. Diese Theorie basiert auf der Annahme, daß jedes Atom im Kristall eine richtende Kraft durch die übrigen Atome des Gitters erfährt, die proportional der Anzahl der bereits gerichteten Atome sein soll. Der Ursprung dieses atomaren Feldes war dagegen völlig unbekannt, und einer Deutung der Weissschen Kräfte auf Grund der klassischen Theorie standen folgende Schwierigkeiten im Wege: Magnetische Wechselwirkungskräfte zwischen den Atomen sind stets um einige Größenordnungen kleiner als die aus den ferromagnetischen Experimenten folgenden atomaren Felder. Elektrische Wechselwirkungen führen zwar zur richtigen Größenordnung; dagegen würde man eher erwarten, daß die elektrischen Wechselwirkungen zweier Atome dem Quadrat des Cosinus ihres gegenseitigen Neigungswinkels als dem Cosinus proportional wären, entgegen den Voraussetzungen der Weissschen Theorie. Andere Schwierigkeiten wurden noch ausführlich von Lenz** diskutiert, und es gelang Ising***, zu zeigen, daß auch die Annahme richtender, hinreichend großer Kräfte zwischen je zwei Nachbaratomen einer Kette nicht genügt, um Ferromagnetismus zu erzeugen.

In ein neues Stadium ist der ferromagnetische Fragenkomplex getreten durch die Uhlenbeck-Goudsmitsche Theorie des Spinelektrons. Insbesondere folgt aus dem bekannten Faktor $g = 2$ beim Einstein-de Haas-Effekt (der ja eben bei ferromagnetischen Substanzen gemessen wurde), daß sich in einem ferromagnetischen Kristall nur die magnetischen Eigenmomente der Elektronen, gar nicht die Atome orientieren. Damit fällt wieder die Möglichkeit fort, die Weissschen Kräfte als elektrische Wechselwirkungen, abhängig von der relativen Spinrichtung der Elektronen,

* P. Weiss, Journ. de phys. (4) **6**, 661, 1907 und Phys. ZS. **9**, 358, 1908.
** W. Lenz, Phys. ZS. **21**, 613, 1920.
*** E. Ising, ZS. f. Phys. **31**, 253, 1925.

zu deuten, da wir wissen, daß solche Kräfte nicht existieren. Ferner hat Pauli* zeigen können, daß bei Vernachlässigung der Wechselwirkung der Elektronen in einem Metall bei Anwendung der Pauli-Fermi-Diracschen Statistik stets Para- oder Diamagnetismus resultiert.

§ 1. Modellmäßige Grundlagen der Theorie. Der Grundgedanke der hier versuchten Theorie ist dieser: Die empirischen Ergebnisse stellen uns beim Ferromagnetismus vor eine ganz ähnliche Sachlage, wie wir sie früher einmal beim Spektrum des Heliumatoms getroffen haben. Aus den Termen des Heliumatoms schien damals zu folgen, daß zwischen den Spinrichtungen zweier Elektronen eine starke Wechselwirkungsenergie herrschte, die zu der Aufspaltung des Termschemas in Singulett- und Triplettsysteme führte. Damals konnte diese Schwierigkeit gelöst werden durch den Nachweis, daß die scheinbare große Wechselwirkung indirekt hervorgebracht wurde durch ein Resonanz- oder Austauschphänomen, das für alle quantenmechanischen Systeme gleicher Partikeln charakteristisch ist. Es liegt also nahe, auch zur Erklärung der ferromagnetischen Erscheinungen dieses Austauschphänomen heranzuziehen. Wir werden zu zeigen versuchen, daß die Coulombschen Wechselwirkungen zusammen mit dem Paulischen Prinzip ausreichen, um die gleichen Wirkungen hervorzubringen wie das von Weiss postulierte molekulare Feld. Erst in der letzten Zeit sind die mathematischen Methoden zur Behandlung eines so komplizierten Problems entwickelt worden in den wichtigen Untersuchungen von Wigner**, Hund***, Heitler und London****.

Bevor ich zur eigentlichen Rechnung übergehe, möchte ich eine kurze Übersicht geben über die Approximationsmethoden, die bei der Behandlung der Elektronenbewegung in Metallen zur Anwendung kommen können.

Methode I. Nach Pauli (l. c.) und Sommerfeld† werden die Elektronen in erster Näherung als völlig frei vorausgesetzt. In zweiter Näherung werden etwa die Wechselwirkungen mit den Gitterpunkten als

* W. Pauli, ZS. f. Phys. **41**, 81, 1927.

** E. Wigner, ebenda **40**, 883, 1927; **43**, 624, 1927.

*** F. Hund, ebenda **43**, 788, 1927.

**** W. Heitler und F. London, ebenda **44**, 455, 1927, im folgenden mit I zitiert; W. Heitler, ebenda **46**, 47, 1927 (mit „II" zitiert); ebenda **47**, 835, 1928 (mit „III" zitiert); F. London, ebenda **46**, 455, 1928.

† A. Sommerfeld, ebenda **47**, 1, 1928; vgl. auch W. V. Houston, ebenda **47**, 33, 1928 und C. Eckart, ebenda **47**, 38, 1928.

Zeitschrift für Physik 49, 619–636 (1928)

Störung hinzugefügt (Houston*). Die Wechselwirkung der Elektronen untereinander wird völlig vernachlässigt.

Methode II. Als erste Näherung wird die Bewegung eines Elektrons in einem (in drei Richtungen) periodischen Kraftfeld (das keineswegs klein zu sein braucht) berechnet. In nächster Näherung werden etwa die Störungen berücksichtigt, die von Abweichungen von der Periodizität im Gitter herrühren. Die Behandlung der Wechselwirkung der Elektronen untereinander begegnet hier den gleichen Schwierigkeiten wie bei Methode I.

Methode III. In der ersten Näherung denkt man sich den Gitterabstand sehr groß und nimmt an, daß noch jedes Elektron zu seinem Atom gehört. In nächster Näherung berücksichtigt man den von Heitler und London (l. c. I.) zuerst behandelten Austausch von Elektronen, die sich im ungestörten System an verschiedenen Orten mit gleicher Energie bewegen. Zustände, bei denen sich mehr Elektronen als im ungestörten System bei einem Atom befinden, werden in dieser Näherung nicht mitberücksichtigt.

Der Unterschied dieser drei Methoden wird klarer, wenn wir ihn an einem anderen Beispiel, an dem von Heitler und London (l. c. I) ausführlich behandelten Wasserstoffmolekül erläutern. In Methode I würden wieder die Elektronen zunächst als frei behandelt, was hier natürlich keine geeignete Ausgangslösung zur Berechnung des Moleküls ergäbe. In Methode II würde man ausgehen von den Lösungen des Zweizentrenproblems (vgl. Hund**). Ein Term, bei dem im Grenzfall unendlichen Kernabstands Elektron 1 um den Kern a eine 1 S-Bahn, Elektron 2 um den Kern b eine 1 S-Bahn beschreibt, würde aufspalten in vier Terme (1 bis 4), etwa charakterisiert durch das Schema:

	Kern a	Kern b
1	1	2
2	2	1
3	1,2	—
4	—	1,2

Die Wechselwirkung der beiden Elektronen würde erst in höheren Näherungen mitberücksichtigt. — Die Methode III wird direkt identisch

* W. V. Houston, ZS. f. Phys. **48**, 449, 1928.
** F. Hund, ebenda **40**, 742, 1927.

mit der von Heitler und London benutzten Methode. Nur Term 1 und 2 werden zu einem ungestörten System zusammengefaßt. Von Term 3 und 4 wird angenommen, daß sie zu erheblich höher liegenden Energiewerten führen. Die Termmannigfaltigkeit des ungestörten Systems ist also bei Methode III geringer als bei Methode I oder II.

Es gibt wohl a priori keine Argumente, irgend eines der drei Annäherungsverfahren vor den anderen zu bevorzugen. Methode I wird am ehesten anwendbar sein bei Metallen sehr großer Leitfähigkeit, Methode III bei Metallen geringer Leitfähigkeit. Methode II steht in der Mitte zwischen diesen beiden Grenzfällen.

Für die folgenden Berechnungen habe ich die Methode III zugrunde gelegt, da nur sie eine quantitative Behandlung der Elektronenwechselwirkung gestattet.

§ 2. Die Verteilung der Termwerte. Die folgenden Rechnungen bilden eine einfache Verallgemeinerung der Heitler-Londonschen Untersuchungen (l. c. I) auf den Fall von $2n$ in Wechselwirkung stehenden Elektronen (die Anzahl der Elektronen ist aus rein formalen Gründen als gerade angenommen). Im ungestörten System befinden sich also $2n$ Elektronen in $2n$ verschiedenen (zwar nicht energetisch, wohl aber örtlich verschiedenen) Quantenzellen.

Über die Quantenzahlen der Elektronen in ihren Atomen machen wir zunächst nur die Annahme, daß sie für alle Atome gleich sind. Andere stationäre Zustände des ungestörten Systems werden nicht mit berücksichtigt, es wird angenommen, daß sie zu sehr viel höheren Energiewerten führen würden.

Es handelt sich dann um die Bestimmung der Energiewerte der stationären Zustände des Gesamtsystems, die zum oben beschriebenen Zustand gehören, wenn als Störung die Coulombsche Wechselwirkung der Ladungen eines Atoms mit den Ladungen irgend eines anderen Atoms betrachtet wird. Es wird uns wegen der sonst auftretenden großen rechnerischen Komplikationen nur möglich sein, die Störungsrechnung bis zur ersten Approximation zu verfolgen. Ob diese erste Annäherung für die in der Natur vorkommenden Fälle wirklich ausreicht, muß dahingestellt bleiben. Für die Eigenfunktionen des ungestörten Systems nehmen wir etwa Produkte der Schrödingerschen Eigenfunktionen des Wasserstoffatoms, oder besser der dem betreffenden Atomrest entsprechenden Eigenfunktionen, ganz wie in der zitierten Arbeit von Heitler und London; — es ist wohl überflüssig, hier diese Ansätze ausführlich zu wiederholen. Diese Eigenfunktionen sind zwar nicht orthogonal, aber die Abweichung von der üblichen Behandlung, die hierdurch erforderlich würde, äußert sich erst in den Gliedern zweiter Ordnung, wir können also hier konsequent die im Falle orthogonaler Eigenfunktionen übliche Behandlungsweise anwenden. Infolge der gegenseitigen Störung können die Elektronen eines Atoms mit denen irgend eines anderen in Austausch treten. Solange von den Störungsgliedern zweiter Ordnung abgesehen wird, kommen nur einfache Transpositionen zwischen zwei benachbarten Atomen vor. Nimmt man als einfachsten Fall an, daß im ungestörten System jedes Atom ein Valenzelektron besessen hat, so reduzieren sich die „Austauschglieder“ der Störungsenergie auf die von Heitler und London angegebenen Ausdrücke

$$J_{(k\,l)} = \frac{1}{2}\int \psi_k^{\varkappa}\,\psi_k^{\lambda}\,\psi_l^{\varkappa}\,\psi_l^{\lambda}\left(\frac{2\,e^2}{r_{k\,l}} + \frac{2\,e^2}{r_{\varkappa\lambda}} - \frac{e^2}{r_{\varkappa k}} - \frac{e^2}{r_{\varkappa l}} - \frac{e^2}{r_{\lambda k}} - \frac{e^2}{r_{\lambda l}}\right) d\tau_k\, d\tau_l. \quad (1)$$

Hier bedeuten k und l die Nummern zweier Elektronen, $\varkappa$ und λ die Nummern derjenigen Atomreste, zu denen k und l im ungestörten System gehörten. Als weitere wichtige Konstante geht in die Störungsrechnung ein die rein „statische“ Wechselwirkung:

$$J_E = \int d\tau_1\, d\tau_2 \ldots d\tau_{2n}\, (\psi_1^1)^2\, (\psi_2^2)^2 \ldots (\psi_{2n}^{2n})^2 \left\{\sum_{\substack{k,\,l\\ k>l}} \frac{e^2}{r_{k\,l}} + \sum_{\substack{\varkappa,\,\lambda\\ \varkappa>\lambda}} \frac{e^2}{r_{\varkappa\lambda}} - \sum_{\substack{k,\,\lambda\\ k\neq\lambda}} \frac{e^2}{r_{k\lambda}}\right\}. \quad (2)$$

Die magnetischen Wechselwirkungen lassen wir wegen ihrer Kleinheit ganz unberücksichtigt. Trotzdem werden sich vermöge der Austauschprozesse die Spinmomente sämtlicher Elektronen teils parallel, teils antiparallel stellen. Nimmt man die grundlegende Annahme des Pauliprinzips hinzu, daß die Eigenfunktionen des Gesamtsystems antisymmetrisch in sämtlichen Elektronen sein sollen, so gehört zu jedem Termwert des gestörten Systems ein ganz bestimmtes magnetisches Gesamtmoment, charakterisiert durch das Drehmoment $s\,\frac{h}{2\pi}$ des Systems. Im ganzen gibt es (abgesehen vom Pauliprinzip und Spin) $(2\,n)!$ Terme des gestörten Systems. Eine statistische Behandlung des Ferromagnetismus wird möglich, wenn sämtliche Energiewerte berechnet sind, die zu einem vorgegebenen Werte von s gehören. Diese Aufgabe ist allerdings nicht in dieser Form lösbar, da $2\,n$ eine sehr große Zahl ist. Wir können nur hoffen, einen allgemeinen Einblick in die Verteilung der Eigenwerte bei vorgegebenem s zu bekommen. Wir werden im folgenden die Anzahl der Terme, den Energieschwerpunkt, also den Mittelwert der Energie bei vorgegebenem s und das mittlere Schwankungsquadrat der Energie um diesen Mittelwert ausrechnen. Dann machen wir die allerdings etwas willkürliche Annahme, daß die Energiewerte in erster Näherung in einer Gaußschen Fehlerkurve um den Mittelwert verteilt sind, wobei die Breite der Fehlerkurve aus dem Schwankungsquadrat zu berechnen ist.

Nach den Untersuchungen von Wigner, Hund und Heitler (l. c.) gehört bei Annahme des Pauliprinzips zu jedem Wert s des Gesamtspinmoments ein Termsystem („σ"), charakterisiert durch eine bestimmte Partitio von $2n$ in Summanden:

$$2n = \underbrace{2 + 2 + \cdots + 2}_{(n-s)\text{-mal}} + \underbrace{1 + \cdots + 1}_{2s\text{-mal}} \tag{3}$$

Die Partitio des „reziproken" Systems heißt dann einfach

$$2n = (n - s) + (n + s). \tag{4}$$

Für den Mittelwert der Energie im System σ, den „Energieschwerpunkt", hat Heitler (l. c. II) die Formel angegeben:

$$E_\sigma = \frac{1}{f_\sigma} \sum_P \chi_\sigma^P J_P. \tag{5}$$

Hierin bedeutet χ_σ^P den Gruppencharakter, der zur Permutation P gehört, $f_\sigma = \chi_\sigma^E$ die Anzahl der Terme des Systems. Die Energie des ungestörten Systems ist als additive Konstante weggelassen. — Wir berechnen weiter das mittlere Schwankungsquadrat $\overline{\Delta E^2}$ der Energie um den Wert E_σ: Die Energiewerte ergeben sich als Wurzel einer Gleichung vom f_σ-ten Grade, die man erhält, wenn man die Determinante

$$\begin{vmatrix} \sum_P b_{11}^P J_P - x & \sum_P b_{12}^P J_P \ldots & \sum_P b_{1f}^P J_P \\ \sum_P b_{21}^P J_P & \sum_P b_{22}^P J_P - x \ldots & \vdots \\ \vdots & & \vdots \\ \sum_P b_{f1}^P J_P & & \sum_P b_{ff}^P J_P - x \end{vmatrix} = 0 \tag{6}$$

setzt. Die Summe der Wurzeln dieser Gleichung $\sum x_n$ ist gegeben durch den Koeffizienten von x^{f-1}, also $\sum_{i,P} b_{ii}^P J_P = \sum_P \chi_\sigma^P J_P$, was zur

Gleichung (5) führt. Die Summe $\sum\limits_{n>m} x_n x_m$ ist gegeben durch den Koeffizienten von x^{f-2} in (6), und es gilt daher

$$\sum_{n>m} x_n x_m = \sum_{i>k} \sum_{P,P'} b_{ii}^P b_{kk}^{P'} J_P J_{P'} - \sum_{i>k} \sum_{P,P'} b_{ik}^P b_{ki}^{P'} J_P J_{P'}$$
$$= \frac{1}{2}\Big(\sum_{i,k} \sum_{P,P'} b_{ii}^P b_{kk}^{P'} J_P J_{P'} - \sum_{i,k} \sum_{P,P'} b_{ik}^P b_{ki}^{P'} J_P J_{P'}\Big). \tag{7}$$

Im letzten Ausdruck ist über alle Werte i, k unabhängig zu summieren. Nun bedeutet $\sum\limits_k b_{ik}^P b_{ki}^{P'}$ das i-te Diagonalglied der Produktmatrix $b^P . b^{P'}$. Da die Matrizen b eine Darstellung der Gruppe bilden, so ist

$$b^P . b^{P'} = b^{P.P'}.$$

Beachtet man noch, daß $\chi^P = \sum\limits_i b_{ii}^P$, so folgt

$$\sum_{n>m} x_n x_m = \frac{1}{2} \sum_{PP'} (\chi^P \chi^{P'} - \chi^{P.P'}) J_P J_{P'}. \tag{8}$$

Setzen wir $x_n = E_\sigma + \Delta E_n$ (der Index σ gehört hier eigentlich auch noch zu x_n, ΔE_n; wir lassen ihn der übersichtlichen Schreibweise halber weg), so wird

$$\sum_{n>m} x_n x_m = \sum_{n>m} (E_\sigma + \Delta E_n)(E_\sigma + \Delta E_m) \quad (n, m = 1, \ldots f_\sigma)$$
$$= \frac{f_\sigma (f_\sigma - 1)}{2} E_\sigma^2 + \sum_{n>m} \Delta E_n \Delta E_m; \tag{9}$$

wegen $\sum \Delta E_n = 0$ gilt

$$\sum_{n=1}^{f_\sigma} \Delta E_n^2 = -2 \sum_{n>m} \Delta E_n \Delta E_m. \tag{10}$$

Aus (5), (8), (9), (10) und der Gleichung $f_\sigma = \chi_\sigma^E$ folgt schließlich

$$\sum_{n=1}^{f_\sigma} \Delta E_n^2 = \frac{1}{f_\sigma} \sum_{PP'} (\chi_\sigma^E \chi_\sigma^{P.P'} - \chi_\sigma^P \chi_\sigma^{P'}) J_P J_{P'} \tag{11}$$

und

$$\overline{\Delta E_\sigma^2} = \frac{1}{f_\sigma^2} \sum_{PP'} (\chi_\sigma^E \chi_\sigma^{P.P'} - \chi_\sigma^P \chi_\sigma^{P'}) J_P J_{P'}. \tag{12}$$

Um diese Formeln anwenden zu können, müssen wir noch die Gruppencharaktere für die Permutationen der verschiedenen Klassen ausrechnen. Da alle J_P verschwinden mit Ausnahme der Fälle $P = E$ und P von der Klasse (12) (Transposition), so kommen für $\chi_\sigma^{P.P'}$ nur folgende Typen in Betracht:

$$\chi_\sigma^E, \ \chi_\sigma^{(12)}, \ \chi_\sigma^{(123)}, \ \chi_\sigma^{(12)(34)}.$$

Diese Gruppencharaktere können nach einer von Heitler (l. c. III) beschriebenen Methode Schurs berechnet werden. Es ergibt sich zunächst für die reziproken Termsysteme*:

$$\left.\begin{aligned}
\chi^{E}_{n-s,\,n+s} &= \frac{(2n)!\,(2s+1)}{(n-s)!\,(n+s+1)!},\\
\chi^{(12)}_{n-s,\,n+s} &= \frac{(2n-2)!\,2\,(2s+1)}{(n-s)!\,(n+s+1)!}\,(s^2+s+n^2-2n),\\
\chi^{(123)}_{n-s,\,n+s} &= \frac{(2n-3)!\,2\,(n-1)}{(n-s)!\,(n+s+1)!}\,\{6s^3+9s^2+s(2n^2-10n+3)\\
&\qquad +n(n-5)\},\\
\chi^{(12)(34)}_{n-s,\,n+s} &= \frac{4\,(2n-4)!}{(n-s)!\,(n+s+1)!}\,\{2s^5+5s^4+4s^3(n^2-5n+4)\\
&\qquad + s^2(6n^2-30n+19)\\
&\qquad + 2s(n^4-6n^3+15n^2\\
&\qquad - 14n+3)+n^4-6n^3\\
&\qquad + 14n^2-9n\}.
\end{aligned}\right\}\qquad (13)$$

Die Charaktere der tatsächlich vorkommenden Termsysteme unterscheiden sich von den hier benutzten Charakteren ihrer reziproken nur durch das Vorzeichen. Und zwar ist der Charakter des reziproken Systems $\frac{\text{gleich}}{\text{entgegengesetzt gleich}}$ dem des Systems selbst, wenn die Permutation P aus einer $\frac{\text{geraden}}{\text{ungeraden}}$ Anzahl von Transpositionen hervorgeht.

Bis hierher gilt alles ganz allgemein, ohne Beziehung zu den speziellen Annahmen, die wir etwa über das Kristallgitter oder den Atombau der ferromagnetischen Substanz machen.

Wir müssen unsere Annahmen nun weiter spezialisieren, um rechnen zu können. Zunächst folgt aus der Formel (1) und aus den Rechnungen von Heitler und London, daß $J_{(12)}$ mit wachsendem Abstand exponentiell abnimmt. Ein Atom in einem Gitter wird also in der Hauptsache nur mit seinen „Nachbarn" in Austausch treten können; der Austausch mit irgend einem Atom, das weiter entfernt liegt als die „Nachbaratome", wird dagegen zu vernachlässigen sein. Die Anzahl der „Nachbarn" eines Atoms ist z. B. in einem Molekülgitter zweiatomiger Moleküle 1, in der linearen Kette 2, in einem quadratischen Flächengitter 4, im

* Die von Heitler [l. c. III, Gleichung (32)] angegebenen Werte für f_σ und $\chi_\sigma^{(12)}$ sind durch Druckfehler entstellt.

einfach kubischen Gitter 6, im kubisch-raumzentrierten Gitter 8, im kubisch-flächenzentrierten Gitter 12.

Weiter machen wir die Annahme, daß alle nicht verschwindenden Austauschglieder J_P gleich sein sollen (wir nennen diesen Wert J_0). Dies muß der Fall sein, wenn die Atomreste unmagnetisch, d. h. zentralsymmetrisch sind. Wir berechnen also jetzt E_σ und $\overline{\Delta E_\sigma^2}$ für ein Gitter, in dem jedes Atom z Nachbarn hat. Dabei werden wir nur die höchste Potenz in n und s berücksichtigen und niedrigere Glieder streichen; dies bedeutet, daß wir von den Effekten an den Grenzflächen des Kristalls absehen.

Die Anzahl der Transpositionen, die zum Werte J_0 führen (d. h. die Anzahl von Atompaaren kleinsten Abstandes), ist $\frac{z \cdot 2n}{2} = z \cdot n$. Also folgt aus (5) und (13)

$$E_\sigma = -z \cdot \frac{s^2 + n^2}{2n} J_0 + J_E. \tag{14}$$

Zur Berechnung des Wertes von $\overline{\Delta E_\sigma^2}$ brauchen wir zunächst den Wert des Ausdrucks

$$A_{P,P'} = \frac{1}{f_\sigma^2}(\chi^E \chi^{P.P'} - \chi^P \chi^{P'}) \tag{15}$$

für die verschiedenen möglichen Kombinationen von P und P'. Es ergibt sich (bis auf niedrigere Potenzen von n und s):

$$\left.\begin{array}{l} 1.\ P = P' \\ \qquad A_{(12)(12)} = \dfrac{(n^2 - s^2)(3n^2 + s^2)}{4n^4}, \\ 2.\ P \text{ und } P' \text{ haben ein Element gemeinsam} \\ \qquad A_{(12)(13)} = \dfrac{(n^2 - s^2)s^2}{4n^4}, \\ 3.\ P \text{ und } P' \text{ haben kein Element gemeinsam} \\ \qquad A_{(12)(34)} = \dfrac{-(n^2 - s^2)s^2}{2n^5}. \end{array}\right\} \tag{16}$$

Wenn ein Atom z Nachbarn hat, so kommt der Typus 1 $z \cdot n$-mal vor, der Typus 2 $z(z-1)n$-mal, der Typus 3 $\frac{z^2 n^2}{2}$-mal.

Es folgt schließlich für $\overline{\Delta E_\sigma^2}$ nach Gleichung (12)

$$\overline{\Delta E_\sigma^2} = J_0^2\left(z n \cdot A_{(12)(12)} + 2z(z-1)n A_{(12)(13)} + 2\frac{z^2 n^2}{2} A_{(12)(34)}\right)$$

und
$$\overline{\varDelta E_\sigma^2} = J_0^2 z \frac{(n^2 - s^2)(3n^2 - s^2)}{4n^3}. \tag{17}$$

Die mittlere Abweichung der Energie vom Mittelwert (14) ist also von der Größenordnung $\varDelta E_\sigma \sim J_0 \sqrt{n}$. — σ bedeutet in den vorhergehenden Formeln stets das Termsystem, das zur Partitio (3) und daher zum Gesamtspinmoment s gehört.

§ 3. Statistik; Anschluß an die Weissschen Formeln. Für die folgenden Überlegungen wird die schon oben erwähnte, allerdings etwas willkürliche Annahme zugrunde gelegt, daß die Verteilung der Energiewerte um den Mittelwert näherungsweise die Form einer Gaußschen Fehlerkurve habe. Da die Gesamtzahl der Terme, die zum Spinmoment s gehören, f_σ beträgt*, so nehmen wir also an, daß

$$\frac{f_\sigma}{\sqrt{2\pi\overline{\varDelta E_\sigma^2}}} \cdot e^{-\frac{\varDelta E^2}{2\overline{\varDelta E_\sigma^2}}} \, d\varDelta E \text{ Terme}$$

zwischen $E_\sigma + \varDelta E$ und $E_\sigma + \varDelta E + d\varDelta E$ liegen.

Nach den bisherigen Rechnungen bleibt die Richtung des Gesamtspinmoments s im Kristall ganz unbestimmt, wenn, wie wir aus dem Ergebnis des Einstein-de Haas-Effektes schließen dürfen, die Bahnmomente der Elektronen im Kristall sich kompensieren. Setzen wir nun den Kristall in ein äußeres Magnetfeld der Stärke H, so tritt zur inneren Energie, die von s abhängt, noch eine äußere Energie, die von der Projektion m von s auf das äußere Feld abhängt. Nach bekannten Formeln gilt für diese Zusatzenergie ($g = 2$ für den Spin!)

$$E' = -\frac{e}{\mu c}\frac{h}{2\pi} H . m \; (s \geqq m \geqq -s) \tag{18}$$

(μ = Elektronenmasse).

Es entsteht jetzt die Aufgabe, für eine gegebene Temperatur und einen gegebenen Wert von H den wahrscheinlichsten Wert von m auszurechnen. Wir führen als Abkürzungen ein:

$$\left.\begin{aligned} \alpha &= \frac{1}{kT} \cdot \frac{e}{\mu c} H \cdot \frac{h}{2\pi}, \\ \beta &= \frac{z J_0}{kT}. \end{aligned}\right\} \tag{19}$$

* Ohne Pauliprinzip würde wegen der f_σ-fachen Entartung die Gesamtzahl der Terme in allen Systemen, die zur Partitio gehören, gleich f_σ^2 sein. Die Entartung wird aber durch das Pauliprinzip aufgehoben.

Dann bildet man die Zustandssumme (bis auf einen unwesentlichen, von α unabhängigen Faktor):

$$\sum_{s=0}^{n} \sum_{m=-s}^{+s} \int_{-\infty}^{+\infty} d\Delta E \frac{f_\sigma}{\sqrt{2\pi \overline{\Delta E_\sigma^2}}} \cdot e^{\alpha m + \beta \frac{s^2}{2n} - \frac{\Delta E}{kT} - \frac{\Delta E^2}{2\overline{\Delta E_\sigma^2}}}$$

$$= \sum_{s=0}^{n} \sum_{m=-s}^{+s} f_\sigma e^{\alpha m + \beta \frac{s^2}{2n} + \frac{\overline{\Delta E_\sigma^2}}{2k^2T^2}}.$$

Schließlich folgt nach (17) (wieder unter Weglassung unwichtiger Faktoren):

$$S = \sum_{s=0}^{n} \sum_{m=-s}^{+s} f_\sigma e^{\alpha m + \beta \frac{s^2}{2n} - \beta^2 \frac{s^2(4n^2 - s^2)}{8n^3 z}} \tag{20}$$

Bezeichnet man den Ausdruck $e^{\beta \frac{s^2}{2n} - \beta^2 \frac{s^2(4n^2-s^2)}{8n^3 z}}$ mit $g(s)$ und berücksichtigt, daß $f_\sigma = \binom{2n}{n+s} - \binom{2n}{n+s+1}$, so wird nach Umkehrung der Summationsreihenfolge:

$$S = \sum_{m=-n}^{+n} \sum_{s=|m|}^{n} e^{\alpha m} \cdot g(s) \left[\binom{2n}{n+s} - \binom{2n}{n+s+1}\right],$$

und durch eine Umformung analog zur partiellen Integration:

$$S = S_1 + S_2 = \sum_{m=-n}^{+n} e^{\alpha m} g(m) \binom{2n}{n+m}$$

$$+ \sum_{s=0}^{n} \sum_{m=-s}^{+s} e^{\alpha m} (g(s+1) - g(s)) \binom{2n}{n+s+1},$$

$$S_2 = \sum_{s=0}^{n} \frac{e^{\alpha(s+1)} - e^{-\alpha s}}{e^\alpha - 1} \cdot (g(s+1) - g(s)) \binom{2n}{n+s+1}.$$

Da $g(s)$ eine gerade Funktion von s ist, kann man in der hier durchweg benutzten Näherung in S_2 das Glied mit $e^{-\alpha s}$ streichen und dafür die Summe von $s = -n$ bis $s = +n$ erstrecken. Man sieht dann, daß bis auf Faktoren der Größenordnung 1 auch S_2 die Form von S_1 annimmt (wenn man den Summationsbuchstaben s in S_2 durch m ersetzt).

Es folgt schließlich

$$S = F \cdot \sum_{-n}^{+n} e^{\alpha m} g(m) \binom{2n}{n+m}, \tag{21}$$

wo F eine Funktion der Größen α und β von der Größenordnung 1 ist (wenn α und β von dieser Größenordnung sind).

Für die weitere Rechnung wird angenommen, daß in S der Ausdruck unter dem Summenzeichen ein steiles Maximum an der Stelle $m = m_0$ ($m_0 \approx$ Mittelwert von m) aufweist, was später verifiziert werden soll. Wir setzen $m = m_0 + (m - m_0)$ und entwickeln im Exponenten von $g(m)$ nach Potenzen von $(m - m_0)$. Es ergibt sich:

$$g(m) \approx e^{\beta \frac{m_0^2}{2n} + \beta \frac{m_0}{n}(m - m_0) - \frac{\beta^2}{8 n^3 z}\left[4 m_0^2 n^2 - m_0^4 + 8 m_0 n^2 (m - m_0) - 4 m_0^3 (m - m_0)\right]}$$

Also wird S wieder bis auf unwesentliche Faktoren, die von α nicht abhängen:

$$S = F \sum_{m=-n}^{m=n} e^{\left(\alpha + \beta \frac{m_0}{n} - \beta^2 \frac{m_0}{n z} + \beta^2 \frac{m_0^3}{2 n^3 z}\right) m} \binom{2n}{n+m}$$

$$= F \left[2\,\mathfrak{Cof} \left(\frac{\alpha + \beta \frac{m_0}{n} - \beta^2 \frac{m_0}{n z} + \beta^2 \frac{m_0^3}{2 n^3 z}}{2} \right) \right]^{2n}. \qquad (22\text{a})$$

Aus S erhalten wir in bekannter Weise den wahrscheinlichsten Wert m_0 nach der Gleichung

$$m_0 = \frac{\partial}{\partial \alpha} \log S = n\,\mathfrak{Tg}\, \frac{\alpha + \beta \frac{m_0}{n} - \beta^2 \frac{m_0}{n z} + \beta^2 \frac{m_0^3}{2 n^3 z}}{2}. \qquad (22)$$

Das von der Differentiation des $\log F$ herrührende Glied ist gegen n zu vernachlässigen. Hierin liegt auch die nachträgliche Rechtfertigung dafür, daß wir im Exponenten in der Zustandssumme nur die höchsten Potenzen in n und s berücksichtigt haben, also Faktoren der Größenordnung 1 vernachlässigten.

Formel (22) ist im wesentlichen identisch mit der bekannten Weissschen Formel für den Ferromagnetismus. Daß in (22) $\mathfrak{Tg}\, x$ an Stelle von $\mathfrak{Cotg}\, x - 1/x$ (wie bei Weiss) steht, rührt davon her, daß für ein Spinmoment nur zwei Stellungen in einem äußeren Felde möglich sind. Setzt man

$$\frac{m_0}{n} = y; \quad \alpha + \beta \frac{m_0}{n} - \beta^2 \frac{m_0}{n z} + \beta^2 \frac{m_0^3}{2 n^3 z} = 2x, \qquad (23)$$

so gelten die Gleichungen:

$$\left.\begin{aligned} &\text{I.} \quad y = \mathfrak{Tg}\, x, \\ &\text{II.} \quad 2x = \alpha + y\left(\beta - \frac{\beta^2}{z}\right) + \frac{\beta^2}{2z} y^3. \end{aligned}\right\} \qquad (24)$$

Diese Formeln unterscheiden sich von den Weissschen Formeln noch durch das kubische Glied in (24 II). Die Lösungen von (24) konstruiert man am einfachsten wieder graphisch. Fig. 1 ist mit den Werten $\alpha = 0{,}2$; $\beta = 2$; $z = 8$ gezeichnet.

Für kleine oder negative Werte der Konstante β ergibt sich Paramagnetismus. Ferromagnetismus tritt ein, wenn die Tangente der Kurve II für $y = 0$ einen kleineren Winkel mit der x-Achse einschließt als die Tangente von I; hierbei ist vom Einfluß des kubischen Gliedes zunächst abgesehen. Die Bedingung für Ferromagnetismus lautet also

$$\beta\left(1 - \frac{\beta}{z}\right) \geqq 2. \tag{25}$$

Diese Bedingung ist nur für hohe Werte von z erfüllbar. Der Maximalwert der linken Seite von (25) ist $\left(\beta_{\max} = \frac{z}{2}\right)$ $\frac{z}{2}\left(1 - \frac{1}{2}\right)$, und es folgt:

$$z \geqq 8. \tag{26}$$

Ferromagnetismus ist hiernach nur für die Gittertypen möglich, bei denen ein Atom mindestens acht Nachbarn hat. Für Fe, Co, Ni ist dies der Fall, die Gitter sind alle kubisch, teils raumzentriert ($z = 8$), teils flächenzentriert ($z = 12$). — Bei Mitberücksichtigung des Gliedes dritten Grades in 24 II kann es etwa eben noch für $z = 7$ vorkommen, daß Kurve II zwar (für $\alpha = 0$) am Nullpunkt steiler ansteigt als I, daß aber weiter oben doch noch zwei Schnittpunkte von I und II auftreten. Da $z = 7$ ohnehin nicht vorkommt, wird man dieser Möglichkeit keine physikalische Bedeutung beimessen.

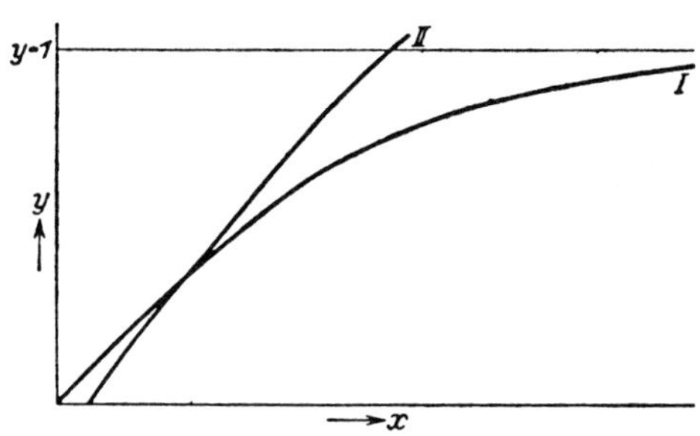

Fig. 1.

Wenn β über den Wert $z/2$ wächst (Anwachsen von β entspricht abnehmender Temperatur), so nimmt nach Gleichung (25) die „Stärke des molekularen Feldes" wieder ab, für $\beta > \frac{z}{2}\left(1 + \sqrt{1 - \frac{8}{z}}\right)$ gibt es nur mehr Paramagnetismus. Wir glauben nicht, daß dieses Resultat physikalische Bedeutung hat. Mathematisch kommt es nämlich durch die angenommene Gaußsche Verteilung der Energiewerte zustande, die zur Folge hat, daß auch für kleine Werte von s Termwerte vorkommen, die (bei positivem J_0) ebenso tief liegen oder tiefer wie die Energie, welche zu $s = n$ gehört. In Wirklichkeit gibt es sicher keinen Energiewert, der tiefer liegt (wenn J_0 positiv) als der für $s = n$. Die

Abweichungen der wirklichen Energieverteilungskurve von der angenommenen machen sich natürlich bei abnehmenden Temperaturen mehr und mehr bemerkbar.

Man wird die hier versuchte vorläufige Theorie dadurch verbessern müssen, daß man die höheren Schwankungsmittelwerte $\overline{\Delta E^3}$, $\overline{\Delta E^4}$ usw. ausrechnet und entsprechend verbesserte Verteilungskurven der Termwerte konstruiert. In dieser verbesserten Theorie werden auf der linken Seite von Gleichung (25) entsprechend höhere Potenzen von β auftreten; die linke Seite von (25) ist also eigentlich eine transzendente Funktion von β; der Wert, den wir hier für die linke Seite von (25) gewonnen haben, wird etwa einer beim zweiten Gliede abgebrochenen Entwicklung dieser Funktion nach Potenzen von $\frac{\beta}{z} = \frac{J_0}{kT}$ entsprechen (unsere Glieder $\beta - \frac{\beta^2}{z}$ können wohl nur um geringe Beträge von den ersten zwei Gliedern der wahren Potenzreihe jener Funktion verschieden sein). Aus dieser Überlegung folgt aber, daß für höhere Werte von $\frac{\beta}{z}$, sagen wir von $\frac{\beta}{z} \gtrsim \frac{1}{2}$ an, eine Untersuchung der höheren Schwankungsgrößen $\overline{\Delta E^n}$ zum Studium des Ferromagnetismus unerläßlich ist. Eine solche genauere Untersuchung der Verteilungskurve wird auch wahrscheinlich den Grenzwert (26) von z noch verschieben. Aber qualitativ wird sich an unseren Ergebnissen wohl nicht viel ändern.

Wenn β erheblich kleiner ist als der durch (25) gegebene Grenzwert, so liefert die Gleichung (24), wie schon erwähnt, Paramagnetismus. Die Rechnung ergibt für kleine Werte von α: $\left(\mathfrak{Tg}\, x \approx x - \frac{x^3}{3} \cdots\right)$:

$$y = \frac{\alpha}{2 - \beta + \frac{\beta^2}{z}} + \frac{\alpha^3}{\left(2 - \beta + \frac{\beta^2}{z}\right)^4} \left(\frac{\beta^2}{2z} - \frac{2}{3}\right) + \cdots \tag{27}$$

Das erste Glied dieser Reihe gibt das Curiesche Gesetz mit einer Modifikation, ähnlich wie in der Weissschen Theorie:

$$m_0 \text{ prop } \frac{1}{T - \Theta} \frac{T}{T\left(1 + \sqrt{1 - \frac{8}{z}}\right) - \Theta\left(1 - \sqrt{1 - \frac{8}{z}}\right)}.$$

Zeitschrift für Physik 49, 619–636 (1928)

Hierin ist die kritische Temperatur

$$\Theta = \frac{2 J_0}{k\left(1 - \sqrt{1 - \frac{8}{z}}\right)}.$$

Bevor ich zur Diskussion des numerischen Wertes von J_0 übergehe, will ich die früher aufgestellte Behauptung rechtfertigen, daß die Glieder der Zustandssumme (21) ein sehr steiles Maximum an der Stelle $m = m_0$ aufweisen. Ich berechne zu diesem Zwecke das mittlere Schwankungsquadrat $\overline{\Delta m^2}$ von m um m_0. Es ist

$$\overline{\Delta m^2} = \overline{m^2} - m_0^2$$

und

$$\overline{m^2} = \frac{1}{S}\frac{\partial^2 S}{\partial \alpha^2}.$$

Aus (22 a):

$$\frac{\partial S}{\partial \alpha} = n(e^x + e^{-x})^{2n-1}(e^x - e^{-x})F + \frac{\partial F}{\partial \alpha}(e^x + e^{-x})^{2n},$$

$$\frac{\partial^2 S}{\partial \alpha^2} = \frac{n(2n-1)}{2}(e^x + e^{-x})^{2n-2}(e^x - e^{-x})^2 F + \frac{n}{2}(e^x + e^{-x})^{2n} F$$
$$+ 2n\frac{\partial F}{\partial \alpha}(e^x + e^{-x})^{2n-1}(e^x - e^{-x}) + \frac{\partial^2 F}{\partial \alpha^2}(e^x + e^{-x})^{2n}.$$

Daraus in erster Näherung:

$$\overline{m^2} = m_0^2 + \frac{n}{2}(1 - \mathfrak{Tg}^2 x) + 2n\,\mathfrak{Tg}\,x \cdot \frac{\partial}{\partial \alpha}\log F.$$

Es wird also

$$\overline{\Delta m^2} = \frac{n}{2}\left(1 - \mathfrak{Tg}^2 x + 4\,\mathfrak{Tg}\,x\,\frac{\partial}{\partial \alpha}\log F\right).$$

Die wahrscheinliche Abweichung des Moments m vom erwarteten Wert m_0 ist also nur von der Größenordnung $\sqrt{n}$, die im Exponenten von $g(m)$ in Gleichung (22) vernachlässigten Glieder sind von der Größenordnung $\frac{\Delta m^2}{n}$, also Größenordnung 1 — in Übereinstimmung mit dem durchweg erstrebten Grad der Annäherung. Die vernachlässigten Glieder sind von der Ordnung der Grenzflächeneffekte.

§ 4. Größe und Vorzeichen des „molekularen Feldes". Die Konstante β muß von der Größenordnung 1 sein, damit Ferromagnetismus möglich ist; also muß $J_0 \sim kT$ sein, wo T bei Fe, Co, Ni Werte der

Ordnung 10^3 Grad annimmt. Es folgt $J_0 \sim 10^{-13}$ erg $\sim \frac{1}{100}$ der Energie des Wasserstoffgrundzustandes. Dies ist eben die Ordnung der Energiebeträge, die für Austauschglieder der Form (1) zu erwarten ist, wenn die Atome nahe nebeneinander liegen. Wenn der Atomabstand größer wird, nehmen die Austauschglieder exponentiell ab. Dies ist der Grund dafür, daß Eisen- oder Nickelsalzlösungen niemals ferromagnetisch sind.

Sehr viel schwieriger ist die Frage nach dem Vorzeichen von J_0 zu beantworten. Heitler und London machen in ihrer Theorie der homöopolaren Bindung die Annahme, daß J_0 ganz allgemein negativ sei, was Ferromagnetismus ausschließen würde. Für den Spezialfall, daß sich die Elektronen ungestört im 1 S-Zustand befinden, folgt nämlich aus allgemeinen Sätzen, daß die Energiewerte so liegen müssen, wie es negativen J_0-Werten entspricht. Eine solche Schlußweise ist indessen nur auf den 1 S-Zustand der Elektronen anwendbar, und man kann zeigen, daß für hohe Hauptquantenzahlen J_0 im allgemeinen positiv wird. Es handelt sich um den Ausdruck:

$$J_0 = \frac{1}{2} \int \psi_k^{\varkappa} \psi_k^{\lambda} \psi_l^{\varkappa} \psi_l^{\lambda} \left(\frac{2e^2}{r_{kl}} + \frac{2e^2}{r_{\varkappa\lambda}} - \frac{e^2}{r_{\varkappa k}} - \frac{e^2}{r_{\varkappa l}} - \frac{e^2}{r_{\lambda k}} - \frac{e^2}{r_{\lambda l}} \right) d\tau_k \, d\tau_l \quad (1)$$

$\varkappa$ und λ waren die Indizes der Atomkerne, k und l die der Elektronen. Die ψ seien zunächst Wasserstoffeigenfunktionen, es wird sich später zeigen, daß die Überlegungen für andere Zentralfelder in der Nähe der Kerne genau so gut gelten. Zunächst kann man bestimmt sagen, daß J_0 für sehr kleine Werte von $r_{\varkappa\lambda}$ positiv wird, da dann das Glied $\frac{1}{r_{\varkappa\lambda}}$ alle anderen überwiegt. Dieses Resultat braucht aber keine physikalische Bedeutung zu haben, da für sehr kleine Werte von $r_{\varkappa\lambda}$ eben die ganze Art der Annäherung illusorisch wird (vgl. eben den Fall des 1 S Terms!). Es kommt also auf den Wert von J_0 für große $r_{\varkappa\lambda}$ an. Wenn J_0 dort positiv ist, wird man annehmen dürfen, daß es im allgemeinen für alle Werte $r_{\varkappa\lambda}$ positiv bleibt. Wir untersuchen also weiter, wie eine Ladungsverteilung der Dichte $\psi_k^{\varkappa} \psi_k^{\lambda}$ bei großen Abständen $r_{\varkappa\lambda}$ aussieht, zunächst etwa für höhere S-Terme. Die Schrödingerfunktionen enthalten als das wichtigste Glied eine e-Funktion, und $\psi_k^{\varkappa} \psi_k^{\lambda}$ enthält daher den Faktor $e^{-\frac{r_{k\varkappa} + r_{k\lambda}}{a_0 n}}$ (a_0 Bohrscher Wasserstoffradius; n Hauptquantenzahl; — eine Verwechslung mit der Elektronenanzahl $2n$ ist wohl nicht zu befürchten). Sieht man von den übrigen Faktoren ab, so ist also die Dichte konstant auf konfokalen Rotationsellipsoiden um die beiden Kerne. Bei wachsendem

Zeitschrift für Physik 49, 619–636 (1928)

Abstand der Kerne artet das Ladungsellipsoid aus in einen Zylinder um die Verbindungslinie der Kerne (dies trifft zu bei jedem Werte der Hauptquantenzahl). Die e-Funktion erscheint weiterhin multipliziert mit einem Polynom in $r_{k\varkappa}$ bzw. $r_{k\lambda}$ vom $(n-1)$-ten Grade. Die Nullstellen dieses Polynoms liegen alle in der Nähe der Kerne, in größerem Abstand von ihnen genügt es, das Polynom durch die höchste Potenz r^{n-1} zu ersetzen. Auch der Verlauf der Zentralkraft in Abständen der Ordnung a_0 vom Kern ist ganz unwesentlich, wenn nur $r_{\varkappa\lambda}$ hinreichend groß ist. — Die Dichteverteilung der Ladung über die Länge des oben beschriebenen Zylinders ist also nicht gleichförmig, sondern näherungsweise proportional $r_{k\varkappa}^{n-1}\, r_{k\lambda}^{n-1}$. Für kleine Werte von n ist diese Verteilung noch ziemlich gleichförmig, es läßt sich leicht einsehen, daß dann in J_0 die negativen Glieder erheblich überwiegen. Mit wachsendem n dagegen erhält die Dichteverteilung ein immer steileres Maximum in der Mitte zwischen beiden Kernen. In der Grenze sehr großer Werte von n strebt daher der Mittelwert der Glieder vom Typus $\frac{1}{r_{k\varkappa}}$, genommen über die oben angegebene Dichteverteilung, dem Werte $\frac{2}{r_{\varkappa\lambda}}$ zu:

$$\overline{\frac{1}{r_{k\varkappa}}} = \overline{\frac{1}{r_{k\lambda}}} = \overline{\frac{1}{r_{l\varkappa}}} = \overline{\frac{1}{r_{l\lambda}}} \rightarrow \frac{2}{r_{\varkappa\lambda}}.$$

Das Glied mit $\frac{1}{r_{kl}}$, das „Selbstpotential" der Dichteverteilung, steigt dagegen mit wachsendem n über alle Grenzen. Für hinreichend hohe Hauptquantenzahlen ist also J_0 sicher positiv. Man kann leicht zeigen, daß an diesem Resultat sich nichts ändert, wenn man die Rechnung für $P, D \ldots$ oder irgendwelche andere höhere Zustände durchführt. Der Grenzwert n, bei dem J_0 zum ersten Male positiv werden kann, ist schwer exakt zu bestimmen. Eine Überschlagsrechnung ergab $n = 3$. Dieser Grenzwert wird eventuell noch vom Werte der übrigen Quantenzahlen mit abhängen. Die Tatsache, daß z. B. das Sauerstoffmolekül empirisch ein magnetisches Moment $2 \cdot \frac{1}{2} \frac{h}{2\pi}$ im Grundzustand besitzt, scheint zu zeigen, daß schon für $n = 2$ J_0 positiv sein kann. Andererseits dürfte aus den vielfach beobachteten negativen kritischen Temperaturen (z. B. bei γ-Eisen) folgen, daß auch für höhere Hauptquantenzahlen J_0 manchmal negativ wird.

Schlußbemerkungen. Die hier beschriebenen Rechnungen führten zu zwei Bedingungen für das Auftreten von Ferromagnetismus:

1. Das Kristallgitter muß von solcher Art sein, daß jedes Atom mindestens 8 Nachbarn hat.

2. Die Hauptquantenzahl der für den Magnetismus verantwortlichen Elektronen muß $n \gtrsim 3$ sein.

Beide Bedingungen zusammen reichen noch lange nicht aus, um Fe, Co, Ni vor allen anderen Stoffen auszuzeichnen; aber Fe, Co, Ni genügen den Bedingungen. Es war ja auch zu erwarten, daß die hier entworfene Theorie einstweilen nur ein qualitatives Schema bieten kann, in das die ferromagnetischen Phänomene später vielleicht eingeordnet werden. Die Theorie bedarf einer Erweiterung für den Fall mehrerer Austauschelektronen pro Atom; ein eingehenderes Studium der $J_{(kl)}$-Werte sowie der Verteilungskurve der Termwerte wird erforderlich sein. Auf diese Fragen wie auf einen eingehenderen Vergleich der Theorie mit den experimentellen Resultaten hoffe ich später eingehen zu können.

Leipzig, Institut für theoretische Physik der Universität.

Zeitschrift für Physik 69, 287–297 (1931)

Zur Theorie der Magnetostriktion und der Magnetisierungskurve.

Von **W. Heisenberg** in Leipzig.

Mit 1 Abbildung. (Eingegangen am 13. März 1931.)

Es wird die Frage diskutiert, welchen Platz die Hypothese der spontan magnetisierten Elementargebiete in der Quantentheorie des Ferromagnetismus einnimmt. Dann wird die genannte Hypothese zur Berechnung der Magnetostriktion von Eisenkristallen unterhalb der Sättigungsmagnetisierung verwendet; ferner wird eine Besonderheit im elastischen Verhalten solcher Kristalle besprochen; schließlich werden die von Becker und Akulov vorgeschlagenen Deutungen der Hysteresiserscheinungen diskutiert.

Die Weisssche Theorie des Ferromagnetismus gibt in ihrer quantentheoretischen Fassung bisher eine befriedigende Beschreibung derjenigen ferromagnetischen Vorgänge, bei denen die magnetischen Wechselwirkungen der Elementarmagnete vernachlässigt werden können; d. h. das Verhalten von ferromagnetischen Körpern in Feldern, die größer sind als etwa 10^4 bis 10^5 Gauß, wird durch die Theorie richtig wiedergegeben. Für die Vorgänge unterhalb der Sättigungsmagnetisierung dagegen ist die Theorie erst anwendbar*, wenn man auch die magnetischen Wechselwirkungen der Spinmagnete und ähnliche Effekte** mit berücksichtigt. Den Terminus „Sättigungsmagnetisierung" werden wir hier so verwenden, wie es die Experimentalphysiker tun, er bedeutet also *nicht* etwa Parallelstellung aller Elektronenspins, sondern bezeichnet die oberhalb 10^5 Gauß nach der Weissschen Theorie herrschende, temperatur*abhängige* Magnetisierung. In der letzten Zeit sind eine Reihe von wichtigen Arbeiten erschienen, in denen gezeigt wird, daß sich die Magnetostriktion von Einkristallen im Falle der Sättigung qualitativ und bei geeigneter Wahl der Konstanten quantitativ erklären läßt durch die magnetischen Wechselwirkungen der Elementarmagnete, wobei unbewußt*** oder bewußt**** angenommen wird, daß die Parallelstellung der Magnete bereits durch die Weissschen Austauschkräfte erfolgt sei und nichts mehr mit den magnetischen Kräften zu tun

* Dieser Punkt wurde übersehen von K. Honda, ZS. f. Phys. **63**, 141, 1930 und N. Akulov, ebenda **64**, 559, 1930.

** Vgl. die ausführlichen Untersuchungen von F. Bloch u. G. Gentile, ZS. f. Phys. (im Erscheinen).

*** N. Akulov, ZS. f. Phys. **52**, 389, 1928; **54**, 582, 1929; **57**, 249, 1929; **29**, 254, 1929; **64**, 817, 1930; **66**, 534, 1930; Phys. ZS. **32**, 107, 1931.

**** G. Mahayani, Phil. Trans. (A) **228**, 63, 1929; R. Becker, ZS. f. Phys. **62**, 553, 1930; **64**, 660, 1930.

habe. Ferner haben Becker (l. c.) und Akulov (l. c.) auf den Zusammenhang zwischen Magnetostriktion und Magnetisierungskurve hingewiesen.

Akulov hat seine Ergebnisse **auch** auf den Fall extrapoliert, wo die Magnetisierung erheblich kleiner ist als die Sättigungsmagnetisierung; hiergegen lassen sich meines Erachtens schwerwiegende theoretische Bedenken geltend machen, und die späteren Rechnungen werden zeigen, daß die Akulovsche Extrapolation an einigen Stellen zu unrichtigen Resultaten geführt hat*.

§ 1. Der Zustand eines ferromagnetischen Einkristalls, dessen Magnetisierung geringer ist als die Sättigungsmagnetisierung, ist in der letzten Zeit von Bloch eingehend untersucht worden. Obwohl die Blochschen Untersuchungen noch nicht abgeschlossen sind, sollen hier einige Resultate erwähnt werden, die als gesichert angesehen werden können und die für das Folgende Bedeutung haben. Der Kristall zerfällt — wie man bisher stets vermutet hat — magnetisch in fadenförmige Elementargebiete, innerhalb deren der Kristall bis zur (temperaturabhängigen) Sättigung magnetisiert ist. Die Ummagnetisierung des Kristalls z. B. bei Änderung des äußeren Feldes vollzieht sich *nicht* durch Drehung der etwa in ihrer Form gegebenen Elementargebiete, sondern durch Verschiebung der Grenzen der Elementargebiete, d. h. Gebiete der einen Spinrichtung wachsen auf Kosten der Gebiete der anderen Spinrichtung. Daraus folgt, daß das Ummagnetisieren im allgemeinen nicht mit Energieänderung verknüpft ist und daß man in Einkristallen die Verteilung der Elementargebiete bei gegebenen äußeren Feldern (und eventuell bei gegebenen Gitterstörungen) als statistischen Gleichgewichtszustand berechnen kann, ohne sich für die *Einstellung* des Gleichgewichts zu interessieren. Auf Grund dieser Vorstellungen sollen im folgenden Magnetostriktion und verwandte Erscheinungen unterhalb der Sättigungsmagnetisierung behandelt werden. Es soll hervorgehoben werden, daß ganz ähnliche Überlegungen schon vielfach in der Literatur angestellt wurden; doch sind mir Anwendungen auf die im folgenden diskutierten Probleme nicht bekannt geworden.

Ich möchte im folgenden zunächst ein spezielles einfaches Beispiel, nämlich das Verhalten von Kobalteinkristallen, besprechen, um die Methode klarzumachen. Wählen wir die hexagonale Achse als X-Achse und be-

* In einer neuerdings erschienenen Arbeit vertritt Akulov (ZS. f. Phys. **67**, 794, 1931) abweichend von seinen bisherigen Untersuchungen die übliche Hypothese der Elementargebiete und gibt damit das sogenannte „Anisotropiegesetz" für Vorgänge *unterhalb* der Sättigungsmagnetisierung auf.

Zeitschrift für Physik 69, 287–297 (1931)

zeichnen die Richtungskosinus des Magnetisierungsvektors mit α_1, α_2, α_3, so wird der magnetische Dipolanteil der freien Energie pro Volumeneinheit des Kristalls *im Falle der Sättigung*

$$F = C_0 - C_1 \alpha_1^2, \tag{1}$$

wie man ohne weiteres aus Symmetriegründen schließen kann. Für Kobalt ist empirisch C_1 positiv. Wir fragen nun nach der freien Energie als Funktion der Magnetisierung J (pro Volumeneinheit) mit den Komponenten J_x, J_y, J_z. Der Einfachheit halber setzen wir $J_z = 0$. Die Sättigungsmagnetisierung nennen wir J_∞.

Die Gleichung (1) läßt sich auch auf die einzelnen Elementargebiete anwenden, da diese bis zur Sättigung magnetisiert sind. Daraus folgt, daß sich im unmagnetischen Zustand sämtliche Elementargebiete parallel oder antiparallel zur Achse (der Richtung leichtester Magnetisierung) einstellen, da diese Anordnung die kleinste Energie liefert. Die wahrscheinlichste Anordnung ist natürlich die, wo ebenso viele Gebiete parallel wie antiparallel zur Achse stehen. Magnetisiert man nun parallel zur Achse, so ist hierzu gar kein merklicher Energieaufwand nötig, da das Umklappen der Elementargebiete nach (1) ohne Energieänderung erfolgt. Nur die freie Energie wird etwas vergrößert, da man die Entropie bei der Orientierung verkleinert. Die hiervon herrührenden Beträge wurden von Bloch (l. c.) abgeschätzt und sollen im folgenden vernachlässigt werden. Magnetisiert man jedoch senkrecht zur Achse, so ist es dazu nötig, sämtliche Elementargebiete aus der X-Achse herauszudrehen, wobei es am wahrscheinlichsten ist, daß sämtliche Gebiete um den gleichen Winkel gedreht werden. Sei die mittlere Anzahl der Gebiete pro Volumeneinheit N, ihr mittleres Volumen also $1/N$, so wird die Energiedifferenz eines Gebiets gegen die Energie im unmagnetischen Zustand im Mittel

$$\frac{1}{N} \cdot C_1 (1 - \alpha_1^2) = \frac{1}{N} C_1 \left(\frac{J_y}{J_\infty}\right)^2,$$

also die Energie pro Volumeneinheit

$$C_1 \left(\frac{J_y}{J_\infty}\right)^2.$$

Die gesamte freie Energie unter Berücksichtigung eines äußeren Magnetfeldes $H\,(H_x, H_y)$ und des Entmagnetisierungsfaktors γ lautet schließlich:

$$F = F_0(J) + C_1 \left(\frac{J_y}{J_\infty}\right)^2 - H_x J_x - H_y J_y + \frac{\gamma}{2} J^2. \tag{2}$$

$F_0(J)$ ist hierbei eine Funktion, die, wie oben erwähnt, für $J < J_\infty$ praktisch konstant ist und die in der Gegend $J = J_\infty$ nach der Weissschen Theorie außerordentlich steil ansteigt, so daß es nicht möglich ist, über J_∞ mit Feldern der Ordnung 10^4 Gauß hinauszukommen. Für die Magnetisierungskurve folgt nach (2)

$$\left.\begin{aligned}\frac{\partial F}{\partial J_x} &= \frac{\partial F_0}{\partial J}\frac{J_x}{J} - H_x + \gamma J_x = 0,\\ \frac{\partial F}{\partial J_y} &= \frac{\partial F_0}{\partial J}\frac{J_y}{J} + \left(\frac{2C_1}{J_\infty^2} + \gamma\right) J_y - H_y = 0.\end{aligned}\right\} \tag{3}$$

Also für

$$J_x^2 + J_y^2 < J_\infty^2$$

ergibt sich

$$\left.\begin{aligned} J_x = \frac{1}{\gamma} H_x;\; J_y = \frac{1}{\gamma + \frac{2C_1}{J_\infty^2}} \cdot H_y\end{aligned}\right\} \tag{4}$$

und für $J_x^2 + J_y^2 = J_\infty^2$ durch Elimination von $\partial F_0/\partial J$:

$$\frac{2C_1}{J_\infty^2} - \frac{H_y}{J_y} + \frac{H_x}{J_x} = 0. \tag{5}$$

Die Gleichungen (3) bis (5) geben eine qualitative und nach geeigneter Wahl der Konstanten C_1 quantitative Beschreibung der Versuche von Kaya*.

§ 2. Das gleiche Verfahren soll nun angewendet werden zur Berechnung der Magnetostriktion von Eiseneinkristallen. Die Richtungen leichtester Magnetisierung sind im Falle des Eisens die Würfelkanten, d. h. die Richtungen 100, 010, 001. Man wird also annehmen müssen, daß im unmagnetischen Zustand die Elementargebiete parallel zu den Würfelkanten orientiert sind, wobei gleichviele Gebiete auf jede der sechs Kantenrichtungen fallen.

Wir fragen zunächst, wie sich die Verteilung ändert, wenn eine bestimmte Magnetisierung parallel zur 100-Richtung vorgeschrieben wird. Zur statistischen Rechnung wird es erlaubt sein, näherungsweise alle Elementargebiete als gleich groß anzunehmen und ihre Anzahl N als gegeben anzusehen. Eine solche Vereinfachung ist strenggenommen unzulässig; man hätte eigentlich eine ins einzelne gehende Statistik der Elementargebiete im Sinne der Blochschen Untersuchung zu treiben. Da ins Endresultat aber weder die Größe, noch die Anzahl der Gebiete eingeht, so

---

* S. Kaya, Sc. Reports Tôhoku Imperial Univ. Tokyo **17**, 1157, 1928.

Zeitschrift für Physik 69, 287–297 (1931)

dürfte die genannte Vereinfachung nicht zu erheblichen Fehlern führen. Es sollen also N_1 Gebiete parallel zur X-Achse, d. h. zur 100-Richtung stehen, N_3 Gebiete antiparallel zu dieser Richtung, und die Anzahl der Gebiete in den übrigen vier Richtungen sei im ganzen N_2. Die Wahrscheinlichkeit einer bestimmten Verteilung N_1, N_2, N_3 ist dann

$$W = \frac{N!}{N_1!\,N_2!\,N_3!}\left(\tfrac{1}{6}\right)^{N_1}\left(\tfrac{4}{6}\right)^{N_2}\left(\tfrac{1}{6}\right)^{N_3}, \tag{6}$$

wofür man näherungsweise schreiben kann

$$\left.\begin{aligned}\log W = N\log N - N_1\log N_1 - N_2\log N_2 - N_3\log N_3\\ + N_1\log\tfrac{1}{6} + N_2\log\tfrac{2}{3} + N_3\log\tfrac{1}{6}.\end{aligned}\right\} \tag{7}$$

Die Nebenbedingungen für die Verteilung sind

$$\left.\begin{aligned}N_1 + N_2 + N_3 &= N\\ \text{und}\qquad N_1 - N_3 &= N\cdot\frac{J}{J_\infty} = N\cdot\delta.\end{aligned}\right\} \tag{8}$$

Für das Maximum von $\log W$ findet man also die Bedingungen (λ, μ sind Lagrangesche Faktoren):

$$\begin{aligned}-\log N_1 - 1 + \log\tfrac{1}{6} + \lambda + \mu &= 0,\\ -\log N_2 - 1 + \log\tfrac{2}{3} + \lambda \qquad &= 0,\\ -\log N_3 - 1 + \log\tfrac{1}{6} + \lambda - \mu &= 0,\end{aligned}$$

und es folgt

$$16\,N_1 N_3 = N_2^2. \tag{9}$$

Eliminiert man N_2 und N_3 aus (8) und setzt in (9) ein, so folgt

$$16\,N_1\,(N_1 - N\delta) = (N - 2\,N_1 + N\delta)^2,$$

oder, wenn $N_1/N = x$ gesetzt wird,

$$16\,x\,(x - \delta) = (1 - 2\,x + \delta)^2$$

und

$$x = \frac{3\delta - 1}{6} + \tfrac{1}{3}\sqrt{3\delta^2 + 1}. \tag{10}$$

Ebenso folgt:

$$\left.\begin{aligned}\frac{N_2}{N} &= \tfrac{4}{3} - \tfrac{2}{3}\sqrt{3\delta^2 + 1},\\ \frac{N_3}{N} &= -\frac{3\delta + 1}{6} + \tfrac{1}{3}\sqrt{3\delta^2 + 1},\\ \delta &= \frac{J}{J_\infty}.\end{aligned}\right\} \tag{11}$$

Um die Magnetostriktion als Funktion der Magnetisierung zu berechnen, ist es jetzt nur noch nötig, die Magnetostriktion der Elementargebiete zu studieren. Da die Elementargebiete bis zur Sättigung magnetisiert sind, können wir die Formeln von Akulov übernehmen. Für die relative Längenänderung eines Gebiets mit der Magnetisierungsrichtung α_1, α_2, α_3, gemessen in der Richtung β_1, β_2, β_3, gilt:

$$\left(\frac{\delta l}{l}\right)_{\alpha,\beta} = c + \varkappa_1 \sum_{i \neq j} \alpha_i^2 \beta_j^2 - \varkappa_2 \sum_{i \neq j} \alpha_i \beta_i \alpha_j \beta_j. \tag{12}$$

Auch diese Formel folgt ohne weiteres aus Symmetriebetrachtungen, es ist völlig überflüssig, zu ihrer Ableitung Gittersummen auszuwerten. Die additive Konstante c in (12) bedeutet, daß nur Unterschiede der Längenänderung in verschiedenen Richtungen bestimmt werden können, da der unmagnetische Zustand des Elementargebiets bei der gegebenen Temperatur nicht realisierbar ist.

Fällt nun die Magnetisierungsrichtung mit einer der Hauptachsen zusammen, so ergibt sich

a) Für den Longitudinaleffekt ($\alpha_i = \beta_i$):

$$\frac{\delta l}{l} = c. \tag{13}$$

b) Für den Transversaleffekt $\sum_i \alpha_i \beta_i = 0$:

$$\frac{\delta l}{l} = c + \varkappa_1. \tag{14}$$

Für die gesamte relative Längenänderung parallel zur X-Achse ergibt sich also schließlich:

$$\begin{aligned} \frac{\delta l}{l} &= c \frac{N_1}{N} + (c + \varkappa_1) \frac{N_2}{N} + c \frac{N_3}{N} \\ &= c + \varkappa_1 \frac{N_2}{N}. \end{aligned} \tag{15}$$

Im unmagnetischen Zustand definieren wir $\delta l/l = 0$ und finden daher

$$0 = c + \varkappa_1 \cdot \tfrac{2}{3}; \tag{16}$$

im Zustande völliger Sättigung ($N_2 = N_3 = 0$) wird

$$\left(\frac{\delta l}{l}\right)_{\text{Sätt.}} = \lambda_0 = c. \tag{17}$$

Die Konstanten c und $\varkappa_1$ bestimmen sich also aus den Gleichungen:

$$c = \lambda_0; \quad \varkappa_1 = -\tfrac{3}{2} \lambda_0. \tag{18}$$

λ_0 hat empirisch bei Eisen ungefähr den Wert $1{,}95 \cdot 10^{-5}$. Übrigens folgt aus $\varkappa_1 = -\frac{3}{2}\lambda_0$ und $c = \lambda_0$ nach (12), daß die Magnetostriktion ohne Volumenänderung erfolgt.

Zeitschrift für Physik *69*, 287–297 (1931)

Für die Magnetostriktion bei mittleren Magnetisierungen folgt aus (15), (11) und (8):

$$\frac{\delta l}{l} = \lambda_0 - \tfrac{3}{2}\lambda_0 \cdot \left(\tfrac{4}{3} - \tfrac{2}{3}\sqrt{3\,\delta^2 + 1}\right)$$
$$= + \lambda_0 \left(-1 + \sqrt{1 + 3\left(\frac{J}{J_\infty}\right)^2}\right). \qquad (19)$$

Wir gehen nun dazu über, die entsprechende Rechnung für die 111-Richtung durchzuführen. Durch die Magnetisierung in der 111-Richtung werden zunächst die drei der betreffenden 111-Richtung nächstliegenden Kantenrichtungen für die Elementargebiete bevorzugt und die Magnetisierung läßt sich ohne merkliche Energieänderung durchführen, bis alle Elementargebiete in diesen Kantenrichtungen eingestellt sind. Die Magnetisierung in der 111-Richtung beträgt dann $J_\infty/\sqrt{3}$. Bis zu diesem Punkt ist die Magnetostriktion exakt Null. Es folgt nämlich aus (12) für

$$\beta_1 = \beta_2 = \beta_3 = \frac{1}{\sqrt{3}}$$
$$\left(\frac{\delta l}{l}\right)_{100;\,111} = \lambda_0 \left(1 - \tfrac{3}{2} \sum_{i \neq j} \alpha_i^2 \beta_j^2\right) = 0. \qquad (20)$$

Von $J = J_\infty/\sqrt{3}$ ab werden die Elementargebiete aus den Würfelkanten herausgedreht und geben alle den gleichen Beitrag zur Magnetostriktion. Sei ε der Kosinus des Winkels, den ein ursprünglich in der 100-Achse orientiertes Gebiet mit der 111-Richtung einschließt, so gilt

$$\alpha_1 = \varepsilon \cdot \frac{1}{\sqrt{3}} + \sqrt{\frac{2}{3}\left(1 - \varepsilon^2\right)} \quad \text{und} \quad \varepsilon = \frac{J}{J_\infty},$$
$$\alpha_2 = \alpha_3 = \sqrt{\frac{1 - \alpha_1^2}{2}}.$$

Für die Magnetostriktion für mittlere Magnetisierungen erhält man also

$$\left.\begin{aligned} \left(\frac{\delta l}{l}\right)_{111;\,111} &= -\frac{2}{3}\varkappa_2 \left(2\,\alpha_1 \sqrt{\frac{1 - \alpha_1^2}{2}} + \frac{1 - \alpha_1^2}{2}\right), \\ \text{wobei} \qquad \alpha_1 &= \frac{J}{J_\infty \cdot \sqrt{3}} + \sqrt{\frac{2}{3}\left[1 - \left(\frac{J}{J_\infty}\right)^2\right]}. \end{aligned}\right\} \qquad (21)$$

Insbesondere gilt für die Sättigung

$$\left(\frac{\delta l}{l}\right)_{111;\,111} = -\frac{2}{3}\varkappa_2 = -\lambda_1; \quad \text{also} \quad \varkappa_2 = \frac{3}{2}\lambda_1 \qquad (22)$$

und schließlich

$$\left.\begin{aligned}\left(\frac{\delta l}{l}\right)_{111;\,111} &= 0 \quad \text{für} \quad 0 \leqslant J \leqslant \frac{J_\infty}{\sqrt{3}} \\ &= -\lambda_1\left(2\alpha_1\sqrt{\frac{1-\alpha_1^2}{2}} + \frac{1-\alpha_1^2}{2}\right) \quad \text{für} \quad \frac{J_\infty}{\sqrt{3}} \leqslant J \leqslant J_\infty.\end{aligned}\right\} \tag{23}$$

λ_1 hat empirisch etwa den Wert $1{,}7 \cdot 10^{-5}$.

Wir wollen endlich die entsprechende Rechnung auch für die 110-Richtung durchführen. Dabei sei N_1 die Anzahl der Gebiete, die nach den beiden der 110-Richtung nächstliegenden Kanten orientiert sind, N_3 die Anzahl für die genau entgegengesetzten Richtungen, N_2 die Anzahl für die zur 110-Richtung senkrechten Kanten. Durch eine einfache Rechnung, die völlig analog ist zu den Gleichungen (6) bis (11), findet man für $J \leqslant \frac{J_\infty}{\sqrt{2}}$:

$$N_1 N_3 = N_2^2 \quad \text{und} \quad N_1 - N_3 = N \cdot \sqrt{2}\,\frac{J}{J_\infty} = N\sqrt{2}\cdot\delta, \tag{24}$$

$$x = \frac{N_1}{N},$$

$$x(x - \delta\sqrt{2}) = (1 - 2x + \delta\sqrt{2})^2, \tag{25}$$

$$\left.\begin{aligned} x = \frac{N_1}{N} &= \frac{3\delta\sqrt{2}+4}{6} - \frac{1}{6}\sqrt{4-6\delta^2}, \\ \frac{N_2}{N} &= -\frac{1}{3} + \frac{1}{3}\sqrt{4-6\delta^2}, \\ \frac{N_3}{N} &= \frac{-3\delta\sqrt{2}+4}{6} - \frac{1}{6}\sqrt{4-6\delta^2}.\end{aligned}\right\} \tag{26}$$

Die senkrecht zur 110-Richtung orientierten Gebiete tragen nach (12) $-\frac{1}{2}\lambda_0$ zur Magnetostriktion bei, die in den anderen Kanten liegenden ergeben $+\frac{1}{4}\lambda_0$. Also folgt für die Magnetostriktion

$$\left(\frac{\delta l}{l}\right)_{110;\,110} = \frac{\lambda_0}{N}\left(\frac{N_1+N_3}{4} - \frac{N_2}{2}\right) = \frac{\lambda_0}{4}\left(2 - \sqrt{4 - 6\left(\frac{J}{J_\infty}\right)^2}\right). \tag{27}$$

Für $J \geqslant J/\sqrt{2}$ werden die Gebiete aus den beiden bevorzugten Kanten heraus auf die 110-Richtung zu gedreht. Sei ε der Kosinus des Winkels, den sie mit der 110-Richtung einschließen, so gilt (für die Gebiete, die ursprünglich in der 100-Richtung standen)

$$\alpha_1 = \frac{\varepsilon + \sqrt{1-\varepsilon^2}}{\sqrt{2}}; \quad \alpha_2 = \sqrt{1-\alpha_1^2}; \quad \alpha_3 = 0.$$

Zeitschrift für Physik *69*, 287–297 (1931)

Also folgt für die Magnetostriktion nach (12):

$$\left(\frac{\delta l}{l}\right) = \lambda_0 - \frac{3}{4}\lambda_0 - \frac{3}{2}\lambda_1 \cdot \alpha_1 \alpha_2 = \frac{\lambda_0}{4} - \frac{3}{4}\lambda_1 (2\varepsilon^2 - 1). \tag{28}$$

Wegen $\varepsilon = J/J_\infty$ gilt also

$$\left.\begin{aligned} \left(\frac{\delta l}{l}\right)_{110;\,110} &= \frac{\lambda_0}{4}\left(2 - \sqrt{4 - 6\left(\frac{J}{J_\infty}\right)^2}\right) \quad \text{für} \quad 0 \leqq J \leqq \frac{J_\infty}{\sqrt{2}}, \\ &= \frac{\lambda_0}{4} - \frac{3\lambda_1}{4}\left[2\left(\frac{J}{J_\infty}\right)^2 - 1\right] \quad \text{für} \quad \frac{J_\infty}{\sqrt{2}} \leqq J \leqq J_\infty. \end{aligned}\right\} \tag{29}$$

Die aus (19), (23) und (29) sich ergebenden Kurven sind in Fig. 1 eingetragen (dabei ist $\lambda_0 = 1{,}95 \cdot 10^{-5}$, $\lambda_1 = 1{,}7 \cdot 10^{-5}$ angenommen, die Punkte bedeuten die von Webster* gemessenen Werte.

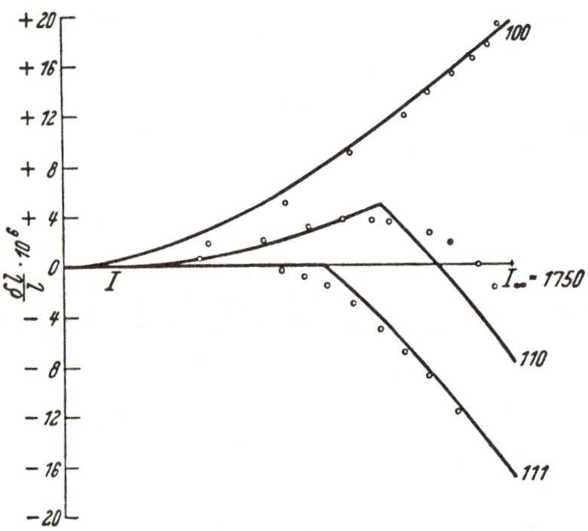

Fig. 1.

Vergleicht man die Formeln für die Längenänderung in den verschiedenen kristallographischen Richtungen, so sieht man, daß das von Akulov vermutete „Gesetz der ferromagnetischen Anisotropie" für Werte von J unterhalb der Sättigung nicht gilt, wie man auch aus den experimentellen Resultaten von Webster schließen kann.

§ 3. Eine weitere interessante Konsequenz der hier durchgeführten Berechnungen ergibt sich für die elastischen Eigenschaften eines Eiseneinkristalls.

Da die Magnetostriktion von Eisen in der 100-Richtung positiv ist, so hat eine Dehnung des Kristalls in dieser Richtung zunächst nur zur

* W. Webster, Proc. Roy. Soc. London (A) **109**, 570, 1925.

Folge, daß die Elementargebiete sich in die Richtung des Zuges (teils parallel, teils antiparallel) einzustellen suchen, daß also so viele Elementargebiete aus den anderen Würfelkanten in die gedehnte Kante wandern, bis ihre Magnetostriktion der Dehnung entspricht. Dieser Prozeß erfolgt nahezu ohne Energieänderung, nur die freie Energie muß sich wegen der Entropieverminderung etwas vergrößern. Erst wenn alle Elementargebiete in der betreffenden 100-Richtung stehen, d. h. wenn die Dehnung gleich der Magnetostriktion bei Sättigung in der betreffenden Richtung geworden ist, ist ein erheblicher Energieaufwand — so, wie er den elastischen Kräften entspricht — nötig, um den Kristall weiter zu dehnen. Bei dem ganzen Prozeß bleibt der Kristall selbstverständlich unmagnetisch, da stets ebenso viele Gebiete parallel wie antiparallel zur Richtung des Zuges stehen. Trägt man die Längenänderung des Kristalls bei Zug in der 100-Richtung als Funktion des Zuges auf, so müßte die Kurve also einen deutlichen Knick aufweisen an der Stelle, wo die Dehnung mit der Sättigungsmagnetostriktion in der 100-Richtung übereinstimmt. Leider sind mir keine Messungen über die elastischen Eigenschaften von Eiseneinkristallen bekannt.

§ 4. Becker* hat durch thermodynamische Überlegungen hergeleitet, daß die Magnetostriktion eines parallel zur Drahtachse magnetisierten Nickeldrahtes unter starkem Zug $1^1/_2$mal so groß sein muß, wie die am unverzerrten Material. Das Beckersche Ergebnis läßt sich anschaulich folgendermaßen deuten: Die Magnetostriktion von Nickel ist negativ. Dehnt man einen Nickeldraht durch starken Zug, so stellen sich also sämtliche Elementargebiete senkrecht zur Drahtachse ein, wobei sich der Draht um die halbe Sättigungsmagnetostriktion [vgl. Gleichung (12)] *ohne* Energieänderung, um den Rest *mit* Änderung der elastischen Energie ausdehnt. Stellt man nun durch ein starkes Magnetfeld sämtliche Elementarmagnete parallel zur Achse, so wird zunächst die genannte Dehnung um die halbe Sättigungsmagnetostriktion rückgängig gemacht, dann kontrahiert sich der Draht um die gewöhnliche Magnetostriktion, so daß im ganzen das 1,5fache der gewöhnlichen Sättigungsmagnetostriktion resultiert. Unterhalb der Sättigung sind die Vorgänge etwas verwickelter, wir wollen sie nicht ausführlich durchrechnen.

§ 5. Schließlich sollen noch kurz die Änderungen besprochen werden, die an den Magnetisierungskurven von Eisen durch Zug z. B. in der 100-Richtung entstehen. Akulov** hat eine Theorie der Hysteresisverluste

* R. Becker, l. c., ZS. f. Phys. **64**, 660, 1930.
** N. Akulov, l. c., ZS. f. Phys. **64**, 817, 1930.

Zeitschrift für Physik 69, 287–297 (1931)

und der Magnetisierungskurven auf die Annahme gegründet, daß z. B. Eisenkristalle (oder wenigstens kleinere Stücke von ihnen) durch starken Zug etwa in der 100-Richtung zu remanenten Magneten werden könnten. Diese Annahme rechtfertigt er durch die Hypothese, daß die Magnetostriktion eines Eisenkristalls von äußeren Spannungen unabhängig sei; macht man diese Hypothese, so folgt in der Tat das Akulovsche Ergebnis auf rein thermodynamischem Wege. Aus der üblichen Vorstellung der Elementargebiete ergibt sich jedoch ohne weiteres, wie oben gezeigt wurde, daß die genannte Hypothese unrichtig ist, wodurch der Akulovschen Theorie der Hysteresis jede Grundlage genommen scheint. Eine einfache Überlegung zeigt unmittelbar, daß Eisenkristalle durch Zug etwa in der 100-Richtung nicht zu permanenten Magneten werden. Der Zug bewirkt ja nur, daß sämtliche Elementargebiete sich in die Achse des Zuges einstellen, und zwar ebenso viele parallel, wie antiparallel. Die Magnetisierungskurven bei Zug sehen also genau so aus, wie die üblichen Magnetisierungskurven in den Richtungen der leichtesten Magnetisierung. Trotz ihrer experimentellen Erfolge dürfte daher die Akulovsche Theorie kaum zur Deutung der Hysteresiserscheinungen geeignet sein.

Eine andere Deutung der Hysteresiserscheinungen ist vorher von Becker (l. c.) vorgeschlagen worden. Nach Becker hindern die durch Bearbeitung oder Verunreinigung entstehenden inneren Spannungen eines ferromagnetischen Materials die Elementargebiete am Umklappen. Solange die Gebiete, innerhalb deren diese Verzerrungen merklich variieren, groß sind gegen die Blochschen Elementargebiete, folgt allerdings aus der hier vertretenen Auffassung, daß die Elementargebiete stets umklappen oder richtiger: sich stets verkleinern und vergrößern können. Wenn jedoch die von Becker diskutierten Verzerrungen innerhalb eines Elementargebiets schon merklich variieren, so läßt sich ein solcher Schluß nicht mehr ziehen und es ist wohl möglich, daß eine eingehendere Diskussion eine befriedigende Deutung der Hysterese geben würde. Allerdings müßte man zu diesem Zweck wohl besonders die Vorgänge bei der Vergrößerung eines Elementargebiets unter Einfluß eines äußeren Magnetfeldes eingehend untersuchen.

aus der L_1 (2 s)- oder L_2 (2 p)-Schale durch das äußere Elektronenbombardement ab. Aber selbst, wenn sich diese beiden Faktoren berücksichtigen ließen, müßten noch weitere Anomalien, wie die Abweichung der Intensitätsverhältnisse der Komponenten irgendeines Multipletts von der Ornstein-Burger-Sommerfeld-Regel erklärt werden (so sollte sich z. B. $\beta_3 : \beta_4$ wie 2 : 1 verhalten, während es bei U gleich 1 : 1 ist). Diese Anomalien sind ähnlich den für die Komponenten der höheren Hauptserienglieder bei den Alkalimetallen beobachteten* und haben wahrscheinlich dieselben Ursachen (Tauchbahnen).

* Vgl. Sambursky, ZS. f. Phys. **49**, 731, 1928, und E. Fermi, ebenda **59**, 680, 1930.

Notiz zur Arbeit des Herrn N. Tunazima: Zum Ferromagnetismus*.

Von **W. Heisenberg** in Leipzig.

(Eingegangen am 18. März 1931.)

Die oben genannte Arbeit des in meinem Institut studierenden Herrn N. Tunazima ist mir niemals vorgelegt worden und wurde ohne mein Wissen publiziert**.

* ZS. f. Phys. **67**, 817, 1931.

** Ich habe den Aufsatz von Herrn Tunazima in gutem Glauben zur Veröffentlichung angenommen, weil mir Herr Tunazima bei der Übersendung des Manuskriptes wörtlich folgendes schrieb: „Meine Arbeit ist bereits von Prof. Heisenberg-Leipzig begutachtet, der meine Gedanken über Ferromagnetismus hoch einschätzt“. Scheel.

Annalen der Physik (5) *10*, 888–904 (1931)

Zum Paulischen Ausschließungsprinzip

Von W. Heisenberg

Nach dem Paulischen Ausschließungsprinzip besteht eine weitgehende Ähnlichkeit zwischen den Termen eines Atomsystems mit n Elektronen und denen eines Systems, bei dem n Elektronen zu einer abgeschlossenen Schale fehlen. Voraussetzung für diese Ähnlichkeit ist die Existenz einer abgeschlossenen Schale und diese hat wieder zur Voraussetzung, daß bei Vernachlässigung der Elektronenwechselwirkung für ein Elektron eine endliche Anzahl energetisch nahe benachbarter Terme existiert, die durch einen (auch gegen die Wechselwirkungsenergie) großen Abstand von den anderen Termen des Systems getrennt sind. Die Äquivalenz des Systems von n Elektronen mit dem System von n „Löchern" ist zunächst vollständig in bezug auf die *Anzahl* der Terme, wie schon von Pauli[1]) festgestellt wurde. Dagegen ist bisher nicht untersucht worden, wie weit die Äquivalenz auch hinsichtlich der *energetischen Lage* der Terme geht. Wenn N die Anzahl der Elektronen in der abgeschlossenen Schale bedeutet, so soll im folgenden gezeigt werden, daß man eine Schrödingergleichung für n Elektronen auch ersetzen kann durch eine ganz ähnlich gebaute, äquivalente Schrödingergleichung für $N - n$ Löcher. Wenn die Wechselwirkung der Elektronen von der Form e^2/r_{lm} war, so wird formal auch die Wechselwirkung der Löcher von der Form e^2/r_{lm}. Allerdings gilt diese Äquivalenz nur so lange, als man die von den Übergangsmöglichkeiten zu angeregten Zuständen (d. h. Zuständen außerhalb der abgeschlossenen Schale) herrührenden Störungen vernachlässigen kann; die Äquivalenz besteht also nur in der sogenannten „ersten Näherung" des Störungsverfahrens, bei dem die Wechselwirkung als Störung betrachtet wird.

1) W. Pauli, Ztschr. f. Phys. **31**. S. 373. 1925.

§ 1. Die Schrödingergleichung des einzelnen Elektrons sei im ungestörten System

$$\left(-\frac{h^2}{8\pi^2\mu}\Delta - eV - E\right)u = 0\,, \tag{1}$$

und führe zu einem System $E_1, E_2 \ldots E_N$ nahe benachbarter Terme mit den Eigenfunktionen $u_1, u_2 \ldots u_N$. Die Störung bestehe aus einer Wechselwirkung je zweier Elektronen H^{lm}

$$\left(\text{z. B.}\quad H^{lm} = \frac{e^2}{r_{lm}}\right)$$

und etwa noch aus einer auf jedes Elektron von außen wirkenden Störung H^l. Die Wellengleichung des Gesamtsystems lautet also

$$\left\{\sum_{l=1}^{n}\left(-\frac{h^2}{8\pi^2\mu}\Delta^l - eV^l + H^l\right) + \sum_{l>m}^{n} H^{lm} - E\right\}\psi = 0. \tag{2}$$

Man kann nun die Eigenfunktion $\psi(\mathfrak{r}_1, \mathfrak{r}_2 \ldots \mathfrak{r}_n)$ entwickeln nach den Eigenfunktionen des ungestörten Systems, die sich als Produkte von Lösungen der Gl. (1) schreiben lassen. Die Koeffizienten S_k dieser Entwicklung hängen dann ab von den Quantenzahlen der einzelnen Elektronen und genügen der Gleichung

$$\sum^k E_{ik}\, S_k - W S_i = 0\,. \tag{2a}$$

Hierbei faßt der Index k die Quantenzahlen sämtlicher Elektronen zusammen und die E_{ik} sind die Matrixelemente des Energieoperators in den ungestörten Eigenfunktionen. Nach dem Vorgang von Jordan, Klein und Wigner[1]) kann man nun statt der Quantenzahlen der Elektronen auch die *Anzahl* der Elektronen N_k in den stationären Zuständen als Variable einführen ($N_k = 0{,}1$). Man sucht also eine Schrödingerfunktion, die nur von den Anzahlen N_k abhängt. Um die zugehörige Schrödingergleichung zu finden, definiert man

$$N_i = a_i^* \, a_i\,, \tag{3}$$

wobei man anschaulich die a_i als die nichtkommutativen Amplituden der dreidimensionalen Wellenfunktion inter-

1) P. Jordan u. O. Klein, Ztschr. f. Phys. **45**. S. 751. 1927; P. Jordan u. E. Wigner, ebenda **47**. S. 631. 1928.

Annalen der Physik (5) *10*, 888–904 (1931)

pretieren kann. Es gelten bei Fermistatistik die Vertauschungsrelationen

$$(4)\qquad \begin{cases} a_i^* a_k + a_k a_i^* = \delta_{ik}\,, \\ a_i a_k + a_k a_i = 0\,, \\ a_i^* a_k^* + a_k^* a_i^* = 0\,. \end{cases}$$

Diese Vertauschungsrelationen sind erfüllbar durch den Ansatz

$$(5)\qquad a_i^* = N_i \Delta_i \nu_i\,; \qquad a_i = \nu_i \Delta_i N_i\,.$$

Δ_i bedeutet einen Operator, der N_i in $1 - N_i$ verwandelt, ν_i eine Vorzeichenfunktion, die der Gleichung

$$(5\,\mathrm{a})\qquad \nu_i = \prod_{k \leq i} (1 - 2 N_k)$$

genügt.

Um die Schrödingergleichung in den Variabeln N_i aufzuschreiben, ist es jetzt noch nötig, den Energieoperator von (2) bzw. (2a) in den Amplituden a_i auszudrücken. Man erhält nach Jordan und Wigner (a. a. O.):

$$(6)\qquad E = \sum_{i,\,k} a_i^* a_k (\delta_{ik} E_i + H_{ik}) + \frac{1}{2} \sum_{i k r s} a_i^* a_k^* a_r a_s H_{ik,\,rs}\,.$$

Hierin haben die Matrixelemente H_{ik} und $H_{ik,\,rs}$ die folgende Bedeutung:

$$(7)\qquad \begin{cases} H_{ik} = \int u_i^*(l) H^l u_k(l)\, d\tau_l\,, \\ H_{ik,\,rs} = \int u_i^*(l) u_k^*(m) H^{lm} u_r(m) u_s(l)\, d\tau_l\, d\tau_m\,. \end{cases}$$

Die E_i sind die Energiewerte der ungestörten Zustände.

Wenn die Summen in Gl. (6) über *sämtliche* Zustände des ungestörten Systems erstreckt werden, so ist (6) mathematisch völlig äquivalent zu Gl. (2). Erstreckt man aber diese Summen nur über die Zustände $1, 2 \ldots N$ der zu betrachtenden Termen*gruppe*, so repräsentiert (6) das zu dieser Termgruppe gehörige Säkularproblem. Seine Lösung liefert eine erste Näherung zur Lösung von (2). Eben dieses Säkularproblem soll nun in ein äquivalentes Säkularproblem für die Löcher transformiert werden. Der Energieoperator für das Säkularproblem der Elektronen lautet also:

(6a) $$E = \sum_{i,k}^{N} a_i^* a_k (\delta_{ik} E_i + H_{ik}) + \frac{1}{2} \sum_{ikrs}^{N} a_i^* a_k^* a_r a_s H_{ik,rs} .$$

Es ist leicht, Gl. (6a) in eine äquivalente Gleichung überzuführen, bei der als Variable nicht die Elektronen, sondern die Löcher erscheinen. Zunächst führen wir ein

(8) $$N_i' = 1 - N_i ,$$

wo demnach N_i' die „Anzahl der Löcher im Zustand i" ($N_i' = 0$ oder 1) bedeutet. In genauer Analogie zu (4) und (5) definieren wir den Operator

(9) $$\Delta_i' = \Delta_i$$

und

(10) $$v_i' = \prod_{k \leq i} (1 - 2N_k') = (-1)^i v_i .$$

Jedoch sollen die a_i' durch die von (5) etwas abweichenden Gleichungen

$$a_i^{*\prime} = N_i' \Delta_i' v_i' (-1)^{i+1} ; \qquad a_i' = v_i' \Delta_i' N_i' (-1)^{i+1}$$

definiert werden. Es gelten dann ungeändert die Vertauschungsrelationen (4) auch für die gestrichenen Amplituden:

(11) $$\begin{cases} a_i'^* a_k' + a_k' a_i'^* = \delta_{ik} , \\ a_i' a_k' + a_k' a_i' = 0 , \\ a_i'^* a_k'^* + a_k'^* a_i'^* = 0 . \end{cases}$$

Aus den Gln. (8) bis (11) folgt jetzt

(12) $$\begin{cases} a_i = v_i \Delta_i N_i = v_i \Delta_i (1 - N_i') = v_i N_i' \Delta_i' \\ \quad = (-1)^{i+1} N_i' \Delta_i' v_i' = a_i'^* , \end{cases}$$

ebenso

(13) $$a_i^* = a_i' .$$

Der Energieoperator (6a) läßt sich also auch in der Form schreiben

(14) $$E = \sum_{ik}^{N} a_i' a_k'^* (\delta_{ik} E_i + H_{ik}) + \frac{1}{2} \sum_{ikrs}^{N} a_i' a_k' a_r'^* a_s'^* H_{ikrs} .$$

Durch Umstellung von Faktoren (die $a_i'^*$ sollen alle nach links gebracht werden) erhält man:

Annalen der Physik (5) *10,* 888–904 (1931)

$$(15)\left\{\begin{aligned} E &= \sum_i^N (H_{ii} + E_i) + \frac{1}{2}\sum_{ik}^N (H_{ik,\,ki} - H_{ik,\,ik}) \\ &- \sum_{ik}^N a_i'^* a_k' \cdot \Big[\delta_{ik} E_i + H_{ki} + \frac{1}{2}\sum_r (H_{kr,\,ri} - H_{rk,\,ri} - H_{kr,\,ir} + H_{rk,\,ir})\Big] \\ &\qquad + \frac{1}{2}\sum_{ikrs}^N a_i'^* \, a_k'^* \, a_r' \, a_s' \, H_{rs,\,ik} \,. \end{aligned}\right.$$

Diese Gleichung für die Amplituden der Löcher hat die gleiche Form, wie die Gl. (6 a) für die Elektronen und man kann von Gl. (15) wieder rückwärts zu einer Wellengleichung in $N - n$ Dimensionen für die Koordinaten der Löcher übergehen. Der wesentliche Unterschied zwischen (6a) und (15) besteht darin, daß der Faktor von $a_i'^* \, a_k'$, der das Störungsfeld des einzelnen Loches (ohne Wechselwirkung) angibt, verschieden ist von dem entsprechenden Faktor für die Elektronen. So ist z. B. im periodischen System das Zentralfeld der Löcher einer abgeschlossenen Schale verschieden von dem Zentralfeld der Elektronen der gleichen Schale, und diese Verschiedenheit rührt von der Wechselwirkung der Elektronen her. Ferner sind die Indizes der Matrix H_{ik} in (15) gegenüber (4) vertauscht. Beim Übergang zur Wellengleichung muß man daher darauf achten, ob die Eigenfunktionen u und die Störungsenergie H^l, H^{lm} reell oder komplex sind. Nehmen wir einmal als Beispiel an, daß $H_{ik} = H_{ki}$ und $H_{ik,\,rs} = H_{rs,\,ik}$ ist. Ferner bezeichnen wir das Zentralfeld der Löcher, dessen Diskussion später folgt, mit $e\,U$ und setzen die Konstante

$$(16)\qquad \sum_i (E_i + H_{ii}) + \frac{1}{2}\sum_{ik} (H_{ik,\,ki} - H_{ik,\,ik}) = A \,.$$

Dann wird die Wellengleichung der Löcher

$$(17)\qquad \left\{\sum_{l=1}^{N-n}\left(\frac{h^2}{8\pi^2\mu}\Delta^l + e\,U^l\right) + \sum_{l>m}^{N-n} H^{lm} + A - E\right\}\psi = 0 \,.$$

Der Übergang von (15) nach (17) hängt allerdings von den speziellen Voraussetzungen ab, die oben gemacht wurden und wird in anderen Fällen etwas verschieden aussehen. Dagegen gilt Gl. (15) ganz allgemein.

§ 2. Die bisherigen allgemeinen Überlegungen sollen nun spezialisiert und ausführlich diskutiert werden für zwei Probleme: Die tiefsten Zustände der Atomspektren im periodischen System und die Leitfähigkeit der Metalle mit anomalem Halleffekt. Wir behandeln zuerst die Atomspektren und verfolgen die Bildung einer abgeschlossenen Schale, deren Elektronen alle die gleiche Azimutalquantenzahl l haben. Die ungestörten Zustände der Elektronen seien also durch ihre magnetische Quantenzahl $m_l = 0, \pm 1, \ldots \pm l$ und ihre Spinrichtung $m_s = \pm \frac{1}{2}$ nummeriert. Die Wirkung der übrigen Elektronen auf die betrachteten werde durch ein geeignetes Zentralfeld V^l repräsentiert. Die Störungsenergie H^l besteht dann in hinreichender Näherung nur aus der Wechselwirkung zwischen Spin und Bahn für jedes einzelne Elektron, ferner gilt

$$H^{lm} = \frac{e^2}{r_{lm}}. \tag{18}$$

Betrachtet man die Säkulargleichung (15) für die in der abgeschlossenen Schale *fehlenden* Elektronen und vergleicht sie mit der Säkulargleichung (6a) für die Elektronen selbst, so findet man folgende Unterschiede: Zunächst tritt in (15) bzw. (17) die Konstante A (Gl. 16) auf, die die Energie der abgeschlossenen Schale darstellt. Dann haben die von H^l herrührenden Bestandteile der Energie des einzelnen Loches das umgekehrte Vorzeichen, wie die des einzelnen Elektrons. Für die Diagonalelemente ist dies trivial; für die zur Spin-Bahnwechselwirkung gehörigen Übergangselemente H^l_{ik} kann man folgendes zeigen: Definiert man als die Eigenfunktionen w_i der Löcher das komplexkonjugierte der Elektroneneigenfunktionen u_i: $w_i = u_i^*$, so geht das Element H^l_{ki} über in

$$\int w_i^* H^l w_k \, d\tau = H'^l_{ik}.$$

Die Multipletts derjenigen Atome, bei denen die Anzahl der Löcher geringer ist, als die der Elektronen, sind also *verkehrt*. Das Resultat, daß die Energie der Löcher genau den entgegengesetzten Wert der Elektronenenergie hat, ist freilich ganz selbstverständlich, solange man von der Wechselwirkung der Elektronen absieht. Aus Gl. (15) folgt nun weiter, daß auch die Wechselwirkung der Löcher die Form e^2/r_{lm} hat, und zwar mit dem gleichen Vorzeichen wie bei den Elektronen, daß aber von der Wechselwirkung noch Zusätze zum Zentralfeld des einzelnen Loches kommen, die im folgenden zu berechnen sind. Wir betrachten zunächst die Terme

$$\frac{1}{2}\sum_r (H_{kr,\,ri} + H_{rk,\,ir}), \tag{19}$$

die zu H_{ki} in (15) hinzutreten.

Aus (7) folgt sofort, daß

$$\sum_r H_{rk,\,ir} = \sum_r H_{kr,\,ri} \tag{20}$$

und es genügt daher, die Summe

$$\sum_r H_{kr,\,ri}$$

zu betrachten. Nach (7) hat sie die Form

$$\sum_r H_{kr,\,ri} = \iint u_k^*(1)\, u_i(1) \sum_r u_r^*(2)\, u_r(2)\, \frac{e^2}{r_{12}}\, d\tau_1\, d\tau_2. \tag{21}$$

Die Eigenfunktionen $u_i(1)$ lassen sich als Produkt einer Funktion $f(r_1)$ des Radius allein mit einer Legendreschen Kugelfunktion $P_l^{m_l}(\cos\vartheta_1)$ und einer Exponentialfunktion $e^{i m_l \varphi_1}$ der Winkel ϑ_1 und φ_1 schreiben; als weiterer Faktor kommt die Spineigenfunktion $v_{m_s}(\sigma_1)$ hinzu ($\sigma_1 = \pm 1$). Wir wollen $f(r_1)$ als normiert annehmen und für die Winkelfunktionen die normierten Funktionen

$$\left\{\begin{aligned} \Pi_l^{m_l}(\vartheta_1) &= \sqrt{\left(l + \frac{1}{2}\right)\frac{(l-m_l)!}{(l+m_l)!}}\; P_l^{m_l}(\cos\vartheta_1) \\ N_{m_l}(\varphi_1) &= \frac{1}{\sqrt{2\pi}}\, e^{i m_l \varphi_1} \end{aligned}\right. \tag{22}$$

einführen.

Es gilt dann

$$u_{m_l,\,m_s}(1) = f(r_1)\, \Pi_l^{m_l}(\vartheta_1)\, N_{m_l}(\varphi_1)\, v_{m_s}(\sigma_1). \tag{23}$$

In der Integration über $d\tau$ in (7) ist die Summation über die Spinvariable mit einbegriffen und diese läßt sich sofort ausführen. Es kommen nämlich nur Übergänge $k \longleftrightarrow i$ vor, bei denen sich die Spinrichtung m_s nicht ändert und für diese gibt die Summation über σ_1 den Faktor 1. Desgleichen gibt die Summation über σ_2 in jedem Glied $u_r^*(2)\, u_r(2)$ den

Faktor 1. Zur Berechnung von (2) muß jetzt noch die Summe

$$S = \sum_{m_l} \Pi_l^{m_l}(\vartheta_2)\, \Pi_l^{m_l}(\vartheta_2)\, \frac{1}{2\pi}$$

ausgeführt werden. Nach dem Additionstheorem der Kugelfunktionen[1]) gilt

$$S = [\Pi_l^0(0)]^2 \frac{1}{2\pi} = \left(l + \frac{1}{2}\right)\frac{1}{2\pi}. \tag{24}$$

Bei der Summation über r in (21) ist aber noch zu beachten, daß es zu jedem Wert von m_l *zwei* Zustände ($m_s = \pm \frac{1}{2}$) gibt. Die Integration über die Winkel ϑ_2 und φ_2 liefert nun für e^2/r_{12} entweder $4\pi e^2/r_1$ oder $4\pi e^2/r_2$ je nachdem $r_1 \gg r_2$ oder $r_2 \gg r_1$. Schließlich erhält man für das zu (19) gehörige Zentralfeld

$$\left\{\begin{aligned} C(r_1) &= \int \sum u_r^*(2) u_r(2) \frac{e^2}{r_{12}}\, d\tau_2 \\ &= 2(2l+1)\cdot\left[\int_0^{r_1} r_2^2\, dr_2 f^2(r_2) \frac{e^2}{r_1} + \int_{r_1}^{\infty} r_2^2\, dr_2 f^2(r_2) \frac{e^2}{r_2}\right]. \end{aligned}\right. \tag{25}$$

Dieses Zentralfeld ist genau das von der Ladungsdichte der abgeschlossenen Schale nach dem Coulombschen Gesetz herrührende elektrostatische Feld.

Wir betrachten ferner die Terme

$$-\frac{1}{2}\sum_r (H_{kr,\,ir} + H_{rk,\,ri})$$

und es genügt wieder, die eine der beiden Summen auszurechnen:

$$\sum_r H_{kr,\,ir} = \int\int u_k^*(1)\, u_i(2) \frac{e^2}{r_{12}} \sum_r u_r^*(2) u_r(1)\, d\tau_1\, d\tau_2. \tag{26}$$

Es kommen wieder nur Übergänge vor, bei denen sich die Spinrichtung nicht ändert, und die Summation über die Spinvariabeln gibt den Faktor 1. Die Eigenfunktionen zerlegen wir nach Gl. (23) und benötigen zur Ausrechnung von (26) die Summe

$$S_{12} = \sum_{m_l} \Pi_l^{m_l}(\vartheta_1)\, \Pi_l^{m_l}(\vartheta_2) \frac{1}{2\pi}\, e^{i(\varphi_2 - \varphi_1) m_l}, \tag{27}$$

---

1) Vgl. etwa A. Sommerfeld, Wellenmechanischer Ergänzungsband zu Atombau u. Spektrallinien, S. 103.

die nach dem Additionstheorem der Kugelfunktionen übergeht in:

$$S_{12} = \Pi_l^0(\chi_{12})\,\Pi_l^0(0)\,\frac{1}{2\pi}, \tag{28}$$

wobei χ_{12} den Winkelabstand von 1 und 2 bedeutet:

$$\chi_{12} = \sphericalangle(\mathfrak{r}_1\,\mathfrak{r}_2)\,.$$

Das Produkt

$$S_{12}\,\frac{e^2}{r_{12}}$$

hängt also nur vom Relativwinkel χ_{12} ab und läßt sich nach gewöhnlichen Kugelfunktionen in $\cos\chi_{12}$ entwickeln. Zur Ausrechnung von (26) braucht man nur den Kooffizienten der Kugelfunktion Π_l^0 in dieser Entwicklung, die anderen Glieder geben keinen Beitrag. Es folgt:

$$\left\{\begin{aligned} S_{12}\,\frac{e^2}{r_{12}} &= \frac{1}{2\pi}\,\Pi_l^0(\chi_{12})\,\Pi_l^0(0)\,\frac{e^2}{\sqrt{r_1^2+r_2^2-2\,r_1\,r_2\cos\chi_{12}}} \\ &= \frac{1}{2\pi}\,\Pi_l^0(\chi_{12})\sqrt{l+\frac{1}{2}}\sum_{l'}\frac{\Pi_{l'}^0(\chi_{12})}{\sqrt{l'+\frac{1}{2}}}\left(\frac{e^2 r_1^{l'}}{r_2^{l'+1}}\ \text{bzw.}\ \frac{e^2 r_2^{l'}}{r_1^{l'+1}}\right), \end{aligned}\right. \tag{29}$$

wobei der erste bzw. zweite Ausdruck in der Klammer gilt, je nachdem $r_2 \geqq r_1$ oder $r_1 \geqq r_2$. Setzt man nun

$$S_{12}\,\frac{e^2}{r_{12}} = \sum a_s\,\Pi_s^0(\chi_{12}) \tag{30}$$

und interessiert sich nur für den Faktor a_l, so folgt:

$$\left\{\begin{aligned} a_l &= \int_{-1}^{+1} S_{12}\,\frac{e^2}{r_{12}}\,\Pi_l^0(\chi_{12})\,d(\cos\chi_{12}) \\ &= \sum_{l'} b_{l\,l'}\left(\frac{e^2 r_1^{l'}}{r_2^{l'+1}}\ \text{bzw.}\ \frac{e^2 r_2^{l'}}{r_1^{l'+1}}\right). \end{aligned}\right. \tag{31}$$

Hierbei ist

$$b_{l\,l'} = \int_{-1}^{+1}\frac{1}{2\pi}\sqrt{\frac{l+\frac{1}{2}}{l'+\frac{1}{2}}}\,\Pi_l^0(\chi_{12})\,\Pi_{l'}^0(\chi_{12})\,\Pi_l^0(\chi_{12})\,d(\cos\chi_{12}). \tag{32}$$

Nach einer Tabelle von Gaunt[1]) ergibt sich: für l' gerade, $2l \geqq l'$

1) J. A. Gaunt, Philos. Trans. of the Roy. Soc. London A. Vol. 228. S. 151. 1929; vgl. besonders S. 195.

$$(33)\qquad b_{l\,l'} = \frac{(l'!)^2\,(2\,l - l')!\left[\left(l + \frac{l'}{2}\right)!\right]^2 \cdot \left(l + \frac{1}{2}\right)^{3/2}}{\pi \cdot (2\,l + l' + 1)!\left[\left(\frac{l'}{2}\right)!\right]^4\left[\left(l - \frac{l'}{2}\right)!\right]^2}\;;$$

für andere Werte von l' und l ist $b_{l\,l'} = 0$. Die Kugelfunktionen Π_s^0 in der Summe (30) ersetzt man nun wieder nach (27) und (28) durch

$$(34)\qquad \frac{1}{2\pi}\,\Pi_s^0(\chi_{12}) = \frac{1}{\Pi_s^0(0)}\sum_m \Pi_s^m(\vartheta_1)\,\Pi_s^m(\vartheta_2)\,N_m^*(\varphi_1)\,N_m(\varphi_2).$$

Man erkennt dann, daß wegen der Orthogonalitätsrelationen die Integration über $\vartheta_1\;\varphi_1\;\vartheta_2\;\varphi_2$ Null ergibt, wenn nicht $i = k$, $s = l$, $m = m_l$. Als Wert des Integrals ergibt sich schließlich:

$$(35)\quad \left\{\begin{aligned} \sum_r H_{kr,\,ir} &= \delta_{ik}\int r_1^2\,d r_1\; r_2^2\,d r_2\, f^2(r_1)\,f^2(r_2) \\ &\cdot \frac{2\pi}{\sqrt{l + \frac{1}{2}}}\sum_{l'=0}^{2l} b_{l\,l'}\left(\frac{e^2\,r_2^{l'}}{r_1^{l'+1}} \text{ bzw. } \frac{e^2\,r_1^{l'}}{r_2^{l'+1}}\right). \end{aligned}\right.$$

Das Zentralfeld, welches von den Austauschwirkungen herrührt, wird also

$$(36)\quad \left\{\begin{aligned} D(r_1) &= \int_0^\infty r_2^2\,d r_2\, f^2(r_2) \\ &\cdot \frac{2\pi}{\sqrt{l + \frac{1}{2}}}\sum_{l'=0}^{2l} b_{l\,l'}\left(\frac{e^2\,r_2^{l'}}{r_1^{l'+1}} \text{ bzw. } \frac{e^2\,r_1^{l'}}{r_2^{l'+1}}\right). \end{aligned}\right.$$

Geht man nun von der Säkulargleichung (15) zur Wellengleichung der Löcher über, so hat man noch zu beachten, daß statt H_{ik} in (6a) stets H_{ki} in (15) steht. Da in den Eigenfunktionen als komplexer Bestandteil nur der Faktor $e^{i\,m_l\,\varphi}$ steht, so genügt es, die Eigenfunktionen der Löcher gegenüber denen der Elektronen mit dem entgegengesetzten Vorzeichen von m_l zu definieren, um eine zu (2) analoge Wellengleichung der Löcher zu erhalten. D. h. ein Loch hat den entgegengesetzten Drehimpuls, wie das entsprechende Elektron.

Die Wellengleichung der Löcher lautet schließlich (hierbei ist $C(r_l) = C^l$, $D(r_l) = D^l$ gesetzt):

$$
(37)\quad \left\{\sum_{l}^{N-n}\left(\frac{h^2}{8\pi^2\mu}\Delta^l - H^l - C^l + D^l + eV^l\right) + \sum_{l>m}^{N-n}\frac{e^2}{r_{lm}} + A - E\right\}\psi = 0.
$$

Wenn man sich auf die erste Näherung beschränkt, kann man also die tiefsten Terme der Atome einer Periode des periodischen Systems auch mit der Wellengleichung (37) statt mit (2) ausrechnen.

Aus diesem Resultat kann man ganz ohne Rechnung folgende an der Erfahrung prüfbaren Konsequenzen ziehen: Der einzige Unterschied zwischen Löchern und Elektronen besteht in dem auf sie wirkenden Zentralfeld. Man wird zweckmäßig zur besseren Konvergenz des Verfahrens dieses Zentralfeld ganz oder teilweise in das ungestörte Problem aufnehmen. Es werden also selbst bei gleicher Kernladung die ungestörten Eigenfunktionen für Löcher und Elektronen zweckmäßig etwas verschieden gewählt. Wir können es aber so einrichten, daß die verschiedenen ungestörten Zentralfelder durch eine einfache Ähnlichkeitstransformation auseinander hervorgehen, wie dies z. B. der Fall ist beim Thomas-Fermischen Feld, oder wenn man das Feld durch eine „effektive Kernladung" annähert. In der durch die Wechselwirkung bedingten Aufspaltung der ungestörten Terme äußert sich dann das verschiedene Zentralfeld nur in einem gemeinsamen konstanten Faktor. Vernachlässigt man die Spinwechselwirkungen, so muß also die Termmannigfaltigkeit eines Atoms mit n Elektronen sich von der eines Atoms mit $N-n$ Elektronen nur durch einen konstanten Faktor und eine additive Konstante unterscheiden. Der konstante Faktor ist stets positiv und bei gleicher Kernladung nicht viel von Eins verschieden. Dagegen haben die Multiplettaufspaltungen in beiden Fällen das entgegengesetzte Vorzeichen.

Als Beispiel[1]) vergleichen wir die tiefsten Terme von Ti, welches zwei Elektronen in $3\,d$-Bahnen enthält, mit denen von Ni, bei dem $10-2=8$ Elektronen in $3\,d$-Bahnen ge-

1) Für wertvolle Hilfe bei der Diskussion des empirischen Materials bin ich Hrn. F. Hund zum größten Dank verpflichtet.

bunden sind.[1]) Es gibt in beiden Fällen die Terme ^3F, ^3P, ^1S, ^1D, ^1G, von denen wir in der folgenden Tabelle die Term*schwerpunkte* eingetragen haben.

Tabelle 1

	Ti	Ni
^3F	223	972
^3P	8547	15696
^1S	15167	—
^1D	7255	13521
^1G	12118	22102

Der Ni-Term ^1S, der zur Konfiguration $(3\,d)^8\,(4\,s)^2$ gehört, ist nicht bekannt. In der nächsten Tabelle sind die Differenzen der Terme gegen den tiefsten eingetragen, ferner enthält die letzte Spalte den Quotienten der betreffenden Termdifferenz bei Ni und bei Ti.

Tabelle 2

	Ti	Ni	Ni/Ti
^3F	0	0	—
^3P	8324	14724	1,769
^1S	14944	—	—
^1D	7032	12549	1,785
^1G	11895	21130	1,776

Man sieht, daß der Quotient der Theorie entsprechend sehr genau konstant ist, wohl genauer, als man nach den Gültigkeitsgrenzen der Rechnung erwarten würde. Immerhin wird man den fehlenden Term $(3\,d)^8\,(4\,s)^2$ ^1S bei Ni nach der obigen Tabelle mit einiger Genauigkeit extrapolieren können. Die Rechnung ergibt

$$\mathrm{Ni}\,(3\,d)^8\,(4\,s)^2\ {}^1S : 27\,500 \pm 150\,.$$

In der gleichen Weise sollen noch die tiefsten Terme von O^{++} (zwei $2p$-Elektronen) verglichen werden mit denen von O (hier fehlen zwei $2p$-Elektronen zur abgeschlossen Schale): Die Tabelle gibt die in Betracht kommenden Terme[2]):

1) Die Angaben sind entnommen aus H. N. Russel, Astrophys. Journal. **66.** S. 335. 1927 und aus einer brieflichen Mitteilung aus dem Institut von Hrn. H. N. Russel an Hrn. F. Hund.

2) Die Termwerte sind entnommen aus: A. Fowler, Proc. Roy. Soc. London A. **117.** S. 317. 1928 und O. Frerichs, Phys. Rev. **36**, 398. 1930.

Tabelle 3

	O^{++}	O
$^3P_0$	444661	109605
$^3P_1$	444545	109672
$^3P_2$	444352	109831
^1D	424385	93962
^1S	401472	76037

In Analogie zu Tab. 2 bilden wir die Differenzen gegen den Schwerpunkt des tiefsten Terms und berechnen den Quotienten:

Tabelle 4

	O^{++}	O	O^{++}/O
^3P	0	0	—
^1D	20066	15791	1,2708
^1S	42979	33716	1,2747

Auch hier sind die Quotienten auffallend genau konstant.

Bevor wir zu einem anderen Beispiel für die Wellengleichung der Löcher übergehen, sollen einige allgemeine Bemerkungen Platz finden. Wenn $X(x\,y\,z\,\sigma)$ ein auf ein einzelnes Elektron bezüglicher Operator ist, also etwa den Strom, die Stromdichte, das Moment des Elektrons bezeichnet, so entspricht diesem Operator im Jordan-Wignerschen Schema der Operator:

(38) $$\sum_{i\,k} a_i^* \, a_k \, X_{ik},$$

wobei

(39) $$X_{ik} = \int u_i^* \, X \, u_k \, d\tau \, .$$

Geht man zu den gestrichenen Amplituden über, so folgt

(40) $$\sum_{ik} a_i^* \, a_k \, X_{ik} = \sum_{ik} (\delta_{ik} - a_k'^* \, a_i') X_{ik} = \sum_i X_{ii} - \sum_{ik} a_i'^* \, a_k' \, X_{ki} .$$

Ein vollständiges Analogon zu Gl. (38) und (39) erhält man aus Gl. (40) nur dann, wenn sich X_{ki} in der Form darstellen läßt:

(41) $$X_{ki} = \int w_i^* \, X \, w_k \, d\tau = X'_{ik},$$

wobei die w_i die Eigenfunktionen der Löcher bedeuten. In den bisherigen Rechnungen haben wir $w_i = u_i^*$ gesetzt (vgl. S. 893), und dann läßt sich in der Tat für eine reelle Koordinatenfunktion X Gl. (41) aus (39) herleiten. Ist X ein andersartiger

Operator, so muß man den Übergang von (39) zu (41) neu diskutieren. Ein allgemeines Resultat läßt sich nur für die *Diagonalelemente* X_{ii} aussprechen: Stromdichte, Moment, Ladungsdichte usw. haben für ein Loch stets das umgekehrte Vorzeichen, wie für das entsprechende Elektron.

§ 3. Als zweites Beispiel für die Verwendung der „Wellengleichung der Löcher" sei die Peierlssche Deutung[1]) des anomalen Halleffektes besprochen. Dieses Problem ist deswegen besonders einfach, weil die Wechselwirkung der Elektronen hier vernachlässigt werden kann: Das ungestörte System besteht aus Elektronen, die in dem dreifach-periodischen Kraftfeld des Kristallgitters (Potential $e\,V_0$) laufen. Als Störungen sind ein äußeres elektrisches Feld und ein äußeres Magnetfeld zu betrachten, ferner die durch die Temperaturbewegung hervorgerufenen Abweichungen von der Gitterperiodizität ($e\,V_1$). Einschließlich aller Störungen wird die Wellengleichung des einzelnen Elektrons demnach:

$$(42)\quad \begin{cases} \Big\{-\dfrac{h^2}{8\pi^2\mu}\Delta - e(V_0+V_1) - eF_x x - \dfrac{h\,e}{4\pi i\mu c}H_z\Big(y\dfrac{\partial}{\partial x} - x\dfrac{\partial}{\partial y}\Big) \\ \qquad + \dfrac{e^2}{8\mu c^2}H_z^2(x^2+y^2) + \dfrac{h}{2\pi i}\dfrac{\partial}{\partial t}\Big\}\psi = 0\,. \end{cases}$$

Hierbei ist das elektrische Feld (F_x) in der x-Richtung angenommen, so daß es die Elektronen in der positiven x-Richtung beschleunigt, das Magnetfeld (H_z) hat die Richtung der Z-Achse. Damit die von F_x und H_z herrührenden Glieder als kleine Störung betrachtet werden können, ist es nötig, ein nicht zu großes Wellenpaket als Ausgangslösung zu betrachten und den Koordinatenanfangspunkt in seine Nähe zu legen. Das Glied mit H_z^2 kann man dann ganz vernachlässigen. Die zu behandelnde Wellengleichung lautet also in dieser Näherung

$$(42\text{a})\quad \begin{cases} \Big\{-\dfrac{h^2}{8\pi^2\mu}\Delta - e(V_0+V_1) - eF_x x \\ \quad -\dfrac{h\,e}{4\pi i\mu c}H_z\Big(y\dfrac{\partial}{\partial x} - x\dfrac{\partial}{\partial y}\Big) + \dfrac{h}{2\pi i}\dfrac{\partial}{\partial t}\Big\}\psi = 0\,. \end{cases}$$

Die Eigenfunktion des ungestörten Problems hat nach Bloch[2]) die Form

---

1) R. Peierls, Ztschr. f. Phys. **53**. S. 255. 1929.

2) F. Bloch, Ztschr. f. Phys. **52**. S. 555. 1928.

$$\psi'_{klm} = e^{2\pi i\left(\frac{kx}{K}+\frac{ly}{L}+\frac{mz}{M}\right)} \cdot u_{klm}(x\,y\,z), \tag{43}$$

wobei K, L, M die Kanten des Elementarquaders sind, der die periodische Randbedingung festlegt, $u_{klm}(x\,y\,z)$ eine Funktion von Gitterperiodizität bedeutet und die k, l, m die Translationsquantenzahlen darstellen. Die Energiewerte des ungestörten Problems ordnen sich bei hinreichend tiefen Potentialmulden V_0 in einzelne Gruppen nahe benachbarter Terme, deren Niveaus durch die Formel

$$E_{klm} = \mathrm{const} - 2\beta\left(\cos\frac{2\pi k a}{K} + \cos\frac{2\pi l a}{L} + \cos\frac{2\pi m a}{M}\right) \tag{44}$$

dargestellt werden (a bedeutet die Gitterkonstante). Eine solche Gruppe kann mit Elektronen aufgefüllt werden und bildet dann eine abgeschlossene Schale. Wir gehen nun von der Wellengleichung (42a) zur entsprechenden Säkulargleichung vom Typus der Gl. (6a) über und von dort zur Säkulargleichung (15) der Löcher. Die Übergangselemente H_{ik} in (6a) sind jetzt in der Form $H_{klm;\,k'l'm'}$ zu schreiben. Um nun von der Säkulargleichung (15) wieder zur Wellengleichung der Löcher überzugehen, müssen wir die Eigenfunktion des einzelnen Loches definieren.

Nach Bloch (a. a. O.) läßt sich die Eigenfunktion (43) des einzelnen Elektrons auch näherungsweise in der Form schreiben:

$$\psi_{klm} = \sum_{g_1 g_2 g_3} e^{2\pi i\left(\frac{g_1 k}{G_1}+\frac{g_2 l}{G_2}+\frac{g_3 m}{G_3}\right)} v_{g_1 g_2 g_3}(x\,y\,z), \tag{45}$$

wobei $v_{g_1 g_2 g_3}$ die Schrödingerfunktion des Atoms mit der Nummer g_1, g_2, g_3 bedeutet, G_1, G_2, G_3 die Anzahlen der Atome in den Kanten K, L, M sind.

Wir setzen nun für die Eigenfunktion des einzelnen Loches

$$\psi'_{klm} = \sum_{g_1 g_2 g_3} e^{-2\pi i\left[g_1\left(\frac{k}{G_1}+\frac{1}{2}\right)+g_2\left(\frac{l}{G_2}+\frac{1}{2}\right)+g_3\left(\frac{m}{G_3}+\frac{1}{2}\right)\right]} v^*_{g_1 g_2 g_3}. \tag{46}$$

Für einen Operator X, der eine reine Koordinatenfunktion ist, gilt dann wieder — allerdings hier nur in erster Näherung:

$$X_{k'l'm';\,klm} = X'_{klm;\,k'l'm'},$$

wie in Gl. (4); diese Beziehung bleibt solange richtig, als man bei der Integration $\int \psi^*_{k'l'm'} X \psi_{klm}\, d\tau$ nur die Glieder der

Form $\int v^*_{g_1 g_2 g_3} v_{g_1 g_2 g_3} X\, d\tau$ berücksichtigt und die Übergangselemente $\int v^*_{g_1 g_2 g_3} v_{h_1 h_2 h_3} X\, d\tau$ vernachlässigt. Diese Vernachlässigung ist bei den hier vorkommenden Koordinatenfunktionen sicher gestattet. Wir erhalten also:

$$[e V_1 + e F_x x]_{k' l' m';\, k l m} = [e V_1 + e F_x x]'_{k l m;\, k' l' m'}. \tag{47}$$

Eine gesonderte Behandlung erfordert jedoch die ungestörte Energie $\left(-\frac{h^2}{8\pi^2\mu}\Delta - e V_0\right)$, sowie das H_z proportionale Glied. Aus Gl. (44) und (46) kann man sofort schließen, daß

$$\left\{\begin{aligned} &\left(-\frac{h^2}{8\pi^2\mu}\Delta - e V_0\right)'_{k l m;\, k l m} \\ &\qquad = \text{const} - \left(-\frac{h^2}{8\pi^2\mu}\Delta - e V_0\right)_{k l m;\, k l m}. \end{aligned}\right. \tag{48}$$

Denn das Glied $\cos\frac{2\pi k a}{K}$ wechselt das Vorzeichen, wenn man k durch $k + \frac{g_1}{2}$ ersetzt. Es bleiben schließlich die Matrixelemente des Operators $x\frac{\partial}{\partial y} - y\frac{\partial}{\partial x}$ zu diskutieren. Bei der Integration über das Volumen fallen die Glieder der Form

$$\int v_{g_1 g_2 g_3}\left(x\frac{\partial}{\partial y} - y\frac{\partial}{\partial x}\right) v_{g_1 g_2 g_3}\, d\tau$$

weg, wenn, wie hier angenommen wird, die Atome ursprünglich in S-Zuständen waren. Es kommt also auf die Übergangselemente an, bei denen eine der Zahlen g_1, g_2, g_3 sich um eins ändert. Beachtet man die hierdurch bedingte Vorzeichenumkehr, so folgt

$$\left\{\begin{aligned} \left(x\frac{\partial}{\partial y} - y\frac{\partial}{\partial x}\right)_{k' l' m';\, k l m} &= \int \psi^*_{k' l' m'}\left(x\frac{\partial}{\partial y} - y\frac{\partial}{\partial x}\right)\psi_{k l m}\, d\tau \\ &\approx -\int \psi'_{k' l' m'}\left(x\frac{\partial}{\partial y} - y\frac{\partial}{\partial x}\right)\psi'^*_{k l m}\, d\tau \\ &= \int \psi'^*_{k l m}\left(x\frac{\partial}{\partial y} - y\frac{\partial}{\partial x}\right)\psi'_{k' l' m'}\, d\tau \\ &= \left(x\frac{\partial}{\partial y} - y\frac{\partial}{\partial x}\right)'_{k l m;\, k' l' m'} \end{aligned}\right. \tag{49}$$

Die Wellengleichung der Löcher, die aus (15) hervorgeht, lautet also schließlich nach (47), (48), (49):

Annalen der Physik (5) *10*, 888–904 (1931)

$$(50)\quad \left\{\begin{array}{l} \left\{-\frac{h^2}{8\pi^2\mu}\Delta - eV_0 + eV_1 + eF_x\cdot x + \frac{h e}{4\pi i\mu c}H_z \right. \\ \left. \quad\cdot\left(y\frac{\partial}{\partial x} - x\frac{\partial}{\partial y}\right) + \frac{h}{2\pi i}\frac{\partial}{\partial t} + \mathrm{const}\right\}\psi = 0\,. \end{array}\right.$$

Auch diese Gleichung gilt nur für Wellenpakete von Löchern, die sich in der Nähe des Koordinatenursprungs bewegen. Vergleicht man (50) mit der Wellengleichung (42a), so erkennt man, daß sich die Löcher genau wie Elektronen *positiver* Ladung unter Einfluß der störenden äußeren Felder bewegen. Andererseits tragen die Löcher auch zum Strom und zur Ladungsdichte genau so bei, wie Elektronen *positiver* Ladung. Die Elektrizitätsleitung in Metallen mit einer geringen Anzahl von Löchern kann also in jeder Beziehung beschrieben werden wie die Leitung in Metallen mit einer geringen Anzahl von *positiven* Leitungselektronen. Daraus folgt unmittelbar der anomale Halleffekt für solche Metalle.

Zusammenfassung

Atomare Systeme, bei denen n Elektronen zu einer abgeschlossenen Schale fehlen, lassen sich näherungsweise behandeln, indem man eine Schrödingergleichung in den Koordinaten der n „Löcher" löst. Als Beispiele für die Methode werden die tiefsten Terme der Atomspektra sowie der anomale Halleffekt besprochen.

(Eingegangen 19. Juni 1931)

PHYSIKALISCHE ZEITSCHRIFT
vereinigt mit dem
JAHRBUCH DER RADIOAKTIVITÄT UND ELEKTRONIK
Nr. 19 — 1. Oktober 1931 — 32. Jahrgang
Redaktionsschluß für Nr. 21 am 14. Oktober 1931.

INHALT:

ORIGINALMITTEILUNGEN.

Über die inkohärente Streuung von Röntgenstrahlen.

Von W. Heisenberg.

Bei der Untersuchung der Struktur mehratomiger Moleküle durch Röntgenstrahlen ist die Kenntnis der Streuung eines einzelnen Atoms von entscheidender Wichtigkeit. Während der kohärente Teil dieser Streuung nach Debye[1]) in hinreichender Näherung aus der Thomas-Fermischen Dichteverteilung bestimmt werden und daher durch Tabulierung einer einzigen Funktion für sämtliche Atome gewonnen werden kann, bleibt für die inkohärente Strahlung zunächst nichts übrig, als die Beiträge der einzelnen Elektronen nach den Formeln von Waller[2]) aufzusummieren; eine Reduktion auf die Thomas-Fermi-Verteilung ist deswegen nicht unmittelbar möglich, weil die inkohärente Strahlung nicht durch die Dichte allein bestimmt ist.

Herr Debye hatte mir die Frage gestellt, ob man nicht durch Methoden, die denen von Thomas[3]) und Fermi[4]) nachgebildet sind, Näherungsformeln für die inkohärente Streustrahlung erhalten könnte.

Für die folgenden Rechnungen werde angenommen, daß die Energie der einfallenden Lichtquanten groß sei gegen die Ionisierungsspannung der innersten Schale des streuenden Atoms und klein gegen μc^2 (μ Elektronenmasse, c Lichtgeschwindigkeit). Für leichte Atome sind beide Forderungen vereinbar. Nach Waller (l. c.) wird dann näherungsweise die einem Übergang des Atoms von Zustand n nach Zustand n' entsprechende Streustrahlung hervorgerufen durch einen schwingenden Dipol der Stärke:

$$d_{nn'} = -\frac{e^2}{8\pi^2\nu^2} E \cdot \gamma_{nn'} \cdot e^{2\pi i\nu' t}. \qquad (1)$$

ν bzw. ν' ist die Frequenz des einfallenden bzw. gestreuten Lichtes, E die Amplitude des elektrischen Vektors im einfallenden Licht. Ferner gilt

$$\gamma_{nn'} = \int \Phi_n^* \Phi_{n'} \sum_{k=1}^{N} e^{-i\mathfrak{s}\mathfrak{r}_k} dV_1 \cdot dV_2 \ldots dV_N. \qquad (2)$$

N ist die Anzahl der Elektronen, Φ_n die Schrödingerfunktion des Atoms im Zustand n, der Vektor $\mathfrak{s}$ ist durch

$$\mathfrak{s} = \frac{2\pi}{c}(\nu\mathfrak{n} - \nu'\mathfrak{n}') \qquad (3)$$

gegeben, wobei $\mathfrak{n}$ einen Einheitsvektor in Richtung des einfallenden Lichtes, $\mathfrak{n}'$ einen in Richtung des gestreuten Lichtquants bedeutet.

Von der Thomsonschen Formel unterscheidet sich Gl. (1) nur durch den Faktor $\gamma_{nn'}$, wir werden also diesen Faktor im folgenden ausschließlich betrachten und den Thomsonschen

1) P. Debye, Physik. Zeitschr. **31**, 419, 1930.
2) I. Waller, Zeitschr. f. Phys. **51**, 213, 1928.
3) L. H. Thomas, Proc. Cambr. Phil. Soc. **23**, 542, 1927.
4) E. Fermi, Zeitschr. f. Phys. **48**, 73, 1928.

Anteil überall weglassen. Die gesamte Streustrahlung des Atoms wird demnach gegeben durch

$$S = \sum_{n'=0}^{\infty} |\gamma_{n n'}|^2 = \tag{4}$$
$$= \sum_{n'} \iint \Phi_n^*(\mathfrak{r}_1, \mathfrak{r}_2 ..)\, \Phi_{n'}(\mathfrak{r}_1 \mathfrak{r}_2 ...)\, \Phi_{n'}^*(\mathfrak{r}'_1, \mathfrak{r}'_2 ...)\, \Phi_n(\mathfrak{r}'_1 \mathfrak{r}'_2 ..) \sum_{k,l=1}^{N} e^{-i\mathfrak{s}(\mathfrak{r}_k - \mathfrak{r}'_l)}\, dV_1 dV_2 .. dV'_1 dV'_2 ..$$

Die Summation über n' läßt sich unter dem Integralzeichen ausführen und gibt nach den Orthogonalitätsrelationen die Diracsche δ-Funktion im Raum der $\mathfrak{r}_1 - \mathfrak{r}'_1$, $\mathfrak{r}_2 - \mathfrak{r}'_2, \ldots$ Es folgt daher

$$S = \int \Phi_n^*(\mathfrak{r}_1 ...)\, \Phi_n(\mathfrak{r}_1 ...) \sum_{k,l=1}^{N} e^{i\mathfrak{s}(\mathfrak{r}_l - \mathfrak{r}_k)}\, dV_1 dV_2 \ldots \tag{5}$$

Aus den Orthogonalitätsrelationen ergibt sich ferner:

$$S = N + \int \Phi_n^* \Phi_n \sum_{k \neq l}^{N} e^{i\mathfrak{s}(\mathfrak{r}_l - \mathfrak{r}_k)}\, dV_1 dV_2 \ldots \tag{6}$$

Bei der Integration über die Volumelemente dV_1, dV_2 usw. in den vorausgehenden Formeln ist genau genommen die Summation über die Spinvariable jedes Elektrons mit inbegriffen; wir haben sie nicht besonders angeschrieben, um die Formeln nicht unnötig zu komplizieren. Bei der Auswertung von (6) haben wir aber jetzt zu beachten, daß sich Φ_n bei Vernachlässigung der Wechselwirkung als antisymmetrische Kombination von Produkten der Eigenfunktionen einzelner Elektronen darstellen läßt, wobei jede Eigenfunktion abhängt von Orts- und Spinvariable des betreffenden Elektrons. Die Zustände numerieren wir mit s und t, die Eigenfunktionen seien $\varphi_s(\mathfrak{r}, \sigma)$, σ bedeutet die Spinvariable.

Aus (6) folgt dann (die Summationen über die Spinvariabeln werden jetzt ausführlich angeschrieben):

$$S = \sum_{s,t} \iint dV dV' \sum_{\sigma} \sum_{\sigma'} \varphi_s^*(\mathfrak{r}, \sigma)\, \varphi_s(\mathfrak{r}, \sigma)\, \varphi_t^*(\mathfrak{r}', \sigma')\, \varphi_t(\mathfrak{r}', \sigma')\, e^{i\mathfrak{s}(\mathfrak{r}-\mathfrak{r}')}$$
$$+ N - \sum_{s,t} \iint dV dV' \sum_{\sigma} \sum_{\sigma'} \varphi_s^*(\mathfrak{r}, \sigma)\, \varphi_t(\mathfrak{r}, \sigma)\, \varphi_t^*(\mathfrak{r}', \sigma')\, \varphi_s(\mathfrak{r}', \sigma')\, e^{i\mathfrak{s}(\mathfrak{r}-\mathfrak{r}')}. \tag{7}$$

Die Summationen über s und t sind hier über alle besetzten Zustände auszuführen. Durch eine einfache Umformung findet man:

$$S = \Big| \sum_t \int dV \sum_{\sigma} \varphi_t^*(\mathfrak{r}, \sigma)\, \varphi_t(\mathfrak{r}, \sigma)\, e^{i\mathfrak{s}\mathfrak{r}} \Big|^2$$
$$+ N - \iint dV dV' \sum_{\sigma} \sum_{\sigma'} \Big| \sum_t \varphi_t^*(\mathfrak{r}, \sigma)\, \varphi_t(\mathfrak{r}', \sigma') \Big|^2 e^{i\mathfrak{s}(\mathfrak{r}-\mathfrak{r}')} \tag{8}$$

Der erste Teil der rechten Seite von (8) stellt die kohärente Streustrahlung dar, deren Amplitude durch

$$a = \sum_t \int dV \sum_{\sigma} \varphi_t^*(\mathfrak{r}, \sigma)\, \varphi_t(\mathfrak{r}, \sigma)\, e^{i\mathfrak{s}\mathfrak{r}} \tag{9}$$

gegeben ist. Durch Abspaltung des Spinanteils (die Spin-Bahnkopplung werde vernachlässigt) läßt sich nämlich $\varphi_t(\mathfrak{r}, \sigma)$ in der Form

$$\varphi_t(\mathfrak{r}, \sigma) = \varphi_t(\mathfrak{r})\, \chi_t(\sigma) \tag{10}$$

schreiben und die Summation über σ in Gleichung (9) gibt jeweils den Faktor 1:

$$a = \sum_t \int dV\, \varphi_t^*(\mathfrak{r})\, \varphi_t(\mathfrak{r})\, e^{i\mathfrak{s}\mathfrak{r}}. \tag{11}$$

Die inkohärente Streustrahlung wird also durch

$$S_i = N - \iint dV\, dV' \sum_{\sigma} \sum_{\sigma'} \Big| \sum_t \varphi_t^*(\mathfrak{r}, \sigma)\, \varphi_t(\mathfrak{r}', \sigma') \Big|^2 e^{i\mathfrak{s}(\mathfrak{r}-\mathfrak{r}')} \tag{12}$$

gegeben und es kommt alles auf die Berechnung der Größe

$$G(\mathfrak{r}\mathfrak{r}', \sigma\sigma') = \sum_t \varphi_t^*(\mathfrak{r}, \sigma)\, \varphi_t(\mathfrak{r}', \sigma') \tag{13}$$

an. Nach Gl. (10) erhält man

$$G(\mathfrak{r}\mathfrak{r}', \sigma\sigma') = \sum_t \varphi_t^*(\mathfrak{r})\, \varphi_t(\mathfrak{r}')\, \chi_t^*(\sigma)\, \chi_t(\sigma'). \tag{14}$$

Wegen des Pauliprinzips können wir nun annehmen, daß zwei Zustände, die sich nur durch die Spinrichtung des Elektrons unterscheiden, stets gleichzeitig besetzt sind; der Fehler, der etwa beim letzten gebundenen Elektron durch diese Annahme entsteht, ist bei Vorhandensein vieler Elektronen gering. Die Summation über t in (14) ergibt deshalb Null, wenn $\sigma \neq \sigma'$. Statt (14) kann man daher schreiben

$$G(\mathfrak{r}\mathfrak{r}', \sigma\sigma') = \delta_{\sigma\sigma'} \cdot \sum_t \varphi_t^*(\mathfrak{r})\, \varphi_t(\mathfrak{r}'), \tag{15}$$

wobei die Summation jetzt nur noch über die besetzten Zustände der Bahnbewegung auszuführen ist und zwar ist (trotz der doppelten Besetzung) jeder besetzte Zustand nur einmal zu zählen.

Die Summe

$$\sum_t \varphi_t^*(\mathfrak{r})\,\varphi_t(\mathfrak{r}')$$

ist bei großer Elektronenanzahl nur für kleine Werte von $|\mathfrak{r}-\mathfrak{r}'|$ merklich von Null verschieden, da sie im Grenzfall, wo die Summe über alle Zustände zu erstrecken ist, in die Diracsche Funktion $\delta(\mathfrak{r}-\mathfrak{r}')$ übergeht. Für $\mathfrak{r}=\mathfrak{r}'$ dagegen stimmt sie bis auf den Faktor 2 mit der Ladungsdichte am Punkt $\mathfrak{r}$ überein.

Thomas und Fermi setzen bekanntlich

$$\sum_t \varphi_t^*(\mathfrak{r})\,\varphi_t(\mathfrak{r}) \approx \iiint\limits_{p\leqq\sqrt{2me\Phi(\mathfrak{r})}} \frac{dp_x\,dp_y\,dp_z}{h^3} = \frac{4\pi}{3h^3}\,[2me\Phi(\mathfrak{r})]^{3/2}. \tag{16}$$

$\Phi(\mathfrak{r})$ bedeutet das Potential am Punkte $\mathfrak{r}$. Zur Begründung von (16) wird angenommen, daß man die Elektronen im Atom praktisch als frei betrachten kann und daß sich ihre Bindung nur darin äußert, daß ihr Translationsimpuls an einer gegebenen Stelle des Raumes einen durch den Energiesatz bestimmten Betrag nicht überschreiten kann. An Stelle der Summation über t tritt also die Integration über p_x, p_y, p_z.

Es liegt nahe, Gl. (16) folgendermaßen zu verallgemeinern:

$$\sum_t \varphi_t^*(\mathfrak{r})\,\varphi_t(\mathfrak{r}') = \iiint\limits_{p\leqq\sqrt{2me\Phi\left(\frac{\mathfrak{r}+\mathfrak{r}'}{2}\right)}} \frac{dp_x\,dp_y\,dp_z}{h^3}\cdot e^{\frac{2\pi i}{h}\mathfrak{p}(\mathfrak{r}-\mathfrak{r}')}. \tag{17}$$

Als Eigenfunktionen sind die der freien Translation eingesetzt, die Summation über t wird wieder ersetzt durch eine Integration über p_x, p_y, p_z, der Höchstwert des Impulses wird aus $\frac{\mathfrak{r}+\mathfrak{r}'}{2}$ berechnet.

Die Berechnung der Streustrahlung ist nun sehr einfach; durch Einsetzen von (17) in (12) erhält man, wenn man gleichzeitig $\frac{\mathfrak{r}+\mathfrak{r}'}{2}=\mathfrak{r}_1$; $\mathfrak{r}-\mathfrak{r}'=\mathfrak{r}_2$ setzt:

$$S_i = N - 2\iint dV_1\,dV_2\,e^{i\mathfrak{s}\mathfrak{r}_2}\iint\limits_{p,p'\leqq\sqrt{2me\Phi(\mathfrak{r}_1)}} \frac{dp_x\,dp_y\,dp_z\,dp'_x\,dp'_y\,dp'_z}{h^6}\cdot e^{\frac{2\pi i}{h}(\mathfrak{p}-\mathfrak{p}')\mathfrak{r}_2}. \tag{18}$$

Der Faktor 2 rührt von der Summation über σ her. Die Integration über $\mathfrak{r}_2$ läßt sich ausführen und ergibt die Diracsche δ-Funktion der Variablen $\mathfrak{p}-\mathfrak{p}'+\frac{\mathfrak{s}h}{2\pi}$, multipliziert mit h^3. Das bedeutet physikalisch, daß beim Streuprozeß die Impulsänderung des Elektrons entgegengesetzt gleich ist der des Lichtquants. Dies folgt auch unmittelbar aus der Annahme freier Elektronen.

Aus (18) wird:

$$S_i = N - 2\int dV_1\cdot\iint\limits_{p,p'\leqq\sqrt{2me\Phi(\mathfrak{r}_1)}} \frac{dp_x\,dp_y\,dp_z\,dp'_x\,dp'_y\,dp'_z}{h^3}\,\delta\!\left(\mathfrak{p}-\mathfrak{p}'+\frac{\mathfrak{s}h}{2\pi}\right). \tag{19}$$

Das sechsfache Integral auf der rechten Seite ist anschaulich das gemeinsame Volumen zweier Kugeln vom Radius $p_0=\sqrt{2me\Phi(\mathfrak{r}_1)}$, deren Mittelpunkte die Strecke $\frac{|\mathfrak{s}|h}{2\pi}$ voneinander entfernt sind. Es folgt daher

$$S_i = N - \frac{8\pi}{3}\int dV_1\left(p_0-\frac{|\mathfrak{s}|h}{4\pi}\right)^2\left(p_0+\frac{|\mathfrak{s}|h}{8\pi}\right)\frac{1}{h^3}, \tag{20}$$

wobei das Integral rechts nur über die Raumgebiete zu erstrecken ist, für die

$$p_0-\frac{|\mathfrak{s}|h}{4\pi}\geqq 0;$$

d. h.

$$S_i = N - \frac{32\pi^2}{3}\int_0^{\varrho} r_1^2\,dr_1\left(p_0-\frac{sh}{4\pi}\right)^2\left(p_0+\frac{sh}{8\pi}\right)\frac{1}{h^3}, \tag{21}$$

wobei

$$\sqrt{2me\Phi(\varrho)}=\frac{sh}{4\pi}.$$

Um die Abhängigkeit dieses Integrals von der Ordnungszahl des betrachteten Atoms zu studieren, setzen wir nach Thomas und Fermi:

$$r_1\cdot N^{1/3}=x;\quad \Phi(r_1)=N^{4/3}\cdot F(x);\quad s=N^{2/3}s_0, \tag{22}$$

und es folgt:

$$S_i = N\left[1-\frac{32\pi^2}{3}\int_0^{x_0}\frac{dx\,x^2}{h^3}\left(\sqrt{2meF(x)}-\frac{s_0h}{4\pi}\right)^2\left(\sqrt{2meF(x)}+\frac{s_0h}{8\pi}\right)\right], \tag{23}$$

wobei

$$\sqrt{2meF(x_0)} = \frac{s_0 h}{4\pi}.$$

Die Funktion $F(x)$ ist bei Thomas und Fermi (l. c.) tabuliert. Die inkohärente Streustrahlung hat nach (23) bis auf den Faktor N stets die gleiche Form der Abhängigkeit von $s_0 = s \cdot N^{-2/3}$. Tabuliert man die Funktion $S_i(s_0)$ nach Gl. (23), so gewinnt man daher die inkohärente Streustrahlung für sämtliche Atome. Es ist interessant, daß die inkohärente Strahlung von $s \cdot N^{-2/3}$ abhängt, während die kohärente bekanntlich von $s \cdot N^{-1/3}$ abhängt. Dieser Umstand zeigt deutlich, daß die inkohärente Streustrahlung im Gegensatz zur kohärenten nicht durch die Ladungsdichte allein bestimmt ist.

Die numerische Auswertung der Gl. (23) wird in einer folgenden Note von Herrn Bewilogua durchgeführt werden.

Die Gültigkeitsgrenzen der Gl. (23) sind etwa folgende: Für sehr kleine Kernladungen versagt Gl. (23), insbesondere wäre beim Wasserstoffatom die Thomas-Fermi-Methode eine sehr schlechte Approximation. Aber auch für sehr schwere Atome ist Gl. (23) kaum anwendbar, da in den praktisch vorkommenden Fällen die Energie des einfallenden Lichtquants dort nicht mehr groß ist gegen die Ionisierungsarbeit der K-Schale. Andererseits ist bei schweren Atomen die inkohärente Strahlung sehr schwach im Vergleich zur kohärenten, so daß erhebliche Fehler in Gl. (23) sich nur wenig bemerkbar machen.

(Eingegangen 9. Juli 1931.)

Über die inkohärente Streuung der Röntgenstrahlen.

Von L. Bewilogua.

Die in der vorhergehenden Arbeit von Heisenberg[1]) entwickelte Theorie gestattet es, aus einer einzigen Kurve oder Tabelle für alle Atome — unter den dort angegebenen Voraussetzungen — die Winkelverteilung der inkohärent gestreuten Strahlung zu entnehmen, wenn man nur für die Abszisse und Ordinate die dem betreffenden Atom entsprechenden Werte wählt. Damit ist es möglich, zusammen mit der bekannten allgemein gültigen Kurve für die kohärente Streustrahlung[2]), die Gesamtstrahlung eines Einzelatoms anzugeben. Von besonderer Bedeutung ist diese Kenntnis für die Auswertung der Winkelabhängigkeit der an Gasen und Dämpfer gestreuten Röntgenstrahlung, die bei mehratomigen Molekeln zur interferometrischen Abstandsbestimmung in der Molekel nach dem Debyeschen Verfahren[1]) und bei einatomigen Molekeln zur Ermittlung der Ladungsverteilung im Atom[2]) führt. Ebenso ist sie unerläßlich für die Diskussion der bei der Kristallreflexion auftretenden diffusen Streustrahlung[3]).

1) W. Heisenberg, Physik. Zeitschr. **32**, 737, 1931.

2) P. Debye, Physik. Zeitschr. **31**, 419, 1930.

1. Es ist für die numerische Rechnung bequem, die von Heisenberg angegebene Schlußformel

$$S_i = N\left[1 - \frac{32\pi^2}{3}\int_0^{x_0}\frac{dx}{h^3}x^2\left(\sqrt{2meF(x)} - \frac{s_0 \cdot h}{4\pi}\right)^2 \left(\sqrt{2meF(x)} + \frac{s_0 \cdot h}{8\pi}\right)\right]$$

etwas anders zu schreiben, indem man von der von Fermi[4]) angegebenen dimensionslosen Schreibweise Gebrauch macht, genau wie das Debye[5]) für die Darstellung der kohärenten Strahlung getan hat. Man mißt alle Längen in Einheiten der für das Atom mit der Elektronenzahl Z charakteristischen Länge $a = \frac{0{,}47}{Z^{1/3}}$ Å und die Potentiale in Einheiten des Grundpotentials $e \cdot Z/a$. Nach Fermi gilt dann $\xi = \frac{x}{a} \cdot N^{1/3}$; $\frac{F(x)}{F_0} = \frac{\varphi(\xi)}{\xi}$; $\varphi(\xi)$ ist bei Fermi tabelliert. Die Gleichung lautet dann:

$$S_i = Z\left[1 - \int_0^{\xi_0}\xi^2\,d\xi\left(\left(\frac{\varphi(\xi)}{\xi}\right)^{1/2} - v\right)^2\left(\left(\frac{\varphi(\xi)}{\xi}\right)^{1/2} + \frac{1}{2}v\right)\right]$$

wo $v = \frac{ks \cdot a}{(6\pi Z)^{1/3}}$ [6]), unter Einsetzung des bekannten Wertes von a ergibt das $v = ks \cdot 0{,}176/Z^{2/3} = ksb$, wo jetzt b die für die inkohärente Streuung

1) P. Debye, Physik. Zeitschr. **28**, 315, 1927; P. Debye, L. Bewilogua, F. Ehrhardt, Physik. Zeitschr. **30**, 84, 1929, Sächs. Akad. Ber. **81**, 29, 1929, P. Debye, Physik. Zeitschr. **30**, 524, 1929; **31**, 142, 348 u. 419, 1930, Zeitschr. f. Elektrochem. **36**, 612, 1930; Proc. Phys. Soc. London **42**, 340, 1930; H. Gajewski, Physik. Zeitschr. **32**, 219, 1931; L. Bewilogua, Physik. Zeitschr. **32**, 114 u. 265, 1931.

2) Vgl. dazu E. O. Wollan, Phys. Rev. **38**, 15, 1931.

3) Vgl. dazu Y. H. Woo, Phys. Rev. **38**, 6, 1931.

4) E. Fermi, Leipziger Vorträge 1928, S. 95.

5) P. Debye, Physik. Zeitschr. **31**, 419, 1930.

6) Heisenberg benutzt die Wallersche Schreibweise $s = \frac{4\pi}{\lambda}\sin\frac{\vartheta}{2}$, während s bei Debye die Winkelvariable $2\sin\frac{\vartheta}{2}$ allein bedeutet. In unserer Formel bedeutet entsprechend ks $\left(\text{mit } k = \frac{2\pi}{\lambda}\right)$ das gleiche wie s bei Heisenberg. (ϑ ist der Winkel zwischen primärem und gestreutem Strahl.)

Bibliographical Citation List

Group 1

Group 2

Group 3

Group 4